Lecture Notes in Computer Science   8834

Commenced Publication in 1973
Founding and Former Series Editors:
Gerhard Goos, Juris Hartmanis, and Jan van Leeuwen

T0212741

Chu Kiong Loo   Keem Siah Yap
Kok Wai Wong   Andrew Teoh
Kaizhu Huang (Eds.)

Neural Information Processing

21st International Conference, ICONIP 2014
Kuching, Malaysia, November 3-6, 2014
Proceedings, Part I

 Springer

Volume Editors

Chu Kiong Loo
University of Malaya, Kuala Lumpur, Malaysia
E-mail: ckloo.um@um.edu.my

Keem Siah Yap
Universiti Tenaga Nasional, Selangor, Malaysia
E-mail: yapkeem@uniten.edu.my

Kok Wai Wong
Murdoch University, Murdoch, WA, Australia
E-mail: k.wong@murdoch.edu.au

Andrew Teoh
Yonsei University, Seoul, South Korea
E-mail: bjteoh@yonsei.ac.kr

Kaizhu Huang
Xi'an Jiaotong-Liverpool University, Suzhou, China
E-mail: kaizhu.huang@xjtlu.edu.cn

ISSN 0302-9743   e-ISSN 1611-3349
ISBN 978-3-319-12636-4   e-ISBN 978-3-319-12637-1
DOI 10.1007/978-3-319-12637-1
Springer Cham Heidelberg New York Dordrecht London

Library of Congress Control Number: 2014951688

LNCS Sublibrary: SL 1 – Theoretical Computer Science and General Issues

Typesetting: Camera-ready by author, data conversion by Scientific Publishing Services, Chennai, India

Printed on acid-free paper

Springer is part of Springer Science+Business Media (www.springer.com)

Preface

This volume is part of the three-volume proceedings of the 21st International Conference on Neural Information Processing (ICONIP 2014), which was held in Kuching, Malaysia, during November 3–6, 2014. The ICONIP is an annual conference of the Asia Pacific Neural Network Assembly (APNNA). This series of ICONIP conferences has been held annually since 1994 in Seoul and has become one of the leading international conferences in the area of neural networks.

ICONIP 2014 received a total of 375 submissions by scholars from 47 countries/regions across six continents. Based on a rigorous peer-review process where each submission was evaluated by at least two qualified reviewers, a total of 231 high-quality papers were selected for publication in the reputable series of *Lecture Notes in Computer Science* (LNCS). The selected papers cover major topics of theoretical research, empirical study, and applications of neural information processing research. ICONIP 2014 also featured a pre-conference event, namely, the Cybersecurity Data Mining Competition and Workshop (CDMC 2014) which was held in Kuala Lumpur. Nine papers from CDMC 2014 were selected for a Special Session of the conference proceedings.

In addition to the contributed papers, the ICONIP 2014 technical program included a keynote speech by Shun-Ichi Amari (RIKEN Brain Science Institute, Japan), two plenary speeches by Jacek Zurada (University of Louisville, USA) and Jürgen Schmidhuber (Istituto Dalle Molle di Studi sull'Intelligenza Artificiale, Switzerland). This conference also featured seven invited speakers, i.e., Akira Hirose (The University of Tokyo, Japan), Nikola Kasabov (Auckland University of Technology, New Zealand), Soo-Young Lee (KAIST, Korea), Derong Liu (Chinese Academy of Sciences, China; University of Illinois, USA), Kay Chen Tan (National University of Singapore), Jun Wang (The Chinese University of Hong Kong), and Zhi-Hua Zhou (Nanjing University, China).

We would like to sincerely thank Honorary Chair Shun-ichi Amari, Mohd Amin Jalaludin, the members of the Advisory Committee, the APNNA Governing Board for their guidance, the members of the Organizing Committee for all their great efforts and time in organizing such an event. We would also like to take this opportunity to express our deepest gratitude to all the technical committee members for their professional review that guaranteed high quality papers.

We would also like to thank Springer for publishing the proceedings in the prestigious LNCS series. Finally, we would like to thank all the speakers, authors, and participants for their contribution and support in making ICONIP 2014 a successful event.

November 2014

Chu Kiong Loo
Keem Siah Yap
Kok Wai Wong
Andrew Teoh
Kaizhu Huang

Organization

Honorary Chairs

Shun-Ichi Amari   RIKEN, Japan
Mohd Amin Jalaludin   University of Malaya, Malaysia

General Chair

Chu Kiong Loo   University of Malaya, Malaysia

General Co-chairs

Yin Chai Wang   University Malaysia Sarawak, Malaysia
Weng Kin Lai   Tunku Abdul Rahman University College,
     Malaysia

Program Chairs

Kevin Kok Wai Wong   Murdoch University, Australia
Andrew Teoh   Yonsei University, Korea
Kaizhu Huang   Xi'an Jiaotong-Liverpool University, China

Publication Chairs

Lakhmi Jain   University of South Australia, Australia
Chee Peng Lim   Deakin University, Australia
Keem Siah Yap   Universiti Tenaga Nasional, Malaysia

Registration Chair and Webmaster

Yun Li Lee   Sunway University, Malaysia

Local Organizing Chairs

Chong Eng Tan   University Malaysia Sarawak, Malaysia
Kai Meng Tay   University Malaysia Sarawak, Malaysia

Workshop and Tutorial Chairs

Chen Change Loy   Chinese University of Hong Kong, SAR China
Ying Wah Teh   University of Malaya, Malaysia
Saeed Reza   University of Malaya, Malaysia
Tutut Harewan   University of Malaya, Malaysia

Special Session Chairs

Thian Song Ong   Multimedia University, Malaysia
Siti Nurul Huda Sheikh
  Abdullah   Universiti Kebangsaan Malaysia, Malaysia

Financial Chair

Ching Seong Tan   Multimedia University, Malaysia

Sponsorship Chairs

Manjeevan Seera   University of Malaya, Malaysia
John See   Multimedia University, Malaysia
Aamir Saeed Malik   Universiti Teknologi Petronas, Malaysia

Publicity Chairs

Siong Hoe Lau   Multimedia University, Malaysia
Khairul Salleh Mohamed
  Sahari   Universiti Tenaga Nasional, Malaysia

Asia Liaison Chairs

ShenShen Gu   Shanghai University, China

Europe Liaison Chair

Wlodzislaw Duch   Nicolaus Copernicus University, Poland

America Liaison Chair

James T. Lo   University of Maryland, USA

Advisory Committee

Lakhmi Jain, Australia
David Gao, Australia
BaoLiang Lu, China
Ying Tan, China
Jin Xu, China.
Irwin King, Hong Kong, SAR China
Jun Wang, Hong Kong, SAR China
P. Balasubramaniam, India
Kunihiko Fukushima, Japan
Shiro Usui, Japan
Minho Lee, Korea
Muhammad Leo Michael Toyad
    Abdullah, Malaysia
Mustafa Abdul Rahman, Malaysia
Narayanan Kulathuramaiyer,
    Malaysia
David Ngo, Malaysia

Siti Salwah Salim, Malaysia
Wan Ahmad Tajuddin Wan Abdullah,
    Malaysia
Wan Hashim Wan Ibrahim, Malaysia
Dennis Wong, Malaysia
Nik Kasabov, New Zealand
Arnulfo P. Azcarraga, Phillipines
Wlodzislaw Duch, Poland
Tingwen Huang, Qatar
Meng Joo Err, Singapore
Xie Ming, Singapore
Lipo Wang, Singapore
Jonathan H. Chan, Thailand
Ron Sun, USA
De-Liang Wang, USA
De-Shuang Huang, China

Technical Committee

Ahmad Termimi Ab Ghani
Mark Abernethy
Adel Al-Jumaily
Leila Aliouane
Cesare Alippi
Ognjen Arandjelovic
Sabri Arik
Mian M. Awais
Emili Balaguer-Ballester
Valentina Emilia Balas
Tao Ban
Sang-Woo Ban
Younès Bennani
Asim Bhatti
Janos Botzheim
Salim Bouzerdoum
Ivo Bukovsky
Jinde Cao
Jiang-Tao Cao
Chee Seng Chan
Long Cheng
Girija Chetty

Andrew Chiou
Pei-Ling Chiu
Sung-Bae Cho
Todsanai Chumwatana
Pau-Choo Chung
Jose Alfredo Ferreira Costa
Justin Dauwels
Mingcong Deng
M.L. Dennis Wong
Hongli Dong
Hiroshi Dozono
El-Sayed M. El-Alfy
Zhouyu Fu
David Gao
Tom Gedeon
Vik Tor Goh
Nistor Grozavu
Ping Guo
Masafumi Hagiwara
Osman Hassab Elgawi
Shan He
Haibo He

Sven Hellbach
Jer Lang Hong
Jinglu Hu
Xiaolin Hu
Kaizhu Huang
Amir Hussain
Kazushi Ikeda
Piyasak Jeatrakul
Sungmoon Jeong
Yaochu Jin
Zsolt Csaba Johanyák
Youki Kadobayashi
Hiroshi Kage
Joarder Kamruzzaman
Shin'Ichiro Kanoh
Nikola Kasabov
Rhee Man Kil
Kyung-Joong Kim
Kyung-Hwan Kim
Daeeun Kim
Laszlo T. Koczy
Markus Koskela
Szilveszter Kovacs
Naoyuki Kubota
Takio Kurita
Olcay Kursun
James Kwok
Sungoh Kwon
Weng Kin Lai
Siong Hoe Lau
Yun Li Lee
Minho Lee
Nung Kion Lee
Chin Poo Lee
Vincent Lemaire
L. Leng
Yee Tak Leung
Bin Li
Yangming Li
Ming Li
Xiaofeng Liao
Meng-Hui Lim
C.P. Lim
Chee Peng Lim
Kim Chuan Lim

Hsuan-Tien Lin
Huo Chong Ling
Derong Liu
Zhi-Yong Liu
Chu Kiong Loo
Wenlian Lu
Zhiwu Lu
Bao-Liang Lu
Shuangge Ma
Mufti Mahmud
Kenichiro Miura
Hiroyuki Nakahara
Kiyohisa Natsume
Vinh Nguyen
Tohru Nitta
Yusuke Nojima
Anto Satriyo Nugroho
Takenori Obo
Toshiaki Omori
Takashi Omori
Thian Song Ong
Sid-Ali Ouadfeul
Seiichi Ozawa
Worapat Paireekreng
Paul Pang
Ying Han Pang
Shaoning Pang
Hyung-Min Park
Shri Rai
Mallipeddi Rammohan
Alexander Rast
Jinchang Ren
Mehdi Roopaei
Ko Sakai
Yasuomi Sato
Naoyuki Sato
Shunji Satoh
Manjeevan Seera
Subana Shanmuganathan
Bo Shen
Yang Shi
Tomohiro Shibata
Hayaru Shouno
Jennie Si
Jungsuk Song

Kingkarn Sookhanaphibarn
Indra Adji Sulistijono
Changyin Sun
Jun Sun
Masahiro Takatsuka
Takahiro Takeda
Shing Chiang Tan
Syh Yuan Tan
Ching Seong Tan
Ken-Ichi Tanaka
Katsumi Tateno
Kai Meng Tay
Connie Tee
Guo Teng
Andrew Teoh
Heizo Tokutaka
Dat Tran
Boris Tudjarov
Eiji Uchino
Kalyana C. Veluvolu
Michel Verleysen
Lipo Wang
Hongyuan Wang

Frank Wang
Yin Chai Wang
Yoshikazu Washizawa
Kazuho Watanabe
Bunthit Watanapa
Kevin Wong
Kok Seng Wong
Zenglin Xu
Yoko Yamaguch
Koichiro Yamauchi
Hong Yan
Wei Qi Yan
Kun Yang
Peipei Yang
Bo Yang
Xu-Cheng Yin
Zhigang Zeng
Min-Ling Zhang
Chao Zhang
Rui Zhang
Yanming Zhang
Rui Zhang
Ding-Xuan Zhou

Table of Contents – Part I

Cognitive Science

Neural Networks and Learning Systems - Theory and Design

Neural Networks and Learning Systems - Applications

Table of Contents – Part II

Kernel and Statistical Methods

Evolutionary Computation and Hybrid Intelligent Systems

Pattern Recognition Techniques

Table of Contents – Part III

Signal and Image Processing

The 2014 Cybersecurity Data Mining Competition and Workshop (CDMC2014)

Intelligent Systems for Supporting Decision-Making Processes: Theories and Applications

Neuroengineering and Neuralcomputing

Cognitive Robotics

Security in Signal Processing and Machine Learning

Learning Systems for Social Network and Web Mining

Transfer Entropy and Information Flow Patterns in Functional Brain Networks during Cognitive Activity

Md. Hedayetul Islam Shovon[1], D (Nanda) Nandagopal[1],
Ramasamy Vijayalakshmi[2], Jia Tina Du[1], and Bernadine Cocks[1]

[1] Cognitive Neuroengineering Laboratory, Division of IT, Engineering and the Environments,
University of South Australia, Adelaide, Australia
[2] Department of Applied Mathematics and Computational Science,
PSG College of Technology, Coimbatore, Tamil Nadu, India
shomy004@mymail.unisa.edu.au

Abstract. Most previous studies of functional brain networks have been conducted on undirected networks despite the direction of information flow able to provide additional information on how one brain region influences another. The current study explores the application of normalized transfer entropy to EEG data to detect and identify the patterns of information flow in the functional brain networks during cognitive activity. Using a mix of signal processing, information and graph-theoretic techniques, this study has identified and characterized the changing connectivity patterns of the directed functional brain networks during different cognitive tasks. The results demonstrate not only the value of transfer entropy in evaluating the directed functional brain networks but more importantly in determining the information flow patterns and thus providing more insights into the dynamics of the neuronal clusters underpinning cognitive function.

Keywords: Transfer entropy, directed functional brain network, EEG, cognitive load, graph theory.

1 Introduction

The human brain is a complex and dense network of billions of interconnected neurons. To quantify the topological features of this network, graph theoretical analysis has been successfully employed by researchers in the recent past [1-3]. Most of this graph theoretical analysis has, however, been applied to undirected networks. Such analysis can, however, be applied to directed networks [2]. This is perhaps, surprising given that directed networks exhibit more prominent features by providing directional interactions between pairwise elements thereby enabling more detailed analysis. Although Granger causality is often used to identify causal relationship in electroencephalogram (EEG) data, it is limited to the linear model of interaction [4]. As a result, it fails to accurately identify causal relationships in highly nonlinear systems such as the human brain. By comparison, the information theoretical measure of transfer entropy (TE) determines the direction and quantifies the information transfer between two processes [5]. TE estimates the amount of activity of a system which is not

C.K. Loo et al. (Eds.): ICONIP 2014, Part I, LNCS 8834, pp. 1–10, 2014.
© Springer International Publishing Switzerland 2014

dependent on its own past activity but on the past activity of another system. It does not require a model of the interaction and inherently non-linear [4]. As a consequence, TE has been used in various applications such as identifying information transfer between auditory cortical neurons using spike train data, investigating the influence of heart rate on breath rate and vice versa, and for the localization of the epileptic focus of epileptic patients on EEG data [5-8]. TE has however not been applied in the construction and analysis of directed functional brain networks (FBN) during different cognitive states. The current study reported in this paper aims to investigate the application of normalized TE to construct a directed FBN. It further explores graph theoretic and statistical analysis to characterize this directed FBN and its patterns of information flow during cognitive tasks. Complex network metrics have also been used to delineate the cognitive tasks and the comparative results are presented. The research literature on TE and graph theoretical analysis are reviewed and described in the following sections.

2   Transfer Entropy

Given two processes x and y, the TE from y to x is shown in Equation 1 [9]:

$$TE_{y \to x} = \sum_{x_{n+1}, x_n, y_n} p(x_{n+1}, x_n, y_n) \log \left(\frac{p(x_{n+1}, x_n, y_n) \cdot p(x_n)}{p(x_n, y_n) \cdot p(x_{n+1}, x_n)} \right) \qquad (1)$$

Here, x_n denotes the status (value) of signal/system x at time n, y_n denotes the status of signal y at time n and x_{n+1} denotes the status of signal x at time $n + 1$. The value of $TE_{y \to x}$ is calculated by summing over all possible combination of x_{n+1}, x_n, and y_n. TE is in the range $0 \le TE_{y \to x} < \infty$. Here, $TE_{y \to x} \ne TE_{x \to y}$. There is another similar equation for $TE_{x \to y}$. In practice, for the calculation of TE, two additional steps were included to improve the calculation accuracy [7, 10]. Due to the finite size and non-stationarity of data, TE matrices usually contain much noise. In the existing literature, noise/bias has been removed from the estimate of TE by subtracting the average transfer entropy from y to x using shuffled version of y denoted by $< TE_{y_{shuffle} \to x} >$, over several shuffles [10]. $y_{shuffle}$ contains the same symbol as in y but those symbols are rearranged in a randomly shuffled order. Then, normalized transfer entropy is calculated from y to x with respect to the total information in sequence x itself. This will represent the relative amount of information transferred by y. The normalized transfer entropy (NTE) is shown in Equation 2 as follows [7]:

$$NTE_{y \to x} = \frac{TE_{y \to x} - < TE_{y_{shuffle} \to x} >}{H(x_{n+1} | x_n)} \qquad (2)$$

In equation 2, $H(x_{n+1} | x_n)$ represents the conditional entropy of process x at time $n + 1$ given its value at time n as shown in equation 3.

$$H(x_{n+1} | x_n) = - \sum_{x_{n+1}, x_n} p(x_{n+1}, x_n) \log \frac{p(x_{n+1}, x_n)}{p(x_n)} \qquad (3)$$

NTE is in the range $0 \leq NTE_{y \rightarrow x} \leq 1$. NTE is 0 when y transfers no information to x, and is 1 when y transfers maximal information to x.

3 Functional Brain Networks and Graph Theoretical Analysis

To capture the dynamic interactions between the neuronal elements of human brain, FBNs can be derived from time series observations of EEG signals [2]. As EEG is cheap non-invasive method with high temporal resolution, it has been used extensively in research, medical diagnosis and brain computer interaction [3, 11]. In the present study, the FBNs are constructed by computing the NTE between EEG channels. According to the graph theory, graph is a mathematical model that consists of vertices (nodes) where the connection between each pairs of vertices is called an edge (link) [1, 2]. In the case of FBNs, scalp electrodes are considered as vertices and the connections/links between electrodes are measured using correlation. The following graph (or complex network) metrics are also used in this study: connectivity density representing the actual number of edges as a proportion to the total number of possible edges [1]; reciprocity representing the ratio of the number of pairs with a reciprocated edges relative to the total number of edges [12]; clustering coefficient quantifying the fraction of triangles around a node [13]; average of shortest path length between all node pairs known as the characteristic path length [14]; small world representing both high clustering and short characteristics path length [14]; node eccentricity representing the maximal shortest path length between any two nodes [12]; and node strength representing the total of all incoming and outgoing link weights [15].

4 Methods

4.1 Participants and EEG Data Acquisition

Six healthy, right handed adults (4 males, 2 females) volunteered for EEG data collection (age range 19-59) at the Cognitive Neuroengineering Laboratory of University of South Australia. The participants were recruited from staff and students cohort of University of South Australia. All reported normal hearing, normal or corrected-to-normal vision, with none reporting any psychological, neurological or psychiatric disorder. EEG data were acquired at a sampling rate of 1000 Hz through a 40 channel Neuroscan Nuamps amplifier using Curry 7 software. The 30 electrode sites used in the current study were based on the international 10-20 convention: FP1, FP2, F7, F3, Fz, F4, F8, FT7, FC3, FCz, FC4, FT8, T3, C3, Cz, C4, T4, TP7, CP3, CPz, CP4, TP8, T5, P3, Pz, P4, T6, O1, Oz and O2. Continuous EEG data were collected as each participant undertook three tasks in a Simuride Driving Simulator. In Task 1 (baseline), data were collected for 2 minutes (each) in both eyes open and eyes closed conditions. In Task 2, participants were asked to drive normally on a virtual winding road for approximately 4 minutes. In Task 3, participants were asked to drive, as per Task 2, while also responding to auditory stimuli. All stimulus onsets and participant responses were time-marked on the EEG record using STIM 2 software.

4.2 EEG Signal Pre-processing

A band pass filter of 1-70 Hz and a notch filter of 50 Hz were applied to the EEG data. Eye blinks were removed using principal component analysis (PCA), with any residual. Bad blocks were removed manually. For Tasks 1 and 2, two seconds epoched averaged data were extracted by applying back-to-back epoching process of Curry software using epoch length of 2 seconds, and then averaging those epochs. For tasks 3, two seconds epochs were extracted by identifying all two seconds epochs from stimulus onsets, where reaction times were between 1.5 and 3 seconds, and then averaging those epochs.

4.3 Proposed Methodology for the Analysis of Directed Functional Brain Network

The pre-processed EEG data during eyes open (EOP), driving only (Drive), and driving with audio distraction (DriveAdo) were used for the construction of TE matrices, where each cell of the TE matrices represents the TE value from one electrode to another. For noise removal, an average shuffled TE matrix (noise matrix) was calculated and subtracted from the original matrix. The data, information processing and associated computational steps are illustrated in Figure 1.

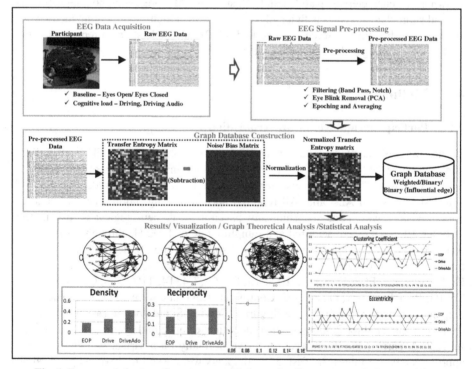

Fig. 1. Framework for transfer entropy and information flow patterns during cognition

The Normalized TE (NTE) matrices were used for the construction of both binary and weighed directed FBNs. The objectives of the proposed methodology are twofold:

1. To quantify the topological features of binary and weighted directed FBNs during different cognitive states; and
2. To determine information flow patterns during different cognitive states.

To investigate the topological features and information flow patterns from constructed FBNs, the following complex network metrics were used. Connectivity density is calculated by counting the actual number of edges in a graph and then diving by all possible number of edges in that graph. For a directed graph with n nodes where there is no self-connections/loop, the total number of possible connections are $n * (n - 1)$. Reciprocity of a graph is calculated by finding the ratio of the total number of bidirectional edges relative to the total number of any directional edges between the node pairs. Clustering coefficient has been calculated using the technique of *directed variant of clustering coefficient* developed by Fagiolo [13]. In a directed graph, 3 nodes can generate up to 8 triangles. The clustering coefficient for node i represents the ratio between all directed triangles actually formed by i and the number of all possible triangles that i could form. The experimental results and analysis are presented in the following section.

5   Results and Discussion

The NTE matrices (each are 30 by 30 in size) during EOP, Drive and DriveAdo states are presented in Figure 2. The increased information flow during cognitive load condition is demonstrated by the appearance of more cluttered brighter pixels (see Figure 2b and 2c). From this experiment, it can be inferred that, during cognitive load, information flow generally increases between the electrodes than in baseline condition (EOP).

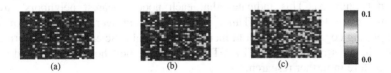

(a)   (b)   (c)

Fig. 2. Normalized transfer entropy matrix during a) EOP, b) Drive and c) DriveAdo

5.1   Analysis of Binary Directed Functional Brain Network

The binary directed FBNs constructed using NTE matrices with threshold=0.002 was utilized in this study and as shown in Figure 3 demonstrating increased connectivity between the electrodes during the cognitive load. Due to space limitations the result of graph metrics analysis is shown only for one participant. As indicated in Figure 4a connectivity density is higher in the cognitive load than the baseline condition, inferring more connections are established to facilitate more active information flow.

Figure 4b shows that reciprocity is higher in the cognitive load condition, which further suggests that most electrode pairs are trying to establish mutual connections to facilitate effective information transmission within the network.

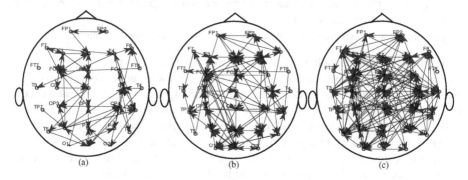

Fig. 3. Binary directed functional brain network during a) EOP, b) Drive and c) DriveAdo

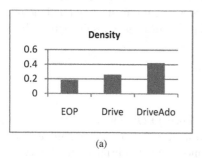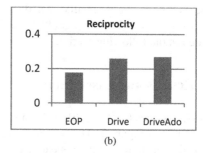

(a)   (b)

Fig. 4. Comparison of a) Connectivity density, and b) Reciprocity of the brain network during EOP, Drive and DriveAdo

Clustering coefficient value increases in almost all of the electrodes during cognitive load (Figure 5). This indicates that each node's nearest neighbors directly communicate and form clusters. This type of segregated neural processing of brain network during cognitive load would increases the local efficiency of the information transfer. The electrodes F3, F8, FT7, FT8 and TP8 are not showing that trend which need to explore in further research.

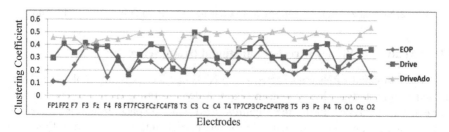

Fig. 5. Clustering coefficient across electrodes during EOP, Drive and DriveAdo

Between the regular lattice and completely random networks there is a class of networks which has the properties of high clustering and short path length. That means, most of the nodes are not neighbors but most of the nodes can traverse each other's by small number of edges. This class of network is called as "small world network" [14]. Small world network can be also characterized as a set of network with both high local and global efficiency [16]. Efficiency estimates how well the information propagates over the network. The small world properties of the directed FBNs during different cognitive tasks are illustrated in Table 1.

Table 1. Small-worldness of directed FBNs during EOP, Drive and DriveAdo

Cognitive Tasks	C	C_{rand}	L	L_{rand}	$\gamma = C/C_{rand}$	$\lambda = L/L_{rand}$	$S = \gamma/\lambda$
EOP	0.4735	0.4708	1.5345	1.5264	1.0056	1.0053	1.0003
Drive	0.5736	0.4882	1.4356	1.5126	1.1749	0.9491	1.2680
DriveAdo	0.7930	0.4905	1.2425	1.5011	1.6167	0.8277	1.9532

Here, C and C_{rand} are the clustering coefficient of a tested and random network respectively; L and L_{rand} are the characteristic path length of a tested and random network respectively. The results reveal that characteristics path length decreases during cognitive load which would suggest a high global efficiency of information transfer in the FBN. The directed FBNs during EOP, Drive and DriveAdo all exhibit small-worldness ($S > 1$), however, the FBN constructed during cognitive load exhibits more small world properties than that of the baseline condition suggesting that the FBN facilitates both high local and global efficiency of information transfer.

5.2 Analysis of Binary Directed Functional Brain Network (Influential Edges Only)

Binary FBNs were constructed by keeping only the influential edges between each electrode pair during different cognitive states; namely, two electrodes a and b, may have two possible edges: $a \rightarrow b$ and $b \rightarrow a$. In the current study, the maximum weight value's edge $a \rightarrow b$ or $b \rightarrow a$ was chosen.

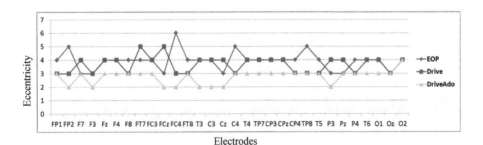

Fig. 6. Comparison of node eccentricity across electrodes during EOP, Drive and DriveAdo

Eccentricity was then calculated for different cognitive states (See Figure 6). The maximum number of edges traversed to reach one node from another during EOP, Drive and DriveAdo are 6, 5 and 4 edges, respectively. It may be concluded that, during cognitive load, information may travel pretty quickly in FBN than in baseline condition.

5.3   Analysis of Weighted Directed Functional Brain Network

NTE matrices without applying any threshold were also used to construct weighted FBN. Figure 7 represents the comparison of the node strengths during different cognitive states computed using the weighted network. As shown, almost all the electrodes have higher strength values during cognitive load, which indicates that each electrode sends and receives more information during cognitive load.

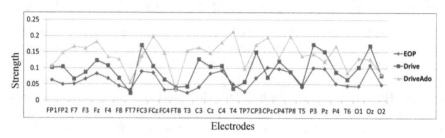

Fig. 7. Comparison of strength across electrodes during EOP, Drive and DriveAdo

5.4   Statistical Analysis

The following experiments were conducted to demonstrate the statistical significance of information flow during various cognitive states. For the three cognitive states, the total information flow of each electrode to all other electrodes was calculated. One-way analysis of variance (ANOVA) was applied to compare the mean information flow of each group and the multi-comparison results are shown in Figure 8.

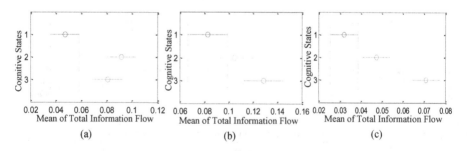

Fig. 8. Multi-comparison of mean total information flow of each electrode to all other electrodes during different cognitive states for 3 participants (a) DP1, (b) DP2 and (c) DP3 (Y-Axis: Cognitive States 1 for EOP, 2 for Drive, and 3 for DriveAdo)

The statistical significance of the differences between means is calculated using t-test at $\alpha = 0.05$ (two-tailed) and the results are shown in Table 2. The results demonstrate that mean of total information flow is significantly different in EOP and DriveAdo experiments for all participants.

Table 2. Statistical validation of mean information flow difference in different cognitive states

States		DP 1		DP 2		DP 3	
		Mean Diff	95% CI	Mean Diff	95% CI	Mean Diff	95% CI
Drive	EOP	0.0244	[0.0244,0.0651]	0.0218	[-0.0033,0.0469]	0.0155*	[0.0053, 0.0258]
DriveAdo	Drive	-0.0108	[-0.0329,0.0113]	0.0239	[-0.0071,0.0550]	0.0235*	[0.0110, 0.0360]
DriveAdo	EOP	0.0339*	[0.0191,0.0487]	0.0457*	[0.0200,0.0715]	0.039*	[0.0285, 0.0495]

*Mean difference is significant at p< .05 level.

6   Conclusion

In this study the information theoretical NTE measure has been applied to construct EEG based directed FBNs in baseline and cognitive load conditions. Using the techniques of signal processing, information and graph theoretic measures, and inferential statistics, information flow patterns during cognition have been detected and identified. The overall results demonstrate that the directed FBNs constructed using NTE are sensitive to cognitive load. This sensitivity of NTE based FBN has the potential to assist the development of quantitative metrics to measure cognition. The NTE approach may have likely application in the clinical diagnosis of cognitive impairments in future. Future research should address the application of various graph mining algorithms on the constructed directed functional brain networks to detect and track possible patterns as well as quantum of information flow during cognitive activity. This may lead to deeper understanding of cognitive function.

Acknowledgement. The authors wish to acknowledge partial support provided by the Defence Science and Technology Organisation (DSTO), Australia. The assistance and technical support provided by fellow researchers Mr Nabaraj Dahal and Mr Naga Dasari are greatly appreciated.

References

1. Rubinov, M., Sporns, O.: Complex network measures of brain connectivity: Uses and interpretations. Neuroimage 52, 1059–1069 (2010)
2. Bullmore, E., Sporns, O.: Complex brain networks: Graph theoretical analysis of structural and functional systems. Nature Reviews Neuroscience 10, 186–198 (2009)
3. Nandagopal, N.D., Vijayalakshmi, R., Cocks, B., Dahal, N., Dasari, N., Thilaga, M., Dharwez, S.: Computational Techniques for Characterizing Cognition Using EEG Data – New Approaches. Procedia Computer Science 22, 699–708 (2013)
4. Vicente, R., Wibral, M., Lindner, M., Pipa, G.: Transfer entropy—a model-free measure of effective connectivity for the neurosciences. Journal of Computational Neuroscience 30, 45–67 (2011)
5. Schreiber, T.: Measuring information transfer. Physical Review Letters 85, 461 (2000)

6. Chávez, M., Martinerie, J., Le Van Quyen, M.: Statistical assessment of nonlinear causality: Application to epileptic EEG signals. Journal of Neuroscience Methods 124, 113–128 (2003)

7. Gourévitch, B., Eggermont, J.J.: Evaluating information transfer between auditory cortical neurons. Journal of Neurophysiology 97, 2533–2543 (2007)

8. Sabesan, S., Narayanan, K., Prasad, A., Iasemidis, L., Spanias, A., Tsakalis, K.: Information flow in coupled nonlinear systems: Application to the epileptic human brain. In: Data Mining in Biomedicine, pp. 483–503. Springer (2007)

9. Kaiser, A., Schreiber, T.: Information transfer in continuous processes. Physica D: Nonlinear Phenomena 166, 43–62 (2002)

10. Neymotin, S.A., Jacobs, K.M., Fenton, A.A., Lytton, W.W.: Synaptic information transfer in computer models of neocortical columns. Journal of Computational Neuroscience 30, 69–84 (2011)

11. Kim, S.P.: A review on the computational methods for emotional state estimation from the human EEG. Computational and Mathematical Methods in Medicine 2013 (2013)

12. Hanneman, R.A., Riddle, M.: Introduction to social network methods. University of California Riverside (2005), published in digital form at http://faculty.ucr.edu/~hanneman/

13. Fagiolo, G.: Clustering in complex directed networks. Physical Review E 76, 026107 (2007)

14. Watts, D.J., Strogatz, S.H.: Collective dynamics of 'small-world' networks. Nature 393, 440–442 (1998)

15. Brain Connectivity Toolbox, https://sites.google.com/site/bctnet/measures/list

16. Latora, V., Marchiori, M.: Economic small-world behavior in weighted networks. The European Physical Journal B-Condensed Matter and Complex Systems 32, 249–263 (2003)

Human Implicit Intent Discrimination
Using EEG and Eye Movement

Ukeob Park[1], Rammohan Mallipeddi[2], and Minho Lee[2,*]

[1] Department of Robot Engineering, Kyungpook National University
[2] School of Electronics Engineering,
Kyungpook National University
1370 Sankyuk-Dong, Puk-Gu, Taegu 702-701, South Korea
uepark@ee.knu.ac.kr, {mallipeddi.ram,mholee}@gmail.com

Abstract. In this paper, we propose a new human implicit intent understanding model based on multi-modal information, which is a combination of eye movement data and brain wave signal obtained from eye-tracker and Electroencephalography (EEG) sensors respectively. From the eye movement data, we extract human implicit intention related to features such as fixation count and fixation duration corresponding to the areas of interest (AOI). Also, we analyze the EEG signals based on phase synchrony method. Combining the eye movement and EEG information, we train several classifiers such as support vector machine classifier, Gaussian Mixture Model and Naïve Bayesian, which can successfully identify the human's implicit intention into two defined categories, i.e. navigational and informational intentions. Experimental results show that the human implicit intention can be better understood using multimodal information.

Keywords: brain-computer interface (BCI), electroencephalographic (EEG), eye movement, phase synchrony, intent recognition, multi-modality.

1 Introduction

Human intent recognition is crucial for an efficient non-verbal human computer interaction. In cognitive science, intention modeling and recognition is considered to create a new paradigm in human computer interface (HCI) and human robot interaction (HRI) [1, 2]. In Psychology, human intention can be divided into 2 types – explicit and implicit **[3]**. In last few years, several attempts have been made to understand and distinctly recognize subject's implicit intention using techniques like eye movement, Electroencephalography (EEG), etc. [4].

In this work, we propose a method to integrate two different modalities, eyeball movement data and EEG data, in order to improve the classification of human's implicit intention in real-world environment. Using the multimodal information, we compare performance of several classifiers that is trained to classify human's implicit

---

* Corresponding author.

C.K. Loo et al. (Eds.): ICONIP 2014, Part I, LNCS 8834, pp. 11–18, 2014.
© Springer International Publishing Switzerland 2014

intention as navigational and informational intentions. Besides, in the same experiment, we compare performances using unimodal (only eyeball movement or EEG signal) with multi-modal (both eyeball movement and EEG signal).

This paper is organized as follows. In section 2, we define the different human implicit intentions and present our methodology based on the multimodal information. In addition, we briefly summarize the eye movement analysis and phase locking value (PLV). In section 3, we present the experiment setup and results of the proposed model. Section 4 concludes the paper.

2   Methods

The proposed method is based on the multimodal features extracted from eyeball movement patterns and brain wave signals. Fig. 1 shows an overall diagram of the proposed method. As the model is based on the multimodal information, to collect the data from different tools, we need to synchronize the starting cue in order to get both the data simultaneously. In this section, we explain how the eye movement data and EEG data can be analyzed to understand the human implicit intention. Once the multimodal data are collected, they are used to train a support vector machine to classify human intention into navigational and informational intentions.

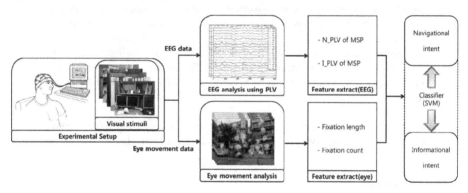

Fig. 1. Overview of experimental procedure

2.1   Definition of User Intention

Human intention can be explicit or implicit in nature. In particular, humans express their explicit intentions using different sequence of actions. In contrast, implicit human intentions are subtle, vague and often difficult to interpret [5, 6]. According to Bernard [3], human implicit intention during a visual search can be classified as navigational and informational intention. The different human implicit intentions during a visual stimulus can be defined as follows.

Navigational intention: it refers to the human's idea to find any object in a visual input without a particular motivation.

Informational intention: it refers to the human's aspiration to find a particular object of interest or to behave with a motivation.

2.2   Eye Movement Analysis

Fig. 2 shows the block diagram of the visual signal analysis for recognizing the human implicit intention based on the eyeball movement patterns [7]. We use the Tobii 1750 eye tracking system to analyze the human eye movement data. In the images, which are presented as visual stimuli, we define the areas of interest (AOI) to analyze the eyeball movement based on the intention of the subject under consideration. Amongst the several features measured by the Tobii 1750 system, we select the dominant features such as fixation length, fixation count to classify the subject's implicit intention into navigational and informational intention in real-time.

Fixation length: duration of each individual's fixations within an AOI

Fixation count: number of times the participant fixates on an AOI

Fig. 2. Block diagram of feature extraction from eye movement data

2.3   EEG Analysis Using Phase Locking Value (PLV)

PLV synchronization measures the synchronization level of EEG at every time instant between any two electrode pairs in the range of (0 - 1) [8, 9]. EEG phase difference is also used to estimate conduction velocity and synaptic integration time [10, 11]. Phase locking value (PLV) is a measure for studying the synchronization phenomena in EEG signals. It is similar to cross spectrum but independent of amplitude of the two signals [8]. Making use of PLV, we can measure the synchronization between all the electrode pairs in EEG collection montage. Synchronization measure (PLV) can be obtained as follows [8]:

$$PLV = \frac{1}{N} \left| \sum_{n=1}^{N} \exp(j\{\Delta\Phi(t,n)\}) \right| \tag{1}$$

Where N is the total number of trials, $\Delta\Phi(t,n)$ is the phase difference $\Phi_1(t,n)-\Phi_2(t,n)$ between the pair of brain nodes, and t is the time of each trial. The range of PLV values varies between 0 and 1. PLV = 1 indicates perfect coupling between the electrode pairs, while PLV = 0 indicates no coupling.

During a given intention (information intention/navigation intention and rest), the most significant pair (MSP) represents the most reactive electrode pairs (electrode pairs/locations) on the EGG montage. After determining the PLV of all the electrode pairs corresponding to each intention, where the MSP defined as the electrode pair that has the most PLV difference between these two events, the MSP electrodes can be identified as follows.

$$MSP - I = \underset{e}{argmax}\{PLV_{information} - PLV_{rest}\} \tag{2}$$

$$MSP - N = \underset{e}{argmax}\{PLV_{navigation} - PLV_{rest}\} \tag{3}$$

Where *MSP - I* and *MSP - N* corresponds to MSP's identified for information intention and navigation intention, respectively. In Eqs. (2) and (3), *e* represents the electrode pairs. Therefore, it is possible to identify both the brain connectivity and the most reactive electrode pairs based on PLV using Eqs. (2) and (3). According to the work in [12], we set the number of MSP as 25.

2.4   Multimodal Fusion

In each modality, there are some useful features to discriminate the human intention. The extracted features can be used to train a classifier. In the current work, we employ three kinds of classifiers, i.e. SVM, GMM and Naïve Bayesian. As mentioned in section 2.2, we can extract eye movement related features as fixation length, fixation count. In case of phase analysis of EEG, we can also extract some feature at each period of user's intention. According to extracted MSP, we can also calculate the PLV in each pair during navigational/informational intention periods. Using the four features, such as fixation length, fixation count, PLV of navigational and informational intents (*PLV − N / I*) from both the modalities, as the inputs of the classifier, the human implicit intention can be classified as navigational and informational intention.

3   Experiment Design and Results

Eight healthy male subjects participated in the experiment. EEG data from 32 channels were recorded with BIOSEMI (www.biosemi.com) amplifier. The stimuli presentation schema of the experiment is shown in Fig. 3. Subjects had to perform the following tasks during each trial. At the same time, to analyze the human eyeball movement patterns, we use the Tobii 1750 eye-tracking system.

Just see a image

| Find cups | Find an opened book | Find wine bottles | Find doughnuts | Find a coke |

Fig. 3. Instructions to participants (Blue box: navigational intention, Red box: informational intention)

3.1 Experiment Design

One session consists of 5 trials. In each trial, various images regarding to navigation/informational intention, which are close to a real-life scenarios, are shown to the subjects as shown in Fig. 3. Besides, each stimulus includes a specific object where the participants need to focus as shown in Fig. 3. So, the subject should keep an eye on the object during the informational intention. In eye movement analysis, AOI can be located around the specified object surroundings.

In total, 5 sessions are conducted and hence a total of 25 trials per subject are performed. Blank images, which are shown during the intention change, are to prevent the confusion for human subjects. Random images in each sequence are presented to avoid a bias effect of human intention in iterative experiments. And also, resting time, which is set for user's normal state without any intention, is considered before the main experiment trials.

3.2 Intent Classification by Eye Movement Data

There exist some useful features in eye movement analysis, which is the fixation length and fixation count corresponding to an AOI [7]. According to previous research about interpreting human's intention based on eye movement, it can be observed that there is significant difference in the distribution of the parameter values in the feature space during the different intent conditions in the stimuli. The difference in the variation of the two parameters values during the different intent conditions can also be observed in Fig. 4.

Due to the significant difference in the distribution of the parameter values, it would be possible to train a classifier which can classify the different human intention conditions. Therefore, the combination of two parameters (fixation length and fixation count in AOI) considered can be assumed to be appropriate for the classification of different intent conditions during a visual task.

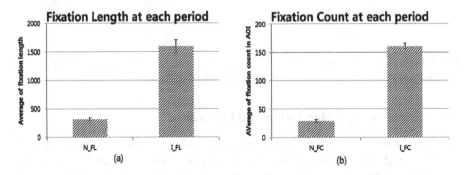

Fig. 4. Variation in the mean and STD values of the parameters (fixation length (FL) and fixation count (FC)) during a different intent conditions; (a) Variation in fixation length and (b) Variation in fixation count

3.3   Intent Classification by EEG Data

We selected 25 *MSP - N* and 25 *MSP - I* identified in theta band to classify navigational and informational intentions. *MSP - N* and *MSP - I* are identified using Eq. (2) in section 2.3. Fig. 5 (a) shows the part of results obtained with 8-th subject as an example case. The average PLV of 25 *MSP - N* and 25 *MSP - I* during both the events for 8-th subject are also illustrated. Although PLV values are difference from each other, all of participants also shown this trend of plot such as an 8th result. The reason of difference of PLV value of MSP in each other is that human's thinking process is different.

One can easily observe the *MSP - N* having higher PLV level compared to *MSP - I* during navigational intention and vice versa during informational intention. The difference in PLV level of navigational and informational intention is crucial for the intent classification. According to Eqs. (4) and (5) in **[13]**, the difference in PLV level of identified MSP's can be calculated. The largest difference of PLV level is evident in theta band **[13]**. Based on this finding, we propose that theta band can be reliably used in phase analysis for intent recognition. In other words, firing intention signal at theta band is larger than the others **[13]**. Therefore, PLV of *MSP - N / I* at each period is characterized to discriminate human's implicit intention based on PLV analysis.

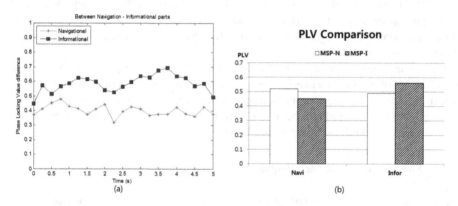

Fig. 5. Result (about Sub #8) of PLV analysis in theta band; (a)   An average of PLV of MSP in each intention part, (b) Comparison of PLV (MSP-I, MSP-N) value

3.4   Intent Classify Using Multimodal Information

To classify a subject's implicit intention in real-world environment, we use three kinds of classifier (SVM, GMM and Naïve Bayesian) using the extracted features during each intention part (navigational intention, informational intention). From the eye movement data, we can get 2 features (fixation length and fixation count) per subjects. However, in case of brain wave signal, we extract PLV of 25 *MSP - N / I* per each subject. That is, the PLV of MSP is used as a feature in EEG analysis. Based on these features, we construct input data by concatenating the eye movement features and PLV features. Then, we train the train data that is selected randomly to classify the data as two classes such as navigational and informational intentions. For the SVM classifier, we use a radial basis kernel. And also, we use a Gaussian distribution in Naïve Bayesian classifier. Table 1 shows the performance of several classifier train/test rate per each modality. We can then observe that the performance of the proposed method that uses both eyeball movements and brain wave signals is higher if compared to the classifier based on unimodality.

Table 1. Performance of three kinds of classifiers between each modality and multi-modality

Modality	Performance of several kinds of classifier; Test rate (%)± STD		
	SVM with RBF	GMM	Naïve Bayesian
EEG features	83.9 ± 0.07	83.95 ± 3.43	84.58 ± 0.16
Eye movement features	85.8 ± 0.06	84.42 ± 5.34	84.15 ± 0.28
Eye movement and EEG features	90.9 ± 0.07	89.08 ± 3.67	72.08 ± 1.15

Since the data was collected from 8 participants during 25 trials, we had a total of 200 samples per intent condition. Out of these 200 samples, 100 randomly selected samples were used to train each classifier model, whereas the remaining 100 samples were used for testing. And, we also use a cross validation. During the cross validation the testing procedure is iterated for 30 times and the average of the 30 iteration is reported. Finally, we get the most performance 90.9% using multi-modality with SVM classifier. After comparing each kinds of state-of-the-art classifier, performance of SVM with RBF is better than other performance.

4   Conclusions

In this paper, we propose a human's implicit intention recognition model based on both eye movement and EEG data analysis. We define and present enhanced classification results of human's implicit intention understanding. In order to recognize the human's implicit intention for given real-visual stimuli, an eye tracking system, Tobii 1750, and BIOSEMI were used. Through the experiment, we can extract some useful features in order to distinguish between navigational and informational intention. From the eye movement analysis, we get the fixation length, fixation count. For the EEG signals, we also get the useful features such as PLV of MSP for classifying two different human implicit intentions. Then, in order to classify the human's implicit intention, we used three kinds of state-of-the-art classifier such as SVM, GMM, Naive Bayesian using the concatenating input features of eye movement and brain signals. The experimental results show that the proposed method has a plausible performance to recognize human implicit intentions, which is better than using just one modality.

In future work, we would like to apply the model mentioned above in web queries to determine the user's implicit intention.

Acknowledgement. This research was supported by the Original Technology Research Program for Brain Science through the National Research Foundation of Korea(NRF) funded by the Ministry of Education, Science and Technology (2014-028219) (50%) and also Basic Science Research Program through the National Research Foundation of Korea(NRF) funded by the Ministry of Science, ICT and future Planning(2013R1A2A2A01068687) (50%).

References

1. Breazeal, C.: Social interactions in HRI: The robot view. IEEE T. Syst. Man. Cy. C: Applications and Reviews 34, 181–186 (2004)
2. Farrah, W., Kwang-Hyun, P., Dae-jin, K., Jin-Woo, J., Zeungnam, B.: Intention reading towards engineering applications for the elderly and people with disabilities. International Journal of ARM 7, 3–15 (2006)
3. Jansen, B.J., Booth, D.L., Spink, A.: Determining the informational, navigational, and transactional intent of Web queries. Inform. Process. Manag. 44, 1251–1266 (2008)
4. Ferreira, A., Celeste, C.W., Cheein, F.A., Bastos-Filho, T.F., Sarcinelli-Filho, M., Carelli, R.: Human-machine interfaces based on EMG and EEG applied to robotic systems. J. Neuroeng. Rehabil. 5, 10 (2008)
5. Goodrich, M.A., Schultz, A.C.: Human-robot interaction: A survey. Foundations and Trends in Human-Computer Interaction 1, 203–275 (2007)
6. Jaimes, A., Sebe, N.: Multimodal human-computer interaction: A survey. Comput. Vis. Image Und. 108, 116–134 (2007)
7. Jang, Y.M., Mallipeddi, R., Lee, S., Kwak, H.W., Lee, M.: Human intention recognition based on eyeball movement pattern and pupil size variation. Neurocomputing 128, 421–432 (2013)
8. Sun, J., Sun, X., Tong, S.: Phase synchronization analysis of EEG signals: An evaluation based on surrogate tests. IEEE Trans. Biomed. Eng. 59, 2254–2263 (2012)
9. Lachaux, J.-P., Rodriguez, E., Martinerie, J., Varela, F.J.: Measuring Phase Synchrony in Brain Signals. Hum. Brain Mapp. 8, 194–208 (1999)
10. Suzuki, H.: Phase relationships of alpha rhythm in man. Jpn. J. Physiol. 24, 569–586 (1974)
11. Nunez, P.: Electrical Fields of the Brain. Oxford University Press, Mass. (1981)
12. Gonuguntla, V., Wang, Y., Veluvolu, K.C.: Phase Synchrony in Subject-Specific Reactive Band of EEG for Classification of Motor Imagery Tasks. In: Proceedings of 35th Annual International IEEE EMBS Conference, pp. 2784–2787 (2013)
13. Park, U., Veluvolu, K.C., Lee, M.: Phase synchrony for human implicit intent differentiation. In: Proceedings of 20th Annual International IEEE ICONIP Conference, pp. 427–433 (2013)

Towards Establishing Relationships between Human Arousal Level and Motion Mass

Sven Nõmm[1], Tiit Kõnnusaar[2], and Aaro Toomela[2]

[1] Department of Computer Science, Tallinn University of Technology,
Akadeemia tee 15a, 12618, Tallinn, Estonia
sven.nomm@ttu.ee
[2] Institute of Psychology, Tallinn University,
Narva mnt, Tallinn, Estonia
Narva mnt. 25, 10120
tiit@print.ee, aaro.toomela@tlu.ee

Abstract. Preliminary results describing the relationship of the human arousal level with the amount and smoothness of their locomotion are reported in this paper. While there is a number of solid results indicating that in many cases arousal level may influence motor activities, measuring the strength, and modeling such relation remains relatively neglected area. The main weakness of the existing results is, that unlike the measurements of the arousal level which are described by the measured value of skin conductance, the locomotion parameters are determined on the basis of human observations and therefore contain certain degree of subjectiveness. Approach proposed in this paper targets to eliminate such subjectiveness. Trajectories of the limb joints will be recorded by the motion capture system. Amount and smoothness of the locomotion will be expressed by means of so-called *motion mass* parameters computed on the basis of recorded trajectories. Then relations between the arousal level and amount of locomotion will be studied.

Keywords: Neural activity, Arousal level, Motion Mass, Modeling, Electrodermal activity, Skin conductance.

1 Introduction

Pilot results, on the establishing relationship between the arousal level on the one side and the locomotion amount of the human limbs on the other side, are reported in this paper. Usually measured value of skin conductance (SC) is used to describe arousal level of a human being. It has been found that electro dermal activity is related to activation of several brain regions including those responsible for locomotion [1],[2]. This relationship is especially remarkable in performing tasks that require effort [3]. In clinical studies of intensive care the skin conductance value of the patients has been found to be related to motor activity [4]. Skin conductance is also related to everyday motor activities, such as drivers' brake pressure during driving [5]. Skin conductance and its relation to the different neuropsychological processes has been extensively studied [6], [7], [8].

C.K. Loo et al. (Eds.): ICONIP 2014, Part I, LNCS 8834, pp. 19–26, 2014.

While the SC were always measured by the special equipment, up to a recent time locomotion amount was assessed on the basis of human made observations and expressed in terms of some kind of motor activity assessment scale. In the best of the author's knowledge there are just a few contributions available where level of locomotion is described by means of objectively measured parameters. For example in [5] angle of the steering wheel and break pressure were measured.

Approach proposed in this paper targets to relate measured values of the skin conductance to the observed locomotion amount of the human limbs. Locomotion amount will be measured using motion capture system which will exclude any possible subjectiveness introduced by human made observations. The *Motion Mass* parameters, which provide numeric measure of the amount and the smoothness of the human limb movements [9], will be calculated on the basis of raw data recorded by motion capture system. Once arousal level and amount of the locomotion are described numerically one may formally describe the strength of the relationship and if possible build the model to estimate one parameter on the basis of the other. In [9] it was demonstrated that values of the *Motion Mass* parameters reflect changes observed during the learning of a new motor activity and in turn describe changes in the quality of motion planning. Human actions are usually target to achieve a certain goal [10], [11]. When motor activity is required, planning of the motions possesses a crucial importance in achieving the goal [12]. Obviously inadequate motions planning may not only be the obstacle to achieve the goal of the action but also lead to unwanted consequences like traumas etc. At the same time measuring the SC usually requires special equipment which in turn limits human movements. In this context ability to estimate arousal level on the basis of measured parameters of the motion may allow to detect higher arousal levels without imposing any limits on human locomotion. In [5] it was pointed out, that higher levels of arousal do not always reflected by the changes of motor activities and *vice versa*. This leads another direction of the research. Namely determine the types of arousal which influence amount and the smoothness of the the human motions.

The organization of the paper is as follows. Main goals of the paper are formulated in Section 2. Mathematical tools and experimental setting are described in Sections 3 and 4 respectively. Analysis of the achieved results is presented in Section 5. Concluding remarques are drawn in the last section.

2   Problem Formalisation

It has been found, as we showed above, that for large groups of individuals there is a relation between the arousal levels and locomotion parameters, at least for the arousals caused by certain stimuli types. It is usually assumed that group-level results apply to individual level as well. If this hypothesis proved true, then for the certain cases it would be possible to develop a model estimating value of the SC based on the measured amount and smoothness of the locomotion and in turn determine the arousal level based on the parameters of captured motions. Pilot study is required to determine which *Motion Mass* parameters provide better ground for modeling of the SC values, cluster individuals by their responsiveness to the different stimuli and analyze which stimuli cause arousal levels to influence locomotion activity. This leads following goals of the present research

- Cluster the group of individuals by the values of correlation coefficient between the chosen *Motion Mass* parameters and the amount of SC changes.
- For the individuals demonstrating strong correlation between the *Motion Mass* parameters and amount of SC change attempt to construct the model to estimate amount of SC change as a function of *Motion Mass* parameters.

3   Mathematical Tools

Main goal of the present research requires one to possess an ability to measure the both arousal level, and locomotion amount. While SC measurements are widely accepted to represent level of arousal, up to now there is no widely accepted technique to measure locomotion amount on the basis of captured motion data. In [9] the notion of the *Motion Mass* was proposed as the measure of the amount and smoothness of the movements associated with the motion or motor activity. In order to make this paper self-sufficient let us briefly remind the definition and meaning of this notion. In [9] *Motion Mass* is defined as the set of four parameters; *Trajectory Mass, Acceleration mass, Combined Euclidean Distance* and length of the motion in time.

$$M_J = \{T_J, A_J, E_J, t\}. \tag{1}$$

Denote $J = \{j_1, j_2, \ldots, j_n\}$ the set of joints describing certain limb or limbs. Let T_{j_i} be the length of the trajectory of the joint j_i, observed during the motion then *Trajectory Mass* is defined as the sum of the trajectory lengthes of each joint of the set J.

$$T_J = \sum_{i=1}^{n} T_{j_i}. \tag{2}$$

Acceleration Mass and *Combined Euclidean Distance* are defined in the similar way as follows

$$A_J = \sum_{i=1}^{n} A_{j_i}. \tag{3}$$

$$E_J = \sum_{i=1}^{n} E_{j_i} \tag{4}$$

Trajectory mass describes amount of the limb movements associated with motion and *acceleration mass* describes their smoothness. It was demonstrated in [9] that the values of the *Motion Mass* reflect changes of human motor functions while individual learning new motor activity. Therefore, those parameters are suitable as the measure of the locomotion amount. To compute numeric values of the *Motion Mass* parameters, actual trajectories of the limb joints are required. Motion capture system was used to record trajectories of the limb joints. In such setting all the measures will be collected by computerized systems, therefore human subjectiveness is excluded.

Unlike the measured value of SC, which is recorded for each instance of time, parameters of the *Motion Mass* are associated with time intervals. In order to compare SC to the *Motion Mass* parameters one has to choose from a two following alternatives. The first one is to introduce the analogue of the *Motion Mass* parameters for each time instance. The second one, is to derive the parameters describing amount and/or smoothness of the SC changes for a given time interval, similarly to those of the *Motion Mass* (1). Present contribution pursues the second alternative. Define the amount of the changes in SC associated with the certain time interval in the similar way to *Trajectory Mass*. Denote C_i amount of the changes of SC during time interval i. In the case of SC the amount of changes and smoothness would strongly correlate, therefore there is no sense to compute the last one.

4   Experimental Setting

For the pilot research, a group of 10 individuals was randomly chosen from a population of eighth grade adolescents. The entire range of the motion activities human may perform is too wide to be considered in a single paper, therefore results reported in the present contribution are narrowed to the studies of upper right limb motions of a seated individual. Performed activity was limited to the manipulation with the computer mouse. Individual was asked to play game and respond on different stimuli by mouse clicks. During the experiments the individual was exposed to the sequence of different stimuli, whereas irritating stimulus was always followed by the stimulus which usually has calming effect. For example one of the irritating stimuli was frightening and it was archived by showing the video of frightened cat. Calming stimulus was usually provided by the video of trees during the autumn. The sequence of stimuli consisted totally of a 21 interval. For each time interval SC, was recorded together with the trajectories of the right hand joints.

Experiment environment is presented in Figure 1. Recording time for each individual took 20 minutes. The hardware setting consists of two devices connected to the PC. Values of SC were measured and recorded by MP150WSW with GSR100C amplifier produced by BIOPAC Systems, Inc. SC was recorded in micro Siemens, which is standard measure for such experiments, whereas frequency was 200Hz. Low pass filter was applied then to smoothen the data and eliminate nonspecific electrodermal reactions (sparks). Within the frameworks of present contribution Kinect sensor was used to capture human motions and record trajectories of the limb joints. In spite of its simplicity it has proved itself to be precise enough to be applied in such delicate area like medicine [13]. After necessary processing, for each of 21 interval amount of SC changes and parameters of the *Motion Mass* were computed.

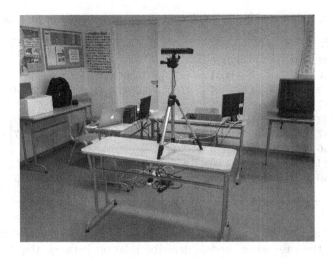

Fig. 1. Hardware setting

5   Analysis

Present research concentrate its attention on *Trajectory Mass* and *Acceleration mass*, which leads the square matrix, with the rows corresponding to the intervals and columns to the amount of SC change, *Trajectory Mass* and *Acceleration mass*. Pearson correlation coefficients r together with corresponding p−values

Table 1. Relation between *Trajectory Mass* and amount of SC changes

ID	r	p-value	Corresponding linear model
41	0.51	0.0272	
3	0.74	0.0001	$C(i) = 1.31T_J(i) + 7.76$
52	0.79	$2.2e - 05$	$C(i) = 0.07T_J(i) + 0.11$
57	0.44	0.0486	
42	0.81	$9.9e - 06$	$C(i) = 0.42T_J(i) + 0.04$
47	0.84	$1.4e - 06$	$C(i) = 0.2T_J(i) + 2.11$
46	0.53	0.0128	
24	0.82	$5.2e - 06$	$C(i) = 0.17T_J(i) + 0.42$
40	0.78	$3e - 05$	$C(i) = 0.09T_J(i) + 0.31$
11	0.74	0.0001	$C(i) = 0.39T_J(i) + 2.79$

computed between the amount of SC change and *Trajectory Mass*, for each individual, together with corresponding linear models wherever applicable, are presented in Table 1. Table 2 describes similar relation between the Acceleration mass and amount of the SC changes.

Let us now turn our attention to the standardized residuals. Figure 2 depicts standardized residuals computed for the models describing relation between the

Table 2. Relation between *Acceleration Mass* and amount of SC changes

ID	r	p-value	Corresponding linear model
41	0.51	0.0176	
3	0.73	0.0001	$C(i) = 0.001A_j(i) + 8.89989656$
52	0.86	$4.9e - 07$	$C(i) = 0.0002A_j(i) + 0.012423054$
57	0.50	0.0203	
42	0.84	$1.5e - 06$	$C(i) = 0.0007A_j(i) - 0.004897169$
47	0.83	$3.0e - 06$	$C(i) = 0.0003A_j(i) + 2.313445807$
46	0.53	0.0119	
24	0.84	$1.6e - 06$	$C(i) = 0.0002A_j(i) + 0.36474467$
40	0.78	$2.1e - 05$	$C(i) = 0.0002A_j(i) + 0.313465413$
11	0.76	$5.7e - 05$	$C(i) = 0.0005A_j(i) + 3.375591976$

Trajectory Mass and the amount of SC change and Figure 3 depicts standardized residuals for the case when models describe relation between the *Acceleration mass* and the amount of SC change.

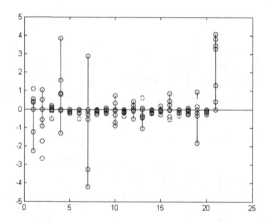

Fig. 2. Standardized residuals for the $C(i) = aT_J(i) + b$ type models

One may easily see that for many models standardized residuals corresponding to the computer task intervals 1, 2, 4, 7 and 21 are in absolute value greater than 2, which indicates that corresponding observation points may be outliers. Remind here that each interval corresponds to the different type of stimulus, therefore analysis of standardized residuals may allow to determine stimuli which cause arousal levels with lesser or greater influence on motor functions.

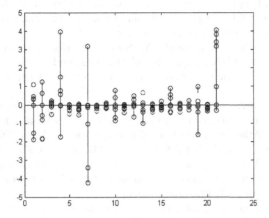

Fig. 3. Standardized residuals for the $C(i) = aA_J(i) + b$ type models

6   Conclusions

Results of the pilot research reported in this paper has clearly demonstrate that amount and smoothness of the locomotion was significantly related to the level of arousal in all cases. The strength of such relationship allows to divide individuals in to two groups. For those who demonstrate higher strengths of such relations models to estimate amounts of SC changes were developed. Studies of the corresponding standardized residuals has revealed that certain types of stimuli cause abnormal arousal levels. In other words, arousal types which either are not related to locomotion or influence locomotion too much. On the one hand, ability to relate machine measured locomotion parameters to the level of arousal provides an alternative approach to measure the last one without limiting human motor activities. On the other hand such ability allows to study in detail influence of the arousal level on the motions planning process.

Acknowledgments. S.Nõmm was supported by the Research Funding Project of the Tallinn University of Technology B37. A. Toomela was supported by the ETAG Institutional Research Funding project IUT 3-3.

References

[1] Frederikson, M., Furmak, T., Olsson, M.T., Fisher, H., Andersson, J., Långström, B.: Functional neuroanatomical correlates of electrodermal activity: A positron emission tomographic study. Psychophysiology 35, 179–185 (1998)

[2] MacIntosh, B.J., Mraz, R., McIlroy, W.E., Graham, S.J.: Brain activity during a motor learning task: An fmri and skin conductance study. Human Brain Mapping 28(12), 1359–1367 (2007)

[3] Mochizuki, G., Hoque, T., Mraz, R., MacIntosh, B., Graham, S., Black, S., Staines, W., McIlroy, W.: Challenging the brain: Exploring the link between effort and cortical activation. Brain Research 1301, 9–19 (2009)

[4] Günther, A., Bottai, M., Schandl, A., Storm, H., Rossi, P., Sackey, P.: Palmar skin conductance variability and the relation to stimulation, pain and the motor activity assessment scale in intensive care unit patients. Critical Care 17(2), 1–7 (2013)

[5] Helander, M.: Applicability of drivers' electrodermal response to the design of the traffic environment. Journal of Applied Psychology 63(4), 481–488 (1978)

[6] Roy, J., North Atlantic Treaty Organization. Scientific Affairs Division: Progress in electrodermal research. NATO ASI series: Life sciences. Plenum Press (1993)

[7] Alvarsson, J.J., Wiens, S., Nilsson, M.E.: Stress recovery during exposure to nature sound and environmental noise. International Journal of Environmental Research and Public Health 7(3), 1036–1046 (2010)

[8] Boucsein, W.: Electrodermal Activity. The Springer series in behavioral psychophysiology and medicine. Springer (2012)

[9] Nõmm, S., Toomela, A.: An alternative approach to measure quantity and smoothness of the human limb motions. Estonian Journal of Engineering 19(4), 298–308 (2013)

[10] Anokhin, P.K.: Ocherki po fiziologii funktsionalnykh sistem. Medicina, Moskow (1975) (in Russian)

[11] Toomela, A.: Biological roots of foresight and mental time travel. Integrative Psychological and Behavioral Science 44(2), 97–125 (2010)

[12] Lauria, A.: Vyshije korkovyje funktsii tsheloveka i ikh narushenija pri lokal'nykh porazenijakh mozga (Higher cortical functions in man and their disturbances in local brain lesions). Moscow University Press, Moscow (1969) (in Russian)

[13] Nomm, S., Buhhalko, K.: Monitoring of the human motor functions rehabilitation by neural networks based system with kinect sensor. In: Proc. of the 12th IFAC Symposium, Analysis, Design, and Evaluation of Human-Machine Systems, Las-Vegas, Nevada, USA, pp. 249–253 (2013)

Estimating Nonlinear Spatiotemporal Membrane Dynamics in Active Dendrites

Toshiaki Omori

Department of Electrical and Electronic Engineering,
Graduate School of Engineering, Kobe University
1–1, Rokkodai-cho, Nada-ku, Kobe 657–8501, Japan
omori@eedept.kobe-u.ac.jp
http://www2.kobe-u.ac.jp/~omoritos/

Abstract. Recent advances in measurement technology enables us to obtain spatotemporal data from neural systems as imaging data. In this study, we propose a statistical method to estimate nonlinear spatiotemporal membrane dynamics of active dendrites. We formulate generalized state space model of active dendrite, based on multi-compartment model. Membrane dynamics and its underlying electrical properties are simultaneously estimated by using sequential Monte-Carlo method and EM algorithm. Using the proposed method, we show that nonlinear spatiotemporal dynamics in active dendritic can be extracted from partially observable data.

Keywords: Multi-compartment model, Dendrite, Spatiotemporal dynamics, Probabilistic time-series analysis.

1 Introduction

Recent findings such as dendritic spikes and backpropagations suggest that dendrite contributes more important roles in neural information processings in our brain [1,2,3,4,5,6]. For example, experimental results showed that dendritic processing plays a key role in directional selectivity for visual stimuli [7,8]. However, the mechanism of dendritic spatiotemporal information processings remain unclear.

Great advances in measurement technology enables us to deal with spatiotemporal data from neural systems including dendrites as imaging data. However, the observable information in the measurements are limited, compared with the complexity of the entire neural system. Some estimation techniques are proposed using the state space modeling approach to extract the spatiotemporal dynamics of the dendrites [9,10,11,12]. In some of previous methods, only membrane potentials are estimated while assuming the parameters underlying spatiotemporal dynamics are known [10], and most of previous methods only consider the estimation of linear dynamics in multi-compartment models or nonlinear dynamics in single-compartment models, although it is important to establish nonlinear dynamics for spatiotemporal membrane evolution in multi-compartment models to reveal dendritic information processings [9,11,12].

C.K. Loo et al. (Eds.): ICONIP 2014, Part I, LNCS 8834, pp. 27–34, 2014.
© Springer International Publishing Switzerland 2014

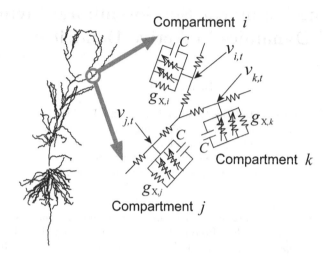

Fig. 1. A schematic diagram of multi-compartment model. Neuron has spatially-extended structure in dendrite. In the multi-compartment model, membrane electrical response at each position is described by using a compartment; each compartment has passive and active channels where the dynamics of membrane potential and channel variables obeys conductance-based model. Each compartment is connected to other compartments according to the neuronal morphology.

In this study, we propose a statistical method to estimate nonlinear spatiotemporal membrane dynamics of active dendrites in order to extract nonlinear dynamics of dendritic membrane. We employ a framework of probabilistic information processing to extract the nonlinear spatiotemporal dynamics from partially observable data. First, we formulate generalized state space model of active dendrite, based on multi-compartment model. Next, sequential estimation algorithm is derived for the generalized state space model. Estimation of membrane dynamics and its underlying electrical properties are simultaneously estimated by using sequential Monte-Carlo method and EM algorithm. Using the proposed method, we show that nonlinear spatiotemporal dynamics in active dendrites can be extracted from partially observable data.

2   State-Space Modeling of Nonlinear Spatiotemporal Dynamics in Dendrite

In this section, we formulate a generalized state-space model of dendritic dynamics in probabilistic manner. We first derive a system model, which describes spatiotemporal nonlinear dynamics of dendrite. Next we formulate an observation model, which reflects partially observable situation seen in imaging experiments.

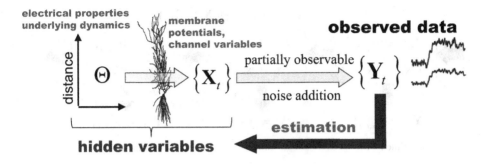

Fig. 2. A framework to estimate the nonlinear spatiotemporal dynamics of dendritic membrane. The hidden variables $\{X_t\}$ including membrane potentials ($\{v_{X,x}\}$) and channel variables ($\{m_{X,x}\}$, $\{h_{X,x}\}$), and those underlying electrical properties Θ are simultaneously estimated from partially observable data $\{Y_t\}$.

2.1 System Model

Dynamics of Membrane Potential. A membrane potential $v_{x,t}$ at compartment x and time t is assumed to obey the following differential equation:

$$C\frac{dV_x}{dt} = -\sum_X g_{X,x}(m_{X,x})^{M_X}(h_{X,x})^{N_X}(V_x - E_X) - g_{\text{axial}}\sum_y (V_x - V_y)$$
$$+ I_{\text{ext},x} + \xi_x^{(V)}(t) \tag{1}$$

where $m_{X,x}$ and $h_{X,x}$ in the first term show the activation and inactivation channel variables, respectively. Each compartment is assumed to have some kinds of membrane currents $\sum_X g_{X,x} m^{M_X} h^{N_X}(V_x - E_X)$, axial currents $g_{\text{axial}}\sum_{y \in N_x}(V_x - V_y)$, external input currents $I_{ext,x}$ and noise current $\xi_x^{(V)}(t)$. The maximal membrane conductances and reversal potentials and membrane capacitance are expressed by $g_{X,x}$, E_X and C, respcetively. By discretizing Eq. (1) with respect to time, we derive the following equation.

$$v_{x,t+1} = v_{x,t} - \Delta\sum_X g_{X,x}(m_{X,x,t})^{M_X}(h_{X,x,t})^{N_X}(v_{x,t} - E_X)$$
$$- \Delta g_{\text{axial}}\sum_y (v_{x,t} - v_{y,t}) + \Delta I_{\text{ext}} + \Delta\xi_{x,t}^{(v)} \tag{2}$$

where time width is set to Δ and we put $C = 1$ without loss of generality.

Based on the statistics of noise, probabilistic density function of membrane potential is described by probabilistic model $p(v_{x,t+1}|\{v_{x,t}\}, m_{X,x,t}, h_{X,x,t})$ which depends on the state at preceding time. If the noise obeys white Gaussian noise, the probabilistic density function can be described by

$$p(v_{x,t+1}|\{v_{x,t}\}, m_{X,x,t}, h_{X,x,t}) = \mathcal{N}(v_{x,t+1}|\mu_{x,t+1}, \sigma_v^2) \tag{3}$$

where the average is expressed by $\mu_{x,t+1} = v_{x,t} - \Delta\sum_X g_{X,x}(m_{X,x,t})^{M_X}(h_{X,x,t})^{N_X}$ $(v_{x,t} - E_X) - \Delta g_{\text{axial}}\sum_y (v_{x,t} - v_{y,t}) + \Delta I_{\text{ext}}$, and the variance by σ_v^2.

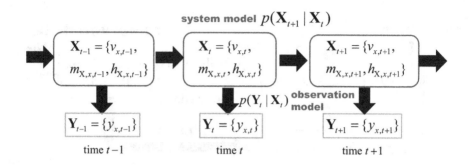

Fig. 3. Graphical model of generalized state-space model of dendritic nonlinear spatiotemporal dynamics in the proposed model. Hidden states $\boldsymbol{X}_t$ including spatiotemporal distribution of membrane potentials $v_{x,t}$ and that of channel variables $m_{X,x,t}$ and $h_{X,x,t}$ show time evolutions from time t to $t+1$, based on the system model $p(\boldsymbol{X}_{t+1}|\boldsymbol{X}_t)$. Within multi-dimensional hidden variables $\boldsymbol{X}_t$, only partial information with lower dimension are observed as $\boldsymbol{Y}_t$, based on the observation model $p(\boldsymbol{Y}_t|\boldsymbol{X}_t)$.

Dynamics of Channel Variables. In the conductance-based models such as the Hodgkin-Huxley model, channel variables $m_{X,x,i}$ and $h_{X,x,i}$ obey the following first-order kinetics:

$$\frac{dm_{X,x}}{dt} = \alpha_{m_X}(V_x)(1 - m_{X,x}) - \beta_{m_X}(V_x)m_{X,x} + \xi_x^{(m_X)}(t) \qquad (4)$$

$$\frac{dh_{X,x}}{dt} = \alpha_{h_X}(V_x)(1 - h_{X,x}) - \beta_{h_X}(V_x)h_{X,x} + \xi_x^{(h_X)}(t) \qquad (5)$$

where $\alpha_{m_X}(V_x)$, $\alpha_{h_X}(V_x)$, $\beta_{m_X}(V_x)$, and $\beta_{h_X}(V_x)$ are functions of membrane potential. By descretizing Eq. (4) and (5) with respect to time, we obtain the probabilistic model of the channel variables: $p(m_{X,x,t+1}|m_{X,x,t}, v_{x,t})$ and $p(h_{X,x,t+1}|h_{X,x,t}, v_{x,t})$. If noise obeys white Gaussian noise, the probabilistic density function can be described by

$$p(m_{X,x,t+1}|m_{X,x,t}, v_{x,t}) = \mathcal{N}(m_{X,x,t+1}|\mu_{m_X,x,t+1}, \sigma_{X_m}^2) \qquad (6)$$

$$p(h_{X,x,t+1}|h_{X,x,t}, v_{x,t}) = \mathcal{N}(h_{X,x,t+1}|\mu_{h_X,x,t+1}, \sigma_{X_h}^2) \qquad (7)$$

Based on probabilistic models (Eqs. (3), (6) and (7)), the system models for all the hidden state vectors $\boldsymbol{X}_t = \{v_{x,t}, m_{X,x,t}, h_{X,x,t}\}$ are summarized as $p(\boldsymbol{X}_{t+1}|\boldsymbol{X}_t)$.

2.2   Observation Model

Each compartment has a multi-dimensional state such as membrane potential $v_{x,t}$ and channel variables $m_{X,x,t}$ and $h_{X,x,t}$ for each type of ion channel X. We assume the situation that only membrane potential can be observed; the observed variable $y_{x,t}$ is assumed to be expressed as follows:

$$y_{x,t} = g(v_{x,t}) + \xi_{x,t}^{(y)} \tag{8}$$

where $g(\cdot)$ is observation function of true membrane potential $v_{x,t}$, and $\xi_{x,t}^{(y)}$ is an observation noise. According to the statistics of the noise, we can derive probabilistic version of observation model $p(y_{x,t}|v_{x,t})$. If the observation noise obeys white Gaussian noise, the probabilistic density function is described by

$$p(y_{x,t}|v_{x,t}) = \mathcal{N}(y_{x,t}|g(v_{x,t}), \sigma_y^2). \tag{9}$$

The observation model for the entire multi-compartment model is expressed as $p(\boldsymbol{Y}_t|\boldsymbol{X}_t)$.

3   Estimation of Hidden Variables

Here we describe the method to estimate latent variables $\{\boldsymbol{X}_t\}$ from observable data $\{\boldsymbol{Y}_t\}$. Hidden variables at time t, $\boldsymbol{X}_t$, is estimated using the observable data up to the same time $\boldsymbol{Y}_{1:t}$ based on the filtering distribution as follows:

$$p(\boldsymbol{X}_t|\boldsymbol{Y}_{1:t}) = \frac{p(\boldsymbol{Y}_t|\boldsymbol{X}_t)p(\boldsymbol{X}_t|\boldsymbol{Y}_{1:t-1})}{\int p(\boldsymbol{Y}_t|\boldsymbol{X}_t)p(\boldsymbol{X}_t|\boldsymbol{Y}_{1:t-1})d\boldsymbol{X}_t} \tag{10}$$

where $p(\boldsymbol{X}_t|\boldsymbol{Y}_{1:t-1})$ shows a predictive distribution given the observable data up to the previous time $\boldsymbol{Y}_{1:t-1}$ as follows:

$$p(\boldsymbol{X}_t|\boldsymbol{Y}_{1:t-1}) = \int p(\boldsymbol{X}_t|\boldsymbol{X}_{t-1})p(\boldsymbol{X}_{t-1}|\boldsymbol{Y}_{1:t-1})d\boldsymbol{X}_{t-1} \tag{11}$$

Since we have assumed that the nonlinearity in the generalized state space model, the integrations in both distributions become analytically intractable. In the present study, we employ sequential Monte-Calro method to tackle this difficulty and perform the filtering and prediction iteratively.

In addition to hidden variables of $\boldsymbol{X}_t$, a set of parameters Θ including maximal membrane conductances $\{g_{X,x}\}$ is unknown. The EM algorithm [13] is employed in order to estimate those parameters underlying the nonlinear spatiotemporal dynamics. In the E-step, expectation of log-likelihood function is calculated

$$Q(\Theta|\Theta_k) = \langle \log p(\{\boldsymbol{X}_t\}, \{\boldsymbol{Y}_t\}|\Theta) \rangle_{p(\{\boldsymbol{X}_t\}|\{\boldsymbol{Y}_t,\},\Theta_k)} \tag{12}$$

where Θ_k shows a set of parameters estimated at step k of the EM algorithm. In the M-step, we obtain the set of parameters Θ which maximizes $Q(\Theta|\Theta_k)$ as Θ_{k+1},

$$\Theta_{k+1} = \arg\max_{\Theta} Q(\Theta|\Theta_k) \tag{13}$$

By performing the E-step and the M-step iteratively, we employ the converged value of Θ_k as estimated parameters.

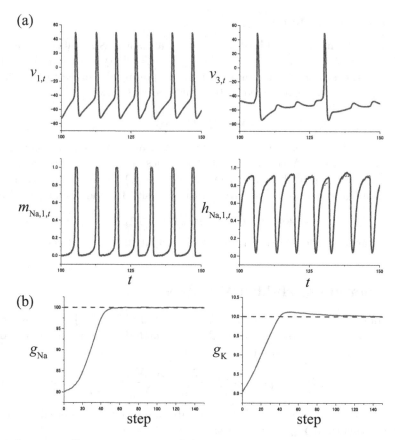

Fig. 4. Latent nonlinear spatiotemporal dynamics extracted by the proposed method. Hidden variables $\{\boldsymbol{Y}_t\} = \{v_{x,t}, m_{\mathrm{X},x,t}, h_{\mathrm{X},x,t}\}$ for membrane potentials and channel variables and Θ for electrical properties are simultaneously estimated. (a) estimated membrane potentials and channel variables. The estimated membrane potentials at compartments 1 and 3 ($v_{1,t}$ and $v_{3,t}$) are shown in top two figures, whereas the estimated sodium activation and inactivation variables ($m_{\mathrm{Na},1,t}$ and $h_{\mathrm{Na},1,t}$) are shown in the bottom two figures. Estimated membrane potentials at multiple compartments (red solid lines) show similar behavior to the true ones (dashed blue lines). Furthermore, estimated channel variables (both activation and inactivation) are similar to the true ones. (b) estimated underlying electrical properties Θ. Maximal membrane conductances of sodium and potassium currents (g_{Na} and g_{K}) are simultaneously estimated from observed data. As estimation step proceeds, estimated conductances (solid red lines) converges to the true conductances (dashed blue lines).

4   Results

In this section, we evaluate the effectiveness of the proposed method by using simulated data of multi-compartment model with active channels. We assume that each compartment has passive and active channels, and only noisy membrane potentials are partially observable. By using the proposed method, we estimate not only membrane potentials but also other hidden variables such as activation and inactivation variables for active channels for every compartments. Furthermore, we also estimate underlying parameters including maximal membrane conductances of active channels, which govern the nonlinear spatiotemporal dynamics of dendritic membrane.

4.1   Estimation of Membrane Potentials and Channel Variables

Here we extract true membrane potentials $v_{x,t}$ and channel variables of sodium current m_{Na}, h_{Na} and potassium current m_K, h_K. The estimated time evolution of these hidden variables are shown in Fig. 4(a). We find that true membrane potential at each compartment x can be estimated accurately. Furthermore, non-observable other hidden variables such as sodium channels can be estimated as shown in the bottom figures in the Fig. 4(a). Hidden variables for potassium channels can be estimated as well (data not shown). These results suggest that proposed method enables us to extract hidden variables under nonlinear spatiotemporal dynamics of dendritic membrane.

4.2   Estimation of Electrical Properties Governing Nonlinear Spatiotemporal Dynamics

Electrical properties such as maximal membrane conductances should be estimated since these properties govern nonlinear spatiotemporal dynamics of dendritic membrane potentials and channel variables. The estimated maximal membrane properties are shown in Fig. 4(b). In this results, not only electrical properties but also membrane potentials and channel variables are simultaneously estimated. In Fig. 4(b), we find that estimated maximal membrane conductances converge to true value (dashed line). These results show that the proposed method can estimate nonlinear spatiotemporal dynamics of dendritic membrane.

5   Concluding Remarks

In this study, we have proposed a statistical method to estimate spatiotemporal membrane dynamics of active dendrites. Generalized state space model of active dendrite has been derived based on multi-compartment model of dendrites. A novel spatiotemporal dynamics extraction technique has been realized by using sequential Monte-Carlo method and EM algorithm. Using the proposed method, we have shown that inner state of neurons such as membrane potential and ion channel variables, and those underlying parameters are simultaneously estimated. These results show that nonlinear spatiotemporal dynamics in active dendritic can be extracted from partially observable data by means of the proposed method.

Acknowledgments. The author is grateful to Prof. Masato Okada, Prof. Koji Hukushima of the University of Tokyo, Prof. Toru Aonishi of Tokyo Institute of Technology, and Prof. Seiichi Ozawa and Prof. Jun Kitazono of Kobe University for valuable comments on this study. This work was partially supported by Grants-in-Aid for Scientific Research on Innovative Areas (No. 25120010) and for Young Scientists (No. 25730147) from the Ministry of Education, Culture, Sports, Science and Technology of Japan.

References

1. Häusser, M.: Diversity and dynamics of dendritic signaling. Science 290, 739–744 (2000)
2. Stuart, G., Spruston, N., Häusser, M.: Dendrites, 2nd edn. Oxford University Press (2007)
3. Spruston, N.: Pyramidal neurons: Dendritic structure and synaptic integration. Nature Reviews Neuroscience 9, 206–221 (2008)
4. Shepherd, G.M.: The synaptic organization of the brain. Oxford University Press (2003)
5. Omori, T., Aonishi, T., Miyakawa, H., Inoue, M., Okada, M.: Estimated distribution of specific membrane resistance in hippocampal CA1 pyramidal neuron. Brain Research 64, 199–208 (2006)
6. Omori, T., Aonishi, T., Miyakawa, H., Inoue, M., Okada, M.: Steep decrease in the specific membrane resistance in the apical dendrites of hippocampal CA1 pyramidal neurons. Neuroscience Research 64, 83–95 (2009)
7. Yates, D.: Dendritic processors. Nature Reviews Neuroscience 15, 815 (2013)
8. Sivyer, B., WIlliams, S.R.: Direction selectivity is computed by active dendritic integration in retinal ganglion cells. Nature Neuroscience 16, 1848–1856 (2013)
9. Huys, Q.J., Paninski, L.: Smoothing of, and parameter estimation from, noisy biophysical recordings. PLoS Computational Biology 5, 1–15 (2009)
10. Paninski, L.: Fast Kalman filtering on quasilinear dendritic trees. Journal of Computational Neuroscience 28, 211–228 (2010)
11. Omori, T., Aonishi, T., Okada, M.: Statistical estimation of non-uniform distribution of dendritic membrane properties. Advances in Cognitive Neurodynamics 3, 649–655 (2013)
12. Kitazono, J., Omori, T., Aonishi, T., Okada, M.: Estimating membrane resistance over dendrite Using Markov Random field. IPSJ Transaction on Mathematical Modeling and Its Applications 5, 89–94 (2012)
13. Bishop, C.M.: Pattern Recognition and Machine Learning. Springer (2006)

Inter Subject Correlation of Brain Activity during Visuo-Motor Sequence Learning

Krishna Prasad Miyapuram[1], Ujjval Pamnani[1], Kenji Doya[2], and Raju S. Bapi[3]

[1] Cognitive Science Program, Indian Institute of Technology, Gandhinagar, India
{kprasad,ujjval.pamnani}@iitgn.ac.in
[2] Neural Computation Unit, Okinawa Institute of Science and Technology, Okinawa, Japan
doya@oist.jp
[3] Center for Neural & Cognitive Sciences,
Department of Computer and Information Sciences, University of Hyderabad, India
bapics@uohyd.ernet.in

Abstract. Brain imaging using functional MRI allows us to understand brain function while participants are engaged in meaningful tasks. Traditionally the experimental paradigms have been limited to repeated presentation of stimuli to participants followed by a model-based analysis of the data. The Inter Subject Correlation (ISC) analysis allows a model-free analysis while participants are presented with naturalistic stimuli such as watching a movie. We extend the ISC approach to a learning paradigm in which participants are repeatedly performing a motor sequence in response to visual stimuli. We qualitatively compare the correlation results across learning sessions. The preliminary result we observe is shift of correlation activity in cerebellum across sessions. A model-based analysis identifying task related activity compared to baseline is also reported.

Keywords: inter subject correlation analysis, visuomotor, sequence learning.

1 Introduction

Functional magnetic resonance imaging (fMRI) allows us to measure brain activations corresponding to specific cognitive phenomena while participants are engaged in a particular task [1]. Design and analysis of such experiments are based on the cognitive subtraction technique, i.e. task-related activation is typically identified by comparing against activation in a baseline condition. The corresponding analysis of fMRI data is a model-based technique. For example, a general linear model (GLM) approach can be used to specify the task conditions that the participant was presented. For detecting the brain activation, i.e. to improve the signal to noise ratio, fMRI experiments require the experimental trials of the task to be repeated a number of times. It is assumed that the repeated trials are similar to each other. Hence in the parlance of signal processing, averaging over multiple experimental trials would yield a good fMRI signal. Further, in order to generalize the inferences from brain imaging experiments to the population, the experiments are collected over a number of subjects.

C.K. Loo et al. (Eds.): ICONIP 2014, Part I, LNCS 8834, pp. 35–41, 2014.
© Springer International Publishing Switzerland 2014

A random effects analysis with the subjects as a random factor is used to make inferences.

There exist several model-free methods for analyzing fMRI data, such as the Independent Component Analysis (ICA) [2]. A more recently developed data driven method is the Inter-Subject Correlation (ISC) analysis [3]. Inter subject correlation (ISC) analysis aims to quantify to what extent brains of different individuals operate in a similar manner [3]. Previous studies in ISC analysis have used naturalistic stimuli identically across participants. These paradigms are not amenable to a model-based analysis such as the General Linear Model because of the complex nature of the stimuli and the experimental conditions are not repeatedly presented as typically done in fMRI experiments. We present a model-free analysis using Inter-Subject Correlation in a visuo-motor sequence learning task [4, 5]. Because ISC is a model-free approach, it has been applied to naturalistic stimuli such as watching a dance [6], watch a movie [7], spoken & written narratives [8], speech comprehension [9], real-life risk communication [10], action observation [11] etc. Through ISC analysis, we can find shared hemodynamic activity in the brain across subjects during the experimental task. Basically, it finds correlation coefficients between fMRI time series of the participants in corresponding brain regions. One study [12] investigated intra-subject correlations by repeatedly presenting stimuli in order to test the reliability of hemodynamic activity in natural viewing. Pajula et al. [13] have validated the ISC approach with that of a stimulus – model based analysis and found the same foci of hemodynamic activity. Moreover, it may also give us a cursory look of co-activation in different brain regions while performing the task.

The present research extends ISC analysis for analyzing a block-design fMRI experiment in which the participants repeatedly performed visuo-motor sequence learning. When analyzing tasks that involve learning, the experimental trials are all not similar to each other. Hence, the assumptions of the GLM do not strictly hold for learning paradigms. It is also known that the corresponding brain activity would shift between different regions as the learning progresses. The inter-session differences in such tasks can be interpreted in the context of learning-related changes in brain activation.

2   Design of Experiment

Eight participants performed a visuomotor sequence learning experiment [5]. The Task condition required participants to learn, by trial and error, the correct order of pressing two keys corresponding to two colored circles presented simultaneously on the screen. Six such sets were presented. There were four possible colors – red, green, blue, and yellow. The stimuli could appear in four possible positions – up, down, left, right. The responses were recorded on a keyboard with similar spatial configuration. The order of keys depended on the color of the stimuli, which remained fixed throughout the experiment. The positions at which the stimuli were presented was randomized every trial (see Fig. 1). The response was made depending on the position of the stimuli. An example is given here. One set of stimuli containing blue and red circles is presented simultaneously at the left and bottom positions on the screen. Participants have to discover the correct order of these two stimuli by trial and error. Let's say the correct order was blue followed by red. Now the participant successively presses the two buttons corresponding to the positions at which the blue and red stimuli were presented i.e. left followed by bottom keys. This is done for six such sets. If an error is made, then a flash appears on the screen and participants repeat the sequence from the first set. Upon completion of the sequence, the trial is repeated in a block until fixed duration of 36 sec. There were a total of four learning sessions, each with six blocks of sequence task. The baseline condition was alternating with the sequence task, in which one colored circle was presented randomly at one of the four positions. The participants had to simply press the corresponding button on the keypad. There were a total of seven baseline blocks per session each of duration 18 sec and every session began and ended with the baseline condition.

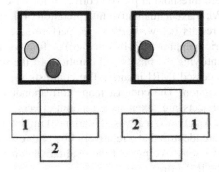

Fig. 1. Color Sequence Learning Task. In one trial, a set of two colored circles blue and red are presented on the screen. In another trial, the same set is displayed at a different position. Participants press the button corresponding to blue circle first followed by the button corresponding to the red circle using a keypad with four keys placed in the same spatial configuration.

3   Data Collection and Analysis

Functional images were collected in a 1.5 T whole-body scanner (Shimadzu-Markoni Magnex Eclipse). A time series of 228 whole-brain scans were obtained for each experiment composed of four sessions. In each scan, a set of 50 axial T2* weighted gradient-echo echo-planar images[repetition time (TR) 6000 ms, echo time (TE) 55 ms, Flip angle (FA) 90, matrix 64 × 64, Field of view (FOV) 192 × 192 mm and slice thickness of 3 mm] covering the whole-brain were collected parallel to the anterior commissure - posterior commissure (AC-PC) line. In addition, a high-resolution T1-weighted anatomical brain image consisting of 191 sagittal slices (TR 12 ms, TE 4.5 ms, FA 20, matrix 256×256, FOV 256×256 mm and slice thickness of 1 mm) was collected for each subject. The ethics committee of the Brain Activity Imaging Center (BAIC), Advanced Telecommunications Research Institute International (ATR), Kyoto, Japan approved the experimental protocol.

Images were preprocessed with SPM8 [14]. The preprocessing for each subject was done using the following procedure. Images were corrected for head movements (realignment). The normalization of images to template was done using the following procedure. First, the structural image was coregistered with the first functional image. The structural image was used to calculate the normalization parameters after segmentation into gray matter, white matter and cerebro-spinal fluid volumes. These normalization parameters were applied for all functional images. An isometric 3D Gaussian kernel with a full-width at half maximum of 8 mm was used for smoothing, as the final step of preprocessing the fMRI data.

We use the toolbox for inter-subject correlation analysis of fMRI developed by Kauppi et. al. [15] .The four sessions of fMRI data were entered separately into a single ISC analysis. The Pearson correlation coefficient (r) of fMRI time series was calculated for all pairs of subjects and the average correlation coefficient was taken at every voxel. The correlation maps were threshold at $p<0.05$ corrected for false discovery rate (FDR) at full frequency band. We make a qualitative interpretation of the correlations observed for each session in the results below. Further, we performed a model-based analysis using statistical parametric mapping (SPM8 software) as follows. First, a general linear model was specified identifying the onsets and durations of the sequence learning task for each subject's preprocessed fMRI images. The contrasts were specified to identify brain activity for each session. The contrast maps corresponding to each of the four sessions from all the subjects are taken to a random-effects group analysis. In this second-level analysis, an ANOVA (Analysis of Variance) model was implemented. The final group-level results are identified at a relatively liberal threshold of $p<0.001$, uncorrected for multiple comparisons. We present the qualitative comparison of the two approaches for analyzing the fMRI time series.

4 Results

In the following, we use the words correlation and activation interchangeably. Inter Subject correlations (see Fig. 2) in session 1 were found in the posterior cerebellum, medial orbitofrontal/ ventromedial prefrontal cortex, anterior striatum, dorsolateral prefrontal cortex, and parts of temporal, parietal and occipital lobes accompanied by extensive correlation in cortical motor areas. In session 2, the cerebellar correlations were found more in the anterior and dorsal regions. Other regions with high correlations were medial orbitofrontal/ ventromedial prefrontal cortex, posterior portions of dorsal striatum, frontopolar areas, and portions of temporal, parietal, and occipital regions concentrated medially. Interestingly, the correlations in cortical motor regions found extensively in session 1 were found to be negative in session 2. In session 3, cerebellar activity continued to be localized in the anterior and dorsal portions. The correlations in orbitofrontal cortex extended to lateral regions also. The ventromedial prefrontal correlations were also present. Correlations in visual areas were more concentrated to medial portion. Correlations in cortical motor areas were observed more laterally. In session 4, the cerebellar correlations were localized more towards anterior region. The correlations in ventral frontal, temporal and parietal areas were extensively found in lateral regions. Striatum and medial prefrontal correlations were also found. The lateral cortical activity in motor areas was persistent in session 4.

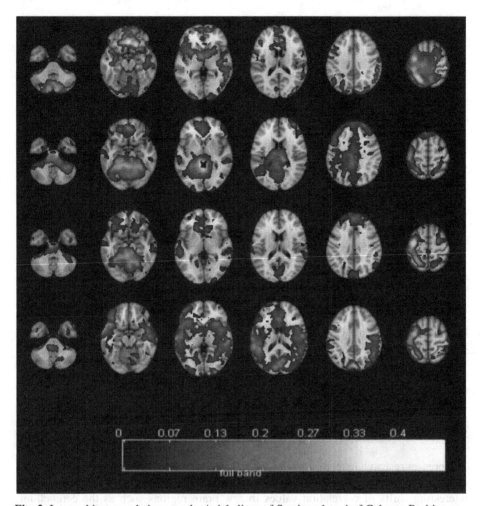

Fig. 2. Inter subject correlations result. Axial slices of Sessions 1 to 4 of Color to Position sequence learning task are shown as rows with slices at Z = -40, -16, 0, 16, 36, 60 mm respectively, shown in columns. Values depicted are the average correlation values (Pearson coefficient) that survived the significance threshold of p<0.05 FDR corrected (full frequency band).

When compared to the ISC analysis, the General linear model analysis revealed fewer locations of brain activity even at a relatively liberal threshold of p<0.001. For a qualitative comparison we have depicted the results at the same brain slices as the results of ISC analysis. We notice that the extent of correlation-based activity was much larger throughout different areas of the brain compared to the model-based analysis. The model based-analysis however did identify more specific brain regions activated (see the sagittal slice in Fig. 3).

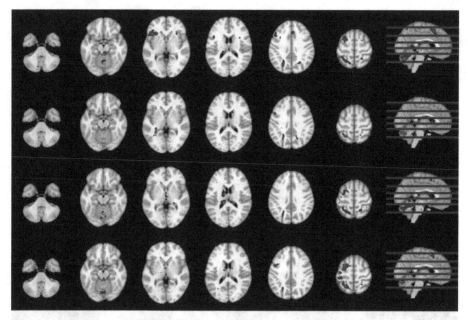

Fig. 3. General linear model result - axial slices of Sessions 1 to 4 of Color to Position sequence learning task are shown as rows at the same locations as Fig. 2. Additionally a mid-sagittal view is included with overlaid activations. Values depicted are the scores of statistical parameter estimated from the GLM analysis that survived threshold of p<0.001, uncorrected for multiple comparisons.

5   Discussion and Conclusion

The present research extends a model-free approach of inter-subject correlation analysis to a visuo-motor sequence learning paradigm. We qualitatively identified some interesting shifts of correlation values in few brain regions such as the cerebellum across different learning sessions. The correlation maps had a greater extent of activations compared to the model-based analysis. This is possibly because of the block design of the experiment with alternating baseline and sequence learning. Future work can aim at delineating the time series from sequence learning blocks alone for ISC analysis. It can be extended to a qualitative comparison of the ISC approach to other model-free approaches such as the Independent Component Analysis.

One limitation of our experimental paradigm is that the task performed required a motor action on behalf of participants. Different participants will have different rate of learning. Hence, unlike paradigms of naturalistic stimuli, our trial and error based learning experiment is not identical across participants. While the model-based analysis is specifically task-related activity corresponding to visuo-motor sequence learning compared to baseline, the ISC analysis performs a time series comparison across the entire experimental session i.e. including both baseline and task. The comparison of the model-free and model-based approaches allows us to conclude that there is information spread across a number of brain regions, as revealed by the ISC analysis. We speculate that this correlation activity represents information that perhaps is sub-threshold in a traditional General Linear Model. With this limitation, this research extends ISC analysis to experiments beyond naturalistic stimuli.

References

1. Huettel, S., Song, A., McCarthy, G.: Functional Magnetic Resonance Imaging 2nd edn. Sinauer, Massachusetts (2009)
2. Brown, G., Yamada, S., Sejnowski, T.: Independent component analysis at the neural cocktail party., 54-63 (January 2001)
3. Hasson, U., Nir, Y., Levy, I., Fuhrmann, G., Malach, R.: Intersubject Synchronization of Cortical Activity During Natural Vision., 1634 (2004)
4. Chandrasekhar Pammi, V., Miyapuram, K., , A., Samejima, K., Bapi, R., Doya, K.: Changing the structure of complex visuo-motor sequences selectively activates the fronto-parietal network. NeuroImage 59 (2011)
5. Miyapuram, K. P.: Visuomotor Mappings and Sequence Learning: A Whole-Brain fMRI Investigation. M.Tech. Thesis, University of Hyderabad, India (2004)
6. Jola, C., McAleer, P., Grosbras, M., Love, S., Morison, G., Pollick, F.: Uni- and multisensory brain areas are synchronised across spectators when watching unedited dance recordings., 265-84 (June 2013)
7. Kauppi, J.-P., Jääskeläinen, I., Sams, M., Tohka, J.: Inter-subject correlation of brain hemodynamic responses during watching a movie: localization in space and frequency. (March 2010)
8. Regev, M., Honey, C., Simony, E., Hasson, U.: Selective and Invariant Neural Responses to Spoken and Written Narratives. The Journal of Neuroscience 33(40), 15978-15988 (October 2013)
9. Wilson, S., Molnar-Szakacs, I., Iacoboni, M.: Beyond Superior Temporal Cortex:Intersubject Correlations in Narrative Speech Comprehension. Cerebral Cortex 18, 230-242 (January 2008)
10. Schmalzle, R., Hacker, F., Renner, B., Honey, C., Schupp, H.: Neural Correlates of Risk Perception during Real-Life Risk Communication. The Journal of Neuroscience 33(25), 10340-10347 (June 2013)
11. Nummenmaa, L., Smirnov, D., Lahnakoski, J., Glearean, E., Jaaskelainen, I., Sams, M., Hari, R.: Mental Action Simulation Synchronizes Action–Observation Circuits across Individuals. The Journal of Neuroscience 34(3), 748-757 (January 2014)
12. Hasson, U., Malach, R., Heeger, D.: Reliability of cortical activity during natural stimulation. Trends Cogn Sci. 14(1), 40 (January 2010)
13. Pajula, J., Kauppi J-P, J.-P., Tohka, J.: Inter-Subject Correlation in fMRI: Method Validation against Stimulus-Model Based Analysis., e41196 (2012)
14. Friston, K., Ashburner, J., Kiebel, S., Nichols, T., Penny, W.: Statistical Parametric Mapping: The Analysis of Functional Brain Images. Academic Press (2007)
15. Kauppi, J.-P., Pajula, J., Tohka, J.: A versatile software package for inter-subject correlation based analysis of fMRI. (2014)

An Agent Response System
Based on Mirror Neuron and Theory of Mind

Kyon-Mo Yang and Sung-Bae Cho

Dept. of Computer Science, Yonsei University
134 Shinchon-dong, Seodaemoon-gu, Seoul 120-749, Korea
kmyang@sclab.yonsei.ac.kr, sbcho@cs.yonsei.ac.kr

Abstract. The applications of service agent have been proliferated. For responding the user intention in a flexible environment, researchers need to incorporate the aspects of biological response method. Especially, the investigations about generating the agent behavior like the human are studied using rule or ontology. However these previous system do not work flexibly. Propose an agent response system based on human brain processes can responds in changeable situation flexibly. There are well known theories to investigate the process of generating response in human brain. First, the mirror neuron investigated the intuitive response process. Second, the theory of mind studied the response process for solving the complicated tasks. In other word, the response process in brain was investigated for the immediate response and the complex response. The proposed system implements this human brain function using a modular behavior selection network and a STRIPS planning. The system applies the home service agent and we evaluate the performance using the data by 7 subjects.

Keywords: Intelligent agent, response model, theory of mind, mirror neuron.

1 Introduction

The service agents have been integrated the human function such as conversational, emotional, brain factor, and so on. Traditionally, these agents facilitated human-computer interaction in many services and helped more natural communication. In this regard, the methods of recognizing the user intention from sensory information and of responding it became the core components in the service agent. However, one of the problems in the previous system is not to respond in changeable situation.

The goal of the proposed response system aims to respond the user intention like human. The system is imitated by the cognitive process of the human brain: The mirror neuron system (MN) and the theory of mind system (ToM) [1]. The mirror neuron system is used for responding to the user intentions intuitively [2]. The theory of mind system makes for responding the complex intentions through the sequence production [3]. These human functions are implemented by the modular behavior selection network and the STRIPS planning.

C.K. Loo et al. (Eds.): ICONIP 2014, Part I, LNCS 8834, pp. 42–49, 2014.

We apply the proposed system to the home service agent and verify the performance to compare of the optimal response which is obtained 7 subjects. In addition, we calculate the performance time to verify the improvement of the response speed of the proposed system.

The rest of the paper is organized as follows. Section 2 presents the related works for intelligent agent, cognition model in the brain, and planning system. Section 3 describes in the details of the proposed system. Section 4 reports the experiments conducted to show the usefulness of the system in the implemented home service.

2   Related Works

The intelligent agents can respond to changes in its environment and interacts with others. These agents have integrated a lot of the fields such as home service, robot, network, mobile, and so on [4]. Tong et al. proposed a three-layer agent-based web service workflow model using ontology [5]. The purpose of this agent is that users only need to focus on what they want rather than how to achieve. Giraffa et al. applied the intelligent agents in tutoring system [6]. Tutoring agents are entities whose ultimate purpose is communicate with the student in order to efficiently fulfill their respective tutoring function, as part of the pedagogical mission of the system. Garvey and Sankaranarayanan applied the intelligent agent to flight search and booking architecture [7]. The system provided the real time viewing of flight arrivals and departures in smartphone. However, these previous agent systems using rule-based or ontology had the limitation in the specific service and the lack of generating flexible response like human. To improve these limitations, the proposed system is implemented considering the human brain process.

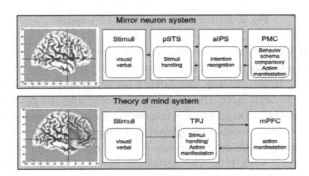

Fig. 1. Cognitive process of human brain

A mirror neuron system and a theory of mind are well known systems in terms of the cognitive process of human brain. Figure 1 shows the cognitive process of human brain. The mirror neuron system consists of three parts. The system relates to respond to user intention intuitively. In the system, superior temporal sulcus (pSTS) handles the stimuli input that is combined the visual and verbal information. Anterior intraparietal sulcus (aIPS) recognizes user intention and premotor cortex (PMC) manifests the action comparing behavior schema [8]. The theory of mind system consists of two parts:

Temporo-parietal junction (TPJ) and medial prefrontal cortex (mPFC). The TPJ is crucial for the representation of goals and intentions, and mPFC plays a role in reflective reasoning about actions and judgments including goals and intentions [9]. The difference of both systems is whether to require the reasoning or not for the response generation. The agent response system like the human brain process requires to the component of making the action intuitively and of solving the complex goals.

3   Agent Response System

The proposed system aims to generate the reasonable action more quickly in the changeable situation and respond about the immediate intention and the complex intention as shown in figure 2.

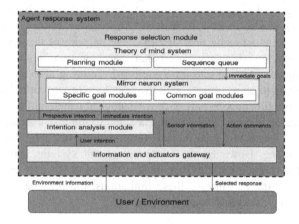

Fig. 2. System architecture for agent response

At the bottom, the user and the environment layer is called as physical layer. That includes simple sensors used to the collect information of interest from the environment, above which there is the gateway that handles the information and sends the action commands to the physical layer. The information processed by the gateway is essential for the intention analysis module which is responsible for recognizing user intention.

The intention analysis module recognizes user intention either prospective or immediate using ontology. We define the two types of user intention with reference to Grafton and Tipper's work [10]. The immediate intention is that the user wants to control one type of object by the command of the direct meaning. The prospective intention involves the control of the various objects by the command in other word, the system need to require the reasoning processes of goal of the intention.

The response selection module consists of the mirror neuron system and the theory of mind system. To implement the similar function as the MN in the system, the system utilizes the behavior selection network [11]. Also, the similar function of the ToM makes the sequence using the STRIPS planning method [12]. The detail of response selection module will be discussed in section 3.1 and 3.2.

3.1   Response Selection Module

The MN system needs to implements the reactive planning system as one of the components of response selection module because it generates the most suitable behavior for an environment using the sensory information and the established goals. The proposed system uses the BSN because it can generate the appropriate action intuitively in the changeable environment.

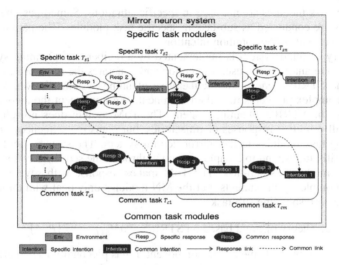

Fig. 3. Mirror neuron system using modular BSN

- Definition 1. Maes' BSN $B_M = \{E, A, G\}$

Here the parameter $E = \{e_1, e_2, \dots, e_n\}$ is the set of environments such as state of light, windows, user, and so on. The parameter $A = \{a_1, a_2, \dots, a_n\}$ is the set of the action nodes such as turn on the device, turn off the device, and so on. The parameter $G = \{g_1, g_2, \dots, g_n\}$ is the set of goals such as controlling windows [2]. We modularize the BSN to control objects as the effect of response because these objects are gathered together for the same purpose and activity that can represent the purpose of objects based on activity theory [13]. When the people take some tasks, the task can be used specific situation or common situation. The BSNs have two types: Specific task module and common task module. Figure 3 shows the mirror neuron system based on BSN.

- Definition 2. Mirror neuron system based on BSN $MN_{BSN} = \{T_s, T_c\}$

The specific BSNs are designed for specific services. The goal of these BSNs is to react in the specific situation. The common BSNs are not used in specific situations. Sometimes, the common BSNs conduct as sub-goal in specific BSNs though the common link. The purpose of the link is to map to the common BSNs.

- Definition 3. Specific task module $T_s = \{E, R_s, R_c, I_s\}$

The specific task module responds to the intention in specific situations. These modules consist of four parameters. The parameter E is information about the environment using sensors or command. The response R_s and R_c are the specific response and the common response, respectively. For instance, turning off light is the response of intention that is to control light only. Sending warning message, however, is the response that can occur in several intentions. The parameter I_s is the specific task called the immediate intention.

- Definition 4. Common task module $T_c = \{E, R_c, I_c\}$

The common task modules are not used in specific situations. Sometimes, the common task modules are conducted as sub-task in specific task modules. The parameter I_c is the common task called the common intention.

The theory of mind system is implemented by a deliberative planning system because the system analyzes the user's intention and generates the sequence to approach the goal. The STRIPS, one of the deliberative planning systems, is well known for solving the complex problems in real environment. We implement the theory of mind system using the STRIPS planning system. The ToM system does not plan the sequences of all primitive responses or trajectories, but plans the sequences of sub-task of conducting task modules in the mirror neuron system. The system should be controlled explicitly to respond to complex intention through the sequence of several independent MN modules as sub-task. The ToM makes response with several sub-tasks correctly in complex environments, but the MN modules only deal with current situations and one corresponding sub-task.

- Definition 5. Theory of mind based on STRIPS $ToM_{STRIPS} = \{P, A\}$

The parameter A represents the action component. It has preconditions and effects. The preconditions must be "true" before that the action can be executed. The effects are "true" in the world after that the action is executed. When an action is executed, the preconditions are removed from the world state and the effects are added. The parameter P is the planer. Pre-defined actions are also organized in plan decompositions, whose detail is how one plan can be executed by performing a sequence of component action.

3.2 Response Selection

In this section, we present how the response is selected depending on the intentions. The flowchart of generating the agent response is shown in figure 4. When the user intention sends to the response module, it is analyzed as the prospective intention or the immediate intention. If the user intention is the prospective intention, the system makes sequence using the planning module like the reasoning process of ToM system. The system checks the action that is satisfied all preconditions. If such action exists, the effect of the action is conducted. Some effect of action pushes a sequence queue to the name of the task module in MN system. If the action that is satisfied all preconditions do not exist, the planner sends an error message to the gateway. When the process of planner is completed, the immediate modules are executed depending on the sequence until there are no more tasks.

When the user intention is single or the sequence is situated, the task modules are conducted. First, the system searches the specific task module that maps the name of the task or single intention. If the task module exists, the system calculates a response level as equation (1). We define the equation of the response level at time t with reference to the activation level of BSN.

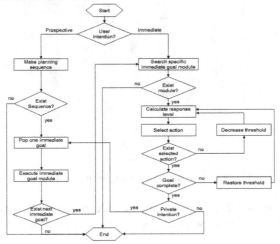

Fig. 4. Flowchart for response generation

$$\alpha(y,t) = input\_from\_environment(y,t) +$$
$$\sum_{x,z}(spread\_bw(x,y,t) + spread\_fw(x,y,z) - takes\_away(z,y,t)), \quad (1)$$

where x ranges over the modules of the task, z ranges over the modules of the task minus that module y. Next, the system selects the response and sends the action commands to the gateway. Finally, when the response node satisfies the task, the response generating process is finished.

4   Experiments

We implemented the home service environment using Unity3D and applied the proposed agent response system. The system has three types of specific goal: "TV management", "Radio management", and "Light management". The "Warning task" is a common goal. In addition, the system has one planner in the ToM system: "Saving energy", "Controlling appliance", and "Controlling temperature".

4.1   Performance of the Proposed System

We conduct the performance test for verifying whether the system offers the appropriate response in a given situation or not. The system obtains the current state of the object and the environment given situation. The state is set to change every time and 7 subjects respond the optimal service before the system offers the services. Each subjects response 10 times per each service and compare with the result of system and the optimal service. They evaluate the system performance that ranged from 1 which means "strongly incongruent" to 5 which means "strongly suitable".

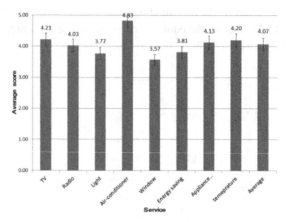

Fig. 5. Performance of each type of intention

The average score is 4.07 as shown in figure 6. As the result, the expected response by subjects is similar in comparison to the service by the proposed system.

4.2   Performance Time of Proposed System

In this section, the comparison results verify the outstanding performance time of the proposed system. It is very important requirement of the smart home agent. As the reason, we conduct the test about three systems: Sequential execution using rule, user command, and proposed method. Sequential execution means the system executes the fixed order for immediate goal. User command means the system require user's command when the system responds to private intention for making a sequence. The experiment is conducted in the configuration: Intel Core™ i7-2600L CPU, 16.0 GB RAM, and Window7. The experiment is conducted ten times.

Table 1. Comparison of performance time

	Sequential execution	User command	Proposed method
Average time (Sec.)	9.725	11.357	7.525

Table 1 shows the average time about the given 20 situations. As the result, the proposed system has the fastest generation time in changeable situation among other methods. We conduct the statistical tests to verify the usefulness of the proposed system. As the result of the t-test, we obtain the significance probability of 0.05, which confirms that the proposed method was more appropriate than the other methods.

5   Conclusion

In this paper, we propose the response system to generate the appropriate action more quickly based on human brain process: Mirror neuron system and theory of mind. The mirror neuron system responds to user intention intuitively and the theory of mind system recognizes prospective user intention and response. We implement the mirror neuron system using the reactive planning system because of the quality of the intuitive generation. The theory of mind system is implemented using deliberative planning system because of the quality of making the sequence. The implemented agent system responds to user intentions in the home service environment and assesses the performance of the relevant response. In addition, we compare the performance time.

Acknowledgements. This research was supported by the Original Technology Research Program for Brain Science through the National Research Foundation of Korea (NRF) funded by the Ministry of Education, Science and Technology (2010-0018948).

References

1. Kuniyoshi, Y., Yorozu, Y., Ohmura, Y., Terada, K., Otani, T., Nagakubo, A., Yamamoto, T.: From Humanoid Embodiment to Theory of Mind. In: Iida, F., Pfeifer, R., Steels, L., Kuniyoshi, Y. (eds.) Embodied Artificial Intelligence. LNCS (LNAI), vol. 3139, pp. 202–218. Springer, Heidelberg (2004)
2. Duijnhoven, D.V.: The Role of the Mirror Neuron System in Action Understanding and Empathy, pp. 1–20 (2010)
3. Amodio, D.M., Frith, C.D.: Meeting of Minds: the Medial Frontal Cortex and Social Cognition. Nature Reviews Neuroscience 7(4), 268–277 (2006)
4. Padgham, L., Winikoff, M.: Developing Intelligent Agent Systems: A Practical Guide, vol. 13. John Wiley & Sons (2005)
5. Tong, H., Cao, J., Zhang, S.: An Agent-based Web Service Workflow Model. In: Networking, Sensing and Control, pp. 1583–1588 (2008)
6. Giraffa, L.M.M., Viccari, R.M.: The Use of Agents Techniques on Intelligent Tutoring Systems. In: Int. Conf. of the Chilean Computer Science Society, pp. 76–83 (1998)
7. Garvey, F., Sankaranarayanan, S.: Intelligent Agent based Flight Search and Booking System. Int. Journal of Advanced Computer Science and Applications 4(1), 12–28 (2012)
8. Cross, E.S., Hamilton, A.F.D., Grafton, S.T.: Building a Motor Simulation de Novo: Observation of Dance by Dancers. Neuroimage 31(3), 1257–1267 (2006)
9. Saxe, R., Powell, L.J.: It's the Thought That Counts Specific Brain Regions for One Component of Theory of Mind. Psychological Science 17(8), 692–699 (2006)
10. Grafton, S.T., Tipper, C.M.: Decoding Intention: A Neuroergonomic Perspective. NeuroImage 59, 14–24 (2012)
11. Maes, P.: How to Do the Right Thing. Connection Science Journal 1(3), 291–323 (1989)
12. Wilkins, D.E.: Practical Planning: Extending the Classical AI Planning Paradigm. Morgan Kaufmann (1988)
13. Engestrom, Y.: Activity Theory and Individual and Social Transformation. In: Perspectives on Activity Theory, pp. 19–38. Cambridge University Press (1999)

Dynamic of Nitric Oxide Diffusion
in Volume Transmission: Model and Validation

Fernández López Pablo[1], García Báez Patricio[2],
and Suárez Araujo Carmen Paz[1]

[1] Institute for Cybernetics, University of Las Palmas de Gran Canaria, Spain
{pfernandez,cpsuarez}@dis.ulpgc.es
[2] Dept. of Statistics, Operations Research and Computation,
University of La Laguna, Spain
pgarcia@ull.es

Abstract. Cellular communiction is one mechanism that connects nerve
cells to cognition. At present it seems that synaptic transmission may not
be the only type of signal processing between cells. *Volume transmission*
(VT) is a process that is performed by means of a gas diffusion process, which is obtained with a diffusive type of signal. This work shows a
Diffusion Model for NO based on Bessel Functions that are valid for homogeneous and isotropic environments for instantaneous generation and
their diffusion occurs while being constrained by cylindrical morphology. The model is validated with experimental data from the dynamic
behaviour of NO in endothelium cells. Capacities are analysed in the
study of the *Diffusion* and *Autoregulation* of NO, and their role in *Fast
Diffusion Neural Propagaton* of NO is observed.

Keywords: Nitric Oxide, Volume transmission, NO dynamics, Fast Diffusion Neural Propagation, Bessel functions.

1   Introduction

The underlying mechanisms of brain activity need to be studied in order to
understand structure and brain function, in addition to computational processes.
Volume Transmission (VT) is one of these mechanisms, and is complementary
to classic neural signal transmission. VT is based on the diffusion of neuro-active
substances such as *Nitric Oxide* (NO) in the Extracellular Space.

NO is a gaseous liposoluble molecule, with a permeable membrane and is char-
acterised by high diffusibility. NO dynamics make up diverse processes: *Generation* or *Synthesis*, which occurs in the framework of the synaptic transmission;
Diffusion, which is controlled by the gradient of its own concentration; and *Autoregulation* and *Recombination* with other substances. NO dynamics, as a brain
messenger, is not experimentally defined.

As opposed to other approaches in this research area [1], [2], [3], [4], [5] and
[6] our aim is to emulate the behaviour of NO with minimal constraint assumptions regarding the characteristics of the specific morphologies present in the
underlying processes in its dynamic.

C.K. Loo et al. (Eds.): ICONIP 2014, Part I, LNCS 8834, pp. 50–58, 2014.
© Springer International Publishing Switzerland 2014

We present a Diffusion Model of NO based on Bessel Functions. It is an analytical model and is based on phenomenological aspects of the transportation of molecular matter in isotropic and homogeneous media [7]. The model is tested and validated with the experimental behavioural data of NO dynamics measured by Tadeusz Malinski et al [2]. An analysis of Fast Diffusion Neural Propagaton (FDNP) [8] and [9] is also carried out.

2 Diffusion Model Based on Bessel Functions

In order to develop this model we present our hypothesis that presynaptic specialisation is not necessary for the generacion or synthesis of NO, that is, the complete surface of the neuron can be seen as a possible place for this generation. This occurrence allows us to consider the volume morphology of the diffusion process of NO, an aspect which heavily depends on the NO dynamic. We consider, in this study, the neuron as a basic emissor module. Hence we consider the construction of a model where the diffusion is defined on the radius of the neuron, and, consequently leads to the study of the expression of the diffusion in a cylindrical environment (figure 1).

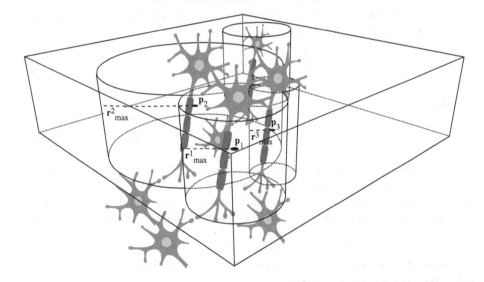

Fig. 1. Environment Ω for the diffusion of NO in cylindrical coordinates

Starting with the General Diffusion Equation with $x = rcos(\theta)$ and $y = rsin(\theta)$, we can obtain the expression for the diffusion based on the cylindrical coordinates r, θ and z, as shown in expression (1).

$$\frac{\partial c}{\partial t} = \frac{1}{r}\{\frac{\partial}{\partial r}(rD\frac{\partial c}{\partial r}) + \frac{\partial}{\partial \theta}(rD\frac{\partial c}{\partial \theta}) + \frac{\partial}{\partial z}(rD\frac{\partial c}{\partial z})\} \tag{1}$$

Given that the dynamic of diffusion is the same for any value of z and θ, an expression for the diffusion equation is (2)

$$\frac{\partial c}{\partial t} = \frac{1}{r}\frac{\partial}{\partial r}(rD\frac{\partial c}{\partial r}) \tag{2}$$

The prior assumption establishes that the diffusion behaves the same on the entire circumference with radius r where the neuron is located at the center. We introduce a term that represents the autoregulation process, and assume that it is proportional to the concentration of the substance that is present in a given moment. We obtain the basic expression for the diffusion of NO with a morphology of cylindrical diffusion, equation (3) in the following form:

$$\frac{\partial c}{\partial t} = \frac{1}{r}\frac{\partial}{\partial r}(rD\frac{\partial c}{\partial r}) - \gamma c \tag{3}$$

This expresion leads to the concentration of NO for a value of r and for some initial conditions and specific environment in a given homogeneous and isotropic medium. Assume a cylinder like the one shown in figure 1, and a value r_{max} where $c(r_{max}, t) = 0$ for all values of t. Then r can take on values in the interval $[0, r_{max}]$, and for $t = 0$ the following initial condition is $c(r, 0) = f(r)$, allowing us to define the form of the concentration initially with $t = 0$. Partial derivatives can then be used to solve the equation with the following infinite series of appropriately weighted Bessel functions:

$$c(r, t) = \sum_{n=1}^{\infty} A_n J_0(\frac{\xi_n}{r_{max}}r)e^{-(\frac{\xi_n^2}{r_{max}^2}+\frac{\gamma}{D})Dt} \tag{4}$$

Where

$$A_n = \frac{2}{r_{max}^2 J_1^2(\xi_n)} \int_0^{r_{max}} f(r)J_0(\frac{\xi_n}{r_{max}}r)rdr \tag{5}$$

and $J_0(x)$ and $J_1(x)$ are first type order zero and order one Bessel functions, respectively (see figure 2), and $f(r)$ is defined as the initial condition $c(r, 0)$, ξ_n are the different roots of the Bessel function $J_0(x)$, D is the diffusion constant associated with the homogeneous and isotropic medium, causing the the diffusion and γ is the autoregulation constant.

The calculation of the A_n coefficients is done as a function of $f(r)$, and when a step function is involved, as shown in figure 2(b), the integral can be directly evaluated, as shown in expression (6),

$$A_n = \frac{2\rho r_0 J_1(\frac{\xi_n}{r_{max}}r_0)}{\xi_n r_{max} J_1^2(\xi_n)} \tag{6}$$

leading to the following infinite series, which defines the concentration proportion of NO in the environment points where it is diffusing at every moment.

$$c(r, t) = \sum_{n=1}^{\infty} \frac{2\rho r_0 J_1(\frac{\xi_n}{r_{max}}r_0)J_0(\frac{\xi_n}{r_{max}}r)}{\xi_n r_{max} J_1^2(\xi_n)}e^{-(\frac{\xi_n^2}{r_{max}^2}+\frac{\gamma}{D})Dt} \tag{7}$$

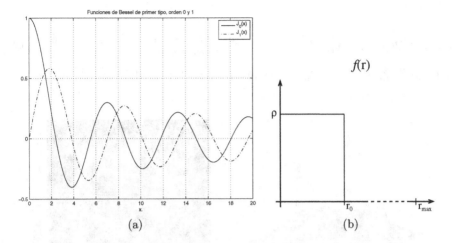

Fig. 2. Figure a) First type Bessel Functions, order 0 and 1, respectively, used in the calculation of A_n. Figure b) Function $f(r)$, where $f(r) = \rho$ when $0 \le r \le r_0$, and 0 when $r_0 < r$. The edge of the cylinder is r_{max}, so that $c(r, t) = 0$ when $r_{max} < r$ is satisfied.

The resultant model allows us to analyze the behaviour of two of the key processes in the dynamics of NO. These processes are NO Diffusion and NO Autoregulation. The process of Generation or Synthesis is constrained by a spontaneous generation of NO at the beginning of the process. As regards the diffusion morphology, and as stated previously, this model only considers cylindrical diffusion.

Once the diffusion process for one case is obtained the next step is to derive the expression for the NO dynamic in a environment Ω such as the one shown in figure 1 where a set of N diffusion processes are present whose individual dynamics for the NO concentration are shown as $\{c_1(r,t), c_2(r,t), c_3(r,t), \ldots, c_i(r,t), \ldots, c_N(r,t)\}$. These diffusion processes are produced at specific points, and belong to the environment $\{p_1, p_2, p_3, \ldots, p_i, \ldots, p_N\} \in \Omega$, and have respective attributes which define their specific diffusion dynamic. A maximum diffusion radius for each process is associated $\{r_{max}^1, r_{max}^2, r_{max}^3, \ldots, r_{max}^i, \ldots, r_{max}^N\}$, and the implied NO is not capable of reaching longer distances. It is represented by $\{f_1(r), f_2(r), f_3(r), \ldots, f_i(r), \ldots, f_N(r)\}$ for the concentration at $t = 0$, with diffusion constants $\{D_1, D_2, D_3, \ldots, D_i, \ldots, D_N\}$ and autoregulation $\{\gamma_1, \gamma_2, \gamma_3, \ldots, \gamma_i, \ldots, \gamma_N\}$.

Using previous developments, the general expression of the dynamics of the NO concentration in a generic point $p_k \in \Omega$, belonging to the environment is seen as:

$$c(p_k, t) = \sum_{i=1}^{N} H(p_k, t, i) \tag{8}$$

Where

$$H(p_k, t, i) = \{ \begin{matrix} c_i(|\ p_k - p_i\ |, t - t_0^i) & If\ |\ p_k - p_i\ | < r_{max}^i \wedge t \geq t_0^i \\ 0 & otherwise \end{matrix} \quad (9)$$

and $\{t_0^1, t_0^2, t_0^3, \ldots, t_0^i, \ldots, t_0^N\}$ are starting times for the diffusion processes.

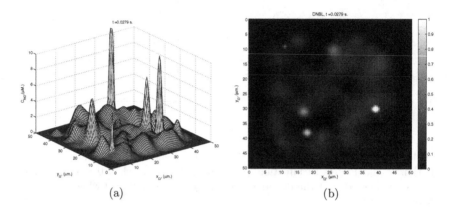

(a)   (b)

Fig. 3. Generation of complex structures

The proposed models allows us to determine the generation of complex structures that occur in these simultaneous NO diffusion processes, (see figure 3).

3   Analysis and Model Validation

The model proposed in this paper reproduces the dynamic behaviour of NO reported in Tadeusz Malinski et al. [2] where a maximum induced value of 1 μM for the NO concentration in the membrane of a endothelial cell of 1 μm. diameter was obtained.

Parameters in the model are set with maximum diffusion radius of 1 μm., establishing a cylindrical diffusion environment with a 2 μm. diameter. An endothelium cell with 1 μm. diamter is assumed to be located in the centre, hence its surface is determined by a $r = 0.5$ μm. radius. In order to determine the shape to maximise NO concentration we consider the funcion $f(r)$, with $r = 0.5$ μm. for 1 μM. concentration. Figure 4(a) reveals that there is a relationship of exponential growth between the strength of the NO source (ρ) and the size of r_0. Smaller r_0 creates a spontaneous increase in the quantity of NO needs in the centre of the cylinder.

This figure also allows us to analyse how much time is necessary to reach the maximum value of the concentration with $r = 0.5$ μm. to converge to $1.8\ 10^{-5}s$. while the size of the source is reduced. This calculation is found to be directly related to the NO concentration profiles which are shown at position $r = 0.5$ μm. for different values of r_0 and ρ (see figure 4(b)).

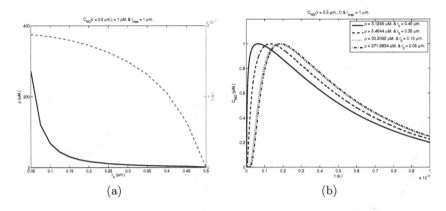

Fig. 4. Behaviour of cylindrical diffusion model for different functions $f(r)$, using $r_{max} = 1$ μm. Figure a shows the relationship between r_0 and ρ (black graph) together with the relationship between r_0 and t (red graph). Figure b) shows increasing NO concentration profiles for different values of r_0 and ρ.

The model has been subjected to a second test, which is used to determine the implications of the amount of NO in the environment. In this case we use parameters for the a maximum diffusion radius of 100 μm., creating a cylindrical environment with 200 μm., and then observe what $f(r)$ should look like under the same conditions previously established (maximum concentration of 1 μM. a distance of $r = 0.5$ μm.). Figures 5(a) and 5(b) reveal that the implied variables (ρ and r_0) react similarly to the ones in the previous case where the maximum diffusion radius was 1 μm. The only observable difference is in the concentration profiles, and they do not appear to be implied in the NO behaviour for $r = 0.5$ μm. when it reaches its maximum value. Figures 4(b) and 5(b) show that the elimination of NO in these profiles takes longer.

Observed behaviour supports the Fast Diffusion Neural Propagation (FDNP) phenomenon, since the dynamic behaviour of NO in locations near its generation process and is not affected by the reach of NO. Hence, the way in which this NO dynamic has an impact in that environment is independent of the reach of NO.

Figure 6 shows the explicit form of the FDNP phenomenon. Figure 6(a) shows the maximum NO concentration values for different values of r_{max}, and figure 6(b) shows the time needed to reach the maximum value, as a function of r. Notice that FDNP is present in the neighbourhood of 0.25 μm. to 0.35 μm., where the influence of NO in reaching maximum concentrations and in the time needed for the same. When $r > 0.35$ μm. as seen in figure (6(a)), we see that the maximum concentration level of NO starts to differ based on the reached value, and the difference in the time variable to reach these maxima are more apparent (see figure 6(b)).

How the model represents autoregulation is analysed in figure 7(a), which shows a comparison of NO concentrations for with radius 0.5 μm. for a NO dynamic with $r_{max} = 1$ μm. and values of $r_0 = 0.25$ μm. and $\rho = 10.8440$ μM.

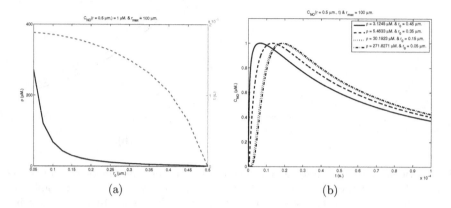

(a)   (b)

Fig. 5. Cylindrical diffusion model behaviour for different function $f(r)$, using $r_{max} = 100 \ \mu m$. Figure a) shows the relationship between r_0 and ρ (black graph), and the relationship between r_0 and t (red graph). Figure b shows different concentration profiles for NO for different values of r_0 and ρ.

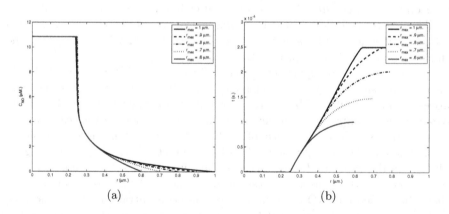

(a)   (b)

Fig. 6. Observed FDNP phenomenon using variation of maximum reached NO concentration levels, figure a),and using time variation to reach these maxima, figure b). $f(r)$ is shown with values of $r_0 = 0.25 \ \mu m.$ and $\rho = 10.8440 \ \mu M$.

When an autoregulation process is not present, the Dynamic of NO reaches a maximum concentration at 1 ρM. when $r = 0.5 \ \mu m$., which corresponds to observed biological behaviour. Concentration profiles during autoregulation is based on a value of $\gamma = 1104 \ s^{-1}$, but does not reach this maximum value, revealing that most of the generated NO is destroyed from the autorregulacion process before it reaches a distance $r = 0.5 \ \mu m$.

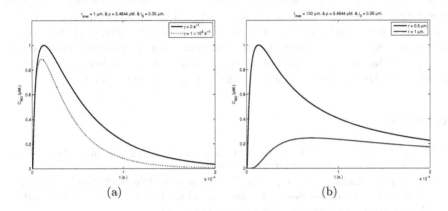

(a)   (b)

Fig. 7. Concentration profiles of NO for distinct parametrized dynamics with the same $f(r)$ but different maximum levels of NO, a) maximum level $r_{max} = 1 \ \mu m$. and b) maximum level $r_{max} = 100 \ \mu m$

The last analysis of NO concentration profiles compares behaviour along the radius r for values of $r = 0.5 \ \mu m$. and $r = 1 \ \mu m$. when the maximum radius is reached at $r_{max} = 100 \ \mu m$. Figure 7(b) shows the importance of the distance variable. For instance, when the radius only moves 0.5 μm. from the area of generation or spontaneous synthesis this causes significant changes in the shape of the NO profile, in addition to the maximum magnitude and the amount of time to reach it.

4   Conclusions

A study of the NO dynamic in biolgical and artificial environments was carried out, leading to an increased understanding of NO and advances in the modeling of NO diffusion.

Observations of NO concentration profiles in different positions at specific distances from the source for a morphology of puntual NO generation and a cylindrical diffusion, as well as in homogenerous and isotropic environments diffusion were performed.

The proposed NO diffusion model was validated biologically based on experimental data of NO behaviour found in Tadeusz Malinski et al. [2]. We analysed the process of Fast Diffusion Neural Propagation (FDNP). This process characterizes the underlying VT transmission scheme in such a way that its influence on the environment depends on the distance to the diffusion process. Furthermore it is independent the maximum radius of NO propagation, that is, the reach of NO.

The generation of complex structures that occur in simultaneous NO diffusion processes have been shown, in addition to the proposed models capacity for their determination. This capability leads to significant implications for neural information transmission and learning, which can be included in future works.

Finally we conclude that in order to consider other diffusion morphologies, as well as generation processes that are different from the puntual generation of NO and non-homogeneous and non-isotropic characteristic environments, further modeling of the NO dynamic must be done using a discrete perspective.

References

1. Edelman, G.M., Gally, J.A.: Nitric oxide: linking space and time in the brain. Proc. Natl. Acad. Sci. USA 89, 11651–11652 (1992)
2. Mailinski, T., Radonski, M.W., Taha, Z., Moncada, S.: Direct Electrochemical Measurement of Nitric Oxide Released from human Platelets. Biochemical and Biophysical Research Communications 194(2), 960–965 (1993)
3. Lancaster Jr., J.R.: Simulation of the diffusion and reaction of endogenously produced nitric oxide. Proc. Natl. Acad. Sci. USA 91, 8137–8141 (1994)
4. Wood, J., Garthwaite, J.: Models of the Diffusional Spread of Nitric Oxide: Implications for Neural Nitric Oxide Signalling and its Pharmacological Properties. Neuropharmacology 33(11), 1235–1244 (1994)
5. Philippides, A., Husbands, P., O'shea, M.: Four-Dimensional Neuronal Signaling by Nitric Oxide: A Computational Analysis. The Journal of Neuroscience 20(3), 1199–1207 (2000)
6. Vaughn, M.W., Kuo, L., Liao, J.C.: Effective diffusion distance of nitric oxide in the microcirculation. Am. J. Physiol., 1705–1714 (1998)
7. Suárez Araujo, C.P., Lopez, P.F., Báez, P.G.: Towards a Model of Volume Transmission in Biological and Artificial Neural Networks: A CAST Approach. In: Moreno-Díaz Jr., R., Buchberger, B., Freire, J.-L. (eds.) EUROCAST 2001. LNCS, vol. 2178, pp. 328–342. Springer, Heidelberg (2001)
8. Suárez Araujo, C.P.: Study and Reflections on the Functional and Organisational Role of Neuromessenger Nitric Oxide in Learning: An Artificial and Biological approach. Computer Anticipatory Systems, AIP 517, 296–307 (2000)
9. Fernández López, P.: Estudio Computacional de la Difusión del Óxido Nítrico y sus Implicaciones en los Procesos de Comunicación y Acoplamiento Celular, Aprendizaje y Memoria: Modelos y Teorías. Ph.D thesis, University of Las Palmas de GC (2014)

A Computational Model of the Relation between Regulation of Negative Emotions and Mood

Altaf Hussain Abro, Michel C.A. Klein, Adnan R. Manzoor,
Seyed Amin Tabatabaei, and Jan Treur

Vrije Universiteit Amsterdam, Artificial Intelligence Section
De Boelelaan 1081, 1081 HV Amsterdam, the Netherlands
{a.h.abro,michel.klein,a.manzoorrajper,
s.tabatabaei,j.treur}@vu.nl

Abstract. In this paper a computational model is presented that describes the role of emotion regulation to reduce the influences of negative events on long term mood. The model incorporates an earlier model of mood dynamics and a model for the dynamics of emotion generation and regulation. Example model simulations are described that illustrate how adequate emotion regulation skills can prevent that a depression is developed.

Keywords: depression, emotion regulation, mood regulation, agent.

1 Introduction

Emotions were traditionally seen as a neural activation states without a function[1]. However, relevant research provides evidence that emotions are functional [2, 3] and provide information about the ongoing fight between a human being and its environment[4]. In addition to the theories that exist in social psychology also in recent neurological literature many contributions (e.g.,[2, 4]) can be found about the relation between emotion and brain functioning. For example, emotional responses relate to activations in the brain within the limbic centers (generating emotions), and cortical centers (regulating emotions); cf.[5, 6]. Previously emotions were often left out of cognitive models; however since the awareness that emotions play a vital role in human lifes is increasing, cognitive models are developed that include the generation and regulation of emotions as well. A useful basic theory for the latter is the one of Gross: on how individuals regulate which emotions they have, when they have them and how they experience and express them[7].

Emotions are different from mood, and emotion regulation is different from mood regulation[7, 8]. Emotions are instantaneous in nature and are specific reactions to a particular event, usually for a short period of time. Emotions help us to set priorities in our lives, taking initiatives in changing situations or making decisions based on how we feel, whether we are happy, angry, frustrated, bored or sad. Emotion regulation describes how a subject can use specific strategies to affect the emotion response levels. Mood, on the other hand, is a more general feeling such as happiness, sadness,

C.K. Loo et al. (Eds.): ICONIP 2014, Part I, LNCS 8834, pp. 59–68, 2014.
© Springer International Publishing Switzerland 2014

frustration, or anxiety that exists for a longer period of time. Mood regulation usually involves the deliberate choice of mood-affecting activities, such as pleasant activities[9]. It has been found that recurring events triggering stressful emotions have a bad influence over time on mood and can easily lead to depression when subjects are vulnerable for that [10, 11].

In this paper a computational model is introduced that combines the short-term emotional reaction on stressful events with the long term dynamics of mood. The model is based on existing model for mood dynamics[12] and the theory for emotion regulation introduced by Gross [7, 8, 13]. In the current paper, it is shown how this process of emotion regulation can help people to maintain a healthy mood in case of the occurrence stressful events.

The paper is organized as follows. First, in Section 2 some background information about the mood model and the process of emotion regulation is presented. In Section 3 the integrated model is explained in detail. In Section 4 simulation results are provided to show the influence of stressful events in different scenario's, thereby providing evidence for the feasibility of the model. Finally, Section 5 concludes the paper.

2   Background on Emotion Regulation and Mood Dynamics

The model presented in this paper adopts Gross' theory of emotion regulation and an existing model of mood dynamics[12]. Both elements are introduced here briefly.

2.1   Emotion Regulation

Controlling emotions or regulating them is often related with the suppression of an emotional response, for example, expressing a neutral poker face. This kind of regulating emotions is sometimes considered not very healthy, and a risk for developing serious kinds of medical problems. However, it has been found that the strategies to regulate emotions are much more varied. For example, closing or covering your eyes when a movie is felt as too scary, or avoiding an aggressive person are other forms of emotion regulation mechanisms[13, 14].

The framework introduced by Gross describes how emotions can be regulated or controlled in different phases of the process during which emotions are generated[7]. Gross distinguishes cognitively regulated emotions, which occurs relatively early on in the emotion generation process (e.g., re-interpretation) and behaviorally regulated emotions, that happen relatively late in the emotion generative process (e.g., suppression).

Over a longer period of time several strategies for emotion regulation have been described in the literature. In general they are classified into two major categories. The first category covers the antecedent focused strategies that can be used before an emotional response has an effect on the behavior. In this category of emotion regulation, emotions may be regulated at four different points in the emotion generation process (a) selection of the situation, (b) modification of the situation, (c) deployment of attention, (d) change of cognition. The second category is formed by the response focused strategies, which can be used in situations where the emotion response already is coming into effect; this is also called modulation of responses[7]. In the current paper the focus is on antecedent focused strategies, in particular re-interpretation of world information by belief change.

2.2   Modeling Emotion Regulation

Based on the theory of emotion generation and regulation described above, a computational model of emotion regulation has been introduced before [15] and applied in the context of contagion and decision making. A detailed discussion of this model is given in that paper; however, here a brief summary is given of these concepts and their dynamics. As illustrated in the dashed box in the upper part of Fig.1 the following concepts play their part in the model: control state (cs), beliefs (bel), feeling (feel), preparation (prep), and sensory representation (srs(x)). The aim of the model is to describe how negative beliefs and feelings are generated and how alternative, more positive beliefs can be generated to regulate the negative feeling. The model is in-spired from various neurological theories [16–19] , from fMRI experiments it has been found that emotion regulation occurs through the interaction between prefrontal cortex and amygdala. Here less interaction or weak connections between amygdala and prefrontal cortex lead to less adequate emotion regulation[16].

In the model, antecedent focused emotion regulation is achieved by the interplay of three states cs(b, c), bel(c), feel(b). Negative weights are assigned to the connec-tions from the control state cs to negative beliefs bel(c) and negative feelings feel(b). Positive weights are assigned to connections in the opposite direction.   In the exam-ple scenario only two beliefs are taken into account: a positive belief which may asso-ciate to good feeling and a negative belief which is related to a stressful feeling (ac-tually for the sake of simplicity there is only one negative feeling state in the scena-rio). A control state is used to determine whether an unwanted emotion through a negative belief has occurred (as a form of monitoring as happens in the prefrontal cortex). If so, by becoming activated the control state suppresses these negative ef-fects. Furthermore, as they concern opposite interpretations of the world information, both beliefs inhibit each other, which is modelled by assigning negative weights to their mutual connections. In the literature (e.g.,[20]) emotion generation and emotion regulation are sometimes considered as overlapping in one process.

In the model introduced here on the one hand both subprocesses (emotion genera-tion and regulation) are clearly distinguished but on the other hand by the cyclic con-nections between them and the dynamics created by these cycles the processes are fully integrated into one process.

The sensory representation srs(w) of a world state w is associated both with a nega-tive and a positive belief, as a basis for two different interpretations of the same world information;, as discussed earlier they suppress each other by a form of inhibition. Only the negative belief has a connection with the preparation for a negative emotion-al response prep(b). The feeling state feel(b) has an impact on this preparation state prep(b), which in turn has an impact on feeling state feel(b) through srs(b) which makes it recursive; this is often called an as-if body loop in the literature (e.g.,[2]).

2.3   Modeling Mood Dynamics and Depression

The model of mood dynamics is depicted in Fig. 1 (lower part). The main concepts include the *mood level, appraisal* and *coping skills* of a person, and how the levels for these states affect the external behavior in the form of selection of situations over time (*objective emotional value of situation*). The model is based upon a number of psychological theories, see [12] for a mapping between the literature and the model itself.

In the model, a number of states are defined, whereby to each state at each point in time a number on the interval [0,1] is assigned. First, the state *objective emotional value of situation* represents the value of the situation a human is in (without any influence of the current state of mind of the human). The state *appraisal* represents the current judgment of the situation given the current state of mind (e.g., when you are feeling down, a pleasant situation might no longer be considered pleasant). The *mood level* represents the current mood of the person, whereas *thought* indicates the current level of thoughts (i.e., the positivism of the thoughts). The *long term prospected mood* indicates what mood level the human is striving for in the long term, whereas the *short term prospected mood level* represents the goal for mood on the shorter term (in case you are feeling very bad, your short term goal will not be to feel excellent immediately, but to feel somewhat better). The *sensitivity* indicates the ability to select situations in order to bring the *mood level* closer to the *short term prospected mood level*. *Coping* expresses the ability of a human to deal with negative moods and situations, whereas *vulnerability* expresses how vulnerable the human is for negative events and how much impact that structurally has on the mood level. Both *coping* and *vulnerability* have an influence on all internal states except the prospected mood levels, but in Figure 1, those arrows are left out for clarity reasons. Finally, *world event* indicates an external situation which is imposed on the human (e.g., losing your job).

3   Integrated Model

The integrated model describes how the emotion generation and regulation mechanism influences the mood dynamics. It describes how specific stressful events generate specific instantaneous negative feelings, which have a negative effect on the (subjective) appraisal (also called sevs – subjective emotional value of the situations of the person) of the more general situations of the person and thus on the mood. When emotion regulation is taking place, the instantaneous feelings will be less negative and thus reduce the influence of the stressful events on the mood. To implement this principle in the model, a connection from the negative feeling in the regulation model to appraisal state in the mood model is introduced. The purpose of this connection is to model the effect of negative but short term feelings on the (longer term) mood. In the model, only negative feelings are considered. For beliefs, there is both a positive and a negative variant. The world(w), sensor(w), srs(w) states may lead to the negative and positive belief as alternative interpretations of the same world information.

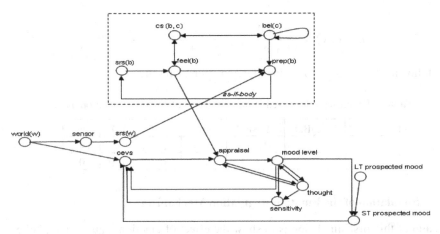

Fig. 1. The integrated model: emotions about stressful events and their influence on mood

4   Simulation Results

In this section, example simulation results are presented that show how emotion regulation can help to change bad beliefs and feelings into more positive beliefs and feelings, and thus protects the mood against stressful events. First, some details of the model design and its implementation and the parameter values used are described.

As mentioned, for the model of mood dynamics (the lower part of Fig. 1) an existing model is used. Due to the lack of space, we have to refer to original article [12] for the numerical details of this part of model.

In the emotion regulation model, the activation level of a state is determined by the impact of all the incoming connections from other states thereby being multiplied by their corresponding connection weights. In the simulations, the connection weights set at the following values: $w_{worldstate\_sensor}$ 1.0, $w_{sensor-srsw}$ 1.0, $w_{srsw-PosBel}$ 0.4, $w_{srsw-NegBel}$ 0.9, w_{NegBel_prep} 0.9, w_{Prep_srsb} 0.9, w_{srsb_feel} 0.9, w_{feel_prep} 0.4, w_{cs_feel} -0.2, w_{cs_negBel} -0.35, w_{NegBel_PosNeg}= -0.3, w_{PosNeg_NegBel}= -0.1. When no emotion regulation takes place w_{feel_cs} and w_{negBel_cs} are taken 0. For scenarios in which emotion regulation takes place, the value of w_{feel_cs} and w_{negBel_cs} change from 0 to 3 and 0.05.

In particular, for a state causally affected by multiple other states, to obtain their combined impact, first the activation levels V_i for these incoming state are weighted by the respective connection strengths w_i thus obtaining $X_i=w_i v_i$ and then, these values X_i are combined, using a combination function $f(X_1,.., X_n)$. In the context of emotion regulation model, the combination function is based on the following function:

$$V_{new} = V_{old} + adapt_{ER} * th(\tau, \sigma, X_1+X_2+...+X_n)$$

Where $adapt_{ER}$ is an adaptation factor, determines the speed with which the value of state changes. The $adapt_{ER}$ for all states of the emotion regulation model is equal to 6. And,

$$th(\tau, \sigma, X) = \left(\frac{1}{1 + e^{-\sigma(X-\tau)}} - \frac{1}{1 + e^{-\sigma\tau}}\right)(1 + e^{-\sigma\tau})$$

The following table shows the value of σ and τ for each state:

Table 1. Parameter values used in the simulation of emotion regulation model

	NegBel	PosBel	prep	srs	feel	cs
τ	9	9	4	3	5	4
σ	0.1	0.1	0.4	0.2	0.10	0.5

4.1   Simulation of the Emotion Regulation Mechanism

The aim of this first simulation is to show the effect of emotion regulation on beliefs and negative feelings. In this experiment, it is assumed that a bad event happened for 3.3 hours. In this case, there are two different beliefs about this event, a negative and one positive one (for example, after losing a match, negative belief is that you played awfully and blame yourself, and positive belief is that your rival was much more powerful and the stadium was full of his fans).

Fig. **3** shows the results of this simulation. As can be seen, when no emotion regulation takes place (Fig. **3a**), the negative event dominates and it leads to a high value of negative feeling. However, when the emotion regulation does take place (Fig. **3b**), the generated negative belief and negative feelings lead to the activation of the control state, and consequently it causes weakening of the negative belief and due to this the positive belief can become dominant. Eventually, activation of the control state also decreases the value of negative feeling (purple line). This substantially reduces the effect of the negative feeling: it can be seen that the integral of the area below the feeling is 2.3 times smaller than in the upper graph.

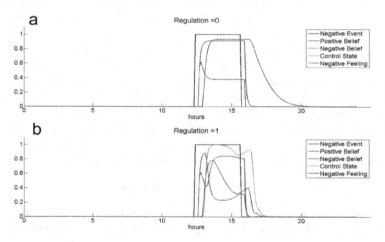

Fig. 2. Simulation results of emotion regulation model when a bad event happens. a) without emotion regulation b) with emotion regulation.

4.2   Simulation of the Mood

The integrated model was used to simulate three types of persons in different situations. The different types are characterized by different values for the parameters *coping, vulnerability* and *LT prospected mood level*. The first type of person is an *emotionally stable person*, defined by having good coping skills that balance out any vulnerability and by having the desire to have a good mood: coping is 0.5, vulnerability 0.5 and LT prospected mood level 0.8. An *emotionally slightly unstable person* is defined by having some vulnerability and bad coping skills and the desire to have a medium mood: settings 0.1, 0.9 and 0.6 respectively. The third type, an *emotionally very unstable person*, is characterized by settings 0.01, 0.99 and 0.6. As start value for OEVS the equilibrium state is used; this needs to be calculated for each type so that when no events occur, the person stays balanced with al variables equal to LT prospected mood level. For type the OEVS is 0.8, for type 2 it is 0.94 and for type 3 the stable OEVS is 0.999.

The six weights between mood, thoughts and appraisal can also be varied to simulate different personal characteristics. However, in these simulations they have been set at the following values: $w_{appraisal\_mood}$ 0.7, $w_{thoughts\_mood}$ 0.3, $w_{appraisal\_thoughts}$ 0.6, $w_{mood\_thoughts}$ 0.4, $w_{mood\_appraisal}$ 0.5, $w_{thoughts\_appraisal}$ 0.5. In each iteration, the value of each state(V_{new}) in the mood model is defined according the weighted sum of its inputs and its old value(V_{old}):

$$V_{new} = V_{old} + adapt_{mood} * (w_1V_1+W_2V_2+..)$$

The adaptation factor for all states in the mood model is 0.1. By comparing the adaptation factors of the mood model and the emotion regulation model, we see that the states of the emotion regulation model are updated 60 times faster than the states of the mood model. This is in line with the background provided in the introduction, which says that the emotions are much more short-time events than mood.

In **the first scenario,** three short (3.3 hours) bad events occur with the time interval of 12 hours. The length of the scenario is three weeks (504 hours). Table 2 shows the value of mood after one, two and three weeks, and the minimum value of mood, for each person when the emotion regulation is on or off.

Table 2. Simulation results when three bad events happen

	Person 1		Person 2		Person3	
	Without ER	With ER	Without ER	With ER	Without ER	With ER
Week 1	0.79008	0.78098	0.43627	0.50024	0.32473	0.39851
Week 2	0.80356	0.79827	0.45075	0.50162	0.27910	0.32961
Week 3	0.79825	0.80059	0.46553	0.50530	0.24847	0.28355

As Table 2 shows, a stable person does not require emotion regulation to handle these bad events (the value of mood does not change significantly when emotion regulation is on or off). However, emotion regulation is critical for person 2 (unstable). In fact, if emotion regulation does not take place, he/she will become depressed after these bad events (a depression is defined as a mood level below 0.5 during at least 336 hours (two weeks)[21]); while if emotion regulation does take place, the value of the mood will not go below 0.5 during this simulation. In contrast, the

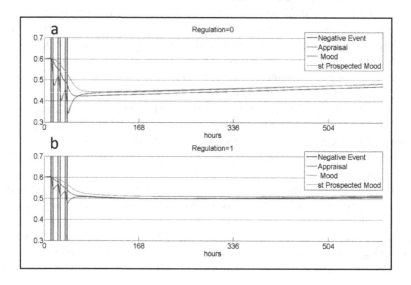

Fig. 3. Simulation results of first scenario for person with unstable characteristics. a) without emotion regulation b) with emotion regulation

emotion regulation does not save a very unstable person from depression. However, even in this case, the emotion regulation postpones the starting point of depression (time which the value mood become less than 0.5) for almost one day (22.8 hours).

Fig. 4 shows the results of the simulation of the first scenario for an unstable person. When ER is off (Fig. **4**a) events lead to a depression. While ER is ON (Fig. **4**b), it decreases the effect of negative events and saves the person from depression.

In **the second scenario**, bad events occur every 3 weeks in one year. Table 3 shows the minimum, average and maximum value of mood in last 3 weeks of this simulation for each person.

Table 3. Simulation results when bad events happen every 3 weeks during one year

	Person 1		Person 2		Person3	
	Without ER	With ER	Without ER	With ER	Without ER	With ER
Minimum	0.7607	0.7817	0.4414	0.4802	0.01060	0.0350
Average	0.7928	0.7914	0.4656	0.4899	0.01443	0.0437
Maximum	0.8017	0.8002	0.4893	0.5014	0.02052	0.5959

Fig. 5 shows the results of this simulation for unstable person. As it can be seen, after each bad event the person tries to recover his situation. However, if ER is off, the mood will not raise to 0.5. In contrast, when ER is ON, after each bad event, moods fall to below 0.5 and again recover to a value higher than 0.5.

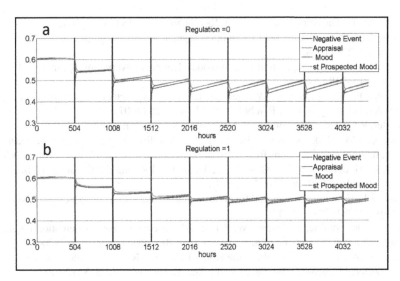

Fig. 4. Simulation results of the second scenario for person with unstable characteristics. a) without emotion regulation b) with emotion regulation

5   Discussion

In this paper, a computational model has been presented for the effect of emotion regulation (cf. [7]) on long term dynamics of mood. This model was used to analyze (by performing simulation experiments) the effect of emotion regulation on the mood level of people with different characteristics. The example simulation results presented have shown how the emotion regulation can prevent the depression in unstable persons, and postpone it in very unstable persons. This is in line with literature addressing the effect of stressful events on depression, such as [10] and [11]

In future work, a focus will be on modeling the effect of learning emotion regulation, i.e., to learn to generate positive beliefs about different events. Such learning can be supported, for example, by training in real or virtual training environments.

References

1. Hebb, D.O.: The organization of behavior: A neuropsychological theory. Psychology Press (2002)
2. Damasio, A.: The feeling of what happens: Body Emot. Mak. Conscious. Vintage, Lond (2000)
3. Oatley, K., Johnson-Laird, P.N.: Towards a cognitive theory of emotions. Cogn. Emot. 1, 29–50 (1987)
4. Schwarz, N., Clore, G.L.: Mood, misattribution, and judgments of well-being: Informative and directive functions of affective states. J. Pers. Soc. Psychol. 45, 513 (1983)
5. Papez, J.W.: A proposed mechanism of emotion. Arch. Neurol. Psychiatry 38, 725–743 (1937)

6. Dalgleish, T.: The emotional brain. Nat. Rev. Neurosci. 5, 583–589 (2004)
7. Gross, J.J.: The emerging field of emotion regulation: An integrative review. Rev. Gen. Psychol. 2, 271 (1998)
8. Gross, J.J.: Emotion regulation in adulthood: Timing is everything. Curr. Dir. Psychol. Sci. 10, 214–219 (2001)
9. Cuijpers, P., Van Straten, A., Warmerdam, L.: Behavioral activation treatments of depression: A meta-analysis. Clin. Psychol. Rev. 27, 318–326 (2007)
10. Kessler, R.C.: The effects of stressful life events on depression (1997)
11. Monroe, S.M., Harkness, K.L.: Life stress, the " kindling" hypothesis, and the recurrence of depression: Considerations from a life stress perspective. Psychol. Rev. 112, 417 (2005)
12. Both, F., Hoogendoorn, M., Klein, M.C., Treur, J.: Modeling the Dynamics of Mood and Depression. In: ECAI, pp. 266–270 (2008)
13. Gross, J.J.: Emotion regulation: Affective, cognitive, and social consequences. Psychophysiology 39, 281–291 (2002)
14. Goldin, P.R., McRae, K., Ramel, W., Gross, J.J.: The neural bases of emotion regulation: Reappraisal and suppression of negative emotion. Biol. Psychiatry 63, 577 (2008)
15. Manzoor, A., Treur, J.: Modelling the Role of Emotion Regulation and Contagion in Socially Affected Decision Making. Procedia-Soc. Behav. Sci. 97, 73–82 (2013)
16. Kim, M.J., Loucks, R.A., Palmer, A.L., Brown, A.C., Solomon, K.M., Marchante, A.N., Whalen, P.J.: The structural and functional connectivity of the amygdala: From normal emotion to pathological anxiety. Behav. Brain Res. 223, 403–410 (2011)
17. Phelps, E.A., Delgado, M.R., Nearing, K.I., Le Doux, J.E.: Extinction learning in humans: Role of the amygdala and vmPFC. Neuron 43, 897–905 (2004)
18. Sotres-Bayon, F., Bush, D.E., LeDoux, J.E.: Emotional perseveration: An update on prefrontal-amygdala interactions in fear extinction. Learn. Mem. 11, 525–535 (2004)
19. Yoo, S.-S., Gujar, N., Hu, P., Jolesz, F.A., Walker, M.P.: The human emotional brain without sleep—a prefrontal amygdala disconnect. Curr. Biol. 17, R877–R878 (2007)
20. Gross, J.J., Barrett, L.F.: Emotion Generation and Emotion Regulation: One or Two Depends on Your Point of View. Emot. Rev. 3, 8–16 (2011)
21. American Psychiatric Association et al.: Diagnostic and statistical manual of mental disorders, 4th edn. Am. Psychiatry. Assoc., Washington, DC (1994)

Low-Cost Representation
for Restricted Boltzmann Machines

Son N. Tran and Artur d'Avila Garcez

Department of Computer Science
City University London, UK
Son.Tran.1@city.ac.uk, aag@soi.city.ac.uk

Abstract. This paper presents a method for extracting a low-cost representation from restricted Boltzmann machines. The new representation can be considered as a compression of the network, requiring much less storage capacity while reasonably preserving the network's performance at feature learning. We show that the compression can be done by converting the weight matrix of real numbers into a matrix of three values $\{-1, 0, 1\}$ associated with a score vector of real numbers. This set of values is similar enough to Boolean values which help us further translate the representation into logical rules. In the experiments reported in this paper, we evaluate the performance of our compression method on image datasets, obtaining promising results. Experiments on the MNIST handwritten digit classification dataset, for example, have shown that a 95% saving in memory can be achieved with no significant drop in accuracy.

Keywords: Restricted Boltzmann Machines, Low-cost Representation, Knowledge Extraction.

1 Introduction

Restricted Boltzmann Machines (RBMs) [5,1] are a generative model which can learn interesting hidden features from data. In many applications, RBMs have been shown advantageous over traditional feature extraction at training classifiers, especially when RBMs are stacked onto a deep network to form, e.g. a Deep Belief Network [2]. However, due to their structural complexity, these feature learning models require a large storage of memory. In this paper, we propose a method for extracting a low-cost representation from RBMs, as a step towards the use of Deep Belief Networks in memory-limited devices. The low-cost representation is expected to require less memory for storage, while reasonably preserving the performance of the RBMs. Furthermore, we show that our low-cost representation can be translated into a logic-like language, thus providing an intuitive understanding of the data and being compatible with boolean circuits.

We are concerned with the use of RBMs as feature extractors whereby the hidden features are generated from a logistic function of the weighted combination of the original features, obtained from a dataset. For the low-cost representation, we use the same logistic function with some changes to the weights. In particular, we convert the weight matrix of real values from an RBM into a matrix

C.K. Loo et al. (Eds.): ICONIP 2014, Part I, LNCS 8834, pp. 69–77, 2014.

where each entry has three possible states $\{-1, 0, 1\}$, and each column vector has a real-valued score associated with it. Since there are only three possible values for each element of the matrix, one needs to use only two bits to represent them. Furthermore, by removing any rows where the low-cost matrix has value zero, a further compression can be achieved. The result, as we shall see, is that the relationships among the input variables in an RBM can be represented by logic rules, similarly to [7]. Experiments on the MNIST handwritten digit classification dataset have shown that a 95% saving in memory can be achieved with no significant drop in accuracy, and up to 99% saving can be achieved with the low-cost feature extraction still offering a significant improvement on the baseline SVM classification applied directly to the input data.

The remainder of the paper is organized as follows. In Section 2, we present background on RBMs. In Section 3, we present the low-cost representation and the network-compression algorithm. In Section 4, we recall the relationship between the compressed representation and logic. Section 5 contains experimental results on the MNIST and related datasets, and Section 6 concludes the paper and discusses directions for future work.

2   Background

A Restricted Boltzmann Machine [5] is a two-layer symmetric connectionist system with no connections between units in the same layer. We use V and H to denote, respectively, the visible and hidden layers of an RBM. We use $W \in \mathbb{R}^{I \times J}$, where I is the number of visible units and J is the number of hidden units, to denote the RBM's weight matrix, with w_{ij} denoting the connection weight from visible unit i to hidden unit j. The energy function of a network with states of visible layer $V = v$ and hidden layer $H = h$ is given by:

$$\mathbf{E}(v, h) = -\sum_{ij} v_i w_{ij} h_j - \sum_i a_i v_i - \sum_j b_j h_j \qquad (1)$$

Here, w_{ij}, a_i, b_j are the connection weights, biases for visible units, and biases for hidden units, respectively. The joint distribution of the network's states is: $P(v, h) = \frac{e^{-\mathbf{E}(v,h)}}{Z}$, with $Z = \sum_{v,h} e^{-\mathbf{E}(v,h)}$. Given the state of a layer, the units in the other layer are conditionally independent and can be sampled from the following distributions:

$$P(v_i|h) = \sigma(\sum_j w_{ij} h_j + a_i)$$
$$P(h_j|v) = \sigma(\sum_i w_{ij} v_i + b_j) \qquad (2)$$

with $\sigma(x) = \frac{1}{1+e^{-x}}$, called a logistic function.

Training RBMs is difficult due to the computational intractability of the partition function Z. However, one can use efficient approximation methods such as Contrastive Divergence [1] to estimate such parameters reasonably well.

In previous work, [7], we have shown that the memory cost of an RBM can be reduced by pruning low-scoring feature detectors. In what follows, we show that the memory cost can be further reduced by converting the weight matrix into a three-valued matrix and a vector of scores.

3   Low-Cost Representation for RBMs

RBMs have been used as a powerful tool for the extraction of features from a dataset. Normally, the RBM is used to perform a non-linear transformation of the data from its original space v to the space of hidden variables h. The probability f_j of unit h_j in the hidden layer being activated given an input v, is given by (from Eq. (2)):

$$f_j = \sigma(\mathbf{w}_j^\top v + b_j) \tag{3}$$

where $\mathbf{w}_j$ is column vector j in the weight matrix W of the RBM, and is also known as a basis vector or feature detector.

We now propose a function to transform the features from the data space to the same hidden space as follows:

$$f_j' = \sigma(c_j \mathbf{s}_j^\top v + b_j) \tag{4}$$

where c_j is a real value and $\mathbf{s}_j \in \{-1, 0, 1\}^I$ is a low-cost vector having the same size as $\mathbf{w}_j$, with element s_{ij} having one of the values -1, 0, or 1. In order to make our proposed features f_j' useful, we need to be able to extract c_j and $\mathbf{s}_j$ from the feature detector $\mathbf{w}_j$ of the RBM such that f_j' approximates f_j. We do this by minimizing the squared Euclidean distance between the basis vector $\mathbf{w}_j$ and the low-cost vector $\mathbf{s}_j$ weighted by c_j, as follow:

$$d(\mathbf{w}_j, c_j \mathbf{s}_j) = \frac{1}{I} \sum_i \|w_{ij} - c_j s_{ij}\|_1^2 \tag{5}$$

Note that (5) is quadratic, the optimal value of c_j can be found by setting the derivatives of the squared Euclidean to zeros, such that:

$$c_j = \frac{\sum_i w_{ij} s_{ij}}{\sum_i s_{ij}^2} \tag{6}$$

Since the value of s_{ij} is in the set $\{-1, 0, 1\}$, we have:

$$\|w_{ij} - c_j s_{ij}\|_1^2 = \|\mathrm{abs}(w_{ij}) - c_j \frac{s_{ij}}{\mathrm{sign}(w_{ij})}\|_1^2 = \begin{cases} (\mathrm{abs}(w_{ij}) + c_j)^2 & \text{if } s_{ij} \neq \mathrm{sign}(w_{ij}) \\ (\mathrm{abs}(w_{ij}) - c_j)^2 & \text{if } s_{ij} = \mathrm{sign}(w_{ij}) \\ \mathrm{abs}(w_{ij})^2 & \text{if } s_{ij} = 0 \end{cases} \tag{7}$$

Here, $\mathrm{abs}(w_{ij})$ and $\mathrm{sign}(w_{ij})$ are functions that return the absolute value and sign of w_{ij}, respectively. Since $(\mathrm{abs}(w_{ij}) + c_j)^2 > (\mathrm{abs}(w_{ij}) - c_j)^2$ and $(\mathrm{abs}(w_{ij}) + c_j)^2 > \mathrm{abs}(w_{ij})^2$, $s_{ij} = 0$ will minimize the Euclidean distance if and only if $\mathrm{abs}(w_{ij})^2 \leq (\mathrm{abs}(w_{ij}) - c_j)^2$ from which $c_j \geq 2 \times \mathrm{abs}(w_{ij})$.

We are now able to describe the procedure to extract c_j and s_{ij} from the RBM, as follows:

Step 1: Initialize s_j so that $s_{ij} = \text{sign}(w_{ij})$ [1]

Step 2: For each hidden node j, compute c_j using Eq. (6)

Step 3: For each connection weight w_{ij}, set $s_{ij} = 0$ if $c_j \geq 2 \times \text{abs}(w_{ij})$

Step 4: Compute c_j; if c_j is unchanged then stop, otherwise go to Step 3.

4   Relation to Logic Representation

In this section, we show that the use of low-cost vectors is similar to the confidence logic representation from [6,7]. A confidence-based logic formula is a logic programming (*if-then*) implication of the form $c : \mathsf{h} \leftarrow \mathsf{b}_1 \wedge \cdots \wedge \mathsf{b}_n$, where h is a logical atom and each b_i, $1 \leq i \leq n$, is a logical literal (an atom or its negation), labelled by a real-valued number c called a *confidence-value*. For example: $1.5 : \mathsf{h} \leftarrow \mathsf{b}_1 \wedge \neg \mathsf{b}_2 \wedge \mathsf{b}_3$ associates hypothesis (hidden unit) h with beliefs (visible units) $\mathsf{b}_1, not\ \mathsf{b}_2, \mathsf{b}_3$ with confidence value 1.5.

If we remove every s_{ij} whose $c_j = 0$ from the low-cost vector s_j then we are able present each function $f'_j = \sigma(c_j s_j^\top v + b_j)$ in the following confidence-logic form:

$$c_j : f'_j \leftarrow \bigwedge_{s_{i'j}=1} v_{i'j} \wedge \bigwedge_{s_{i''j}=-1} \neg v_{i''j} \tag{8}$$

In what follow, we present two examples, using the XOR function and the MNIST images data set, to illustrate the above translation.

Example 1. XOR function

We trained an RBM with 10 hidden units on the truth-table of the XOR function with 3 variables X, Y, Z. Suppose that we would like Z to be our target variable (notice that we could have equally chosen X or Y without retraining the model). Below, we present the confidence-logic rules in which literal h_j appears together with target literal z or $\neg \mathsf{z}$. By combining rules of the form $\mathsf{h} \leftarrow \mathsf{z}$ and $\mathsf{z} \leftarrow \mathsf{h}$ into $\mathsf{h} \leftrightarrow \mathsf{z}$, we obtain the set of rules below; treating h_j as an intermediate concept and combining each pair of rules to obtain rules relating x, y and z directly, and ignoring the confidence-values, one obtains the four rules for XOR, e.g., from $\mathsf{h}_2 \leftarrow \neg \mathsf{x} \wedge \neg \mathsf{y}$ and $\neg \mathsf{z} \leftrightarrow \mathsf{h}_2$, one obtains $\neg \mathsf{z} \leftarrow \neg \mathsf{x} \wedge \neg \mathsf{y}$.

$6.843 : \mathsf{h}_2 \leftarrow \neg \mathsf{x} \wedge \neg \mathsf{y}; 4.008 : \neg \mathsf{z} \leftrightarrow \mathsf{h}_2;$	$5.342 : \mathsf{h}_3 \leftarrow \mathsf{x} \wedge \mathsf{y}; 4.008 : \neg \mathsf{z} \leftrightarrow \mathsf{h}_3$
$3.984 : \mathsf{h}_5 \leftarrow \neg \mathsf{x} \wedge \neg \mathsf{y}; 4.008 : \neg \mathsf{z} \leftrightarrow \mathsf{h}_5;$	$2.668 : \mathsf{h}_6 \leftarrow \mathsf{x} \wedge \mathsf{y}; 4.008 : \neg \mathsf{z} \leftrightarrow \mathsf{h}_6$
$4.611 : \mathsf{h}_7 \leftarrow \neg \mathsf{x} \wedge \mathsf{y}; 4.008 : \mathsf{z} \leftrightarrow \mathsf{h}_7;$	$2.389 : \mathsf{h}_8 \leftarrow \mathsf{x} \wedge \mathsf{y}; 4.008 : \neg \mathsf{z} \leftrightarrow \mathsf{h}_8$
$3.847 : \mathsf{h}_9 \leftarrow \mathsf{x} \wedge \neg \mathsf{y}; 4.008 : \mathsf{z} \leftrightarrow \mathsf{h}_9;$	$4.015 : \mathsf{h}_{10} \leftarrow \neg \mathsf{x} \wedge \mathsf{y}; 4.008 : \mathsf{z} \leftrightarrow \mathsf{h}_{10}$

Example 2. Handwritten Characters

We have applied the same process from the previous example to the MNIST dataset. We trained a sparse RBM[4] with 500 hidden units and 794 visible units (784 pixel variables and 10 softmax class variables). Below, we present a

---

[1] $s_{ij} = 0$ if $w_{ij} = 0$.

0.765 : Zero ← ▨ ∧ ▨ ∧ ▨ ∧ ▨ ∧ ▨ ∧ ▨

0.831 : One ← ▨ ∧ ▨ ∧ ▨ ∧ ▨ ∧ ▨ ∧ ▨

0.524 : Two ← ▨ ∧ ▨ ∧ ▨ ∧ ▨ ∧ ▨ ∧ ▨

0.687 : Three ← ▨ ∧ ▨ ∧ ▨ ∧ ▨ ∧ ▨ ∧ ▨

0.348 : Four ← ▨ ∧ ▨ ∧ ▨ ∧ ▨ ∧ ▨ ∧ ▨

visualization of the confidence-logic rules whereby a positive literal v_i is shown as a white pixel, a negative literal $\neg v_i$ is shown as a black pixel, and a literal that does not appear in the rule is shown as a grey pixel. Normally, the rules are organized, by the way in which they are obtained, into two levels: one with relations between pixel variables and hidden variables, and another with relations between hidden variables and target variables. For ease of presentation, we omit the scores (confidence-values) from the first level, and also omit the negative literals from the second level, before we replace the hidden literals (intermediate concepts) with the visible literals for visualization. Because of space restrictions, we only show 6 images per rule for 5 (out of 10) rules.

5 Experimental Results

We performed experiments with the MNIST handwritten digits dataset[2], TiCC handwritten characters dataset[3] and YALE face dataset[4]. In each dataset, we divide the data into training, validation and test set. For the MNIST dataset, we use a subset of the training data with 10,000 samples ($MNIST_{10K}$), 2000 validation samples, and 10,000 test samples for a digits recognition task (from 0 to 9). We also use the same test set to test the representation extracted from RBMs trained on the entire training set with 60,000 samples[5] ($MNIST_{60K}$). The TiCC dataset consists of 18,189 training samples, 1,250 validation samples, and 18,177 test samples for a person's letter recognition task (from A to Z). We divide the YALE dataset into a training set with 135 samples, thus 9 samples per person, and the test set with 30 samples. We used an SVM with Gaussian kernel as a classifier to measure the performance of the extracted low-cost representation in comparison with the RBMs. Model selection is performed by running a grid-like search (except for the YALE dataset) over the learning rates for the RBMs (between 0.001 and 1), cost (between 0.0001 and 100), and gamma (between 0.0001 and 100)) for the SVM, all on a log-scale. We did not select the number of hidden units in the RBMs, instead we tested RBMs with 500 and 1000 hidden units only, simply to investigate whether the size of the network

---

[2] http://yann.lecun.com/exdb/mnist/
[3] http://algoval.essex.ac.uk:8080/icdar2005/index.jsp?page=ocr.html
[4] http://vision.ucsd.edu/content/yale-face-database
[5] Here, we re-use the hyper-parameters from the experiment with 10,000 training samples.

affects the quality of the extracted low-cost representation. All the results using the MNIST dataset can be reproduced using the MATLAB code provided at https://github.com/sFunzi/Low-costRBM/. The reader can contact the authors directly if interested in the results obtained using the other datasets.

The memory needed by each type of representation, i.e. RBMs and our low-cost representation, can be defined as follows:

$$M_{RBM} = T \times C_{word} \times I \times J$$
$$M_{low-cost} = (2 \times I \times J) + (T \times C_{word} \times J) \tag{9}$$

where C_{word} is the number of bits of a computer word in a device, and T is the number of computer words of a real-valued data type. For example, in a 32-bit machine, a RBM with 784 visible units and 500 hidden units will cost $2 \times 32 \times 784 \times 500 = 25,088,000$ bits for a double precision floating point type. In the case of an implementation of low-cost vectors in a computing device which needs 2 bits to represent an element in the vector then the memory cost should be $(2 \times 784 \times 500) + (2 \times 32 \times 500) = 816,000$ bits. The ratio of memory saved by the low-cost representation over the RBM can be measured by:

$$r_{save} = \frac{M_{RBM} - M_{low-cost}}{M_{RBM}} \times 100\% \tag{10}$$

Table 1. The expected memory saving ratios for an RBM with 784 visible units and 500 hidden units using standard floating point data types in a 32-bit computer; *pruning* refers to the percentage of low-scoring hidden nodes removed

	float	double
r_{save} no pruning	93.622%	96.747%
r_{save} 20% pruning	94.898%	97.398%
r_{save} 40% pruning	96.173%	98.048%
r_{save} 60% pruning	97.449%	98.699%
r_{save} 80% pruning	98.724%	99.349%

In our experiments, we have trained RBMs using double-precision floating point weight matrices on a 32-bit computer in order to evaluate the performance of the low-cost representation in comparison with that of the original RBMs at performing feature extraction. Hence, our purpose is to compare the accuracy and ratio of memory saved of the RBMs and their low-cost representation. We assume the existence of a feasible hardware implementation of the low-cost representation. We also investigate how accuracy drops as the RBMs are pruned, in comparison with pruning of their low-cost representation with respect to the ratio of memory saved. Pruning of x% of a network (RBM or its low-cost counterpart) means that the x% lowest-scoring vectors s_j, i.e. with the smallest values of c_j, are removed, as done in [7].

Table 2 contains the accuracies of the RBMs with 500 and 1000 hidden nodes trained on four datasets, and the accuracies of their low-cost counterparts, all

Table 2. Average test set performance of RBMs in comparison with their low-cost representation on four different datasets

	TiCC	MNIST$_{10K}$	MNIST$_{60K}$	YALE face
RBM (J=500)	94.851% ± 0.033	97.198 ± 0.060	98.553% ± 0.031	95.000% ± 2.833
Low-cost	94.711% ± 0.072	97.240 ± 0.089	98.530% ± 0.040	94.333% ± 3.865
RBM (J=1000)	94.928% ± 0.016	97.245% ± 0.031	98.680% ± 0.024	97.000% ± 2.919
Low-cost	94.729% ± 0.070	97.219% ± 0.056	98.562% ± 0.035	96.667% ± 1.757

on the held-out test sets. We have run each experiment 10 times and report the mean accuracy, along with standard deviation. The results show that the performance of the low-cost representation can be almost identical to that of the RBMs, with high consistency.

Next, we evaluate the effectiveness of the low-cost representation in comparison with pruning the RBM. For both the RBM and its low-cost counterpart, one can rank and remove the low-scoring vectors s_j, for which c_j is relatively low. For the sake of comparison, we prune 20%, 40%, 60% and 80% of both the RBMs and the low-cost representation, and evaluate performance. As expected, the average test set error increases with the pruning. However, results show that more than 98% memory saving can be achieved by the low-cost representation with the feature extraction still offering a significant improvement on the baseline SVM classification obtained from the input data directly.

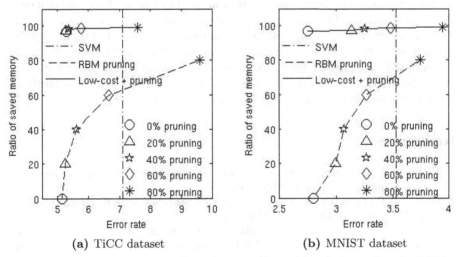

(a) TiCC dataset   (b) MNIST dataset

Fig. 1. Error rate progression in comparison with memory capacity gains for RBMs and low-cost RBMs pruned by 0, 20, 40, 60 and 80%

In order to show the usefulness of the compressed representation at feature extraction, we use the classification accuracy obtained by an SVM on the original input data as baseline. We found that for the MNIST$_{60K}$ and YALE face datasets, the features extracted by either the RBM or the low-cost representation produced only a slight improvement on the original data trained using an SVM. In the experiments with the TICC and $MNIST_{10K}$ datasets, however, feature extraction outperformed the SVMs. Therefore, we have chosen the latter two datasets to visualize and evaluate the effect of pruning, as shown in Figure 1 for RBMs containing 500 hidden units only.

In Figure 1, the SVM line indicates the test set error on the raw input data without using RBMs at all. This line separates the space into an area where the use of an RBM, low-cost or otherwise, can improve performance (on the left hand side) and an area where feature extraction, whichever the memory capacity gains, is not warranted (on the right hand side). Notice that, in the case of the MNIST dataset, since a 0.2% increase in accuracy is generally accepted as a significant improvement [3], Figure 1 shows that approximately 98% of memory capacity gains can be obtained from storing a low-cost RBM for feature extraction, while preserving a significant improvement over the baseline SVM classification applied to the raw input data.

6   Conclusions and Future Work

We have presented a method for the extraction of a low-cost representation from restricted Boltzmann machines, which may be seen as a step towards the integration of deep networks in memory limited devices. The new representation offers a compression of the network, which theoretically requires less storage memory, while preserving to some extent most of the network's performance at feature learning. In the experiments reported in this paper, it is shown that the low-cost representation proposed here is advantageous over RBMs in terms of memory efficiency. The experiments also indicate that the performance of the low-cost RBMs is almost identical, in practice, or acceptably lower than that of the full RBMs. As future work, we intend to consider details of hardware implementation and a real application using lower-memory devices.

References

1. Hinton, G.E.: Training products of experts by minimizing contrastive divergence. Neural Comput. 14(8), 1771–1800 (2002)
2. Hinton, G.E., Osindero, S., Teh, Y.-W.: A Fast Learning Algorithm for Deep Belief Nets. Neural Comp. 18(7), 1527–1554 (2006)
3. Larochelle, H., Bengio, Y.: Classification using discriminative restricted boltzmann machines. In: Proceedings of the 25th International Conference on Machine Learning, ICML 2008, pp. 536–543. ACM, New York (2008)
4. Lee, H., Ekanadham, C., Ng, A.Y.: Sparse deep belief net model for visual area v2. In: Advances in Neural Information Processing Systems. MIT Press (2008)

5. Smolensky, P.: Information processing in dynamical systems: Foundations of harmony theory. In: Parallel Distributed Processing: Volume 1: Foundations, pp. 194–281. MIT Press, Cambridge (1986)
6. Tran, S., Garcez, A.: Logic extraction from deep belief networks. In: ICML 2012 Representation Learning Workshop, Edinburgh (July 2012)
7. Tran, S.N., d'Avila Garcez, A.: Knowledge extraction from deep belief networks for images. In: IJCAI-2013 Workshop on Neural-Symbolic Learning and Reasoning (2013)

Add-if-Silent Rule for Training Multi-layered Convolutional Network Neocognitron

Kunihiko Fukushima

Fuzzy Logic Systems Institute,
680-41 Kawazu, Iizuka, Fukuoka 820-0067, Japan
fukushima@m.ieice.org
http://personalpage.flsi.or.jp/fukushima/index-e.html

Abstract. The neocognitron is a multi-layered convolutional network that can be trained to recognize visual patterns robustly. This paper discusses a new neocognitron, which uses the *add-if-silent* rule for training intermediate layers and the method of *interpolating-vector* for classifying patterns at the highest stage of the hierarchical network. By the add-if-silent rule, a new cell is generated when all postsynaptic cells are silent. The generated cell learns the activity of the presynaptic cells in one-shot, and its input connections will never be modified afterward. Thus the training process is very simple, and does not require time-consuming calculation such as the gradient descent process. This paper analyzes how the size of training set affects the performance of the neocognitron and show that the add-if-silent rule can produce feature-extracting cells efficiently even with a small number of training patterns.

Keywords: add-if-silent, neocognitron, convolutional network, pattern recognition, size of training set.

1 Introduction

Multi-layered neural networks show a large power for robust recognition of visual patterns. Training intermediate layers of a multi-layered network, however, is a difficult problem, because it is not easy to know intuitively the desired response of cells of intermediate layers. Various methods have been proposed for training intermediate layers of a multi-layered network [1,2].

The author proposed previously another learning rule, named *add-if-silent*, by which a new cell is generated if all postsynaptic cells are silent in spite of non-silent presynaptic cells [3]. The generated cell learns the activity of the presynaptic cells in one-shot, by setting its input connections to be proportional to the activity of the presynaptic cells. Once a cell is generated, its input connections will never be modified afterward. Thus the training process is very simple, and does not require time-consuming calculation such as the gradient descent process. For example, the supervised backpropagation for fine-tuning, which is used after pre-training in many deep-learning algorithms, is not required.

The neocognitron is a multi-layered convolutional network that can be trained to recognize visual patterns robustly [4,5]. This paper discusses a neocognitron,

C.K. Loo et al. (Eds.): ICONIP 2014, Part I, LNCS 8834, pp. 78–85, 2014.

which uses an improved add-if-silent rule [6] for training intermediate layers and the method of *interpolating-vector* (*IntVec*) [7] for classifying patterns at the highest stage of the network. We analyze how the size of training set affects the performance of the neocognitron and show that the add-if-silent rule can produce feature-extracting cells efficiently even with a small training set.

2 Neocognitron

2.1 Network Architecture

The neocognitron consists of layers of S-cells, which resemble simple cells in the visual cortex, and layers of C-cells, which resemble complex cells. As shown in Fig. 1, layers of S-cells and C-cells are arranged alternately in a hierarchical manner. U_{Sl}, for example, indicates the layer of S-cells of the lth stage.

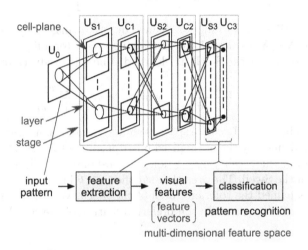

Fig. 1. The architecture of the neocognitron

Each layer of the network is divided into a number of sub-layers, called *cell-planes*, depending on the feature to which cells respond preferentially. A cell-plane is a group of cells that are arranged retinotopically and share the same set of input connections. All cells in a cell-plane have receptive fields of an identical characteristic, but the locations of the receptive fields differ from cell to cell.

The stimulus pattern is presented to the input layer, U_0. An S-cell of U_{S1} responds selectively to an edge of a particular orientation. The input connections to S-cells, except in U_{S1}, are variable and are modified through learning. After having finished the learning, S-cells come to work as feature-extracting cells.

In each stage of the network, the output of layer U_{Sl} is fed to layer U_{Cl}. Each C-cell has fixed excitatory connections from a group of S-cells of the corresponding cell-plane. Through these connections, each C-cell averages the responses of S-cells whose receptive field locations are slightly deviated.

2.2   Feature Extraction by S-cells

S-cells work as feature extractors. As illustrated in Fig. 2(a), each S-cell of layer U_{Sl} ($l \geq 2$) receives excitatory connections directly from a group of C-cells, which are cells of the preceding layer U_{Cl-1}. Let a_n be the strength of the excitatory connection from the nth C-cell, whose output is x_n. We now use vector notation $\boldsymbol{x} = (x_1, \cdots, x_n, \cdots)$ to represent the input signal to the S-cell, namely, the response of the presynaptic C-cells. The S-cell also receives an inhibitory connection of strength $-\theta$ ($0 \leq \theta < 1$) through a V-cell. The V-cell receives fixed excitatory connections from the same group of C-cells as does the S-cell and calculate the norm of $\boldsymbol{x}$. Namely, $v = \|\boldsymbol{x}\|$.

The output u of the S-cell is given by

$$u = \frac{1}{1-\theta} \cdot \varphi \left[\sum_n a_n x_n - \theta v \right] \tag{1}$$

where $\varphi[\ ]$ is a rectified linear function defined by $\varphi[x] = \max(x, 0)$. Connection $\boldsymbol{a}$ is given by $\boldsymbol{a} = \boldsymbol{X}/\|\boldsymbol{X}\|$, where $\boldsymbol{X}$ is the training vector (or a linear combination of training vectors) that the S-cell has learned. Then, (1) reduces to

$$u = \|\boldsymbol{x}\| \cdot \frac{\varphi[s-\theta]}{1-\theta} \qquad \text{where} \quad s = \frac{(\boldsymbol{X}, \boldsymbol{x})}{\|\boldsymbol{X}\| \cdot \|\boldsymbol{x}\|} \tag{2}$$

In the multi-dimensional feature space, s shows a kind of similarity between $\boldsymbol{X}$ and $\boldsymbol{x}$, which is defined by the normalized inner product of $\boldsymbol{X}$ and $\boldsymbol{x}$. If similarity s is larger than θ, the S-cell yields a non-zero response [5]. Thus θ determines the threshold of the S-cell. The area that satisfies $s > \theta$ in the multi-dimensional feature space is named the tolerance area of the S-cell. We call $\boldsymbol{X}$ the reference vector of the S-cell. It represents the preferred feature of the S-cell.

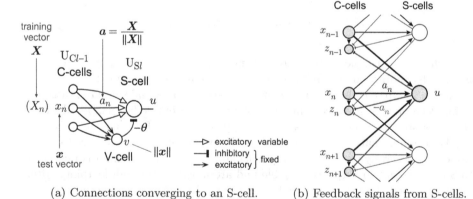

(a) Connections converging to an S-cell.   (b) Feedback signals from S-cells.

Fig. 2. Connections feed-forward to and feedback from an S-cell

2.3   Add-if-Silent Rule for Training the Intermediate Layer

The *add-if-silent* rule is used for unsupervised learning of the intermediate layer U_{S2} [3,5]. During learning, training patterns from a training set are presented one by one to the input layer U_0, and the response of layer U_{C1} works as a training stimulus for U_{S2}.

If all postsynaptic S-cells are silent for a training stimulus, a new S-cell is generated and added to the layer. The strength of the input connections of the generated S-cell is determined to be proportional to the response of the presynaptic C-cells at this moment. The learning is done in one-shot: once an S-cell is generated and added to the network, the input connections to the S-cell do not change any more. This means that the learning by the add-if-silent is a process of choosing reference vectors from the large set of training vectors.

Thus the training process is very simple, and does not require time-consuming calculation such as the gradient descent process. Presenting each training pattern only once is enough to complete the learning process.

Under the add-if-silent rule, no cell can be generated any more within the tolerance areas of existing S-cells, whose size is determined by the threshold θ of S-cells. As a result, reference vectors of generated S-cells come to distribute uniformly in the multi-dimensional feature space after presentation of a high enough number of training vectors.

If the threshold of S-cells during recognition is kept to the same value as in the learning, however, S-cells behave like grandmother cells. Namely, each test vector elicits a response from only one (or a small number of) S-cell. This is not desirable for robust recognition of deformed patterns. To make these S-cells respond in such a way that the input pattern be represented by a population coding, we use *dual threshold* for S-cells [3]. After the finish of the learning, the threshold of S-cells is set to a lower value than the threshold for the learning.

To apply the add-if-silent rule to a neocognitron, a slight modification is required because the neocognitron is a convolutional network. Each layer of the neocognitron consists of a number of cell-planes. In a cell-plane, all cells are arranged retinotopically, and share the same set of input connections. This condition of shared connections has to be kept even during learning.

In the neocognitron, generation of a new S-cell means a generation of a new cell-plane. To keep the condition of shared connections, all S-cells in the generated cell-plane are organized so as to have the same input connections as the added S-cell. Since the added S-cell thus works like a seed in crystal growth, we call it a *seed-cell*.

Suppose a training stimulus is presented to U_{S2}. If there is a retinotopic location where all S-cells are silent in spite of non-zero training stimulus, a new S-cell is generated and is added to the layer. The newly added S-cell learns this training stimulus, and plays the role of a seed-cell.

After the generation of the new cell-plane, if there still remains any area in which all postsynaptic S-cells are silent in spite of non-zero training stimulus, the same process of generating a cell-plane is repeated until the whole area is covered by non-silent S-cells.

If there are a number of retinotopic locations where postsynaptic S-cells are silent but presynaptic C-cells are not silent, we have to decide which location should be chosen first. Since presynaptic cells that do not contribute yet for eliciting responses from existing postsynaptic cells convey important information, we look for the location of such presynaptic cells.

We use negative feedbacks for this purpose [6]. Fig. 2(b) illustrates this process. Negative feedback signals from postsynaptic cells suppress the activity of the presynaptic cells, where the strength of the feedback connections is $-\boldsymbol{a}$. If a certain feature is extracted correctly by a postsynaptic cell, the negative feedback signals from the cell will suppress the activity of presynaptic cells that constitute the feature. If a cell remains unsuppressed by the negative feedback, it means that there still remains a feature that has not been extracted by any postsynaptic cells yet. Hence a new cell is generated at the retinotopic location where the activity of the presynaptic cells remain unsuppressed.

Let the feedback signal from the postsynaptic S-cell to the nth presynaptic C-cell be $y_n = a_n u$. Then, from (2),

$$\boldsymbol{y} = \boldsymbol{a}\,u = \frac{\boldsymbol{X}}{\|\boldsymbol{X}\|} \cdot u = \frac{\boldsymbol{X}}{\|\boldsymbol{X}\|} \cdot \|\boldsymbol{x}\| \cdot \frac{\varphi[s - \theta]}{1 - \theta} \tag{3}$$

The suppressed activity of the nth presynaptic cell is given by $z_n = \varphi[x_n - y_n]$. If $\boldsymbol{x} = \boldsymbol{X}$, we have $s = 1$ and then $\boldsymbol{y} = \boldsymbol{X}$. This means that, if the feature of the stimulus vector $\boldsymbol{x}$ is correctly extracted by the postsynaptic S-cell, we have $z_n = 0$. On the other hand, if a feature fails to be extracted correctly by the S-cell, z_n keeps a large value.

Actually, there are a number of postsynaptic S-cells, and feedback signals come, not only from a single S-cell, but from a number of postsynaptic S-cells. If z_n is large, it means that the stimulus feature centered at location n has not been extracted correctly by any of the existing S-cells yet.

We then choose the retinotopic locations sequentially from the place at which $\sum z_n$ is the largest, where the sum is taken for all cells that have the same retinotopic location, and generate a new S-cell to extract the feature located there. It should be noted here that $\boldsymbol{z}$ is used only for determining the retinotopic location where the add-if-silent rule is to be applied, and that the reference vector of the generated S-cell is set to be proportional, not to $\boldsymbol{z}$, but to $\boldsymbol{x}$.

2.4 Interpolating-Vector for the Highest Stage

We use the method of *interpolating-vector* (*IntVec*) for U_{S3} (the highest stage) to classify input patterns based on the features extracted by the intermediate layers [3,5,7]. The input signals to an S-cell of U_{S3} is the response of C-cells of U_{C2}. The response of the S-cell is given by $u = \|\boldsymbol{x}\| \cdot s$ from (2), because $\theta = 0$ for U_{S3}. For the economy of computational cost, analysis of the response of C-cells of U_{C2} is actually performed only at the retinotopic location where the response of the V-cell is the largest.

Before explaining the method of learning, we first explain the basic idea of classification by the IntVec. Each S-cell (hence its reference vector) has a label indicating the class to which it belongs. In the multi-dimensional feature space, we assume plane segments spanned by every trio of reference vectors of the same label. Every plane segment is assigned the same label as the reference vectors that span it. We then measure distances (based on similarity) to these plane segments from the test vector. The label of the nearest plane segment indicates the result of pattern recognition. In an initial version of the IntVec, we used line segments spanned by pairs of reference vectors, instead of plane segments spanned by trios. In this paper, we use the IntVec from trios, because it can produces a smaller recognition error than the IntVec from pairs.

The IntVec is used also for the learning. Every time a training vector x is presented during learning, we try to classify it using the IntVec. If the result of classification is wrong, or if the similarity to the nearest plane segment is smaller than a certain threshold Θ, a new S-cell is generated. The training vector x is adopted as the reference vector of the generated S-cell, which is assigned the label of the class name.

The learning is carried out in two steps. In each step, the same set of training vectors is presented. In the first step, once new reference vectors are generated, they are not modified. In the second step, however, if the result of classification is correct, the three reference vectors, that span the nearest plane segment, learn the training vector by adjusting their values toward the training vector.

3   Computer Simulation

With computer simulation, we test how the training set affect the performance of the neocognitron. For the simulation, we use handwritten digits (free writing) of the ETL1 database [8]. In each run of the simulation, training and test sets are made of patterns randomly sampled from the ETL1 database. There is no overlap between the two sets. The size of a test set is always 5,000 (500 patterns for each digit), but the size of a training set differs depending on the experiment. In each experiment shown below, we repeated the simulation seven times and averaged the results. Error bars show the standard deviation.

3.1   Effect of the Size of Training Set

To see how the size of training set affects the recognition error, we measure the recognition error under different sizes of training set. Sizes of the training set are chosen independently for layers U_{S2} and U_{S3}. Since the recognition error is affected, not only by the size of training set, but also by threshold Θ for the IntVec, we first compared the effect of the size of training set under the same value of $\Theta = 0$. It should be noted here that the recognition error can be reduced further by choosing a larger value of Θ (See 3.2).

Fig. 3 show how N_{L2} (the size of training set for U_{S2}) affects the recognition error. The curves with thin dotted lines show the cases where N_{L3} (the size of training set for U_{S3}) is fixed. The curve with thick line shows the case with $N_{L3} = N_{L2}$ (the case where the same training set is used for both U_{S2} and U_{S3}).

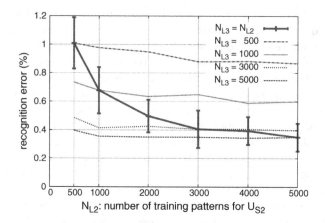

Fig. 3. Recognition error vs. N_{L2}. N_{L2}: the size of training set for U_{S2}. N_{L3}: the size of training set for U_{S3}. Here, threshold for the IntVec is $\Theta = 0$.

As can be seen from these curves, the recognition error is affected largely by N_{L3}, but not so much by N_{L2}. In other words, the effect of N_{L2} is very small, and it is almost sufficient to use $N_{L2} = 1,000$. Even a smaller size of $N_{L2} = 500$ does not increase the recognition error so much.

The reason for this can be explained as follows. The intermediate stages of the neocognitron are trained to extract local features. These local features are chosen during the learning, based on the difference in shape between local features, independently of the class name of the pattern that contains these local features. During the recognition, the same local features can be used in common for classifying entire patterns of different classes. Hence the use of a large set of training patterns is not necessarily required for training intermediate layers.

On the other hand, it is very important to present varieties of training patterns during the supervised learning for U_{S3}. The recognition error is mainly affected by N_{L3} (the size of the training set for U_{S3}).

3.2   Effect of Threshold for the Interpolating-Vector

The recognition error of the neocognitron is also affected by the value of threshold Θ during the training of U_{S3} by the IntVec. A higher value of Θ usually reduces the recognition error but increases K_{S3}, the number of cell-planes generated in U_{S3}. A larger value of K_{S3} requires a larger computational cost for classification by the IntVec. Fig. 4 shows how the recognition error changes with Θ. In the figure, the error rate is plotted as the ordinate against K_{S3} as the abscissa.

The curves in the figure show the cases with $N_{L3} = 3,000$ and $5,000$. Although each curve shows the case of $N_{L2} = N_{L3}$, we can have almost the same result even if N_{L2} is fixed to a smaller number, say, around $N_{L2} \approx 1,000$, because the error rate is affected mostly by N_{L3}, and little by N_{L2}.

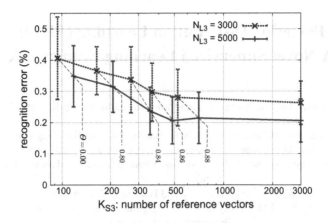

Fig. 4. Recognition error vs. K_{S3} under different values of threshold Θ during the training by the IntVec. The curves show the case of $N_{L2} = N_{L3}$.

4 Discussion

This paper has shown that the add-if-silent is a powerful and efficient rule for the learning of intermediate layers of multi-layered convolutional networks. The add-if-silent is an unsupervised learning method with very small computational cost. Different from the gradient-descent algorithms, for example, it does not require any repetitive calculation.

References

1. Hinton, G.E., Osindero, S., Teh, Y.: A Fast Learning Algorithm for Deep Belief Nets. Neural Computation 18, 1527–1554 (2006)
2. Bengio, Y.: Learning deep architectures for AI. Foundations and Trends in Machine Learning 2(1), 1–127 (2009)
3. Fukushima, K.: Training Multi-layered Neural Network Neocognitron. Neural Networks 40, 18–31 (2013)
4. Fukushima, K.: Neocognitron: A Self-organizing Neural Network Model for a Mechanism of Pattern Recognition Unaffected by Shift in Position. Biol. Cybernetics 36(4), 193–202 (1980)
5. Fukushima, K.: Artificial Vision by Multi-layered Neural Networks: Neocognitron and its Advances. Neural Networks 37, 103–119 (2013)
6. Fukushima, K.: One-shot Learning with Feedback for Multi-layered Convolutional Network. In: Wermter, S., Weber, C., Duch, W., Honkela, T., Koprinkova-Hristova, P., Magg, S., Palm, G., Villa, A.E.P. (eds.) ICANN 2014. LNCS, vol. 8681, pp. 291–298. Springer, Heidelberg (2014)
7. Fukushima, K.: Interpolating Vectors for Robust Pattern Recognition. Neural Networks 20(8), 904–916 (2007)
8. ETL1 database, http://projects.itri.aist.go.jp/etlcdb/

Posterior Distribution Learning (PDL): A Novel Supervised Learning Framework

Enmei Tu[1], Jie Yang[1,*], Zhenghong Jia[2], and Nicola Kasabov[3]

[1] Institute of Image Processing and Pattern Recognition, Shanghai Jiao Tong University, China
hellotem@hotmail.com, jieyang@sjtu.edu.cn
[2] School of Information Science and Engineering, Xinjiang University, Urumqi, 830046, China
jzhh@xju.edu.cn
[3] The Knowledge Engineering and Discovery Research Institute,
Auckland University of Technology, Auckland, New Zealand
nkasabov@aut.ac.nz

Abstract. In order to obtain a robust supervised model with good generalization ability, traditional supervised learning method has to be trained with sufficient well labeled and uniformly distributed samples. However, in many real applications, the cost of labeled samples is generally very expensive. How to make use of ample easily available unlabeled samples to remedy the insufficiency of labeled samples to train a supervised model is of great interest and practical significance. In this paper we propose a new supervised learning framework, Posterior Distribution Learning (PDL), which could train a robust supervised model with very a few labeled samples by including those unlabeled samples into training stage. Experimental results on both synthetic and real world data sets are presented to demonstrate the effectiveness of the proposed framework.

Keywords: distribution learning, nonlinear regression, manifold classification.

1 Introduction

Supervised learning method is widely used in various areas, because once it is well trained, it builds a model in the whole input space and thus can predict any unseen sample with high speed and good accuracy. However, in order to obtain such a model with good generalization ability, one needs to train a supervised classifier with sufficient well labeled and uniformly distributed samples. But in many real applications, the cost of labeled samples is generally very expensive, as the labeling process usually takes much time and resource. This poses an obstacle to applying supervised learning method to those applications in which one needs to classify large amount of unlabeled samples with a very few labeled samples. How to make use of the large quantity of easily available unlabeled samples to train a supervised learning classifier and, meanwhile, to reduce the demand and requirement upon the labeled samples is still an interesting problem and of great practical significance.

---

* Corresponding author.

C.K. Loo et al. (Eds.): ICONIP 2014, Part I, LNCS 8834, pp. 86–94, 2014.

The problem of incorporating unlabeled samples to remedy the insufficiency of labeled samples to improve the performance of supervised learning method has been studied for many years and previous works generally can be casted into two categories. One is the co-training strategy [1-4]. These algorithms either require the data set has two or more distinct views, each of which is sufficient to make good classification alone, or require different classifiers can discover the diversity in the data set. But most of real world data sets do not meet these requirements. The other category is self-training [5-8]. These algorithms are restricted to binary classification tasks, in which only labeled samples of one class (called positive samples) are known and labeled samples of the other class (called negative samples, possibly a mixture of several classes) are unknown. A theoretical study of this strategy can be found in [9].

In this paper we propose a novel supervised learning framework which contains two steps to incorporate the distribution of unlabeled samples to train a robust supervised model with a few labeled samples: the posterior probability estimation and the posterior distributions regression. The first step estimates the posterior probabilities of each sample in a part of (or the whole) the data set and the second step fit a single multivariate posterior distribution function for all classes in the data space. When a new sample comes, the posterior distribution function can give directly its posterior probability to each class and thus the class label can be obtained using minimum Bayes error rule. Unlike previous works, the new supervised learning framework has the following characteristics: (1) it does not put constraints on the data set, such as multi-view property; (2) it is multi-class and more time efficient because it does not require training several classifiers iteratively. Instead, it fits a single model in the input space; (3) most importantly, it can greatly improve classification results by incorporating the distribution of unlabeled sample. Experimental results on both synthetic and real world data sets are presented to demonstrate validity and effectiveness of the proposed framework and algorithm.

2   Posterior Distribution Learning (PDL)

Let us consider a data set $X = \{x_1, x_2, ..., x_{t-1}, x_t, x_{t+1}, ..., x_n\}$. The first t samples are labeled by $\omega_i \in \{1, 2, ...C\}, i = 1...t$ and the rest samples are unlabeled. We use $X_T = \{x_1, x_2, ..., x_{t-1}, x_t\}$ and $X_U = \{x_{t+1}, x_{t+2}, ..., x_n\}$ to denote the labeled sample set and unlabeled sample set, respectively. With a little abuse of notation, we also use X (similarly, X_T and X_U) to denote the data matrix $(x_1, x_2, ..., x_n) \in R^{d \times n}$. We define a learning set $X_L = \{x_1, x_2, ..., x_{t-1}, x_t, x_{t+1}, ..., x_l\}$, which contains all the labeled samples and $l - t$ ($l \leq n$) unlabeled samples in X.

2.1   A New Supervised Learning Framework

The diagram of the proposed supervised learning framework is shown in Figure 1.

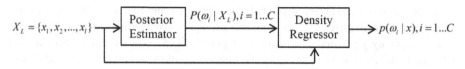

Fig. 1. Diagram of the proposed supervised learning framework that includes distribution of unlabeled samples into training stage

The posterior estimator computes $P(\omega_i \mid x_j)$, the posterior probability of $x_j, j = 1..l$ coming from class $\omega_i, i = 1..C$ .We require the output of the Estimator to have the following properties:

- $\forall x_j \in X_L, P(\omega_j = i \mid x_j) \geq 0; i = 1..C$.

- $\forall x_j \in X_L, \sum_{i=1}^{C} P(\omega_j = i \mid x_j) = 1$.

- $x_j \in X_L$, $P(\omega_j = i \mid x_j) = 1$ if x_j is labeled to class i .

- $x_j \in X_L$, $P(\omega_j = i \mid x_j) > P(\omega_j = k \mid x_j)$, $k \neq i$ and $k = 1..C$, if x_j is unlabeled and more similar to labeled samples in class i than other classes.

Let $F(x_j) = \left(P(\omega_j = 1 \mid x_j), P(\omega_j = 2 \mid x_j), ..., P(\omega_j = C \mid x_j) \right) \in R^C$, then $F(x_j)$ can be deemed as a sample point of a continuous function $F(x)$ and it can be regressed by the Density Regressor. After this, for a new sample x , its posterior probability can be obtained directly with $F(x)$ and the class label is determined by the minimum Bayes error rule.

2.2   Posterior Estimator: Propagate Posterior Probability from Labeled Samples to Unlabeled Samples

The posterior probabilities of a labeled sample are known, i.e. $P(\omega_j = i \mid x_j) = 1$ if x_j is from class i and $P(\omega_j = k \mid x_j) = 0, k \neq i$. We set $P(\omega_j = i \mid x_j) = 0, i = 1..C$ if x_j is unlabeled. Define an initial posterior probability matrix $F_T \in R^{l \times C}$ over X_L as $(F_T)_{ij} = P(\omega_j = i \mid x_j)$. Then we construct a full connected graph $G(X_L, A)$ on learning set X_L , where the adjacency matrix $A = \{A_{ij} \mid i, j = 1..l\}$ is computed by Gaussian kernel. Then the posterior probability matrix F_L is computed as follows

1. Compute $S = D^{-1/2} A D^{-1/2}$, where D is a diagonal matrix with $D_{ii} = \sum_{j=1}^{l} A_{ij}$.

2. Evaluate equation $\tilde{F} = I_{rate} S \tilde{F} + (I - I_{rate}) F_T$ repeatedly until convergence.

3. Normalize $F_L = G^{-1} \tilde{F}$, where G is a diagonal matrix with $G_{ii} = \sum_{j=1}^{C} \tilde{F}_{ij}$.

where I is the identity matrix and I_{rate} is a diagonal matrix defining the propagation rates on different directions. The method to determine I_{rate} will be described in Section 2.4. We treat the posterior probability of the labeled samples as information sources (or activation sources) and propagate this information iteratively over the graph [10, 11]. In the first iteration $\tilde{F} = F_T$.

2.3   Density Regressor: A Robust Multivariate Nonlinear Model to Regress the Posterior Distribution

Denote $F(x_j) = \left(P(\omega_j = 1 \mid x_j), P(\omega_j = 2 \mid x_j),..., P(\omega_j = C \mid x_j) \right)$, i.e. $F(x_j)$ is the jth row of F_L. We build a model $F(x) = \left(p(\omega = 1 \mid x), p(\omega = 2 \mid x),..., p(\omega = C \mid x) \right) \in R^C$, where the posterior density function of class i is $p(\omega = i \mid x) = w_i^T \varphi_i(x) + b_i$ and $\varphi_i(\cdot)$ is a unknown nonlinear mapping function which maps the input space to the so-called feature space, whose dimensionality are unknown and can be very large (possibly infinite). Without loss of generality, we assume all mapping functions are same, written as $\varphi(x)$. So $F(x) = W^T \varphi(x) + b$, where $W = (w_1, w_2,..., w_C)^T$, whose row dimension is same as that of the feature space. Note that $F(x)$ is a vector-valued function.

We compute $F(x)$ by solving the following optimization problem

$$\min J(W, b, E) = \frac{1}{2}\|W\|^2 + \frac{1}{2}\gamma\sum_{j=1}^{l} v_j \|r_j\|^2, \ s.t. \ F(x_j) = W^T \varphi(x_j) + b + r_j, j = 1..l \quad (1)$$

where $E = (r_1, r_2,..., r_l) \in R^{C \times l}$ is the error matrix and γ is a regularization parameter. It is worth noting that because the mapping function is unknown and the dimensionality of the feature space can be arbitrary large, problem (1) is quite different from ridge regression (or linear regression or Tikhonov regularization) and cannot be solved directly in primal form. $v = (v_1, v_2,..., v_l)$ is a weight vector to reduce the sensitivity of the sum of squared error (SSE) to noise and outliers. Method to obtain v will be given in Section 2.4. One can also simply set v_k to 1. In this case, problem (1) becomes a regular least squares support vector machine [12].

It can be shown that problem (1) is a convex optimization problem. According to KKT conditions, the optimal solution meets the following linear equations

$$\begin{bmatrix} 0 & e^T \\ e & K + \eta V \end{bmatrix} \begin{bmatrix} b^T \\ \Lambda^T \end{bmatrix} = \begin{bmatrix} 0 \\ F_L^T \end{bmatrix} \quad (2)$$

where K is a kernel matrix $K_{ij} = \varphi(x_i)^T \varphi(x_j)$ and $\Lambda = (\lambda_1, \lambda_2,..., \lambda_l) \in R^{c \times l}$ is the Lagrange multiplier matrix. V is a diagonal matrix with $V_{ii} = v_i$ and $e = (1,1,...,1)^T \in R^l$. After obtaining b and Λ, we can evaluate $F(x)$ at any point x by

$$F(x) = \Lambda \hat{K} + b \qquad (3)$$

where $\hat{K} = \left(\varphi(x_1), \varphi(x_2), ..., \varphi(x_l)\right)^T \varphi(x)$. Note that to compute $F(x)$ one only needs to know $\varphi(x)^T \varphi(x)$ even the concrete form of $\varphi(x)$ is unknown. This is the so-called kernel trick. We define $K_{ij} = \varphi(x_i)^T \varphi(x_j) = \exp(-\|x_i - x_j\|^2 / 2\sigma^2)$, the Gaussian kernel which performs well for most situations.

2.4   The Propagation Rate Matrix and the Weight Vector

The propagation rate matrix I_{rate} has to be designed so that the propagation speed alone high sample density direction is large while speed along sparse sample density direction is small, because sparse region usually means boundaries or overlapping region of classes and low speed can effectively suppress incorrect propagation. Here I_{rate} is computed by $(I_{rate})_{ii} = \exp\left(-\bar{d}_i^2 / 2\sigma^2\right)$, where $\bar{d}_i$ is the mean distances between x_i and its N nearest neighbors (N is empirically set to 20). Following a similar approach as in [11], it is easy to demonstrate the convergence of the algorithm.

The weight vector of $x_j, j = 1..l$ is set to $v_j = \max_{i=1..C} F(x_j) - \max_{k=1..C, k \neq i} F(x_j)$, the difference between its largest and the second largest posterior probability. Because equal posterior probabilities indicate that the sample is ambiguous to all classes and thus the posterior probability is less informative and unreliable, so the weight should be small.

3   Experimental Results

3.1   Synthetic Data Sets

We conduct experiments on two synthetic data sets: two-moon data set and two-circle data set. The data sets are shown in Fig 2, in which the black dots are unlabeled samples and the color shapes are labeled samples, one labeled sample for each class.

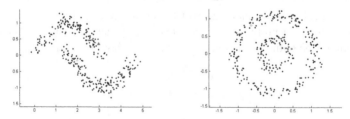

Fig. 2. Two synthetic data sets. The black dots are unlabeled samples. The green rounds and the red triangles are labeled samples, one for each class.

The experimental results of PDL are shown in Fig 3 and results of SVM in Fig 4. For SVM, we use radial basis kernel with $\sigma = 0.1$. From these results we can see that given a very few labeled samples and many unlabeled samples, PDL can train a supervised model with a very good generalization ability. In contrast, traditional supervised learning method fails to obtain a reliable supervised model because the labeled samples contain very limited information to train the model with good generalization.

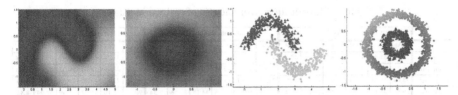

Fig. 3. Experimental results of PDL. Top: posterior probability distribution $F(x)$ learnt in the input space; bottom: classification result of out-of-sample data generated independently.

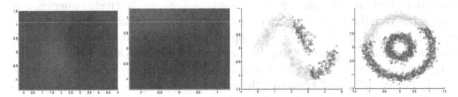

Fig. 4. Experimental results of SVM. Top: posterior probability distribution $F(x)$ learnt in the input space; bottom: classification result of out-of-sample data generated independently.

3.2   Comparison with Traditional Supervised Learning Method

We conduct experiments on 6 real world data sets: USPS hand-writing digit image and UCI repository data sets. The information of the data sets is listed in Table 1.

Table 1. Inforamtion of the experimental data sets

	USPS	segmentation	banknote	pendigits	skin	miniBoo
n	9298	2086	1348	10992	245057	130064
d	256	19	4	16	3	50
C	10	6	2	10	2	2

The split of X_L and X_U is random and the size of them are $0.7n$ and $0.3n$, respectively. The baseline algorithms are: support vector machine (SVM), k nearest neighbors (kNN), artificial neural networks (ANN), Naïve Bayes (NB) and decision tree (DT). We use radial basis kernel for SVM and three-layer back-propagation networks with $\lceil d/2 \rceil_{10}$ neurons in the hidden layer for ANN, where $\lceil x \rceil_{10}$ is the smallest integer greater than or equal to the $\max(x,10)$. Baseline algorithms are trained with X_T. Each algorithm runs 10 times and the final result is the average error rate of classifying X_U over 10 runs. We adopt the grid-search strategy to tune the parameter and the one producing lowest error rate is selected. Experimental results are shown in Table 2.

We can see that by taking the unlabeled samples into training stage, PDL outperforms traditional supervised learning algorithms. It is worth noting that for traditional supervised learning algorithms trained with a very few labeled samples, the error rate does not always decrease as the number of labeled samples increase. This indicates

Table 2. Classification error rate (%) of real world data set

	1 labeled sample / class						3 labeled samples / class						5 labeled samples / class					
	SVM	kNN	ANN	NB	DT	PDL	SVM	kNN	ANN	NB	DT	PDL	SVM	kNN	ANN	NB	DT	PDL
usps	59.77	63.66	65.61	45.97	84.73	**45.41**	49.79	41.95	42.20	97.85	74.77	**23.53**	43.16	31.08	37.87	85.47	69.14	**18.76**
segmentation	38.21	56.76	55.35	82.97	71.58	**30.27**	29.60	32.88	42.73	50.00	63.58	**20.10**	25.88	28.43	25.05	30.53	29.82	**18.21**
banknote	38.69	45.11	38.27	38.69	45.11	**8.72**	27.75	32.81	17.01	41.19	44.52	**4.94**	16.10	20.52	12.20	29.33	16.79	**2.89**
pendigits	89.63	67.19	47.21	66.59	89.08	**18.47**	89.57	28.17	21.68	44.18	80.56	**7.34**	90.39	21.16	19.77	35.99	61.04	**8.82**
skin	71.63	71.61	44.07	83.13	71.61	**15.95**	28.62	28.43	16.17	18.62	71.40	**12.69**	71.47	25.37	20.63	23.28	27.37	**7.96**
miniBoo	71.80	71.80	29.97	95.11	71.80	**22.27**	71.85	29.93	31.96	71.92	71.85	**19.11**	71.87	18.56	35.21	27.68	26.83	**18.31**

that given a very small number of labeled samples, some of the labeled samples may bring negative influence to the supervised model and the impact can be very large because of insufficiency of training samples. But for PDL, this negative effect is overcome because the posterior propagation algorithm performs a kind of graph diffusion, so even a few and not well positioned labeled samples can be utilized correctly. This is the key difference between PDL and traditional supervised learning method for training a supervised model.

3.3   Comparison with the State-of-Art Algorithms

We also compare the proposed PDL with two recently reported algorithms, the Tri-training (TriT) [13] and the virtual label regression (VLR) [14], which also train supervised learning classifiers using both labeled and unlabeled samples. For the two algorithms, we use the parameters provided in the paper. These algorithms and PDL are all trained with X_L and then used to classify X_U. The average results over 10 runs are reported in Table 3.

Table 3. Classification error rate (%) of real world data set

	1 labeled sample / class			3 labeled samples / class			5 labeled samples / class		
	TriT	VLR	PDL	TriT	VLR	PDL	TriT	VLR	PDL
usps	68.12	55.22	**45.41**	50.22	41.45	**23.53**	33.44	36.16	**18.76**
segmentation	74.49	31.08	**30.27**	40.35	28.43	**20.10**	25.58	24.46	**18.21**
banknote	41.88	22.81	**8.72**	22.07	4.91	4.94	6.47	1.38	2.89
pendigits	84.57	40.76	**18.47**	41.28	29.20	**7.43**	23.49	28.91	**8.82**
skin	54.33	32.96	**15.95**	34.21	17.46	**12.69**	21.42	17.92	**7.96**
miniBoo	39.00	32.82	**22.27**	20.72	**18.11**	19.11	**17.22**	27.65	18.31

We can see that PDL can achieve better results for most of the cases. For tri-training or co-training algorithms, it is easy to introduce noise labels as the training set grows and the impact of these noise labels can be very large. For VLR, it regresses a linear model using the discrete class-indicator vector of each sample, so it can hardly capture the nonlinearity and continuousness of the posterior density functions. In contrast, PDL first propagates posterior information from labeled samples to unlabeled samples on graph and then regresses a nonlinear model in the input space. But graph has a tight relationship with the manifold structure, thus the underlying manifold information is thus also encoded in the distribution function $F(x)$. Therefore PDL shows a promising potential for classifying manifold-distributed data set.

4   Discussions and Conclusions

In this paper we developed a novel two-step framework to learn a robust supervised model using a very few training samples and plenty of unlabeled samples. The framework first propagates posterior information from labeled samples to unlabeled samples and then regresses a multivariate posterior function in input space. As the experiments demonstrated, the proposed method can greatly reduce the number of training samples and thus reduce human burden to obtain labeled samples. This has not only theoretical interest but also great practical value. In the future we will focus on theoretical analysis and extending the framework to other existing supervised learning algorithms.

Acknowledgements: This research is partly supported by NSFC, China (No: 61273258, 31100672, 61375048), Ph.D. Programs Foundation of Ministry of Education of China (No.20120073110018).

References

1. Blum, A., Mitchell, T.: Combining labeled and unlabeled data with co-training. In: COLT, pp. 92–100. ACM (1998)
2. Nigam, K., McCallum, A.K., Thrun, S., Mitchell, T.: Text classification from labeled and unlabeled documents using EM. Machine Learning 39, 103–134 (2000)
3. Amini, M.-R., Gallinari, P.: The use of unlabeled data to improve supervised learning for text summarization. In: SIGIR, pp. 105–112. ACM (2002)
4. Goldman, S., Zhou, Y.: Enhancing supervised learning with unlabeled data. In: ICML, pp. 327–334. Citeseer (2000)
5. Denis, F.: PAC learning from positive statistical queries. In: Richter, M.M., Smith, C.H., Wiehagen, R., Zeugmann, T. (eds.) ALT 1998. LNCS (LNAI), vol. 1501, pp. 112–126. Springer, Heidelberg (1998)
6. Li, X., Liu, B.: Learning to classify texts using positive and unlabeled data. In: IJCAI, vol. 3, pp. 587–592 (2003)
7. Liu, B., Lee, W.S., Yu, P.S., Li, X.: Partially supervised classification of text documents. In: ICML, vol. 2, pp. 387–394. Citeseer (2002)
8. Chi, M., Bruzzone, L.: A semilabeled-sample-driven bagging technique for ill-posed classification problems. IEEE Geoscience and Remote Sensing Letters 2, 69–73 (2005)
9. Lee, W.S., Liu, B.: Learning with positive and unlabeled examples using weighted logistic regression. In: ICML, vol. 3, pp. 448–455 (2003)

10. Shrager, J., Hogg, T., Huberman, B.A.: Observation of phase transitions in spreading activation networks. Science 236, 1092–1094 (1987)
11. Zhou, D., Bousquet, O., Lal, T.N., Weston, J., Schölkopf, B.: Learning with local and global consistency. In: NIPS, pp. 595–602 (2004)
12. Suykens, J.A., Vandewalle, J.: Least squares support vector machine classifiers. Neural Processing Letters 9, 293–300 (1999)
13. Zhou, Z.-H., Li, M.: Tri-training: Exploiting unlabeled data using three classifiers. IEEE Transactions on Knowledge and Data Engineering 17, 1529–1541 (2005)
14. Nie, F., Xu, D., Li, X., Xiang, S.: Semisupervised dimensionality reduction and classification through virtual label regression. IEEE Transactions on Systems, Man, and Cybernetics, Part B: Cybernetics 41, 675–685 (2011)

Computational Model of Neocortical Learning Process: Prototype

Jing Xian Teo and Henry Lee Seldon[*]

Faculty of Information Science & Technology, Multimedia Univ.
75450 Melaka, Malaysia
lee@mmu.edu.my

Abstract. The balloon model of neocortical growth and learning claims that learning starts with larger groups of functional units (neuron columns) responding to a signal, but with training and lateral cortical expansion and inhibition, the number of units responding to a particular signal decreases as the units become better able to differentiate similar inputs. This process is different from most Artificial Neural Networks, but has some similarities with Temporal Organizing Maps (TOM). This paper describes the architecture and testing of a prototype computational model, a variation on TOMs, which seeks to emulate the anatomical and physiological behavior. Preliminary results indicate that it is consistent with predictions.

Keywords: Balloon Model, Neocortex, Temporal Organizing Map TOM.

1   Introduction

Starting at birth, the human neocortex starts to learn to differentiate speech sounds. According to the "balloon model" developed by Seldon (Harasty et al., 2003; Seldon, 2005, 2006), several neuroanatomical and neurophysiological processes involving the cortical neuronal columns are occurring.

1. The cortex is anatomically a sheet of neuronal columns, each of which is a "perceptual unit" (Fig. 1). The columns start with "inter-column spacings" or "center-center distances". The input from subcortical centers to the auditory cortex is widely spread through afferent axon arbors.
2. The first speech signals excite diffuse and overlapping ranges of neuronal columns, resulting in fuzzy or blurry perception. The horizontal green line in Fig. 2 shows that 5 cortical units (blue) are excited (red arrows) by each input (purple).
3. With activity and training, a couple of things happen.
   (a) The subcortical and intracortical myelin grows, thereby causing the cortex to expand laterally and become thinner like a balloon. The neuronal columns move physically apart. Compare the distances between blue units in Fig. 3 with those in Fig. 2. The total mass of the cortex does not increase so much.

---

[*] Corresponding author.

C.K. Loo et al. (Eds.): ICONIP 2014, Part I, LNCS 8834, pp. 95–102, 2014.
© Springer International Publishing Switzerland 2014

(b) The range of excited neuronal columns for each stimulus shrinks, due partly to the increasing lateral inhibition within the cortex (the dark yellow dashes in Figs. 2-3 show the lateral inhibition from unit X4[4]) and to the stretching. The horizontal green line in Fig. 3 covers only 3 cortical units (blue).

4. Finally, the range of columns excited by a particular stimulus is small. This allows the same number of columns to differentiate many more signals. Also, the map units (neuronal columns) are arranged non-linearly, with all the spacings dependent on the local activity and growth.

These are summarized in figures 2-3. Physiological studies support it (Yvert et al., 2001; Mäkelä et al., 2003).

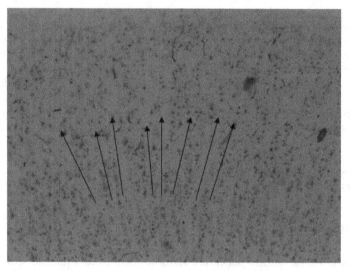

Fig. 1. Columnar organization of neurons in Layer 3 of human auditory cortex (arrows). Layer 4 is below. Microscope image, Nissl stain.

In contrast, most artificial neural networks (ANNs) work differently. Firstly, the human cortex has no blueprint of relevant speech sounds when it starts, so ANNs like perceptrons are not the same. Self-organizing maps (SOMs, Kohonen, 1990) and Temporal-organizing maps (TOM, Durand & Alexandre, 1995; Sarlin, 2013) are similar in that they are "feed-forward" networks which do not require predefined categories. However, they develop according to a "winner neuron" algorithm and do not include non-linear stretching. The balloon model starts with "many winners" and reduces their number with training and lateral cortical expansion.

We are experimenting with a computational model which tries to mimic the processes described by the balloon model. In Step 1 we create a simple, 1-layer model and observe whether it can expand in accordance with the balloon model, and whether the number of responding neuronal columns shrinks with training.

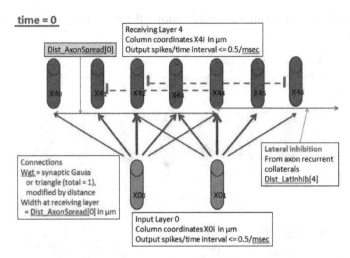

Fig. 2. Anatomical / physiological Balloon model at the start

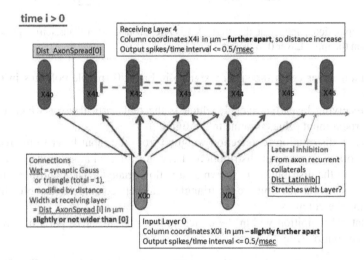

Fig. 3. Anatomical / physiological Balloon model after activity and training

2   Methodology

The Stimuli: Many publications have indicated that Voice Onset Time (VOT) is an important feature in distinguishing several consonants (Schwartz & Tallal, 1980). To correctly perceive speech the cortex must differentiate between voiced and unvoiced consonants - ba versus pa, da versus ta, ga versus ka. Therefore, we have chosen ba and pa for our sample stimuli; they were each spoken with a low voice and a high voice.

The Model: The computational model follows the neuroanatomical one, as shown in figures 4-6.

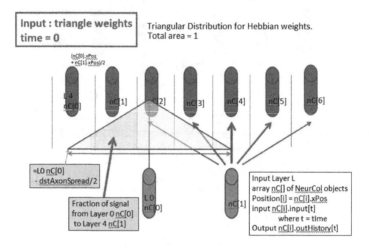

Fig. 4. Computational model at the start. A triangle (instead of a Gaussian) represents the weights from the input Layer 0 to the receiving Layer 4.

- The input layer (with the speech signal) is Layer 0 (purple columns in the figures).
- The receiving layer is Layer 4, which is the real cortical layer which receives subcortical input (blue columns in the figures).
- The axon distribution (connection weights) from the input layer to the receiving layer follow a gaussian distribution. They are all excitatory (red arrows). For simplicity the distribution is shown as a yellow triangle. The contributions of input to each receiving unit are the triangle areas between the vertical blue lines separating the columns.
- The lateral inhibition within the cortex follows the negative parts of a Mexican Hat distribution (Fig. 6).

With training the cortical units (blue) move further apart (Fig. 4 vs Fig. 5), so the stimulus from each input layer unit (purple) reaches fewer units. The lateral inhibition behaves similarly (Fig. 6); as the cortex expands laterally, the distribution of inhibition over the units changes (yellow dashes and triangles).

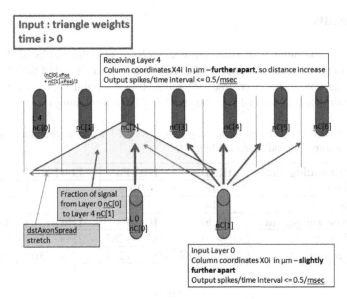

Fig. 5. Computational model after activity and training. Receiving units have spread apart.

The intracortical lateral inhibition behaves similarly. The next figure shows the inhibition emanating from cortical unit nC[4].

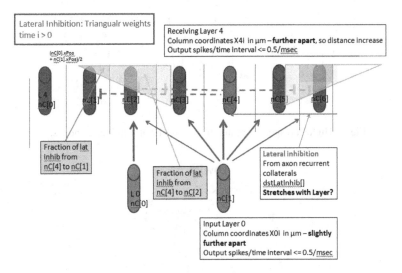

Fig. 6. Lateral inhibition after activity and training

3   Results

For each trial the number of receiving nodes (20 or 60), the widths of the gaussian input distribution (standard deviation, "afferent axon spread") and of the lateral inhibition, and the unit integration time (for integrating temporal signals) were set. The results are shown at iteration 1 (i.e. the start) and 10 (after training).

Table 1. Number and width of responding nodes, with 20 nodes, integration time 20 msec, afferent axon spread 45 and lateral inhibition 15

No. of responding nodes	HighBa	HighPa	LowBa	LowPa
Iteration 1	18	20	17	15
Iteration 10	14	14	12	12
Response range, μm	**HighBa**	**HighPa**	**LowBa**	**LowPa**
Iteration 1	-255→+285	-285→+285	-225→+255	-165→+255
Iteration 10	-288→+287	-288→+257	-287→+286	-226→+226

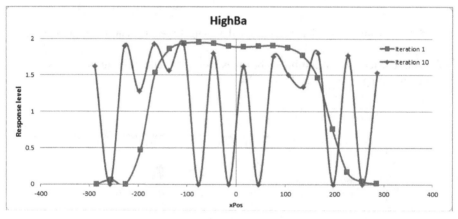

Fig. 7. HighBa responses at iteration 1 and 10 of training. 20 nodes.

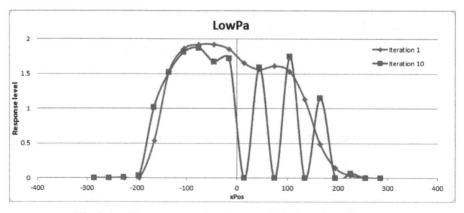

Fig. 8. LowPa responses at iteration 1 and 10 of training. 20 nodes.

Table 2. Number and width of responding nodes, with 60 nodes, integration time 20 msec, afferent axon spread 120 and lateral inhibition 30

No. responding nodes	HighBa	HighPa	LowBa	LowPa
Iteration 1	34	34	30	28
Iteration 10	32	30	26	22
Response range, μm	HighBa	HighPa	LowBa	LowPa
Iteration 1	-495→+495	-495→+495	-435→+435	-345→+465
Iteration 10	-618→+617	-618→ +557	-497→+496	-346→+406
[main response]	[-468→+497]	[-498→+507]	[-347→+496]	[-346→+406]

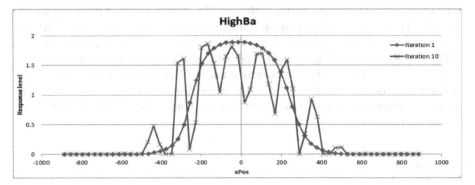

Fig. 9. HighBa responses at iteration 1 and 10 of training. 60 nodes.

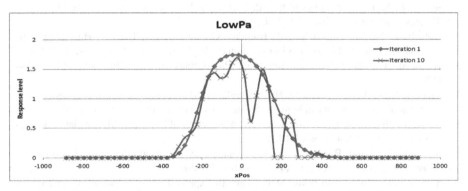

Fig. 10. LowPa responses at iteration 1 and 10 of training. 60 nodes.

Overall the test results can be summarized:

- Number of responding nodes shrinks with training for all inputs; neuronal columns move further apart as a function of input activity.
- Narrower lateral inhibition causes stronger inhibition of neighboring columns by the excited columns. The model is indicating that this parameter is critical for the final distribution of responding nodes.
- The syllables (ba and pa) and the pitches (high and low) are differentiated by the number of responding columns and the widths of the response regions.

4   Discussion

The neuroanatomical model predicts that initially stimuli will each excite a wide range of neuronal columns, and that the excited range will shrink (in numbers of columns) with training. With several parameter settings the computational model has shown this behavior up to now (April 2014). The neuroanatomical model also predicts that the spatial intervals between excited columns will increase with training, due to myelination; this has also been borne out by the computational model.

Up to now the computational model has differentiated ba and pa signals some, but not very well. This is due partly to the use of a very small number of identical input frequencies for all test inputs, partly to the use of a small number of receiving nodes (initially 20), partly to the 1-layer model (whereas the real cortex uses several layers), partly to uniform integration times for all nodes (in contrast to the real cortex), etc.

Also, the balloon model introduces many additional parameters such as distances, lateral expansion rates, widths and expansion rates of Gaussian input versus Mexican Hat lateral inhibition distributions, etc. These problems are being addressed in the ongoing development of this model, from both the computational and the neuroanatomical perspectives.

References

1. Durand, S., Alexandre, F.: Learning speech as acoustic sequences with the unsupervised model, TOM. In: Proceedings of the 7th International Conference on Neural Networks and Their Applications, pp. 267–273 (1995)
2. Harasty, J., Seldon, H.L., Chan, P., Halliday, G., Harding, A.: The Left Human Speech-Processing Cortex is Thinner but Longer than the Right. Laterality 8(3), 247–260 (2003)
3. Kohonen, T.: The Self-Organizing Map. Proceedings of the IEEE 78(9), 1464–1480 (1990)
4. Mäkelä, A.M., Alku, P., Tiitinen, H.: The auditory N1m reveals the left-hemispheric representation of vowel identity in humans. Neuroscience Letters 353(2), 111–114 (2003)
5. Sarlin, P.: Self-Organizing Time Map: An Abstraction of Temporal Multivariate Patterns. Neurocomputing 99, 496–508 (2013)
6. Schwartz, J., Tallal, P.: Rate of acoustic change underlie hemispheric specialization for speech perception. Science 207(4437), 1380–1381 (1980)
7. Seldon, H.L.: Does Brain White Matter Growth Expand the Cortex like a Balloon? Hypothesis and Consequences. Laterality 10(1), 81–95 (2005)
8. Seldon, H.L.: Cortical Laminar Thickness and Column Spacing in Human Temporal and Inferior Parietal Lobes: Intra-Individual Anatomical Relations. Laterality 11(3), 226–250 (2006)
9. Yvert, B., Crouzeix, A., Bertrand, O., Seither-Presiler, A., Christo, P.: Multiple Supratemporal Sources of Magnetic and Electric Auditory Evoked Middle Latency Components in Humans. Cereb. Cortex 11(5), 411–423 (2001)

Active Learning with Maximum Density and Minimum Redundancy

Yingjie Gu[1,2], Zhong Jin[1], and Steve C. Chiu[2]

[1] Computer Science and Engineering,
Nanjing University of Science and Technology, Nanjing 210094, China
csyjgu@gmail.com, zhongjin@njust.edu.cn
[2] Department of Electrical Engineering,
Idaho State University, Pocatello 83209-8060, USA
chiustev@isu.edu

Abstract. Active Learning is a machine learning technique that selects the most informative examples for labeling so that the classification performance would be improved to its maximum possibility. In this paper, a novel active learning approach based on Maximum Density and Minimum Redundancy (MDMR) is proposed. The objective of MDMR is to select a set of examples that have large density and small redundancy with others. Firstly, we propose new methods to measure the density and redundancy of examples. Then a model is built to select examples by combining density and redundancy and dynamic programming algorithm is applied to solve the problem. The results of the experiment on terrain classification have demonstrated the effectiveness of the proposed approach.

Keywords: active learning, classification, density, redundancy.

1 Introduction

In many real-world applications, there are large numbers of unlabeled data while the labels are expensive and difficult to get. And much redundant data, which slows down the training process without improving the classification result, also exist in the training set. Active learning [1] was proposed to select the most informative examples for labeling and training a classifier, thus the labels of testing examples can be predicted most precisely. The kernel problem of active learning is how to measure the value of each example and how to select the most informative examples from the unlabeled data set.

There are many criteria in active learning for examples selection. Uncertainty sampling is one of the most widely used criterion that queries the examples whose labels are most uncertain under the current trained classifier. The most popular uncertainty sampling is SVM_{active} [2], which selects the examples nearest to the current decision boundary. Other criteria like variance reduction [3], density [4], and diversity [5] also have been widely applied to active learning.

Optimum Experimental Design (OED) [6], which refers to the problem of selecting examples for labeling in statistics, has attracted an increasing amount

C.K. Loo et al. (Eds.): ICONIP 2014, Part I, LNCS 8834, pp. 103–110, 2014.

of attention [7] [8]. The example **x** is referred to as experiment and its label y is referred to as measurement. OED tries to select examples so that the variances of a parameterized model are minimized. OED has two types of criteria. One is D, A, and E-Optimal Design that choose data points to minimize the variance of the model's parameters. The other is I and G-optimal Design that minimize the variance of the prediction value.

Active learning based on OED selects the most informative points while it is unable to exploit the redundancy between selected points. In this paper, we proposed an active learning algorithm called MDMR to select a set of points with maximum density and minimum redundancy. By combining examples' density and redundancy, every selected example is informative and the redundancy between selected examples are small.

The rest of this paper is organized as follows: In Section 2, we elaborate the proposed active learning approach MDMR. The experimental settings and results are presented in Section 3. Finally, we discuss the conclusion and future work in Section 4.

2   Active Learning with Maximum Density and Minimum Redundancy

The general problem of active learning can be described as follows. Given a set of points $\mathcal{X} = \{\mathbf{x}_1, \mathbf{x}_2, ..., \mathbf{x}_n\}$, where each $\mathbf{x}_i$ is an instance of d-dimensional vector, find a subset $\mathcal{Z} = \{\mathbf{z}_{s_1}, \mathbf{z}_{s_2}, ..., \mathbf{z}_{s_k}\} \subseteq \mathcal{X}$, which contains the most informative points. In other words, if the points $\mathbf{z}_{s_i} (i = 1, 2, ..., k)$ are labeled and used as training data, the labels of testing data can be predicted most precisely.

In this section, a novel active learning algorithm is proposed to select examples by considering examples' density and redundancy.

2.1   Density and Redundancy

Information density is an important criterion for active learning since examples in dense regions are expected to be representative and informative. Thus we aim to select a set of examples that have large density. Firstly, we use Gaussian kernel to construct a complete graph with all unlabeled examples. The weight W_{ij} between $\mathbf{x}_i$ and $\mathbf{x}_j$ is defined as

$$W_{ij} = exp(-\frac{\|\mathbf{x}_i - \mathbf{x}_j\|^2}{2\sigma^2}) \tag{1}$$

where σ is the parameter of gaussian kernel. As shown in (1), W_{ij} is large if $\mathbf{x}_i$ and $\mathbf{x}_j$ are very close to each other. The large weight W_{ij} means $\mathbf{x}_i$ and $\mathbf{x}_j$ are highly connected, or they have large similarity.

For an example in dense region, it should be very close to its neighbors, which means the weight between the example and its neighbors should be large. Therefore, the average weight between an example and its p-nearest neighbors is able to measure the density of the example. The density of $\mathbf{x}_i$ is defined as follows:

$$den(\mathbf{x}_i) = \frac{1}{p} \sum_{\mathbf{x}_j \in N_p(\mathbf{x}_i)} W_{ij} \tag{2}$$

Where $N_p(\mathbf{x}_i)$ is the p-nearest neighbors of $\mathbf{x}_i$.

Density-based active learning is able to select the most representative examples, but it is unable to exploit the redundancy between the selected examples. In other words, some selected examples may have similar information. Hence each example has maximum information can't guarantee the global information is maximum. Here we exploit the redundancy among the selected examples.

The examples have large weight are usually highly connected to each other. They probably have more redundant information than the examples whose weight is small. So the selected examples are required to have small weight with each other. Here, the maximum weight between an example and other selected examples are used to measure the redundancy of the example. If the maximum weight is very small, the example has little redundancy with other selected examples.

Suppose we have selected a set of k examples $Z_k = \{\mathbf{x}_{s_1}, \mathbf{x}_{s_2}, ..., \mathbf{x}_{s_k}\}$ from $\mathcal{X}$. The redundancy between example $x_{s_i} (i > k)$ and Z_k can be described as follows:

$$red(\mathbf{x}_{s_i}, Z_k) = \max_{\substack{1 \leq j \leq k \\ j \neq i}} W_{s_i s_j} \tag{3}$$

2.2 The Proposed Approach

In this work, we aim to select k examples $(\mathcal{Z})$ with maximum density and minimum redundancy from $\mathcal{X}$. Suppose Z_k is an arbitrary subset of $\mathcal{X}$ that contains k examples and $Z_k = \{\mathbf{x}_{s_1}, \mathbf{x}_{s_2}, ..., \mathbf{x}_{s_k}\}$. The final selected k examples $\mathcal{Z}$ can be obtained by solving the following problem:

$$\mathcal{Z} = \arg\max_{Z_k \subseteq \mathcal{X}} \sum_{i=1}^{k} (den(\mathbf{x}_{s_i}) - \lambda \, red(\mathbf{x}_{s_i}, Z_k)) \tag{4}$$

where λ is the tradeoff parameter that can determine the importance of density and redundancy.

Unfortunately, the optimization problem (4) is a highly complicated problem. To get the optimal subset $\mathcal{Z}$, we would have to search over all possible sets to determine the unique optimal $\mathcal{Z}$. It is impossible to finish in short time with the number of examples increased.

However, it should be noted that $den(\mathbf{x}_{s_i})$ is only dependent on s_i while $red(\mathbf{x}_{s_i}, Z_k)$ is related with $\{\mathbf{x}_{s_1}, \mathbf{x}_{s_2}, ..., \mathbf{x}_{s_k}\}$. Suppose $X_u = \{\mathbf{x}_1, \mathbf{x}_2, ..., \mathbf{x}_u\}(u = 1, ..., n)$ and $Z(u, v)(v \leq u)$ denotes the optimal solution of selecting v examples from X_u. We transform the problem (4) into a relatively simple form:

$$\mathcal{Z} = \arg\max_{Z_k \subseteq \mathcal{X}} \sum_{i=1}^{k} (den(\mathbf{x}_{s_i}) - \lambda \, red(\mathbf{x}_{s_i}, Z(s_i - 1, i - 1))) \tag{5}$$

where $red(\mathbf{x}_{s_i}, Z(s_i - 1, i - 1))$ is the redundancy between $\mathbf{x}_{s_i}$ and the selected $i - 1$ examples from $X_{s_i - 1} = \{\mathbf{x}_1, \mathbf{x}_2, ..., \mathbf{x}_{s_i - 1}\}$.

Obviously, $red(\mathbf{x}_{s_i}, Z(s_i - 1, i - 1))$ is relevant with $\{s_1, ..., s_{i-1}\}$ but irrelevant with $\{s_{i+1}, .., s_k\}$. This means that when we select the $i - th$ example, it is required to have small redundancy with the selected examples $\{\mathbf{x}_{s_1}, ..., \mathbf{x}_{s_{i-1}}\}$. This guarantees that the next selected example must be different from the previous selected examples. This idea is in accord with the process of sequential examples selection.

2.3   The Dynamic Programming Approach

The problem (5) can be solved by dynamic programming that breaks it down into simpler subproblems. Suppose $F(u, v)(u \geq v)$ denotes the maximum volume of information of selecting v examples from X_u. As defined above, $Z(u, v)$ denotes the optimal solution of selecting v examples from X_u, hence $\mathcal{Z} = Z(n, k)$. $F(u, v)$ and $Z(n, k)$ can be described as follows:

$$F(u, v) = \max_{Z_v \subseteq X_u} \sum_{i=1}^{v} (den(\mathbf{x}_{s_i}) - \lambda \, red(\mathbf{x}_{s_i}, Z(s_i - 1, i - 1))) \qquad (6)$$

$$Z(u, v) = \arg\max_{Z_v \subseteq X_u} \sum_{i=1}^{v} (den(\mathbf{x}_{s_i}) - \lambda \, red(\mathbf{x}_{s_i}, Z(s_i - 1, i - 1))) \qquad (7)$$

where $u \in \{1, 2, ..., n\}$, $v \in \{1, 2, ..., k\}$, and $u \geq v$.

Our final goal is to find $Z(n, k)$ that decides which k examples should be selected from the n unlabeled examples.

It should be noted that there are two special situations: $v = 1$ and $u = v$. If $v = 1$, there are no redundancy since only one example is selected. Therefore, the example with maximum density should be selected. If $u = v$, obviously, all of the examples in X_u should be selected. So

$$F(u, v) = \begin{cases} \max\limits_{\mathbf{x}_i \in X_u} den(\mathbf{x}_i) & if \ v = 1 \\ \sum_{i=1}^{u} den(\mathbf{x}_i) - \lambda \, red(\mathbf{x}_i, X_{i-1}) & if \ u = v \end{cases} \qquad (8)$$

Suppose we have already obtained the optimal solution of selecting $v - 1$ and v examples from X_{u-1}, now we consider the optimal solution of selecting v examples from u unlabeled examples. If the example $\mathbf{x}_u$ has small density and large redundancy with $Z(u - 1, v - 1)$, obviously we will not select the example $\mathbf{x}_u$. Hence the optimal solution of selecting v examples from X_u should be the same as selecting v examples from X_{u-1}. On the contrary, if the example $\mathbf{x}_u$ has large density and small redundancy with $Z(u - 1, v - 1)$, we prefer to select it for labeling. In this situation, since $Z(u - 1, v - 1)$ is the optimal solution of selecting $v - 1$ examples from X_{u-1}, the optimal solution of selecting v examples from X_u is $Z(u, v) = Z(u - 1, v - 1) \cup \mathbf{x}_u$.

Table 1. The process of the proposed active learning algorithm

---

Input:

    Initial unlabeled data set $\mathcal{X} = \{x_1, x_2, .., x_n\}$, the gaussian parameter (σ), the number of nearest neighbor (p), the tradeoff parameter (λ), the number of examples need to select (k)

Output:

    $Z(n, k)$: the k selected examples

Procedure:

    compute weight W, $den(x_i), (i = 1, 2, ..., n)$

    Initilize X_u, $F = \mathbf{0}_{nk}$, $Z(u, v) = \emptyset$

    For $u = 1 : n$

        $F(u, 1) = \max\limits_{x_i \in X_u} den(x_i)$

        $Z(u, 1) = \arg\max\limits_{x_i \in X_u} den(x_i)$

    End

    For $u = 2 : n$

        For $v = 2 : k$

            $C(u) = den(x_u) - \lambda\, red(x_u, Z(u - 1, v - 1))$

            $F(u, v) = \max(F(u - 1, v), F(u - 1, v - 1) + C(u))$

            $Z(u, v) = \begin{cases} Z(u - 1, v) & if\ F(u, v) = F(u - 1, v) \\ Z(u - 1, v - 1) \cup x_u & else \end{cases}$

        End

    End

Return $Z(n, k)$

In general, the relationships between $F(u - 1, v - 1)$, $F(u - 1, v)$, and $F(u, v)$ can be described as follows:

$$C(u) = den(x_u) - \lambda\, red(x_u, Z(u - 1, v - 1)) \tag{9}$$

$$F(u, v) = \max(F(u - 1, v), F(u - 1, v - 1) + C(u)) \tag{10}$$

where $2 \leq u \leq n$, $2 \leq v \leq k$ and $v \leq u$. $Z(u, v)$ can be computed as follows:

$$Z(u, v) = \begin{cases} Z(u - 1, v) & if\ F(u, v) = F(u - 1, v) \\ Z(u - 1, v - 1) \cup x_u & else \end{cases} \tag{11}$$

Since $Z(1, 1), Z(2, 1)$ is easy to obtain, the global optimal solution $Z(n, k)$ can be obtained by iteration. The dynamic programming approach to solve the examples selection problem is summarized in Table 1. As can be seen from Table 1, the proposed active learning algorithm is easy to perform and the computational cost is low.

3   Experiments

In this section, experiments of terrain classification are performed with different active learning algorithms. In order to demonstrate the effectiveness of our proposed algorithm, we evaluate and compare four active learning methods:

- **Random Sampling** (Rand) method, which selects examples randomly from unlabeled data set.
- **D-Optimal Design** (DOD) as described in Section 2.1.
- **Manifold Adaptive Experimental Design** (MAED) Algorithm [9], which performed convex TED in manifold adaptive kernel space.
- **Active Learning with Maximum Density and Minimum Redundancy** (MDMR), which is proposed in this paper.

3.1   Data and Experimental Settings

Terrain image dataset used in the experiment was constructed by us from the Outex Database [10], which is consisted of two data sets: Outex-0 and Outex-1. Each of them includes 20 outdoor scene images and the size of each image is 2272×1704. The images are marked as one type of bush, grass, tree, sky, road, and building. The marked area of each image is cut into patches with size 64×64 and each patch is regarded as an example. In this work, we extract 50 patches of each class (totally 300 patches) to construct a pool of unlabeled data set for examples selection. The testing examples, which are predicted to evaluate active learning algorithms' performance, are also consisted of 300 patches (50 patches from each class).

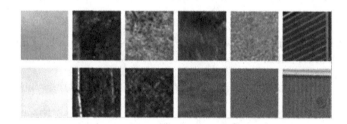

Fig. 1. Patch examples of Outex: sky, tree, bush, grass, road, and building

Two examples of each class are shown in Figure 1. It is difficult to classify these terrains directly in image color space. Thus color histogram feature [11] and texture feature with the rotation-invariant operators $LBP^{riu2}_{8,1+16,3}$ [12] are extracted and combined together. As a result, each example is represented by a 43-dimensional feature vector.

Logistic regression with l_2 regularization is used as classifier and the regularization parameter is set to be 0.5. The parameters in our proposed active learning algorithms are set as follows: the gaussian kernel parameter ($\sigma = 0.1$), the number of nearest neighbor ($p = 15$), and the tradeoff parameter ($\lambda = 10$).

3.2   Results

The process of the experiments are as follows: Firstly, we select $k(k = 5, 10, 15, ..., 50)$ examples from unlabeled data set for labeling and training a classifier.

Then we perform classification on testing data set and the accuracy is defined as Correct Classification Rate (CCR). The experiments are repeated 20 times and the average accuracy is computed as the final result.

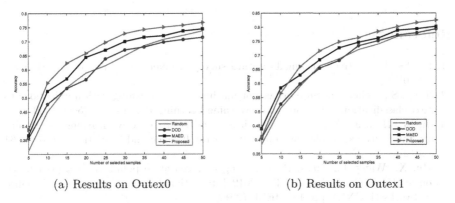

(a) Results on Outex0   (b) Results on Outex1

Fig. 2. Classification performance on Outex0 and Outex1 dataset using Rand, DOD, MAED, and the proposed active learning algorithm

Figure 2 shows the average classification accuracy versus the number of training (selected) examples. As can be seen, our proposed MDMR algorithm significantly outperforms the other active learning algorithms in most cases. The MAED algorithm outperforms Random Sampling and DOD method in most cases. DOD and Random Sampling perform comparably to each other. When only five examples are selected, there exists at least one class that does not have any labeled examples. Therefore, in this case, all of the algorithms yield low classification accuracy. As the number of selected examples increases, the classification accuracy of all of the algorithms increases. As shown in 2, with only 40 selected examples, MDMR algorithms performs comparably to or even better than the other algorithms with 50 selected examples. Our MDMR algorithm yields the highest classification accuracy.

4 Conclusion

In this paper, we have introduced a novel active learning algorithm called MDMR, which selects the examples with maximum density and minimum redundancy. The experimental results on terrain classification demonstrate that it is better than other popular active learning algorithms.

The disadvantage of this proposed algorithm is that it is not global optimal since the examples are sequentially selected. The redundancy of selected example $\mathbf{x}_{s_t}$ is measured by the redundancy between $\mathbf{x}_{s_t}$ and previous selected $t-1$ examples $\{\mathbf{x}_{s_1}, \mathbf{x}_{s_2}, ..., \mathbf{x}_{s_{t-1}}\}$. Thus the redundancy among all the selected examples may not be minimum. Moreover, combining different criteria such as density and redundancy is a significant problem in active learning. There is a lot of work that needs to be explored.

Acknowledgements. This work is partially supported by National Natural Science Foundation of China under Grant Nos. 61373063, 61233011, 61125305, 61375007, 61220301, and by National Basic Research Program of China under Grant No. 2014CB349303.

References

1. Settles, B.: Active learning literature survey. University of Wisconsin, Madison (2010)
2. Tong, S., Koller, D.: Support vector machine active learning with applications to text classification. The Journal of Machine Learning Research 2, 45–66 (2002)
3. Ji, M., Han, J.: A variance minimization criterion to active learning on graphs. In: International Conference on Artificial Intelligence and Statistics, pp. 556–564 (2012)
4. Hu, X., Wang, L., Yuan, B.: Querying representative points from a pool based on synthesized queries. In: The 2012 International Joint Conference on Neural Networks (IJCNN), pp. 1–6. IEEE (2012)
5. Chakraborty, S., Balasubramanian, V., Panchanathan, S.: Dynamic batch mode active learning. In: 2011 IEEE Conference on Computer Vision and Pattern Recognition (CVPR), pp. 2649–2656. IEEE (2011)
6. Atkinson, A.C., Donev, A.N., Tobias, R.D.: Optimum experimental designs, with SAS. Oxford University Press, Oxford (2007)
7. He, X.: Laplacian regularized d-optimal design for active learning and its application to image retrieval. IEEE Transactions on Image Processing 19(1), 254–263 (2010)
8. Gu, Y., Jin, Z.: Neighborhood preserving d-optimal design for active learning and its application to terrain classification. Neural Computing and Applications 23(7-8), 2085–2092 (2013)
9. Cai, D., He, X.: Manifold adaptive experimental design for text categorization. IEEE Transactions on Knowledge and Data Engineering 24(4), 707–719 (2012)
10. University of oulu texture database, http://www.outex.oulu.fi/temp/
11. Procopio, M.J., Mulligan, J., Grudic, G.: Learning terrain segmentation with classifier ensembles for autonomous robot navigation in unstructured environments. Journal of Field Robotics 26(2), 145–175 (2009)
12. Ojala, T., Pietikainen, M., Maenpaa, T.: Multiresolution gray-scale and rotation invariant texture classification with local binary patterns. IEEE Transactions on Pattern Analysis and Machine Intelligence 24(7), 971–987 (2002)

One-to-Many Association Ability of Chaotic Quaternionic Multidirectional Associative Memory

Takumi Okutsu and Yuko Osana

Tokyo University of Technology,
1401-1 Katakura, Hachioji, Tokyo, Japan
osana@stf.teu.ac.jp

Abstract. In this paper, we investigate one-to-many association ability of multi-valued patterns in the Chaotic Quaternionic Multidirectional Associative Memory (CQMAM). The CQMAM is based on the Multidirectional Associative Memory and composed of quaternionic neurons and chaotic quaternionic neurons, it can realize one-to-many associations of M-tuple multi-valued patterns. Although the conventional Chaotic Complex-valued Multidirectional Associative Memory with variable scaling factor (CCMAM) can realize one-to-many associations of M-tuple multi-valued patterns, the one-to-many association ability of the CQMAM is better than that of the conventional CCMAM.

1 Introduction

Recently, many associative memories have been proposed in the field of neural networks, most of these models can not deal with multi-valued patterns and one-to-many associations. As models which can deal with multi-valued patterns and one-to-many associations, we have proposed the Chaotic Complex-valued Bidirectional Associative Memory (CCBAM)[1][2], the Chaotic Complex-valued Multidirectional Associative Memory (CCMAM)[3] and the Chaotic Complex-valued Multidirectional Associative Memory (CCMAM) with variable scaling factor[4]. These models are composed of complex-valued neurons[5] and chaotic complex-valued neurons[6] and the association of multi-valued patterns is realized by complex-valued neurons, and one-to-many association is realized by chaotic complex-valued neurons. However, one-to-many association ability decreases when the number of states (S) increases in these models. We have also proposed the Chaotic Quaternionic Multidirectional Associative Memory (CQMAM)[7]. This model is composed of quaternionic neurons[8] and chaotic quaternionic neurons[9], and it can realize one-to-many associations of M-tuple multi-valued patterns. Since this model is composed of quaternionic and chaotic quaternionic neurons, it is more likely to have better one-to-many association ability.

In this paper, we investigate one-to-many association ability of multi-valued patterns in the Chaotic Quaternionic Multidirectional Associative Memory (CQMAM).

C.K. Loo et al. (Eds.): ICONIP 2014, Part I, LNCS 8834, pp. 111–118, 2014.

2   Chaotic Quaternionic Neuron Model

Here, we examine the chaotic quaternionic neuron model[9] which is used in the Chaotic Quaternionic Multidirectional Associative Memory. The chaotic quaternionic neuron model is the extended chaotic neuron model[10] in order to deal with internal states and output of neurons which are represented in quaternions.

The output of the chaotic quaternionic neuron model at the time t is calculated by

$$x(t+1) = f\left(A(t) - \alpha(t)\sum_{d=0}^{t} k^d x(t-d) - \theta\right) \tag{1}$$

$$(A(t), x(t), \theta \in \mathbb{H}, \quad k, \alpha(t) \in \mathbb{R})$$

where $A(t)$ is the external input at the time t, k is the damping factor $(0 < k < 1)$ and θ is the threshold of the neuron. And $\mathbb{H}$ shows the set of quaternions, and $\mathbb{R}$ shows the set of real numbers. $\alpha(t)$ is the scaling factor of the refractoriness at the time t, and it is given by

$$\alpha(t) = a + b \cdot \sin(c \cdot t) \tag{2}$$

where a, b and c are coefficients. In the conventional chaotic neuron model[10], the scaling factor of refractoriness α is constant. In contrast, we have proposed the chaotic neural network with variable scaling factor in order to improve the dynamic association ability[11]. In the chaotic quaternionic neuron model, by introducing the variable scaling factor, we can except that the dynamic association ability improves.

And, $f(\cdot)$ is the output function which is given by

$$f(u) = f^{(e)}(u^{(e)}) + f^{(i)}(u^{(i)})i + f^{(j)}(u^{(j)})j + f^{(k)}(u^{(k)})k \tag{3}$$

$$f^{(e)}(u) = f^{(i)}(u) = f^{(j)}(u) = f^{(k)}(u) = \tanh\left(\frac{u}{\varepsilon}\right) \tag{4}$$

where ε is the steepness parameter, and i, j and k are imaginary units.

3   Chaotic Quaternionic Multidirectional Associative Memory

Here, we explain the Chaotic Quaternionic Multidirectional Associative Memory (CQMAM)[7].

3.1   Structure

The CQMAM has more than two layers and each layer has Key Input Part and Context Part. Fig.1 shows the structure of the 3-layered CQMAM. In this model, quaternionic neurons[8] and chaotic quaternionic neurons[9] are used, and the Key Input Part are composed of quaternionic neurons and the Context Part are composed of chaotic quaternionic neurons.

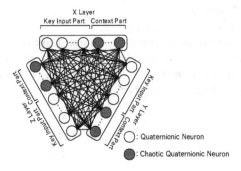

Fig. 1. Structure of CQMAM

3.2   Learning Process

In the CQMAM, the connection weights are trained by the orthogonal learning. However, the orthogonal learning can not deal with the training pattern set including one-to-many relations because the stored common data cause superimposed patterns. In the CQMAM, the patterns with its own contextual information are memorized by the orthogonal learning as similar as the conventional CCMAM[3].

The connection weights from the layer y to the layer x, $\boldsymbol{w}^{xy}$ and the connection weights from the layer x to the layer y , $\boldsymbol{w}^{yx}$ are determined as follows:

$$\boldsymbol{w}^{xy} = \boldsymbol{X}_y(\boldsymbol{X}_x^*\boldsymbol{X}_x)^{-1}\boldsymbol{X}_x^* \tag{5}$$

$$\boldsymbol{w}^{yx} = \boldsymbol{X}_x(\boldsymbol{X}_y^*\boldsymbol{X}_y)^{-1}\boldsymbol{X}_y^* \tag{6}$$

where * shows the conjugate transpose, and -1 shows the inverse. And, $\boldsymbol{X}_x$ and $\boldsymbol{X}_y$ are the training pattern matrix which are memorized in the layer x and the layer y, and are given by

$$\boldsymbol{X}_x = \{\boldsymbol{X}_x^{(1)}, \cdots, \boldsymbol{X}_x^{(p)}, \cdots, \boldsymbol{X}_x^{(P)}\} \tag{7}$$

$$\boldsymbol{X}_y = \{\boldsymbol{X}_y^{(1)}, \cdots, \boldsymbol{X}_y^{(p)}, \cdots, \boldsymbol{X}_y^{(P)}\} \tag{8}$$

where $\boldsymbol{X}_x^{(p)}$ is the pattern p which is stored in the layer x, $\boldsymbol{X}_y^{(p)}$ is the pattern p which is stored in the layer y and P is the number of the training pattern sets.

3.3   Recall Process

Since contextual information is usually unknown for users, in the recall process, only the Key Input Part receives input. In the CQMAM, since the chaotic quaternionic neurons in the Context Part change their states by chaos, one-to-many associations can be realized.

The recall process of the CQMAM has the following procedures when the input pattern is given to the layer x.

Step 1 : Input to Layer x

The input pattern is given to the layer x.

Step 2 : Propagation from Layer x to Other Layers

When the pattern is given to the layer x, the information is propagated to the Key Input Part in the other layers. The output of the neuron k in the Key Input Part of the layer y ($y \neq x$), $x_k^y(t)$ is calculated as

$$x_k^y(t) = f\left(\sum_{j=1}^{N^x} w_{kj}^{yx} x_j^x(t)\right) \tag{9}$$

where N^x is the number of neurons in the layer x, w_{kj}^{yx} is the connection weight from the neuron j in the layer x to the neuron k in the layer y, and $x_j^x(t)$ is the output of the neuron j in the layer x at the time t. And $f(\cdot)$ is the output function which is given by Eq.(4).

Step 3 : Propagation from Other Layers to Layer x

The output of the neuron j in the Key Input Part of the layer x, $x_j^x(t+1)$, is calculated as

$$x_j^x(t+1) = f\left(\sum_{y \neq x}^{M}\left(\sum_{k=1}^{n^y} w_{jk}^{xy} x_k^y(t)\right) + v A_j\right) \tag{10}$$

where M is the number of layers, n^y is the number of neurons in the Key Input Part of the layer y, w_{jk}^{xy} is the connection weight from the neuron k in the layer y to the neuron j in the layer x, v is the connection weight from the external input, and A_j is the external input (See **3.4**) to the neuron j in the layer x.

The output of the neuron j of the Context Part in the layer x, $x_j^x(t+1)$ is calculated as

$$x_j^x(t+1) = f\left(\sum_{y \neq x}^{M}\left(\sum_{k=1}^{n^y} w_{jk}^{xy} \sum_{d=0}^{t} k_m^d x_k^d(t-d)\right)\right.$$
$$\left. -\alpha(t)\sum_{d=0}^{t} k_r^d x_j^x(t-d)\right) \tag{11}$$

where k_m and k_r are damping factors. And, $\alpha(t)$ is the scaling factor of the refractoriness at the time t, and is given by Eq.(2).

Step 4 : Repeat

Steps 2 and **3** are repeated.

3.4   External Input

In the CQMAM, the external input A_j is always given so that the key pattern does not change into other patterns. If the pattern is given to the layer x and the initial input does not include noise, we can use the initial input pattern $x_j^x(0)$ as the external pattern. However, since the initial input pattern sometimes includes noise, so we use the following pattern $\hat{\boldsymbol{x}}_j^x(t_{in})$ when the network becomes stable t_{in} as an external input. Here, t_{in} is given by

$$
t_{in} = \min\left\{ t \left| \sum_{j=1}^{n^x} (\hat{\boldsymbol{x}}_j^x(t) - \hat{\boldsymbol{x}}_j^x(t-1)) = 0 \right. \right\} \tag{12}
$$

where n^x is the number of neurons in the Key Input Part of the layer x. And $\hat{\boldsymbol{x}}_j^x(t)$ is the quantized output of the neuron j in the layer x at the time t.

4   Computer Experiment Results

4.1   One-to-Many Association Ability Comparison with Conventional CCMAM

Here, we compared one-to-many association ability of the CQMAM with the conventional CCMAM[3]. In this experiment, 16-valued random patterns in oneto- many relations were memorized in the 3-layered proposed CQMAM whose parameters are shown in Table 1 and common pattern was given to the network as an initial input and investigated how many patterns to be recalled appeared

Table 1. Experimental Conditions

The Number of Neurons (Key Input Part)		400
The Number of Neurons (Contextual Part)		100
Parameter in Output Function	ε	0.02
Decay Parameters	k_m	0.89
	k_r	0.96
Weights from External Input	v	50
Parameters for Scaling Factor	a	1.2
for Refractoriness	b	1.0
	c	π

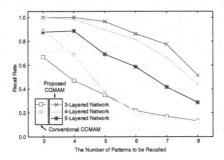

Fig. 2. One-to-Many Association Ability of Proposed CQMAM and Conventional CC-MAM

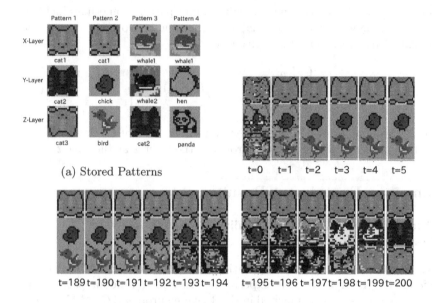

(a) Stored Patterns

t=0   t=1   t=2   t=3   t=4   t=5

t=189 t=190 t=191 t=192 t=193 t=194   t=195 t=196 t=197 t=198 t=199 t=200

(b) Association Result

Fig. 3. Association Result for Noisy Input

during t = 0 ∼ 999. Fig.2 shows the one-to-many association ability of the proposed CQMAM and the conventional CCMAM. This is the average result of 100 trials. As shown in this figure, we can confirmed that the one-to-many association ability of the CQMAM is better than that of the conventional CCMAM.

4.2   Robustness for Noisy Input

Here, we examined the robustness for noisy input in the CQMAM. In this experiment, four 16-valued pattern sets shown in Fig.3(a) were memorized in the 3-layered proposed CQMAM and the common pattern *cat1* with 20 % noise was given to X-Layer as an initial input at $t = 0$. Fig.3(b) shows an association result of the CQMAM for noisy input. As shown in this figure, at $t = 0$, the superimposed patterns composed of the patterns 1 and 2 corresponding to the input pattern *cat1* appeared in Y-Layer and Z-Layer. And then, the chaotic quaternionic neurons in the Contextual Part changed their states by chaos, as a result, the patterns *cat2* and *cat3* (pattern 1) and the patterns chick and bird (pattern 2) were recalled correctly.

Fig.4 shows the robustness for noisy input of the CQMAM. As shown in this figure, the CQMAM has superior robustness for noise.

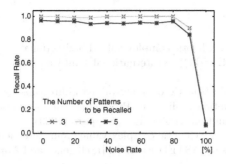

Fig. 4. Robustness for Noisy Input

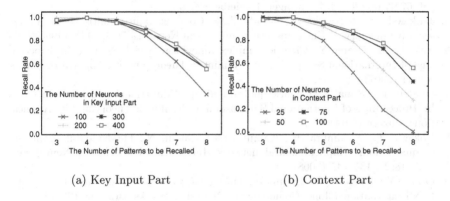

(a) Key Input Part   (b) Context Part

Fig. 5. One-to-Many Association Ability in Various Size CQMAM

4.3   One-to-Many Association Ability in Various Size CQMAM

Here, we examined one-to-many association ability of the CQMAM in various size CQMAM. Fig.5 shows the one-to-many association ability of the CQMAM in various size CQMAM. As shown in this figure, one-to-many association ability of the larger size CQMAM is better than that of the smaller size CQMAM.

5   Conclusion

In this paper, we have investigated one-to-many association ability of multivalued patterns in the Chaotic Quaternionic Multidirectional Associative Memory (CQMAM). We carried out a series of computer experiments and confirmed that (1) its one-to-many association ability is better than that of the conventional Chaotic Complex-valued Multidirectional Associative Memory (CCMAM) with variable scaling factor, (2) it has has superior robustness for noisy input, and (3) one-to-many association ability of the larger size CQMAM is better than that of the smaller size CQMAM.

References

1. Yano, Y., Osana, Y.: Chaotic complex-valued bidirectional associative memory. In: Proceedings of IEEE and INNS International Joint Conference on Neural Networks, Atlanta (2009)
2. Yano, Y., Osana, Y.: Chaotic complex-valued bidirectional associative memory – one-to-many association ability. In: Proceedings of International Symposium on Nonlinear Theory and its Applications, Sapporo (2009)
3. Shimizu, Y., Osana, Y.: Chaotic complex-valued multidirectional associative memory. In: Proceedings of IASTED Artificial Intelligence and Applications, Innsbruck (2010)
4. Yoshida, A., Osana, Y.: Chaotic complex-valued multidirectional associative memory with variable scaling factor. In: Honkela, T. (ed.) ICANN 2011, Part I. LNCS, vol. 6791, pp. 266–274. Springer, Heidelberg (2011)
5. Jankowski, S., Lozowski, A., Zurada, J.M.: Complex-valued multistate neural associative memory. IEEE Transactions on Neural Networks 7(6), 1491–1496 (1996)
6. Nakada, M., Osana, Y.: Chaotic complex-valued associative memory. In: Proceedings of International Symposium on Nonlinear Theory and its Applications, Vancouver (2007)
7. Okutsu, T., Osana, Y.: Chaotic quaternionic multidirectional associative memory. In: Proceedings of International Symposium on Nonlinear Theory and its Applications, Luzern (2014)
8. Isokawa, T., Nishimura, H., Kamiura, N., Matsui, N.: Fundamental properties of quaternionic Hopfield neural network. International Journal of Neural Systems 18(2), 135–145 (2008)
9. Osana, Y.: Chaotic quaternionic associative memory. In: Proceedings of IEEE and INNS International Joint Conference on Neural Networks, Brisbane (2012)
10. Aihara, K., Takabe, T., Toyoda, M.: Chaotic neural networks. Physics Letter A 144(6 & 7), 333–340 (1990)
11. Osana, Y.: Recall and separation ability of chaotic associative memory with variable scaling factor. In: Proceedings of IEEE and INNS International Joint Conference on Neural Networks, Hawaii (2002)
12. Hagiwara, M.: Multidirectional associative memory. In: Proceedings of IEEE and INNS International Joint Conference on Neural Networks, Washington, D.C., vol. 1, pp. 3–6 (1990)

An Entropy-Guided Adaptive Co-construction Method of State and Action Spaces in Reinforcement Learning

Masato Nagayoshi[1], Hajime Murao[2], and Hisashi Tamaki[3]

[1] Niigata College of Nursing, 240 Shinnan-cho, Joetsu 943-0847, Japan
nagayosi@niigata-cn.ac.jp
[2] Kobe University, 1-2-1, Tsurukabuto, Nada-ku, Kobe 657-8501, Japan
[3] Kobe University, 1-1, Rokkodai-cho, Nada-ku, Kobe 657-8501, Japan

Abstract. Engineers and researchers are paying more attention to reinforcement learning (RL) as a key technique for realizing computational intelligence such as adaptive and autonomous decentralized systems. In general, it is not easy to put RL into practical use. In previous research, Nagayoshi et al. have proposed an adaptive co-construction method of state and action spaces. However, the co-construction method needs two parameters for sufficiency of the number of learning opportunities. These parameters are difficult to set. In this paper, first we propose an entropy-guided adaptive co-construction method with and index using the entropy instead of the parameters for sufficiency of the number of learning opportunities. Then, the performance of the proposed method and the efficiency of interactions between state and action spaces were confirmed through computational experiments.

Keywords: reinforcement learning, interactions between state and action spaces, co-construction of state and action spaces, entropy.

1 Introduction

Engineers and researchers are paying more attention to reinforcement learning (RL)[1] as a key technique in developing autonomous systems. In general, however, it is not easy to put RL to practical use. Such issues as satisfying the requirements of learning speed, resolving the perceptual aliasing problem, and designing reasonable state and action spaces for an agent, etc., must be resolved. Our approach mainly deals with the problem of designing state and action spaces. By designing suitable state and action spaces adaptively, it can be expected that the other two problems will be resolved simultaneously. Here, the problem of designing state and action spaces involves the following two requirements: (i) to keep the characteristics of the original search space as much as possible in order to seek strategies that lie close to the optimal, and (ii) to reduce the search space as much as possible in order to expedite the learning process. In general, these requirements are in conflict.

C.K. Loo et al. (Eds.): ICONIP 2014, Part I, LNCS 8834, pp. 119–126, 2014.

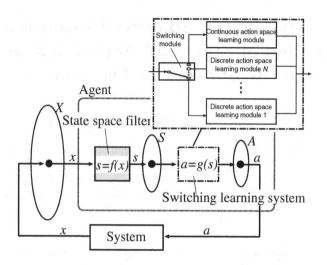

Fig. 1. Developing reinforcement learning model

Previously, some possible solutions for the state space design have been proposed[2–5]. On the other hand, some possible solutions for the action space design have been proposed[6, 7] also. Here, Nagayoshi et al. [8] reconstitute these two construction methods[5, 7] as one co-construction method by treating them as a combined method for mimicking an infant's perceptual and motor developments. We can not find other methods to co-construct the state and the action spaces. However, this co-construction method needs two parameters for sufficiency of the number of learning opportunities. These parameters are difficult to set.

In this paper, first we propose an entropy-guided adaptive co-construction method based on an index using a trend of the entropy instead of the parameters for sufficiency of the number of learning opportunities. Then, through computational experiments by a path planning problem with two-dimensional continuous state space and one-dimensional continuous action space, the performance of the proposed method and the efficiency of interactions between state and action spaces are confirmed .

2   Developing Reinforcement Learning

2.1   Outline of a Computational Model

Nagayoshi et al. have proposed an developing RL model to mimic the processes of an infant's perceptual and motor developments. The model is constructed by "state space filter[5]" which is a mapping between an input state space $\mathcal{X}$ and the state space $\mathcal{S}$, and a "switching learning system[7]" as shown in Fig. 1.

The state space filter mimics the process of perceptual development, in which perceptual differentiation progresses by differentiating the state space gradually from the undifferentiated state space, as the infant becomes older and more experienced. In parallel, the switching learning system also mimics the process of motor development, in which gross motor skills develop before fine motor skills by switching discrete action space learning modules (hereafter called "DA modules") from a more coarse-grained DA module to a more fine-grained DA module, and finally switching to a continuous action space learning module (hereafter called a "CA module").

2.2   Co-construction Method of State and Action Spaces

Basic Idea. A variety of methods can be considered to acquire the state space filter and the switching learning module. Here, we have proposed a method based on introducing and referring to the entropy which is defined by action selection probability distributions in a state. It is expected that the method (i) is able to learn in parallel the state space filter and the switching learning system, and (ii) does not required specific RL methods for the learning module.

The entropy of action selection probability distributions using Boltzmann selection in a state $H_D(s)$ is defined by

$$H_D(s) = -(1/\log|\mathcal{A}_D|) \sum_{a \in \mathcal{A}_D} \pi(a|s) \log \pi(a|s) \tag{1}$$

where $\pi(a|s)$ specifies the probabilities of taking each action a in each state s, $\mathcal{A}_D$ is the action space and $|\mathcal{A}_D|$ is the number of available actions of the DA module.

In this paper, sufficiency for the number of learning opportunities is judged using the following properties of the entropy $H_D(s)$.

- The entropy $H_D(s)$ becomes smaller with the number of learning opportunities.
- On the other hand, if the number of learning opportunities is sufficient, then the entropy does not become smaller any more and remains unchanged or becomes larger temporarily.

Thus, after the entropy becomes smaller, if the entropy remains unchanged or becomes larger, the number of learning opportunities is judged as sufficient.

Then, if the number of learning opportunities in a state s is sufficient, then the state space filter is adjusted by dividing the state s or the learning module is switched to the more fine-grained DA module, and finally ends with the CA module.

Here, we propose an adaptive entropy-guided co-construction method in which the adjustment of the state space filter is carried out first, and the adjustment of the state space filter or the switching of learning modules is carried out alternately upon the number of learning opportunities being sufficient.

Also note that, the effectiveness of the adjustment of the state space filter and the switching of learning modules are judged by the following: When the number of learning opportunities is sufficient $H_D^+(s)$ gets smaller than the pre-sufficient entropy $H_D^-(s)$ immediately after adjusting the state space filter or switching the learning module. When both adjusting the state space filter and switching learning modules are continuously ineffective, adjusting the state space filter and switching learning modules in the state are no longer carried out.

2.3 Sufficiency of the Number of Learning Opportunities

In previous research, the co-construction method of Nagayoshi et al.[8] needs θ_H and θ_L to judge sufficiency of the number of learning opportunities where θ_H is a threshold value of the entropy, and θ_L is a threshold value of the number of learning opportunities. These parameters are difficult to set. In this paper, sufficiency of the number of learning opportunities is judged using a trend of the entropy $H_D(s)$ (Eq.1).

However, preliminary computational experiments indicate that the entropy of the action selection probability in the state s shows a range of fluctuations. In order to decrease wrong judgments by the influence of the fluctuations, the time of the occurrence of the reversal of a downward trend is detected using MACD (Moving Average Convergence / Divergence)[9], which is one of the most popular tools in technical analysis trading. In addition, the entropy $H_D^*(s)$, after updating the Q-value, is used to refine the detection only when the agent selects an action with the maximal Q-value, if the learning module is Q-learning. Then, a short-term ($n_T = \theta_{EMAS}$) EMA (Exponential Moving Average) value and a long-term ($n_T = \theta_{EMAL}$) EMA value of the entropy are calculated according to the following equation after every update of $H_D^*(s)$.

$$\text{EMA}^{n_T}(s(t)) = (1 - \alpha_{EMA}) \times \text{EMA}_{old}^{n_T}(s(t)) + \alpha_{EMA} \times H_D^*(s(t)) \qquad (2)$$

where $\text{EMA}_{old}^{n_T}(s(t))$ is the latest known value of the n_T-term EMA in $s(t)$, $\alpha_{EMA} = 2/(n_T + 1)$ and n_T are constant numbers expressing the smoothing constant and the average amount of time respectively.

A MACD (Moving Average Convergence / Divergence) value is calculated according to the following equation, after updating the short-term and the long-term EMA values.

$$\text{MACD}(s) = \text{EMA}^{\theta_{EMAS}}(s) - \text{EMA}^{\theta_{EMAL}}(s) \qquad (3)$$

In addition, a 'signal' value is a moving average for the latest series of θ_{MACD} values of MACD(s). In particular, when the 'signal' value increases from a value equal to or smaller than 0, to larger than or equal to 0, the number of learning opportunities in the state is judged as sufficient.

The usual parameters of MACD ($\theta_{EMAS} = 12, \theta_{EMAL} = 26$ and $\theta_{MACD} = 9$) are used in the following experiments.

Adjustment of a State Space Filter. If the number of the learning opportunities in a state s is sufficient, then the state space filter is adjusted by dividing the range of the input state mapped to the state s into two parts for each dimension, and mapping each part to a different state. Through this operation, the size of the state space increases by $(2^M - 1)$ after being divided, where M is the number of dimensions. Also note that the values of the new $2M$ states are the value of the state before it is divided.

Table 1. Parameters for experiments

Parameter	Description	Value
$\alpha_Q, \alpha_C, \alpha_\mu, \alpha_\sigma$	learning rate of Q-learning, critic, μ ,σ	0.1
γ	discount rate	0.9
τ	temperature used by Boltzmann selection	0.1

Switching of Learning Modules. In this paper, Q-learning(hereafter called "QL") and Actor-critic[10](hereafter called "AC") are applied to the DA module and the CA module, respectively. The learning module is switched in the order of the DA module with an action space divided evenly into $n, 2n, \cdots, 2^{(N-1)}n$, and finally ending with the CA module, where N is the number of DA modules.

If the number of learning opportunities in a state is sufficient, then the learning module is switched to the more fine-grained DA module, and finally ends with the CA module. The Q-values of newly added actions a_i at this time are set according to the following formula:

$$Q(s, i) = \max_{j \in i-1, i+1} Q(s, j) \tag{4}$$

where action $i - 1$ and $i + 1$ are adjacent to action i. This formula is set in consideration of a more efficient search as well as the idea of the optimistic initial valuess[1].

In the procedure to switch learning modules, the result of QL is succeeded by AC and the following procedure is carried out. 1. The state value of the critic, $V(s)$, is initialized by

$$V(s) = \sum_{a \in \mathbf{A}_Q} \pi(a|s) \cdot Q(s, a) \tag{5}$$

2. The normal probability distribution used by the actor is calculated by

$$\mu(s) = \arg \max_{a \in \mathbf{A}_Q} Q(s, a), \tag{6}$$

$$\sigma(s) = |A_Q(\arg \max_{a \in \mathbf{A}_Q} Q(s, a))|/4 \tag{7}$$

where μ is the mean, σ is standard error of the mean, $|A_Q(i)|$ is the range of the action space which represents action i of QL, and $\sigma(s)$ is set at $|A_Q(i)|/4$ in order to select an action in the range $|A_Q(i)|$ with a probability of 0.95.

3 Computational Examples

3.1 Path Planning Problem

The proposed method is applied to a so-called "path planning problem" where an agent is navigated from a start point to a goal area in a continuous space as shown in Fig. 2. Here, the agent has a circular shape (diameter = 50[mm]), and the continuous space is 500[mm] × 500[mm] bounded by the external wall with internal walls as shown in black. The agent can observe the center position of the agent: (x_A, y_A) as the input, and move 25[mm] in a direction, i.e., decide the direction: θ_A as the output.

The positive reinforcement signal $r_t = 10$ (reward) is given to the agent only when the center of the agent arrives at the goal area and the reinforcement signal $r_t = 0$ at any other steps. The period from when the agent is located at the start point to when the agent is given a reward or 10,000 steps pass away, labeled as 1 episode, is repeated.

3.2 Comparison to Various Co-construction Methods

In this section, the proposed method (hereafter called method "A") is compared with the following two co-construction methods: (1) method B, in which the adjustment of the state space filter is carried out first and continues until the effectiveness of the adjustment is lost. Then, the switching of the learning module is carried out until the effectiveness of the switching is lost. These procedures are carried out alternately. (2) method C, in which the adjustment of the state space filter and the switching of the learning module are carried out simultaneously. Also note that the adjustment of the state space filter is only carried out after switching to the CA module. When continuously ineffective in both adjusting the state space filter and switching learning modules, adjusting the state space filter and switching learning modules in the state are no longer carried out in the same way as method A. After switching to the CA module, $\max \pi(a|s)$ is used for judging the effectiveness of the adjustment of the state space filter instead of $H_D(s)$. Here, the initial state space filter, the switching learning system, the range of $\sigma(x)$, initial Q-values, the number of adjustments of the state space filter and parameters for experiments are set at the same as above. The average number of steps, the average number of adjustments of the state space filter and the average number of switching to DA modules and the CA module, that is QL

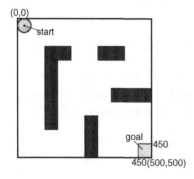

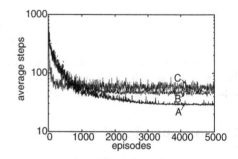

Fig. 2. Path planning problem

Fig. 3. Required steps: comparison to various co-construction methods

and AC, required to accomplish the task were observed during learning over 20 simulations with various methods, as described in Figs. 3,4,5 respectively.

Learning speed and obtained control rule: It can be seen from Fig. 3 that, (1) method A shows a better performance than any other methods with regard to the control rule obtained. (2) method C shows a better performance than any other methods with regard to the learning speed.

The number of adjustments of the state space filter, that is the number of divisions of the state space: It can be seen from Fig. 4 that, (1) method C is smaller than any other methods. (2) method B is larger than any other methods. This is regarded as the result of adjusting the state space filter first until the effectiveness of the adjustment is lost.

The number of switching learning modules: It can be seen from Fig. 5 that, (1) with regard to the number of switching to QL, method C is smaller than any other methods, and method B is larger than any other method, but it is also considered to be a result of switching learning modules until the effectiveness of switching is lost. (2) with regard to the number of switching to AC, method B is smaller than any other methods. Thus, it is thought that method B was mostly ineffective in switching to QL. Then, method A is larger than any other methods. Thus, it is considered that method A has more effective for switching learning modules than any other methods. Therefore, we have confirmed that method A demonstrates better performances than any other method for the path planning problem. In addition, because method A, which constructs the state and the action spaces mutually, showed better performance than the method C, which constructs the spaces simultaneously, we have confirmed that interactions between the state and the action spaces improve the performance. However, it could be considered that these experiments happen to mutually suit grain sizes of the state and action spaces mutually. For this reason, we need to re-create the experiments with various grain sizes of the state and action spaces.

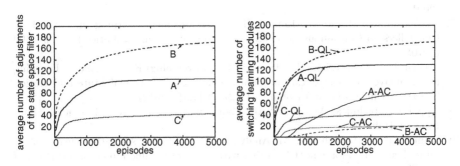

Fig. 4. Average number of adjustments of the state space filter

Fig. 5. Average number of switching learning modules

4   Conclusion

In this paper, we proposed an entropy-guided adaptive co-construction method based on an MACD, which is one of the most popular tools in technical analysis trading, using an entropy instead of two parameters for sufficiency of the number of learning opportunities. Then, through computational experiments by a path planning problem with two-dimensional continuous state space and one-dimensional continuous action space, the performance of the proposed method and the efficiency of interactions between state and action spaces were confirmed.

Our future projects include to re-create the experiments with various grain sizes of a state and an action spaces, etc.

References

1. Sutton, R.S., Barto, A.G.: Reinforcement Learning. A Bradford Book. MIT Press (1998)
2. Sutton, R.S.: Generalization in Reinforcement Learning: Successful Examples Using Sparse Coarse Coding. In: Advances in Neural Information Processing Systems: Proc. of the 1995 Conf., pp. 1038–1044 (1996)
3. Doya, K.: Temporal Difference Learning in Continuous Time and Space. In: Proc. of the 1995 Conf. on Advances in Neural Information Processing Systems, pp. 1073–1079 (1996)
4. Murao, H., Kitamura, S.: QLASS: An enhancement of Q-learning to generate state space adaptively. In: Proc. of Fourth European Conf. on Artificial Life, ECAL 1997 (1997)
5. Nagayoshi, M., Murao, H., Tamaki, H.: A State Space Filter for Reinforcement Learning. In: Proc. of the Eleventh International Symposium on Artificial Life and Robotics, pp. 615–618 (2006)
6. Morimoto, J., Doya, K.: Acquisition of stand-up behavior by a real robot using hierarchical reinforcement learning. Robotics and Autonomous System 36, 35–51 (2001)
7. Nagayoshi, M., Murao, H., Tamaki, H.: A Reinforcement Learning with Switching Controllers for Continuous Action Space. Artificial Life and Robotics 15(1), 97–100 (2010)
8. Nagayoshi, M., Murao, H., Tamaki, H.: Developing Reinforcement Learning for Adaptive Co-construction of Continuous High-dimensional State and Action Spaces. Artificial Life and Robotics 17(2), 204–210 (2012)
9. Appel, G.: Technical Analysis Power Tools for Active Investors. Financial Times Prentice Hall (2005)
10. Kimura, H., Kobayashi, S.: An Analysis of Actor/Critic Algorithms using Eligibility Traces: Reinforcement Learning with Imperfect Value Function. In: Proc. of the Fifteenth International Conference on Machine Learning, pp. 278–286 (1998)

A Nodes Reduction Procedure
for RBFNDDA through Histogram

Pey Yun Goh, Shing Chiang Tan, and Wooi Ping Cheah

Multimedia University, Jln. Ayer Keroh Lama, 75450 Melaka
{pygoh,sctan,wpcheah}@mmu.edu.my

Abstract. This paper presents a two-stage learning algorithm to reduce the hidden nodes of a radial basis function network (RBFN). The first stage involves the construction of an RBFN using the dynamic decay adjustment (DDA) and the second stage involves the use of a modified histogram algorithm (HIST) to reduce hidden neurons. DDA enables the RBFN to perform constructive learning without pre-defining the number of hidden nodes. The learning process of DDA is fast but it tends to generate a large network architecture as a result of its greedy insertion behavior. Therefore, an RBFNDDA-HIST is proposed to reduce the nodes. The proposed RBFNDDA-HIST is tested with three benchmark medical datasets. The experimental results show that the accuracy of the RBFNDDA-HIST is compatible with to that of RBFNDDA but with less number of nodes. This proposed network is favorable in a real environment because the computation cost can be reduced.

Keywords: radial basis network, nodes reduction, histogram, dynamic decay adjustment.

1 Introduction

The Radial Basis Function Network (RBFN) is one of the popular artificial neural networks (ANNs) being used due to its fast learning characteristic through the use of locally-tuned neurons [1, 2]. It learns the non-linear relationship among the input and output data with a simple topological structure [3, 4]. An RBFN is also an universal approximator with the use of continuous functions (usually the Gaussian functions) in the hidden units for information processing [1] [4] [5]. An RBFN can be trained by using a dynamic decay adjustment (DDA) algorithm [24]. The DDA algorithm can speed up the training process of a network to construct an RBFNDDA with a satisfactory performance in prediction. A note is, users should set manually the number of hidden nodes in a conventional RBFN but an RBDNDDA requires no such a parameter setting because the network topology can be built up automatically. However, such constructive-learning approach could result in large network architecture. A huge network architecture is fast in convergence and also favorably used for solving problems [6]. However, the complexity of a huge network is high and over-fitting is likely to occur during its training process. The network may give a poor performance on presence of spurious information such as outliers, noise and overlapping nodes [7, 8].

C.K. Loo et al. (Eds.): ICONIP 2014, Part I, LNCS 8834, pp. 127–134, 2014.
© Springer International Publishing Switzerland 2014

On the other hand, a small network has a better generalization ability but to search for a small yet suitable network architecture could be a time-consuming task [8, 9]. There is also a possibility where the small network may not be able to solve problem and trapped in local minima [8]. Various methods have been proposed to improve generalization capability of an ANN. One of the common methods is node pruning. Sietsma and Dow [13] proposed a two-stage pruning algorithm. In the first stage, a hidden node is removed if it has a constant output over all the training inputs and if the outputs of any two hidden nodes are the same. In the second stage, nodes are removed when they are linearly independent from other nodes in the same layer. As commented by [9], the method [13] will cause the network to take a long training time. Liang [14] proposed a node pruning method with the use of orthogonal projection and weight crosswise propagation (CP) calculation. The pruning method consists of two stages. In Stage 1, the node with shortest distance in its orthogonal projection is removed. In Stage 2, the author used the weight CP to propagate the information of deleted hidden nodes to the other hidden nodes. As such, information loss in stage 1 can be reduced. Zhang and Qiao [15] proposed the pruning algorithm based on the neural complexity (PBNC). The PBNC algorithm calculates the network complexity after deleting a hidden node. All hidden nodes must be selected once. The node with the highest neural complexity is removed. The learning process of the network is terminated when the average error of the adjustment process is less or equal to the objective error. Indeed, these methods [13, 14, 15] involved a complicated calculation process to prune the network.

As simpler and more efficient methods are favorably adopted to formulate solutions to many real-world problems [10]. The research in this study is aimed to reduce the network size of an RBFN with the use of histogram. Histogram is a traditional statistical approach that enables the visualization of statistical approximation [11]. In this work, it is employed to reduce hidden nodes of an RBFN due to its simplicity and computational efficiency [12]. Histogram is commonly used to approximate the distribution of data. From the literature, histogram is vastly applied in the field of image processing and computer vision, especially in the usage of summarizing the characteristics of images [11]. There are researchers who used histogram for pruning in speech recognition. Researchers [22, 23] proposed to use a histogram-based pruning method to further limit the search space in automatic speech recognition system. In this work, we proposed a node reduction method by absorbing a histogram algorithm [16] into the RBFNDDA learning process. Notably, the histogram algorithm was proposed by Shimazaki and Shinomoto [16] to optimize the distribution of neuronal spike signals. To our best knowledge, the use of a histogram approach for reducing the number of hidden nodes of an ANN is new.

The organization of this paper is as follows. In Section 2, the details of the RBFNDDA, the histogram algorithm and the proposed RBFN are described. In Section 3, the results from an experimental study using several benchmark data sets from the UCI machine-learning repository [17] are presented, analyzed and compared. Lastly, the conclusion of the paper and future work are described in Section 4.

2   The Methods

2.1   Overview of RBFNDDA

The DDA algorithm is applied to build an RBFNDDA. During training, new nodes are inserted into the hidden layer of RBFNDDA to encode new information from the data samples. The dynamics of RBFNDDA are governed by two user-defined parameters, i.e., the positive threshold θ^+ and the negative threshold θ^-. They regulate the width of a node (prototype) and correspond to distinguish a prototype from its neighbors (prototypes) of other classes. In this regard, θ^+ represents the lowest correct-classification probability for the correct class, and θ^- represents the highest probability tolerable to an incorrect class. In [24], the default settings of θ^+ and θ^- are 0.4 and 0.2 respectively. The training algorithm of RBFNDDA in a single epoch is as follows.

Step 1: Initialize the weight $W_i = 0$

Step 2: Consider a training input x of a class k, assume that $\boldsymbol{P}_i^k$ denote an RBF hidden node of class k has been inserted in the network. Increase the weight $W_i^k = W_i^k + 1$ if the $\boldsymbol{P}_i^k$ has the Gaussian activation $R_i^k \geq \theta^+$ (in other words, the input is correctly classified)

Otherwise, insert a new hidden node $\boldsymbol{P}_{m_k}^k$ (m_k denote number of nodes for class k) and perform the following actions.

Set $m_k = m_k + 1$

$W_{m_k}^k = 1$

Center of neuron $\boldsymbol{z}_{m_k}^k = x$

Width $\sigma_{m_k}^k = \min\limits_{\substack{1 \leq b \leq m_j}}^{j \neq k} \left\{ \sqrt{-\dfrac{||z_b^j - z_{m_k}^k||^2}{\ln \theta^-}} \right\}$

Step 3: Shrink the width of all the conflicting nodes where $j \neq k, 1 \leq b \leq m_j$

$$\sigma_b^j = \min\left\{\sigma_b^j, \sqrt{-\dfrac{||x - z_b^j||^2}{\ln \theta^-}}\right\}$$

Step 4: Repeat steps 2 and 3 for the next training inputs.

2.2   Overview of Histogram

Finding a suitable bin size is important to construct a histogram for representing the actual data distribution as close as possible [16]. As such, Shimazaki and Shinomoto [16] proposed a histogram algorithm to determine a suitable bin size with equal width and number of sequences (trials) required when representing the time-dependent spike rate. The optimal bin size can be obtained by minimizing $C_n(\Delta)$. Throughout the process, the number of bin N, the width Δ and the n sequences will change accordingly to generate the cost function. In [16], N is a setting from a range between 2 and 50 whereas n is set as 30. The optimum bin size N with the most minimum cost function will be selected. The algorithm is as below:

Step 1: Observation period T is divided into N bins of width Δ. The frequency of spikes k_i from all n sequences that enter the i-th bin is computed.

Step 2: Compute the mean, $\bar{k}$ and variance, v of the number of spikes, as follows.

$$\bar{k} \equiv \frac{1}{N}\sum_{i=1}^{N} k_i \tag{1}$$

$$v \equiv \frac{1}{N}\sum_{i=1}^{N}(k_i - \bar{k})^2 \tag{2}$$

Step 3: Calculate the cost function (C_n)

$$C_n(\Delta) = \frac{2\bar{k}-v}{(n\Delta)^2} \tag{3}$$

Steps 4: Repeat steps 1 to 3 by varying different numbers of bin to find the corresponding bin width Δ that minimizes $C_n(\Delta)$.

2.3   The Proposed Method

The proposed method is a two-stage learning process. Inputs are trained through RBFNDDA in the first stage. The generated RBFNDDA hidden nodes (or RBF centers) are sent to the proposed method, which we called HIST algorithm. This HIST algorithm is an extension of Shimazaki and Shinomoto [16]'s histogram algorithm. Our experiment does not need the number of sequences, as the purpose of HIST is not to train the nodes but to identify unneeded nodes. Therefore, n sequence is set as 1. The minimum bin number N is set as minimum 3 instead of 2. The reason is too less number of bins will cause difficulty in nodes reduction and degrade the performance of classification. In order to decide which bins contain unneeded nodes, steps $7 - 11$ are proposed. The hidden nodes have multiple dimensions but HIST accepts one dimension inputs. Therefore, each hidden node is transformed to one dimension by aggregating sum of all its attributes and enter the HIST according to class label. The details of the HIST algorithm are as below:

Step 1: All hidden nodes H are categorized according to class c and transformed to one dimension (eq. 4, D = number of dimension),

$$Input\ X_i = \sum_{j=1}^{D} x_j \qquad i=1,...,H \tag{4}$$

Step 2: All Input Xs of class c are divided into N bins of width Δ (eq. 5). Count the frequency of Input X, k_i which enters the i-th bin.

$$\Delta = \frac{Input\ X_{max}-Input\ X_{min}}{N} \tag{5}$$

Step 3: Generate the mean, $\bar{k}$ (eq. (1)) and variance, v (eq. 2) of k_i

Step 4: Calculate the C_n (eq. 3).

Step 5: Step 1 through 4 are repeated by varying the setting of N and Δ to search for an optimum bin size and width that minimize $C(\Delta)$.

Step 6: Histogram is constructed using the optimum bin size and width from Step 5.

Step 7: Compute the magnitude $p(a)$ (eq. 6), which is the probability of the RBFNDDA weight for each bin, where a = value of RBFNDDA weightage. Assume that a bin contains 7 Input X, with 3 of them are having the weightage a, which is 1, another 3 Input X with weightage which is 3 and 1 Input X with weightage which is 5. Therefore, the probability $p(a)$ are $\frac{3}{7}, \frac{3}{7}$, and $\frac{1}{7}$ respectively. The number of weight category m for this bin is 3.

$$p(a) = \frac{number\ of\ Input\ X\ of\ a\ RBFNDDA\ weight\ category}{total\ number\ of\ Input\ X\ per\ bin} \tag{6}$$

Step 8: Compute the expected value $E(a)$ for each bin, where $E(a) = \sum_1^m a\,p(a)$. ($E(a)$ reflects the score of weightage for each bin. Continue with the example in Step 7, $E(a) = (1)\left(\frac{3}{7}\right) + (3)\left(\frac{3}{7}\right) + (5)\left(\frac{1}{7}\right) = 2.43$. The highest the score, the more important is the bin.)

Step 9: Normalize $E(a)$ between 0 and 1. Assume that the histogram has 3 bins, each bin with the $E(a)$ score is 2.43, 0.38 and 4.50. After normalization, the obtained scores are 0.50, 0 and 1.00 respectively.

Step 10: Bins with normalized expected value lowered than a threshold (we set it as 0.2) are deleted. As such, based on the assumption in Step 9, bin 2 is deleted.

Step 11: Repeat Steps 2 to 10 for all the classes.

Step 12: Send retains nodes back to RBFNDDA and proceed to testing phase.

Normalization of $E(a)$ is to generalize the proposed method to all data sets as each data set will generate different number of DDA weights with different aggregate sum of input X. Without generalization, the threshold used to delete the bin need to change accordingly. With normalization, one threshold value can be used for different data sets.

3   Experiment and Discussion

The results of RBFNDDA-HIST is benchmarked with three medical data sets from the UCI Machine Learning Repository [17]: Diabetes, Cancer and Heart. All of the data sets consist of two classes with the number of samples 768, 699, 270, respectively, and number of attributes 8, 9 and 13 respectively. In this experiment, we used a two-fold cross validation where 50% of the data was used for training and the remaining data was used for testing. The involved parameter settings were $\theta^+ = 0.4$, $\theta^- = 0.2$, $n = 1$, N = minimum 3, maximum 50, and threshold for $E(a) = 0.2$. RBFNDDA was trained in multi epochs before the execution of the HIST algorithm. The maximum training epoch of RBFNDDA was set as 6. The proposed RBFNDDA-HIST was run on a computer having the following specifications: operating system Windows 7, Intel Core (TM) CPU i5-2410M and 4.0 GB RAM. The performance of RBFNDDA-HIST is then compared with the original RBFNDDA and RBFNDDA-T [18].

RBFNDDA-T is an online pruning technique where node is marked as temporary once it is inserted into the network. The status of the node will change to permanent if more than two samples are covered by the node. After each epoch training, nodes with temporary status are deleted and are not used in the next training epoch. Paetz [18] conducted an experiment in 8 runs for both RBFNDDA and RBFNDDA-T. In our work, the experiment was repeated for 30 runs, and the average results in terms of the number of hidden nodes and the accuracy rates of RBFNDDA-HIST were computed. The results are listed in Table 1.

Table 1. Experiment results (Acc. = average test accuracy; #nodes = average number of hidden nodes; standard deviation in round brackets)

| Data Set | Paetz [18] | | | | RBFNDDA-HIST | |
| | RBFNDDA | | RBFNDDA-T | | | |
	Acc.	# nodes	Acc.	# nodes	Acc.	# nodes
Diabetes	74.35	288.5	73.5	65.6	72.85	202.9
	(0.97)	(6.1)	(2.38)	(4.2)	(1.24)	(26.2)
Cancer	96.86	70.6	96.9	38.3	96.80	19.1
	(0.99)	(13.4)	(0.39)	(4.6)	(0.97)	(10.3)
Heart	79.26	83.6	79.82	32.6	80.10	36.95
	(2.83)	(3.3)	(3.54)	(2.1)	(2.55)	(1.9)

As compared to RBFNDDA, RBFNDDA-HIST manages to reduce the hidden nodes significantly with the percentages of reduction are 29.68%, 73.00% and 55.80% in Diabetes, Cancer and Heart respectively. As compared to RBFNDDA-T, the accuracy of RBFNDDA-HIST in Diabetes is lower with a higher number of hidden nodes; whereas in Cancer and Heart, RBFNDDA-HIST is competitive to give high accuracy rates and small number of hidden nodes. The Diabetes data set consists of data samples that are highly overlapped [18]. In Diabetes, RBFNDDA-T achieves better result in node reduction, i.e. 77.26% because RBFNDDA-T implements node pruning for unwanted information from its network whereas RBFNDDA-HIST re-organizes all existing nodes of RBFNDDA into a more compact structure without removing such overlapping information intensively. On the other hand, for Cancer data sets, RBFNDDA-HIST achieves better result where the percentage of node reduction for RBFNDDA-HIST is 73.00% and RBFNDDA-T is 45.75% from RBFNDDA. The percentages of node reduction for both RBFNDDA-T and RBFNDDA-HIST are almost similar when dealing with Heart data set. In other words, the performances of RBFNDDA-HIST and RBFNDDA-T in node reduction are problem-dependent. But RBFNDDA-HIST shows a significant advantage in generating a more compact network structure when compared to RBFNDDA.

Further comparison was performed with different experiment setup to see how the reduction of nodes will affect the accuracy and computational cost of RBFNDDA-HIST. We run this experiment by using the Cancer data set and compare with the reported methods in [19]. The experiment excluded all the missing values. The remaining 400 instances were used in training and 283 instances were used in testing. Table 2 presents the comparison of the results. The numbers of hidden nodes for MPANN are lower than those of RBFNDDA-HIST, which are 4.125 ± 1.360 and

21.567 ± 7.655 respectively. This is expected as we aim to have a simple and yet a fast learning neural network. While MPANN is having a complicated construction of neural network that match this reason: a successive smaller networks can be time consuming [8]. The computational cost of RBFNDDA-HIST is much lower than the methods in [19, 20, 21]. The accuracy is about the same when compared with these methods.

Table 2. Performance comparison (accuracy and number of epoch)

Methods	Accuracy	Number of epoch
RBFNDDA-HIST	0.976±0.313	3
MPANN [19]	0.981±0.005	5100
EAA [20]	0.981±0.464	200000
C-net [21]	0.975±1.800	10000

4   Conclusion

In this study, a RBFNDDA-HIST network is proposed to reduce hidden nodes to constitute a more compact network structure than the original RBFNDDA. The process is simple and fast in handling superfluous nodes. This is further shown when the proposed method is compared with MPANN, EAA and C-net. In the future, additional experiments will be carried out by using other benchmark and real data sets to examine the effectiveness of RBFNDDA-HIST.

References

1. Meng Joo, E., Shiqian, W., Juwei, L.: Face recognition using radial basis function (RBF) neural networks. In: Proceedings of the 38th IEEE Conference on Decision and Control, vol. 3, pp. 2162–2167 (1999)
2. Moody, J., Darken, C.J.: Fast learning in networks of locally-tuned processing units. Neural Computation 1(2), 281–294 (1989)
3. Guang-Bin, H., Saratchandran, P., Sundararajan, N.: A generalized growing and pruning RBF (GGAP-RBF) neural network for function approximation. IEEE Transactions on Neural Networks 16(1), 57–67 (2005)
4. Kang, L., Jian-Xun, P., Er-Wei, B.: Two-stage mixed discrete-continuous identification of Radial Basis Function (RBF) Neural models for nonlinear systems. IEEE Transactions on Circuits and Systems I: Regular Papers 56(3), 630–643 (2009)
5. Freeman, J.A.S., Saad, D.: Learning and generalization in radial basis function networks. Neural Computation 7(5), 1000–1020 (1995)
6. Yu, H.: Network complexity analysis of multilayer feedforward artificial neural networks. In: Schumann, J., Liu, Y. (eds.) Appl. of Neural Networks in High Assur. Sys. SCI, vol. 268, pp. 41–55. Springer, Heidelberg (2010)
7. Zhang, G.P.: Avoiding pitfalls in neural network research. IEEE Transactions on Systems, Man, and Cybernetics, Part C: Applications and Reviews 37(1), 3–16 (2007)
8. Reed, R.: Pruning algorithms-a survey. IEEE Transactions on Neural Networks 4(5), 740–747 (1993)

9. Karnin, E.D.: A simple procedure for pruning back-propagation trained neural networks. IEEE Transactions on Neural Networks 1(2), 239–242 (1990)
10. Augasta, M.G., Kathirvalavakumar, T.: A novel pruning algorithm for optimizing feed-forward neural network of classification problems. Neural Processing Letters 34(3), 241–258 (2011)
11. Ioannidis, Y.: The histogry of histogram. In: Proceedings of the 29th International Conference on Very Large Data Bases, vol. 29, pp. 19–30 (2003)
12. Legg, P.A., Rosin, P.L., Marshall, D., Morgan, J.E.: Improving accuracy and efficiency of registration by mutual information using Sturges' histogram rule. In: Medical Image Understanding and Analysis, pp. 26–30 (2007)
13. Sietsma, J., Dow, R.J.F.: Neural net pruning-why and how. In: IEEE International Conference on Neural Networks, vol. 1, pp. 325–333 (1988)
14. Liang, X.: Removal of hidden neurons in multilayer perceptrons by orthogonal projection and weight crosswise propagation. Neural Computing and Applications 16(1), 57–68 (2007)
15. Zhang, Z., Qiao, J.: A node pruning algorithm for feedforward neural network based on neural complexity. In: 2010 International Conference on Intelligent Control and Information Processing (ICICIP), pp. 406–410 (2010)
16. Shimazaki, H., Shinomoto, S.: A method for selecting the bin size of a time histogram. Journal of Neural Computation 19(6), 1503–1527 (2007)
17. Asunction, A., Newman, D.J.: No Title. University of California, School of Information and Computer Science, Irvine, CA (2007)
18. Paetz, J.: Reducing the number of neurons in radial basis function networks with dynamic decay adjustment. Neurocomputing 62, 79–91 (2004)
19. Abbass, H.A.: An evolutionary artificial neural networks approach for breast cancer diagnosis. Artificial Intelligence in Medicine 25(3), 265–281 (2002)
20. Fogel, D.B., Wasson III, E.C., Boughton, E.M.: Evolving neural networks for detecting breast cancer. Cancer Letters 96(1), 49–53 (1995)
21. Abbass, H.A., Towsey, M., Finn, G.: C-Net: A method for generating non-deterministic and dynamic multivariate decision trees. Knowledge and Information Systems 3(2), 184–197 (2001)
22. Pylkkonen, J.: New pruning criteria for efficient decoding. In: Proceedings of the 9th European Conference on Speech Communication and Technology, pp. 581–584 (2005)
23. Steinbiss, V., Tran, B.H., Ney, H.: Improvement in beam search. In: International Conference on Spoken Language Processing, pp. 2143–2146 (1994)
24. Berthold, M.R., Diamond, J.: Constructive training of probabilistic neural networks. Neurocomputing 19, 167–183 (1998)

Toroidal Approximate Identity Neural Networks Are Universal Approximators

Saeed Panahian Fard* and Zarita Zainuddin

School of Mathematical Sciences, Universiti Sains Malaysia
11800 USM, Pulau Pinang, Malaysia
saeedpanahian@yahoo.com
zarita@cs.usm.my
http://math.usm.my/

Abstract. The approximation of a continuous function on the torus $\mathbb{T}^2$ is an important problem in approximation theory of artificial neural networks. In this work, we investigate the universal approximation capability of one-hidden layer feedforward toroidal approximate identity neural networks. To this end, we present notions of toroidal convolution and toroidal approximate identity. Using these notions, we apply a convolution linear operator approach to prove uniform converges in terms of continuous functions on the torus $\mathbb{T}^2$. Using this result, we also prove a main theorem. The main theorem shows that one-hidden layer feedforward toroidal approximate identity neural networks are universal approximators in the space of continuous functions on the torus $\mathbb{T}^2$.

Keywords: Toroidal approximate identity, Toroidal approximate identity neural networks, Toroidal activation functions, Toroidal convolution, Two dimensional torus, Universal approximation.

1 Introduction

The approximation of a continuous function on the torus $\mathbb{T}^2$ is area of burgeoning interest with applications to bioinformatics, meteorology, and oceanography [1], [2]. A few efforts has been done for approximation of continuous functions on the torus $\mathbb{T}^2$ [3], [4]. To the best of our knowledge, there exist noting notable contributions focused on the approximation of continuous functions on the torus $\mathbb{T}^2$ by using feedforward artificial neural networks. Thus, we are motivated to investigate the universal approximation capability of a one-hidden layer feedforward neural networks in the space of continuous functions on the torus $\mathbb{T}^2$.

The universal approximation capability of feedforward artificial neural networks is one of the most important problems in neural networks theory [5]. That is, under what conditions one-hidden layer feedforward neural networks can approximate arbitrary function belong to a certain set of functions with arbitrary accuracy on a compact domain? Some authors (Artega and Marrero,

---

* Corresponding author.

C.K. Loo et al. (Eds.): ICONIP 2014, Part I, LNCS 8834, pp. 135–142, 2014.
© Springer International Publishing Switzerland 2014

2013; Costarelli, 2014, Lin et al., 2014) surveyed this topic.

In the other direction, one-hidden layer feedforward approximate identity neural networks has been constructed based on the notion of approximate identity [9]. The different types of one-hidden layer feedforward approximate identity neural networks have been introduced [10], [11], [12], [13], [14], [15], [16], [17]. These networks have been constructed based on the modification of approximate identity neural networks.

The main purpose of this work is to develop a theory of the universal approximation capability of one-hidden layer feedforward approximate identity neural networks. In fact, we are concerned with the universal approximation capability of one-hidden layer feedforward toroidal approximate identity neural networks. The networks are called toroidal approximate identity neural networks, because the networks are constructed based on the idea of toroidal approximate identity as the modification of the notion of approximate identity. Note that, we restrict our analyses to the space of continuous functions on the torus $\mathbb{T}^2$.

In this work, we mainly discuss the following problems: first, we introduce the notions of toroidal convolution and toroidal approximate identity. Using these notions, we prove a theorem that shows the convolution of every continuous function f on the torus $\mathbb{T}^2$ with toroidal approximate identity converges uniformly to f on the space of continuous functions on the torus $\mathbb{T}^2$. Using this result, we also prove a main theorem. The main theorem shows that one-hidden layer feedforward toroidal approximate identity neural networks are universal approximators in the space of continuous functions on the torus $\mathbb{T}^2$. The proof of the main theorem is constructed based on a fact concerning theory of ϵ-net, and it is a straightforward modification of the proof of Theorem 1 [18].

The work is organized as follows: in Section 2, we will recall some standard notations and present preliminaries required for the next section. In Section 3, we will prove a theorem. The theorem shows that the convolution of every continuous function f on the torus $\mathbb{T}^2$ with toroidal approximate identity converges uniformly to f on the space of continuous functions on the torus $\mathbb{T}^2$. In Section 4, we will present a main result of this work. The main result shows the analysis of the universal approximation capability of one-hidden layer feedforward toroidal approximate identity neural networks in the space of continuous functions on the $\mathbb{T}^2$ embedded in Euclidean space $\mathbb{R}^4$. In Section 5, discussion is given. Finally, we will conclude the work with some remarks in Section 6.

2   Notations and Preliminaries

In this section, we are going to introduce two basic notions required for the analyses that follows in the next sections. The torus $\mathbb{T}^2$ is as the Cartesian product of the two unit circle $\mathbb{S}^1 \times \mathbb{S}^1$.

Now, we present the notion of toroidal convolution.

Definition 1. *Let $f, g : \mathbb{T}^2 \to \mathbb{R}$, be integrable functions. The toroidal convolution $f * g$ is defined as follows:*

$$f * g(x) := \frac{1}{4\pi^2} \int_{\mathbb{T}^2} f(x - y)g(y)dy, x \in \mathbb{T}^2.$$

Then, we give the notion of toroidal approximate identity.

Definition 2. *Let $\{\phi_n\}_{n=1}^{\infty}$, $\phi_n : \mathbb{T}^2 \to \mathbb{R}$, be a sequence of functions. The sequence is called a toroidal approximate identity if it satisfies the following conditions:*

1) $\dfrac{1}{4\pi^2} \displaystyle\int_{\mathbb{T}^2} \phi_n(x - y)dy = 1, \forall n \in \mathbb{N}, x \in \mathbb{T}^2$;

2) for arbitrary $\epsilon > 0$, $\delta > 0$, and $\mathbb{T}_\delta^2(x) := \{y \in \mathbb{T}^2 | dist(x, y) < \delta\}$, there exists a number N such that if $n \geqslant N$ it results

$$\frac{1}{4\pi^2} \int_{\mathbb{T}^2 \backslash \mathbb{T}_\delta^2(x)} |\phi_n(x - y)|dy \leq \epsilon.$$

In the next section, we will show an approximation result in the space of continuous functions on the torus $\mathbb{T}^2$.

3   The Approximation Result in the Space of Continuous Functions on the Torus $\mathbb{T}^2$

In this section, we prove Theorems 1 that shows the convolution of every continuous function f on the torus $\mathbb{T}^2$ with toroidal approximate identity converges uniformly to f on the space of continuous functions on the torus $\mathbb{T}^2$.

Theorem 1. *Let $C(\mathbb{T}^2)$ be the real linear space of all continuous functions on the torus $\mathbb{T}^2$. Let $\{\phi_n\}_{n \in \mathbb{N}}$, $\mathbb{T}^2 \to \mathbb{R}$ be a toroidal approximate identity. Then for every $f \in C(\mathbb{T}^2)$, $\phi_n * f$ converges uniformly to f on $C(\mathbb{T}^2)$.*

Proof. Let f be a continuous function in $C(\mathbb{T}^2)$, let $x \in \mathbb{T}^2$ be a points of its continuity, and let $\epsilon > 0$. Choose a $\delta > 0$ such that $\|f(x) - f(y)\|_{C(\mathbb{T}^2)} < \epsilon$ when y is in the set $\mathbb{T}_\delta^2(x) := \{y \in \mathbb{T}^2 | dist(x, y) < \delta\}$. Let us define $\{\phi_n * f\}_{n \in \mathbb{N}}$ by $\phi_n(x - y) = n\phi(x - ny)$, we consider

$$\phi_n * f(x) - f(x)$$

$$= \frac{1}{4\pi^2} \int_{\mathbb{T}^2} \phi_n(x - y)f(y)dy - \frac{1}{4\pi^2} \int_{\mathbb{T}^2} \phi_n(x - y)dy f(x)$$

$$= \frac{1}{4\pi^2} \int_{\mathbb{T}^2} \phi_n(x - y)\{f(y) - f(x)\}dy$$

$$= \frac{1}{4\pi^2} \int_{\mathbb{T}^2} n\phi(x - ny)\{f(y) - f(x)\}dy$$

$$= \frac{1}{4\pi^2} \Big(\int_{\mathbb{T}_\delta^2(x)} + \int_{\mathbb{T}^2 \backslash \mathbb{T}_\delta^2(x)}\Big)n\phi(x - ny)\{f(y) - f(x)\}dy$$

$$= I_1 + I_2.$$

Subsequently we calculate I_1, I_2 as follows;

$$\|I_1\|_{C(\mathbb{T}^2)} \leq \frac{1}{4\pi^2} \int_{\mathbb{T}^2_\delta(x)} |n\phi(x-ny)| \|f(y) - f(x)\|_{C(\mathbb{T}^2)} dy$$

$$< \frac{\epsilon}{2\frac{1}{4\pi^2}\int_{\mathbb{T}^2} |\phi(x-t)|dt} \frac{1}{4\pi^2} \int_{\mathbb{T}^2_\delta(x)} |n\phi(x-ny)| dy$$

$$= \frac{\epsilon}{2\frac{1}{4\pi^2}\int_{\mathbb{T}^2} |\phi(x-t)|dt} \frac{1}{4\pi^2} \int_{\mathbb{T}^2_{n\delta}(x)} |\phi(x-t)|dt$$

$$\leq \frac{\epsilon}{2\frac{1}{4\pi^2}\int_{\mathbb{T}^2} |\phi(x-t)|dt} \frac{1}{4\pi^2} \int_{\mathbb{T}^2} |\phi(x-t)|dt = \frac{\epsilon}{2}.$$

For I_2, we have

$$\|I_2\|_{C(\mathbb{T}^2)} \leq \frac{1}{4\pi^2} \int_{\mathbb{T}^2\setminus\mathbb{T}^2_\delta(x)} |n\phi(x-ny)| \|f(y) - f(x)\|_{C(\mathbb{T}^2)} dy$$

$$\leq 2\|f\|_{C(\mathbb{T}^2)} \frac{1}{4\pi^2} \int_{\mathbb{T}^2\setminus\mathbb{T}^2_\delta(x)} n|\phi(ny)| dy$$

$$= 2\|f\|_{C(\mathbb{T}^2)} \frac{1}{4\pi^2} \int_{\mathbb{T}^2\setminus\mathbb{T}^2_{n\delta}(x)} |\phi(t)| dt.$$

Since

$$\lim_{n\to\infty} \frac{1}{4\pi^2} \int_{\mathbb{T}^2\setminus\mathbb{T}^2_{n\delta}(x)} |\phi(t)| dt = 0,$$

there exists an $n_0 \in \mathbb{N}$ such that for all $n \geq n_0$,

$$\frac{1}{4\pi^2} \int_{\mathbb{T}^2\setminus\mathbb{T}^2_{n\delta}(x)} |\phi(t)| dt < \frac{\epsilon}{4\|f\|_{C(\mathbb{T}^2)}}.$$

Combining I_1 and I_2 for $n \geq n_0$, we have

$$\|\phi_n * f(x) - f(x)\|_{C(\mathbb{T}^2)} < \epsilon.$$

□

Using Theorem 1, we shall prove the main result of this work in the next section.

4   The Main Result

In this section, we prove Theorem 2 as the main result of this work. Theorem 2 shows that one-hidden layer feedforward toroidal approximate identity neural networks are universal approximators in the space of continuous functions on the torus $\mathbb{T}^2$.

Theorem 2. *Let $C(\mathbb{T}^2)$ be the real linear space of all continuous functions on the torus $\mathbb{T}^2$, and $V \subset C(\mathbb{T}^2)$ a compact set. Let $\{\phi_n\}_{n\in\mathbb{N}}$, $\phi_n : \mathbb{T}^2 \to \mathbb{R}$ be a toroidal approximate identity. Let the family of functions $\{\sum_{j=1}^{M} \lambda_j \phi_j(x - y_j) | \lambda_j \in \mathbb{R}, x \in \mathbb{T}^2, y_j \in \mathbb{T}^2, M \in \mathbb{N}\}$, be dense in $C(\mathbb{T}^2)$, and given $\epsilon > 0$. Then there exists an $N \in \mathbb{N}$ which depends on V and ϵ but not on f, such that for arbitrary $f \in V$, there exist weights $c_k = c_k(f, V, \epsilon)$ satisfying*

$$\left\| f(x) - \sum_{k=1}^{N} c_k \phi_k(x - y_k) \right\|_{C(\mathbb{T}^2)} < \epsilon.$$

Moreover, every c_k is a continuous function of $f \in V$.

Proof. Since V is compact, for arbitrary $\epsilon > 0$, there is a finite $\frac{\epsilon}{2}$-net $\{f^1, ..., f^M\}$ for V. Thus, for arbitrary $f \in V$, there is an f^j such that $\| f - f^j \|_{C(\mathbb{T}^2)} < \frac{\epsilon}{2}$. For arbitrary f^j, by assumption of the theorem, there are $\lambda_i^j \in \mathbb{R}, N_j \in \mathbb{N}$, and $\phi_i^j(x - y_i)$ such that

$$\left\| f^j(x) - \sum_{i=1}^{N_j} \lambda_i^j \phi_i^j(x - y_i) \right\|_{C(\mathbb{T}^2)} < \frac{\epsilon}{2}. \tag{1}$$

For arbitrary $f \in V$, we define

$$F_-(f) = \{j | \ \| f - f^j \|_{C(\mathbb{T}^2)} < \frac{\epsilon}{2}\},$$

$$F_0(f) = \{j | \ \| f - f^j \|_{C(\mathbb{T}^2)} = \frac{\epsilon}{2}\},$$

$$F_+(f) = \{j | \ \| f - f^j \|_{C(\mathbb{T}^2)} > \frac{\epsilon}{2}\}.$$

Therefore, $F_-(f)$ is not empty according to the definition of $\frac{\epsilon}{2}$-net. If $\widetilde{f} \in V$ approaches f such that $\| \widetilde{f} - f \|_{C(\mathbb{T}^2)}$ is small enough, then we have $F_-(f) \subset F_-(\widetilde{f})$ and $F_+(f) \subset F_+(\widetilde{f})$. Thus, $F_-(\widetilde{f}) \bigcap F_+(f) \subset F_-(\widetilde{f}) \bigcap F_+(\widetilde{f}) = \emptyset$, which implies $F_-(\widetilde{f}) \subset F_-(f) \cup F_0(f)$. We finish with the following:

$$F_-(f) \subset F_-(\widetilde{f}) \subset F_-(f) \cup F_0(f). \tag{2}$$

Define

$$d(f) = \left[\sum_{j \in F_-(f)} (\frac{\epsilon}{2} - \| f - f^j \|_{C(\mathbb{T}^2)}) \right]^{-1}$$

and

$$f_h = \sum_{j \in F_-(f)} \sum_{i=1}^{N_j} d(f) \left(\frac{\epsilon}{2} - \| f - f^j \|_{C(\mathbb{T}^2)} \right) \lambda_i^j \phi_i^j(x - y_i) \tag{3}$$

then, $f_h \in \left\{ \sum_{j=1}^{M} \lambda_j \phi_j(x - y_j) \right\}$ approximates f with accuracy ϵ :

$$\| f - f_h \|_{C(\mathbb{T}^2)}$$

$$= \left\| \sum_{j \in F_-(f)} d(f) \left(\frac{\epsilon}{2} - \| f - f^j \|_{C(\mathbb{T}^2)} \right) \right.$$

$$\left. \left(f - \sum_{i=1}^{N_j} \lambda_i^j \phi_i^j(x - y_i) \right) \right\|_{C(\mathbb{T}^2)}$$

$$= \left\| \sum_{j \in F_-(f)} d(f) \left(\frac{\epsilon}{2} - \| f - f^j \|_{C(\mathbb{T}^2)} \right) \right.$$

$$\left. \left(f - f^j + f^j - \sum_{i=1}^{N_j} \lambda_i^j \phi_i^j(x - y_i) \right) \right\|_{C(\mathbb{T}^2)}$$

$$\leq \sum_{j \in F_-(f)} d(f) \left(\frac{\epsilon}{2} - \| f - f^j \|_{C(\mathbb{T}^2)} \right)$$

$$\left(\| f - f^j \|_{C(\mathbb{T}^2)} + \left\| f_j - \sum_{i=1}^{N_j} \lambda_i^j \phi_i^j(x - y_i) \right\|_{C(\mathbb{T}^2)} \right)$$

$$\leq \sum_{j \in F_-(f)} d(f) \left(\frac{\epsilon}{2} - \| f - f^j \|_{C(\mathbb{T}^2)} \right) \left(\frac{\epsilon}{2} + \frac{\epsilon}{2} \right)$$

$$= \epsilon. \tag{4}$$

In the next step, we prove the continuity of c_k. For the proof, we use (2) to obtain

$$\sum_{j \in F_-(f)} \left(\frac{\epsilon}{2} - \| \widetilde{f} - f^j \|_{C(\mathbb{T}^2)} \right)$$

$$\leq \sum_{j \in F_-(\widetilde{f})} \left(\frac{\epsilon}{2} - \| \widetilde{f} - f \|_{C(\mathbb{T}^2)} \right)$$

$$\leq \sum_{j \in F_-(\widetilde{f})} \left(\frac{\epsilon}{2} - \| \widetilde{f} - f^j \|_{C(\mathbb{T}^2)} \right) + \sum_{j \in F_0(f)} \left(\frac{\epsilon}{2} - \| \widetilde{f} - f^j \|_{C(\mathbb{T}^2)} \right). \tag{5}$$

Let $\widetilde{f} \to f$ in (5), then we have

$$\sum_{j \in F_-(\widetilde{f})} \left(\frac{\epsilon}{2} - \| \widetilde{f} - f^j \|_{C(\mathbb{T}^2)} \right) \to \sum_{j \in F_-(f)} \left(\frac{\epsilon}{2} - \| f - f^j \|_{C(\mathbb{T}^2)} \right). \tag{6}$$

This obviously demonstrates that $d(\widetilde{f}) \to d(f)$. Thus, $\widetilde{f} \to f$ results

$$d(\widetilde{f}) \left(\frac{\epsilon}{2} - \| \widetilde{f} - f^j \|_{C(\mathbb{T}^2)} \right) \lambda_i^j \to d(f) \left(\frac{\epsilon}{2} - \| f - f^j \|_{C(\mathbb{T}^2)} \right) \lambda_i^j. \tag{7}$$

Let $N = \sum_{j \in F_-(f)} N_j$ and define c_k in terms of

$$f_h = \sum_{j \in F_-(f)} \sum_{i=1}^{N_j} d(f) \left(\frac{\epsilon}{2} - \| f - f^j \|_{C(\mathbb{T}^2)} \right) \lambda_i^j \phi_i^j (x - y_i)$$

$$\equiv \sum_{k=1}^{N} c_k \phi_k (x - y_k)$$

From (6), c_k is a continuous functional of f. □

5   Discussion and Conclusion

By toroidal functions, we mean functions whose supports are on the torus $\mathbb{T}^2$. Toroidal functions occur frequently in the study of the wind directions. As the main result of this work, we have found that toroidal continuous functions can be approximate by one-hidden layer feedforwad toroidal approximate identity neural networks. In this context, Theorem 2 indicates that one-hidden layer feedforwad toroidal approximate identity neural networks are universal approximators in the space of continuous functions on the torus $\mathbb{T}^2$. The importance of our finding is that this finding extends the universal approximation capability of feedforward neural networks to dimension higher than three. However, a restriction is worth noting. In fact, we have imposed the dimensional restriction on the torus. We have demonstrated that there are no notable contributions focused on the approximation of a continuous function on the torus $\mathbb{T}^2$ by using feedforward neural networks. In the current work, we have developed the theory of the universal approximation capability of one-hidden layer feedforward toroidal approximate identity neural networks. To this end, we have presented the notion of toroidal convolution and the notion of toroidal approximate identity. Using these notions, we have derived two theorems. In Theorem 1, we have proved that the convolution linear operators of every continuous function f on the torus $\mathbb{T}^2$ with toroidal approximate identity converges uniformly to f on the space of continuous functions the torus $\mathbb{T}^2$. Using this result, we have also proved Theorem 2 as the main result of the study. In Theorem 2, we have proved that one-hidden layer feedforward toroidal approximate identity neural networks are universal approximators in the space of the continuous functions on the torus $\mathbb{T}^2$.

Acknowledgements. The authors would like to thank the reviewers for their valuable comments and suggestions. This work was supported by Universiti Sains Malaysia under Grant No. 1001/PMATHS/811161.

References

1. Marzio, M.D., Panzera, A., Taylor, C.C.: Kernel density estimation on the torus. Journal of Statistical Planning and Inference 141, 2156–2173 (2011)
2. Taylor, C.C., Mardia, K.V., Marzio, M.D., Panzera, A.: Validating protein structure using kernel density estimates. Journal of Applied Statistics 39, 2379–2388 (2012)

3. Potts, D.: Approximation of scattered data by trigonometric polynomials on the torus and the 2-sphere. Advances in Computational Mathematics 21, 21–36 (2004)
4. Kushpel, A., Grandison, C., Ha, M.D.: Optimal sk-splines approximation and reconstruction on the torus and sphere. International Journal of Pure and Applied Mathematics 29, 469–490 (2006)
5. Nong, J.: Conditions for RBF neural networks to universal approximation and numerical experiments. In: Zhao, M., Sha, J. (eds.) ICCIP 2012, Part II. CCIS, vol. 289, pp. 299–308. Springer, Heidelberg (2012)
6. Arteaga, C., Marrero, M.: Universal approximation by radial basis function networks of Delsarte translates. Neural Networks 46, 299–305 (2013)
7. Costarelli, D.: Interpolation by neural network operators activated by ramp function. Journal of Mathematical Analysis and Applications 419, 574–598 (2014)
8. Lin, S., Rong, Y., Xu, Z.: Multivariate Jackson-type inequality for a new type neural network approximation. Applied Mathematical Modelling (2014), http://dx.doi.org/10.1016/j.apm.2014.05.018
9. Cao, L., Xi, L., Zhang, Y.: L^p estimate of convolution transform of singular measure by approximate identity. Nonlinear Analysis 94, 148–155 (2014)
10. Zainuddin, Z., Panahian, F. S.: Double approximate identity neural networks universal approximation in real Lebesgue spaces. In: Huang, T., Zeng, Z., Li, C., Leung, C.S. (eds.) ICONIP 2012, Part I. LNCS, vol. 7663, pp. 409–415. Springer, Heidelberg (2012)
11. Panahian Fard, S., Zainuddin, Z.: On the universal approximation capability of flexible approximate identity neural networks. In: Wong, W.E., Ma, T. (eds.) Emerging Technologies for Information Systems, Computing, and Management. LNEE, vol. 236, pp. 201–207. Springer, Heidelberg (2013)
12. Panahian Fard, S., Zainuddin, Z.: The universal approximation capabilities of Mellin approximate identity neural networks. In: Guo, C., Hou, Z.-G., Zeng, Z. (eds.) ISNN 2013, Part I. LNCS, vol. 7951, pp. 205–213. Springer, Heidelberg (2013)
13. Panahian Fard, S., Zainuddin, Z.: The universal approximation capability of double flexible approximate identity neural networks. In: Wong, W.E., Zhu, T. (eds.) Computer Engineering and Networking. LNEE, vol. 277, pp. 125–133. Springer, Heidelberg (2014)
14. Panahian Fard, S., Zainuddin, Z.: Analyses for $L^p[a, b]$-norm approximation capability of flexible approximate identity neural networks. Neural Computing and Applications 24, 45–50 (2014)
15. Panahian Fard, S., Zainuddin, Z.: The universal approximation capabilities of "2π-periodic approximate identity" neural networks. In: 2013 International Conference on Information Science and Cloud Computing, pp. 793–798. IEEE Press (2014)
16. Panahian Fard, S., Zainuddin, Z.: The universal approximation capabilities of "double 2π-periodic approximate identity" neural networks. Soft Computing, doi: 10.1007/s00500-014-1449-8
17. Zainuddin, Z., Panahian Fard, S.: A study on the universal approximation capability of 2-spherical approximate identity neural networks. In: 5th International Conference on Mathematical Model for Engineering Science, pp. 23–27. WSEAS Press (2014)
18. Wu, W., Nan, D., Li, Z., Long, J., Wang, J.: Approximation to compact set of functions by feedforward neural networks. In: Proceedings of the 20th International Joint Conference on Neural Networks, pp. 1222–1225 (2007)

Self-organizing Neural Grove

Hirotaka Inoue

Department of Electrical Engineering and Information Science,
Kure National College of Technology,
2-2-11 Agaminami, Kure, Hiroshima, 737-8506 Japan
hiro@kure-nct.ac.jp

Abstract. Recently, multiple classifier systems have been used for practical applications to improve classification accuracy. Self-generating neural networks (SGNN) are one of the most suitable base-classifiers for multiple classifier systems because of their simple settings and fast learning ability. However, the computation cost of the multiple classifier system based on SGNN increases in proportion to the numbers of SGNN. In this paper, we propose a novel pruning method for efficient classification and we call this model a self-organizing neural grove (SONG). Experiments have been conducted to compare the SONG with bagging and the SONG with boosting, and support vector machine (SVM). The results show that the SONG can improve its classification accuracy as well as reducing the computation cost.

1 Introduction

Classifiers need to find hidden information within a large amount of given data effectively and classify unknown data as accurately as possible [1]. Recently, to improve the classification accuracy, multiple classifier systems such as neural network ensembles, bagging, and boosting have been used for practical data mining applications [2]. In general, base classifiers of multiple classifier systems use traditional models such as neural networks (backpropagation network and radial basis function network) [3] and decision trees (CART and C4.5) [4].

Neural networks have great advantages such as adaptability, flexibility, and universal nonlinear input-output mapping capability. However, to apply these neural networks, it is necessary for the network structure and some parameters to be determined by human experts, and it is quite difficult to choose the right network structure suitable for a particular application at hand. Moreover, they require a long training time to learn the input-output relation of the given data. These drawbacks prevent neural networks from being the base classifier of multiple classifier systems for practical applications.

Self-generating neural networks (SGNN) [5] have a simple network design and high speed learning. SGNN are an extension of the self-organizing maps (SOM) of Kohonen [6] and utilize the competitive learning which is implemented as a self-generating neural tree (SGNT). The abilities of SGNN make it suitable for the base classifier of multiple classifier systems. In order to improve in the accuracy of SGNN, we proposed ensemble self-generating neural networks (ESGNN) for classification [7] as one of multiple classifier systems. Although the accuracy of ESGNN improves by using various

C.K. Loo et al. (Eds.): ICONIP 2014, Part I, LNCS 8834, pp. 143–150, 2014.
© Springer International Publishing Switzerland 2014

SGNN, the computation cost, that is the computation time and the memory capacity increases in proportion to the increase in numbers of SGNN in multiple classifier systems.

In an earlier paper [8], we proposed a pruning method for the structure of the SGNN in multiple classifier systems to reduce the computation cost. In this paper, we propose a novel pruning method for more effective processing and we call this model a self-organizing neural grove (SONG). This pruning method is constructed in two stages. In the first stage, we introduce an on-line pruning algorithm to reduce the computation cost by using class labels in learning. In the second stage, we optimize the structure of the SGNT in multiple classifier systems to improve the generalization capability by pruning the redundant leaves after learning. In the optimization stage, we introduce a threshold value as a pruning parameter to decide which subtree's leaves to prune and estimate with 10-fold cross-validation [9]. After the optimization, the SONG improve its classification accuracy as well as reducing the computation cost. We use bagging [10] and boosting [11] as a resampling technique for the SONG.

We investigate the improvement performance of the SONG by comparing it with support vector machine (SVM) [12] using ten problems in the UCI machine learning repository [13].

The rest of the paper is organized as follows. The next section shows how to construct the SONG. Section 3 shows the experimental results. Then section 4 is devoted to some experiments to investigate the incremental learning performance of SONG.

2   Constructing Self-organizing Neural Grove

In this section, we describe how to prune redundant leaves in the SONG. First, we mention the on-line pruning method in the learning of SGNT. Second, we show the optimization method in constructing the SONG.

2.1   On-line Pruning of Self-generating Neural Tree

SGNN are based on SOM and are implemented as an SGNT architecture. The SGNT can be constructed directly from the given training data without any intervening human effort. The SGNT algorithm is defined as a tree construction problem of how to construct a tree structure from the given data which consists of multiple attributes under the condition that the final leaves correspond to the given data.

Before we describe the SGNT algorithm, we denote some notations.

- input data vector: $e_i \in \mathbb{R}^m$.
- root, leaf, and node in the SGNT: n_j.
- weight vector of n_j: $w_j \in \mathbb{R}^m$.
- the number of the leaves in n_j: c_j.
- distance measure: $d(e_i, w_j)$.
- winner leaf for e_i in the SGNT: n_{win}.

The SGNT algorithm is a hierarchical clustering algorithm. The pseudo C code of the SGNT algorithm is given as follows:

Algorithm (SGNT Generation)

Table 1. Sub procedures of the SGNT algorithm

Sub procedure	Specification
$copy(n_j, e_i/w_{win})$	Create n_j, copy e_i/w_{win} as w_j in n_j.
$choose(e_i, n_1)$	Decide n_{win} for e_i.
$leaf(n_{win})$	Check n_{win} whether n_{win} is a leaf or not.
$connect(n_j, n_{win})$	Connect n_j as a child leaf of n_{win}.
$prune(n_{win})$	Prune leaves if the leaves have the same class.

```
Input:
  A set of training examples E = {e_i},
  i = 1, ... , N.
  A distance measure d(e_i,w_j).
Program Code:
  copy(n_1,e_1);
  for (i = 2, j = 2; i <= N; i++) {
    n_win = choose(e_i, n_1);
    if (leaf(n_win)) {
      copy(n_j, w_win);
      connect(n_j, n_win);
      j++;
    }
    copy(n_j, e_i);
    connect(n_j, n_win);
    j++;
    prune(n_win);
  }
Output:
  Constructed SGNT by E.
```

In the above algorithm, several sub procedures are used. Table 1 shows the sub procedures of the SGNT algorithm and their specifications.

In order to decide the winner leaf n_{win} in the sub procedure choose(e_i,n_1), competitive learning is used. This sub procedure is recursively used from the root to the leaves of the SGNT. If an n_j includes the n_{win} as its descendant in the SGNT, the weight w_{jk} ($k = 1, 2, \ldots, m$) of the n_j is updated as follows:

$$w_{jk} \leftarrow w_{jk} + \frac{1}{c_j} \cdot (e_{ik} - w_{jk}), \quad 1 \leq k \leq m. \tag{1}$$

In the SGNT, the input vector x_i corresponds to e_i, and the desired output y_i corresponds to the network output o_i which is stored in one of the leaf neurons, for $(x_i, y_i) \in D$. Here, D is the training data set which consists of data $\{x_i, y_i | i = 1, \ldots, N\}$, $x_i \in \mathbb{R}^m$ is the input and y_i is the desired output. After all training data are inserted into the SGNT as the leaves, the leaves each have a class label as the outputs and the weights of each node are the averages of the corresponding weights of all its leaves. The whole network of the SGNT reflects the given feature space by its topology.

We explain the SGNT generation algorithm using an simple example. In this example, m is one and the four training data (x_i, y_i) is (1,1), (2,2), (3,3), and (4,4). Hence, $e_{11} = 1, e_{21} = 2, e_{31} = 3$, and $e_{41} = 4$. Fig. 1 shows an example of the SGNT generation. First, e_{11} is just copied to a neuron n_1 as the root, and e_{11} is substituted to w_{11} (Fig. 1 (a)). In Fig. 1, the circle is the neuron, the integer in the circle is the number of neuron j, the integer of left-upper of the circle is c_j, and the integer of under the circle is w_{j1}. Next, n_2 and n_3 are generated as the children of n_1 with $w_{21} = 1, w_{31} = 2$. w_{11} is updated by e_{21} to $1 + 1/2(2-1) = 1.5$ (Fig. 1 (b)). Next, the winner in $\{n_1, n_2, n_3\}$ is n_3 since $d(e_3, w_1) = 1.5, d(e_3, w_2) = 2$, and $d(e_3, w_3) = 1$; and thus, n_4 and n_5 are generated as the children of n_3 with $w_{41} = 2, w_{51} = 3$. w_{31} is updated by e_{31} to $2 + 1/2(3-2) = 2.5$ and w_{11} is updated by e_{31} to $1.5 + 1/3(3-1.5) = 2$ (Fig. 1 (c)). Finally, n_6 and n_7 are generated as the children of n_5 with $w_{61} = 3, w_{71} = 4$. w_{51} is updated by e_{41} to $3 + 1/2(4-3) = 3.5$, w_{31} is updated by e_{41} to $2.5 + 1/3(4-2.5) = 3$, and w_{11} is updated by e_{41} to $2 + 1/4(4-2) = 2.5$ (Fig. 1 (d)).

Note, to optimize the structure of the SGNT effectively, we remove the threshold value of the original SGNT algorithm in [5] to control the number of leaves based on the distance because of the trade-off between the memory capacity and the classification accuracy. In order to avoid the above problem, we introduce a new pruning method in the sub procedure `prune(n_win)`. We use the class label to prune leaves. For leaves that have the n_{win}'s parent node, if all leaves belong to the same class, then these leaves are pruned and the parent node is given to the class.

2.2 Optimization of the SONG

The SGNT has the capability of high speed processing. However, the accuracy of the SGNT is inferior to the conventional approaches, such as nearest neighbor, because the SGNT has no guarantee to reach the nearest leaf for unknown data [14]. Hence, we construct the SONG by taking the majority of multiple SGNT's outputs to improve the accuracy.

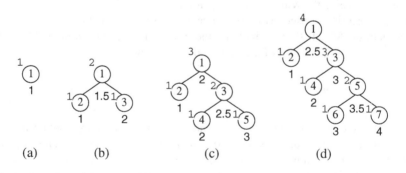

Fig. 1. An example of the SGNT generation

Although the accuracy of the SONG is superior or comparable to the accuracy of conventional approaches, the computational cost increases in proportion to the increase in the number of SGNTs in the SONG. In particular, the huge memory requirement prevents the use of SONG for large datasets even with the latest computers.

In order to improve the classification accuracy, we propose an optimization method of the SONG for classification. This method has two parts, the merge phase and the evaluation phase. The merge phase is performed as a pruning algorithm to reduce dense leaves (Fig. 2).

```
1 begin    initialize j = the height of the SGNT
2    do for each subtree's leaves in the height j
3      if the ratio of the most class ≥ α,
4      then merge all leaves to parent node
5      if all subtrees are traversed in the height j,
6      then j ← j − 1
7    until j = 0
8 end.
```

Fig. 2. The merge phase

This phase uses the class information and a threshold value α to decide which subtree's leaves to prune or not. For leaves that have the same parent node, if the proportion of the most common class is greater than or equal to the threshold value α, then these leaves are pruned and the parent node is given the most common class.

The optimum threshold values α of the given problems are different from each other. The evaluation phase is performed to choose the best threshold value by introducing 10-fold cross validation (Fig. 3).

3 Experimental Results

We investigate the computational cost (the memory capacity and the computation time) and the classification accuracy of the SONG with bagging for ten benchmark problems

```
1 begin initialize α = 0.5
2    do for each α
3      evaluate the merge phase with 10-fold CV
4      if the best classification accuracy is obtained,
5      then record the α as the optimal value
6      α ← α + 0.05
7    until α = 1
8 end.
```

Fig. 3. The evaluation phase

in the UCI machine learning repository [13]. We evaluate how the SONG is pruned using 10-fold cross-validation for the ten benchmark problems. In this experiment, we use a modified Euclidean distance measure for the SONG. Since the performance of the SONG is not sensitive to the threshold value α, we set the different threshold values α to vary from 0.5 to 1; $\alpha = [0.5, 0.55, 0.6, \ldots, 1]$. We set the number of SGNT K in the SONG as 25 and execute 100 trials by changing the sampling order of each training set. All experiments in this section were performed on an UltraSPARC workstation with a 900MHz CPU, 1GB RAM, and Solaris 8.

Table 2 shows the average classification accuracy of 10 trials for the SONG with bagging and boosting. On boosting, we implement AdaBoost [11] to the SONG. Since original AdaBoost algorithm have been proposed for binary classification problems, we use four binary classification problems in Table 2. In comparison with boosting, bagging is superior to boosting on all of the 4 datasets. In short, bagging is better than boosting in terms of the classification accuracy.

To show the advantages of the SONG, we compare it with SVM on the same problems. In the SONG, we choose the best classification accuracy of 100 trials with bagging. In SVM, we use C-SVM in libsvm [12] with radial basis function kernel. We select the parameters of SVM, the cost parameters C and the kernel parameters γ, from $15 \times 15 = 225$ combinations by 10-fold cross validation; $C = [2^{12}, 2^{11}, 2^{10}, \ldots, 2^{-2}]$ and $\gamma = [2^4, 2^3, 2^2, \ldots, 2^{-10}]$. We normalize the input data from 0 to 1 for all problems in SONG and SVM. All methods are compiled by using gcc with the optimization level -O2 on the same workstation.

Table 3 shows the classification accuracy, the memory requirement, and the computation time achieved by the SONG and SVM. Next, we show the results for each category. First, in view point of the classification accuracy, the SONG superior to SVM 3 of the 10 datasets and degrade 1.7% in the average. Second, in terms of the memory requirement, even though the SONG includes the root and the nodes which are generated by the SGNT generation algorithm, this is less than SVM for 8 of the 10 datasets. Although the memory requirement of the SONG is totally used K times in Table 3, we release the memory of SGNT for each trial and reuse the memory for effective computation. Therefore, the memory requirement is suppressed by the size of the single SGNT. Finally, in view of the computation time, although the SONG consumes the cost of K times of the SGNT to construct the model and test for the unknown dataset, the average computation time is faster than SVM. The SONG is slower than SVM for small

Table 2. The average classification accuracy of 10 trials for the SONG with bagging and boosting. The standard deviation is given inside the bracket ($\times 10^{-3}$.)

Dataset	SONG with bagging			SONG with boosting		
	SGNT	SONG	ratio	SGNT	SONG	ratio
breast-cancer-w	0.96(4.74)	0.975(2.86)	+1.5	0.96(6.47)	0.957(4.13)	-0.3
ionosphere	0.847(19.3)	0.89(8.23)	+4.3	0.854(18.26)	0.773(17.4)	-8.1
liver-disorders	0.571(21.4)	0.636(11.0)	+6.5	0.588(17.0)	0.572(24.3)	-1.6
pima-diabetes	0.705(9.8)	0.754(4.96)	+4.9	0.696(12.2)	0.722(6.82)	+2.6
Average	0.771	0.814	+4.3	0.775	0.756	-1.9

Table 3. The classification accuracy, the memory requirement, and the computation time of ten trials for the best pruned SONG and SVM

Dataset	classification acc.		memory requirement		computation time (s)	
	SONG	SVM	SONG	SVM	SONG	SVM
balance-scale	0.885	**0.992**	109.93	**60.6**	**0.82**	4.77
breast-cancer-w	**0.976**	0.973	**26.8**	79.6	**1.18**	**0.64**
glass	**0.758**	0.738	**91.33**	132.4	**0.36**	0.61
ionosphere	0.912	**0.954**	**51.38**	147.9	1.93	**1.25**
iris	**0.973**	0.96	**11.34**	51.3	0.13	**0.06**
letter	0.958	**0.977**	6208.03	7739.7	**208.52**	2359.39
liver-disorders	0.685	**0.73**	**134.17**	214.5	**0.54**	2.07
new-thyroid	0.972	**0.977**	45.74	**44.1**	0.23	**0.22**
pima-diabetes	0.764	**0.766**	**183.57**	363.5	**1.72**	5.63
wine	0.983	**0.989**	**11.8**	62.2	0.31	**0.15**
Average	0.887	**0.904**	**687.41**	889.58	**21.57**	236.88

datasets such as glass, ionosphere, and iris. However, the SONG is faster than SVM for large datasets such as balance-scale, letter, and pima-diabetes. Especially, in letter, the computation time of the SONG is faster than SVM about 11 times. We need to repeat 10-fold cross validation many times to select the optimum parameter for α, k, C, and γ. This evaluation consumes much computation time for large datasets such as letter. Therefore, the SONG based on the fast and compact SGNT is useful and practical for large datasets. Moreover, the SONG has the ability of parallel computation because each classifier behaves independently. In conclusion, the SONG is a practical method for large-scale data mining compared with SVM.

4 Conclusions

In this paper, we proposed a new pruning method for the multiple classifier system based on SGNT, which is called SONG, and evaluated the computation cost and the accuracy. We introduced an on-line and off-line pruning method and evaluated the SONG by 10-fold cross-validation. Experimental results showed that the memory requirement reduced remarkably, and the accuracy increased by using the pruned SGNT as the base classifier of the SONG. The SONG is a useful and practical multiple classifier system to classify large datasets. In future work, we will study a parallel and distributed processing of the SONG for large scale data mining.

Acknowledgment. The author would like to thank the anonymous referees for their helpful comments.

References

1. Han, J., Kamber, M.: Data Mining: Concepts and Techniques. Morgan Kaufmann Publishers, San Francisco (2000)
2. Quinlan, J.R.: Bagging, Boosting, and C4.5. In: Proceedings of the Thirteenth National Conference on Artificial Intelligence, Portland, OR, August 4-8, pp. 725–730. AAAI Press, The MIT Press (1996)
3. Bishop, C.M.: Neural Networks for Pattern Recognition. Oxford University Press, New York (1995)
4. Duda, R.O., Hart, P.E., Stork, D.G.: Pattern Classification, 2nd edn. John Wiley & Sons Inc., New York (2000)
5. Wen, W.X., Jennings, A., Liu, H.: Learning a neural tree. In: The International Joint Conference on Neural Networks, Beijing, China, November 3-6, vol. 2, pp. 751–756 (1992)
6. Kohonen, T.: Self-Organizing Maps. Springer, Berlin (1995)
7. Inoue, H., Narihisa, H.: Improving generalization ability of self-generating neural networks through ensemble averaging. In: Terano, T., Liu, H., Chen, A.L.P. (eds.) PAKDD 2000. LNCS (LNAI), vol. 1805, pp. 177–180. Springer, Heidelberg (2000)
8. Inoue, H., Narihisa, H.: Optimizing a multiple classifier system. In: Ishizuka, M., Sattar, A. (eds.) PRICAI 2002. LNCS (LNAI), vol. 2417, pp. 285–294. Springer, Heidelberg (2002)
9. Stone, M.: Cross-validation: A review. Math. Operationsforsch. Statist., Ser. Statistics 9(1), 127–139 (1978)
10. Breiman, L.: Bagging predictors. Machine Learning 24, 123–140 (1996)
11. Freund, Y., Schapire, R.E.: Boosting: Foundations and Algorithms. MIT Press, Cambridge (2012)
12. Chang, C.-C., Lin, C.-J.: LIBSVM: A library for support vector machines. ACM Transactions on Intelligent Systems and Technology 2, 27:1–27:27 (2011), Software available at http://www.csie.ntu.edu.tw/~cjlin/libsvm
13. Frank, A., Asuncion, A.: UCI machine learning repository (2010), http://archive.ics.uci.edu/ml
14. Inoue, H.: Self-Organizing Neural Grove: Efficient Multiple Classifier System with Pruned Self-Generating Neural Trees. In: Franco, L., Elizondo, D.A., Jerez, J.M. (eds.) Constructive Neural Networks. SCI, vol. 258, pp. 281–291. Springer, Heidelberg (2009)

Transfer Learning Using the Online FMM Model

Manjeevan Seera[1,*], Chee Peng Lim[2], and Chu Kiong Loo[1]

[1] Faculty of Computer Science and Information Technology,
University of Malaya, Malaysia
[2] Centre for Intelligent Systems Research, Deakin University, Australia
mseera@um.edu.my

Abstract. In this paper, we present an analysis on transfer learning using the Fuzzy Min-Max (FMM) neural network with an online learning strategy. Transfer learning leverages information from the source domain in solving problems in the target domain. Using the online FMM model, the data samples are trained one at a time. In order to evaluate the online FMM model, a transfer learning data set, based on data samples collected from real landmines, is used. The experimental results of FMM are analyzed and compared with those from other methods in the literature. The outcomes indicate that the online FMM model is effective for undertaking transfer learning tasks.

Keywords: Data classification, fuzzy min-max neural network, online learning, transfer learning.

1 Introduction

Transfer learning focuses on utilizing data from the one (source) domain to solve problems in a different, but related domain, or known as the target domain [1]. Traditional machine learning models typically assume that the training samples collected have the same distribution and characteristics as the incoming data samples during operation [2]. In real-world situations, this assumption may not always be true. As an example, in time-varying environments, the training data samples are unlikely to have the same distribution as the actual incoming data samples during the operation phase, since the environment is non-stationary.

In order to have an effective machine learning model, a large number of data samples from a particular task are required during the training phase [3]. This can be a problem when it is not feasible to collect a large number of data samples, or it is too costly to do so in real environments. As such, in this paper, transfer learning using a machine learning model that is able to learn incrementally is studied. The rationale behind transfer learning is although data distribution in the source and target domains is different, common knowledge in both domains can be adapted during learning [4].

In transfer learning, the key issues to solve include differences in the task and type of knowledge to be transferred [1]. Transfer learning can be utilized in various real-world applications. It has been shown to be useful for learning manufacturing conditions and requirements of high-volume products and for adapting them in improving the processes of low-volume products [5]. In another case study, the location of an object can be found

---

* Corresponding author.

C.K. Loo et al. (Eds.): ICONIP 2014, Part I, LNCS 8834, pp. 151–158, 2014.

using access points with machine learning models with transfer learning [6]. As standard methods could not adapt to changes in the environment, an indoor location estimation model based on transfer learning and feature relevance network-based method was introduced [6]. For the detection of facial landmark and facial expression recognition, a regression-based transfer learning technique was used, where only data samples from one of the two classes were used in the target domain [7]. The results indicated increased accuracy rates in facial-based detection and recognition applications [7].

We aim to tackle transfer learning tasks using the Fuzzy Min-Max (FMM) supervised neural network [8] in this paper. The online learning strategy of FMM allows it to incrementally learn new knowledge from new data samples without the need to retrain the network with previously learned knowledge [9]. The main issue in online learning is on how the system can adapt to new information in an incremental manner without forgetting previously learned knowledge, which is known as the stability-plasticity dilemma [10]. Here, FMM possesses salient features to overcome this stability-plasticity dilemma in undertaking online learning problems. Since FMM is able to avoid the need of retraining, learning can be performed on-the-fly by training one data sample after another [8]. This is important to tackle large data sets in time-varying environments.

This study is an extension of [11], whereby learning is conducted using individual data samples, instead of blocks of data samples. Each data sample is fed to FMM on-the-fly for evaluation and training. As such, the main contribution of this study is a truly online learning model for tackling transfer learning problems. The resulting model eliminates the periodic en-bloc learning cycle as adopted in [11]. Its efficacy is clearly shown based on the benchmark problems presented in Section 3.

The organization of this paper is as follows. In Section 2, the online learning model of FMM is detailed. This is followed by an explanation on the transfer learning experiment and the results in Section 3.

2   Online FMM Model

2.1   An Overview of the Online FMM Model

In this section, the procedure of online FMM is detailed. Table 1 shows the overall online FMM procedure for transfer learning.

Table 1. The procedure of online FMM for transfer learning

1.	Initialize the minimum and maximum points FMM hyperbox points
2.	Load a data sample, and evaluate (test) the performance of FMM with the data sample
3.	Use the data sample with the following steps of online learning
	a) Identify the closest hyperbox for the input data sample for expansion, otherwise add new hyperbox
	b) Check whether the expansion process causes any overlaps among hyperboxes of different classes
	c) If overlaps exist, eliminate overlaps by contracting hyperboxes
4.	Repeat steps 2 and 3 until all data samples have been used for evaluation (testing) and followed by training

By creating the hyperboxes online and incrementally, FMM is able to absorb knowledge autonomously when it is provided with increasing number of data samples. As such, an online learning strategy is devised such that FMM can be exploited to tackle transfer learning problems. A set of data samples is presented on-the-fly, one at a time. For each data sample, FMM is first used to provide a predicted target class for the data sample (i.e. evaluation/testing); therefore producing a performance metric. Then, FMM training with the data sample is initiated, with online, one-pass learning.

The testing and then training process is repeated for the remaining data samples. This online learning strategy allows FMM to adapt to data distribution that can change over time, or data samples that are drawn from noisy and/or non-stationary environments. In other words, FMM handles changes in the target domain by learning the new characteristics embedded in the incoming data samples in an incremental manner; therefore transfer learning takes place. The dynamics and structure of FMM are detailed next.

2.2   Dynamics of the FMM Model

The FMM classification model [8] consists of an input layer (F_A), a hidden (hyperbox) layer (F_B), and an output layer (F_C). The number of input nodes equals the number of dimensions of the input pattern, while the number of nodes in the output layer equals the number of target classes. F_A to F_B connections are made through matrices V and W, which store the minimum and maximum points of the associated hyperboxes, respectively. The connections between nodes F_B and F_C are binary-valued, and are stored in matrix U.

The FMM model learns by forming a knowledge base comprising a set of multi-dimensional hyperboxes incrementally. The size of each hyperbox is controlled by a user-defined parameter, θ, set between 0 and 1. The number of hyperboxes is large when θ is small, and vice versa. The fuzzy membership function is measured with respect to the minimum and maximum points of the hyperbox, and to the extent in which an input pattern fits within the hyperbox. For an input pattern X with n-dimension in a unit cube space, I^n, each hyperbox with a fuzzy set B_j is defined as:

$$B_j = \left\{ X, V_j, W_j, f\left(X, V_j, W_j \right) \right\} \ \forall X \in I^n, \tag{1}$$

where V_j and W_j is the minimum and maximum points of the jth hyperbox. The combined fuzzy set that is used to classify the Kth target class, C_k, is

$$C_k = \bigcup_{j \in K} B_j, \tag{2}$$

where K is the index of class k associated hyperboxes.

In FMM, the learning process is concerned with finding and fine-tuning the hyperbox boundaries. The learning algorithm of FMM allows hyperboxes of the same class to overlap one another, while overlapping among hyperboxes of different target classes needs to be eliminated. The membership function for the jth hyperbox, $B_j(X_h)$, $0 \leq B_j(X_h) \leq 1$, measures the extent to which the hth input pattern, X_h, falls outside hyperbox B_j. This can be considered as a measurement of the extent of each component is lower (or higher) than the minimum (or maximum) point along each dimension that extends over the minimum and maximum bounds of the hyperbox. The function that meets all these criteria is the sum of two complements, i.e., the average number of the maximum point violation and the average number of the minimum point violation. The resulting membership function is:

$$b_j(X_h) = \frac{1}{2n} \sum_{i=1}^{n} \begin{bmatrix} \max\left(0, 1 - \max\left(0, \gamma \min\left(1, x_{hi} - w_{ji}\right)\right)\right) \\ + \max\left(0, 1 - \max\left(0, \gamma \min\left(1, x_{hi} - w_{ji}\right)\right)\right) \end{bmatrix}, \tag{3}$$

where, $X_h = (x_{h1}, x_{h2}, \ldots, x_{hn}) \in I^n$ is the hth input pattern, and γ is the sensitivity parameter that controls how quickly the membership values decrease when the distance between X_h and B_j increases, $V_j = (v_{j1}, v_{j2}, \ldots, v_{jn})$ is the minimum point of B_j and $W_j = (w_{j1}, w_{j2}, \ldots, w_{jn})$ is the maximum point of B_j. The F_B to F_C connection is determined as follows.

$$u_{jk} = \begin{cases} 1 & \text{if } b_j \text{ is a hyperbox for class } C_k \\ 0 & \text{otherwise} \end{cases}, \tag{4}$$

where b_j is the jth node and C_k is the kth node. Each F_C node represents a target class. The output of the F_C node constitutes the degree to which input pattern X_h fits within class k. The transfer function of each F_C node performs the fuzzy union operation of the appropriate hyperbox fuzzy set values, and is defined as:

$$c_k = \max_{j=1}^{m} b_j u_{jk}, \tag{5}$$

Note that the F_C class nodes can be utilized in two ways. The outputs are used directly when a soft decision is required. On the other hand, if a hard decision is required, the F_C node with the highest value is determined, and its node value is set to 1 to indicate that it is the closest target class, while the remaining F_C node values are set to 0, i.e., the principle of winner-takes-all [12].

The learning algorithm of FMM consists on a series of expansion and contraction processes of the hyperboxes. The training set, D, consists of M ordered pairs, $\{X_h, C_h\}$, where, $X_h = (x_{h1}, x_{h2}, \ldots, x_{hn}) \in I^n$ is the hth input pattern, and $C_h \in \{1, 2, \ldots, m\}$ is the index of one of the m target classes. The learning process of FMM begins by selecting an ordered pair from D and finding a hyperbox of the same class that can be expanded. The expansion criterion has a constraint to be met, and is defined as:

$$n\theta \geq \sum_{i=1}^{n} \left(\max\left(w_{ji}, x_{hi}\right) - \min\left(v_{ji}, x_{hi}\right) \right), \tag{6}$$

where θ is the hyperbox size. When the expansion criterion cannot be found, a new hyperbox is formed in the hidden layer of the network.

There is a possibility of overlaps among the existing hyperboxes when the hyperboxes expand. An overlap test is introduced to check if the overlap occurs among the same or different classes. A number of cases exist to identify the overlap between two hyperboxes from different classes. During the search process, if an overlap is found, the index of the dimension and the smallest overlap value is identified and used for the contraction process. Details of FMM are available in [8].

3   Experimental Study

In this section, we present an empirical evaluation of the FMM model to tackle transfer learning problems using the Landmine data set [13]. To comprehensively evaluate the efficacy of FMM in tackling transfer learning problems, a series of experiments with both offline and online learning was conducted. In offline learning experiments, the data samples were divided into training and test sets. FMM was first trained using the training data set. The trained network was then evaluated using the test set. The experiment was repeated ten times. On the other hand, in online learning experiments, the data samples were fed one at a time. Each data sample was first used as a test datum to produce the performance metrics of FMM, and then used as the training datum to reinforce the knowledge base of FMM in tackling changes or drifts in knowledge contained in the incoming data samples. In the offline experiments, the average results of 10 runs were computed while for the online experiments, the averages in moving window of 100 samples were computed. All experiments were conducted using MATLAB® R2013a.

3.1   Landmine Data Set

The Landmine data set was collected from real landmines. A total of 29 data sets were available [13], which were collected from different landmine fields. In each data sample, a nine-dimensional feature vector was formed based on features extracted from radar images, and the target classes were either true or false mines. As the data sets were collected from various regions with different types of ground surface conditions, they were dominated by different distributions [15]. As explained in [14], the first ten data sets were collected from foliated regions, while data sets 20 to 24 were from bare earth or desert. Following the experiments in [15], the first five data sets were combined as the source data samples, while data sets 20 to 24 were used as the target data samples. In addition, data sets 6 to 10 were combined as they exhibited similar distributions with that of the first five data sets [15].

Shi *et al.* [15] applied an active learning algorithm (i.e., AcTraK or Actively Transfer Knowledge) with the concept of using out-of-domain data in prediction of domain data, and applied it to the Landmine data set. The results were compared with those from TrAdaBoost [15]. Table 2 shows the FMM accuracy rates and those reported in [15]. The online FMM scores are the highest for all five experiments, which is followed by offline FMM. The results from both online and offline experiments are close to AcTraK [15], and higher (by up to 5%) than those from TrAdaBoost [15]. It should be noted that while the difference in accuracy between online FMM and AcTraK (as well as offline FMM) is small, online FMM achieves these results with only one-pass learning through the data samples without retraining.

Based on the experiments in Table 2, the average computational durations using an Intel Core™ i5 1.70GHz processor with 4GB RAM for a single run of offline and online FMM experiment were 3.5 and 4.3 seconds, respectively. For the online learning performance of FMM, the average accuracy was calculated using a 100-moving window. Figs. 1 to 5 show the average accuracy for groups 1 to 5 *vs* the individual groups from data sets 20 to 24. In general, all accuracy rates are above 80%, indicating good performances of online FMM. The numbers of hyperboxes increased with increasing number of data subsets. This signified that FMM was able to absorb knowledge online when new data samples were presented incrementally on an online learning setting.

Table 2. Accuracy rates from different models

1-5 *vs*	TrAdaBoost [15]	AcTraK [15]	Offline FMM	Online FMM
20	89.76%	94.49%	94.72%	95.15%
21	86.04%	94.48%	94.68%	95.08%
22	90.50%	94.49%	94.75%	95.23%
23	88.42%	94.49%	94.72%	95.18%
24	90.70%	94.49%	94.78%	95.25%

From Figs. 1 to 5, the performance goes up and down, but all above 80%. Besides that, online FMM has a dip in its performance when the data distribution changes. As an example, there was a reduction in performance when moving from one group of data samples to the next, e.g. from data sample 800 to 1000 in Fig. 1. It can also be seen the largest decrease in performance occurred when a new group of data samples was applied, e.g. from data sample 2600 onwards when data set 20 was first used. Being a truly online learning system, FMM quickly builds up knowledge from the incoming data samples; therefore stabilizing its accuracy rate.

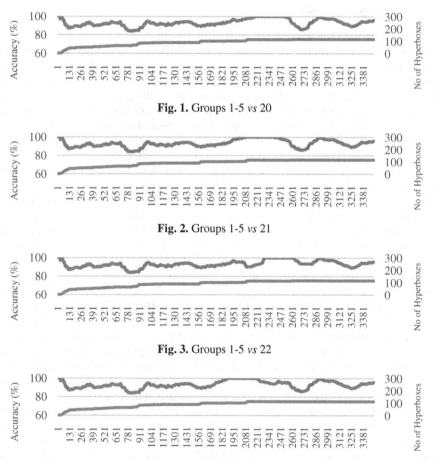

Fig. 1. Groups 1-5 *vs* 20

Fig. 2. Groups 1-5 *vs* 21

Fig. 3. Groups 1-5 *vs* 22

Fig. 4. Groups 1-5 *vs* 23

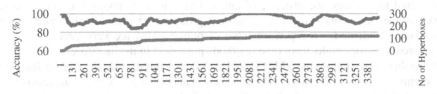

Fig. 5. Groups 1-5 *vs* 24

In addition to the individual experiments, a group-based experiment between 1 to 5 *vs* 6 to 10 as in [15] was conducted. The accuracy rates of all experiments were close (above 94%). The accuracy rate was achieved by online FMM, followed by offline FMM. The online learning performance of FMM is shown in Fig. 6. Similarly, the accuracy rates became higher with increasing number of hyperboxes when more and more data samples were presented, indicating a truly online learning system.

Table 3. Accuracy rates of the group-based experiment

1-5 *vs*	TrAdaBoost [15]	AcTraK [15]	Offline FMM	Online FMM
6-10	94.76%	94.49%	94.88%	95.32%

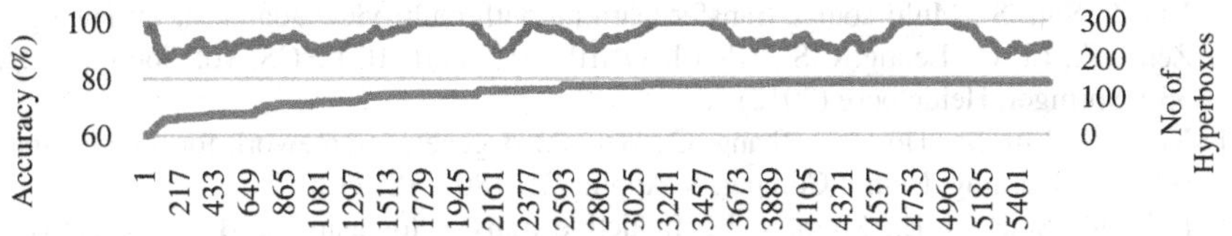

Fig. 6. Groups 1-5 *vs* 6-10

From Fig. 6, it can be seen that the accuracy rates are all above 80%, similar to those in previous experiments. In Fig. 6, a number of sharp dips can be noticed, e.g. from data samples 2000 to 2400 owing to the change in the incoming data from groups 1 to 5 to groups 6 to 10. In all figures, peaks and troughs in the online performance curves can be observed. The decrease in accuracy rates occurs when the data samples drawn come from different sources. However, FMM is able to absorb new information from the new incoming samples, as indicated by the increase in the number of hyperboxes; therefore regaining its performance. Both offline and online FMM models produce better results than those reported in [15]. It should be noted that the online FMM model has the advantage of learning on-the-fly and in one pass through the data samples without the need of re-training. The performance curves clearly demonstrate its effectiveness in handling transfer learning problems in an online learning environment. This is the key benefit of the online FMM model, which constitutes the main contribution of this study.

4   Conclusions

The use of FMM in tackling transfer learning tasks has been presented in this paper. An online learning strategy that allows transfer learning to take place in FMM has been formulated. A series of experiments using the Landmine data samples has been conducted to evaluate the efficacy of online FMM in undertaking transfer learning tasks. Six experiments using individual and group-based data samples comparison have been conducted. The performances of online FMM are very encouraging, as compared with the results from other methods published in the literature.

In further work, the online FMM model will be explored to tackle transfer learning problems in various domains. One such application is on condition monitoring of machine faults, where faults from different, but related, machines can be easily tackled. In addition, hardware implementation of the online FMM model can be accomplished to handle real-time transfer learning problems.

Acknowledgment. This project is supported by UMRG Research Subprogram (Project Number RP003D-13ICT).

References

[1] Boutsioukis, G., Partalas, I., Vlahavas, I.: Transfer learning in multi-agent reinforcement learning domains. In: Sanner, S., Hutter, M. (eds.) EWRL 2011. LNCS, vol. 7188, pp. 249–260. Springer, Heidelberg (2012)

[2] Kocer, B., Arslan, A.: Genetic transfer learning. Expert Systems with Appl. 37, 6997–7002 (2010)

[3] Xu, Z., Sun, S.: Multi-source transfer learning with multi-view adaboost. In: Huang, T., Zeng, Z., Li, C., Leung, C.S. (eds.) ICONIP 2012, Part III. LNCS, vol. 7665, pp. 332–339. Springer, Heidelberg (2012)

[4] Yang, S., Lin, M., Hou, C., Zhang, C., Wu, Y.: A general framework for transfer sparse subspace learning. Neural Comput. & Applic. 21, 1801–1817 (2012)

[5] Luis, R., Sucar, L.E., Morales, E.: Inductive transfer for learning Bayesian networks. Machine Learning 79, 227–255 (2010)

[6] Seok, H.S., Hwang, K.B., Zhang, B.: Feature Relevance Network-Based Transfer Learning for Indoor Location Estimation. IEEE Trans. Systems, Man, and Cybernetics, Part C: Appl. and Reviews 41, 711–719 (2011)

[7] Chen, J., Liu, X.: Transfer Learning with One-Class Data. Pattern Recognition Letters (2013), doi:10.1016/j.patrec.2013.07.017

[8] Simpson, P.K.: Fuzzy Min-Max Neural Networks-Part 1: Classification. IEEE Trans. Neural Networks 3, 776–786 (1992)

[9] Shen, F., Yu, H., Sakurai, K., Hasegawa, O.: An incremental online semi-supervised active learning algorithm based on self-organizing incremental neural network. Neural Comput. & Applic. 20, 1061–1074 (2011)

[10] Carpenter, G.A., Grossberg, S.: The art of adaptive pattern recognition by a self-organizing neural network. IEEE Comput. 21, 77–88 (1988)

[11] Seera, M., Lim, C.P.: Transfer learning using the online Fuzzy Min–Max neural network. Neural Comput. & Applic., 1–12 (2013)

[12] Kohonen, T.: Self-Organization and Associative Memory. Springer, Berlin (1984)

[13] Landmine data, http://www.ee.duke.edu/lcarin/LandmineData.zip

[14] Xue, Y., Liao, X., Carin, L., Krishnapuram, B.: Multi-task learning for classification with Dirichlet process priors. The Journal of Machine Learning Research 8, 35–63 (2007)

[15] Shi, X., Fan, W., Ren, J.: Actively transfer domain knowledge. In: Daelemans, W., Goethals, B., Morik, K. (eds.) ECML PKDD 2008, Part II. LNCS (LNAI), vol. 5212, pp. 342–357. Springer, Heidelberg (2008)

A Supervised Methodology to Measure the Variables Contribution to a Clustering

Oumaima Alaoui Ismaili[1,2], Vincent Lemaire[2], and Antoine Cornuéjols[1]

[1] AgroParisTech 16, rue Claude Bernard 75005 Paris
[2] Orange Labs, 2 av. Pierre Marzin, 22300 Lannion

Abstract. This article proposes a supervised approach to evaluate the contribution of explanatory variables to a clustering. The main idea is to learn to predict the instance membership to the clusters using each individual variable. All variables are then sorted with respect to their predictive power, which is measured using two evaluation criteria, i.e. accuracy (*ACC*) or Adjusted Rand Index (*ARI*). Once the relevant variables which contribute to the clustering discrimination have been determined, we filter out the redundant ones thanks to a supervised method. The aim of this work is to help end-users to easily understand a clustering of high-dimensional data. Experimental results show that our proposed method is competitive with existing methods from the literature.

1 Introduction

Everyday, huge amounts of data are generated by users via the web, social networks, etc. Clustering algorithms are a tool of choice to explore these high-dimensional data sets. However, their use is often hampered by the lack of understandability of the results. End-users would like to identify the most relevant variables that suffice to explain the observed clusters, but these are not easily detectable once a clustering has been performed. It is therefore crucial to be able to evaluate the contribution of each descriptive variable to the clustering process. Indeed, not all variables are relevant to the clustering: some may be irrelevant, some may be noisy and some may be redundant or (and) correlated.

The purpose of this study is to find a simple way to assist the analysts in their interpretation of a clustering result. The idea is to sort variables according to their contribution to a clustering using a supervised approach. The importance of a variable is evaluated as its power to predict the membership of each object to a cluster. In this paper, we restrict ourselves to an univariate classifiers to obtain an univariate weight for each variable.

The paper is organized as follows: Section 2 describes briefly some related work. Then, Section 3 presents the proposed method to score the contribution of variables to a clustering. This section also presents an alternative method to eliminate redundant variables among the relevant variables. The experimental results are presented Section 4. Finally, the perspectives and the further research are presented as a conclusion in the last section.

2 Related Work

Recently, the measure of the importance of the variables has been increasingly studied in the unsupervised learning. The methods proposed in this context can mainly be divided into two categories: *features selection* and *validation indices*.

C.K. Loo et al. (Eds.): ICONIP 2014, Part I, LNCS 8834, pp. 159–166, 2014.

Features selection methods can be grouped either as *wrapper* or as *filter* approaches. The *wrapper* approach aims to incorporate the feature selection in the clustering process, whereas, the idea of the *filter* approach is first to pre-select the features and then to use the selected features in the clutering process. In the unsupervised context, the *wrapper* methodology was initially proposed by Brodley in [1].

Inspired by the idea given in [1], Zhu et al. presented in [2] a novel method called ULAC. This method is essentially based on the analysis of the correlation among the variables. Moreover, some methods aim at removing the redundancy among variables. Accordingly, they rely on estimations of mutual information or of correlation ([3],[4],[5]). Mitra et al. proposed in [3] a method based on a measure of similarity between variables after elimination of the redundant variables. This measure is defined as the lowest eigenvalue of the correlation matrix. In [4], Vesanto et al. used a visualization tool (SOM-based approach) to detect the correlation between features. The same approach is used by Guerif et al in [5]. The difference between the two approaches is that Guerif et al. integrate a weight criterion in the SOM algorithm to reduce the effect of redundancy.

Other approaches have been presented to evaluate the clustering performance introducing criteria such as validation indices which can be adapted to evaluate the variables importance. Those approaches are divided in two main types: *external* and *internal* [6]. The *external* approaches exploit the supervised information given by the ID-cluster (identification given to each discovered cluster that can be subsequently used as a "label"). Among these approaches, we can cite: Adjusted Rand index [7], F-measure [8] and MMI [9]. The internal approaches use unsupervised criteria like the inertia. Among these methods, we can cite: Davies-Bouldin [10], Silhouette [11], Dunn-index [12], SD [13], XB-index [14], I-index [14] and BIC [15] indices.

3   Contribution

In this section, we propose two supervised approaches which fall within the context of the external validation indices. These approaches allow an interpretation of the clustering output based on relevant variables in case where the clsutering does not suffer from a very bad quality (otherwise there is no sense to interpret the result). In the remainder of this paper, we call this output (or the clustering result) *'the reference clustering'*. The first supervised approach consists in measuring the variables importance with respect to their predictive power regarding the cluster Ids. The second one aims at detecting the redundant variables.

3.1   Variables Importance

The objective of this work is to propose a simple way to identify the most relevant features from the output of a clustering. In order to retain all variables, we rank the variables according to their importance without doing a selection. The main idea is to turn this problem into a supervised classification problem where the cluster membership (ID-cluster) is used as a target class. Then, for each variable, we use a supervised classification algorithm to predict the ID-cluster. We define the importance of variables as their power to predict the ID cluster: a variable is relevant only if it is able to predict correctly the ID cluster obtained from the reference clustering (i.e. clustering using all variables). To measure the importance of each variable, we use two evaluation criteria: Accuracy and Adjusted Rand Index:

- *Accuracy* (ACC) criterion: a variable is considered relevant if the associated accuracy value is high.
- *Adjusted rand index (or ARI)* is a popular cluster validation index proposed by Hubert and Arabie [7]. It can be used to evaluate the performance of the classification as in [16]. In this work, we calculate the ARI between: (i) the reference clustering (ii) the predicted membership (ID-cluster) associated to the variable of which we want to measure the importance. The idea behind this is to compare the reference clustering with each predictive membership associated to each variable. So, a variable is important if the associated predictive ID-cluster is highly similar to the reference clustering, i.e. the ARI value is close to 1.

The algorithm 1 presented below provides a summary of our approach:

An interesting measure of importance must allow us to sort variables according to their relevance in a clustering process and the least influent variables should only contain little or irrelevant information to create the clusters. Consequently, the quality of the obtained clustering which is deprived of these variables remains substantially the same or even slightly better (less noise). In contrast, the removal of an important variable deprives the algorithm of important information and leads to a poor clustering result.

Notations:

X: The training database constituted of N examples and d explanatory variables, (X_{ab} is the value of the variable b for the example a)

M: A supervised classifier

CLU: A clustering algorithm

M_{ref}: The reference clustering model

$IdClusters$: A vector of the N memberships

R : Ranking of the d explanatory variables

$XPRE \leftarrow$ preprocessing (X)
$M_{ref} \leftarrow$ train $(CLU, XPRE)$
$IdClusters \leftarrow$ Membership$(XPRE, M_{ref})$
for *i=1 to d* **do**
    $M_i \leftarrow$ train $(XPRE_{.i}, IdClusters)$
    $ACC_i \leftarrow$ computeAccuracy (M_i)
    $ARI_i \leftarrow$ computeAdjustedRandIndex (M_i)
end
$R_{ACC} \leftarrow$ sortInDescendingOrder $(ACC_i, i=1$ to d$)$
$R_{ARI} \leftarrow$ sortInDescendingOrder $(ARI_i, i=1$ to d$)$

Algorithm 1. Algorithm for ranking

To compare our proposed method to other existing methods from the literature, the curve of the ARI values versus the number of variables used will be plotted. This curve is obtained as follows:

For each iteration until one reaches the number of variables:

- Eliminate the less relevant variable with respect to the chosen criterion;
- New partition: run the clustering algorithm without this variable;
- Calculate the ARI value between the reference clustering and the new partition.

The review of the results can be visually made by observing the curve evolution (for example, see Figure 1).

3.2 Redundant Variables

Once the variables that are the most informative for the clustering have been identified, it is important to filter out the redundant ones in order to improve the understandability of the result. To solve this problem, we propose a supervised approach.

The concept of redundancy is based on the similarity between partitions obtained using the "predicted ID-Clusters" (using Algorithm 1) for each variable. The assumption is: X_i and X_j are redundant if they produce similar partitions when considering their "predicted ID-Clusters" (using M_i and M_j). A way to measure the similarity between these two partitions is to use the ARI criterion. For example, the ARI criterion will be close to 1 when it calculated between two partitions containing same "predicted ID-Clusters" or between two partitions containing symetric "predicted ID-Clusters". The resulting algorithm is presented below (see Algorithm 2).

Notations:
X: The training database constituted of N examples and d explanatory variables
$M_\cdot$: d supervised classifier models coming from the Algorithm 1
$PredId$: A vector of size N of the predicted ID-Cluster for a given explanatory variable
RE: A matrix of size dxd values

$XPRE \leftarrow$ preprocessing (X)
for $i=1$ to d **do**
  | $PredId(d) \leftarrow$ PredictionOfTheMembership $(M_i, XPRE_{\cdot i})$
end
for all $pairs$ of $variable$ (l, m) **do**
  | $RE(l, m) \leftarrow$ computeAdjustedRandIndex $(PredId(l), PredId(m))$
end

Algorithm 2. Algorithm for redundant variables

4 Experimental Results

4.1 Protocol

To evaluate the behavior of our approach, we have selected 3 different datasets from the UCI [17]: WINE, PIMA and WAVEFORM datasets. The two first datasets are used to illustrate the competitiveness of the proposed method to measure the variables importance comparing to two other methods from the literature. Among these methods, we decide to use efficient and often used indexes from the literature: Davies-Bouldin [10] and SD indexes [13]. The last dataset is used to illustrate the behavior of our approach to detect the redundant variables.

We proceed as follows to evaluate the performance of our approach:

- the pre-processing used is standardization[1];
- to obtain the reference clustering, the K-means algorithm [18] has been used where:
  - K is equal to the number of target class for each used datasets (as in [19]);
  - the method used to initialize the centroids is K-means++ algorithm [20];
  - the number of replicates is 25 [2].
- a decision tree (CART) [21] has been used to predict the ID-cluster[3].

4.2 Variables Contribution

The first experimentation to test our approach is made using the WINE dataset which is constituted of $N = 178$ sample points described with $d = 13$ variables and associated with three different classes. The ARI obtained between the reference clustering (using K-means algorithm, where K=3) and the target class is equal to 0.91. Figure 1 presents the evolution of the ARI curve for the three approaches (SD, DB and the supervised approach using ARI or ACC to measure the contribution of variables in the clustering results) versus the number of variables. The table 1 (left part) presents the list of the ranked variables (from the most important to the least important) for the three approaches.

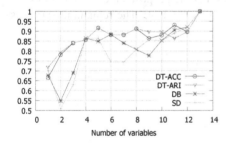
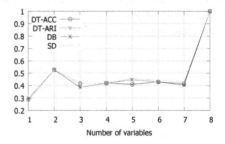

Fig. 1. Evolution of the ARI criterion for the 4 methods (K=3)

Fig. 2. Evolution of the ARI criterion for the four methods (K=2)

---

[1] All the experimentation have been realized using R (http://www.r-project.org/) and are easily reproducible.

[2] The initialization process and the nature of the K-means algorithm does not guarantee to reach a global minimum. Therefore the algorithm has to be run several times.

[3] To evaluate the importance of the variables for the clustering, we need to choose a classifier which does not modify the representation used to elaborate the reference clustering; i.e the data after the pre-processing step.

Table 1. Ranking of the variables

Index	Wine													Pima							
DB	V7	V6	V10	V1	V12	V13	V9	V8	V4	V5	V3	V2	V11	V8	V2	V1	V3	V6	V5	V4	V7
SD	V7	V6	V10	V1	V12	V9	V8	V13	V5	V4	V3	V2	V11	V8	V2	V1	V3	V6	V7	V4	V5
ARI-Tree	V7	V13	V12	V1	V10	V6	V11	V2	V9	V4	V8	V5	V3	V8	V2	V1	V3	V5	V6	V4	V7
ACC-Tree	V7	V13	V12	V1	V10	V6	V11	V2	V9	V4	V5	V8	V3	V8	V2	V1	V3	V5	V6	V4	V7

The PIMA data dataset contains $N = 768$ sample points described with $d = 8$ variables which are associated with two different classes. The ARI obtained between the reference clustering (using K-means algorithm, where $K = 2$) and the target class is equal to 0.11. Table 1 (right part) and Figure 2 present respectively the list of the ranked variables (from the most important to the least important) and the evolution of ARI curve for the three approaches (DB, SD and the proposed approach).

The results obtained on PIMA and WINE show that the proposed method is competitive with regards to DB and SD approaches on these two datasets.

4.3 Redundant Variables

To test the ability of our approach to detect the redundant variables, we use the WAVE-FORM dataset. This dataset consists of $n = 5000$ sample points described with 40 variables and associated with three different classes: only the first 21 variables are real attributes for this database and most of these are relevant to a classification problem whereas the last 19 variables are noisy standard centered Gaussian variables (for more details see [21], page 43 - 49). Figure 3 shows that the proposed method identifies the irrelevant set of variables $W = V1, V21 - V37, V39, V40$. The remaining variables are all relevant variables for the clustering.

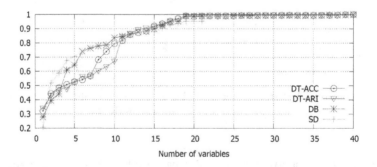

Fig. 3. Evolution of ARI criterion for the three methods (K=3)

To identify the redundant variables, we use the method described in Section 3.2. Table 2 presents the ARI values calculated between two pairs of relevant variables (the 20 variables identified by the proposed method using the ACC criterion). In this table, if we consider only the values above 0.96 to focus the attention on the high values of redundancy. The set of redundant variables is then : $R = V38, V2, V20, V19, V3$. Finally the set of relevant variables is $V = V4 - V18$. These obtained results are similar to those obtained using RD-MCM selection features method (see [19]). The ARI value obtained between the predicted ID-cluster using all variables (41 variables) and the predicted ID-cluster using the relevant variable (18 variables) is equal to 0.935.

Table 2. ARI values between pairs of relevant variables

	V7	V15	V8	V14	V16	V6	V13	V12	V17	V9	V5	V10	V4	V18	V11	V3	V19	V20	V2	V38
V7	1,00	0,51	0,42	0,42	0,39	0,41	0,41	0,38	0,37	0,36	0,35	0,34	0,34	0,34	0,32	0,32	0,32	0,32	0,32	0,32
V15		1,00	0,66	0,62	0,59	0,59	0,58	0,54	0,52	0,50	0,49	0,46	0,46	0,46	0,44	0,44	0,44	0,44	0,44	0,44
V8			1,00	0,76	0,72	0,70	0,67	0,62	0,59	0,56	0,54	0,51	0,51	0,51	0,49	0,48	0,48	0,48	0,48	0,48
V14				1,00	0,81	0,78	0,75	0,67	0,62	0,59	0,57	0,54	0,53	0,53	0,51	0,50	0,51	0,50	0,50	0,50
V16					1,00	0,84	0,77	0,71	0,66	0,61	0,60	0,56	0,56	0,56	0,53	0,53	0,52	0,53	0,52	0,52
V6						1,00	0,86	0,75	0,68	0,63	0,62	0,58	0,57	0,58	0,55	0,54	0,54	0,54	0,54	0,54
V13							1,00	0,79	0,72	0,66	0,65	0,60	0,60	0,60	0,58	0,57	0,57	0,57	0,57	0,57
V12								1,00	0,88	0,81	0,78	0,74	0,73	0,73	0,69	0,69	0,68	0,68	0,68	0,68
V17									1,00	0,89	0,84	0,81	0,79	0,79	0,75	0,74	0,74	0,74	0,74	0,74
V9										1,00	0,91	0,86	0,85	0,84	0,80	0,79	0,79	0,79	0,79	0,79
V5											1,00	0,89	0,87	0,86	0,83	0,82	0,81	0,82	0,82	0,82
V10												1,00	0,94	0,91	0,88	0,87	0,86	0,86	0,86	0,86
V4													1,00	0,94	0,89	0,88	0,87	0,87	0,87	0,87
V18														1,00	0,92	0,89	0,88	0,88	0,89	0,88
V11															1,00	0,95	0,93	0,92	0,92	0,92
V3																1,00	0,96	0,95	0,93	0,94
V19																	1,00	0,97	0,95	0,96
V20																		1,00	0,97	0,97
V2																			1,00	1,00
V38																				1,00

5   Conclusion

This paper has presented a supervised method to measure the importance of the variables used in a clustering. This method turned the problem into a supervised classification problem to sort variables according to their importance at the end of the clustering convergence. The experimental results corroborated the competitiveness of the method comparing to other methods from the literature. It has been incorporated successfully in the process of marketing service in the french Orange company. Future works will be done to incorporate the method in the convergence of the clustering algorithm and to measure the variables importance as a multivariate supervised classification problem.

References

1. Dy, J.G., Brodley, C.E.: Feature selection for unsupervised learning. J. Mach. Learn. Res. 5, 845–889 (2004)
2. Liu, P., Zhu, J., Liu, L., Li, Y., Zhang, X.: Application of feature selection for unsupervised learning in prosecutors' office. In: Wang, L., Jin, Y. (eds.) FSKD 2005. LNCS (LNAI), vol. 3614, pp. 35–38. Springer, Heidelberg (2005)

3. Mitra, P., Murthy, C.A., Pal, S.K.: Unsupervised feature selection using feature similarity. IEEE Trans. Pattern Anal. Mach. Intell. 24(3), 301–312 (2002)
4. Vesanto, J., Ahola, J.: Hunting for correlations in data using the self-organizing map. In: Proceeding of the International ICSC Congress on Computational Intelligence Methods and Applications (CIMA 1999), pp. 279–285. ICSC Academic Press (1999)
5. Guérif, S., Bennani, Y., Janvier, E.: μ-SOM: Weighting features during clustering. In: Proceeding of the 5th Workshop on Self-organizing Maps (WSOM 2005), pp. 397–404 (2005)
6. Halkidi, M., Batistakis, Y., Vazirgiannis, M.: On clustering validation techniques. J. Intell. Inf. Syst. 17(2-3), 107–145 (2001)
7. Hubert, L., Arabie, P.: Comparing partitions. Journal of Classification 2(1), 193–218 (1985)
8. Larsen, B., Aone, C.: Fast and effective text mining using linear-time document clustering. In: Proceedings of the Fifth ACM SIGKDD International Conference on Knowledge Discovery and Data Mining (ICDM), pp. 16–22. ACM, New York (1999)
9. Alok, A.K., Siparna, S., Ekbal, A.: A min-max distance based external cluster validity index: MMI. In: HIS, pp. 354–359. IEEE (2012)
10. Davies, D.L., Bouldin, D.W.: A cluster separation measure. IEEE Transactions on Pattern Analysis and Machine Intelligence PAMI-1(2), 224–227 (1979)
11. Rousseeuw, P.: Silhouettes: A graphical aid to the interpretation and validation of cluster analysis. J. Comput. Appl. Math. (1), 53–65 (1987)
12. Dunn, J.C.: A Fuzzy Relative of the ISODATA Process and Its Use in Detecting Compact Well-Separated Clusters. Journal of Cybernetics 3(3), 32–57 (1973)
13. Halkidi, M., Vazirgiannis, M., Batistakis, Y.: Quality scheme assessment in the clustering process. In: Zighed, D.A., Komorowski, J., Żytkow, J.M. (eds.) PKDD 2000. LNCS (LNAI), vol. 1910, pp. 265–276. Springer, Heidelberg (2000)
14. Xie, X.L., Beni, G.: A validity measure for fuzzy clustering. IEEE Trans. Pattern Anal. Mach. Intell. 13(8), 841–847 (1991)
15. Raftery, A.: A note on Bayes factors for log-linear contingency table models with vague prior information. Journal of the Royal Statistical Society, 249–250 (1986)
16. Santos, J.M., Embrechts, M.: On the use of the adjusted rand index as a metric for evaluating supervised classification. In: Alippi, C., Polycarpou, M., Panayiotou, C., Ellinas, G. (eds.) ICANN 2009, Part II. LNCS, vol. 5769, pp. 175–184. Springer, Heidelberg (2009)
17. Blake, C.L., Merz, C.J.: Uci repository of machine learning databases (1998)
18. MacQueen, J.B.: Some methods for classification and analysis of multivariate observations. In: Cam, L.M.L., Neyman, J. (eds.) Proc. of the Fifth Berkeley Symposium on Mathematical Statistics and Probability, vol. 1, pp. 281–297. University of California Press (1967)
19. Celeux, G., Martin-Magniette, M.L., Maugis, C., Raftery, A.E.: Comparing model selection and regularization approaches to variable selection in model-based clustering. Journal de la Société Française de Statistique (2014)
20. Arthur, D., Vassilvitskii, S.: K-means++: The advantages of careful seeding. In: Proceedings of the Eighteenth Annual ACM-SIAM Symposium on Discrete Algorithms, SODA 2007, pp. 1027–1035 (2007)
21. Breiman, L., Friedman, J.H., Olshen, R.A., Stone, C.J.: Classification and regression trees. Wadsworth International Group (1984)

Coupling between Spatial Consistency of Neural Firing and Local Field Potential Coherence: A Computational Study

Naoyuki Sato

Department of Complex and Intelligent Systems,
School of Systems Information Science, Future University Hakodate,
116-2 Kamedanakano-cho, Hakodate-shi, Hokkaido 041-8655, Japan
satonao@fun.ac.jp
http://www.fun.ac.jp/~satonao/

Abstract. Computational modeling with biologically-plausible parameters has shown an association between the coherence of local field potentials (LFP) in two cortical areas and the consistency of their spatial firing patterns. In this study, the dynamic properties of this association were evaluated. The association was strong for each input sequence, step change in input, repetitive alternation of input, and continuous change of input. The time constant of the spatially consistent firing was ~100 ms, and this was thought to be related to a winner-take-all process among the neuronal subpopulations in each area. Furthermore, the results of the phase-coupling analysis suggested that waves slower than 10 Hz in the LFPs produced a time window for transmitting the spatial firing patterns between areas.

Keywords: network model, cortical column, information transfer, long-range synchronization, electroencephalogram.

1 Introduction

A local field potential (LFP) is an electrical potential of a neural population that occurs within ~250 μm. Similar to electroencephalograms that reflect neural activity within >1 cm^2, the LFPs are known to change with changes in the cognitive process, such as visual perception and memory. These signals are thought to provide clues for understanding information processing in large-scale neural networks.

Computational studies have shown associations between neural firing and gamma-band LFPs by using neural models with biologically plausible parameters [1–4] and such associations have been supported by physiological experiments [5]. Mazzoni et al. (2008) [2] evaluated the data for the temporal patterns of LFP and showed that channels with different frequency can transmit information independently. The author evaluated the association between LFP coherence and spatial firing patterns between two cortical areas [4] and showed a strong association between them. This finding is thought to be important for bridging the

C.K. Loo et al. (Eds.): ICONIP 2014, Part I, LNCS 8834, pp. 167–174, 2014.
© Springer International Publishing Switzerland 2014

LFP phenomena and the functional neural processes implemented in canonical networks consisting of multiple cortical columns with lateral inhibition. However, these evaluations have only been performed when the network was in a stable state (~1 s). It remains unclear whether the association between spatial firing patterns and LFP coherence is stable for temporally changing input signals.

In this study, the relationship between the spatial patterns of firing and LFP coherence in two cortical areas was evaluated by using a network model used in the previous study [4] under the condition of temporally changing input signals.

2   Model

The model was originally proposed by Mazzoni et al. (2008) [2] and modified in a previous study [4] with the newly introduced structure of two areas consisting of two cortical columns. The network of an area consisted of 4,000 pyramidal cells and 1,000 inhibitory cells. Each neuron was randomly connected to other cells with fixed connectivity (the values of which are shown in Fig. 1) and each received strong excitatory synaptic currents of background activity and external input. All external currents were given by a random Poisson spike train.

Each neuron k is described by its membrane potential V_k that evolves according to

$$\tau_m \dot{V}_k = -V_k + I_{Ak} - I_{Gk}$$

where I_{Ak} is the AMPA-type excitatory synaptic current and I_{Gk} is the GABA-type inhibitory current received by neuron k. When the membrane potential crosses the threshold of 18 mV the neuron potential is reset at a value of 11 mV. The synaptic currents are given by

$$\tau_{dA} \dot{I}_{Ak} = -I_{Ak} + x_{Ak}, \quad \tau_{dG} \dot{I}_{Gk} = -I_{Gk} + x_{Gk}$$

$$\tau_{rA} \dot{x}_{Ak} = -x_{Ak} + \tau_m \left(J_k^{pyr} \sum_{pyr} \delta(t - t_k^{pyr} - \tau_L) + J_k^{ext} \sum_{ext} \delta(t - t_k^{ext} - \tau_L) \right)$$

$$\tau_{rG} \dot{x}_{Gk} = -x_{Gk} + \tau_m \left(J_k^{int} \sum_{int} \delta(t - t_k^{int} - \tau_L) \right)$$

where $\delta(t)$ denotes the delta function and $t_k^{pyr/int/ext}$ is the time that the spikes are received from the pyramidal neurons/ interneurons connected to neuron k or from external input that consists of background activity (1.2 spikes/ms), input from inter-cortical connections, and additional input stimuli (ranging from 0 to ~0.8 spikes/ms). The values of the model parameters are τ_m=20 ms (pyramidal) or 10 ms (interneuron), the refractory period is 2 ms (pyramidal) or 1 ms (interneuron), τ_{dA}= 2 ms (pyramidal) or 1 ms (interneuron), τ_{dG}= 5 ms, τ_{rA}=0.4 ms (pyramidal) or 0.2 ms (interneuron), τ_{rG}=0.25 ms, τ_L=1 ms (within the same area) or 4 ms (otherwise), J_k^{pyr}=0.42 (pyramidal) or 0.7 (inter.), J_k^{int}=1.7 (pyramidal) or 2.7 (interneuron), and J_k^{ext}=0.55 (pyramidal) or 0.95 (interneuron).

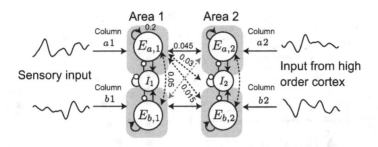

Fig. 1. Basic structure of the model. Each oval represents the cell population, and E and I denote pyramidal and interneuron subpopulations in an area, respectively. The neurons in area 1 and 2 are sparsely connected, and their connection probabilities are denoted along the arrow between the subpopulations. Areas 1 and 2 were assumed to receive sensory input and input from higher order cortex, respectively.

For the instantaneous firing rate in a subpopulation in Area i at time t, $E_{a/b,i}(t)$ was calculated with a 5-ms bin, except for that described in Section 3.3. The spatial pattern of neural firing in Area i was defined by a vector $\mathbf{E}_i(t) = (\mathbf{E}_{a,i}(t), \mathbf{E}_{b,i}(t))$, and the spatial consistency of the firing in the two areas was given by the inner product $\mathbf{E}_1(t)\mathbf{E}_2(t)$ that was associated with the connection structure shown in Fig. 1.

The time-frequency power of the LFP was analyzed with a Morlet wavelet transformation (width=5) and the LFP coherence between areas 1 and 2 was calculated. The relationships between the LFP coherence and the spatial consistency of firing, $\mathbf{E}_1(t)\mathbf{E}_2(t)$, were evaluated with Spearmans correlation coefficients.

3   Results

3.1   Response to a Step Change in Input Signal

The first simulation evaluated the response of the model to a step change in input signal, which was done to show the dynamic properties of the spatial consistency of firing and the LFP coherence in the network (Fig. 2a). Figure 2b shows the temporal evolution of the model to a step input of 0.2 spikes/ms. Figure 2c shows the LFP gamma coherence (30–80 Hz) between the two areas for each condition of bias input to Area 2 (0, 0.1 and 0.2 spikes/ms), where the rise time of the LFP coherence was found to be shorter for the larger bias input. A similar tendency was found for the spatial consistency of firing (Fig. 2d). When the delay time of LFP coherence and spatial consistency was defined by a time point that was half of the average value in the period of 100–200 ms (Fig. 2e), the delay times were strongly correlated (R=0.85). According to the regression line, the spatial consistency preceded the LFP coherence by ∼8 ms. Figure 2f shows the correlation between the spatial consistency of firing and the LFP coherence. The correlation was high (R=0.98) and were in agreement with the results obtained with a stable input [4].

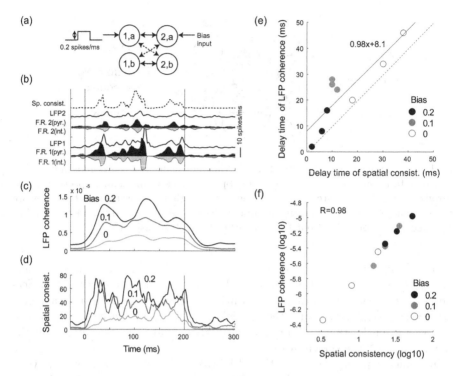

Fig. 2. Response to a step change in the input signal. (a) Parameters of the input sequence. (b) Temporal evolution of the model. The black and gray areas indicate the instantaneous firing rate of the subpopulations of pyramidal neurons and inhibitory neurons in each area, respectively. The solid and dotted lines indicate the local field potential (LFP) in each area and the spatial consistency of the firing, respectively. Two vertical lines denote onset and offset of the step input signal. (c, d) Temporal evolution of the LFP coherence (30–80 Hz) and the spatial consistency of the firing. The black, gray and pale gray lines denote the condition of the bias input 0, 0.1 and 0.2, respectively. (e) The relationship between the delay time of the spatial consistency and those of LFP coherence. (f) The relationship between the spatial consistency and the LFP coherence.

3.2   Response to the Repetitive Alternation of the Input Pattern

The second simulation investigated the temporal context effects of the model by using input signals that repetitively changed to result in different spatial patterns. In this simulation, two pairs of subpopulations, $E_{1,a} - E_{2,a}$ and $E_{1,b} - E_{2,b}$, were alternated with a cycle of 20, 100, 200 or 300 ms (Fig. 3a). Bias input was added to subpopulation E_{2a} to disturb the formation of spatially consistent firing patterns (0, 0.1 or 0.2 spikes/ms).

Figure 3b shows the temporal evolution of the instantaneous firing rate, the LFP and the spatial consistency of firing (cycle=400 ms). The firing rates in subpopulations a and b were shown to alternate according to the change in the input pattern. The rise time of the firing rate was shown at ∼100 ms, which

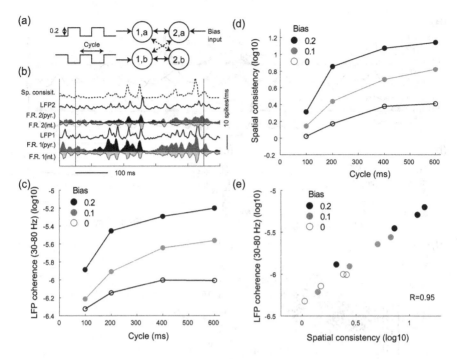

Fig. 3. Response to the repetitive alternation of the input pattern. (a) The parameters of the input signals. (b) The temporal evolution of the instantaneous firing rate (the black, gray, and pale gray areas indicate the pyramidal cells in subpopulations a and b and the inhibitory cells, respectively), the LFP (solid lines), and the spatial consistency of the firing (dashed line). (e, d) The LFP coherence and the spatial consistency of the firing for each condition of the input cycle. (c) Correlation between the spatial consistency of the firing and the LFP coherence.

was longer than the delay time found in the previous section. The relationship between the LFP coherence and the input cycle is shown in Fig 3c. The LFP coherence increased for the longer cycle and reached a plateau at ∼200 ms. This was thought to related to the time for formation of spatially consistent firing patterns with temporal context. Interestingly, this profile was not affected by bias input. The results were similar for the spatial consistency of firing (Fig. 3d).The correlation between the spatial consistency of firing and LFP coherence was again strong in these conditions (R=0.95) (Fig. 3e).

3.3   Response to Continuously Changing Input Signals

In this section, the relationship between the spatial consistency of firing and LFP coherence was evaluated with continuously changing input signals in which four independent input signals were introduced to subpopulations in each area (Fig. 1). Each input was given by random sequence with low-pass filtering (a fourth order Butterworth filter with a cut-off frequency of 20 Hz) and regulated

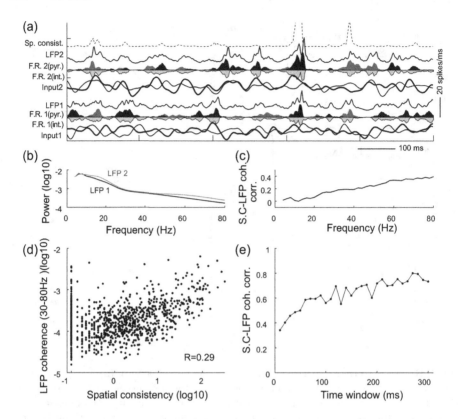

Fig. 4. Response to the countinuously changing input sequence. (a) Temporal evolution of the input sequence (the bold and thin lines indicate the input to subpopulations a and b, respectively), the firing rate, the LFP and the spatial consistency of firing. (b) The LFP power in each area. (c) Correlation between the spatial consistency of firing and the LFP coherence at each h frequency band. (d) Correlation between the spatial consistency and the LFP coherence (30–80 Hz). (e) The correlation coefficient between spatial consistency and the LFP coherence in relationship to the time window used in the calculation of the firing rate and LFP coherence.

to have a mean value of 1.4 spikes/ms including background activity, and a standard deviation of 1.0 spikes/ms.

Figure 4a shows the temporal evolution of the model. In different to results in the previous sections (Figs. 2b and 3c), the spatial patterns of the firing in the two areas were not always consistent. The lower frequency components in the LFP powers in each area were strong according to the input signals (Fig. 4b), but the strong correlation between the spatial consistency of firing and the LFP coherence still appeared in the gamma band (Fig. 4c). This result agrees with the findings of the previous study using stable input signals [4].

Figure 4d shows the relationship between the spatial consistency and the LFP gamma coherence (30-80 Hz). The correlation was weaker (R=0.29) than those

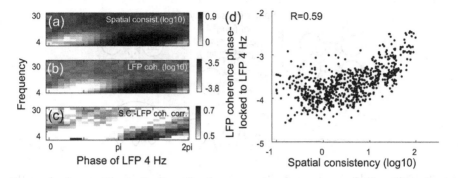

Fig. 5. Results of the phase-coupling analysis. The phase of the average LFPs in the two areas was used in the analysis. (a, b) Phase locking of the spatial consistency of the firing and the LFP coherence (30–80 Hz). (c) The phase-dependent correlation between the spatial consistency of firing and the LFP coherence. (d) Correlation between the spatial consistency and the LFP coherence during the phase value of $\pi \sim 1.5\pi$ of the average LFP of 4 Hz.

reported in the previous section. This was thought to be due to the narrow time window for the calculation of the firing rate (5 ms). When the time window increased, the correlation coefficient increased and reached a plateau at a window of ~100 ms.

The phase coupling analysis was performed to examine the relationship of the experimental evidence for the phase coupling betewen neural firing and the slow waves in the LFP [5]. Here, the LFP phase was defined by the average of the LFPs in the two areas. According to Mazzoni et al. (2010) [3], the large LFP was defined to correspond to the negative phase. Figs. 5a and 5b show the phase locking of the spatial consistency of the firing and the LFP gamma coherence (30–80 Hz) at each frequency band. Both of them were found to be phase locked to a slow wave of the LFP taht had a range of 4–16 Hz. The correlation between the spatial consistency and the LFP coherence was strong in these conditions (Fig. 5c). When the particular phase ($\pi \sim 1.5\pi$) of a LFP of 4 Hz was analyzed, the correlation between the spatial consistency and the LFP gamma coherence became large (a time window of 50 ms was used for the calculation of the firing rate).

4   Discussion

This study evaluated the relationship between the spatial consistency of firing and the LFP coherence between two areas under conditions of temporally changing input signals. The results demonstrated for using step changes in the input (Section 3.1), repetitive alternations in the spatially input pattern (Section 3.2), and continuously changing input (Section 3.3) showed the strong correlation between the spatial consistency of firing and the LFP gamma coherence, which agreed with previous results obtained with a stable input signal [4]. The time

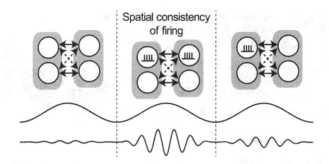

Fig. 6. Time window of the spatially consistent firing given by the slow wave in the LFP

scale in the formation of the spatially consistent firing pattern was suggested as ~100 ms (Figs. 3c, 3d and 4e). According to the results reported in Sections 3.1 and 3.2, this time scale was thought to be associated with a winner-take-all process within the subpopulations of neurons in each area. This result also agreed with the phase-coupling analysis that showed that the firing and LFP gamma were locked to the LFP components below ~10 Hz. Interestingly, this time scale appeared similar to the experimental observations obtained with a phase-coupling analysis [6]. The current results supported the idea that the slow oscillations produced units for transmitting spatial firing patterns between cortical areas (Fig. 6).

Acknowledgments. This work was supported by JSPS KAKENHI Grant Number 26540069.

References

1. Brunel, N., Wang, X.J.: What determines the frequency of fast network oscillations with irregular neural discharges? I. Synaptic dynamics and excitation-inhibition balance. Journal of Neurophysiology 90, 415–430 (2003)
2. Mazzoni, A., Panzeri, S., Logothetis, N.K., Brunel, N.: Encoding of naturalistic stimuli by local field potential spectra in networks of excitatory and inhibitory neurons. PLoS Computational Biology 4, e1000239 (2008)
3. Mazzoni, A., Whittingstall, K., Brunel, N., Logothetis, N.K., Panzeri, S.: Understanding the relationships between spike rate and delta/gamma frequency bands of LFPs and EEGs using a local cortical network model. NeuroImage 52, 956–972 (2010)
4. Sato, N.: Spatial consistency of neural firing regulates long-range local field potential synchronization: A computational study. Neural Networks (2014), doi:10.1016/j.neunet.2014.07.004
5. Whittingstall, K., Logothetis, N.K.: Frequency-band coupling in surface EEG reflects spiking activity in monkey visual cortex. Neuron 64, 281–289 (2009)
6. Canolty, R.T., Knight, R.T.: The functional role of cross-frequency coupling. Trends in Cognitive Sciences 14, 506–515 (2010)

Fading Channel Prediction Based on Self-optimizing Neural Networks

Tianben Ding and Akira Hirose

Department of Electrical Engineering and Information Systems,
The University of Tokyo, 7-3-1 Hongo, Bunkyo-ku, Tokyo 113-8654, Japan
ding_tei@eis.t.u-tokyo.ac.jp, ahirose@ee.t.u-tokyo.ac.jp
http://www.eis.t.u-tokyo.ac.jp/

Abstract. Channel prediction is an important technique for compensating fading channel in mobile communications. We proposed a channel prediction method based on complex-valued neural networks as a previous work. In this paper, we introduce a penalty function to the weight update in the complex-valued neural network to realize a learning dynamics that can self-optimize network structures according to fast changing communication environments. This presents an adaptive and highly accurate channel prediction method. We demonstrate the ability of the proposed method in a series of simulations.

Keywords: Fading, channel prediction, adaptive transmission, complex-valued neural network, sparse representation.

1 Introduction

Performance of mobile communications often suffers from various fading phenomena. To reduce the adverse effect, there are some techniques such as pre-equalization and transmission power control. Such adaptive techniques require channel prediction for the channel changes in time [7,5]. To realize a low-cost and high-precision prediction, we proposed a new prediction method that separates multipaths in chirp z-transform (CZT) [7] according to Jakes model [11], and predicts them separately based on complex-valued neural networks (CVNN) [3,4].

The learning ability of neural networks including CVNN is greatly affected by the size of network structure[6]. For instance, a too large network for a problem results in high calculation cost and a large generalization error. On the other hand, a too small network structure cannot assure sufficient flexibility to express a problem, and its learning convergence at a learning point is low.

In our proposed channel prediction method, the structure of the CVNN (the number of input terminals and neurons in the hidden layer) was set according to experience. However, in a real communication environment, communications is forced to work in a variety of communication situations, and those situations fluctuate complicatedly. The most suitable network structure in such situations may also change. This is why we need an adaptive network structure in channel

C.K. Loo et al. (Eds.): ICONIP 2014, Part I, LNCS 8834, pp. 175–182, 2014.

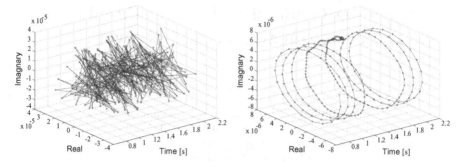

Fig. 1. Variation of channel characteristic on user terminal in time in complex plane

Fig. 2. Variation of channel characteristic of a separated single path in time in complex plane

prediction. There are various studies to get optimized structure[9]. For example, pruning neural networks start learning with a large structure, and then prune redundant connection weights and neurons to obtain an optimum network [8,12], whereas growing neural networks raise the size of a structure from a small network to a larger one [1]. In this paper, we propose to apply a penalty function to connection weights update in the CVNN for realizing a network structure with adaptive connection depending on the communication situations. A new channel prediction method with proposed CVNN structure presents highly accurate predictions under fluctuating communication environments.

2   Channel Prediction Based on CVNN and Conventional Network Structure

According to the Jakes model, a fading channel can be modeled as the summation of sinusoids at a receiver, which are the multipath rays caused by scattering and reflection. Each sinusoid can be characterized by a set of path parameters such as amplitude a_m, Doppler frequency f_m, and phase shift ϕ_m. The channel characteristic $c(t)$ as a function of time t is the summation of M complex signal paths and expressed as

$$c(t) = \sum_{m=1}^{M} c_m(t) = \sum_{m=1}^{M} a_m e^{j(2\pi f_m t + \phi_m)}. \tag{1}$$

Fig. 1 shows a fading channel observed at a receiver that changes irregularly in the complex plane, and has difficulty in prediction. We decomposed this channel characteristic into multiple channel characteristics of respective multipaths by detecting peaks of Doppler frequency presented by CZT [13,10]. We also focused that the separated channel characteristics $c_m(t)$ have rotary motion in the complex plane as shown in Fig. 2, and proposed a new channel prediction method based on CVNNs [6]. The CVNNs is a suitable framework for treating rotation

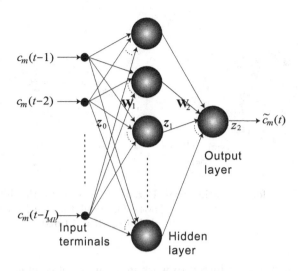

Fig. 3. Layered complex-valued neural network

in the complex plane, and can deal with channel characteristics of respective paths as complex entities.

Fig. 3 shows the structure of the multilayer (ML) CVNN used in our previous work. It has input terminals, a hidden-neuron layer and an output-neuron layer. The input terminals of the layered CVNN distribute input signals, $c_m(t-1)$, ..., $c_m(t - I_{ML})$, to the hidden-layer neurons as their inputs z_0. In the same way, the outputs of the hidden-layer neurons z_1 are passed to the output-layer neuron as its inputs. The outputs z_l in layer l are given by adopting an amplitude-phase-type activation function f_{ap} to weighted inputs z_{l-1} ($z_l = f_{ap}(z_{l-1} \cdot W_l)$).

We updated the connection weights W_l in the CVNN as follows. The ML-CVNN regards past channel characteristics of respective paths $\hat{c}_m(t)$ estimated by CZT as the teacher signal, and the previous channel characteristics $\hat{c}_m(t - 1), ..., \hat{c}_m(t - I_{ML})$ as the input signals. The output z_2 was used to update the weights under the steepest descent method. The steepest descent method updates the weights to minimize the difference

$$E_l \equiv \frac{1}{2}|z_l - \hat{z}_l|^2 \qquad (2)$$

where z_l are the output signals and $\hat{z}_l$ are the teacher signals in layer l. The teacher signals in the hidden layer $\hat{z}_1$ are the signals obtained through the backpropagation of the teacher signal $\hat{z}_2$.

In the previous work, the structure of the CVNN represented as the number of input terminals and neurons in the hidden layer was set according to results of numerical experiments. The combination of the input terminals and the hidden layer was decided to maximize the prediction accuracy in various communication situations. The final combination set also provided high prediction ability in actual experiments [4]. However, in a real communication environment,

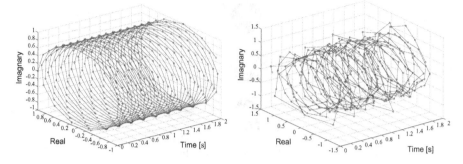

Fig. 4. Model signal for evaluating proposed method

Fig. 5. Model signal with white Gaussian
noise for simulating hard predicting condition (SNR = 14 dB)

a communication system is forced to work in a variety of communication situations, and those situations fluctuate complicatedly. The most suitable network
structure in such situations may also change.

3   CVNN Dynamics for Self-optimization

In this paper, we propose a new network dynamics which can prune or grow
connections depending on communication situations.

Theoretically, if a network of a given size can express a channel characteristic,
a larger network can also express it. All the redundant connections or neurons
are expected to have zero strengths in an ideal structure. However, an actual
neural network spreads non-zero weights all over the network, and may cause
larger generalization error.

To overcome this effect, we introduce a penalty function to restrict the connection of networks in a suitable small size. The l_0 norm to $\boldsymbol{W}_l$ is the basic penalty
to restrict the number of the connection weights, but we can use l_1 norm as a
substitute to l_0 norm for a practical penalty function, and the difference at layer
l is expressed as

$$E_l \equiv \frac{1}{2}|\boldsymbol{z}_l - \hat{\boldsymbol{z}}_l|^2 + \alpha \|\boldsymbol{W}_l\|_1 \tag{3}$$

where α is a coefficient to express the degree of the penalty function. This is
equivalent to (2) with l_1 norm of $\boldsymbol{W}_l$, and minimizing this difference means the
restriction of non-zero weights to minimal number in the connections. In other
words, the penalty function introduces sparsity to weights update. This is can
also considered as a problems similar to the sparse representation [2]. We can
also use the steepest descent method to update the weights here.

Table 1. Prediction Parameters of CVNN

Prediction Parameter	Value
ML-CVNN input terminal number I_{ML}	30
ML-CVNN hidden neural number J_{ML}	30
Iteration of ML-CVNN weight update R_{ML}	10 times

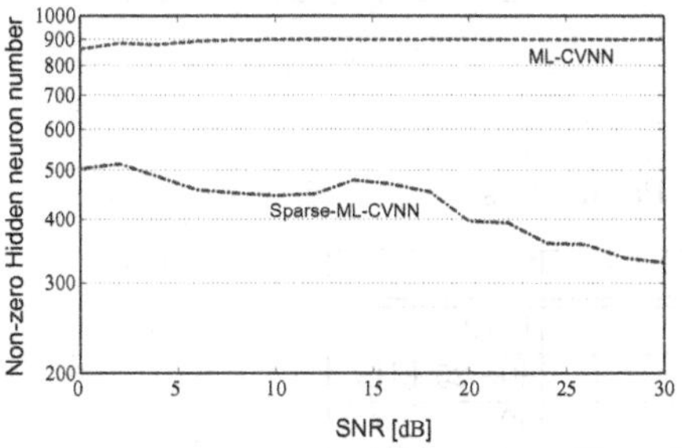

Fig. 6. Number of non-zero-amplitude weights in the hidden layer versus signal to noise ratio

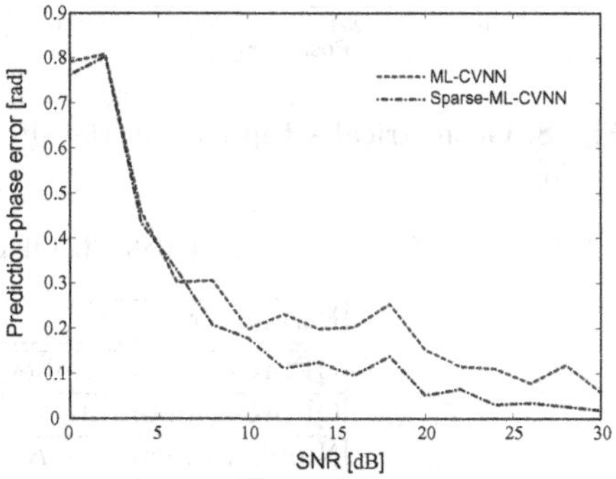

Fig. 7. Prediction-phase error versus signal to noise ratio

4 Numerical Experiments

4.1 Model Test for Evaluating Effectiveness of Sparse CVNN

The performance of the proposed channel prediction based on the CVNN with the penalty function is discussed in this section. For the first, we test accuracy of the proposed method in a modeled numerical experiment. We generate a signal rotating in the complex plane in time shown in Fig. 4, and add white Gaussian noise with various magnitude to get hard situation for prediction. We evaluate and compare the accuracy of prediction based on the conventional ML-CVNN and the ML-CVNN with proposed dynamics (Sparse-ML-CVNN). Table 1 represents the network parameters used in the experiment.

Fig. 6 shows the number of non-zero weights ($> \max(\boldsymbol{W}_1)/1000$) in the hidden layer versus signal to noise ration. In the ML-CVNN, almost all connection weights play a part to represent input signals. On the other hand, in Sparse-ML-CVNN, more connections are pruned for any SNR signals than in ML-CVNN. This results show the added penalty term restricts the connections to small number and the non-zero connection weights are sparse. Moreover, pruning in the Sparse-ML-CVNN becomes stronger depending on a rise of the SNR. This means that the network dynamics prunes more redundant connections automatically according to the penalty function in easy situations for the prediction methods.

Fig. 7 shows the prediction accuracy versus signal to noise ratio. Here, we evaluate the accuracy by using prediction-phase error that is an important value in

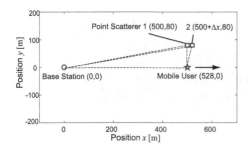

Fig. 8. Geometrical setup used in the simulation

Table 2. Simulation Parameters

Parameter	Value
Carrier frequency f_c	2 GHz
User speed v	14 m/s
Number of paths M	3

Table 3. OFDM Parameters

Parameter	Value
QPSK symbol number	1612
QPSK symbol rate F	406.25 kHz
Number of carriers K	52
FFT Size	64
Carrier spacing F/K	7812.5 Hz
Spacing between carriers	3.98 MHz
Size of guard interval	16
TDD frame length	5 ms
TDD symbol number in a frame s	2500 symbol
Sampling rate	500 kHz

channel prediction. The prediction errors of ML-CVNN and Sparse-ML-CVNN decrease with the increase of SNR, because the white noise decreases and the prediction becomes easier in the larger SNR situations. Sparse-ML-CVNN presents lower prediction error than ML-CVNN over 5 dB. By thinking with the results shown in Fig. 6, Sparse-ML-CVNN can self-optimize its structure according to prediction situations and improve the prediction accuracy.

4.2 Performance of Proposed ML-CVNN in Fading Channel Prediction

In this section, we evaluate the fading channel prediction performance based on combination of Sparse-ML-CVNN and CZT. The geometrical setup is shown in Fig. 8 and the simulation parameters are shown in Table 2. We assume an orthogonal frequency-division multiplexing (OFDM) with QPSK modulation and a time division duplex (TDD) as the communication scheme. Table 3 lists the system parameters. We assume the same prediction parameters as those in Table 1. Here, we name the prediction method based on combination of Sparse-ML-CVNN and CZT as CZT-Sparse-ML-CVNN method, same structure without the penalty function as CZT-ML-CVNN method, and different structure with the

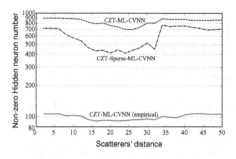

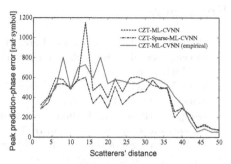

Fig. 9. Number of non-zero-amplitude weights in the hidden layer versus scatterers' distance Δx

Fig. 10. Prediction-phase error versus scatterers' distance Δx

empirically decided input terminals $I_{\mathrm{ML}} = 11$ and the hidden neural number $J_{\mathrm{ML}} = 10$ as CZT-ML-CVNN (empirical).

Fig. 9 shows the number of non-zero weights ($> \max(\boldsymbol{W}_1)/1000$) in the hidden layer versus scatterers' distance Δx. In CZT-Sparse-ML-CVNN, more connections are pruned for any Δx in comparison with CZT-ML-CVNN. In addition, the structure size of the Sparse-ML-CVNN changes greatly with the change of Δx.

Fig. 10 shows prediction-phase error curves accumulated in a unit TDD frame versus scatterers' distance Δx. This results show CZT-ML-CVNN is too large to predict channel characteristics, but we can improve the prediction performance by introducing the sparsity. The effectiveness of the sparsity especially appears in the decrease of the maximum prediction error in Fig. 10. In addition, CZT-Sparse-ML-CVNN has better prediction accuracy than CZT-ML-CVNN (empirical) in the almost all Δx conditions. This result shows that the proposed CZT-Sparse-ML-CVNN can adapt its network structure automatically even in prediction conditions difficult for empirical structure. These results present that the proposed method has a higher performance in fading channel prediction.

5 Conclusion

We proposed a new adaptive method for optimizing structure of complex-valued neural networks according to fading channel changes. The proposed CVNN restricts its connections to small number based on the added penalty function. Therefore, the proposed CVNN can automatically change its network structure depending on the change of communication environment, and the channel prediction method based on it presents high prediction accuracy. A series of simulations demonstrated that the proposed method has better performance than the conventional methods.

References

1. Barakat, M., Druaux, F., Lefebvre, D., Khalil, M., Mustapha, O.: Self adaptive growing neural network classifier for faults detection and diagnosis. Neurocomputing 74(18), 3865–3876 (2011)
2. Baraniuk, R.G.: Compressive sensing (lecture notes). IEEE Signal Processing Magazine 24(4), 118–121 (2007)
3. Ding, T., Hirose, A.: Fading channel prediction based on complex-valued neural networks in frequency domain. In: Proceedings of URSI International Symposium on Electromagnetic Theory (EMTS), pp. 640–643 (May 2013)
4. Ding, T., Hirose, A.: Fading channel prediction based on combination of complex-valued neural networks and chirp z-transform. IEEE Transactions on Neural Networks and Learning Systems 25(9), 1685–1695 (2014)
5. Duel-Hallen, A.: Fading channel prediction for mobile radio adaptive transmission systems. Proceedings of the IEEE 95(12), 2299–2313 (2007)
6. Hirose, A.: Complex-Valued Neural Networks, 2nd edn. SCI, vol. 400. Springer, Heidelberg (2012)
7. Jakes, W.C. (ed.): Microwave Mobile Communications, 2nd edn. Wiley-IEEE Press (1994)
8. Karnin, E.D.: A simple procedure for pruning back-propagation trained neural networks. IEEE Transactions on Neural Networks 1(2), 239–242 (1990)
9. Lu, T.C., Yu, G.R., Juang, J.C.: Quantum-based algorithm for optimizing artificial neural networks. IEEE Transactions on Neural Networks and Learning Systems 24(8), 1266–1278 (2013)
10. Ozawa, S., Tan, S., Hirose, A.: Errors in channel prediction based on linear prediction in the frequency domain: A combination of frequency-domain and time-domain techniques. URSI Radio Science Bulletin 337, 25–29 (2011)
11. Rabiner, L.R., Schafer, R.W., Rader, C.: The chirp z-transform algorithm. IEEE Transactions on Audio and Electroacoustics 17, 86–92 (1969)
12. Reed, R.: Pruning algorithms - a survey. IEEE Transactions on Neural Networks 4(5), 740–747 (1993)
13. Tan, S., Hirose, A.: Low-calculation-cost fading channel prediction using chirp z-transform. Electronics Letters 45(8), 418–420 (2009)

Invariant Multiparameter Sensitivity
of Oscillator Networks

Kenzaburo Fujiwara, Takuma Tanaka, and Kiyohiko Nakamura

Department of Computational Intelligence and Systems Science,
Interdisciplinary Graduate School of Science and Engineering,
Tokyo Institute of Technology, Yokohama, 226-8502, Japan
kenzaburo.fujiwara@gmail.com

Abstract. The behavior of neuronal and other biological systems is determined by their parameter values. We introduce a new metric to quantify the sensitivity of output to parameter changes. This metric is referred to as invariant multiparameter sensitivity (IMPS) because it takes on the same value for a class of equivalent systems. As a simplification of neuronal membrane, we calculate, in parallel resistor circuits, the values of IMPS and a previously studied metric of parameter sensitivity. Furthermore, we simulate phase oscillator models on complex networks and clarify the property of IMPS.

Keywords: Parameter Sensitivity, Complex Network, Phase Oscillator, Synchronization.

1 Introduction

A large number of mathematical models have been proposed in order to explain complex phenomena in brain including learning, chaotic behavior and synchronization [1, 2]. Because these models have many parameters, it would be desirable to know how changes of parameter values influence the output of a model. Information about the relationship between parameter changes and output of models is indispensable in designing models, fitting parameters and understanding the dynamics of systems [3, 4].

In this paper, we investigate parameter sensitivity, that is, the response of output to small changes of parameters. Parameter sensitivity has been intensively studied in circuit theory, particularly in resistor-capacitor networks [4–7]. In biochemical modeling, metrics of sensitivity are also used in quantifying robustness of systems [8]. In neuronal modeling and machine learning, it is important to estimate how sensitively output, such as firing rate and generalization error, changes in response to small parameter changes.

Several metrics of parameter sensitivity have been proposed in previous studies. Single parameter sensitivity (SPS) allows us to quantify the output changes in response to small change of a single parameter. Multiparameter sensitivity (MPS) is a generalization of SPS to multiple parameters [9]. MPS is defined as the square root of the sum of the square of SPSs. However, as will be shown later,

C.K. Loo et al. (Eds.): ICONIP 2014, Part I, LNCS 8834, pp. 183–190, 2014.

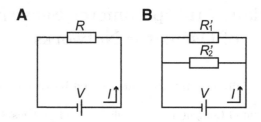

Fig. 1. Circuit A and circuit B are equivalent if $R = R_1'R_2'/(R_1' + R_2')$

MPS gives different values for such equivalent electric circuits as in Fig. 1(A) and Fig. 1(B).

We propose in this paper a new metric of sensitivity, which we call invariant multiparameter sensitivity (IMPS). This paper is organized as follows. In section 2, we introduce sensitivity metrics previously proposed and define IMPS. Then we derive basic properties of IMPS. In section 3, we examine properties of IMPS by applying it to a simple circuit. In section 4, we further investigate IMPS for nonlinearly coupled oscillators. Since it was reported that networks in brain are scale-free networks [10], in which the number of connections of each vertex obeys a power-law distribution, we examine the system of oscillators on a scale-free network. In section 5, we summarize our results and discuss potential applications.

2   Parameter Sensitivity

Dynamical systems are expressed by first-order differential equations

$$\dot{\boldsymbol{x}} = F\left(t, \boldsymbol{x}, \boldsymbol{p}\right), \tag{1}$$

where t is time, $\boldsymbol{x} = [x_1, x_2, x_3, \ldots, x_n]$ is the state variable vector and $\boldsymbol{p} = [p_1, p_2, p_3, \ldots, p_m]$ is the parameter vector.

For the output q of the system, single parameter sensitivity for parameter p_i, which we call SPS_i, is defined as

$$\mathrm{SPS}_i = \frac{p_i}{q}\frac{\partial q}{\partial p_i} = \frac{\partial \ln q}{\partial \ln p_i}. \tag{2}$$

SPS_i is the ratio of the change of output q to the change of parameter p_i. However, SPS_i does not quantify sensitivity to the change of other parameters. MPS, which is defined as

$$\mathrm{MPS}^2 = \sum_{i=1}^{m} \mathrm{SPS}_i^2, \tag{3}$$

is known as a metric to estimate sensitivity to the change of the whole parameter set of the system [7, 9]. As shown in the next section, MPS often gives different values for two equivalent models, and thereby MPS is not appropriate for comparing sensitivities between models. Thus, we introduce a new metric, invariant multiparameter sensitivity (IMPS). IMPS is defined by the sum of absolute values of SPSs as

$$\text{IMPS} = \sum_{i=1}^{m} |\text{SPS}_i| . \tag{4}$$

IMPS gives the same values for equivalent models in many cases. Assuming that $q(p_1, p_2, p_3, \ldots, p_m)$ is a homogeneous function of degree k and that SPSs in equation (4) have the same sign, we obtain

$$\text{IMPS} = \sum_{i=1}^{m} \left| \frac{p_i}{q} \frac{\partial q}{\partial p_i} \right|$$
$$= \left| \sum_{i=1}^{m} \frac{p_i}{q} \frac{\partial q}{\partial p_i} \right|$$
$$= |k| , \tag{5}$$

where we used Euler's theorem

$$p_1 \left(\frac{\partial q}{\partial p_1} \right) + p_2 \left(\frac{\partial q}{\partial p_2} \right) + \cdots + p_m \left(\frac{\partial q}{\partial p_m} \right) = kq(p_1, p_2, p_3, \ldots, p_m). \tag{6}$$

Hence IMPS is constant. IMPS is invariant for all models satisfying the following conditions: (1) the outputs are expressed by homogeneous functions of parameters; and (2) SPSs take on the same sign.

3   Circuit Toy Models

In this section we examine circuit toy models. Consider that there is one resistor R in a circuit as in Fig. 1(A). We denote the electric energy consumption by W and the voltage of the voltage source by V. We assume that W is the output. MPS of this circuit equals 1. The same current-voltage relationship as the circuit shown in Fig. 1(A) can be realized by the circuits equivalent to it such as that in Fig. 1(B) if $R = R_1' R_2'/(R_1' + R_2')$. MPS of the circuit in Fig. 1(B) is given by

$$\text{MPS}^2 = \sum_{i=1}^{n} \text{SPS}_i^2$$
$$= \sum_{i=1}^{2} \left(\frac{R_i'}{W} \frac{\partial W}{\partial R_i'} \right)^2$$
$$= \left(\frac{R_2'}{R_1' + R_2'} \right)^2 + \left(\frac{R_1'}{R_1' + R_2'} \right)^2$$
$$< 1. \tag{7}$$

Thus, MPS of the circuit in Fig. 1(B) is less than MPS of that in Fig. 1(A). In contrast, IMPS of the circuit in Fig. 1(B) is given by

$$
\begin{aligned}
\text{IMPS} &= \sum_{i=1}^{n} |\text{SPS}_i| \\
&= \frac{R_2'}{R_1' + R_2'} + \frac{R_1'}{R_1' + R_2'} \\
&= 1,
\end{aligned}
\tag{8}
$$

which equals IMPS of that in Fig. 1(A). It can be easily shown that IMPS is the same for the energy consumption of the equivalent RC circuits, by which the electric properties of neuronal membrane have been modeled [1].

4   Nonlinear Model

In this section, we investigate the IMPS of the system of phase oscillators on a Barabási–Albert network as an example of neuronal networks. Barabási–Albert model is the most thoroughly studied scale-free network model [11]. We generate Barabási–Albert networks with average degree of 4. We start from 2 vertices and add a vertex with 2 edges in each step until we have N vertices.

We assume that N oscillators are connected to each other by the adjacency matrix $\mathbf{A_{BA}}$ of a Barabási–Albert network. The dynamics of oscillator i are described by

$$
\frac{d\theta_i}{dt} = \omega_i + \sum_{j=1}^{N} K_{ij} \sin(\theta_j - \theta_i),
\tag{9}
$$

where K_{ij} is the (i, j)-element of the connection weight matrix defined by $\mathbf{K} = \alpha \mathbf{A_{BA}}$ and the natural frequency ω_i is drawn from the Gaussian distribution with unit variance. We assume that $\sum_{i=1}^{N} \omega_i = 0$ without loss of generality. Here, α is the connection strength. We use the circular variance V of the oscillators

$$
V = 1 - r = 1 - \frac{1}{N} \sqrt{C^2 + S^2}
\tag{10}
$$

in the phase-locked state as an output, where r is the Kuramoto order parameter, $C = \sum_{i=1}^{N} \cos \theta_i$ and $S = \sum_{i=1}^{N} \sin \theta_i$.

In the phase-locked state, the right-hand side y_i' of equation (9) is 0, that is,

$$
\mathbf{y}' = \mathbf{0}.
\tag{11}
$$

Here we derive the relationship between the connection weights and the phases under the condition that equation (11) is satisfied. Assuming that $\Delta \mathbf{K}$ is small, we obtain

$$y_i' + \Delta y_i' = \omega_i + \sum_{j=1}^{N}(K_{ij} + \Delta K_{ij}) \sin(\theta_j + \Delta\theta_j - \theta_i - \Delta\theta_i)$$

$$\approx \omega_i + \sum_{j=1}^{N}(K_{ij} + \Delta K_{ij}) \left[\sin(\theta_j - \theta_i) + \cos(\theta_j - \theta_i)(\Delta\theta_j - \Delta\theta_i)\right].$$

$$(12)$$

Subtracting y_i' from both sides yields

$$\Delta y_i' \approx \sum_{j=1}^{N} K_{ij} \cos(\theta_j - \theta_i)(\Delta\theta_j - \Delta\theta_i)$$

$$+ \sum_{j=1}^{N} \Delta K_{ij} \left[\sin(\theta_j - \theta_i) + \cos(\theta_j - \theta_i)(\Delta\theta_j - \Delta\theta_i)\right]. \quad (13)$$

Thus we obtain

$$\frac{\partial y_i'}{\partial \theta_j} \equiv J_{ij}', \tag{14}$$

$$\frac{\partial y_i'}{\partial K_{lm}} = \begin{cases} \sin(\theta_m - \theta_l) & i = l \\ 0 & i \neq l \end{cases}, \tag{15}$$

where

$$J_{ij}' = \begin{cases} -\sum_{s=1}^{N} K_{is} \cos(\theta_s - \theta_i) & i = j \\ K_{ij} \cos(\theta_j - \theta_i) & i \neq j \end{cases}. \tag{16}$$

$\mathbf{J'}$ is of $N-1$ rank, because Laplacian matrices of connected graphs are of $N-1$ rank [12]. Adding the same value to all θ_i's of a phase-locked solution results in another phase-locked solution, and the latter cannot be distinguished from the former in terms of V. Thus we cannot determine the unique phase-locked solution for this model. However, we can set the average phase to 0, which will not ruin the generality of our argument, because we are interested only in circular variance V of the oscillators. Assuming $\sum_{i=1}^{N} \theta_i = 0$, we can replace equation (11) with

$$y_i \equiv \omega_i + \sum_{j=1}^{N} K_{ij} \sin(\theta_j - \theta_i) + \sum_{j=1}^{N} \theta_j = 0. \tag{17}$$

Hence $\partial y_i / \partial \theta_j$ can be derived as

$$\frac{\partial y_i}{\partial \theta_j} = J_{ij}' + 1 \equiv J_{ij}. \tag{18}$$

J is full rank. Thus we have

$$\frac{\partial \theta_i}{\partial K_{lm}} = -\sum_{j=1}^{N} (\mathbf{J}^{-1})_{ij} \delta_{jl} \sin(\theta_m - \theta_l)$$

$$= -(\mathbf{J}^{-1})_{il} \sin(\theta_m - \theta_l). \quad (19)$$

Hence the derivative of V with respect to K_{lm} is given by

$$\frac{\partial V}{\partial K_{lm}} = -\frac{1}{2N} \left(C^2 + S^2\right)^{-1/2} \frac{\partial \left[\left(\sum_{i=1}^{N} \cos\theta_i\right)^2 + \left(\sum_{i=1}^{N} \sin\theta_i\right)^2\right]}{\partial K_{lm}}$$

$$= \frac{1}{N^2 r} \left(S \sum_{i=1}^{N} \cos\theta_i (\mathbf{J}^{-1})_{il} - C \sum_{i=1}^{N} \sin\theta_i (\mathbf{J}^{-1})_{il}\right) \sin(\theta_m - \theta_l). \quad (20)$$

From the above analysis, we numerically obtain IMPS as

$$\text{IMPS} = \sum_{\langle lm \rangle} |\text{SPS}_{lm}| = \sum_{\langle lm \rangle} \left|\frac{K_{lm}}{V} \frac{\partial V}{\partial K_{lm}}\right|, \quad (21)$$

where $\langle \rangle$ is the summation over the connected oscillator pairs. In the initial state, all phases are uniformly distributed. When α is sufficiently large, the oscillators are phase locked as shown in Fig. 2. We calculate the IMPS for various values of α under phase-locked conditions.

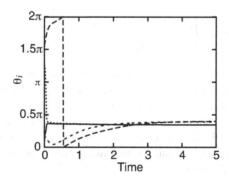

Fig. 2. Synchronization of 1000 oscillators on a Barabási–Albert network with average degree of 4. Phases of 4 out of 1000 oscillators are shown. The connection strength α is set to 2.

The IMPS of this model is shown in Fig. 3. IMPS gives similar values for system size $N = 1000$ (Fig. 3A) and 10000 (Fig. 3B), whereas MPS exhibits system-size dependency (Fig. 3C). Unlike the toy circuit model in section 3,

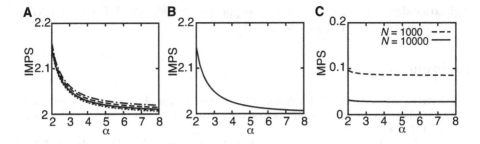

Fig. 3. IMPS of oscillator networks with $N = 1000$ (A) and $N = 10000$ (B). Panel A shows the results of networks generated by 5 different random seeds. Panel C shows the MPS for $N = 1000$ (dashed line) and the MPS for $N = 10000$ (solid line).

IMPS for this system of phase oscillators does not take on a constant value because of the nonlinearity of this system. If α is sufficiently large, the nonlinearity of sine coupling is ignorable. Therefore, as α increases, IMPS of this oscillator system converges to 2 because the circular variance V of this model converges to a homogeneous function of degree -2.

5   Discussion

In this paper, we have reviewed the previously proposed metrics, SPS and MPS, which quantify parameter sensitivity. We have formulated an improved metric, IMPS. IMPS gives the same value for equivalent models in many cases. This gives IMPS a significant advantage over MPS, which gives different values for equivalent systems. In the analysis of the simple circuits, IMPS has given the same value for equivalent parallel circuits. Then we have applied IMPS to nonlinear complex systems. As a first step for applying IMPS to neuronal systems, we have used the phase oscillator model and the Barabási–Albert model because those two models are widely used in the previous researches [11, 13, 14].

Formerly, invariance of IMPS was reported only for RC network circuits [4–7]. In this paper, we have shown its invariance in a wider setting than previous studies. In the system of phase oscillators, IMPS is not always invariant because of the nonlinearity. Our results suggest that IMPS is a metric reflecting both structure and dynamics of the systems. Thus IMPS would allow us to estimate the dynamics and excitability of individual neurons and synaptic connectivity between neurons.

In a future work, it should be examined how structure and nonlinearity of systems are reflected in the value of IMPS. In particular, we will apply IMPS to the system of neurons on a Watts–Strogatz small-world network [15]. Furthermore, the relation between IMPS and previously proposed network metrics, such as cluster coefficient and average path length [15, 16], should be investigated. The application of IMPS to the real neuronal networks is also of interest.

Acknowledgment. This work was supported by MEXT/JSPS KAKENHI Grant Numbers 24651184 and 25115710.

References

1. Hodgkin, A.L., Huxley, A.F.: A quantitative description of membrane current and its application to conduction and excitation in nerve. The Journal of Physiology 117, 500–544 (1952)
2. Aoyagi, T., Kang, Y., Terada, N., Kaneko, T., Fukai, T.: The role of Ca^{2+}-dependent cationic current in generating gamma frequency rhythmic bursts: Modeling study. Neuroscience 115, 1127–1138 (2002)
3. Csete, M.E., Doyle, J.C.: Reverse engineering of biological complexity. Science 295, 1664–1669 (2002)
4. Leeds, J.V., Ugron, G.: Simplified multiple parameter sensitivity calculation and continuously equivalent networks. IEEE Transactions on Circuit Theory 14, 188–191 (1967)
5. Roska, T.: Summed-sensitivity invariants and their generation. Electronics Letters 4, 281–282 (1968)
6. Goddard, P., Spence, R.: Efficient method for the calculation of first- and second-order network sensitivities. Electronics Letters 5, 351–352 (1969)
7. Rosenblum, A., Ghausi, M.: Multiparameter sensitivity in active RC networks. IEEE Transactions on Circuit Theory 18, 592–599 (1971)
8. Maeda, K., Kurata, H.: Quasi-multiparameter sensitivity measure for robustness analysis of complex biochemical networks. Journal of Theoretical Biology 272, 174–186 (2011)
9. Goldstein, A., Kuo, F.: Multiparameter sensitivity. IRE Transactions on Circuit Theory 8, 177–178 (1961)
10. Eguiluz, V.M., Chialvo, D.R., Cecchi, G.A., Baliki, M., Apkarian, A.V.: Scale-free brain functional networks. Physical Review Letters 94, 018102 (2005)
11. Barabási, A.L., Albert, R.: Emergence of scaling in random networks. Science 286, 509–512 (1999)
12. Olfati-Saber, R., Murray, R.M.: Consensus problems in networks of agents with switching topology and time-delays. IEEE Transactions on Automatic Control 49, 1520–1533 (2004)
13. Kuramoto, Y.: Chemical oscillations, waves and turbulence. Springer, Berlin (1984)
14. Teramae, J., Tanaka, D.: Robustness of the noise-induced phase synchronization in a general class of limit cycle oscillators. Physical Review Letters 93, 204103 (2004)
15. Watts, D.J., Strogatz, S.H.: Collective dynamics of 'small-world' networks. Nature 393, 440–442 (1998)
16. Albert, R., Barabási, A.L.: Statistical mechanics of complex networks. Reviews of Modern Physics 74, 47–97 (2002)

Spatial-Temporal Saliency Feature Extraction for Robust Mean-Shift Tracker*

Suiwu Zheng, Linshan Liu, and Hong Qiao

State Key Laboratory of Management and Control for Complex Systems
Institute of Automation, Chinese Academy of Sciences ZhongGuanCunDong Road,
HaiDian District, Beijing, China
{suiwu.zheng,linshan.liu,hong.qiao}@ia.ac.cn

Abstract. Robust object tracking in crowded and cluttered dynamic scenes is a very difficult task in robotic vision due to complex and changeable environment and similar features between the background and foreground. In this paper, a saliency feature extraction method is fused into mean-shift tracker to overcome above difficulties. First, a spatial-temporal saliency feature extraction method is proposed to suppress the interference of the complex background. Furthermore, we proposed a saliency evaluation method by fusing the top-down visual mechanism to enhance the tracking performance. Finally, the efficiency of the saliency features based mean-shift tracker is validated through experimental results and analysis.

Keywords: Saliency Feature, Mean-Shift, Object Tracking.

1 Introduction

Mean-shift was first proposed by Fukunaga[1] as a mode seeking method for data clustering purpose in 1975. It was enriched by importing kernel function and weight coefficient, and was applied into image processing field[2]. Comaniciu[3] proposed kernel based mean-shift tracker, which has been quite successfully applied in many areas.

Mean-shift tracking algorithm has been proved to be efficient in many scenarios due its simplicity and and fast convergence property attributes. However, there still have some problems needed to be solved to adapt to the complex dynamic environment. For example, the original weighted color histogram feature in mean-shift tracker may lead to failure in target tracking when similar color feature appears in background or the target appearance changes drastically.

Aiming to encode more information to enhance the robustness of visual representation, multiple cues are fused in mean-shift tracking methods recently[4]. Wang[7] integrated color and texture-shape cues and embedded them into kernel

---

* This work is supported in part by the National Key Technology R&D Program of China #2012BAI34B02, National Natural Science Foundation of China(NNSF) Grants #61101221, #60725310, #61033011.

C.K. Loo et al. (Eds.): ICONIP 2014, Part I, LNCS 8834, pp. 191–198, 2014.

based tracking with adaptive confidential coefficient. Leichter[8] focus on multiple reference integrated histogram. Jia[6] introduced the histogram of gradient (HOG) feature into kernel based tracking method. Multi-cues methods tend to produce higher dimension combined histogram to represent target, which may prevent these approaches from real-time applications. How to combine these cues into mean-shift tracker framework[9][10]in a simple and efficient form to achieve good performance is remaining an unsolved problem.

In this paper, we present an improved mean-shift tracker which embedded spatial-temporal saliency feature extraction and evaluation methods. The spatial-temporal saliency feature, composed of the spatial saliency feature and motion saliency feature, is proposed to find the most distinguished features between the target and background. The saliency evaluation mechanism could enhance the tracking performance greatly in the cluttered and dynamic environment. Another important benefit of the proposed method is that the saliency feature map could be easily fused with the traditional color histogram feature and embedded into the mean-shift framework. By introducing the spatial-temporal saliency feature, our mean-shift tracker will be robust to the targets and scenes which has identical texture feature and with different spatial and motion patterns.

The flowchart of proposed tracking method is given as follows.

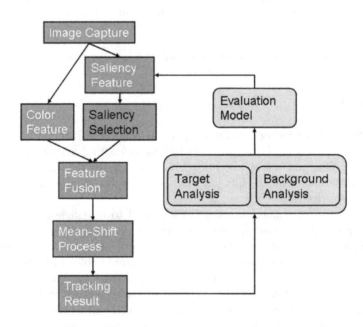

Fig. 1. The flowchart of the proposed tracker. The blue part is the Saliency Evaluation Mechanism.

2   Spatial-Temporal Saliency Feature

Saliency attention feature have been extended to the applications in computer vision, including: object detection[11] and object tracking[15]. There are many methods to produce the saliency feature map. Itti[11] computed the saliency under several features separately, and saliency is defined as the relative center-surround contrast between object region and its neighborhood surroundings in Gaussian kernel. Though the result of saliency maps meet the biological mechanism in some degree, it is generally blurry and has high computation complexity. Histogram based saliency[16] is based on global contrast and may cause unnecessary error.

Inspired by the superiority of saliency attention above and the characteristic of mean-shift tracker, we propose a spatial-temporal saliency feature extraction method to get the most distinguished saliency map for mean-shift tracker. Differ from previous approaches, our saliency attention extraction method is closely related to the tracking and take full advantages of the spatial structure and motion feature. The spatial-temporal saliency map will be fused with the traditional color saliency map in mean-shift tracker by using an adaptive weight method.

2.1   Spatial Saliency Feature

According to the difference of feature domain, the spatial saliency feature extraction methods are divided into spatial domain based and frequency domain based. Spectral residual[12] and image label[13] are fast by using the frequency methods. In this section, we use discrete cosine transform to extract the frequency feature[14]. It is simple and fast.

The pulse discrete cosine transform of the target area X is given as follows.

$$\hat{X} = sign(DCT(X)) \tag{1}$$

Inverse transform to the spatial domain:

$$\boldsymbol{X} = IDCT(sign(DCT(X))) \tag{2}$$

The spatial saliency map of the target could be computed through equ.3. The example result could be seen in fig.2.

$$S = G * (\boldsymbol{X} \circ \boldsymbol{X}) \tag{3}$$

where G is the two dimension Gaussian filter, and it is used to smooth the result. S is the spatial saliency map of target. $*$ is the Hadamard product. $\circ$ is the convolution.

2.2   Motion Saliency Feature

The motion feature is very important to visual tracking. In this paper, the motion feature is obtained from the difference between consecutive frames, which is represented as equ.4.

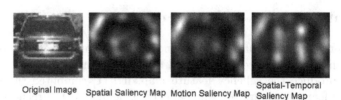

Original Image   Spatial Saliency Map   Motion Saliency Map   Spatial-Temporal Saliency Map

Fig. 2. Extraction of Spatial-Temporal Saliency Map

$$T = S_t - S_{t-1} \tag{4}$$

where T is the motion saliency map. S_t and S_{t-1} are the spatial saliency map in the t_{th} and $(t-1)_{th}$ frame respectively. Therefore, the result of motion saliency could be computed by using equ.3 to equ.4. The example temporal feature could be seen in fig.2.

2.3   Spatial-Temporal Saliency Feature Fused with Color Feature

In the traditional color based mean-shift tracking method, color feature distribution of the target(the center is x_c) is often represented as $H_c = \{q_u^c\}_{u=1,\cdots,m}$.

$$q_u^c = C \sum_{i=1}^{n} k(\|\frac{x_i - x_c}{a_c}\|^2)\delta[b^c(x_i) - u] \tag{5}$$

while $\sum_{u=1}^{m} q_u^c = 1$, function $b_c : R^2 \rightarrow \{i = 1, \cdots, m\}$ makes $b_c(x_i)$ the corresponding color index of x_i. $\{x_i\}_{i=1,\cdots,n}$ denotes the coordinate of each pixel, n is the number of pixels in the target, $\delta(\cdot)$is $Dirac$ function, $a_c = \sqrt{w^2 + h^2}$ represents size of the area, $u = 1, \cdots, m$ means the arbitrary color index from 1 to m, C is the normalization coefficient, $k(x)$ is the Epanechnikov kernel function.

The spatial-temporal saliency feature histogram $H_s = \{q_u^s\}_{u=1,\cdots,m}$ could be gotten through the spatial-temporal saliency map by using the histogram computing method.

$$q_u^s = C_s \sum_{i=1}^{n} k(\|\frac{x_i - x_s}{a_s}\|^2)\delta[b^s(x_i) - u] \tag{6}$$

The representation of the target could be described by fusing the spatial-temporal saliency feature and the color feature(see equ.7).

$$H_{cs} = w_c H_c + w_s H_s = \{q_u^{cs}\}_{u=1,\cdots,m} \tag{7}$$

w_c and w_s are the weights of the two feature. w_c and w_s could be set in a fix weight or adaptively adjusted in the tracking process according to the type of the environment.

3 The Improved Mean-Shift Tracker and Saliency Evaluation Mechanism

3.1 Similarity

The fused feature model of the target is $H_{cs} = \{q_u^{cs}\} u = 1, \cdots, m$, see euq.8.

$$q_u^{cs} = C_{cs} \sum_{i=1}^{n} k(\|\frac{x_i - x_{cs}}{a}\|^2) \delta[b^{cs}(x_i) - u], \sum_{u=1}^{m} q_u^{cs} = 1 \qquad (8)$$

The fused feature model of the candidate could be represented as $H_{csc} = \{p_u^{csc}(y)\}, u = 1, \cdots, m$.

The similarity between the candidate and the target could be computed through Bhattacharyya coefficient.

$$\rho_{csc} = \sum_{u=1}^{m} \sqrt{p_u^{csc}(y) q_u^{cs}} \qquad (9)$$

3.2 Mean-Shift Process

The principle of mean-shift tracker could be described as follows.

Set y_0 is the position of the target in the last frame. The Taylor expansion of Bhattacharyya coefficient in y_0 is:

$$\rho(csc) = \frac{1}{2} \sum_{u=1}^{m} \sqrt{p_u^{csc}(y_0) q_u^{cs}} + \frac{C_{csc}}{2} \sum_{i=1}^{n} w_i k(\|\frac{y - x_i}{a_{csc}}\|^2) \qquad (10)$$

while C_{csc} is the texture normalization coefficient, and

$$w_i = \sum_{u=1}^{m} \sqrt{\frac{q_u^{cs}}{p_u^{csc}(y_0)}} \delta[b_{csc}(x_i) - u] \qquad (11)$$

Using the mean-shift iteration method to reach the maximum similarity ρ, we can get the local optimized target position:

$$y_1 = \frac{\sum_{i=1}^{n} x_i w_i g(\|\frac{y_0 - x_i}{a_{csc}}\|^2)}{\sum_{i=1}^{n} w_i g(\|\frac{y_0 - x_i}{a_{csc}}\|^2)} \qquad (12)$$

where $g(x) = -k'(x)$ is the kernel density estimation.

3.3 Saliency Evaluation Mechanism

Though the spatial-temporal saliency features are extracted by using the intrinsic frequency characteristic of the input image and the motion feature of the tracking task, it could not be proved that the selected saliency features are all important to the target. Some part of the extracted saliency feature may be unrelated and even bad to search the target due to the complexity and dynamics of the tracking task.

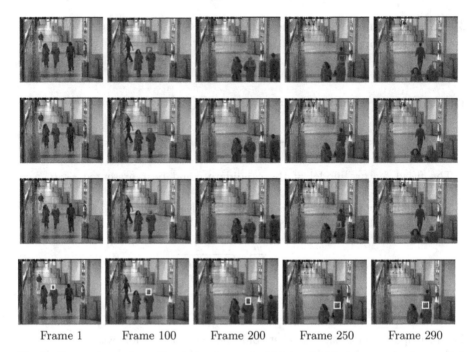

Frame 1   Frame 100   Frame 200   Frame 250   Frame 290

Fig. 3. Comparison of results of classic mean-shift tracker[3], improved color model tracker[5], multi-cues tracker[7] and the proposed method(From top to down) by using Caviar video in PETS 2004

To improve the robust of our tracker, a saliency evaluation mechanism is introduced into the tracking process(see blue part in fig.1). The principle of the evaluation is judging the importance of each pixel in the selected saliency feature by using a criterion, and remove the unimportant pixels or give a small weight. In this paper, we build the evaluation criterion by analysis the appearance of the target and background in the early frames, and give a new weight to each saliency point by the criterion. The proposed evaluation mechanism improve the tracker greatly through our experiments.

4   Experimental Results and Analysis

To validate the performance of the improved mean-shift tracker, the proposed algorithm is tested on the benchmark videos and actual videos with complex scene, multi-Persons with mutual intersection. Four mean-shift based trackers are implemented to make the comparison of the results. They are classic mean-shift tracker[3], improved color model tracker[5], multi-cues tracker[7] and the proposed method. The experimental results and discussion are given as follows.

4.1   WalkByShop1cor Video in CAVIAR Datasets

The frame rate of videos in CAVIAR is 25fps, and the image size is 384*288. The scene of WalkByShop1cor Video is multi-persons with clutter background in a building, and there exists some instant occlusions. One person tracking result and the comparisons are given in fig.3. The results show that the classic mean-shift tracker failed in the one person tracking process. Our proposed method could localize the object accurately.

4.2 Multi-persons with Mutual Intersection

In this part, the four algorithms are tested on the video with multi-person intersection. The results and comparisons are given in fig.4. It shows that the classic mean-shift tracker and the improved color model tracker[5] are fails when the intersection occurs, while our proposed method and the multi-cues tracker[7] succeed.

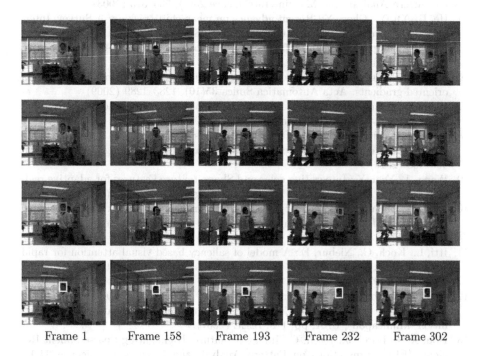

Frame 1   Frame 158   Frame 193   Frame 232   Frame 302

Fig. 4. Comparison of results of classic mean-shift tracker[3], improved color model tracker[5], multi-cues tracker[7] and the proposed method(From top to down) by using the video with multi-person intersection

5   Conclusion

In this paper, spatial-temporal saliency feature and a saliency evaluation method are introduced into mean-shift framework for reliable tracking. The spatial saliency feature and motion saliency feature are fused to find the most distinguished features between the target and background. The saliency evaluation mechanism could enhance the tracking performance greatly in the cluttered and dynamic environment. Finally, the efficiency of the proposed method is validated by some contrast experiments and analysis.

References

1. Fukunaga, K., Hostetler, L.: The estimation of the gradient of a density function with application in pattern recognition. IEEE Trans. on Information Theory 21(1), 32–40 (1975)
2. Cheng, Y.: Mean Shift, Mode seeking, and cluttering. IEEE Trans. on Pattern Analysis and Machine Intelligence 17(8), 790–799 (1995)
3. Comaniciu, D., Ramesh, V., Meer, P.: Kernel-based object tracking. IEEE Trans. on Pattern Analysis and Machine Intelligence 25(5), 564–575 (2003)
4. He, L., Xu, Y., Chen, Y.: Recent advance on mean shift tracking: A Survey. International Journal of Image and Graphics 13(3), 84–113 (2013)
5. Li, P.: An improved mean shift algorithm for object tracking. Acta Automatica Sinica 33(4), 347–354 (2009)
6. Jia, H., Zhang, Y.: Multiple kernels based object tracking using histograms of oriented gradients. Acta Automatica Sinica 35(10), 1283–1289 (2009)
7. Wang, Y., Liang, Y., Zhao, C.: Kernel-based tracking based on adaptive fusion of multiple cues. Acta Automation Sinica 34(4), 393–399 (2008)
8. Leichter, I., Lindenbaum, M., Rivlin, E.: Mean shift Tracking with multiple reference color histograms. Computer Vision and Image Understanding (CVIU) 114(3), 400–408 (2010)
9. Wang, J., Yagi, Y.: Integrating color and Shape-texture Features for adaptive real-time tracking. IEEE Trans. on Image Processing 17(2), 235–240 (2008)
10. Ning, J., Zhang, L.: Robust object tracking using joint color-texture histograms. International Journal of Pattern Recognition and Artificial Intelligence 23(7), 1245–1263 (2009)
11. Itti, L., Koch, C., Niebur, E.: A model of saliency based visual attention for rapid scene analysis. IEEE Trans. on Pattern Analysis and Machine Intelligence 20(11), 1254–1255 (1998)
12. Hou, X.D., Zhang, L.Q.: Saliency detection. A Spectral Residual Approach. In: Proceeding of the 2007 IEEE Conference on Computer Vision and Pattern Recognition, Minneapolis, USA, pp. 1–8. IEEE (2007)
13. Hou, X.D., Harel, J., Koch, C.: Image Signature: Highlighting Sparse Salient Regions. IEEE Transactions on Pattern Analysis and Machine Intelligence 34(1), 194–201 (2012)
14. Yu, Y., Wang, B., Zhang, L.M.: Bottom-up attention: Pulsed PCA transform and pulsed cosine transform. Cogn. Neurodyn. 5(4), 321–332 (2011)
15. Mahadevan, V., Vasconcelos, N.: Saliency-based discriminant tracking. In: IEEE Conference on Computer Vision and Pattern Recognition, pp. 1007–1013 (2009)
16. Cheng, M., Zhang, G., Mitra, N., et al.: Global contrast based salient region detection. In: IEEE Conference on Computer Vision and Pattern Recognition, pp. 409–416 (2011)

BOOSTRON:
Boosting Based Perceptron Learning

Mirza M. Baig[1], Mian. M. Awais[1], and El-Sayed M. El-Alfy[2,⋆]

[1] School of Science and Engineering (SSE)
Lahore University of Management Sciences (LUMS)
Lahore 54792, Pakistan
[2] College of Computer Sciences and Engineering
King Fahd University of Petroleum and Minerals
Dhahran 31261, Saudi Arabia
{mirza,awais}@lums.edu.pk, alfy@kfupm.edu.sa

Abstract. A novel boosting based perceptron learning algorithm is presented that uses AdaBoost along with a new representation of decision stumps using homogenous coordinates. The new representation of decision stumps makes perceptron an instance of boosting based ensemble. As Boostron minimizes an exponential cost function instead of the mean squared error minimized by the perceptron learning algorithm, it gives improved performance for classification problems. The proposed method is compared to the perceptron learning algorithm using several classification problems of varying complexity.

1   Introduction

Single-node perceptron [1] is a simple mathematical model of neurons in the human brain that stores the learned information as weights of the connections and has been frequently used as a building block for more complex multi-layer feed-forward neural networks [2, 3]. A single-node perceptron, as shown in Figure 1(a), has an $(n+1)-$dimensional vector $\bar{x} = [x_0, x_1, x_2, \ldots, x_n]$ as the input and produces its output by taking a dot product of the input vector with a $(n+1)-$dimensional weight vector, $\bar{W} = [w_0, w_1, w_2, \ldots, w_n]$. The component x_0 is permanently set to -1 and represents an external bias element. The perceptron output, y, is mostly computed using a non-linear activation function. For example, the *sign* of the product is the predicted output if the target output is ± 1 (binary classification). Mathematically, such output is given as,

$$y = sign\left(\bar{W}.\bar{x}^t\right) = sign\left(\sum_{i=0}^{n} w_i.x_i\right). \tag{1}$$

where $\bar{x}^t$ is the transpose of $\bar{x}$.

To learn the weight vector $\bar{W}$, a perceptron is provided with a set of p labeled training examples of the form $(\bar{x}_i, y_i);\ i = 1 \ldots p$. The weight vector elements

---

⋆ On leave from the College of Engineering, Tanta University, Egypt.

C.K. Loo et al. (Eds.): ICONIP 2014, Part I, LNCS 8834, pp. 199–206, 2014.
© Springer International Publishing Switzerland 2014

are initialized to 0's and are iteratively modified for each misclassified training example $(\bar{x}_i, y_i)$ using the perceptron learning law,

$$\bar{W}_{new} = \bar{W}_{old} + \eta.(y_i - y_i').\bar{x}_i. \tag{2}$$

The perceptron learning law adapts the weight vector by minimizing the mean-squared error over the training data [4]. The perceptron structure to handle a multiclass learning problem is typically as shown in Figure 1(b); it has $(n + 1)$ input units directly connected to m output units corresponding to the m classes. The weights of each output unit are learned independently using one-vs-remaining encoding of the classes.

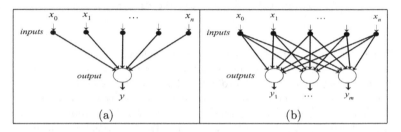

Fig. 1. Structure of a perceptron with (a) Single output, (b) Multiple outputs

The least mean squared criteria might fail to produce an optimal decision boundary for linearly non-separable problems [5]. To solve this problem a novel method, *boostron*, for learning a perceptron is introduced. The boostron minimizes an exponential cost function [5] and hence results into a more accurate classifier.

The remaining paper is organized as follows. Section 2 introduces the proposed method. Section 3 provides the experimental setup, results and comparison of the proposed method with the standard perceptron learning algorithm, and Section 4 concludes the paper.

2 Proposed Method

A short review of the AdaBoost algorithm followed by the proposed representation of decision stumps and the method of combining the decision stumps [7] with AdaBoost to learn a perceptron is presented in this section.

We assume the standard supervised learning setting and hence the learning algorithm is provided with p labeled training examples $(\bar{x}_i, y_i), i = 1...p$ each consisting of n *real-valued* attributes x_i^j and a binary class label $y_i \in \{-1, +1\}$.

2.1 The AdaBoost Algorithm

The AdaBoost algorithm [6], shown as Algorithm 1, is often used to construct highly accurate classifier ensembles from a moderately accurate base classifier. It takes p labeled training examples as input and iteratively selects T classifiers h_k by modifying a weight distribution, D_k, maintained on the training examples. The final ensemble is constructed by taking a linear combination of the selected classifiers using,

$$H(\bar{x}) = sign\left(\sum_{k=1}^{T} \alpha_k . h_k(\bar{x})\right), \tag{3}$$

where the weight of each classifier, α_k, is computed by using the error of the classifier w.r.t. the running distribution used to select the classifier.

Algorithm 1. AdaBoost [6]

Require: Examples $(\bar{x}_1, y_1) \ldots (\bar{x}_p, y_p)$ where
    $\bar{x}_i$ is the i-th training instance features and $y_i \in \{-1, +1\}$, and
    parameter $T =$ number of base learners in the ensemble
1: Set $D_1(i) = 1/p$ for $i = 1 \ldots p$
2: **for** $k = 1$ to T **do**
3:     Select a classifier h_k using the weights D_k
4:     Compute $\epsilon_k = Pr[h_k(\bar{x}_i) \neq y_i]$ w.r.t. D_k
5:     Set $\alpha_k = \frac{1}{2}log(\frac{1-\epsilon_k}{\epsilon_k})$
6:     Set $D_{k+1}(i) = \frac{D_k(i)\exp(-\alpha_k y_i . h_k(\bar{x}_i))}{Z_k}$; where Z_k is the normalization factor
7: **end for**
8: Output classifier $H(\bar{x}) = sign\left(\sum_{k=1}^{T} \alpha_k . h_k(\bar{x})\right)$

The AdaBoost algorithm uses a binary classifier and has been extended by Schapire and Singer [8] to handle confidence-rated outputs of a base classifier. They presented a different criterion of selecting the base classifier and a new method for computing the weight of the selected classifier by using,

$$\alpha_k = \frac{1}{2}ln\left(\frac{1+r_k}{1-r_k}\right), \tag{4}$$

where r_k is the difference of the weights of correctly classified instances and incorrectly classified instances w.r.t. the running distribution D_k.

2.2 Boostron

Decision stumps have been frequently used as base classifiers in AdaBoost to construct highly accurate classifier ensembles [8, 9]. A decision stump is a single node decision tree that makes predictions based on only one of the feature values. It consists of a feature index, say j, and a threshold, δ, such that all instances are partitioned into two sets using an **if-then-else** rule of the following form:

if $x_i^j \leq \delta$ **then**
    Class = +ve/-ve
else
    Class = -ve/+ve
end if

For $\bar{x}_i \in R^n$ the above decision stump can be converted into an equivalent classifier by using the inner product defined over R^n. For example, in the case where +1 label is assigned if $x_i^j \leq \delta$, we can represent the decision stump as,

$$s(\bar{x}_i) = -\left(\bar{w}.\bar{x}_i^t - \delta\right). \tag{5}$$

In equation 5, all components of the row vector $\bar{w} = [w_1, w_2, ..., w_n]$ are 0's except the component w_i . The sign of $s(\bar{x}_i)$ is the classification decision and its magnitude is the confidence of prediction. Such a classifier can be represented as a single dot product by representing instance $\bar{x}_i$ and the vector $\bar{w}$ using the homogenous coordinates,

$$s(x_i) = \bar{W}.\bar{X}_i^{\;t}, \tag{6}$$

where the vector $\bar{X}_i = [x_i^1, x_i^2, ..., x_i^n, 1]$ is obtained from the instance $\bar{x}_i$ by adding a 1 as the $(n+1)^{st}$ component and the vector $\bar{W} = -[w_1, w_2, ..., w_n, -\delta]$ obtained from $\bar{w}$ by adding $-\delta$ as the $(n+1)^{st}$ component. The final form of boosted classifier, as given in Equation 3, using the new representation of decision stump becomes,

$$H(\bar{x}) = sign\left(\sum_{k=1}^{T} \alpha_k.h_k(\bar{x})\right) = sign\left(\sum_{k=1}^{T} \alpha_k.\left(\bar{W}_k.\bar{x}^t\right)\right). \tag{7}$$

By using simple arithmetic manipulation, the above equation can be written as,

$$H(\bar{x}) = sign\left(\bar{W}.\bar{x}^t\right) \tag{8}$$

where the $(n+1)$−dimensional vector, $\bar{W} = \sum_{k=1}^{T} \left(\alpha_k.\bar{W}_k\right) = [w_1, w_2, ..., w_{n+1}]$, is the weighted sum of the selected decision stumps. The classifier given by Equation 8 is equivalent to a perceptron as given in Equation 1.

3   Evaluation

This section provides a detailed description of the datasets, experimental settings and results obtained for comparing the *boostron* with the perceptron learning algorithm.

Table 1 shows detailed statistics and method of computing error rate estimates for 7 binary and 12 multiclass learning datasets. These datasets cover a wide variety of classification problems including a very small lung cancer dataset that has only 32 training examples, and larger datasets including the letters recognition and pen-digit recognition datasets.

Table 1. Description of Datasets

Dataset Name	Dimensions	Training Instances	Test Instances	Total Classses	Error Estimate
Balance Scale	4	625		2	Cross Validation
Breast Cancer	30	569		2	Cross Validation
Spambase	57	4601		2	Cross Validation
EEG	14	14979		2	Cross Validation
Two Norm	20	2000	2000	2	Training/Test
Three Norm	20	1000	2000	2	Training/Test
Ring Norm	20	2479	2442	2	Training/Test
Iris	4	150		3	Cross Validation
Forest Fire	5	500		4	Cross Validation
Glass	10	214		7	Cross Validation
Vowels	10	528	372	11	Cross Validation
Land State	37	4435	2000	8	Training/Test
Wine	13	178		3	Cross Validation
Waveform	21	5000		3	Cross Validation
Pen digits	16	7494	3498	10	Training/Test
Letters	16	16000	4000	26	Training/Test
Segmentation	20	210	2100	8	Training/Test
Yeast	8	98 0	504	10	Training/Test
Lung Cancer	56	32		3	Cross Validation

In all our experiments the extension of AdaBoost algorithm [8] that uses confidence rated predictions has been used along with decision stumps to create the classifiers. The output of each decision stump has been normalized to obtain confidence rated predictions in the range $[-1, 1]$.

The first set of experiments compares the two algorithms for the 7 binary classification problems. Table 2 summarizes the training and test accuracies of the two perceptron learning algorithms for these datasets. The boostron based learning algorithm considerably improved the performance of the resulting perceptron for the six linearly non-separable binary classification problems including the breast cancer detection, spambase recognition, heart disease detection(EEG), balance scale, and the simulated two-norm and three-norm problems. In the case of ring norm dataset the boostron converged to a degraded decision boundary.

Table 2. Error Rate Comparison for 7 binary classification Datasets

Dataset	Training Error%		Test Error%	
	Boostron	Perceptron	Boostron	Perceptron
Balance Scale	**6.56**	7.68	**7.32**	9.73
Breast Cancer	**9.72**	16.95	**9.14**	16.87
Spambase	**24.73**	39.34	**24.67**	39.43
EEG	**41.53**	51.27	**44.53**	55.27
Two Norm	**2.25**	5.05	**2.65**	5.0
Three Norm	**26.81**	35.6	**29.05**	35.7
Ring Norm	45.25	**30.83**	46.71	**31.58**

Table 3. Error Rate Comparison for 12 Multiclass Datasets

Dataset	Training Error%		Test Error%	
	Boostron	**Perceptron**	**Boostron**	**Perceptron**
Iris	**7.36**	34.07	**7.66**	40.27
Forest Fire	**16.62**	22.71	**17.73**	24.24
Glass	**18.22**	37.34	**18.82**	39.52
Vowel	71.02	**66.29**	73.59	**73.38**
Land State	**28.88**	76.80	**31.05**	77.65
Wine	**29.40**	55.25	**30.08**	56.76
Waveform	**18.08**	27.43	**18.24**	27.62
Pen Digits	26.90	**8.57**	31.33	**13.84**
Letters	62.81	**54.87**	64.90	**56.00**
Segmentation	**16.19**	46.19	**16.48**	48.76
Yeast	**43.88**	58.33	**48.02**	69.08
Lung Cancer	3.7	**1.25**	38.33	**36.67**
KDD-CUP 99	**6.35**	11.81	**11.22**	18.31

The second set of experiments compares the two algorithms over the 12 multiclass learning datasets taken from the UCI machine learning repository [10]. In these experiments, each multiclass problem has been decomposed into a number of binary classification problems using one-vs-remaining encoding and, similar to Figure 1(b), a perceptron has been learned for each of the resulting binary classification problems. The class corresponding to the most confident positive label has been the predicted class of an instance $\bar{x}$.

The summary of results, given in Table 3, shows that boostron mostly converged to a significantly improved decision boundary than the standard perceptron. However, for the three larger learning problems including letters, pen digits and vowel recognition, the boostron converged to an inferior decision boundary.

3.1   Boostron Based Intrusion Detection System

The last experiment reports the performance of a multiclass boostron for a 23-class Intrusion Detection System (IDS) used to monitor a TCP/IP network traffic. This dataset has been adopted from the KDD Cup 99 (KDD99) dataset [11] prepared and managed by MIT Lincoln Labs and is the most dominant intrusion detection dataset used in the machine learning community [12–16]. The adopted dataset has 494011 connections; each described using 41 attributes and a label identifying the type of the connection (either normal or one of the attacks). The dataset contains 3 dominant classes comprising of more than 98% of the total training examples. A detailed summary of the dataset is shown in Table 4. To measure the performance of boostron based system, the the dataset has been partitioned randomly into a training set (3% of data) and a test set (remaining 97% data). The experiment has been repeated 10 times to obtain an average performance of the intrusion detection system.

Table 5 gives the confusion matrix and the values of different measures for the three dominant classes covering most of the dataset. The IDS achieved a

Table 4. KDD99 Class Frequency

Class	Frequency	Class	Frequency	Class	Frequency
1	2203	9	7	17	10
2	30	10	107201	18	1589
3	8	11	231	19	280790
4	53	12	97278	20	2
5	12	13	3	21	979
6	1247	14	4	22	1020
7	21	15	264	23	20
8	9	16	1040		

Table 5. Test Performance of Intrusion Detection System for Three Dominant Classes

NEPTUNE			NORMAL			SMURF		
	+ve	-ve		+ve	-ve		+ve	-ve
+ve	100006	3622	+ve	82205	11824	+ve	271298	134
-ve	154	373761	-ve	6381	377133	-ve	69	206042
Accuracy: 0.9921			Accuracy: 0.9619			Accuracy: 0.9996		
Precision: 0.9985			Precision: 0.9280			Precision: 0.9997		
Recall: 0.9650			Recall: 0.8743			Recall: 0.9995		

test accuracy of 96.19% for the class representing normal TCP/IP traffic where as the precision and recall rates are around 92% and 87%, respectively. For the other two classes, the values of the accuracy and precision have been higher than 99% whereas the value of recall has been better than 96%. We also found that on average the proposed IDS attained 99.6% accuracy, 95.34% precision and 95.34% recall for the entire KDD99 dataset.

4   Conclusions

A boosting-based perceptron learning algorithm has been presented and compared to the standard perceptron learning algorithm using several classification tasks of varying complexity. The proposed method produced significantly accurate decision boundaries for most of the considered problems.

Acknowledgments. The authors would like to acknowledge the support provided by the Lahore University of Management Science (LUMS),the Higher Education Commission of Pakistan(HEC) and the support provided by King Abdulaziz City for Science and Technology (KACST) through the Science & Technology Unit at King Fahd University of Petroleum & Minerals (KFUPM) through project No. 11-INF1658-04 as part of the National Science, Technology and Innovation Plan.

References

1. Rosenblatt, F.: The perceptron: A probabilistic model for information storage and organization in the brain. Psychological Review 65(6), 386 (1958)
2. Hertz, J., Anders, K., Richard, G.P.: Introduction to the theory of neural computation. Addison Wesley (1990)
3. Nielson, H.R.: Neurocomputing. Addison Wesley (1990)
4. Rumhlhart, D.E., McClelland, J.L.: Parallel and Distributed Processing: Explorations in Microstructure of Cognition. Foundations, vol. 1. MIT Press, Cambridge (1990)
5. Friedman, J., Hastie, T., Tibshirani, R.: Additive logistic regression: A statistical view of boosting. The Annals of Statistics 38(2), 337–374 (2000)
6. Freund, Y., Schapire, R.E.: A Decision Theoretic Generlization of Online Learning. Journal of Computer and System Sciences 55(1), 119–139 (1997)
7. Iba, W., Langley, P.: Induction of One-Level Decision Trees. In: Proceedings of the Ninth International Conference on Machine Learning (1992)
8. Schapire, R.E., Singer, Y.: Improved Boosting algorithm using confidence rated predictions. Machine Learning 37(3), 297–336 (1999)
9. Viola, P., Jones, M.: Rapid object detection using a boosted cascade of simple features. In: Proceedings of the 2001 IEEE Computer Society Conference on Computer Vision and Pattern Recognition, CVPR 2001, vol. 1, p. 511 (2001)
10. Frank, A., Asuncion, A.: UCI Machine Learning Repository. School of Information and Computer Science. University of California, Irvine, CA (2010), http://archive.ics.uci.edu/ml
11. KDD Cup 1999 dataset for network-based intrusion detection systems (1999), http://kdd.ics.uci.edu/databases/kddcup99/kddcup99.html
12. Feng, W., Zhang, Q., Hu, G., Huang, J.X.: Mining network data for intrusion detection through combining SVMs with ant colony networks. Future Generation Computer Systems (2013)
13. Li, W., Liu, Z.: A method of SVM with normalization in intrusion detection. Procedia Environmental Sciences 11, 256–262 (2011)
14. Altwaijry, H., Algarny, S.: Bayesian based intrusion detection system. Journal of King Saud University, Computer and Information Sciences 24(1), 1–6 (2012)
15. Amiri, F., Yousefi, M.R., Lucas, C., Shakery, A., Yazdani, N.: Mutual information-based feature selection for intrusion detection systems. Journal of Network and Computer Applications 34(4), 1184–1199 (2011)
16. Bolón-Canedo, V., Sánchez-Maroño, N., Alonso-Betanzos, A.: Feature selection and classification in multiple class datasets: An application to KDD cup 99 dataset. Expert Systems with Applications 38(5), 5947–5957 (2011)

G-Stream: Growing Neural Gas
over Data Stream

Mohammed Ghesmoune, Hanene Azzag, and Mustapha Lebbah

University of Paris 13, Sorbonne Paris City
LIPN-UMR 7030 - CNRS
99, av. J-B Clément – F-93430 Villetaneuse, France
`firstname.secondname@lipn.univ-paris13.fr`

Abstract. Streaming data clustering is becoming the most efficient way
to cluster a very large data set. In this paper we present a new approach,
called G-Stream, for topological clustering of evolving data streams. G-
Stream allows one to discover clusters of arbitrary shape without any
assumption on the number of clusters and by making one pass over the
data. The topological structure is represented by a graph wherein each
node represents a set of "close" data points and neighboring nodes are
connected by edges. The use of the *reservoir*, to hold, temporarily, the
very distant data points from the current prototypes, avoids needless
movements of the nearest nodes to data points and therefore, improving
the quality of clustering. The performance of the proposed algorithm is
evaluated on both synthetic and real-world data sets.

Keywords: Data Stream Clustering, Topological Structure, Growing
Neural Gas.

1 Introduction

Clustering is the problem of partitioning a set of observations into clusters such
that observations assigned in the same cluster are similar (or close) and the inter-
cluster observations are dissimilar (or distant). The other objective of clustering
is to quantify the data by replacing a group of observations (cluster) with one
representative observation (or prototype). A data stream is a sequence of potentially infinite, non-stationary (i.e., the probability distribution of the unknown
data generation process may change over time) data arriving continuously (which
requires a single pass through the data) where random access to data is not feasible and storing all arriving data is impractical. Mining data streams can be
defined as the process of finding a complex structure in a large data. Clustering
data streams requires a process capable of partitioning observations continuously
with restrictions of memory and time. In literature, many data stream algorithms
have been adapted from clustering algorithms, e.g., the density-based method
DbScan [6,8], the partitioning method k-means [1], or the message passing-based
method AP [14]. In this paper, we propose G-Stream (Growing Neural Gas over
Data Stream), a novel algorithm for discovering clusters of arbitrary shape in an
evolving data stream, whose main features and advantages are described below:

C.K. Loo et al. (Eds.): ICONIP 2014, Part I, LNCS 8834, pp. 207–214, 2014.
© Springer International Publishing Switzerland 2014

- The topological structure is represented by a graph wherein each node represents a cluster, which is a set of "close" data points and neighboring nodes (clusters) are connected by edges. The graph size is not fixed but may evolve.
- We use an exponential fading function to reduce the impact of old data whose relevance diminishes over time. For the same reason, links between nodes are also weighted by an exponential function.
- Unlike many other data stream algorithms that start by taking a significant number of data points for initializing the model (these data points can be seen several times), G-Stream starts with only two nodes. Several nodes (clusters) are created in each iteration, the opposite of traditional GNG [7].
- All aspects of G-Stream (including creation, deletion and fading of nodes, edges management, and reservoir management) are performed online.
- A reservoir is used to hold, temporarily, the very distant data points, compared to the current prototypes.

The remainder of this paper is organized as follows: Section 2 is dedicated to related works. Section 3 describes the G-Stream algorithm. Section 4 reports the experimental evaluation on both synthetic and real-world datasets. Section 5 concludes this paper.

2   Related Works

This section discusses previous works on data stream clustering problems, and highlights the most relevant algorithms proposed in literature to deal with this problem. Most of existing algorithms divided the clustering process in two phases: (1) *Online*, the data will be summarized, (2) *Offline*, final clusters will be generated. Both *CluStream* [2] and *DenStream* [6] use a temporal extension of the *Clustering Feature vector* [13] (called *micro-clusters*) to maintain statistical summaries about data locality and timestamps during the online phase. By creating two kinds of micro-clusters (*potential* and *outlier micro-clusters*), *DenStream* overcomes one of the drawbacks of *CluStream*, its sensitivity to noise. In the offline phase, the micro-clusters found during the online phase are considered as *pseudo-points* and will be passed to a variant of k-means in the *CluStream* algorithm (resp. to a variant of DbScan in the *DenStream* algorithm) in order to determine the final clusters. *StreamKM++* [1] maintains a small sketch of the input data using the *merge-and-reduce* technique. The merge step is performed by a means of data structure, named *bucket set*. The reduce step is performed by a significantly different summary data structure, the *coreset tree*. *SOStream* [8] is a density-based clustering algorithm inspired by both the principle of DbScan algorithm and that of the self-organizing maps (SOM) [9]. *E-Stream* [12] classifies the evolution of data into five categories: appearance, disappearance, self evolution, merge, and split. It uses another data structure for saving summary statistics, named α-bin histogram. *StrAP* [14], an extension of the Affinity Propagation algorithm for data stream, uses a reservoir for saving potential outliers. *AING* [5], an incremental GNG that learns automatically the distance thresholds of nodes based on its neighbors and data points assigned to the concerned

Table 1. Comparison between algorithms (WL: weighted links, 2 phases : online+offline)

Algorithms	based on	topology	WL	phases	remove	merge	split	fade
G-Stream	NGas	✓	✓	online	✓	✗	✗	✓
AING	NGas	✓	✗	online	✗	✓	✗	✗
CluStream	k-means	✗	✗	2 phases	✓	offline	✗	✗
DenStream	DbScan	✗	✗	2 phases	✓	offline	✗	✓
SOStream	DbScan, SOM	✗	✗	online	✓	✓	✗	✓
E-Stream	k-means	✗	✗	2 phases	✓	✓	✓	✓
StreamKM++	k-means	✗	✗	2 phases	✓	✓	✓	✓
StrAP	AP	✗	✗	2 phases	✓	✗	✗	✓

node. It merges nodes when their number reaches a given *upper bound*. Table 1 summarizes the main features offered by each algorithm in terms of: basic clustering algorithm, whether the algorithm identifies a topological structure or not, whether links (if those exists) between clusters (nodes) are weighted, how many phases does it adopt (online and offline), operations for updating clusters (remove, merge, and split cluster), and the *fading* function.

3   Growing Neural Gas over Data Stream

In this section we introduce Growing Neural Gas over data Stream (G-Stream) and highlight some of its novel features. G-Stream is based on Growing Neural Gas (GNG), which is an incremental self-organizing approach that belongs to the family of topological maps such as Self-Organizing Maps (SOM) [9] or Neural Gas (NG) [10]. It is an unsupervised algorithm capable of representing a high dimensional input space in a low dimensional feature map. Typically, it is used for finding topological structures that closely reflect the structure of the input distribution. We assume that the data stream consists of a sequence $\mathcal{DS} = \{\mathbf{x}_1, \mathbf{x}_2, ..., \mathbf{x}_n\}$ of n (potentially infinite) data streams arriving in times $T_1, T_2, ..., T_n$, where $\mathbf{x}_i = (x_i^1, x_i^2, ...x_i^d)$ is a vector in $\Re^d$. At each time, G-Stream is represented by a graph $\mathcal{C}$ where each node represents a cluster. Each node $c \in \mathcal{C}$ has a prototype $\mathbf{w}_c = (w_c^1, w_c^2, ...w_c^d)$ (resp. a distance threshold δ_c) representing its position (resp. the distance from the node to the farthest data point assigned to it). Starting with two nodes, and as a new data point is reached, the nearest and the second-nearest nodes are identified, linked by an edge, and the nearest node with and topological neighbors are moved toward the data point. Each node has an accumulated error variable and has weight, which varies over time using Fading function. Using Edge management, one, two or three nodes are inserted into the graph between the nodes with the largest error values. Nodes can also be removed if they are identified as being superfluous. Figure 1 represents a general diagram of the algorithm.

   Fading Function: In most data stream scenarios, more recent data can reflect the emergence of new trends or changes in data distribution [3]. There

Fig. 1. Diagram of G-Stream algorithm.

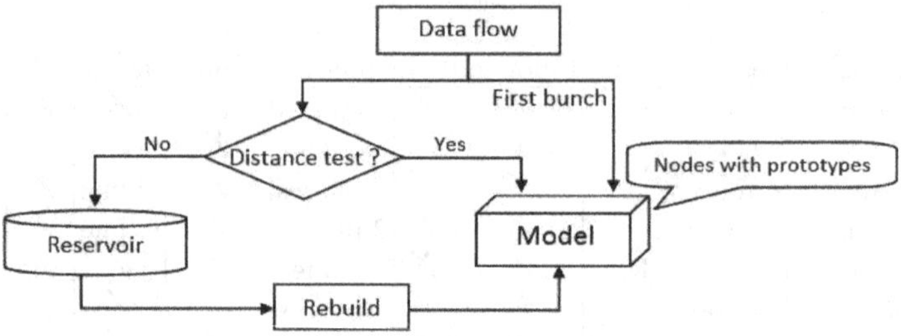

are three models of window commonly studied in the data stream: landmark window, sliding window and damped window. We consider, like many others, the damped window model, in which the weight of each data point decreases exponentially with time t via a fading function $f(t) = 2^{-\lambda_1(t-t_0)}$, where $\lambda_1 > 0$, defines the rate of decay of the weight over time. t denotes the current time and t_0 is the timestamp of the data point. The weight of a node is based on data points associated therewith: $weight(c) = \sum_{i=1}^{m} 2^{-\lambda_1(t-t_{i_0})}$, where m is the number of points assigned to the node c in the current time t. If the weight of a node is less than a parameter value then this node is considered as outdated and then deleted (with its links).

Edge Management: The edge management procedure performs operations related to updating graph edges, as illustrated in steps 13-14 of the algorithm. The way to increase the age of edges is inspired by the fading function in the sense that the creation time of a link is taken into account. Contrary to the *fading* function, the age of the links will be strengthened by the exponential function $2^{\lambda_2(t-t_0)}$, where $\lambda_2 > 0$, defines the rate of growth of the age over time. t denotes the current time and t_0 is the creation time of the edge. The next step is to add a new edge that connects the two closest nodes. The last step is to remove each link exceeding a maximum age, since these links are no longer useful because they were replaced by younger and shorter edges that were created during the graph refinement in steps 15-20.

Reservoir Management: The aim of using the reservoir is to hold, temporarily, the far data points. As mentioned before, each node has a threshold distance. The first bunch of data is assigned to nearest nodes without comparing distances thresholds. The distance threshold of each node is learned by taking the maximum distance of the node to the farthest point that it has been assigned. When the reservoir is full, its data is re-passed to the learning. They are placed in the heap of the data stream, $\mathcal{DS}$, to be dealt with first and the distance thresholds of nodes are updated accordingly.

4 Experimental Evaluations

In this section, we present an experimental evaluation of G-Stream algorithm. We compared our algorithm with the GNG algorithm and two relevant

---

Algorithm 1. G-Stream

    Data: $\mathcal{DS} = \{\mathbf{x}_1, \mathbf{x}_2, ..., \mathbf{x}_n\}$
    Result: set of nodes $\mathcal{C} = \{c_1, c_2, ...\}$ and their prototypes $\mathbf{W} = \{\mathbf{w}_{c_1}, \mathbf{w}_{c_2}, ...\}$

1 Initialize node set $\mathcal{C}$ to contain two nodes, c_1 and c_2: $\mathcal{C} = \{c_1, c_2\}$;
2 **while** *there is a data point to proceed* **do**
3   Get the next data point in the data stream, $\mathbf{x}_i$;
4   Find the nearest node $bmu_1 \in \mathcal{C}$ and the second nearest node $bmu_2 \in \mathcal{C}$;
5   **if** $\|\mathbf{x}_i - \mathbf{w}_{bmu_1}\| > \delta_{bmu_1}$ **then**
6   put $\mathbf{x}_i$ in the reservoir;
7   **if** *the reservoir is full* **then** Reservoir management
8   **else**
9   Increment the number of points assigned to bmu_1;
10        $error(bmu_1) = error(bmu_1) + \|\mathbf{x}_i - \mathbf{w}_{bmu_1}\|^2$;
11   Move bmu_1 and its topological neighbors towards $\mathbf{x}_i$:
          $\mathbf{w}_{bmu_1} = \mathbf{w}_{bmu_1} + \alpha_1.\|\mathbf{x}_i - \mathbf{w}_{bmu_1}\|$;
12        $\mathbf{w}_c = \mathbf{w}_c + \alpha_2.\|\mathbf{x}_i - \mathbf{w}_c\|$ for all direct neighbors c of node bmu_1;
13   Increment the age of all edges emanating from bmu_1 and weight them;
14   **if** *bmu_1 and bmu_2 are connected by an edge* **then** set the age of this edge to zero **else** create an edge between bmu_1 and bmu_2, and mark its time stamp Remove edges whose age is greater than age_{max};
15   **if** *the number of points passed is multiple of a parameter β* **then**
16   **for** *i=1 to 3* **do**
17   Find node q with the maximum accumulated error;
18   Find the neighbor f of q with the largest accumulated error;
19   Add the new node, r, half-way between nodes q and f;
20   Insert edges connecting the new node r with nodes q and f, and remove the original edge between q and f;
21   Application of *fading*, delete outdated and isolated nodes;
22   Finally, decrease the error of all units;

---

data stream algorithms. Our experiments were performed on MATLAB platform using real-world and synthetic datasets. Table 2 overviews all the datasets used. The real-world databases were taken from the UCI repository [4]. DS1 is generated by `http://impca.curtin.edu.au/local/software/synthetic-data-sets.tar.bz2`. Uniform is generated with matlab code. The letter4 dataset is generated by a Java code `https://github.com/feldob/Token-Cluster-Generator`. The algorithms are evaluated using three performance measures: Rand, Normalized Mutual Information and Accuracy (Purity) with the aim of maximizing each measure [11]. As explained in section 3, GNG and G-Stream algorithms start with two nodes. We used an online version of GNG but without the parameters that we added and this, precisely, to show the interest and contribution of these parameters in G-Stream. Therefore, we did experiments by initializing two nodes randomly among the first 20 points and we repeated this 10 times. We used the same initialization for both algorithms (G-Stream and GNG) and the average

Table 2. Overview of all datasets

Datasets	size	#features	#classes
DS1	9153	2	14
Uniform	24000	2	4
letter4	9344	2	7
Shuttle	43500	9	7
L-recognition	20000	16	26
KddCup1	49402	34	18

value with its standard deviation is reported in Table 3. These results show that G-Stream algorithm outperforms the GNG algorithm on almost all the datasets. For comparison purpose, we used CluStream with **stream** R package `http://cran.r-project.org/web/packages/stream/index.html`. Comparison is also performed with StreamKM++ [1]. Results are reported in Table 3. Again, with reference to Table 3, it is noticeable that G-Stream's Accuracies (Acc) are higher for all datasets as compared to GNG, StreamKM++, and CluStream. The NMI values are also higher than the other algorithms except for CluStream in DS1, and for the two data stream algorithms in Uniform. The Rand values are also higher than the other algorithms except in Uniform and shuttle. We remind that G-Stream proceeds in one single phase whereas CluStream and StreamKM++ proceed in two phases (online and offline phase).

Table 3. Comparing G-Stream with different algorithms

Datasets		GNG	**G-Stream**	StreamKM++	CluStream
DS1	Acc	0.511±0.251	**0.993±0.006**	0.675±0.018	0.701±0.028
	NMI	0.491±0.132	0.712 ±0.004	0.702±0.021	0.723±0.022
	Rand	0.621±0.122	**0.846±0.001**	0.844±0.004	0.845±0.007
Uniform	Acc	1 ± 0	**1 ± 0**	0.998±0.004	0.995±0.012
	NMI	0.492±0	0.568±0.003	0.777±0.007	0.787±0.015
	Rand	0.754±0	0.765±0	0.855±0.003	0.868±0.011
letter4	Acc	0.577±0.201	**0.991±0**	0.687±0.026	0.934±0.026
	NMI	0.529±0.074	**0.607±0**	0.553±0.022	0.264±0.034
	Rand	0.686±0.084	**0.812±0**	0.794±0.014	0.341±0.004
Shuttle	Acc	0.963±0.002	**0.973±0.004**	0.822±0.003	0.899±0.017
	NMI	0.355±0	**0.362±0.007**	0.258±0.015	0.340±0.035
	Rand	0.378±0.001	0.376±0.001	0.753±0.039	0.559±0.059
L-recognition	Acc	0.077±0.068	**0.408±0.019**	0.161±0.009	0.181±0.009
	NMI	0.046±0.096	**0.437±0.015**	0.239±0.015	0.267±0.011
	Rand	0.541±0.139	**0.956±0.001**	0.861±0.006	0.851±0.007
KddCup1	Acc	0.929±0.085	**0.998±0.001**	0.768±0	0.998±0
	NMI	0.655±0.319	**0.602±0.032**	0.012±0.003	0.022±0.002
	Rand	0.824±0.206	**0.655±0.045**	0.623±0.003	0.369±0.083

Figure 2a (resp. Figure 2b) compares G-Stream (red line with circle) with GNG (blue line with cross) with respect to accuracy (resp. RMS error, number of nodes). On almost all times, the accuracy value (resp. RMS error) of G-Stream is higher (resp. is less) than the one of GNG. Figure 2c compares the two algorithms in terms of number of nodes creating the graph. Despite this we create several nodes at each iteration (against a single node for GNG), the number of nodes created by G-Stream becomes steady (against a continuously increase for GNG) due to the application of the fading function. The same result can be shown on the rest of the datasets. The second row of Figure 2 shows the

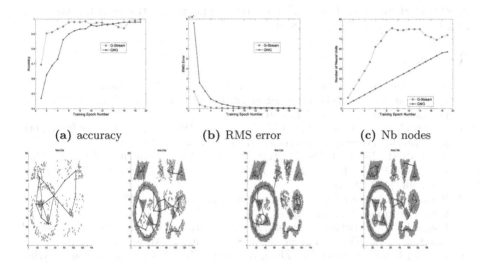

(a) accuracy   (b) RMS error   (c) Nb nodes

Fig. 2. DS1 Experimentation. Accuracy, RMS error, and number of nodes for G-Stream on DS1. The second row shows visual result of G-Stream on DS1 (dataset and topological result).

evolution of the creation of nodes by applying G-Stream on DS1 (green points represent data points of the data stream and blue ones are nodes of the graph with edges in blue lines). It illustrates that G-Stream manages to recognize the structures of the data stream and can separate these structures with the best visualization. Due to space limitations, we omitted the visual results about the other datasets since they have the same interpretation as that we have shown.

5   Conclusion

In this paper, we have proposed G-Stream, an efficient method for topological clustering an evolving data stream in an online manner. In G-Stream, nodes are weighted by a fading function as well as edges by an exponential function. Starting with two nodes, G-Stream confronts the arriving data points to the current prototypes, storing the very distant ones in a reservoir, learns the threshold distances automatically, and many nodes are created in each iteration. Experimental evaluation over a number of real and synthetic data sets demonstrates the effectiveness and efficiency of G-Stream in discovering clusters of arbitrary shape. Our experiments show that G-Stream outperformed the GNG algorithm in terms of visual results and criteria as accuracy, rand and NMI. G-Stream is also compared, in terms of clustering quality, to two relevant data stream algorithms, results are promising. We plan in future to make adaptive windows, make our algorithm as autonomous as possible and develop it in Spark or Storm for testing on large datasets with other data stream algorithms.

Acknowledgments. This work has been supported by the French foundation PIA Grant Big data "Square Predict".

References

1. Ackermann, M.R., Märtens, M., Raupach, C., Swierkot, K., Lammersen, C., Sohler, C.: Streamkm++: A clustering algorithm for data streams. ACM Journal of Experimental Algorithmics 17(1) (2012)
2. Aggarwal, C.C., Watson, T.J., Ctr, R., Han, J., Wang, J., Yu, P.S.: A framework for clustering evolving data streams. In: VLDB, pp. 81–92 (2003)
3. de Andrade Silva, J., Faria, E.R., Barros, R.C., Hruschka, E.R., de Carvalho, A.C.P.L.F., Gama, J.: Data stream clustering: A survey. ACM Comput. Surv. 46(1), 13 (2013)
4. Bache, K., Lichman, M.: UCI machine learning repository (2013), http://archive.ics.uci.edu/ml
5. Bouguelia, M.R., Belaïd, Y., Belaïd, A.: An adaptive incremental clustering method based on the growing neural gas algorithm. In: ICPRAM, pp. 42–49 (2013)
6. Cao, F., Ester, M., Qian, W., Zhou, A.: Density-based clustering over an evolving data stream with noise. In: SDM, pp. 328–339 (2006)
7. Fritzke, B.: A growing neural gas network learns topologies. In: NIPS, pp. 625–632 (1994)
8. Isaksson, C., Dunham, M.H., Hahsler, M.: SOStream: Self organizing density-based clustering over data stream. In: Perner, P. (ed.) MLDM 2012. LNCS (LNAI), vol. 7376, pp. 264–278. Springer, Heidelberg (2012)
9. Kohonen, T., Schroeder, M.R., Huang, T.S. (eds.): Self-Organizing Maps, 3rd edn. Springer-Verlag New York, Inc., Secaucus (2001)
10. Martinetz, T., Schulten, K.: A "Neural-Gas" Network Learns Topologies. In: Artificial Neural Networks I, pp. 397–402 (1991)
11. Strehl, A., Ghosh, J.: Cluster ensembles — a knowledge reuse framework for combining multiple partitions. Journal of Machine Learning Research 3, 583–617 (2002)
12. Udommanetanakit, K., Rakthanmanon, T., Waiyamai, K.: E-stream: Evolution-based technique for stream clustering. In: Alhajj, R., Gao, H., Li, X., Li, J., Zaïane, O.R. (eds.) ADMA 2007. LNCS(LNAI), vol. 4632, pp. 605–615. Springer, Heidelberg (2007)
13. Zhang, T., Ramakrishnan, R., Livny, M.: Birch: An efficient data clustering method for very large databases. In: SIGMOD Conference, pp. 103–114 (1996)
14. Zhang, X., Furtlehner, C., Sebag, M.: Data streaming with affinity propagation. In: Daelemans, W., Goethals, B., Morik, K. (eds.) ECML PKDD 2008, Part II. LNCS (LNAI), vol. 5212, pp. 628–643. Springer, Heidelberg (2008)

Combining Active Learning and Semi-supervised Learning Using Local and Global Consistency

Yingjie Gu[1,2], Zhong Jin[1], and Steve C. Chiu[2]

[1] Computer Science and Engineering, Nanjing University of Science and Technology,
Nanjing 210094, China
csyjgu@gmail.com, zhongjin@njust.edu.cn
[2] Department of Electrical Engineering, Idaho State University,
Pocatello 83209-8060, USA
chiustev@isu.edu

Abstract. Semi-supervised learning and active learning are important techniques to solve the shortage of labeled examples. In this paper, a novel active learning algorithm combining semi-supervised Learning with Local and Global Consistency (LLGC) is proposed. It selects the example that can minimize the estimated expected classification risk for labeling. Then, a better classifier can be trained with labeled data and unlabeled data using LLGC. The experiments on two datasets demonstrate the effectiveness of the proposed algorithm.

Keywords: Active learning, semi-supervised learning, image classification.

1 Introduction

In traditional machine learning approaches to classification, only labeled examples are used to train the classifier. But in many real-world applications, there is a large number of unlabeled examples. Whereas labeled examples are usually difficult and expensive to obtain. Two typical methods to address this problem are semi-supervised learning [1] and active learning [2]. Semi-supervised learning combines both labeled examples and unlabeled examples to train a better classifier. Active learning usually selects a set of unlabeled instances for experts labeling, a better classifier can be trained by labeled examples afterwards.

The kernel of active learning is how to measure examples' value and which examples should be selected for labeling. There are many criteria in active learning to instruct examples selection. Uncertainty sampling is one of the most widely used criterion that queries the examples whose labels are most uncertain under the current classifier. Other criteria like variance reduction [3], Expected Model Change [4], Expected Error Reduction [5][6], and diversity [7] have also been widely applied to active learning.

With the same number of labeled examples, both active learning and semi-supervised learning usually perform better than supervised learning. It may make sense to utilize active learning in conjunction with semi-supervised learning.

C.K. Loo et al. (Eds.): ICONIP 2014, Part I, LNCS 8834, pp. 215–222, 2014.
© Springer International Publishing Switzerland 2014

Specifically, we firstly select a set of unlabeled examples to be labeled by experts. Then, both labeled examples and unlabeled examples are used to train classifiers. In [5], Zhu et al. combined active learning and semi-supervised learning using Gaussian Fields and Harmonic Functions (GFHF). Active learning is performed on top of the semi-supervised learning scheme by selecting examples to minimize the estimated expected classification risk.

Since Learning with Local and Global Consistency (LLGC) [8] presents a promising performance in semi-supervised learning, we explore the combination of active learning and LLGC in this paper. In active learning process, the example which can minimize the estimated expected classification risk is selected to be labeled. Then, a classifier is learned by LLGC with labeled data and unlabeled data. The experiments of image classification on two datasets demonstrate the effectiveness of the proposed algorithm.

The rest of this paper is organized as follows: In Section 2, we review semi-supervised Learning with Local and Global Consistency. The combination of active learning and LLGC is introduced in Section 3. In Section 4, we present the experimental settings and results. Finally, the conclusion and future work are discussed in Section 4.

2   Semi-supervised Learning with Local and Global Consistency

We begin by briefly describing the semi-supervised learning method LLGC [8]. Suppose there are l labeled examples $(x_1, y_1), ..., (x_l, y_l)$ and u unlabeled examples $x_{l+1}, ..., x_{l+u}$; usually $l \ll u$. y_i is the label of example x_i. For a c-class classification problem, $y_i \in \{1, 2, ..., c\}, i = 1, ..., l$. The labeled set and unlabeled set are denoted by $\mathcal{L}$ and $\mathcal{U}$, and $n = l + u$. The goal is to predict the labels of the unlabeled examples.

Let $\mathcal{F}$ denote the set of $n \times c$ matrices with nonnegative entries. Define a $n \times c$ matrix $Y \in \mathcal{F}$ with $Y_{ij} = 1$ if x_i is labeled as $y_i = j$ and $Y_{ij} = 0$ otherwise. A matrix $F \in \mathcal{F}$ is a matrix that labels all examples x_i with a label $y_i = argmax_{j \leq c} F_{ij}$. If F is defined as $F = [F_1^T, ..., F_n^T]^T$, F can be understander as a vectorial function which assigns a vector F_i to each example x_i. The LLGC algorithm is as follows:

1. Constrcut the affinity matrix W defined by $W_{ij} = exp(-||x_i - x_j||^2/2\sigma^2)$ if $i \neq j$ and $W_{ij} = 0$ if $i = j$.
2. Compute $S = D^{-1/2}WD^{-1/2}$ where D is a diagonal matrix with $D_{ii} = \sum_{j=1}^{n} W_{ij}$.
3. Iterate $F(t+1) = \sigma S F(t) + (1-\sigma)Y$ until convergence, where σ is a parameter in $(0, 1)$.
4. Define $F^* = \lim_{t \to \infty} F(t)$. The label of x_i is predicted as $y_i = \arg\max_{j \leq c} F_{ij}^*$.

We firstly construct a graph $G = (V, E)$ on $\mathcal{L} \cup \mathcal{U}$, where the vertex set V is the set of all examples and the edges E are weighted by W. Then, the weight matrix W is normalized symmetrically. In the iteration, each examples receives information from its neighbors (first term), and retains its initial information (second term). The information is spread symmetrically since S is a symmetric matrix. Finally, the label of each unlabeled examples is predicted as the class of which it has received most information during the iteration process.

By computing the limit of the sequence $\{F(t)\}$, we can obtain

$$F^* = (1 - \alpha)(I - \alpha S)^{-1}Y \tag{1}$$

for classification, which is equivalent to

$$F^* = QY \tag{2}$$

where $Q = (I - \alpha S)^{-1}$. Since S is fixed, Q is also fixed in the learning process.

A regularization framework was also proposed by Zhou et al. for this method. The cost function associated with F with regularization parameter $\mu > 0$ is defined as

$$\mathcal{Q}(F) = \frac{1}{2}(\sum_{i,j=1}^{n} W_{ij} \| \frac{1}{\sqrt{D_{ii}}}F_i - \frac{1}{\sqrt{D_{jj}}}F_j \|^2 + \mu \sum_{i=1}^{n} \|F_i - Y_i\|^2) \tag{3}$$

The optimal decision function is $F^* = \arg\min_{F \in \mathcal{F}} \mathcal{Q}(F)$. More on this semi-supervised learning framework can be found in [8].

3   Active Learning

In this section, we propose to perform active learning with LLGC. The basic idea of the proposed active learning is to select the example that can minimize the classification risk of the examples.

With both labeled examples and unlabeled examples, we can train a classifier (decision function F) using LLGC. The class of unlabeled example x_i is predicted as $y_i = \arg\max_{j \leq c} F^*_{ij}$. Suppose $P(y_i|x_i)$ is the probability distribution of the examples' labels. We assume that the distribution $P(y_i|x_i)$ can be estimated based on decision function F.

$$P(y_i = j|x_i) = \frac{F_{ij}}{\sum_{t=1}^{c} F_{it}} \tag{4}$$

We define the true risk $\mathcal{R}(P)$ of the classification based on labels' distribution P. Thus

$$\mathcal{R}(P) = \sum_{i=1}^{n}(1 - \max_{j=1,...,c} P(y_i = j|x_i)) \tag{5}$$

If we perform active learning to select an unlabeled example x_k for experts labeling, we will receive an answer y_k^* ($y_k^* \in \{1, ..., c\}$). Before we selecting x_k for labeling, $Y_{kj} = 0$ ($j = 1, ..., c$). After labeling x_k and adding (x_k, y_k^*) to labeled set, the matrix Y should be updated and denoted by $Y^{+(x_k, y_k^*)}$ where

$Y_{k,y_k^*}^{+(x_k,y_k^*)} = 1$. The decision function F and the probability distribution P will also change

$$F^{+(x_k,y_k^*)} = QY^{+(x_k,y_k^*)} \tag{6}$$

$$P^{+(x_k,y_k^*)}(y_i = j|x_i) = \frac{F_{ij}^{+(x_k,y_k^*)}}{\sum_{t=1}^{c} F_{it}^{+(x_k,y_k^*)}} \tag{7}$$

If (x_k, y_k^*) is added to the labeled set, the estimated classification risk is

$$\mathcal{R}(P^{+(x_k,y_k^*)}) = \sum_{i=1}^{n}(1 - \max_{j=1,\dots,c} P^{+(x_k,y_k^*)}(y_i = j|x_i)) \tag{8}$$

Before we querying experts about the label of x_k, the true label y_k^* is unknown. But we can obtain the labels' distribution $P(y_i|x_i)$ from decision function F. Therefore, the expected classification risk after querying x_k is estimated as

$$\mathcal{R}(P^{+x_k}) = \sum_{j=1}^{c} P(y_k = j|x_k)\mathcal{R}(P^{+(x_k,j)}) \tag{9}$$

We aim to select the example that can minimize the expected estimated risk. Therefore, the index of the selected example is

$$s = \underset{k \in \{l+1,\dots,n\}}{\arg\min} \mathcal{R}(P^{+x_k}) \tag{10}$$

Once the label y_s^* of the example x_s is queried from experts, (x_s, y_s^*) will be added to the labeled set. The label matrix Y will be updated to $Y^{+(x_s,y_s^*)}$ and the decision function will be retrained by equation (6). In fact, the update operation of label matrix Y is only to change one element in Y, namely set Y_{s,y_s^*} to be 1. The retraining step $F^{+(x_s,y_s^*)} = QY^{+(x_s,y_s^*)}$ is equivalent to update the y_s^*-th column of the matrix F.

$$F_{\cdot y_s^*}^{+(x_s,y_s^*)} = F_{\cdot y_s^*} + Q_{\cdot y_s^*} \tag{11}$$

where $F_{\cdot y_s^*}$ and $Q_{\cdot y_s^*}$ denote the y_s^*-th column of matrices F and Q. Of course $F_{\cdot j}^{+(x_s,y_s^*)} = F_{\cdot j}$ if $j \neq y_s^*$. It is easy to prove that the equation (6) is equivalent to equation (11). But the computation of equation (11) is much faster than equation (6).

The process of the proposed active learning combining LLGC is concluded in Table 1. It is the procedure of selecting one example for experts labeling. In applications, the examples selection often repeats many times until the stop criterion is reached.

4   Experiment

In order to assess the effectiveness of the proposed technique, we evaluate and compare five active learning methods:

Table 1. The process of the proposed active learning algorithm

---

Input:
    Initial labeled data set $(x_1, y_1), ..., (x_l, y_l)$, unlabeled data set $x_{l+1}, ..., x_{l+u}$,
    the guassian kernel parameter σ, the tradeoff parameter α
Output:
    The selected example
Procedure:
    Construct label matrix Y, compute weight matrix W and S, Q, F
    For $k = l + 1 : n$
        For $y_k = 1 : c$
$$F^{+(x_k, y_k)} = F, \quad F^{+(x_k, y_k)}_{\cdot y_k} = F_{\cdot y_k} + Q_{\cdot y_k}$$
$$P^{+(x_k, y_k)}(y_i = j | x_i) = \frac{F^{+(x_k, y_k)}_{ij}}{\sum_{t=1}^{c} F^{+(x_k, y_k)}_{it}}$$
$$\mathcal{R}(P^{+(x_k, y_k)}) = \sum_{i=1}^{n}(1 - \max_{j=1,...,c} P^{+(x_k, y_k)}(y_i = j | x_i))$$
        End
$$\mathcal{R}(P^{+x_k}) = \sum_{j=1}^{c} P(y_k = j | x_k)\mathcal{R}(P^{+(x_k, j)})$$
    End
$$s = \underset{k \in \{l+1,...,n\}}{\arg\min} \mathcal{R}(P^{+x_k})$$
Return x_s

- Random Sampling with LLGC classifier (RS+LLGC), which randomly selects examples for labeling and uses LLGC classifier.
- Most Uncertain with LLGC classifier (MU+LLGC), which selects the most uncertain example from LLGC classifier for labeling. The index of the most uncertain example is

$$s = \underset{i=l+1,...,n}{\arg\min} F_{ij_1} - F_{ij_2} \tag{12}$$

where $j_1 = \underset{j=1,...,c}{\arg\max} F_{ij}$, $j_2 = \underset{j=1,...,c, j \neq j_1}{\arg\max} F_{ij}$.

- Multiclass-level uncertainty with SVM classifier (MCLU+SVM), which was proposed in [9].
- MinRisk+GFHF, which was proposed in [5].
- MinRisk+LLGC, which is proposed in this paper. The parameter α is set to 0.99 and σ is set to 0.1.

In the following sections, we carry out classification experiments on two real-world data sets to compare different active learning algorithms quantitatively.

4.1 Handwritten Digits Recognition

The USPS handwritten digits data set is used in this experiment. The data set contains 8-bit gray-scale images of '0' through '9'. The size of each image is 16×16 pixels. Thus, each digit image is represented as a 256-dimensional vector.

On this data set, we used digits 1, 2, 3, and 4 in our experiments as the four classes. 500 examples from each class are randomly selected so there are totally 2000(500 × 4) examples. Only 1 example from each class is randomly selected as initial labeled example. Thus there are 4 labeled examples and 1996 unlabeled examples. We apply each active learning algorithm to select k ($k = 1, 2, ..., 10$) examples for labeling. A classifier can be trained with LLGC or SVM method. Lastly, we predict the labels of the rest unlabeled examples and compute the classification accuracy. The experiments are repeated for 30 times and the average accuracy is obtained.

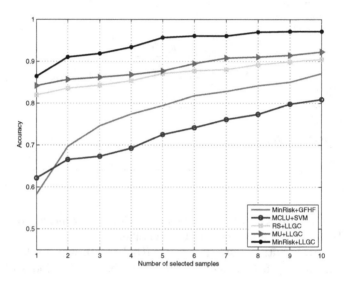

Fig. 1. The average classification accuracy on usps dataset

Fig.1 shows the average classification accuracy versus the number of examples selected by active learning methods. As can be seen, our MinRisk+LLGC algorithm significantly outperforms the other active learning algorithms. MU+LLGC performs the second best. The active learning combining GFHF (MinRisk+ GFHF) is not better than our proposed method. MCLU+SVM is worse than others since it is unable to use unlabeled examples to train a classifier.

4.2   Terrain Classification

In this section, we apply active learning algorithms to terrain classification problems. Terrain image dataset used in the experiment was constructed by us from the Outex Database [10], which consists of two data sets: Outex-0 and Outex-1. Each of them includes 20 outdoor scene images and the size of each image is 2272 × 1704. The images are marked as one type of bush, grass, tree, sky, road, and building. The marked area of each image is cut into patches with size 64 × 64 and each patch is regarded as an example. Two examples of each class are shown in Fig. 2. Both color histogram feature and LBP feature are extracted and combined to represent each example. We extract 100 patches of each class (totally 600 patches) to construct a pool of unlabeled data set for examples selection. Firstly, only 1 example of each class is labeled as initial labeled set. Then, active learning is used to select k ($k = 1, 2, ..., 10$) examples for labeling. Lastly, a classifier is trained and the labels of the unlabeled examples are predicted.

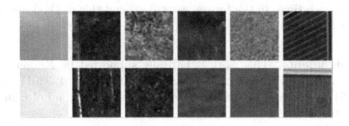

Fig. 2. Examples of Outex from categories: sky, tree, bush, grass, road, and building

The average classification accuracies on Outex-0 and Outex-1 are shown in Fig.3. As can been seen, our MinRisk+LLGC outperforms the other algorithms in most of the cases. MinRisk+GFHF performs the second best on Outex-0 while worse than MU+LLGC on Outex-1. MCLU+SVM performs the worst on two datasets since it is a supervised learning method that does not use unlabeled data in learning.

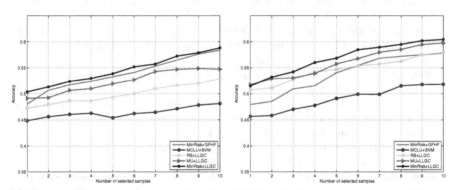

(a) The classification accuracy on Outex-0 (b) The classification accuracy on Outex-1

Fig. 3. The average classification accuracy on Outex-0 and Outex-1

To sum up, semi-supervised learning (LLGC, GFHF) performs better than supervised learning (SVM) with the same number labeled examples. Our proposed MinRisk+LLGC outperforms MinRisk+GFHF, MU+LLGC, and RS+LLGC in most of the cases.

5 Conclusion

In this paper, a novel active learning algorithm which combining semi-supervised learning with LLGC is proposed. The example that can minimize the estimated expected classification error is selected for labeling. Experiments on two datasets indicate that the proposed algorithm can be highly effective.

MinRisk+LLGC is a single-mode active learning algorithm that selects only one example each time. In the future, we will expend this method into a batch-mode active learning.

Acknowledgements. This work is partially supported by National Natural Science Foundation of China under Grant Nos. 61373063, 61233011, 61125305, 61375007, 61220301, and by National Basic Research Program of China under Grant No. 2014CB349303.

References

1. Zhu, X.: Semi-supervised learning literature survey. Computer Science 2, 3 (2006)
2. Settles, B.: Active learning literature survey. University of Wisconsin, Madison (2010)
3. Ji, M., Han, J.: A variance minimization criterion to active learning on graphs. In: International Conference on Artificial Intelligence and Statistics, pp. 556–564 (2012)
4. Settles, B., Craven, M., Ray, S.: Multiple-instance active learning. In: Advances in Neural Information Processing Systems, NIPS, pp. 1289–1296. MIT Press (2008)
5. Zhu, X., Lafferty, J., Ghahramani, Z.: Combining active learning and semi-supervised learning using gaussian fields and harmonic functions. In: ICML 2003 Workshop on the Continuum from Labeled to Unlabeled Data in Machine Learning and Data Mining, pp. 58–65 (2003)
6. Guo, Y., Greiner, R.: Optimistic active-learning using mutual information. In: IJ-CAI, vol. 7, pp. 823–829 (2007)
7. Chakraborty, S., Balasubramanian, V., Panchanathan, S.: Dynamic batch mode active learning. In: 2011 IEEE Conference on Computer Vision and Pattern Recognition (CVPR), pp. 2649–2656. IEEE (2011)
8. Zhou, D., Bousquet, O., Lal, T.N., Weston, J., Schölkopf, B.: Learning with local and global consistency. In: NIPS, vol. 16, pp. 321–328 (2003)
9. Demir, B., Persello, C., Bruzzone, L.: Batch-mode active-learning methods for the interactive classification of remote sensing images. IEEE Transactions on Geoscience and Remote Sensing 49(3), 1014–1031 (2011)
10. University of oulu texture database, http://www.outex.oulu.fi/temp/

Complex-Valued Neural Networks – Recent Progress and Future Directions

(Invited Paper)

Akira Hirose

Department of Electrical Engineering and Information Systems, The University of Tokyo
7-3-1 Hongo, Bunkyo-ku, Tokyo 113-8656, Japan
ahirose@ee.t.u-tokyo.ac.jp
http://www.eis.t.u-tokyo.ac.jp/

Abstract. This invited paper presents and discusses the recent progress, present and prospective applications, and the future directions of complex-valued neural networks (CVNNs) including hypercomplex-valued neural networks (HVNNs).

1 Introduction

Complex-valued neural networks (CVNNs), including hypercomplex-valued networks (HVNNs), have been making extensive progress in these years. Correspondingly some special issues have also been planned and published in, e.g., IEEE Transactions on Neural Networks and Learning Systems [1]. Special sessions are also held constantly in conferences such as ICONIP and IJCNN. In addition, many books have been published [2] [3] [4] [5] [6] [7]. This invited paper presents the recent progress and the future directions of CVNNs by describing two technological points: compatibility with the wave nature and the sparsity existing essentially in a complex number. Besides, we review one of the most recent advanced applications in quaternion neural networks (QNNs), a type of HVNNs, applied in the field of radar imaging to contribute to solving environmental issues [8]

2 Features Specific to CVNNs: Two Key Points among Others

2.1 Compatibility with the Wave Nature

Fig. 1 is a diagram showing the specific features and application fields of CVNNs. CVNNs show excellent generalization characteristics in particular to deal with waves such as electromagnetic, sonic and generally quantum waves as well as to process wave-related information [9] [10]. Focusing on the wave nature, we can trace back the CVNN history to the middle of the 20th century. The first introduction of phase information into computational systems was made by Eiichi Goto in 1954 in his invention of "Parametron" [11] [12] [13]. He utilized the phase of a high-frequency carrier to represent binary or multivalued information. However, the computational principle employed there was "logic" of Turing type, or von Neumann type, based on symbol processing, so that he was indifferent to making further extensive use of the phase.

C.K. Loo et al. (Eds.): ICONIP 2014, Part I, LNCS 8834, pp. 223–230, 2014.

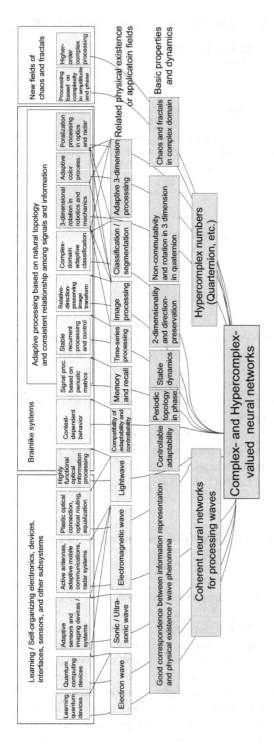

Fig. 1. Specific features and application fields of complex- and hypercomplex-valued neural networks classified by physical existence as well as neural dynamics (Ref. [3], modified)

In the present CVNN fields, contrarily, researchers extend the world of computation to pattern processing based on a novel use of the structure of complex-amplitude (phase and amplitude) information. This is an important technological point also in optical and/or quantum computing to develop new algorithms where the physical field exists as the superimposition of quantum wave [14] [15] [16] [17] [18] [19] [20].

The application fields are very wide in a diverse range of wave engineering such as wireless communications [21] [22] [23] [24] [25] [26] [27] [28] [29], electromagnetic-wave imaging [30] [31] [32] [33] [34], in particular, in ground penetrating radar systems [35] [36] [37] [38] [39] [40] and ultrasonic sensing [41] [42] [43]. Even in the processing of ordinary images and signals, the CVNNs work very effectively when we employ a periodic representation of information in frequency or real domains [44] [45]. Wave informatics will remain the most important application field.

2.2   The Sparsity Existing Essentially in the Number Itself

The origin of the wave compatibility of complex number is "sparsity in itself," i.e., sparsity lying in the number representation. We follow the discussion in Ref.[9]. As we focus on multiplication out of the four arithmetic operations of complex numbers, we can represent a complex number as a real 2×2 matrix. That is, with every complex number $c = a + jb$, where a and b are real numbers and j is imaginary unit, we associate a C-linear transformation T_c of $z = x + jy$ as

$$T_c : C \to C, \quad z \mapsto cz = ax - by + j(bx + ay) \tag{1}$$

If we identify C with R^2 by

$$z = x + jy = \begin{pmatrix} x \\ y \end{pmatrix} \tag{2}$$

it follows that

$$T_c \begin{pmatrix} x \\ y \end{pmatrix} = \begin{pmatrix} ax - by \\ bx + ay \end{pmatrix} = \begin{pmatrix} a & -b \\ b & a \end{pmatrix} \begin{pmatrix} x \\ y \end{pmatrix} \tag{3}$$

In other words, the linear transformation T_c determined by $c = a + jb$ is expressed by a matrix that means phase rotation and magnitude amplification or attenuation as

$$\begin{pmatrix} a & -b \\ b & a \end{pmatrix} = r \begin{pmatrix} \cos\theta & -\sin\theta \\ \sin\theta & \cos\theta \end{pmatrix} \tag{4}$$

where $r \equiv \sqrt{a^2 + b^2}$ and $\theta \equiv \arctan b/a$ denote amplification or attenuation of amplitude and rotation angle applied to the complex signal z, respectively.

Let us consider a set of weights in a neural network. In comparison with the fact that a general 2×2 matrix has four independent parameters (complonents), the above relationship means the reduction of the parameter number to two. This is the sparsity that the complex representation possesses in itself. The sparsity brings about the excellent generalization ability. The utilization of this essential sparsity is one of the key points in the future development.

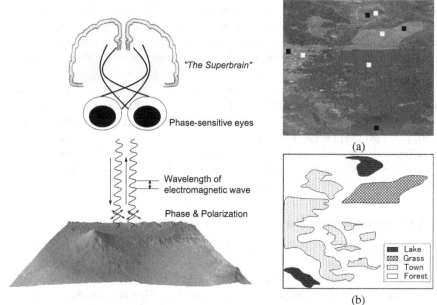

Fig. 2. Geographical profile acquisition using phase- and polarization-sensitive eyes and a *superbrain* brought up with the special eyes [3] (modified)

Fig. 3. (a) Google photo of Fujisusono area with teacher areas indicated by rectangles, and (b) its sketch [8]

3   Quaternion Neural Networks in Adaptive Polarimetric Synthetic Aperture Radar Systems

As described above, radar imaging is one of the most important technological fields, to which the CVNNs contribute actively and widely, since microwave and millimeter-wave sense and carry information of imaging targets not only in its amplitude but also in its phase and polarization. Adaptive complex-valued filters play a significantly important role. Among others, CVNN framework is extremely useful for elimination of interference noise and distortion [46] [47] [48].

Here, in this paper, we briefly review an recently developed adaptive land-vegetation classification technique based on a quaternion neural network (QNN) in a polarimetric synthetic aperture radar [8] [49] [50] [51] . This type of airborne/satellite-borne radar system realizes the so-called full polarimetric observation by transmitting horizontally and vertically polarized waves one after the other, and by receiving waves scattered at earth surface with horizontal and vertical polarization antennas simultaneously [52]. The state of polarization (SoP) is represented on/in Poincare sphere space without excess or deficiency. The change or difference of SoP is presented by the motion/difference of vectors mostly on the Poincare sphere, or on a spherical shell inside, in three dimension. This fact leads to a high compatibility of SoP with QNNs since QNNs deal with three-dimensional rotations with excellent generalization ability.

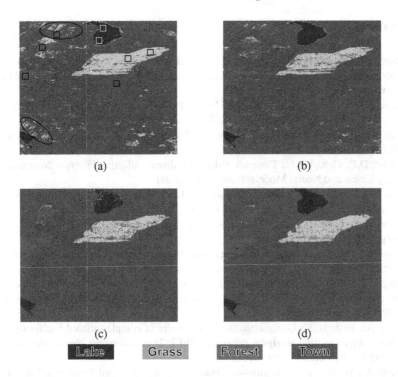

Fig. 4. Classification results for the Fujisusono area generated by (a) Coherency matrix with Wishart classifier (C-Wishert, conventional); (b) Coherency matrix with real-valued neural network (C-RVNN); (c) Poincare vectors with real-valued neural network (P-RVNN); and (d) Pincare vectors with quaternion neural network (P-QNN) [8]

Fig. 2 is an example of observation area. A QNN classifies the earth vegetation adaptively based on the full polarimetric imaging data with supervised learning. Fig. 4 compares the result with conventional ones. That is, Fig. 4(a) presents the result of conventional method combining coherency matrix representation and Wishert classification, while (b) the result of coherency matrix with a real-valued neural network (RVNN), (c) the result of Poincare sphere representation with a RVNN, and finally (d) the result of Poincare representation with a QNN. The QNN result is found clearly superior to others. This technology will contribute to solving CO_2 and hydrological cycle issues as well as to disaster monitoring. Please refer to, e.g., Ref. [8] for details.

4   Summary

We briefly presented and discussed the recent progress and the future directions of the complex-valued neural networks including the hypercomplex-valued networks. We also reviewed a quaternion neural-network based polarimetric radar imaging system that shows excellent vegetation classification ability.

Acknowledgment. This work was partly supported by KDDI Foundation.

References

1. Hirose, A., Aizenberg, I., Mandic, D.P. (Guest-eds.): Special Issue on Complex- and Hypercomplex-Valued Neural Networks. IEEE Transactions on Neural Networks and Learning Systems (September 2014)
2. Hirose, A (ed.): Complex-Valued Neural Networks: Advances and Applications. IEEE Press Series on Computational Intelligence. IEEE Press and Wiley, New Jersey, U.S.A (2013)
3. Hirose, A.: Complex-Valued Neural Networks, 2nd edn. SCI, vol. 400. Springer, Heidelberg (2012)
4. Mandic, D.P., Goh, V.S.L.: Complex Valued Nonlinear Adaptive Filters – Noncircularity, Widely Linear and Neural Models. Wiley (April 2009)
5. Nitta, T.: Complex-Valued Neural Networks: Utilizing High-Dimensional Parameters. Information Science Reference, Pennsylvania (2009)
6. Aizenberg, I.: Complex-Valued Neural Networks with Multi-Valued Neurons. SCI, vol. 353. Springer, Heidelberg (2011)
7. Hirose, A.: Complex-Valued Neural Networks: Theories and Applications. Series on Innovative Intelligence, vol. 5. World Scientific Publishing, Singaore (2003)
8. Shang, F., Hirose, A.: Quaternion neural-network-based PolSAR land classification in Poincare-sphere-parameter space. IEEE Transactions on Geoscience and Remote Sensing 52(9), 5693–5703 (2014)
9. Hirose, A., Yoshida, S.: Generalization characteristics of complex-valued feedforward neural networks in relation to signal coherence. IEEE Transactions on Neural Networks and Learning Systems 23, 541–551 (2012)
10. Hirose, A., Yoshida, S.: Comparison of complex- and real-valued feedforward neural networks in their generalization ability. In: Lu, B.-L., Zhang, L., Kwok, J. (eds.) ICONIP 2011, Part I. LNCS, vol. 7062, pp. 526–531. Springer, Heidelberg (2011)
11. Goto, E.: The parametron – A new circuit element which utilizes non-linear reactors. Paper of Technical Group of Electronic Computers and Nonlinear Theory, IECE (July 1954) (in Japanese)
12. Goto, E.: On the application of parametrically excited non-linear resonators. The Journal of the Institute of Electrical Communication Engineers of Japan (IECE) 38(10), 2761 (1955) (in Japanese)
13. Takahasi, H.: An exerimental decimal calculator. Paper of Technical Group of Electronic Computers, IECE (March 1956) (in Japanese)
14. Hirose, A., Kiuchi, M.: Coherent optical associative memory system that processes complex-amplitude information. IEEE Photon. Tech. Lett. 12(5), 564–566 (2000)
15. Kawata, S., Hirose, A.: A coherent optical neural network that learns desirable phase values in frequency domain by using multiple optical-path differences. Optics Letters 28(24), 2524–2526 (2003)
16. Kawata, S., Hirose, A.: Frequency-multiplexed logic circuit based on a coherent optical neural network. Applied Optics 44(19), 4053–4059 (2005)
17. Kawata, S., Hirose, A.: Frequency-multiplexing ability of complex-valued Hebbian learning in logic gates. International Journal of Neural Systems 12(1), 43–51 (2008)
18. Hirose, A., Eckmiller, R.: Behavior control of coherent-type neural networks by carrier-frequency modulation. IEEE Transactions on Neural Networks 7, 1032–1034 (1996)
19. Tazuke, K., Muramoto, N., Matsui, N., Isokawa, T.: An application of quantum-inspired particle swarm optimization to function optimization problems. In: Proceedings of the International Joint Conference on Neural Networks (IJCNN), Dallas, pp. 1234–1239 (August 2013)

20. Takata, T., Isokawa, T., Matsui, N.: Performance analysis of quantum-inspired evolutionary algorithm. Journal of Advanced Computational Intelligence and Intelligent Informatics 15(8), 1095–1102 (2011)

21. Ding, T., Hirose, A.: Fading channel prediction based on combination of complex-valued neural networks and chirp Z-transform. IEEE Transactions on Neural Networks and Learning Systems 25(9), 1685–1695 (2014)

22. Ding, T., Hirose, A.: Fading channel prediction based on complex-valued neural networks in frequency domain. In: Proceedings of International Symposium on Electromagnetic Theory (EMTS), Hiroshima, May 20-24, pp. 640–643 (2013)

23. Matsui, H., Hirose, A.: Nonlinear prediction of frequency-domain channel parameters for channel prediction in fading and fast Doppler-shift change environment. In: Proceedings of International Symposium on Antennas and Propagation (ISAP), pp. 1132–1135 (October 2012)

24. Yoshida, H., Hirose, A.: Beamforming for impulse-radio uwb communication systems based on complex-valued spatio-temporal neural networks. In: Proceedings of International Symposium on Electromagnetic Theory (EMTS), Hiroshima, pp. 848–851 (May 2013)

25. Hong, X., Chen, S.: Modeling of complex-valued Wiener systems using B-spline neural network. IEEE Transactions on Neural Networks 22(5), 818–825 (2011)

26. Savitha, R., Suresh, S., Sundararajan, N.: A fully complex-valued radial basis function network and its learning algorithm. International Journal of Neural Systems 19(4), 253–267 (2009)

27. Suksmono, A.B., Hirose, A.: Beamforming of ultra-wideband pulses by a complex-valued spatio-temporal multilayer neural network. International Journal of Neural Systems 15(1-2), 85–91 (2005)

28. Chang, A.C., Jen, C.W., Su, I.J.: Robust adaptive array beamforming based on independent component analysis with regularized constraints. IEICE E90-B(7), 1791–1800 (2007)

29. Wei, Z.Q., Xu, F., Jin, Y.Q.: Phase unwrapping for SAR interferometry based on an ant colony optimization algorithm. International Journal of Remote Sensing 29(3), 711–725 (2008)

30. Suksmono, A.B., Hirose, A.: Adaptive noise reduction of InSAR images based on a complex-valued MRF model and its application to phase unwrapping problem. IEEE Transactions on Geoscience and Remote Sensing 40(3), 699–709 (2002)

31. Suksmono, A.B., Hirose, A.: Interferometric sar image restoration using monte-carlo metropolis method. IEEE Trans. on Signal Processing 50(2), 290–298 (2002)

32. Peng, Z., Wang, H., Zhang, G., Yang, S.: Spotlight SAR images restoration based on tomography model. In: Asia-Pacific Conference on Synthetic Aperture Radar (APSAR), Xi'an, pp. 1060–1063 (2009)

33. Onojima, S., Arima, Y., Hirose, A.: Millimeter-wave security imaging using complex-valued self-organizing map for visualization of moving targets. Neurocomputing 134, 247–253 (2014)

34. Onojima, S., Hirose, A.: One-dimensional-array millimeter-wave imaging of moving targets for security purpose based on complex-valued self-organizing map (CSOM). In: Huang, T., Zeng, Z., Li, C., Leung, C.S. (eds.) ICONIP 2012, Part V. LNCS, vol. 7667, pp. 229–236. Springer, Heidelberg (2012)

35. Ejiri, A., Hirose, A.: Landmine visualization system utilizing multiple complex-valued SOMs integrating multimodal information. In: World Congress on Computational Intelligence (WCCI)/International Joint Conference on Neural Networks (IJCNN), Brisbane, pp. 1233–1239 (June 2012)

36. Nakano, Y., Hirose, A.: Adaptive identification of landmine class by evaluating the total degree of conformity of ring-SOM. Australian Journal of Intelligent Information Processing Systems 12, 23–28 (2010)

37. Nakano, Y., Hirose, A.: Improvement of plastic landmine visualization performance by use of ring-CSOM and frequency-domain local correlation. IEICE Transactions on Electronics E92-C(1), 102–108 (2009)
38. Masuyama, S., Hirose, A.: Walled LTSA array for rapid, high spatial resolution, and phase sensitive imaging to visualize plastic landmines. IEEE Transactions on Geoscience and Remote Sensing 45(8), 2536–2543 (2007)
39. Masuyama, S., Yasuda, K., Hirose, A.: Multiple mode selection of walled-ltsa array elements for high resolution imaging to visualize antipersonnel plastic landmines. IEEE Geoscience and Remote Sensing Letters 5(4), 745–749 (2008)
40. Yang, C.C., Bose, N.: Landmine detection and classification with complex-valued hybrid neural network using scattering parameters dataset. IEEE Transactions on Neural Networks 16(3), 743–753 (2005)
41. Fujimoto, M., Hori, T.: Sub-band processing for DOA estimation of UWB signal. IEICE Communications Express 1(1), 23–27 (2012)
42. Terabayashi, K., Hirose, A.: Proposal of ultra-short-pulse acoustic imaging using complex-valued spatio-temporal neural-network null-steering. In: Lee, M., Hirose, A., Hou, Z.-G., Kil, R.M. (eds.) ICONIP 2013, Part III. LNCS, vol. 8228, pp. 217–224. Springer, Heidelberg (2013)
43. Terabayashi, K., Hirose, A.: Ultra-wideband direction-of-arrival estimation using complex-valued spatiotemporal neural networks. IEEE Transactions on Neural Networks and Learning Systems (to appear, 2014)
44. Aizenberg, I.: Periodic activation function and a modified learning algorithm for the multi-valued neuron. IEEE Transactions on Neural Networks 21(12), 1939–1949 (2010)
45. Wong, W.K., Loo, C.K., Lim, W.S., Tan, P.N.: Thermal condition monitoring system using log-polar mapping, quaternion correlation and max-product fuzzy neural network classification. Neurocompu. 74, 164–177 (2010)
46. Yamaki, R., Hirose, A.: Singular unit restoration in interferograms based on complex-valued Markov random field model for phase unwrapping. IEEE Geoscience and Remote Sensing Letters 6(1), 18–22 (2009)
47. Natsuaki, R., Hirose, A.: Circular property of complex-valued correlation learning in CMRF-based filtering for synthetic aperture radar interferometry. Neurocomputing 134, 165–172 (2014)
48. Natsuaki, R., Hirose, A.: Phase property in complex-correlation and real-imaginary-correlation filtered SAR interferograms and its influence on DEM quality. In: Proceedings of Asia-Pacific Conference on Synthetic Aperture Radar (APSAR), Tsukuba, pp. 218–221 (September 2013)
49. Shang, F., Hirose, A.: Classification features in phase components of mechanism vectors in PolInSAR optimization. In: Proceedings of International Symposium on Antennas and Propagation (ISAP), Nagoya, pp. 114–117 (October 2012)
50. Shang, F., Hirose, A.: Polarimetric-basis transformation for land classification in PolInSAR. Electronics Letters 49(1), 69–71 (2013)
51. Shang, F., Hirose, A.: PolSAR land classificatoin by using quaternion-valued neural networks. In: Proceedings of Asia-Pacific Conference on Synthetic Aperture Radar (APSAR), Tsukuba, pp. 593–596. IEEE Geoscience and Remote Sensing Society Japan Chapter/IEICE Electronics Society (September 2013)
52. Yamaguchi, Y., Sato, A., Boerner, W.M., Sato, R., Yamada, H.: Four-component scattering power decomposition with rotation of coherency matrix. IEEE Transactions on Geoscience and Remote Sensing 49(6), 2251–2258 (2011)

A Cascade System of Simple Dynamic Binary Neural Networks and Its Sparsification

Jungo Moriyasu and Toshimichi Saito

Hosei University, Koganei, Tokyo, 184-8584 Japan
tsaito@hosei.ac.jp

Abstract. This paper studies a cascade system of two simple dynamic binary neural networks characterized by signum activation function and ternary connection parameters. In order to store a desired binary periodic orbit, we present a simple method based on the correlation learning. In order to sparsify the network connection, we present a simple method based on the genetic algorithm. The sparsification can be effective to reinforce stability of the stored periodic orbit.

Keywords: Digital dynamical systems, stability, power electronics.

1 Introduction

This paper presents a cascade system of two simple dynamic binary neural networks (SDNNs [1]-[3]). The SDNN is a two-layer network with delayed feedback. It is characterized by signum activation functions [4] [5], ternary connection parameters, and integer threshold parameters. The cascade system (CSDN) can generate a variety of binary periodic orbits (BPOs) and is basic to develop deep systems of the SDNNs [3]. The CSDN is an example of digital dynamical systems such as cellular automata [6]. Such digital systems can output various phenomena and are applicable to engineering systems, e.g., information compressors [7], image processors [8], communication systems [9], and switching circuits [10]-[12]. This paper studies learning and sparsification of the CSDN. As a basic learning problem, we consider storage of one desired BPO. The BPO is applicable to a control signal of switching circuits [12]. For the storage, we present a simple method based on the correlation (CL) learning [13] [14]. The CL-based learning guarantees storage of a class of BPOs. Next, we consider sparsification of the CDNN and stabilization of the stored BPO. For the sparsification, we present a simple method based on the genetic algorithm (GA). The GA-based method can sparsify the connection of the CSDN. The sparsification can be effective to reinforce stability of the stored BPO. Performing fundamental numerical experiments for two typical examples, the algorithm efficiency is confirmed.

Note that this is the first paper of sparsification and the CSDN. In Ref. [2], sparsification is applied to single SDNN (not to CSDN) and the algorithm is different from this paper. In Ref. [3], the deep system is not sparsified and the stability of the stored BPO is weak. In Ref. [12], the three-layer dynamic binary neural network employs neither CL-based learning nor sparsification.

C.K. Loo et al. (Eds.): ICONIP 2014, Part I, LNCS 8834, pp. 231–238, 2014.
© Springer International Publishing Switzerland 2014

2   The Cascade System

The dynamics of the CSDN is described by

$$\text{1st layer: } x_i^{t+1} = \text{sgn}\left(\sum_{j=1}^{N} w_{ij}^1 x_j^t - \theta_i^1\right) \quad \text{ab. } \boldsymbol{x}^{t+1} = F_1(\boldsymbol{x}^t)$$

$$\text{2nd layer: } x_i^{t+2} = \text{sgn}\left(\sum_{j=1}^{N} w_{ij}^2 x_j^{t+1} - \theta_i^2\right) \quad \text{ab. } \boldsymbol{x}^{t+2} = F_2(\boldsymbol{x}^{t+1}) \quad (1)$$

$$\text{sgn}(x) = \begin{cases} +1 \text{ for } x \geq 0 \\ -1 \text{ for } x < 0 \end{cases} \quad i = 1 \sim N$$

where $\boldsymbol{x}^t \equiv (x_1^t, \cdots, x_N^t)$, $x_i^t \in \{-1, 1\} \equiv \boldsymbol{B}$, is a binary state vector at discrete time t. The signum activation function realizes the binarization. The weighting parameters are ternary $w_{ij}^l \in \{-1, 0, 1\}$ and the threshold parameters are integer $\theta_i^l \in \boldsymbol{Z}$ ($i = 1 \sim N$, $j = 1 \sim N$, $l = 1, 2$). As an initial state vector $\boldsymbol{x}^1$ is given, the first layer outputs $\boldsymbol{x}^2$, the second layer outputs $\boldsymbol{x}^3$. The $\boldsymbol{x}^3$ is fed back to the first layer. Repeating in this manner, the CSDN can generate various binary sequences. As shown in Fig. 1, the CSDN is a cascade system of two different SDNNs F_1 and F_2:

$$\text{SDNN} : \boldsymbol{x}^{t+1} = F_1(\boldsymbol{x}^t), \ \boldsymbol{x}^{t+1} = F_2(\boldsymbol{x}^t)$$
$$\text{CSDN} : \boldsymbol{x}^{t+2} = F_D(\boldsymbol{x}^t) \equiv F_2(F_1(\boldsymbol{x}^t)) \quad (2)$$

where t takes integer values for the SDNN and t takes odd integer values for the CSDN. The SDNN can generate various BPOs and the CSDN can generate richer BPOs than SDNN. Since the number of lattice points in the domain of F_D is 2^N, direct memory of all the inputs/outputs of F_D becomes hard/impossible as N increases. However, in the CSDN, the number of parameters is polynomial $2 \times (N^2 + N)$. Since the number of inputs/outputs of the F_D is finite, the steady state of the CSDN is a BPO.

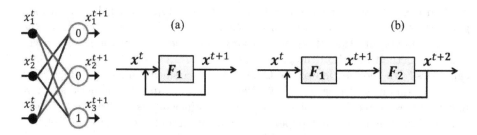

Fig. 1. (a) SDNN. Red and blue segments represent $w_{ij} = +1$ and $w_{ij} = -1$, respectively. $w_{ij} = 0$ means no connection. The threshold parameters θ_i are shown in the circles. (b) The cascade system of two SDNNs.

3 Learning of Binary Periodic Orbit

We consider a basic learning problem: storage of one BPO with period $2T$:

$$z^1, z^2, \cdots, z^{2T}; \ z^{2T+1} = z^1; \ z^t = (z^t_1, \cdots, z^t_N) \in B^N \tag{3}$$

The learning problem is determination of the parameters w^l_{ij} and θ^l_i to store the BPO where $i = 1 \sim N$, $j = 1 \sim N$, and $l = 1, 2$. This BPO is the teacher signal and its storage is divided by the 1st and 2nd layers as the following:

1st layer: $(\boldsymbol{\xi}^1_1, \boldsymbol{\xi}^1_2) \equiv (z^1, z^2), (\boldsymbol{\xi}^2_1, \boldsymbol{\xi}^2_2) \equiv (z^3, z^4), \cdots, (\boldsymbol{\xi}^T_1, \boldsymbol{\xi}^T_2) \equiv (z^{2T-1}, z^{2T})$
2nd layer: $(\boldsymbol{\xi}^1_3, \boldsymbol{\xi}^1_4) \equiv (z^2, z^3), (\boldsymbol{\xi}^2_3, \boldsymbol{\xi}^2_4) \equiv (z^4, z^5), \cdots, (\boldsymbol{\xi}^T_3, \boldsymbol{\xi}^T_4) \equiv (z^{2T}, z^1)$.

where $\boldsymbol{\xi}^\tau_j \equiv (\xi^\tau_{j1}, \cdots, \xi^\tau_{jN})$, $j = 1 \sim 4$ and $\tau = 1 \sim T$. The CL-based learning of the 1st layer is defined as the following:

$$w^1_{ij} = \begin{cases} +1 \text{ for } c_{ij} > 0 \\ 0 \quad \text{ for } c_{ij} = 0 \\ -1 \text{ for } c_{ij} < 0 \end{cases}, \ c_{ij} = \sum_{\tau=1}^{T} \xi^\tau_{2i} \xi^\tau_{1j}, \ \theta^1_i = \frac{R_i + L_i}{2}$$

$$R_i = \min_\tau \sum_{j=1}^{N} w_{ij} \xi^\tau_{1j} \text{ for } \xi^\tau_{2i} = +1, \quad L_i = \max_\tau \sum_{j=1}^{N} w_{ij} \xi^\tau_{1j} \text{ for } \xi^\tau_{2i} = -1 \tag{4}$$

The weighting parameters w^1_{ij} are given by ternarising the correlation matrix elements c_{ij}. After w^1_{ij} are given, the threshold parameters θ^1_i are determined by the quantities R_i and L_i. Note that R_i (L_i) exists if $\xi^\tau_{2i} = +1$ $(\xi^\tau_{2i} = -1)$ for some τ. If $\xi^\tau_{2i} = -1$ $(\xi^\tau_{2i} = +1)$ for all τ then R_i (L_i) does not exist and let $\theta^1_i = N + 1$ $(\theta^1_i = -N - 1)$. Replacing $(\boldsymbol{\xi}^1_1, \boldsymbol{\xi}^1_2)$ with $(\boldsymbol{\xi}^1_3, \boldsymbol{\xi}^1_4)$, we obtain the CL-based learning of the 2nd layer. Storage of the BPO is guaranteed if

$$R_i > L_i \text{ is satisfied for } i \text{ such that both } R_i \text{ and } L_i \text{ exist.} \tag{5}$$

In order to consider the CL-based learning, we consider two teacher signals of $N = 9$. The first one is the BPO1 with period 6 in Table 1. Applying the CL-based learning, the BPO1 can be stored into the SDNN in Fig. 2 (a).

In order to visualize the dynamics of the SDNN, we introduce the Grey-code-based return map (Gmap). As shown in Eq. (2), the SDNN is described by the mapping F_1 from B^N to itself. Applying the Grey code to express 2^N elements of B^N, F_1 can be expressed by a mapping from a set of lattice points $L \equiv (C_1, \cdots, C_{2^N})$ to itself. This is the Gmap from L to L. Figure 3 (a) shows an Gmap of the SDNN where storage of BPO1 can be confirmed.

For the Gmap, we give basic definitions. For a Gmap $G : L \to L$, we will say $p \in L$ is a binary periodic point (BPP) with period k if $G^k(p) = p$ and $G^l(p) \neq p$ for $0 < l < k$ where G^k is the k-fold composition of G. A sequence of BPPs $(G(p), \cdots, G^k(p)$ is referred to as a binary periodic orbit (BPO). We will say $q \in L$ is an eventually periodic point (EPP) of a BPO if it is not a BPP and falls into the BPO. BPO and EPP characterize the steady and transient states, respectively.

The second example is the BPO2 with period 12 in Table 2. Note that BPO2 corresponds to control signal of a basic ac-ac converter in power electronics [11].

Table 1. Teacher signal BPO1

z^1	$(-1,+1,-1,-1,+1,-1,-1,-1,+1)$
z^2	$(+1,-1,-1,-1,+1,-1,-1,+1,-1)$
z^3	$(+1,-1,-1,+1,-1,-1,-1,+1,-1)$
z^4	$(-1,-1,+1,+1,-1,-1,+1,-1,-1)$
z^5	$(-1,-1,+1,-1,-1,+1,+1,-1,-1)$
z^6	$(-1,+1,-1,-1,-1,+1,-1,-1,+1)$

Table 2. Teacher signal BPO2

z^1	$(-1,-1,+1,-1,+1,-1,-1,-1,+1)$
z^2	$(+1,-1,-1,-1,+1,-1,-1,+1,-1)$
z^3	$(+1,-1,-1,+1,-1,-1,-1,+1,-1)$
z^4	$(-1,-1,+1,+1,-1,-1,-1,-1,+1)$
z^5	$(-1,-1,+1,+1,-1,-1,+1,-1,-1)$
z^6	$(-1,-1,+1,-1,-1,+1,-1,+1,-1)$
z^7	$(-1,+1,-1,-1,-1,+1,-1,+1,-1)$
z^8	$(-1,+1,-1,+1,-1,-1,+1,-1,-1)$
z^9	$(-1,+1,-1,-1,+1,-1,+1,-1,-1)$
z^{10}	$(+1,-1,-1,-1,-1,+1,+1,-1,-1)$
z^{11}	$(+1,-1,-1,-1,-1,+1,-1,-1,+1)$
z^{12}	$(-1,+1,-1,-1,+1,-1,-1,-1,+1)$

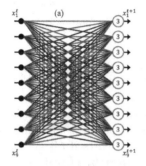 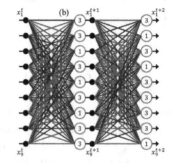

Fig. 2. (a) SDNN and (b) CSDN after the CL-based learning

Applying the CL-based learning, BPO2 can be stored into the CSDN in Fig. 2 (b) (cannot be stored into the SDNN). Figure 3 (c) shows Gmaps of the 1st and 2nd layers F_1 and F_2. Applying the two Gmaps alternately, we obtain the dynamics of the CSDN. In the figure, we can confirm that BPO2 with period 12 is stored. Figure 3 (b) shows composition of the two Gmaps ($F_D = F_1 \circ F_2$).

4 Sparsification and Stability

Here, we consider sparsification of the connection of the CSDN (SDNN) and stability of the stored BPO. In order to control the sparsity, we present a simple method based on the genetic algorithm (GA). The GA has the chromosome $\{c_{ij}^l\}$ which correspond to the weighting parameter $\{w_{ij}^l\}$ where $i = 1 \sim N$, $j = 1 \sim N$, and $l = 1, 2$. Each chromosome consists of $N \times N \times 2$ genes. In the update of the chromosomes, the following two fitness functions are used.

Convergence rate (CR):

$$F_1(c_{ij}^l) = (\ (\#\text{EPPs} + \#\text{BPPs}) \text{ of teacher signal})/2^N \times 100 \qquad (6)$$

Sparsity rate (SR): $F_2(c_{ij}^l) = (\#\text{zeros in } \{c_{ij}^l\})/(N \times N \times 2) \times 100$

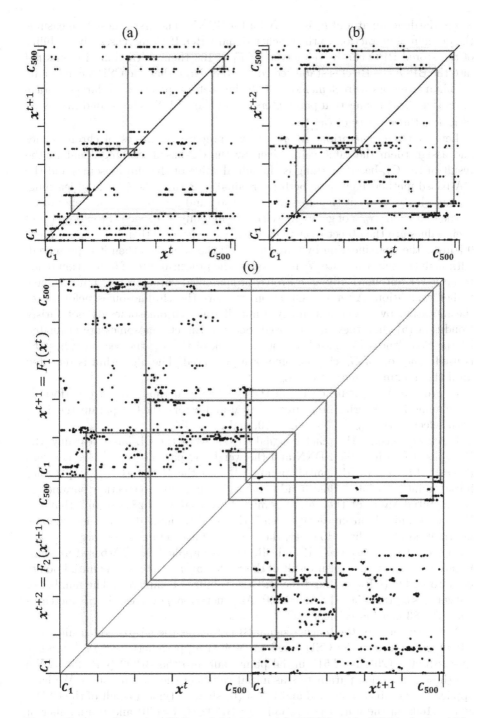

Fig. 3. Gmaps after the CL-based learning. (a) SDNN with BPO1 (red orbit). (b) CSDN with BPO2 (red orbit). (c) 1st and 2nd layers of CSDN with BPO2.

where the denominator of F_2 is $N \times N$ for the SDNN. The first fitness F_1 measures the domain of attraction to the teacher signal BPO. If F_1 increases, the stability of the stored BPO can be reinforced. If $F_1 = 100$ then all the initial points fall into the BPO: the BPO is stabilized completely. Note that the CSDN can handle 2^N binary vectors. Direct memory of all the vectors spends much larger storage units than $2N^2$ connection parameters w_{ij}^l of the CSDN. The second fitness F_2 measures the sparsity of (w_{ij}^l).

Let g be the generation. At $g = 0$, we prepare G_e pieces of chromosomes and assign them the same value given by the CL-based learning. That is, the result of the CL-based learning is the initial value of the chromosomes and the GA-based method must give better (or equal) CR than the CL-based learning. For simplicity, we omit the crossover operation and use the mutation operation. In the mutation, $N_g\%$ of genes are selected randomly from each chromosome and their values are the object to change. If the value is not 0 then it is changed into 0 (this is the sparsification by 0-insertion). If the value is 0 then it is preserved. Although the selection rate N_g is constant, the practical rate of 0-insertion tends to be lower automatically as g evolves if sparsified chromosomes tend to have higher evaluation. After the mutation, we sort the chromosomes before/after the mutation by the elite strategy. First, if some chromosomes do not satisfy Condition (5) then they are removed. Second, the chromosomes are evaluated by the first fitness F_1. Third, in the tie-break of the F_1, the second fitness F_2 is used. The top of G_e chromosomes are preserved. The algorithm is repeated until the maximum generation G_{max}.

In order to investigate effects of the GA-based method, we have performed basic numerical experiments. After the trial-and-errors, the GA parameters have been selected as $G_{max} = 500$, $G_e = 20$, and $N_g = 4\%$.

First, we consider the teacher signal BPO1. After the CL-based learning, the BPO1 is stored into the SDNN and the weighting parameters w_{ij}^1 are not zero (SR=0). 144 out of 512 initial points fall into the BPO1 (CR=28%). Other initial points fall into other steady states. We use this connection parameters as an initial value of the chromosomes. Note that CR=28% means that the BPO is stabilized automatically even if the teacher signal does not include any information of stability. However, the domain of attraction is not large.

In order to increase the CR and SR, we have applied the GA-based method. Figure 4 (a) shows a typical result of the SDNN at $g = 100$: all the initial points fall into BPO1 (CR=100) and the BPO is stabilized completely. The number of the zeros is 56 (SR=56/81 × 100 ≈ 69). As generation g evolves, the SR increases and SR≈ 83 is achieved at $g = G_{max} = 500$.

Next, we consider the teacher signal BPO2. After the CL-based learning, the BPO2 is stored into the CSDN and the weighting parameters w_{ij}^1, w_{ij}^2 are not zero (SR=0). 120 out of 512 initial points fall into the BPO2 (CR=23%). We use these connection parameters as initial values of the chromosomes. We have applied the GA-based method and Fig. 4 (b) shows a typical result of the CSDN at $g = 100$: all the initial points fall into BPO2 (CR=100) and the number of

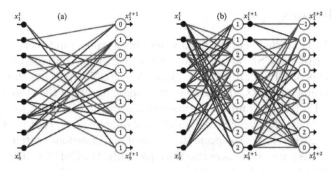

Fig. 4. (a) SDNN. (b) CSDN after the GA-based method.

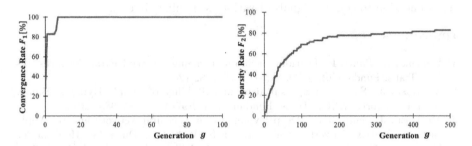

Fig. 5. Evolution process of the best chromosome for the teacher signal BPO1. Left: F_1(convergence rate), right: F_2(sparsity rate).

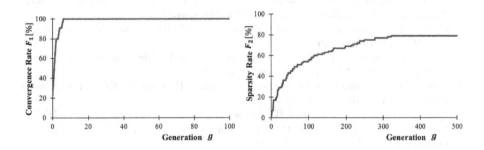

Fig. 6. Evolution process of the best chromosome for BPO2. Left: F_1, right: F_2

zeros is 90 (SR=$90/162 \times 100 \approx 55$). As g evolves, the SR increases and SR≈ 79 is achieved at $g = G_{max} = 500$.

Figures 5 and 6 show the evolution process of CR and SR, respectively. In the figure, the initial generation $g = 0$ means the result of the CL-based learning. As the generation g evolves, the first fitness F_1 of CR increases rapidly in the early stage. After the CR is saturated, the second fitness F_2 of SR can increase.

5   Conclusions

Learning and sparsification of the CSDN have been studied in this paper. The 1st and 2nd layers of the CSDN share the output of BPOs. Applying the CL-based learning, a class of BPO can be stored into the CSDN. Applying the GA-based method, the connection of CSDN is sparsified and stability of the stored BPO can be reinforced. Performing numerical experiments for typical two examples, the algorithm performance has been investigated. It should be noted that the CSDN can realize various BPOs that correspond to control signal of various switching converters. Although this paper considers two examples only, the CSDN can be developed into reconfigurable and robust control circuits for various switching circuits.

Future problems include analysis of learning and sparsification process, analysis of effects of sparsitication on the stability, analysis of effects of GA parameters, and application to control signals of various switching circuits.

References

1. Kouzuki, R., Saito, T.: Learning of Simple Dynamic Binary Neural Networks. IEICE Trans. Fundamentals E96-A(8), 1775–1782 (2013)
2. Moriyasu, J., Saito, T.: Sparsification and Stability of Simple Dynamic Binary Neural Networks. IEICE Trans. Fundamentals E97-A(4), 985–988 (2014)
3. Moriyasu, J., Saito, T.: A Deep Dynamic Binary Neural Network and Its Application to Matrix Converters. In: Wermter, S., Weber, C., Duch, W., Honkela, T., Koprinkova-Hristova, P., Magg, S., Palm, G., Villa, A.E.P. (eds.) ICANN 2014. LNCS, vol. 8681, pp. 611–618. Springer, Heidelberg (2014)
4. Gray, D.L., Michel, A.N.: A Training Algorithm for Binary Feed Forward Neural Networks. IEEE Trans. Neural Networks 3(2), 176–194 (1992)
5. Chen, F., Chen, G., He, Q., He, G., Xu, X.: Universal Perceptron and DNA-like Learning Algorithm for Binary Neural Networks: Non-LSBF Implementation. IEEE Trans. Neural Networks 20(8), 1293–1301 (2009)
6. Chua, L.O.: A Nonlinear Dynamics Perspective of Wolfram's New Kind of Science, I, II. World Scientific (2005)
7. Wada, W., Kuroiwa, J., Nara, S.: Completely Reproducible Description of Digital Sound Data with Cellular Automata. Physics Letters A 306, 110–115 (2002)
8. Rosin, P.L.: Training Cellular Automata for Image Processing. IEEE Trans. Image Process. 15(7), 2076–2087 (2006)
9. Iguchi, T., Hirata, A., Torikai, H.: Theoretical and Heuristic Synthesis of Digital Spiking Neurons for Spike-Pattern-Division Multiplexing. IEICE Trans. Fundamentals E93-A(8), 1486–1496 (2010)
10. Bose, B.K.: Neural Network Applications in Power Electronics and Motor Drives - an Introduction and Perspective. IEEE Trans. Ind. Electron. 54(1), 14–33 (2007)
11. Rodriguez, J., Rivera, M., Kolar, J.W., Wheeler, P.W.: A Review of Control and Modulation Methods for Matrix Converters. IEEE TIE 59(1), 58–70 (2012)
12. Nakayama, Y., Kouzuki, R., Saito, T.: Application of the Dynamic Binary Neural Network to Switching Circuits. In: Lee, M., Hirose, A., Hou, Z.-G., Kil, R.M. (eds.) ICONIP 2013, Part II. LNCS, vol. 8227, pp. 697–704. Springer, Heidelberg (2013)
13. Hopfield, J.J.: Neural Networks and Physical Systems with Emergent Collective Computation Abilities. Proc. of the Nat. Acad. Sci. 79, 2554–2558 (1982)
14. Araki, K., Saito, T.: An Associative Memory Including Time-Variant Self-Feedback. Neural Networks 7(8), 1267–1271 (1994)

A Model of V4 Neurons Based on Sparse Coding

Hui Wei, Zheng Dong, and Qiang Li

Department of Computer Science, Laboratory of Cognitive Model and Algorithm
Shanghai Key Laboratory of Data Science,
Fudan University Shanghai, China
weihui@fudan.edu.cn

Abstract. Area V4 lies in the middle of the ventral visual pathway in primate brains. It is an intermediate stage in the visual processing for object discrimination. V4 neurons exhibit selectivity to complex boundary conformation. In this paper, we propose a novel model of V4 neurons based on sparse coding. The model is a multi-layer neural network of which the output layer consists of laterally connected V4 units. We provide an informal proof for sparse coding with intra-layer inhibitory connections and show experimentally that this model successfully reproduces shape selectivity observed in V4 neurons. The model provides clues to the high level representation of visual stimuli in the brain.

Keywords: Visual pathway, V4, sparse coding, receptive field, shape selectivity, image representation.

1 Introduction

Primate brains possess two distinct visual systems. As visual information exits the occipital lobe, it follows two main pathways [7,14]. The dorsal pathway terminates in the parietal lobe and is involved with processing the object's spatial location relevant to the viewer. The ventral pathway travels to the temporal lobe and is involved with object discrimination and recognition. Cortical area V4 lies in the middle of the ventral pathway. At lower levels of the pathway (V1 and V2), objects are represented in terms of local orientation [11,12]. At the final stages in IT, neurons tend to selectively respond to complex objects like faces and body parts [1,2]. Area V4 is an intermediate stage in which local orientation signals from lower levels are transformed into complex object selectivity at the final stages. V4 plays a crucial role in the hierarchy of visual shape perception. Understanding the mechanism and constructing models of V4 help to reveal the object recognition mechanism of the ventral visual pathway.

Neurophysiological studies have not produced consistent descriptions of V4 selectivity. V4 neurons are known to be selective about color, shape, depth and even motion [21]. In this paper, we focus on the V4 selectivity for shapes. Early experiments examined the selectivity of cells in V4 with classical stimuli including bars and sinusoidal gratings [6]. Similar to earlier processing stages,

C.K. Loo et al. (Eds.): ICONIP 2014, Part I, LNCS 8834, pp. 239–246, 2014.

V4 cells are tuned for orientation and spatial frequency of edges and linear sinusoidal gratings. However, V4 neurons are also sensitive to more complex shape
properties. Later experiments reported that V4 neurons display a clear bias in
their responses in favor of non-Cartesian gratings and they show a significant
degree of invariance in their selectivity across changes in stimulus position [10].
More recent experiments showed that V4 neurons can be strongly selective about
curvature of contours and angular position of acute curvatures [18,19]. These experiments also suggested that V4 neurons are not sensitive to small displacement
of the stimulus within the receptive field.

In section 2, we review some previous models of area V4. In section 3, we
propose a novel model of V4 shape selectivity based on sparse coding. This is
a multi-layer neural network model. The layer of V4 units gets input from the
afferent layer of complex cells and achieves sparse coding by intra-layer inhibitory
connections. We give an informal proof for the emergence of sparse codes from
lateral inhibition and show experimentally that complex cells' output provides
sufficient information for the emergence of V4 shape selectivity. In section 3.4,
we demonstrate that the sparse codes obtained with this model form a novel
kind of representation of local image structures, which provides important clues
to the high level presentation of visual information in the ventral visual pathway.
The conclusion is summarized in section 4.

2   Previous Models of Area V4

Several models have been proposed to explain the shape selectivity and invariance
of V4 neurons.

The spectral receptive field (SRF) is one model for shape selectivity of V4
neurons [5]. It explains many observations of V4 response patterns. The SRF
describes tuning in terms of the orientation and spatial frequency spectrum.
The model is based on the fact that V4 neurons have large orientation and
spatial frequency bandwidth. They respond selectively to stimuli such as contour
conformations and non-Cartesian gratings, which generally consist of multiple
orientations and spatial frequencies. The spectral model is also invariant to small
changes in stimulus position and thus should explain the invariance property of
V4 response patterns. The model is powerful in describing the shape selectivity
of V4 neurons. However, it provides little explanation for the emergence of the
selectivity.

The hierarchical MAX-pooling (HMAX) model is a generic model for object
recognition in the visual cortex [20]. It was also adopted as a model for V4
shape selectivity and invariance [4]. The selectivity of earlier stages in the visual
pathway can be well modeled as linear filters such as Gabor filters [9] for simple
cells. The HMAX model processes complex stimuli by concatenating multiple
layers of linear filters and inserts maximum operations between these layers in
order to achieve non-linearity. This model conforms to the neuronal connectivity
in the visual pathway but provides no evidence for the maximum operation in
neuroscience.

3   Shape Selectivity Model Based on Sparse Coding

3.1   Sparse Coding

Neural codes in mammalian visual cortex are sparse [22]. Computational models of primary visual cortex have demonstrated that sparse coding can explain the emergence of neurons with localized oriented receptive fields [8,17].Sparse coding even replicates complex neuroscience phenomena that are not well explained by simple linear models of primary visual cortex [13].

Sparse coding can be seen as recovering the code h^* associated with an input x via:

$$h^* = \underset{h}{\operatorname{argmin}} ||x - Wh||_2^2 + \lambda||h||_1, \tag{1}$$

where $x \in \mathbb{R}^{d_x}$, $h \in \mathbb{R}^{d_h}$ and λ is a penalty to enforce the sparsity of the code. Learning the dictionary W can be accomplished by minimizing $\sum_t ||x^{(t)} - Wh^{*(t)}||_2^2$ over the training input $x^{(t)}$ and corresponding sparse code $h^{*(t)}$ determined by equation (1).

Sparse coding can be achieved by a layer of competitive neurons with intra-layer lateral inhibitory connections which are found in visual cortex [3].We use a two-layer neural network model (Fig. 1) to explain how sparse coding is achieved via recurrent inhibitory connections in the output layer. The network consists of an input layer and an output layer. The input layer has d_x units: $x_i, i = 1 \ldots d_x$. The output layer has d_h units: $h_j, j = 1 \ldots d_h$. The weight of the excitatory feedforward connection from the i-th input unit to the j-th output unit is $W_{E_{i,j}}$. The weight of the inhibitory recurrent connection from the j-th output unit to the k-th output unit is $W_{I_{j,k}}$. Suppose the output units are linear, the output vector h can be obtained by solving the following equation.

$$h = W_E x - W_I h, \tag{2}$$

where x is the input vector and h is the output vector.

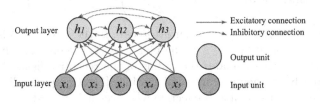

Fig. 1. The neural network model for sparse coding.

We show that iteratively solving the equation (2) minimizes the objective function defined in equation (1). Let $\mathcal{F} = ||x - Wh||_2^2 + \lambda||h||_2^2$ be the objective function to be minimized. The L1 penalty is replaced with a squared Euclidean norm for simplicity. We then calculate the partial derivative of $\mathcal{F}$ with respect to h_k for $k = 1 \ldots d_h$.

$$\frac{\partial \mathcal{F}}{\partial h_k} = \frac{\partial}{\partial h_k} \sum_i \left(x_i - \sum_j W_{i,j} \cdot h_j \right)^2 + \frac{\partial}{\partial h_k} \lambda \sum_j h_j^2$$

$$= 2 \sum_i \left(\sum_j W_{i,j} \cdot h_j - x_i \right) W_{i,k} + 2\lambda h_k$$

$$= 2 \left[W^\top (Wh - x) + \lambda h \right]_k. \tag{3}$$

After each iteration, the change in h is $\Delta h = (W_E x - W_I h - h)$ and thus the change in the objective function $\mathcal{F}$ is approximated as follows.

$$\Delta \mathcal{F} = \sum_k \Delta h_k \frac{\partial \mathcal{F}}{\partial h_k}$$

$$= 2(W_E x - W_I h - h)^\top \cdot \left[W^\top (Wh - x) + \lambda h \right]$$

$$= 2 \left[W_E x - (W_I + 1)h \right]^\top \cdot \left[(W^\top W + \lambda) h - W^\top x \right]. \tag{4}$$

The change in $\mathcal{F}$ is a scalar product of two vectors. Since the weight matrices and sparse bases consist of non-negative values, the two vectors point to opposite directions and thus the scalar product remains negative over every iteration. This explains that the output h of the recurrent network is the sparse code of the network input x.

3.2   Multi-layer Model of V4

According to the hierarchy of the ventral visual pathway, area V4 receives input from the lower levels including area V1 and V2. These areas have been well studied since 1960s by the Nobel Prize winners [11,12] and successive researchers.

Neurons in V1 and V2 respond to local orientations. They fall into two categories, simple cells and complex cells. The receptive fields of simple cells can be understood as linear filters modeled as Gabor functions [9],

$$g_\theta(x, y; \lambda, \sigma_s) = \exp\left(-\frac{x'^2 + y'^2}{2\sigma_s^2} \right) \cos\left(2\pi \frac{x'}{\lambda} + \psi \right), \tag{5}$$

where $x' = x \cos\theta + y \sin\theta$, $y' = -x \sin\theta + y \cos\theta$. In the equation, λ is the wavelength of the sinusoidal factor, ψ is the phase offset, θ represents the preferred orientation and σ_s approximates the radius of the receptive fields. Simple cells respond primarily to oriented edges and gratings [11]. Complex cells differ from simple cells in that a stimulus is effective wherever it is placed in the receptive field, provided that the orientation is appropriate [11]. The complex cells receive input from simple cells with the same preferred orientation and thus have larger receptive fields [16]. They are modeled as Gaussian filters,

$$f(x, y; \sigma_c) = \frac{1}{2\pi\sigma_c^2} \exp\left(-\frac{x^2 + y^2}{2\sigma_c^2} \right), \tag{6}$$

where σ_c represents the radius of the complex receptive fields.

We construct a two-layer neural network to imitate simple cells and complex cells and the network provides input for the V4 units in our model (Fig. 2). Given I as an input image, the output of complex cells with the preferred orientation θ is defined as the following convolution.

$$C_\theta = |I \otimes g_\theta| \otimes f. \tag{7}$$

The output of complex cells with different preferred orientations are fed into V4 units as input.

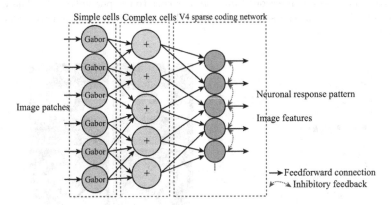

Fig. 2. Multi-layer neural network model of V4

3.3 V4 Sparse Coding Network

An image patch that is covered by a V4 receptive field is fed into the multi-layer neural network model (Fig. 2) and is processed by the first two layers of simple cells and complex cells. The output of complex cells of the same preferred orientation is a matrix defined by equation (7). The values in the matrices corresponding to different preferred orientations are lined-up in a single vector. Sparse codes of such vectors are obtained by the V4 layer by feeding the vectors as input to the V4 sparse coding network.

We examine experimentally that the complex cells' output is sufficient for the emergence of neuronal response pattern of V4. A perceptron model is used in the examination. The perceptron takes the same input with the V4 sparse coding network. It is trained with shapes presented to V4 neurons in [19]. The two shapes are shown in Fig. 3a. The shapes are moved randomly within the receptive field before they are fed to the network so that the perceptron shall be invariant with respect to small changes in stimulus positions. The trained perceptron exhibits strong bias towards the shape with convex to the top right and shares the same preference with the actual V4 neuron. The response map is shown in Fig. 3a where dark colors indicate stronger responses. The weight of the afferent connections to the perceptron is shown in Fig. 3b. The weight of connections from complex cells of different orientations is plotted respectively in different blocks.

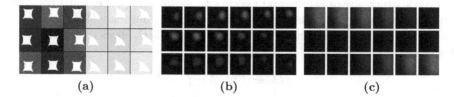

Fig. 3. Model of V4 neurons. (a) A perceptron is trained to simulates the response pattern of a V4 neuron that selective responds to sharp convex to the top right. (b) The weight of afferent connections from complex cells to the perceptron. (c) The weight matrices obtained from sparse coding, which show similar distribution to that of the perceptron.

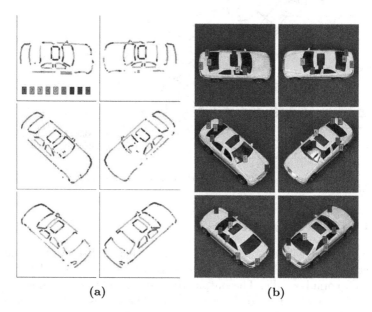

Fig. 4. Clusters of V4 sparse codes

The experiment demonstrates that complex cells provide sufficient information for V4 neurons to show selectivity observed in neurobiological experiments.

The sparse coding network of V4 is trained with Hebbian rules for excitatory connections and anti-Hebbian rules for inhibitory connections [8]. Images patches of natural images are used in the training. One of the sparse bases is shown in Fig.3c. Sub-vectors of the basis corresponding to complex cells' output of different orientations are reshaped into matrices and plotted in different blocks.

3.4   Image Representation in V4

We demonstrate the property of V4 sparse codes in processing images of real objects. We use images of a car from the ETH-80 datasets [15]. In this experiment, the receptive field radii of simple cells and complex cells are 4 and 9 pixels respectively. The preferred orientations that are sampled by these cells start from 0° to 170° in steps of 10°. The V4 sparse codes at the position of edge points are clustered by the K-means algorithm. These clusters are plotted in different colors as shown in Fig. 4a.

The clusters of adjacent points are labeled in Fig. 4b. The labels show clear matching of the same physical position in different views of the car. For instance, clusters labeled 6 appear at the rear of the car and clusters labeled 9 appear at the side of the front window. This experiment demonstrates that similar sparse codes correspond to similar image structures, which implies that the V4 sparse codes form an intermediate representation of the object or shape conformation for object discrimination in the ventral visual pathway.

4   Conclusion

In this paper, we propose a novel model of V4 neurons based on sparse coding. This multi-layer neural network model conforms to the neuronal connectivity in the visual cortex and explains the shape selectivity observed in V4 neurons. We give an informal proof for the emergence of sparse coding in area V4 and demonstrates that the sparse codes provide important clues to the high level representation of visual stimuli in the brain.

Acknowledgments. This work was supported by the 973 Program (Project No. 2010CB327900), the NSFC project (Project No. 61375122, 81373556), and the National "Twelfth Five-Year Plan" for Science and Technology (Project No. 2012BAI37B06).

References

1. Bell, A.H., Hadj-Bouziane, F., Frihauf, J.B., Tootell, R.B.H., Ungerleider, L.G.: Object Representations in the Temporal Cortex of Monkeys and Humans as Revealed by Functional Magnetic Resonance Imaging. Journal of Neurophysiology 101(2), 688–700 (February 2009)
2. Bruce, C., Desimone, R., Gross, C.G.: Visual properties of neurons in a polysensory area in superior temporal sulcus of the macaque. Journal of Neurophysiology 46(2), 369–84 (1981)
3. Budd, J., Kisvrday, Z.: Local lateral connectivity of inhibitory clutch cells in layer 4 of cat visual cortex (area 17). Experimental Brain Research 140(2), 245–250 (2001)
4. Cadieu, C., Kouh, M., Pasupathy, A., Connor, C.E., Riesenhuber, M., Poggio, T.: A Model of V4 Shape Selectivity and Invariance. Journal of Neurophysiology 98(3), 1733–1750 (September 2007)
5. David, S.V., Hayden, B.Y., Gallant, J.L.: Spectral Receptive Field Properties Explain Shape Selectivity in Area V4. Journal of Neurophysiology 96(6), 3492–3505 (2006)
6. Desimone, R., Schein, S.J.: Visual properties of neurons in area V4 of the macaque: sensitivity to stimulus form. Journal of Neurophysiology 57(3), 835–868 (1987)

7. Ettlinger, G.: "Object vision" and "spatial vision": the neuropsychological evidence for the distinction. Cortex 26(3), 319–341 (Sep 1990)
8. Földiák, P.: Forming sparse representations by local anti-Hebbian learning. Biological cybernetics 64(2), 165–170 (1990)
9. Gabor, D.: Theory of communication. Institution of Electrical Engineering (1946)
10. Gallant, J.L., Connor, C.E., Rakshit, S., Lewis, J.W., Van Essen, D.C.: Neural responses to polar, hyperbolic, and Cartesian gratings in area V4 of the macaque monkey. Journal of Neurophysiology 76(4), 2718–2739 (1996)
11. Hubel, D.H., Wiesel, T.N.: Receptive fields, binocular interaction and functional architecture in the cat's visual cortex. The Journal of physiology 160, 106–154 (jan 1962)
12. Hubel, D.H., Wiesel, T.N.: Receptive Fields and Functional Architecture in Two Nonstriate Visual Areas (18 and 19) of the Cat. Journal of Neurophysiology 28(2), 229–289 (1965)
13. Lee, H., Battle, A., Raina, R., Ng, A.Y.: Efficient sparse coding algorithms. In: Neural Information Processing Systems. pp. 801–808 (2007)
14. Lehky, S.R., Sereno, A.B.: Comparison of Shape Encoding in Primate Dorsal and Ventral Visual Pathways. Journal of Neurophysiology 97(1), 307–319 (January 2007)
15. Leibe, B., Schiele, B.: Analyzing appearance and contour based methods for object categorization. In: Computer Vision and Pattern Recognition, 2003. Proceedings. 2003 IEEE Computer Society Conference on. vol. 2, pp. II–409–15 vol.2 (2003)
16. Movshon, J.A., Thompson, I.D., Tolhurst, D.J.: Receptive field organization of complex cells in the cat's striate cortex. The Journal of Physiology 283(1), 79–99 (1978)
17. Olshausen, B.A., Field, D.J.: Sparse coding with an overcomplete basis set: A strategy employed by V1? . Vision Research 37(23), 3311 – 3325 (1997)
18. Pasupathy, A., Connor, C.E.: Responses to Contour Features in Macaque Area V4. Journal of Neurophysiology 82(5), 2490–2502 (1999)
19. Pasupathy, A., Connor, C.E.: Shape Representation in Area V4: Position-Specific Tuning for Boundary Conformation. Journal of Neurophysiology 86(5), 2505–2519 (2001)
20. Riesenhuber, M., Poggio, T.: Hierarchical models of object recognition in cortex. Nature Neuroscience 2, 1019–1025 (1999)
21. Roe, A., Chelazzi, L., Connor, C., Conway, B., Fujita, I., Gallant, J., Lu, H., Vanduffel, W.: Toward a Unified Theory of Visual Area V4 . Neuron 74(1), 12 – 29 (2012)
22. Willmore, B.D.B., Mazer, J.A., Gallant, J.L.: Sparse coding in striate and extrastriate visual cortex. Journal of Neurophysiology 105(6), 2907–2919 (2011)

A Fast and Memory-Efficient Hierarchical Graph Clustering Algorithm*

László Szilágyi[1], Sándor Miklós Szilágyi[2], and Béat Hirsbrunner[3]

[1] Dept. of Control Engineering and Information Technology,
Budapest University of Technology and Economics, Hungary
lazacika@yahoo.com
[2] Dept. of Informatics, Petru Maior University of Tîrgu-Mureş, Romania
[3] University of Fribourg, Switzerland

Abstract. In this paper we propose a quick and memory-efficient implementation of the TRIBE-MCL clustering algorithm, suitable for accurate classification of large-scale protein sequence data sets. A symmetric sparse matrix structure is introduced that can efficiently handle most operations of the main loop. The reduction of memory requirements is achieved by regrouping the operations performed during the expansion matrix squaring. The proposed algorithm is tested on synthetic protein sequence data sets of up to 250 thousand items. The validation process revealed that the proposed method performs in 30% less time than previous efficient Markov clustering algorithms, without losing anything from the partition quality. This novel implementation makes it possible for the user of an ordinary PC to process protein sequences sets of 100,000 items in reasonable time.

Keywords: Protein sequence clustering, Markov clustering, Markov processes, efficient computing, sparse matrix.

1 Introduction

Markov clustering performs a hierarchical grouping of input data based on a graph structure and its associated connectivity matrix. When the input data consists of protein sequences, each sequence will be associated to a node of the graph, and edge weights will be the pairwise similarity values computed with existing alignment methods like: Needleman-Wunsch [5], Smith-Waterman [6], and BLAST [1]. In case of large-scale data sets, the BLAST similarity measures are preferred due to its sparse nature, which allows for quick and memory-efficient processing.

TRIBE-MCL is a clustering method based on Markov chain theory [3], which assigns a graph structure to the protein set such a way that each protein has a

---

* Research supported by the Hungarian National Research Funds (OTKA), Project no. PD103921. The work of S.M. Szilágyi was supported by the TÁMOP-4.2.2.C-11/1/KONV-2012-0001 project, which is funded by the European Union, co-financed by the European Social Fund.

C.K. Loo et al. (Eds.): ICONIP 2014, Part I, LNCS 8834, pp. 247–254, 2014.

corresponding node. Edge weights are stored in the so-called similarity matrix S, which acts as a stochastic matrix. At any moment, edge weight s_{ij} reflects the posterior probability that protein i and protein j have a common evolutionary ancestor. TRIBE-MCL is an iterative algorithm, performing in each loop two main operations on the similarity matrix: inflation and expansion. Inflation raises each element of the similarity matrix to power $r > 1$, which is a previously established fixed inflation rate, favoring higher similarity values in the detriment of lower ones. Expansion, performed by raising matrix S to the second power, is aimed to favor longer walks along the graph. Further operations like column or row normalization, and matrix symmetrization are included to serve the stability and robustness of the algorithm, and to enforce the probabilistic constraint. Similarity values that fall below a previously defined threshold value ε are rounded to zero. Clusters are obtained as connected subgraphs in the graph.

Handling matrices of tens or hundreds of thousand rows and columns is prohibitively costly in both runtime and storage space. Recent fast TRIBE-MCL implementations (e.g. [10]) significantly reduced runtime, but the memory limitations still exist. The main goal of this paper is to introduce a novel fast TRIBE-MCL approach that uses only sparse matrices to store similarity values and a one-dimensional array to store intermediate values of a single row during expansion. This change can upgrade the size of processable data sets by an order of magnitude, and may also improve processing speed. The proposed method will be validated using the protein sequences of the SCOP95 database [7,4,2], and larger synthetic protein data sets [8] derived from SCOP95.

The remainder of this paper is structured as follows. Section 2 presents the details of the proposed efficient TRIBE-MCL algorithm. Section 3 thoroughly evaluates the behavior of the proposed method and discusses the achieved results and outlines the role of each parameter, while section 4 concludes this study.

2   Methods

In this paper we introduce a highly efficient implementation of the TRIBE-MCL algorithm, which also focuses on requiring reduced amount of memory storage. Any kind of TRIBE-MCL needs two instances of the similarity matrix: inflation, normalization, and symmetrization can be performed in a single matrix, but the expansion needs separate matrix instance for the input and the output data. Quick solutions existing so far use a sparse matrix and a two-dimensional array, which is not quite effective in reducing memory needs. The solution introduced here employs two instances of sparse matrix and an extra array that stores a single line of the similarity matrix during expansion.

The data structure employed for sparse matrix representation, shown in Fig. 1, is similar to the sparse supermatrix introduced in our previous work [10]. Each nonzero element of the sparse matrix (s_{ij}) is stored together with an approximated value of its symmetrically situated element in the matrix $(t_{ij} \approx s_{ji})$. Rows are stored concatenated in an array of records, each such record describing a nonzero element in the matrix. The starting address of each row is stored

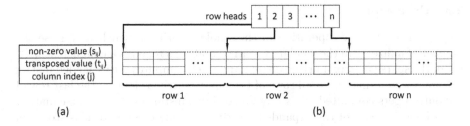

Fig. 1. The employed sparse matrix data structure: (a) the record for storing a single nonzero element; (b) array containing concatenated rows of the matrix

in a separate array of row heads. This whole data structure assures easy parsing of the matrix for all operations. With the exception of expansion, none of the operations increases the nonzero elements in any row. Normalization and inflation keeps the amount of nonzero elements constant, while symmetrization eliminates values under the chosen threshold ε. During the expansion, the sparse matrix is completely rewritten, as presented in Section 2.3.

2.1 Inflation

During inflation the sparse matrix is parsed record by record and both the s_{ij} similarity value and its approximated transposed value t_{ij} are raised to the r-th power. New values overwrite old ones in the same sparse matrix.

2.2 Normalization and Symmetrization

Normalization requires parsing the sparse matrix twice. In a first step, the sum of the similarity values is computed in each row. Let us denote by σ_i the sum of values in row i, $\forall i = 1 \ldots n$, computed as: $\sigma_i = \sum_{j \in \text{row}_i} s_{ij}$. In the second step, each element in the sparse matrix is divided by the corresponding sum: $s_{ij}^{(\text{new})} = s_{ij}\sigma_i^{-1}$ and $t_{ij}^{(\text{new})} = t_{ij}\sigma_j^{-1}$, $\forall i = 1 \ldots n$ and $\forall j \in \text{row}_i$. New similarity values overwrite the old ones in the same instance of sparse matrix.

The approximate symmetry of the similarity matrix S is assured by an iterative process in the main loop, situated between inflation and expansion. In each main loop, the symmetrization step is performed q times, and each symmetrization step is followed by a normalization. One symmetrization step requires a single parsing of the sparse matrix, and computes the following:

$$s_{ij}^{(\text{new})} = \begin{cases} \sqrt{s_{ij}t_{ij}} & \text{if } s_{ij}t_{ij} \geq \varepsilon^2 \\ 0 & \text{otherwise} \end{cases} \qquad \begin{matrix} \forall i = 1 \ldots n, \\ \forall j \in \text{row}_i. \end{matrix} \qquad (1)$$
$$t_{ij}^{(\text{new})} = s_{ij}^{(\text{new})}$$

Parameter q determines how accurate is the approximation of matrix symmetry, while ε is responsible for neglecting unimportant low values of similarity. Neglected values are not just overwritten by zero but completely eliminated from the sparse matrix, so that the time consuming expansion can operate on as few data as possible.

2.3 Expansion

Expansion is the only operation in the whole algorithm, which may raise the number of nonzeros in the similarity matrix, and thus requires a complete rewriting of the sparse matrix. Further on, our approach computes the expanded matrix row by row, and may need all rows of the input matrix as long as the last row of the output gets computed. This way we need two instances of the sparse matrix.

As long as a row of the expanded matrix is getting computed, it is stored in an n-element array. When the row is ready, nonzeros are transferred into the output sparse matrix. Let us denote the elements of this output matrix by $\bar{s}_{ij}$, $i, j = 1 \ldots n$, while $\mathbf{s}_i$ and $\bar{\mathbf{s}}_i$ will stand for row i of the input and output matrix, respectively. Since $\forall i, j = 1 \ldots n$, $\bar{s}_{ij} = \sum_{k \in \text{row}_i} s_{ik} s_{kj}$, we may compute row with index i $(i = 1 \ldots n)$ as

$$\bar{\mathbf{s}}_i = \left(\sum_{k \in \text{row}_i} s_{ik} s_{k1} \quad \sum_{k \in \text{row}_i} s_{ik} s_{k2} \cdots \sum_{k \in \text{row}_i} s_{ik} s_{kn} \right) = \sum_{k \in \text{row}_i} s_{ik} \mathbf{s}_k . \quad (2)$$

Thus by parsing row i and computing a linear combination of rows with index k, where $k \in \text{row}_i$ we obtain a whole row of the expansions' output matrix. Even after having parsed all rows and obtained $\bar{\mathbf{s}}_i$ $(\forall i = 1 \ldots n)$, this is only half the job of expansion, because the new $\bar{t}_{ij}$ values are also needed in the next iteration. These $\bar{t}_{ij}$ values are obtained from the output sparse matrix. A pointer to the current element in each row is needed, initially set to the first element of the row. Then the sparse matrix is parsed row by row, and for each $\bar{s}_{ij}$ existing nonzero in the sparse matrix, the transposed value should be the current element pointed in row j. If there is a correspondence in coordinates, $\bar{t}_{ij}$ gets the pointed $\bar{s}_{ji}$ value, and the pointer in row j steps to the next element. When the row parsing gets to the end of the last row, each current element pointer reaches the end of its own row, and all nonzeros have received their transposed values.

2.4 Algorithm

Let us summarize the proposed approach of the TRIBE-MCL algorithm:

1. Initialize the parameters of the algorithm with the desired values: inflation rate r, similarity threshold ε, and symmetrization steps in each loop q.
2. Load initial similarity matrix and store it as a sparse matrix.
3. Normalize the values in the sparse matrix as described in section 2.2.
4. Perform inflation as described in section 2.1.
5. Normalize and symmetrize the sparse matrix in q steps as indicated in section 2.2, beginning and ending with normalization.
6. Perform expansion as presented in section 2.3.
7. Go back to step 4 unless convergence is achieved.
8. Clusters are obtained as isolated subgraphs in the final similarity graph.

At the convergence point, all isolated subgraphs are complete with approximately equal edge weights within the group.

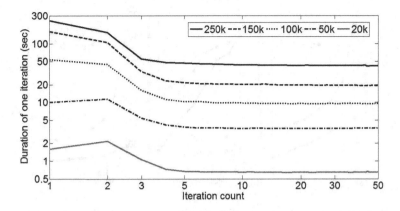

Fig. 2. Duration of first 50 iterations for various test data sizes at $r = 1.5$ and $\varepsilon = 10^{-3}$. The test with 250,000-item data set used $r = 1.7$.

3 Results and Discussion

The proposed algorithm was engaged in a series of numerical tests using synthetic protein data sets of various sizes in the range of 20-250 thousand items, created with the method given in [8]. For each test data size, 15 instances were created and the one with median number of nonzero values was chosen for the test.

Figure 2 exhibits the duration of each of the first fifty iterations in case of various matrix sizes, at inflation rate fixed at $r = 1.5$ and similarity threshold $\varepsilon = 10^{-3}$. As long as most nodes of the graph are connected together, namely in the first 5-6 loops, the computational load is somewhat higher, but it considerably falls thereafter, being virtually constant and low from the 10th loop.

Figure 3 exhibits the effect of the main parameters on the computational load of the algorithm. The input data here consisted of 50 thousand items having a similarity matrix of median density. Figure 3(a) indicates the total runtime of clustering performed in 50 iterations. As the inflation rate grows, the similarity matrix becomes sparser and thus the total runtime and also the length of late iterations is shorter. A lower value of the similarity threshold keeps small similarity values longer in the matrix, and consequently the processing needs more time (Fig. 3(b)).

The ratio between total runtime (duration of 50 iterations) and the length of a late iteration, exhibited in Fig. 3(c) shows us how much longer the first iterations are compared to late ones. This ratio would be 50 if the first loops were not at all computationally harder than the late ones. At $\varepsilon = 10^{-3}$ this ratio stays below 60, indicating that the algorithm quickly gets rid of unnecessary edges in the graph. At $\varepsilon = 10^{-4}$ this ratio can become 100, indicating that lots of computations are done in the first iterations to tear the graph into isolated subgraphs. Considering the fact that final clusters hardly differ from $\varepsilon = 10^{-4}$ to $\varepsilon = 10^{-3}$, choosing a low similarity threshold proves to be a waste of time.

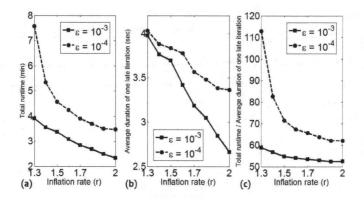

Fig. 3. Benchmark figures for a median density data set of 50k items, showing the influence of r and ε: (a) total runtime plotted vs. inflation rate; (b) average duration of one late iteration plotted vs. inflation rate; (c) the ratio between total runtime and duration of a late iteration plotted vs. inflation rate

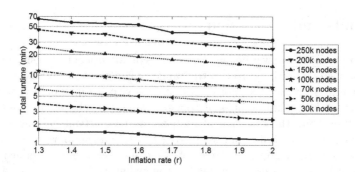

Fig. 4. Total runtime on various test data sizes, plotted against inflation rate, at constant similarity threshold value $\varepsilon = 10^{-3}$

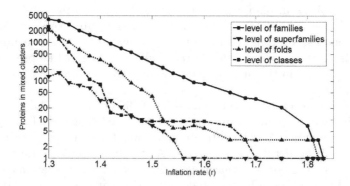

Fig. 5. The amount of proteins in mixed clusters found in a median density data set of 50k items, plotted against inflation rate, at constant $\varepsilon = 10^{-3}$

Figure 4 exhibits total runtime values plotted against inflation rate for input data set sizes ranging from 30,000 to 250,000 items, having the similarity threshold fixed at $\varepsilon = 10^{-3}$. Clustering at $r = 1.3$ takes 50-100% more time than at $r = 2$. Data sets of up to 250k items can be processed in an hour. Doubling the data size seemingly leads to four times longer processing.

Runtime benchmarks indicate that the proposed algorithm performs clustering in 30% less time than our previous high speed solution [10]. All efficiency tests were carried out on a notebook with quad core Haswell i7 processor running at 2.2GHz frequency and 24GB RAM memory, using a single core of the microprocessor.

The main goal of protein clustering is to reveal hidden similarities among proteins. When evaluating the output partition quality, one can count the number of mixed clusters (those which contain proteins from two or more different families) and their cardinality. We have shown in the previous work [9,10], that the inflation rate is the main factor to influence the amount of mixed clusters. The approach proposed here computes the same partitions as the conventional TRIBE-MCL, in a seriously more efficient way. Having drastically reduced the execution time of conventional TRIBE-MCL enabled us to perform a very detailed evaluation of the amount and nature of the mixed clusters in the output partition using synthetic data sets of several tens of thousand items.

Our synthetic data sets inherited the most important properties of SCOP95 proteins database, in which the proteins are organized in four-level hierarchy: classes, folds, superfamilies and families. Test data sets contain up to two dozens of classes, each of them containing several folds, the majority of which contain several superfamilies that are structured into families of proteins. Clusters established by TRIBE-MCL are called pure if all members belong to the same family. Alternately, there can be mixed clusters, which contain proteins of various families. Further on, mixed clusters may be mixtures of any of the four levels: e.g. if a mixed cluster contains proteins of different folds of a single class, then the mixture is called of the level of folds. The partition quality of a clustering algorithm may be characterized by the number of mixtures it produces, and their distribution among the four levels.

Figure 5 gives a detailed report on mixed clusters found in a 50,000-item data set of median density, for various inflation rates and a similarity threshold $\varepsilon = 10^{-3}$. The amount of mixed proteins highly depends on the inflation rate r, and in a very modest way on the similarity threshold ε. As the inflation rate rises, the number of mixed clusters and their cardinality drops quickly. At $r = 1.7$ there are no more mixtures at the level of classes and above $r = 1.83$ there are no more mixtures of any kind. As an effort to extract biologically useful information, it is recommendable to choose such an inflation rate, which produces mixtures and reveals hidden similarities among proteins of various known or unknown origins. Depending on the level and amount of mixtures we are looking for, the ideal inflation rate should be chosen in the range $1.4 - 1.7$.

The upper limit of processable data size is constrained by the memory of the used computer. The main determining factor is the number of nonzero values in the similarity matrix after the first expansion. After 2-3 iterations, the memory necessity stabilizes at a much lower level. With the current version of the algorithm, an ordinary PC with 2GB RAM can process a graph of 30,000 nodes, while 250,000 nodes require an upper class PC with 24GB RAM.

4   Conclusions

In this paper we have proposed an efficient approach to the graph-based TRIBE-MCL clustering method, a useful tool in protein sequence classification. The proposed approach proved extremely quick, and its memory needs are strongly reduced since previous versions. This novel implementation represents a major step of TRIBE-MCL towards handling huge data sets in reasonable time. We have shown that in case of the SCOP95 database and synthetic data sets derived from it, the most useful biological information can be extracted in case of inflation rates in range of $1.4 - 1.7$. Further enhancement of the algorithm's efficiency may be achieved via parallel implementation in CPUs or GPUs.

References

1. Altschul, S.F., Madden, T.L., Schaffen, A.A., Zhang, J., Zhang, Z., Miller, W., Lipman, D.J.: Gapped BLAST and PSI-BLAST: A new generation of protein database search program. Nucl. Acids Res. 25, 3389–3402 (1997)
2. Andreeva, A., Howorth, D., Chadonia, J.M., Brenner, S.E., Hubbard, T.J.P., Chothia, C., Murzin, A.G.: Data growth and its impact on the SCOP database: New developments. Nucl. Acids Res. 36, D419–D425 (2008)
3. Enright, A.J., van Dongen, S., Ouzounis, C.A.: An efficient algorithm for large-scale detection of protein families. Nucl. Acids Res. 30, 1575–1584 (2002)
4. Lo Conte, L., Ailey, B., Hubbard, T.J., Brenner, S.E., Murzin, A.G., Chothia, C.: SCOP: A structural classification of protein database. Nucl. Acids Res. 28, 257–259 (2000)
5. Needleman, S.B., Wunsch, C.D.: A general method applicable to the search for similarities in the amino acid sequence of two proteins. J. Mol. Biol. 48, 443–453 (1970)
6. Smith, T.F., Waterman, M.S.: Identification of common molecular subsequences. J. Mol. Biol. 147, 195–197 (1981)
7. Structural Classification of Proteins database,
   http://scop.mrc-lmb.cam.ac.uk/scop
8. Szilágyi, L., Kovács, L., Szilágyi, S.M.: Synthetic test data generation for hierarchical graph clustering methods. In: Loo, C.K., Yap, K.S., Wong, K.W., Teoh, A., Huang, K. (eds.) ICONIP 2014, Part II. LNCS, vol. 8835, pp. 303–310. Springer, Heidelberg (2014)
9. Szilágyi, L., Medvés, L., Szilágyi, S.M.: A modified Markov clustering approach to unsupervised classification of protein sequences. Neurocomput. 73, 2332–2345 (2010)
10. Szilágyi, S.M., Szilágyi, L.: A fast hierarchical clustering algorithm for large-scale protein sequence data sets. Comput. Biol. Med. 48, 94–101 (2014)

Hopfield-Type Associative Memory with Sparse Modular Networks

Gouhei Tanaka[1,*], Toshiyuki Yamane[2], Daiju Nakano[2], Ryosho Nakane[1]
and Yasunao Katayama[2]

[1] Graduate School of Engineering, The University of Tokyo, Tokyo 113-8656, Japan
gouhei@sat.t.u-tokyo.ac.jp
nakane@cryst.t.u-tokyo.ac.jp
[2] IBM Research - Tokyo, Kawasaki, Kanagawa 212-0032, Japan
{tyamane,dnakano,yasunaok}@jp.ibm.com

Abstract. Modular structures are ubiquitously found in the brain and
neural networks. Inspired by the biological networks, we explore Hopfield-
type recurrent neural networks with sparse modular connectivity for associative memory. We first show that an iterative learning algorithm,
which determines the connection weights depending on the network
topology, yields better performance than the one-shot learning rule. We
then examine the topological factors which govern the memory capacity
of the sparse modular neural network. Numerical results suggest that
the uniformity of the number of connections per neuron is an essential condition for good performance. We discuss a method to design an
energy-efficient neural network.

Keywords: Associative memory, Modular structures, Sparse networks,
Iterative learning algorithms, Bio-inspired computing.

1 Introduction

Artificial neural networks are the computational intelligence based on a simplified mathematical model of biological neural networks in the brain. They are
suitable for a wide variety of computational tasks that are not well handled by
ordinary rule-based programs, such as pattern recognition, classification, data
fitting and prediction, and feature extraction [1]. Although artificial neural networks are expected as a powerful framework of machine learning [2], it is still
challenging to achieve a good balance between high performance and energy
efficiency for their practical use. At present, a hardware implementation of numerous interconnections is particularly hard due to the limitation of technology
and its high power consumption [3]. A possible approach for overcoming this
problem is to reduce the number of individual neuron units and/or the connections between the neuron units while maintaining the computational ability.
In this context, an attention has been paid to the gap between the network
architecture of the computing system and the connectome of the brain [4].

---

* Corresponding author.

C.K. Loo et al. (Eds.): ICONIP 2014, Part I, LNCS 8834, pp. 255–262, 2014.
© Springer International Publishing Switzerland 2014

The Hopfield neural network is a class of recurrent artificial neural networks, which have been widely applied to associative memory, pattern recognition, and combinatorial optimization problems [5]. The Hopfield-type neural network has been highly developed by the proposal of new learning algorithms [6] and the extension to the complex domain [7, 8]. In associative memory, the Hopfield network preliminarily stores the information of a set of patterns (memory patterns) in the connection weights using a learning algorithm. When receiving an input pattern close to a memory pattern, the network produces an output pattern after repeated updates of the neuronal states. If the network output matches the corresponding memory pattern, then the memory association is regarded to be successful. The maximum number of patterns, which can be successfully memorized in the Hopfield network with N neurons, is approximately given by $0.14N$ [9]. Since the original Hopfield model is designed with a fully connected (all-to-all) structure without self-connection, the number of unidirectional connections is given by $N(N-1) \sim N^2$. This means that the number of connections approximately increases with the square of the system size. Particularly in a large-scale network, the dense connectivity is a serious problem in terms of energy efficiency as well as difficulty in hardware implementation [10].

Modifications of the network structure of the original Hopfield network have been often made to seek its biological plausibility. The performance of associative memory with randomly diluted Hopfield networks has been intensively studied [11–14]. It is shown that, for random dilution of connections, the storage capacity decreases in proportional to the percentage of dilution [11]. Namely, the storage capacity is proportional to the number of connections. Following the development of network science and complex network theory, much attention has been paid to the Hopfield model with small-world [15], scale-free [16], and other complex topologies [17–19]. These networks can be regarded as non-randomly diluted networks. In these studies, the effect of the network structure on the computational performance is examined with the one-shot Hebbian learning rule [20]. However, other learning algorithms have not yet been fully considered for the diluted Hopfield networks.

Recent studies on the brain and neuronal networks have revealed that the connectivity of such networks can be characterized by sparseness and modularity [21–25]. There is a possibility that modular structures arise in biological networks due to a selection pressure to reduce the number of connections between network nodes [26]. Motivated by these findings, we explore an energy-efficient Hopfield-type neural network for associative memory using sparse modular structures. Instead of the one-shot Hebbian learning rule assuming a full connection, we use an iterative local learning algorithm [27] which determines the connection weights depending on the network structure. We show that the iterative learning algorithm outperforms the one-shot learning in the storage capacity of the associative memory. We also examine the effect of different topologies of modular networks on the associative memory performance.

In Sec. 2, we describe the learning algorithm and the retrieval process in the Hopfield-type associative memory with a sparse modular structure. In Sec. 3, we show numerical results of the associative memory tests. In Sec. 4, we summarize this study and mention future works.

2   Methods

2.1   Learning Algorithms

We consider a Hopfield-type neural network consisting of N neurons which are interconnected in a sparse modular structure. The total number of unidirectional connections is represented as DN^2 with $0 \leq D \leq 1$, where D is the dilution factor denoting the proportion of the number of existing connections to the possible maximum number of connections [4]. For the original Hopfield network with a full connection, $D = 1$ in the limit $N \to \infty$. The average number of connections per neuron is given by DN. The weight of the connection between neurons i and j is denoted by w_{ij}, which is assumed to be symmetric, i.e. $w_{ij} = w_{ji}$. If neurons i and j are not connected, then we set $w_{ij} = 0$.

An associative memory model stores a set of training patterns by embedding the patten information into the connection weights. This is called a learning phase. The set of training patterns to be stored are given by binary vectors $\mathbf{s}^{(k)} = (s_1^{(k)}, \ldots, s_N^{(k)})^T$ for $k = 1, \ldots, P$, where P is the number of stored patterns and $s_i^{(k)} = 1$ or -1 for $i = 1, \ldots, N$.

In the one-shot Hebbian learning rule [20], the connection weights are determined as follows [5]:

$$w_{ij} = \begin{cases} \frac{1}{N} \sum_{k=1}^{P} s_i^{(k)} s_j^{(k)} & \text{for } i \neq j, \\ 0 & \text{for } i = j. \end{cases} \tag{1}$$

The distribution of the weights is given by a Gaussian distribution including negative values [6]. Since the full connection is assumed in this learning algorithm, non-zero values can be assigned even to the non-existing connections, for which the corresponding weights are neglected in the retrieval phase. This is often referred to as a one-shot algorithm because the weights are calculated in a single step from the training set.

In order to set the connection weights in a topology-dependent way, we adopt the perceptron-style iterative learning algorithm based on Hebb's rule [27]. The local fields for each training pattern are adjusted to be appropriate for a correct retrieval. Initially we set $w_{ij} = 0$ for all the connections. For a randomly chosen training pattern k, the weight corresponding to the existing connection between neurons i and j is updated as follows [27]:

$$w'_{ij} = w_{ij} + \frac{1}{N} s_i^{(k)} s_j^{(k)}, \tag{2}$$

until the following condition is satisfied:

$$h_i^{(k)} s_i^{(k)} \geq T \quad \text{for } i = 1, \ldots, N \text{ and } k = 1, \ldots, P. \tag{3}$$

The local field $h_i^{(k)}$ is defined as $h_i^{(k)} = \sum_{j \in V_i} w_{ij} s_j^{(k)}$, where V_i denotes the set of neurons connected to neuron i, i.e., $V_i = \{j \mid w_{ij} \neq 0\}$. This is the perceptron-style learning with a fixed margin T and a learning rate $1/N$. The weight will converge to an appropriate value if it exists, for which pattern k is guaranteed to be locally stable. We repeat the above procedure for all the stored patterns until inequality condition (3) is satisfied or the number of updates reaches $P \times t_L$.

2.2 Memory Retrieval

Once the connection weights are determined, a randomly chosen stored pattern (the target pattern) is modified by adding noise, or flipping a proportion n of the components, and then given to the network as an input. The states of the neurons are represented by a real vector $\mathbf{x} = (x_1, \ldots, x_N)^T$ with $-1 \leq x_i \leq 1$. The state of neuron i is updated as follows:

$$x_i' = f\left(\sum_{j \in V_i} w_{ij} x_j - \theta_i\right), \tag{4}$$

where the activation function f is given by the sigmoid function $f(x) = 2/(1 + \exp(-x/\epsilon)) - 1$ and θ_i stands for the threshold for firing. The discretized state of neuron i is 1 if $x_i > 0$ and 0 otherwise, for $i = 1, \ldots, N$. The state update is repeated asynchronously until the network state converges to the target pattern or the number of update steps reaches t_R.

3 Results

We generate a sparse modular network consisting of M modules, in each of which the neurons have the same number of connections. The number of neurons in module l is denoted by N_l for $l = 1, \ldots, M$. The size of the modules is assumed to be different according to biological networks [21–25]. The number of connections per neuron in module l is given by $D_l N$, satisfying $\sum_{l=1}^{M} D_l N_l / N = D$. The connections are separated into intramodule ones and intermodule ones, i.e., $D_l = D_l^{\text{intra}} + D_l^{\text{inter}}$. The adjacency matrix for an example of sparse modular networks consisting of 4 modules is shown in Fig. 1(a). The stored patterns with binary components are randomly generated to avoid strong correlations between the patterns. The parameter values are set at $N = 1024$, $D \sim 0.05$, $\epsilon = 0.1$, $\theta_i = 0$ for all i, $T = 1$, and $t_L = t_R = 50000$.

3.1 Comparison of the Learning Algorithms

First, we compare the one-shot and iterative leaning algorithms in associative memory tests using the sparse modular Hopfield networks. Figure 1(b) shows the final overlap m between the retrieved pattern $\mathbf{s}^r$ and the target pattern $\mathbf{s}^t$, evaluated as $m = \sum_{i=1}^{N} s_i^r s_i^t / N$. We observe that the iterative algorithm is much

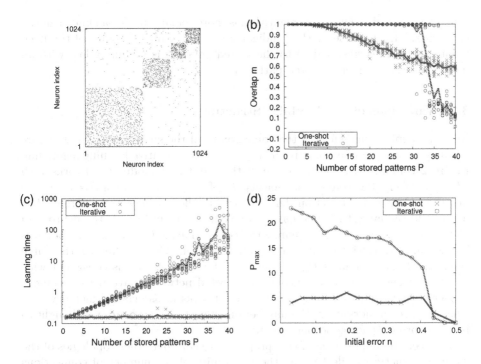

Fig. 1. (a) A representation of the adjacency matrix of a modular network with $M = 4$ modules, where a black point is plotted at (i, j) if $w_{ij} \neq 0$. (b) Overlap m versus the number P of the stored patterns for the one-shot learning (red crosses) and the iterative learning (blue open circles) in the modular network where $M = 4$, $(N_1, N_2, N_3, N_4) = (512, 256, 128, 128)$, $D_l^{\mathrm{intra}} \sim 0.04$ and $D_l^{\mathrm{inter}} \sim 0.01$ for $l = 1, \ldots, 4$, and $n \sim 0.03$. The red solid and blue dashed curves correspond to the average over 10 simulations. (c) Learning time versus the number of stored patterns. (d) The maximum number $P_{\max}$ of stored patterns with variation of the initial error rate n.

better than the one-shot algorithm in terms of the maximum number $P_{\max}$ of correctly retrieved patterns, i.e., $m = 1$ for $P < P_{\max}$. Whereas the final overlap for the one-shot algorithm starts decreasing at around $P \sim 5$, the iterative algorithm can achieve the perfect retrieval until the performance degradation at around $P \sim 30$. The rapid performance degradation suggests that inequality condition (3) is not satisfied before the maximum step is reached, because a successful memory association is guaranteed for a small initial error once a solution of the inequality condition is found. As shown in Fig. 1(c), the learning time for the one-shot algorithm is short and linearly dependent on P, while that for the iterative algorithm seems to grow exponentially with P. The latter is related to the time to find a solution of inequality condition (3). These results show that there is a trade-off between the performance of the successful memory retrieval and the computation time. Figure 1(d) shows the maximum number $P_{\max}$ of stored patterns with variation of the initial error rate n.

The advantage of the iterative algorithm is more remarkable for a smaller initial error rate. The decrease in $P_{\max}$ indicates that the global stability of the fixed point attractors corresponding to the stored patterns is gradually lost with an increase in the number of stored patterns.

3.2 Variations of the Modular Structure

Next, we explore the effect of the connectivity of the sparse modular structure on the performance of associative memory with the iterative learning algorithm. Figure 2 compares the performance of three different modular networks with $M = 4$, where the connection density D of the whole network is the same but the balance of connection densities within the modules are different. The neurons in the largest module have more connections in Case 1, the neurons in all the modules have the same number of connections in Case 2, and the neurons in the smallest modules have more connections in Case 3. Among the three cases, the largest value of $P_{\max}$ is obtained for Case 2. This is consistent with the consequence that homogeneously connected networks have better memory capacity than heterogeneously connected networks in associative memory [17]. Therefore, the uniformity of the number of connections per neuron is thought to be a key property for good performance.

To check the importance of this property, we consider other topologies of the modular structure while keeping the uniformity of the number of connections per neuron. Figure 3(a) shows the effect of the number of modules on the final overlap. The result indicates that the performance is not altered by the number of modules as long as the connection uniformity is maintained. Figure 3(b) shows the effect of the balance between the number of intermodule and intramodule connections per neuron. Even if the network structure is varied from a highly

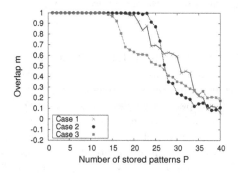

Fig. 2. The parameters are fixed at $M = 4$ and $(N_1, N_2, N_3, N_4) = (512, 256, 128, 128)$, and $D_i^{\text{inter}} \sim 0.01$. (Case 1) The larger module is more densely connected. $(D_1, D_2, D_3, D_4) \sim (0.06, 0.03, 0.15, 0.15)$. (Case 2) The number of connections per neuron is uniform. $(D_1, D_2, D_3, D_4) \sim (0.04, 0.04, 0.04, 0.04)$. (Case 3) The smaller module is more densely connected. $(D_1, D_2, D_3, D_4) \sim (0.02, 0.04, 0.08, 0.08)$.

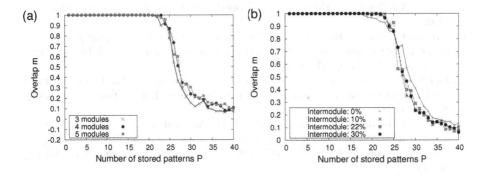

Fig. 3. (a) The effect of the number of modules. (b) The effect of the balance between the number of intermodule and intramodule connections.

modular structure to a random structure by changing the ratio D_l^{inter}/D_l of the intramodule connections, the performance is almost the same for the wide range of P. These results support the consequence that the uniformity of the number of connections per neuron is an essential factor for good performance of the Hopfield-type associative memory with iterative learning algorithm.

4 Conclusions

As a first step towards a realization of bio-inspired energy-efficient computing systems, we have examined the Hopfield-type neural associative memory using sparse modular structures. We have found that the uniformity of the number of connections per neuron governs the performance of the associative memory based on the iterative learning algorithm. This property is considered to be related to the solvability of inequality condition (3). The generality of the property can be checked, for instance, using correlated patterns like real images. Through such an effort, we aim to establish a design method of neural network architectures, which are energy-efficient, cost-effective, and amenable to hardware implementation.

References

1. Bishop, C.M.: Pattern Recognition and Machine Learning, vol. 1. Springer, New York (2006)
2. Hinton, G.E., Salakhutdinov, R.R.: Reducing the dimensionality of data with neural networks. Science 313(5786), 504–507 (2006)
3. Misra, J., Saha, I.: Artificial neural networks in hardware: A survey of two decades of progress. Neurocomputing 74(1), 239–255 (2010)
4. Katayama, Y., Yamane, T., Nakano, D.: An energy-efficient computing approach by filling the connectome gap. In: Ibarra, O.H., Kari, L., Kopecki, S. (eds.) UCNC 2014. LNCS, vol. 8553, pp. 229–241. Springer, Heidelberg (2014)
5. Hopfield, J.J.: Neural networks and physical systems with emergent collective computational abilities. Proc. Natl. Acad. Sci. USA 79(8), 2554–2558 (1982)
6. Davey, N., Hunt, S.P., Adams, R.: High capacity recurrent associative memories. Neurocomputing 62, 459–491 (2004)

7. Hirose, A.: Complex-Valued Neural Networks. SCI, vol. 32. Springer, Heidelberg (2006)

8. Tanaka, G., Aihara, K.: Complex-valued multistate associative memory with non-linear multilevel functions for gray-level image reconstruction. IEEE Trans. Neural Networks 20(9), 1463–1473 (2009)

9. McEliece, R.J., Posner, E.C., Rodemich, E.R., Venkatesh, S.S.: The capacity of the Hopfield associative memory. IEEE Transactions on Information Theory 33(4), 461–482 (1987)

10. Beiu, V., Madappuram, B.A.M., Kelly, P.M., McDaid, L.J.: On two-layer brain-inspired hierarchical topologies – A rent's rule approach –. In: Stenström, P. (ed.) Transactions on HiPEAC IV. LNCS, vol. 6760, pp. 311–333. Springer, Heidelberg (2011)

11. Sompolinsky, H.: Neural networks with nonlinear synapses and a static noise. Physical Review A 34(3), 2571 (1986)

12. Derrida, B., Gardner, E., Zippelius, A.: An exactly solvable asymmetric neural network model. Europhys. Lett. 4(2), 167 (1987)

13. Treves, A., Amit, D.J.: Metastable states in asymmetrically diluted hopfield networks. Journal of Physics A: Mathematical and General 21(14), 3155 (1988)

14. Gardner, E.: Optimal basins of attraction in randomly sparse neural network models. Journal of Physics A: Mathematical and General 22(12), 1969 (1989)

15. Bohland, J.W., Minai, A.A.: Efficient associative memory using small-world architecture. Neurocomputing 38, 489–496 (2001)

16. Stauffer, D., Aharony, A., da Fontoura Costa, L., Adler, J.: Efficient hopfield pattern recognition on a scale-free neural network. The European Physical Journal B-Condensed Matter and Complex Systems 32(3), 395–399 (2003)

17. McGraw, P.N., Menzinger, M.: Topology and computational performance of attractor neural networks. Physical Review E 68(4), 047102 (2003)

18. Torres, J.J., Munoz, M.A., Marro, J., Garrido, P.: Influence of topology on the performance of a neural network. Neurocomputing 58, 229–234 (2004)

19. Kim, B.J.: Performance of networks of artificial neurons: The role of clustering. Physical Review E 69(4), 045101 (2004)

20. Hebb, D.O.: The organization of behavior: A neuropsychological theory. Psychology Press (2005)

21. Scannell, J.W., Blakemore, C., Young, M.P.: Analysis of connectivity in the cat cerebral cortex. The Journal of Neuroscience 15(2), 1463–1483 (1995)

22. Rubinov, M., Sporns, O.: Complex network measures of brain connectivity: uses and interpretations. Neuroimage 52(3), 1059–1069 (2010)

23. Hagmann, P., Cammoun, L., Gigandet, X., Meuli, R., Honey, C.J., Wedeen, V.J., Sporns, O.: Mapping the structural core of human cerebral cortex. PLoS Biology 6(7), e159 (2008)

24. Bullmore, E., Sporns, O.: Complex brain networks: graph theoretical analysis of structural and functional systems. Nature Reviews Neuroscience 10(3), 186–198 (2009)

25. Varshney, L.R., Chen, B.L., Paniagua, E., Hall, D.H., Chklovskii, D.B.: Structural properties of the caenorhabditis elegans neuronal network. PLoS Computational Biology 7(2), e1001066 (2011)

26. Clune, J., Mouret, J.B., Lipson, H.: The evolutionary origins of modularity. Proceedings of the Royal Society B: Biological Sciences 280(1755), 20122863 (2013)

27. Diederich, S., Opper, M.: Learning of correlated patterns in spin-glass networks by local learning rules. Physical Review Letters 58, 949–952 (1987)

Concept Drift Detection Based on Anomaly Analysis

Anjin Liu, Guangquan Zhang, and Jie Lu

Decision Systems & E-Service Intelligence Research Laboratory,
Center for Quantum Computing and Intelligent System, School of Software,
Faculty of Engineering and Information Technology, University of Technology Sydney,
Australia
anjin.liu@student.uts.edu.au;
{guangquan.zhang,jie.lu} @uts.edu.au

Abstract. In online machine learning, the ability to adapt to new concept quickly is highly desired. In this paper, we propose a novel concept drift detection method, which is called Anomaly Analysis Drift Detection (AADD), to improve the performance of machine learning algorithms under non-stationary environment. The proposed AADD method is based on an anomaly analysis of learner's accuracy associate with the similarity between learners' training domain and test data. This method first identifies whether there are conflicts between current concept and new coming data. Then the learner will incrementally learn the non-conflict data, which will not decrease the accuracy of the learner on previous trained data, for concept extension. Otherwise, a new learner will be created based on the new data. Experiments illustrate that this AADD method can detect new concept quickly and learn extensional drift incrementally.

Keywords: Adaptive Intelligent Systems, Online Machine Learning, Incremental Learning, Concept Drift.

1 Introduction

In the real world, there are a growing number of applications generating data continuously and requiring efficient machine learning algorithms to cope with this data. For example, personal assistance applications dealing with information filtering, macroeconomic forecasting, bankruptcy prediction or individual credit scoring [1]. Moreover, the fast pace of preference changing of the target customers (*concept drift*) is also a challenge to existing learning algorithms. As a result, conventional machine learning algorithms, which hold a stationary distribution assumption, will be replaced by more efficient online learning algorithms, which have the ability to adapt to new environment quickly, sooner or later.

The issue of *concept drift* refers to the change of the distribution underlying the data at different time steps [2], in which the term *concept* refers to the distribution of a problem at a certain time step. Concept drift will lead to the predictions of well-trained classifiers become less accurate as time passes. More formally, lets denote the feature vector as x and the class label as y, then an infinite sequence of (x, y) presents

C.K. Loo et al. (Eds.): ICONIP 2014, Part I, LNCS 8834, pp. 263–270, 2014.
© Springer International Publishing Switzerland 2014

the data stream, p_t(x, y) is the distribution of data chuck at time step t, the term concept drift means that p_t(x, y) ≠ p_{t+1}(x, y) [3]. Recall the Bayesian Probability Theory, p(x, y) can be decompose as p(x, y) = p(x) × p(y | x). In Kelly, et al. [4] publication, they concluded that concept drift can be caused by the drifting of p(x) over time t (it can also be written as p(x | t)), or the drifting of p(y | x), which is the conditional probability of feature x, or both. Virtual concept drift is neither the change of p(x | t) nor p(y | x). It is caused by the sampling shift of current p(x | t) or p(y | x), or both.

Concept drift can be categorized into different types based on different criteria as shown in the literatures. Minku, et al. [4] proposed that concept drift could be categorized into 14 types based on the drifting speed, severity, predictability, frequency and recurrence. In the real-world applications, the types of concept drift can be varied and mixed. In addition, in some special cases, virtual drift may have the same effect on learning model as concept change. For example, p_{t+1}(x | t) is the sampling shift of p_t(x | t) and they are not equal, they might be treated as two different concepts and assigned to two learners separately. If the data chuck at t+2 with distribution of mixed p_{t+1}(x | t) and p_t(x | t), neither classifier$_t$ nor classifier$_{t+1}$ would achieve a high accuracy. These issues have made concept drift even difficult to be solved. Current ensemble drift detection and handling approaches treats virtual concept drift and real concept drift as the same problem. These approaches detect drifts based on the outputs of learner at each time step without considering whether the drift is a sampling shift or a new one. As a result, their performances are limited.

Motivated by these issues, we propose a novel drift detection method for online learning algorithms, which runs anomaly analysis on the accuracy associate with the similarity between training domain and test data. In the anomaly analysis, we focus on the data that was correctly classified by existing learners. We compare the similarity of the distribution between the old correctly classified data and the new data. Under normal circumstances, including virtual concept drift, the similarity and the accuracy should stay at a stable ratio. Otherwise, it can be either a concept change or noise. Our approach is capable to high light the data with unknown distribution and identifies the conflicts instances. Therefore, both virtual drift and real concept drift could be handled well.

The organization of this paper is as follow: In the next section, we survey the state of art drift detection methods based on learners' outputs. Section 3 explains and presents the details of our new proposed AADD method. Section 4 presents the preliminary and the evaluation results of the proposed AADD method. Section 5 concludes this paper and discusses some future works.

2   Related Work

This section formally presents the problem of concept drift, and analyzes the advantages and drawbacks of established literature with regard to concept dirt detection based on learners' outputs.

2.1   Problem Description: Concept Drift

We reference the definition from a most recent concept drift review which was given by Zliobaite [1], to present concept drift. In classification problems, at every time step t we have historical data (labelled) available $x^H = (x_1, \ldots , x_t)$. For each time increasing t+1, a new instance x_{t+1} arrive. The task is to predict a label c_i, where $c_i, c_2, \ldots , c_k$ is the set of class labels, for every new coming instance x_{t+1}. The optimal classifier to classify $x \rightarrow c_i$ is completely determined by a prior probabilities for the classes $p(c_i)$ and the class-conditional probability density functions (pdf) $p(x|c_i)$, $i = 1, \ldots, k$. They define a set of prior probabilities of the classes and class-conditional pdfs as concept or data source, denote it S:

$$S = (p(c_1), p(x|c_1)), (p(c_2), p(x|c_2)), \ldots ,(p(c_k), p(x|c_k)) \tag{1}$$

Every instance x_t is generated by a source S_t. If all the data is sampled from the same source, i.e. $S_1 = S_2 = \ldots = S_{t+1} = S$ we say that the concept is stable. If for any two time points i and j exists $S_i \neq S_j$, we say that there is a concept drift.

However, for some special situation S_{new}, $S_i \neq S_j$ and $S_i \cup S_j = S_{new}$, if we treat them as three different concepts, we will have to train three different learners. Moreover, S_{new} would not be able to take the advantages of previous concept or data source S_i and S_j. Therefore, we suggest learning this type of drift incrementally.

2.2   Drift Detection Methods by Outputs of Learners

To the best of our knowledge, explicit drift detection can be categorized into three groups, detecting drift by data distribution [5], learner outputs [6, 7] and competence model [3]. A more comprehensive literature review can be found in [3]. Comparing with other types of drift detection methods, drift detection by learners' outputs is the most intuitive and has a relative low computational cost. On the other hand, this type of drift detection method can only take reactions after drift. In this section, we will give a literature review on the drift detection by learners' outputs.

Drift Detection Method (DDM), which proposed by Gama et al. [7], detects concept drift by tracing the online error rate of the learning algorithm. It treats the error of a set of examples as a Bernoulli trail random variable. The number of errors in a sample set should follow Binomial distribution. The changes in the errors of the algorithm indicate the changes of the class distribution. Since DDM assesses a learner through its overall performance, it is more suitable for identifying concept change and rebuilding models rather than updating an existing learner, which means that it cannot handle slowly gradual drift [6]. Hence, Baena-García et al. [6] proposed a upgraded version of DDM, Early Drift Detection Method (EDDM), to improve the detection in the presence of gradual concept drift. The difference is that EDDM considers the distance between two consecutive erroneous classifications instead of the overall error rate. They assume that the change of the distance reflects the changes of the current concept. Moreover, they applied a warning system to reduce the error detections caused by noise. In spite of that, EDDM is still very sensitive to noisy examples [8].

3   Anomaly Analysis Drift Detection Methods

This section presents the AADD method. In Section 3.1, we give an intuitive explanation of the proposed method. In Section 3.2, a detailed description of the AADD method is explained.

3.1   The Idea of AADD

The main reason why concept drift would lead to well-trained learner becoming inaccurate is either the test data distribution is not learned sufficiently or the class labels changed with the same distribution. If the environment is not noise free, it can also cause the same problem. Therefore, from this point of view, we believe that if we could monitor the performance of a learner on its confident data distribution on which the learner always have a high accuracy, we would be able to quickly identify what caused the drop of accuracy. For example, if a batch of data chuck is from a new distribution and has no conflict with current learner, the similarity would be low and the accuracy would be low as well. We use a table to illustrate the differences.

Table 1. Concept drift similarity & accuracy analysis

Drift or Noise	Similarity Change	Accuracy Change
Noise	No change	Fluctuating
Concept extension (Distribution extended)	Decreasing	Decreasing
Concept change (Conflicts under the same distribution)	No change	Decreasing

The core idea of AADD method is that monitoring the similarity and accuracy of the new available sample data at each time step. By taking the similarity into consideration, AADD would be able to identify whether the incorrect predictions is caused by the conflicts between current learning model and new concept or they are caused by unlearned distribution or noise. In addition, with similarity functions, data can be only trained and analyzed once in an online manner.

3.2   The Method Description

At each time step of the data stream, we assume that there are 2 batches of data, $D_{train}(t)$, $D_{test}(t)$. The $D_{train}(t)$ can be a subset of $D_{test}(t)$ with known labels or they share similar distributions. After utilizing $D_{train}(t)$ to create and update a learner L (step 3, 8 and 11), we change the label of an instance in $D_{train}(t)$ to *TRUE* if it be classified correctly by L, else to *FALSE* to form a new dataset $D`_{train}(t)$. And then we use $D`_{train}(t)$ to create or update learner L's similarity function (step 4, 9 and 12). For each new coming train data chuck, we run anomaly analysis to verify the ratio of the similarity and the accuracy, incremental learning it if no conflict with current learner, otherwise, build new learner base on it, Step 6. The AADD method is described as follows:

Anomaly Analysis Drift Detection:		
Input:		
	Noise sensitive parameter θ	
	Updateable learning algorithm	
	For each time step D_{train}, D_{test}	
Output:		
	Predictions of D_{test}	
1.	**For** t = 0: numBatch	
2.	**If** t = 0	
3.	buildNewLearner(D_{train}(t))	
4.	buildSimilarityFunctions($D`_{train}$ (t))	
5.	**Else**	
6.	conflictDetection(D_{train} (t))	
7.	**If** no conflict	
8.	incrementalLearning(currentLearner, D_{train} (t))	
9.	incrementalLearning(similarityFunction, $D`_{train}$ (t))	
10.	**Else**	
11.	buildNewLearner(D_{train} (t))	
12.	buildSimilarityFunctions($D`_{train}$ (t))	
13.	**End**	
14.	**End**	
15.	**End**	
16.	classification(currentLearner, D_{test}(t))	

The conflict between active learner and new coming data is identified by the following function, which indicates the abnormal data batch:

$$acc_{Dtrain} \times similarity_{Dtrain} < acc_{learner} \times similarity_{Dtrain} - \theta \times (1 - similarity_{Dtrain}) \qquad (2)$$

where acc_{Dtrain} is the accuracy of the training data, $similarity_{Dtrain}$ is the similarity of the training data, $acc_{learner}$ is the stable accuracy of current learner, θ is a parameter to control the sensitive to noise, the smaller value the θ is, the more sensitive to noise.

4   Experiments and Result Analysis

In this section, we present our experiment results of AADD method. First, in Section 4.1, we give the configuration details of the experiments. Secondly, in Section 4.2, we show the accuracy change caused by concept drift and plot the anomaly points at each time step on a graph.

4.1   Experiment Setup

In order to test AADD method, we applied it on the SEA Concepts [9], which has been used by many researchers as a standard to test algorithms for concept drift. This dataset have two class values and three features, with only two features being relevant and the third one being noise. Class values are assigned based on the sum of the two relevant features. If the sum of these two features of one instance is lower than a given threshold, this instance will be assigned to class 1. Otherwise, it will be assigned to class 2. The threshold will be updated after a predefined time step to simulate an abrupt shift in the class boundary. The values of the three features are uniformly distributed between 0 and 10, and the threshold is changed three times throughout the experiment with increasing severity 8→9→7.5→9.5. For example:

Table 2. SEA Concepts

attribute 1(0 - 10)	attribute 2 (0 - 10)	attribute 3 (noise) (0 - 10)	Class {1, 2}
Threshold = 8 (if attribute1 + attribute2 < 8 then class 1, otherwise class2)			
8.498129	1.243221	5.675182	class 2
...	...	...	...
Threshold = 9.5 (if attribute1 + attribute2 < 9.5 then class 1, otherwise class2)			
1.406376	0.738125	2.598439	class 1
...	...	...	...

Our testing procedure is identical to that described in [9]: 50000 instances training data, 250 instances each time step. In our experiment, we used the weka naive bayes updateable classifier as the base learner and the similarity functions. We run the test 10 times and calculate the average accuracy as the final result. The θ we used here is 0.05. Meanwhile, we also run a test with the same classifier but manually discard old learners and create new one at each drift time step. At last, we put our implementation of NSE++ [10] to demonstrate how a concept drift may affect the performance of learning model. The parameters of NSE are same as it was suggested in their paper, a = 10, b = 0.5 data size = 250, base learner is weka naive bayes updateable classifier.

4.2   Experiment Results

Fig 1 is the performance of three concept drift algorithms on SEA concept. As shown above, NB with known drift point was forced to drop old learner and create new one at each drift time step. Therefore, it is barely affected by the change of concept. The difference of accuracy between each concept period is caused by the concept itself (some concepts are easier to be predicted correctly). By contrast, NSE++ with same base classifier, which does not have active drift detection method, have a significant accuracy drop at each drift time step. After that, it recovers from drift gradually. Regarding to AADD, there is no significant accuracy drop at drift time step. However, indeed, there is a recovery period for AADD to get back to normal accuracy by incremental learning. This is because of that the AADD method can identify the conflict learners once there is a drift and abandon them before they could cause any negative effect on new concept predictions.

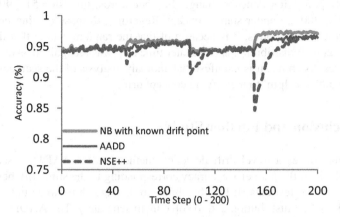

Fig. 1. Compare AADD method with NB and NSE++

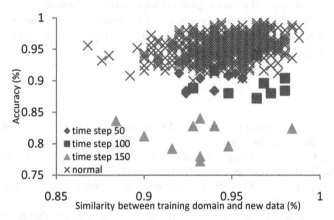

Fig. 2. Anomaly analysis of the accuracy associate with similarity

Fig 2 shows the distribution of the accuracy of the current learner on new coming data and the similarity returned by its similarity function. We put all the 10 runs into one figure so that we could have enough drift points to present the differences between stable points and drifting points. As described in section 4.1, time step 50 is the time when threshold change from 8→9 (presented with diamond), time step 100 is 9→7.5 (presented with square) and time step 150 is 7.5→9.5 (presented with triangle). The points at time step 150 are easy to be understood. As it has the biggest concept change $absolute(7.5 - 9.5) = 2$, the accuracy is much lower than the others with the same similarity. Regarding to time step 50 and 100, both of them have an accuracy drop compare to the normal data points. However, as time 50 has a relatively small change $absolute(8 - 9) = 1$, the shifts of similarity of these points are relatively stable. By contrast, time step 100 has a larger change $absolute(9 - 7.5) = 1.5$, so it would spread into a larger region. From fig 2, we can also see that after quick reaction to adapt to new concept, which is creating new classifier at time step 50, 100 and 150, the rest time step of these new concepts have go back to normal position. There is no other anomaly point after concept change. In other words, time step 51...99, 101 ... 159 and 151... 200 are under stable concept. Regarding to those drifting points that are mixed with normal points, it is because of that the random data at that time does not have a clear drift distribution and the later time step was recognized as drift point.

From the results above, it is manifest that anomaly analysis of the accuracy and the similarity would be helpful to identifying concept drift.

5   Conclusion and Further Study

This paper introduces a novel drift detection method, called AADD, based on the anomaly analysis of the learner's accuracy corresponding to the similarity between its training domain and test data. It has the ability to distinguish between unknown distribution and conflict distribution and to solve them separately. The AADD method is capable to detect concept extension and change efficiently and highlights the conflict instances at the same time. It offers a great convince to drift handling.

Our next attempt will aim to combine some drift handling approaches with AADD to research how to take the advantages of AADD for improving the performance of learning under non-stationary environment.

References

1. Zliobaite, I.: Learning under concept drift: An overview. Technical report, Vilnius University, 2009 techniques, related areas, applications Subjects: Artificial Intelligence (2009)
2. Zliobaite, I., Bifet, A., Pfahringer, B., Holmes, G.: Active Learning With Drifting Streaming Data. IEEE Transactions on Neural Networks and Learning Systems 25, 27–39 (2014)
3. Lu, N., Zhang, G., Lu, J.: Concept drift detection via competence models. Artificial Intelligence 209, 11–28 (2014)
4. Kelly, M.G., Hand, D.J., Adams, N.M.: The impact of changing populations on classifier performance. In: Proceedings of the Fifth ACM SIGKDD International Conference on Knowledge Discovery and Data Mining, pp. 367–371 (1999)
5. Dasu, T., Krishnan, S., Venkatasubramanian, S., Yi, K.: An information-theoretic approach to detecting changes in multi-dimensional data streams. In: Proc. Symp. on the Interface of Statistics, Computing Science, and Applications (2006)
6. Baena-García, M., del Campo-Ávila, J., Fidalgo, R., Bifet, A., Gavaldà, R., Morales-Bueno, R.: Early drift detection method (2006)
7. Gama, J., Medas, P., Castillo, G., Rodrigues, P.: Learning with drift detection. In: Bazzan, A.L.C., Labidi, S. (eds.) SBIA 2004. LNCS (LNAI), vol. 3171, pp. 286–295. Springer, Heidelberg (2004)
8. Nishida, K., Yamauchi, K.: Detecting concept drift using statistical testing. In: Corruble, V., Takeda, M., Suzuki, E. (eds.) DS 2007. LNCS (LNAI), vol. 4755, pp. 264–269. Springer, Heidelberg (2007)
9. Street, W.N., Kim, Y.: A streaming ensemble algorithm (SEA) for large-scale classification. Presented at the Proceedings of the Seventh ACM SIGKDD International Conference on Knowledge Discovery and Data Mining, San Francisco, California (2001)
10. Elwell, R., Polikar, R.: Incremental learning of concept drift in nonstationary environments. IEEE Trans. Neural Netw. 22, 1517–1531 (2011)

Online Learning for Faulty RBF Networks
with the Concurrent Fault

Wai Yan Wan, Chi-Sing Leung, Zi-Fa Han, and Ruibin Feng

Dept. of Electronic Engineering, City University of Hong Kong, Hong Kong
eeleungc@cityu.edu.hk, vianwanwai@yahoo.com.hk
rfeng4-c@my.cityu.edu.hk, zifahan@gmail.com

Abstract. Although there are some batch model learning algorithms for handling the weight fault situation, there are few results about online learning for handling this situation. Besides, a recent article showed that the objective function of the weight fault injection algorithm is not equal to the training error of faulty radial basis function (RBF) networks. This paper proposes an online learning algorithm for handling faulty RBF networks with two types of weight failure. We prove that the trained weight vector converges to the batch mode solution. Our experimental results show that the convergent behavior of the proposed algorithm is better than the conventional online weight decay algorithm.

Keywords: RBF networks, Weight Failure, Online Learning.

1 Introduction

Weight failure is unavoidable because a neural network cannot be implemented in a perfect way. For instance, the finite precision in the trained weights introduces multiplicative weight noise [1]. Physical faults introduce open weight fault [2]. Although many training methods have been developed, most of them focus on one kind of weight failure only [3, 4]. In the real situation, different kinds of weight failure could happen in a singe neural network. There are some results related to the weight noise and weight fault based on the weight decay approach [4]. In this approach, an appropriate weight decay parameter is very important. In [4], a systematic method for selecting the weight decay parameter was proposed. Since the training objective of this approach is not equal to the training set error of faulty networks, the performance of this approach is not optimal. Also, the selection method [4] is suitable for the batch mode learning only.

The weight failure regularization (WFR) [5] is a batch mode approach. In terms of training set error, it is optimal. However, the disadvantage of using the batch mode learning is that it needs to store the entire input-output history. One way to solve this problem is to develop an online algorithm, in which a key issue is the convergent condition. For the RBF model, the online weight fault injection algorithm does not achieve the desired result as we expect [6]. As shown in Figure 1, the performance of the online weight fault injection algorithm is almost similar to that of the standard LMS algorithm.

This paper presents an online version for the WFR. One of the important issues in online learning is the convergent behavior. The common requirement is that the converged solution should be equal to the corresponding batch mode solution. We establish

C.K. Loo et al. (Eds.): ICONIP 2014, Part I, LNCS 8834, pp. 271–278, 2014.

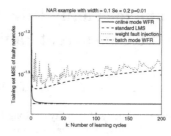

Fig. 1. Convergent behavior of various online algorithms. The details of the two data sets are provided in Section 4.

the convergent condition for the online mode WFR algorithm. The organization of this paper is as follows. Section 2 reviews the concept of RBF networks. Section 3 presents the online mode WFR algorithm and the convergent analysis. The simulation results are presented in Section 4. Section 5 concludes the paper.

2 Background

Denote the training data set as $\mathcal{D}_t = \{(\boldsymbol{x}_i, y_i) : \boldsymbol{x}_i \in \Re^K, y_i \in \Re, i = 1, 2, ..., N\}$, where x_i and y_i are the input and the target output of the i-th sample, respectively. We assume that the output y_i is generated by an unknown system, given by $y_i = f(\boldsymbol{x}_i) + \epsilon_i$, where $f(\cdot)$ is the unknown nonlinear function and ϵ_i's are the independent zero-mean Gaussian random variables with variance of σ_ϵ^2.

In the RBF approach, the unknown system is approximated by $f(\boldsymbol{x}) \approx \tilde{f}(x, w) = \sum_{j=1}^{M} w_j \phi_j(\boldsymbol{x}) = \boldsymbol{\phi}^T(\boldsymbol{x})\boldsymbol{w}$, where M is the number of RBFs, $\boldsymbol{w} = [w_1, \cdots, w_M]^T$ is the weight vector, $\phi_j(\boldsymbol{x}) = \exp\left(-\frac{\|\boldsymbol{x}-\boldsymbol{c}_j\|^2}{s}\right)$ is the j-th basis function, $\boldsymbol{\phi}(\boldsymbol{x}) = [\phi_1(\boldsymbol{x}), \cdots, \phi_M(\boldsymbol{x})]^T$, and $\boldsymbol{c}_j$ is the center of the j-th basis function. We assume that all the RBFs are with the same width s. For the fault-free case, the training set error is given by $\frac{1}{N}\sum_{i=1}^{N}(y_i - \boldsymbol{\phi}^T(\boldsymbol{x}_i)\boldsymbol{w})^2$. The optimal weight vector for minimizing the training set error is given by $\boldsymbol{w}_{ls} = \boldsymbol{H}^{-1}\frac{1}{N}\sum_{i=1}^{N}\boldsymbol{\phi}(\boldsymbol{x}_i)y_i$, where $\boldsymbol{H} = \frac{1}{N}\sum_{i=1}^{N}\boldsymbol{\phi}(\boldsymbol{x}_i)\boldsymbol{\phi}^T(\boldsymbol{x}_i)$.

In the real situation, an RBF network may suffer from multiplicative weight noise and open weight fault at the same time [6]. This fault situation can be modelled as

$$\tilde{w}_{j,b,\beta} = (w_j + b_j w_j)\beta_j \ \forall \ j = 1, \cdots, M, \tag{1}$$

where $\tilde{w}_{j,b,\beta}$ is the faulty weight. In (1) b_j's are independent zero-mean random variables with variance σ_b^2 and their density function are symmetric. Also, in (1), the open fault factors β_j's are identical independent binary random variables. The probability mass function of β_j is given by

$$\text{Prob}(\beta_j = 0) = p \text{ and } \text{Prob}(\beta_j = 1) = 1 - p. \tag{2}$$

Given a faulty weight vector $\tilde{\boldsymbol{w}}_{b,\beta}$, from (1), the training set error is given by

$$\mathcal{E}(\mathcal{D}_t)_{b,\beta} = \frac{1}{N}\sum_{i=1}^{N}(y_i - \boldsymbol{\phi}^T(\boldsymbol{x}_i)\tilde{\boldsymbol{w}}_{j,b,\beta})^2. \tag{3}$$

From the definition of the fault model, we obtain $\langle \beta_j \rangle = \langle \beta_j^2 \rangle = 1 - p$, $\langle \beta_j \beta_{j'} \rangle = (1 - p)^2 \; \forall j \neq j'$, $\langle b_j \rangle = \langle b_j b_{j'} \rangle = 0 \; \forall j \neq j'$ and $\langle b_j^2 \rangle = \sigma_b^2$, where $\langle \cdot \rangle$ is the expectation operation. Taking the expectation over b's and β's, we obtain the expected training error of faulty networks, given by

$$\bar{\mathcal{E}}(\mathcal{D}_t)_{b,\beta} = \frac{1-p}{N} \sum_{i=1}^{N} (y_i - \phi^T(x_i)w)^2 + \frac{p}{N} \sum_{i=1}^{N} y_i^2 + (1-p)w^T[(p+\sigma_b^2)G - pH]w, \quad (4)$$

where $H = \frac{1}{N} \sum_{i=1}^{N} \phi(x_i)\phi^T(x_i)$ and $G = \mathbf{diag}(H)$. Equation (4) shows the training error of faulty RBF networks. Since the first term $\frac{1-p}{N} \sum_{i=1}^{N} (y_i)$ in (4) is not a function of w, we can define the training objective function as

$$\bar{\mathcal{L}}(w) = \frac{1}{N} \sum_{i=1}^{N} (y_i - \phi^T(x_i)w)^2 + w^T \left[(p + \sigma_b^2)G - pH \right] w \quad (5)$$

In (5), the second term can be considered as a regularization term. Considering $\frac{\partial \bar{\mathcal{L}}(w)}{\partial w} = 0$, we obtain the optimal weight vector, given by

$$w^* = \left[(1-p)H + (p + \sigma_b^2)G \right]^{-1} \frac{1}{N} \sum_{i=1}^{N} \phi(x_i)y_i. \quad (6)$$

3 Online Mode and Convergence

This paper considers the cyclic learning scheme. That is, in an iteration cycle, a sample is learned exactly once according to a fixed order. Considering the i-th training sample, from (5) the instantaneous objective function for this sample is given by

$$\bar{\mathcal{L}}^i(w) = (y_i - \phi^T(x_i)w)^2 + w^T((p + \sigma_b^2)G^{(i)} - pH^{(i)})w, \quad (7)$$

where $H^{(i)} = \phi(x_i)\phi^T(x_i)$ and $G^{(i)} = \mathbf{diag}(\phi(x_i)\phi^T(x_i))$. Based on the gradient $\frac{\partial \bar{\mathcal{L}}^i(w)}{\partial w}$, the updating for this sample is given by

$$w^{(k,i)} = (I - \mu_k S^{(i)})w^{(k,i-1)} + \mu_k \phi(x_i)y_i \quad (8)$$

where $S^{(i)} = (p + \sigma_b^2)G^{(i)} + (1-p)H^{(i)})$ and μ_k is the learning rate.

For online learning, one of the important issues is the convergent behavior. The basic requirement is that the algorithm should converge to a solution. Also, the converged solution should be equal to the corresponding batch mode solution. Before we investigate the convergence condition of (8), we state an important lemma [7] that was used in the proofs of the convergence in the standard stochastic gradient and the standard least mean square (LMS). Our proof will also use this lemma.

Lemma 1. *Let $\{z_k\}$ is a sequence of real numbers, given by $z_{k+1} = (1 - \nu_k)z_k + O(\nu_k^2)$, where $z_1 = r_0$, r_0 is an arbitrary number, $O(\nu_k^2)$ is a function of ν_k and its order is greater than ν_k^2, and $\{\nu_k : 0 \leq \nu_k, k \geq 1\}$ is a decreasing sequence. If $\sum_{k=1}^{\infty} \nu_k = \infty$, and $\sum_{k=1}^{\infty} \nu_k^2 < \infty$, then $\lim_{k \to \infty} z_k = 0$.*

The convergent behavior of (8) is given by Theorem 1.

Theorem 1. *For the online algorithm* (8), *let* $\{\mu_k\}$ *be a decreasing sequence and* $\lim_{k \to \infty} \mu_k = 0$. *If* $\sum_{k=1}^{\infty} \mu_k = \infty$ *and* $\sum_{k=1}^{\infty} \mu_k^2 < \infty$, *then* $w^{(k,i)} \longrightarrow w^*$ *as* $k \longrightarrow \infty$, *for all* i, *where* $w^* = \left[(1-p)H + (p + \sigma_b^2)G\right]^{-1} \frac{1}{N} \sum_{i=1}^{N} \phi(x_i) y_i$ *and* w^* *is the batch mode optimal solution.*

(Proof:) We use the concept of the proof in [7] to prove our Theorem 1. Without loss of generality, we consider that $i = 1$. The online rule is

$$w^{(k,1)} = (I - \mu_k S^{(1)}) w^{(k-1,N)} + \mu_k \phi(x_1) y_1. \tag{9}$$

Expressing (9) for one iteration cycle, we obtain

$$w^{(k,1)} = (I - \mu_k \sum_{i=1}^{N} S^{(i)}) w^{(k-1,1)} + \mu_k \sum_{i=1}^{N} \phi(x_i) y_i + \Xi(\mu_k^2), \tag{10}$$

where $\Xi(\mu^2)$ is a vector valued function and the order of each element of $\Xi(\mu^2)$ is greater than μ^2. Define

$$S \equiv \sum_{i=1}^{N} S^{(i)} = \sum_{i=1}^{N} (p + \sigma_b^2) G^{(i)} + (1-p) H^{(i)} = N((p + \sigma_b^2) G + (1-p) H). \tag{11}$$

Equation (10) can be written as

$$w^{(k,1)} = (I - \mu_k S) w^{(k-1,1)} + \mu_k \sum_{i=1}^{N} \phi(x_i) y_i + \Xi(\mu_k^2). \tag{12}$$

Since H is a symmetric positive/semi-positive definite matrix and G is a diagonal matrix with positive elements, S is a positive definite matrix. Applying the eigen decomposition on S, we obtain $S = \Theta^T D \Theta = \Theta^{-1} D \Theta$, where Θ is an orthonormal matrix, D is diagonal matrix and $[D]_{jj} = d_j > 0$ for all j. With the orthonormal matrix, (12) can be rewritten as

$$\dot{w}^{(k,1)} = (I - \mu_k D) \dot{w}^{(k-1,1)} + \mu_k \sum_{i=1}^{N} \dot{\phi}(x_i) y_i + \dot{\Xi}(\mu_k^2). \tag{13}$$

where $\dot{w}^{(k,1)} = \Theta w^{(k-1,1)}$, $\dot{\phi}(x_i) = \Theta \phi(x_i)$ and $\dot{\Xi}(\mu_k^2) = \Theta \Xi(\mu_k^2)$.

Denote $\dot{w}_j^{(k,1)}$ as the j-th element of $\ddot{w}_j^{(k,1)}$, $\dot{\phi}_j(x_i)$ as the j-th element of $\ddot{\phi}(x_i)$ and $\dot{\Xi}_j(\mu_k^2)$ as the j-th element of $\ddot{\Xi}(\mu_k^2)$. Equation (13) can be rewritten in the element-wise form, given by

$$\dot{w}_j^{(k,1)} = (1 - \mu_k d_j) \dot{w}_j^{(k-1,1)} + \mu_k \sum_{i=1}^{N} \dot{\phi}_j(x_i) y_i + \dot{\Xi}_j(\mu_k^2) \text{ for all } j. \tag{14}$$

Furthermore, equation (14) can be expressed as

$$z_{j,k} = (1 - \mu_k d_j) z_{j,k-1} + \ddot{\Lambda}_j(\mu_k^2) \tag{15}$$

for all j, where $z_{j,k} = \dot{w}_j^{(k,1)} - \frac{1}{d_j} \sum_{i=1}^N \dot{\phi}_j(\boldsymbol{x}_i) y_i$. Applying Lemma 1, we have $\lim_{k \to \infty} z_{j,k} = 0$. Therefore, as $k \longrightarrow \infty$,

$$\dot{w}_j^{(k,1)} = \frac{1}{d_j} \sum_{i=1}^N \dot{\phi}_j(\boldsymbol{x}_i) y_i . \tag{16}$$

Therefore, as $k \longrightarrow \infty$, $\dot{w}_j^{(k,1)} = \boldsymbol{D}^{-1} \sum_{i=1}^N \dot{\phi}_j(\boldsymbol{x}_i) y_i$. That means, $\boldsymbol{w}_j^{(k,1)} = [(p + \sigma_b^2)\boldsymbol{G} + (1-p)\boldsymbol{H})]^{-1} \frac{1}{N} \sum_{i=1}^N \phi_j(\boldsymbol{x}_i) y_i$ and $\boldsymbol{w}_j^{(k,1)} = \boldsymbol{w}^*$. The proof is complete. ∎

4   Simulations

Two common data sets, the nonlinear autoregressive (NAR) time series [8] and the abalone (ABA) data set [9], are considered. The NAR series is generated by

$$z(t) = \big(0.8 - 0.5\exp(-z^2(t-1))\big)\, z(t-1)$$
$$- \big(0.3 + 0.9\exp(-z^2(t-1))\big)\, z(t-2) + 0.1\sin(\pi z(t-1)) + \epsilon(t) , \tag{17}$$

where $\epsilon(t)$ is a zero-mean Gaussian random variable with variance $\sigma_\epsilon^2 = 0.01$. One thousand samples, with $z(0) = z(-1) = 0$, are generated. The first 500 samples are selected as the training set. The remaining 500 samples form the test set. The RBF network is used to estimate $y_t = z(t)$ from the past two observations $\boldsymbol{x}_t = [z(t-1), z(t-2)]^T$. The RBF width is equal to 0.1. The RBF centers are selected from the training set based on the concept of the orthogonal least squares (OLS) algorithm [8]. Based on the sorted RBF centers from the OLS algorithm, we can obtain the test set MSE versus the number of RBF nodes, shown in Figure 2(a). It can be seen that 45 RBF nodes are good enough. In the ABA data set, each sample has 7 input features and one output (age). The data set has 4177 samples. The first two thousands samples were used as the training set. The other samples form the test set. The RBF width is equal to 0.1. We use the OLS algorithm to select the centers. From Figure 2(b), it can be seen that around 60 RBF nodes are good enough.

We test the convergent behavior of the online WFR algorithm under the fault situation: $\sigma^2 = 0.4$ and $p = 0.1$. The results are shown in Figure 3. The first row of the

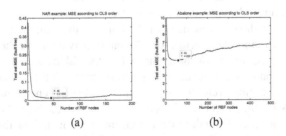

(a)   (b)

Fig. 2. Using the OLS algorithm to select RBF nodes. (a) NAR example. (b) Abalone example. It is seen that the number of the selected nodes in the NAR example should be 45. The number of the selected nodes in the Abalone example should be 60.

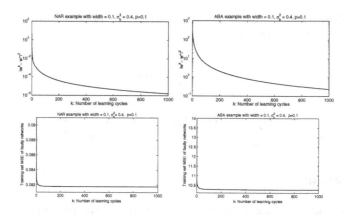

Fig. 3. Convergent Behavior. The first row is the convergence of weight vector. The second row is the convergence of MSE.

figure shows the difference between the online WFR weight vector $w^{k,1}$ and the batch mode optimal solution w^*. Besides, we include the convergence of the mean square error (MSE) of faulty networks in the second row of the figure. From the first row of the figure, it is seen that as we increase the number of the learning cycles, the trained weight vector is more close to the batch mode optimal solution w^*. This behavior agrees with the expectation of Theorem 1. The convergent speed of the adaptive learning rate case is very fast at the beginning phase of learning. It is because that the learning rate is large at this phase of learning. For the NAR case, within hundreds of the learning cycles, the trained weight vector is very close the optimal weight vector. For the ABA case, it is seen that we need more training cycles to approach the optimal weight. In fact, from the second row of Figure 3, in terms of MSE, the online WFR algorithm can settle down in two hundred training cycles.

As a comparison, we also consider the online weight decay algorithm [10]. In the online weight decay algorithm, one important parameter is the weight decay parameter. Since there is no systematic way to set the weight decay parameter, we try 50 different values in the range of $[10^{-5}, 10^{-1}]$ in the logarithmic scale and then we select the best weight decay parameter. The convergence of MSE is shown in Figure 4. It is seen that the online WFR algorithm can settle down around one hundred learning cycles. Also, the MSE performance of WFR algorithm is better than that of the online weight decay algorithm (with optimized weight decay parameter).

We also investigate the performance distribution over the faulty networks. For each trained network, we randomly generate 10, 000 faulty networks. We then measure the test set and training set errors of the faulty networks. Table1 summarizes the mean and standard deviation of the faulty network performances. For the online algorithm we limit the number of learning cycles to 1,000. From the table, in general, the performance of the online WFR learning algorithm is very close to that of the batch WFR algorithm. Also, in terms of mean and standard deviation, the performance of the online WFR algorithm is better than that of the online mode weight decay method.

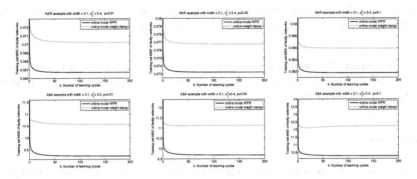

Fig. 4. Comparison between online mode WFR and online mode weight decay

Table 1. MSE of 10,000 faulty neural networks, where number of nodes is 200

	weight noise level σ_b^2	open fault p	batch WFR	online WFR	online weight decay
			NAR example		
Training Set MSE	0.2	0.01	0.0221(0.0041)	0.0221(0.0041)	0.0227(0.0043)
		0.10	0.0270(0.0060)	0.0271(0.0353)	0.0317(0.0086)
	0.4	0.01	0.0299(0.0071)	0.0300(0.0071)	0.0309(0.0075)
		0.10	0.0352(0.0089)	0.0353(0.0090)	0.0396(0.0114)
Test Set MSE	0.2	0.01	0.0246(0.0043)	0.0247(0.0043)	0.0251(0.0046)
		0.10	0.0294(0.0062)	0.0294(0.0062)	0.0338(0.0086)
	0.4	0.01	0.0322(0.0074)	0.0322(0.0073)	0.0329(0.0078)
		0.10	0.0373(0.0094)	0.0374(0.0093)	0.0416(0.0118)
	weight noise level σ_b^2	open fault p	batch WFR	online WFR	online weight decay
			ABA example		
Training Set MSE	0.2	0.01	6.1900(0.6709)	6.2096(0.6708)	7.0061(1.1164)
		0.10	6.7203(0.8572)	6.7530(0.8674)	8.3747(1.9833)
	0.4	0.01	6.9941(0.9475)	7.0286(0.9526)	8.0965(1.6255)
		0.10	7.5101(1.1532)	7.5287(1.1830)	9.3563(2.4420)
Test Set MSE	0.2	0.01	6.7183(0.6995)	6.7532(0.7227)	7.6156(1.1436)
		0.10	7.1832(0.8864)	7.2436(0.9238)	9.0379(1.9713)
	0.4	0.01	7.4337(0.9731)	7.4965(1.0085)	8.6709(1.6667)
		0.10	7.9075(1.1708)	7.9624(1.2352)	9.9685(2.4302)

5   Conclusion

The online WFR algorithm is proposed in this paper to handle faulty RBF networks
with the weight noise and weight fault. The convergent condition is also given. Let the
learning rates $\{\mu_k\}$ be a decreasing sequence and $\lim_{k \to \infty} \mu_k = 0$. We show that if
$\sum_{k=1}^{\infty} \mu_k = \infty$ and $\sum_{k=1}^{\infty} \mu_k^2 < \infty$, then the trained weight vector tends to the solution

of the batch mode WFR algorithm. We then analyze the convergent behavior of the algorithm under different fault levels based on some simulations.

Acknowledgment. The work presented in this paper is supported by a research grant (CityU 115612) from the Research Grants Council of the Government of the Hong Kong Special Administrative Region.

References

1. Burr, J.: Digital Neural Network Implementations. In: Neural Networks, Concepts, Applications, and Implementations, vol. III, pp. 237–285. Prentice Hall, Englewood Cliffs (1995)
2. Bolt, G., Austin, J., Morgan, G.: Fault Tolerant Multi-layer Perceptron Networks. Department of Computer Science, York University, Hestington, York, UK, Tech. Rep:YCS180 (1992)
3. Leung, C.S., Sum, J.: A Fault Tolerant Regularizer for RBF Networks. IEEE Trans. Neural Netw. 19(3), 493–507 (2008)
4. Leung, C.S., Sum, J.: RBF Networks under the Concurrent Fault Situation. IEEE Trans. Neural Netw. and Learning Syst. 23(7), 1148–1155 (2012)
5. Leung, C.-S., Sum, J.P.-F.: Regularizer for Co-existing of Open Weight Fault and Multiplicative Weight Noise. In: Lu, B.-L., Zhang, L., Kwok, J. (eds.) ICONIP 2011, Part III. LNCS, vol. 7064, pp. 276–283. Springer, Heidelberg (2011)
6. Ho, K.J., Leung, C.S., Sum, J.: Convergence and Objective Functions of Some Fault/Noise-injection-based Online Learning Algorithms for RBF Networks. IEEE Trans. Neural Netw. 21(6), 938–947 (2010)
7. Luo, Z.Q.: On the Convergence of the LMS Algorithm with Adaptive Learning Rate for Linear Feedforward Networks. Neural Comput. 3(2), 226–245 (1991)
8. Chen, S.: Local Regularization Assisted Orthogonal Least Squares Regression. Neurocomputing 69(4-6), 559–585 (2006)
9. Sugiyama, M., Ogawa, H.: Optimal Design of Regularization Term and Regularization Parameter by Subspace Information Criterion. Neural Netw. 15(3), 349–361 (2002)
10. Leung, C.S., Wang, H.J., Sum, J.: On the Selection of Weight Decay Parameter for Faulty Networks. IEEE Trans. Neural Netw. 21(8), 1232–1244 (2010)

The Performance of the Stochastic DNN-kWTA Network

Ruibin Feng[1], Chi-Sing Leung[1], Kai-Tat Ng[1], and John Sum[2]

[1] Dept. of Electronic Engineering, City University of Hong Kong, Hong Kong
[2] Institute of Technology Management, National Chung Hsing University Taichung
rfeng4-c@my.cityu.edu.hk, {eeleungc,eektng}@cityu.edu.hk
pfsum@dragon.nchu.edu.tw

Abstract. Recently, the dual neural network (DNN) model has been used to synthesize the k-winners-take-all (kWTA) process. The advantage of this DNN-kWTA model is that its structure is very simple. It contains $2n + 1$ connections only. Also, the convergence behavior of the DNN-kWTA model under the noise condition was reported. However, there is no an analytic expression on the equilibrium point. Hence it is difficult to study how the noise condition affects the model performance. This paper studies how the noise condition affects the model performance. Based on the energy function, we propose an efficient method to study the performance of the DNN-kWTA model under the noise condition. Hence we can efficiently study how the noise condition affects the model performance.

1 Introduction

The winner-take-all (WTA) process is used to find out the largest number out of a list of n numbers [1–3]. The generalized version of the WTA process is the kWTA process, which is used for finding out the k largest inputs from n inputs. Among many models, the dual neural network (DNN)-based kWTA [4] is with the simplest structure. It contains $2n + 1$ connections only, while conventional structures usually require n^2 connections. Owing to the model simplicity, various studies have been done regarding its stability [4,5] and the bounds on its convergence time [6]. In [7], an analytical equation for the convergence time was derived.

Due to its simple structure, the DNN-kWTA model is suitable for hardware implementation. However, circuit implementation has certain limitation [8–11]. For instance, the thermal noise [10] may be injected into the model. Hence, we have studied the dynamic behavior of the DNN-kWTA model under the noise condition in [12]. However, the effect of noise on the performance of the model is still not known.

This paper proposes a computationally efficient method to check whether the noisy model works properly or not. With the method, we can efficiently study the probability that the model produces the correct outputs. Hence we can know how the noise affects the model performance. Section 2 presents the background information. Section 3 studies the performance of the model when the inputs do not have a minimum separation. The paper is then concluded in Section 4.

C.K. Loo et al. (Eds.): ICONIP 2014, Part I, LNCS 8834, pp. 279–286, 2014.

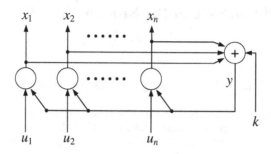

Fig. 1. Structure of a DNN-kWTA network

2 Background

2.1 DNN-kWTA

The DNN-kWTA model, shown in Figure 1, consists of n inputs $\{u_1, \cdots, u_n\}$, n outputs $\{x_1, x_2, \cdots, x_n\}$ and one hidden state y. Without loss of generality, we assume that the inputs are all distinct and bounded by 0 and 1. The operation of the model is given by

$$\epsilon \frac{dy(t)}{dt} = \sum_{i=1}^{n} x_i(t) - k, \tag{1}$$

$$x_i(t) = g(u_i - y(t)) \tag{2}$$

for $i = 1, \cdots, n$, where ϵ is the characteristic time, and $g(s)$ is a threshold comparator:

$$g(s) = \begin{cases} 1 \text{ if } s > 0 \\ 0 \text{ otherwise.} \end{cases} \tag{3}$$

Without loss of generality, we assume that $\epsilon = 1$. Let $\{u_{\kappa_1}, \cdots, u_{\kappa_n}\}$ be the sorted inputs in ascending order and $\{\kappa_1, \cdots, \kappa_n\}$ be the corresponding sorted index list. Furthermore, let $\{x_{\kappa_1}, \cdots, x_{\kappa_n}\}$ be the corresponding outputs. By (1) and (2), $y(t)$ converges in finite time [5, 7]. For this model, at the equilibrium point, only k outputs $\{x_{\kappa_{n-k+1}}, \cdots, x_{\kappa_n}\}$ are equal to 1. Other $n - k$ outputs are equal to zero.

2.2 Noisy DNN-kWTA and Its Existing Results

As hardware implementation can never be perfect [10, 11], there are some stochastic behaviors. For instance, the thermal noise may be introduced into the inputs. Under this situation, the ith input is given by

$$\tilde{u}_i(t) = u_i + \delta_i(t). \tag{4}$$

It consists of a constant value u_i and a random noise $\delta_i(t)$, where the probability density function of $\delta_i(t)$ is given by

$$P(\tilde{u}_i(t)|u_i) = \sqrt{\frac{\alpha_I}{2\pi}} \exp\left(-\frac{\alpha_I(\tilde{u}_i(t) - u_i)^2}{2}\right). \tag{5}$$

The parameter α_I corresponds to the inverse of the noise power.

Besides, the output nodes may have the stochastic behavior. This behavior may come from the random drift in the offset voltage of comparators [11]. Let $\tilde{x}_i(t) \in \{0, 1\}$ be the ith output at time t. The dynamics are given by

$$\frac{dy}{dt} = \sum_{i=1}^{n} \tilde{x}_i(t) - k, \tag{6}$$

where the conditional probability mass function of the output (given $\tilde{u}_i(t)$) is equal to

$$\text{Probit}(\tilde{x}_i(t) = 1 | y(t), \tilde{u}_i(t))$$
$$= \sqrt{\frac{\alpha_O}{2\pi}} \int_{y(t)}^{\infty} \exp\left(-\frac{\alpha_O(z - \tilde{u}_i(t))^2}{2} \right) dz \tag{7}$$

for $i = 1, \cdots, n$. The parameter α_O in (7) controls the shape of the probability mass function. If the stochastic behavior comes from the random drift of the offset voltage, the parameter α_O corresponds to the inverse of the random drift's variance.

In [12] we show that with the two defects, the dynamics of $y(t)$ can be written in the integral form, given by

$$y(t + \tau) - y(t) = \sum_{i=1}^{n} \int_{t}^{t+\tau} \tilde{x}_i(\tau) d\tau - k\tau, . \tag{8}$$

For τ is small,

$$y(t + \tau) = y(t) + \tau \left\{ \sum_{i=1}^{n} \text{Prob}(\tilde{x}_i(t) = 1 | y(t), u_i) - k \right\}. \tag{9}$$

where

$$\text{Prob}(\tilde{x}_i(t) = 1 | y(t), u_i)$$
$$= \int_{-\infty}^{\infty} \text{Probit}(\tilde{x}_i(t) = 1 | y(t), \tilde{u}_i) P(\tilde{u}_i | u_i) d\tilde{u}_i. \tag{10}$$

In [12], we showed that the conditional probability mass function of the output (given u_i) is equal to

$$\text{Prob}(\tilde{x}_i(t) = 1 | y(t), u_i) = \sqrt{\frac{\alpha_O \alpha_I}{2\pi(\alpha_O + \alpha_I)}} \int_{y(t)}^{\infty} \exp\left(-\frac{\alpha_O \alpha_I (z - u_i)^2}{2(\alpha_O + \alpha_I)} \right) dz.$$

Let

$$f(y(t), u_i) = \text{Prob}(\tilde{x}_i(t) = 1 | y(t), u_i) = \frac{\alpha_O \alpha_I}{\alpha_O + \alpha_I}. \tag{11}$$

Then the conditional probability mass function (given u_i) can be rewritten as

$$f(y(t), u_i) = \sqrt{\frac{\alpha}{2\pi}} \int_{y(t)}^{\infty} \exp\left(-\frac{\alpha(z - u_i)^2}{2} \right) dz \tag{12}$$

where the parameter α describes the overall effect of the input noise and the output node stochastic behavior. To sum up, we obtain

$$y(t + \tau) = y(t) + \tau \left\{ \sum_{i=1}^{n} f(y(t), u_i) - k \right\} \tag{13}$$

for small τ. It should be noticed that $f(y(t), u_i)$ is the firing rate of the ith output node. In [12], we showed the following theorems.

Theorem 1 (Existence of equilibrium). *For any constant α, there exists a unique equilibrium point y^* such that $\sum_{i=1}^{n} f(y^*, u_i) - k = 0$.*

Theorem 2 (Convergence). *For any constant α, $\lim_{t \to \infty} y(t) = y^*$. Furthermore, if α is sufficiently large, then $\lim_{t \to \infty} y(t) = (u_{\kappa_{n-k}} + u_{\kappa_{n-k+1}})/2$.*

Theorem 3 (Energy Function). *The dynamics are rewritten as*

$$y(t + \tau) = y(t) - \tau \left. \frac{\partial V}{\partial y} \right|_{y=y(t)} . \tag{14}$$

where

$$V(y) = (k - n)y + \sum_{i=1}^{n} \int \Phi \left(\sqrt{\alpha}(y - u_i) \right) dy \tag{15}$$

is the energy function. $\Phi(s)$ can be expressed in term of the error function $erf(s/\sqrt{2})$ as follows :

$$\Phi(s) = \frac{1}{2} + \frac{1}{2} erf \left(\frac{s}{\sqrt{2}} \right) .$$

where $erf(s) = \frac{1}{\sqrt{\pi}} \int_0^s \exp(-\eta^2) d\eta$.

One can easily show that $V(y)$ is an U-shaped function. Figure 2 shows some examples of this energy function for the inputs equal to 0.5, 0.7, 0.8, 0.4, 0.1 and 0.3, respectively. For finite α, $V(y)$ has one global minimum. For $\alpha \to \infty$, the model becomes the conventional DNN-kWTA model. In this situation, $V(y)$ is flat for $y \in (0.5, 0.7)$.

In [12], we do not discuss how the noise condition affects the performance of the network. In the rest of this paper, we address how the noise condition affects the performance of the network.

3 New Results: Inputs with Continuous Distribution

In the noisy DNN-kWTA model, at the equilibrium point y^* the outputs are not exact equal to 0 or 1. The output $\tilde{x}_i(t)$'s are with firing rate, given by $f(y^*, u_i)$'s. That means, we need some new definitions about winners and losers.

Definition 1 (Winner and Losers). *Given a set of inputs $\{u_1, \cdots, u_n\}$, the ith output node is called winner, if $f(y^*, u_i) \geq 0.5$. Otherwise, the ith output node is called loser.*

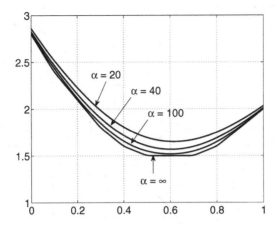

Fig. 2. The energy function $V(y)$ against y for the cases that $\alpha = 20, 40, 100$ and ∞. Here, the number of winners is 2 (i.e. $k = 2$) and the inputs are 0.5, 0.7, 0.8, 0.4, 0.1 and 0.3 respectively.

Besides, we use the following definition to define whether a noisy DNN-kWTA network works properly or not.

Definition 2 (Correct Operation). *Given a set of inputs $\{u_1, \cdots, u_n\}$, the network works properly, if the firing rates $\{f(y^*, u_{\kappa_{n-k+1}}), \cdots, f(y^*, u_{\kappa_{n-k+1}}\}$ are greater than or equal to 0.5, and $\{f(y^*, u_{\kappa_1}), \cdots, f(y^*, u_{\kappa_{n-k}})\}$ are less than 0.5.*

From Definition 2, to study the performance of the noisy DNN-kWTA, we need to simulate the network dynamics. Then we can know the firing rates at the equilibrium point y^*. Hence, intensive network simulations on the network dynamics are required. Instead of measuring the firing rates, we can use the following theorem to check whether the network works properly or not.

Theorem 4 (Correct Operation). *The network works properly if*

$$u_{\kappa_{n-k}} < y^* \leq u_{\kappa_{n-k+1}}. \tag{16}$$

Proof: From (12), we know that $f(y^*, u_i) = \sqrt{\frac{\alpha}{2\pi}} \int_{y^*}^{\infty} \exp\left(-\frac{\alpha(z-u_i)^2}{2}\right) dz$ is a Q-function (the complement of the cdf of the normal distribution). Hence, if $u_{\kappa_{n-k}} < y^*$, then $\{f(y^*, u_{\kappa_1}), \cdots, f(y^*, u_{\kappa_{n-k}})\}$ are less than 0.5. And if $y^* \leq u_{\kappa_{n-k+1}}$, then $\{f(y^*, u_{\kappa_{n-k+1}}), \cdots, f(y^*, u_{\kappa_n})\}$ are greater than 0.5. The proof is complete. ∎

Although we have Theorem 4, it is difficult to verify that the network can work properly or not. It is because there is no an analytic expression on $y*$. This difficulty can be solved based on the property of the energy function $V(y)$.

Since $V(y)$ is a U-shaped function and it has one minimum point only, we can use the values of $\frac{dV}{dy}\big|_{y=u_{\kappa_{n-k}}}$ and $\frac{dV}{dy}\big|_{y=u_{\kappa_{n-k+1}}}$ to check whether $u_{\kappa_{n-k}} < y^* \leq u_{\kappa_{n-k+1}}$ or not.

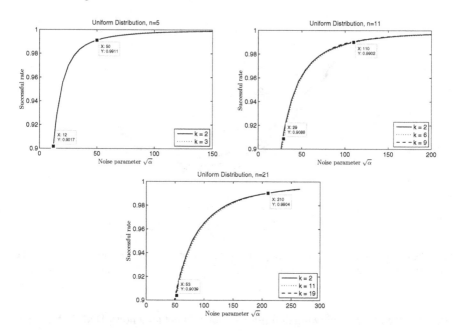

Fig. 3. Successful rates of working properly for the uniform distribution. For each n and each α, we generate 100,000 sets of inputs. We then use Theorem 5 to check the performance of the noisy DNN-kWTA model.

Theorem 5. *For a noisy DNN-kWTA network, if*

$$\left.\frac{dV}{dy}\right|_{y=u_{\kappa_{n-k}}} < 0 \ and \ \left.\frac{dV}{dy}\right|_{y=u_{\kappa_{n-k+1}}} \geq 0, \tag{17}$$

then

$$u_{\kappa_{n-k}} < y^* \leq u_{\kappa_{n-k+1}} . \tag{18}$$

Proof: From Theorems 1-3, we know that $\frac{dV}{dy}$ is a strictly monotonically increasing function of y. Also, $\left.\frac{dV}{dy}\right|_{y=y^*} = 0$. Hence, if $\left.\frac{dV}{dy}\right|_{y=u_{\kappa_{n-k}}} < 0$, then $u_{\kappa_{n-k}} < y^*$. Similarly, if $\left.\frac{dV}{dy}\right|_{y=u_{\kappa_{n-k+1}}} \geq 0$, then $y^* \leq u_{\kappa_{n-k+1}}$. The proof is complete. ∎

Theorem 5 gives us a quick way to study the performance of the network instead of simulating the network dynamics. With Theorem 5, we can studies how the noise level value affects the performance of the model. We consider two cases. The first case is that the inputs are uniformly distributed over 0 an 1. The second case is that the inputs are with the Beta distribution: $\text{Beta}_{a,b}(u) = \frac{\Gamma(a+b))}{\Gamma(a)\Gamma b}u^{a-1}(1-u)^{b-1}$, where $\Gamma(\cdot)$ is the Gamma function. The Beta distribution is a family of distribution functions defined on the interval between 0 and 1. In this experiment, a and b are set to 1.

For each n and each α, we generate 100,000 sets of inputs. We then use Theorem 5 to check the performance of the noisy DNN-kWTA model. The performance of the

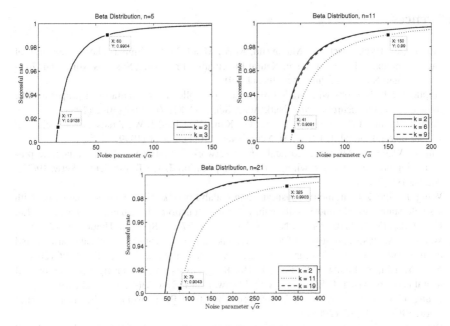

Fig. 4. Successful rates of working properly for the beta distribution. For each n and each α, we generate 100,000 sets of inputs. We then use Theorem 5 to check the performance of the noisy DNN-kWTA model.

noisy DNN-kWTA model is shown in Figures 3 and 4. For the uniform distribution with $n = 5$, if the standard deviation of the noise is equal to $\frac{1}{50}$, the chance of working properly is equal to 99%. For the Beta distribution with $n = 5$, if the standard deviation of the noise is equal to $\frac{1}{60}$, the chance of working properly is equal to 99%.

4 Conclusion

This paper proposed a method (Theorem 5) to check whether the noisy DNN-kWTA model produces the correct firing rate or not. The advantage of the proposed method is that we do not need to simulate the network dynamics. When a number of the input samples are given, we can use this method to study the chance that the network works properly. This method is suitable to predict the performance of the network when there is no minimum separation in the inputs. One of the further directions is to study the performance of the model when there is a minimum separation in the inputs.

Acknowledgment. The work is supported by a grant from the Research Grants Council of the Hong Kong SAR, China (CityU115612) and the Taiwan NSC research grant (NSC 102-2221-E-005-084).

References

1. Lazzaro, J., Ryckebusch, S., Mahowald, M.A., Mead, M.A.: Winner-take-all networks of O(N) complexity. In: Advances in Neural Information Processing Systems, vol. 1, pp. 703–711. Morgan Kaufmann Publishers Inc. (1989)
2. Hopfield, J.J.: Neurons with graded response have collective computational properties like those of two-state neurons. Proc. Natl. Acad. Sci. USA 81(10), 3088–3092 (1984)
3. Sum, J., Leung, C.S., Tam, P., Young, G., Kan, W., Chan, L.W.: Analysis for a class of winner-take-all model. IEEE Trans. Neural Netw. 10(1), 64–71 (1999)
4. Hu, X., Wang, J.: An improved dual neural network for solving a class of quadratic programming problems and its -winners-take-all application. IEEE Trans. Neural Netw. 19(22), 2022–2031 (2008)
5. Wang, J., Guo, Z.: Parametric sensitivity and scalability of k-winners-take-all networks with a single state variable and infinity-gain activation functions. In: Zhang, L., Kwok, J., Lu, B.-L. (eds.) ISNN 2010, Part I. LNCS, vol. 6063, pp. 77–85. Springer, Heidelberg (2010)
6. Wang, J.: Analysis and design of a -winners-take-all model with a single state variable and the heaviside step activation function. IEEE Trans. Neural Netw. 21(9), 1496–1506 (2010)
7. Xiao, Y., Liu, Y., Leung, C.S., Sum, J., Ho, K.: Analysis on the convergence time of dual neural network-based. IEEE Trans. Neural Netw. Learn. Syst. 23(4), 676–682 (2012)
8. Leung, C.S., Sum, J.: A fault tolerant regularizer for RBF networks. IEEE Trans. Neural Netw. 19(3), 493–507 (2008)
9. Wang, L.: Noise injection into inputs in sparsely connected Hopfield and winner-take-all neural networks. IEEE Trans. on System Man and Cybernetics Part B: Cybernetics 27(5), 868–870 (1997)
10. Hu, M., Li, H., Wu, Q., Rose, G.S., Chen, Y.: Memristor crossbar based hardware realization of BSB recall function. In: Proc. the 2012 International Joint Conference on Neural Networks, IJCNN (2012)
11. He, J., Zhan, S., Chen, D., Geiger, R.L.: Analyses of static and dynamic random offset voltages in dynamic comparators. IEEE Trans. on Circuits and Systems I 56(5), 911–919 (2009)
12. Sum, J., Leung, C.S., Ho, K.: Effect of Input Noise and Output Node Stochastic on Wang's kWTA. IEEE Trans. Neural Netw. Learn. Syst. 24(9), 1472–1478 (2013)
13. Bowling, S.R., Khasawneh, M.T., Kaewkuekool, S., Cho, B.R.: A logistic approximation to the cumulative normal distribution. Journal of Industrial Engineering and Management 2(1), 114–127 (2009)

Modularity Maximization Adjusted by Neural Networks

Desiree Maldonado Carvalho, Hugo Resende, and Mariá C.V. Nascimento

Instituto de Ciência e Tecnologia, Universidade Federal de São Paulo,
Rua Talim, 330, São José dos Campos, SP
{dmcarvalho,hresende,mcv.nascimento}@unifesp.br

Abstract. Graphs are combinatorial structures suitable for modelling various real systems. The high clustering tendency observed in many of these graphs has led a large number of researches, among them, we point out the modularity maximization-based community detection algorithms. Although very effective a few studies suggest that, for some networks, this approach does not find the expected communities due to a resolution limit of the measure. In this paper, we propose a way to automatically choose the value of the resolution parameter considered in the modularity by using neural networks. In the computational experiments, we observed that the proposed strategy outperformed another strategies from the literature for hundreds of artificial graphs considering the expected communities.

Keywords: community detection in networks, graph clustering, modularity maximization, neural networks.

1 Introduction

A wide variety of real systems can be modeled into graphs or networks. Consequently, one can perform their pattern analysis by identifying vertex partitions in which the groups, also known as communities, are composed of highly related vertices that are sparsely connected with the rest of the network. Moreover, the community detection problem plays an important role in wide-ranging applications [5].

Efficiently detecting communities in networks poses as a challenge in pattern recognition. To approach this issue, also known as graph clustering, different algorithms were developed, among them we can underline: the *edge betweenness* algorithm [5], *Infomap* [13], *spin glass* community detection strategy [12] and *modularity* maximization based heuristics [5]. Modularity emerged as a breakthrough on the graph clustering subject and the literature presents many modularity maximization based heuristics for detecting communities in networks [4].

Despite the reasonable mathematical foundation behind the modularity definition, for certain types of graphs it is subject to a bottleneck known as *resolution limit*. This scaling problem causes a dependence between the size of the groups of a clustering and the total vertex number [4]. Nevertheless, Newman [10] attested

C.K. Loo et al. (Eds.): ICONIP 2014, Part I, LNCS 8834, pp. 287–294, 2014.

that even though modularity in its original form may be a biased measure, its adjusted version, proposed in [12], does not depend on scaling matters.

In this paper, we propose a strategy to adjust the unbiased modularity formulation proposed in [12] by using neural networks. In the computational experiments, we employed a wide variety of graphs artificially generated by the software described in [7]. Additionally, some topological analysis of these graphs concerning their distribution model were performed and used for adjusting the neural network. The results of this experiment showed a good efficiency of the neural network for determining the λ values, outperforming the results achieved by two other strategies widely employed to this end from the literature.

2   Modularity Maximization Problem

In this paper, a graph $G = (V(G), E(G))$ is represented by a set of vertices, $V(G)$, and a set of edges, $E(G)$, where each edge $e := (v_i, v_j) \in E(G)$ is associated with a non ordered pair of vertices of G. Additionally, a given edge $(v_i, v_j) \in E(G)$ has as ends the vertices v_i and v_j, where $i, j \in \{1, 2, \ldots, |V(G)|\}$. The number of vertices and edges of G are denoted in this paper by $n(G)$ and $m(G)$, respectively. The degree of a vertex v_i from G, here called $d_G(v_i)$, corresponds to the number of times that a vertex v_i appears as an edge end in graph G. A graph induced by a set of vertices $X \subseteq V$ is denoted by $G[X]$.

The definition of a clustering relies on the k-way partition of the vertex set. Let $\mathcal{C} = \{V_1, V_2, \ldots, V_k\}$, with $1 \leq k \leq n$, be a k-way partition of $V(G)$. The induced graph $G[\mathcal{C}] = (V(G), E(G[\mathcal{C}]))$, where $E(G[\mathcal{C}]) := \bigcup_{i=1}^{k} E(G[V_i])$ defines a graph clustering.

A substantial number of studies have been published on this subject, primarily with the purpose of refining the existing graph clustering. Major progress towards in developing graph clustering methods was established by Newman et al. [11]. The authors described an assessment measure, named modularity, that ranks the quality of a graph clustering. Of the many ways for defining this measure, we present a formulation for modularity in Equation (1).

$$q(\mathcal{C}) = \frac{1}{m(G)} \sum_{i=1}^{k} [m(G[V_i]) - p(G[V_i])] \tag{1}$$

where $p(G[V_i]) = \sum_{\forall v_r \neq v_j \in C_i} \frac{d_G(v_r) d_G(v_j)}{2m(G)} + \sum_{\forall v_r \in V_i} \frac{d_G(v_r)^2}{4m(G)}$ is the expected number of edges between vertices from V_i in a random graph with the same degree sequence as G. This difference provides an assessment measure that the higher its value, the better the quality of the partition evaluated.

As it has been proven that the decision version of the modularity maximization problem is $\mathcal{NP}$-complete [2], heuristics are primarily used to tackle large scale graphs [4,1]. In particular, a simple and efficient strategy developed to address large scale graphs is the Louvain method [1]. Due to its good potential for identifying communities in networks with over thousands of vertices [1], in this paper, for attesting the quality of the proposed adjustment strategy by neural network results, we employed this method in the computational experiments.

2.1   Parameterized Modularity

Reichardt et al. [12] derived an assessment community measure that draws on the exploration of the ground state of an infinity spin glass. In this sense, to evaluate network communities, this measure states, as indicated in Equation (2), that existing edges between vertices from distinct groups must be penalized as well as the nonexistence of edges between vertices from the same group.

$$
\begin{aligned}
\mathcal{E}(\mathcal{C}) = &-\sum_{i \neq j} \sigma_{ij} a_{ij} \delta(v_i, v_j) + \sum_{i \neq j} \beta_{ij}(1 - a_{ij})\delta(v_i, v_j) \\
&+ \sum_{i \neq j} \alpha_{ij} a_{ij}[1 - \delta(v_i, v_j)] - \sum_{i \neq j} \gamma_{ij}(1 - a_{ij})[1 - \delta(v_i, v_j)]
\end{aligned}
\tag{2}
$$

where $\mathcal{C}$ is the spin system; a_{ij} is the number of edges between v_i and v_j; $\delta(v_i, v_j)$ is a function that receives 1 if vertices v_i e v_j are in the same spin state, and 0, otherwise; σ_{ij} is the positive weight corresponding to the contribution of edge (v_i, v_j); β_{ij} is the penalty of a pair of non adjacent vertices from a same group; α_{ij} is the penalty with regard to the existence of an edge between vertices v_i and v_j from different groups; and, γ_{ij} is the weight related to the nonexistence of an edge between a pair of vertices v_i and v_j which does not belong to a same group.

Taking into account each existing and nonexistent edge with the same contribution, i.e., $\alpha_{ij} = \sigma_{ij}$ and $\gamma_{ij} = \beta_{ij}$, one may observe that to solve the community detection problem is sufficient to examine the number of existing and nonexistent internal edges. Moreover, Reichardt et al. [12] found as suitable alternatives for the parameters' weights σ_{ij} and β_{ij}, respectively, the values $1 - \lambda p_{ij}$ and λp_{ij}. The additional parameter, λ, corresponds to the contribution of an edge between v_i and v_j whereas p_{ij} concerns the probability of existing an edge between two vertices v_i and v_j. Assuming these weights' choice, Equation (2) can be rewritten as indicated in Equation (3).

$$
\mathcal{E}(\mathcal{C}) = -\sum_{i \neq j} (a_{ij} - \lambda p_{ij})\delta(v_i, v_j)
\tag{3}
$$

The probability p_{ij} can be set up according to the vertex degree distribution of the studied graph. In particular, if we assume the null model proposed by Girvan and Newman [5], inducing to the edge probability $d_G(v_i)d_G(v_j)/2m(G)$. By observing Equation (3), we can notice the resemblance of its average with Equation (1) unless the parameter λ. These assumptions lead to a modified version of modularity, here called parameterized modularity, detailed in Equation (4).

$$
q(\mathcal{C}) = \frac{1}{2m(G)} \sum_{i \neq j} \left(a_{ij} - \lambda \frac{d_G(v_i)d_G(v_j)}{2m(G)} \right) \delta(c_i, c_j)
\tag{4}
$$

These results are considerably important since, depending on the graph topology, the original modularity has drawbacks regarded to the communities size.

In line with these considerations, the main target of this paper is to attempt to overcome the flaws in community identification by employing the parameterized modularity. Additionally, the principal purpose of this study is to set out an automatic way for providing the best fit for the parameter λ from Equation (4) through the neural network Multi-Layer Perceptron (MLP).

Among the wide range of possibilities to evaluate the graph structure, we found the most suitable for this evaluation, one which enumerates the four sized motifs within a network. We thoroughly describe this invariant in next section.

3   Structural Graph Analysis

In this paper, we designed our neural network to deal with complex networks, which are graphs characterized by their large size and particular topological traits that reveal a high clustering tendency. For appropriately investigating structural patterns in a graph in order to provide the best fit for λ, in this paper, we used the fact that some specific structures, the so-called *network motifs*, are more recurrent in complex networks than in random networks [9]. Some authors regard motifs as properties of a network, since these structures, that are induced subgraphs of $k < n$ nodes, have a statistically significant superior frequency of occurrence in a given network than in a random graph. Consequently, according to Milo et al. [9], they might be used to differentiate network classes.

Given a graph G, a network motif can be defined as a ν-node connected subgraph of G induced by some set of edges. For example, if G is a K_5, all possible network four-sized motifs are illustrated in Figure 1.

Fig. 1. All possible four-sized motifs

Therefore, the graph analysis amounts to comparing the number of each ν-sized motif and the average number of these motifs inside random graphs with the same degree sequence as the original one. In this paper, we use the four-sized motifs for extracting information about the networks [8]. Then, we use the occurrence histogram of each of the 6 four-sized network motif patterns feeding MLP with this information. Section 4 shows the proposed learning process employed for detecting suitable λ values depending on this graph topology analysis.

4   Artificial Neural Networks

Artificial Neural Networks (ANNs), also regarded in literature as connectionist networks, are computational models inspired in the central nervous system (CNS) of the brain. In this sense, for achieving our goal of automatically determining values for the parameter λ, we make use of an established ANN, the Multilayer Perceptron Network (MLP) [14].

4.1   Design of the MLP Algorithm

MLP is a well known supervised machine learning strategy, more specifically, a neural network, which requires the training step for setting up its parameters. To this end, we employed a widely adopted method, the Resilient backpropagation (Rprop), which is performed in two phases: feed-forward and feed-backward. These phases are responsible for propagating and updating the network parameters according to the training dataset for which the outputs (clustering) are known. Thus, for these steps, in our strategy we give for each network input, the information about the graph topology. As expected solution (target), we provide the most suitable λ for the input found by an efficient algorithm guided by the parameterized modularity using a finite set of possible λ values. As the most suitable λ we mean the λ which guided to the closest solution to the expected output we had the knowledge. The Normalized Mutual Information (NMI) [3] was the similarity measure between the clustering we used for assessing the closeness between the partitions.

We turned discrete the interval $[1, 5]$ for λ by considering the set $\{1, 1.1, 1.2, \ldots, 4.9, 5\}$. The main reason we fixed these values is that some preliminary tests indicated the employed clustering algorithm achieved better results in the case in which the algorithm was guided with these λ values. Additionally, as an attempt to let the results more robust, we defined classes of λ for the training step of the algorithm as it will be shown in the next section.

5   Computational Experiments

In order to attest the performance of the proposed strategy which aims at defining the most suitable λ values, we used an adapted version of a Louvain method [1] for efficiently detecting clustering. Regarding the MLP architecture, we experimentally set up the following configuration: two hidden layers with 3 neurons each an input with 6 neurons and 8 neurons at the output layer. As aforementioned, the output values were organized into classes of values. These classes were labeled from 1 to 8 and correspond to, respectively, the following intervals of λ: $[1, 1.5]$, $[1.6, 2]$, $[2.1, 2.5]$, $[2.6, 3]$, $[3.1, 3.5]$, $[3.6, 4]$, $[4.1, 4.5]$ and $[4.6, 5]$. Hence, we named our strategy GMOD and employed the following steps to provide the desired results:

GMOD. According to the class yielded by MLP, randomly choose a value within the corresponding class (interval) for specifying the λ for this instance. Find the clustering through the Louvain method guided by the parameterized modularity with this λ.

A total of 800 graphs were artificially generated by using the benchmark network generator software introduced in [7] for the experiments. The generated graphs can be divided into two different classes: one composed by 400 of small-sized clusters and another with 400 networks, but characterized by medium to large-sized clusters. These graph features were chosen due to the intention of exploring mainly the first group of graphs for which pure modularity has a higher probability of flaws.

In line with this, we designed two neural networks, the first for the small-sized graphs and the second taking into account the remaining graphs. It is worth mentioning that, for unlabeled graphs, we can easily differentiate these two types of graphs by analyzing the proportion of vertices with degree lower than their average degree. If the proportion is higher than 50%, it means that the handled graph may have a significant number of small-sized clusters, otherwise, the graph can be discussed as medium to large-sized clusters.

In addition to the analysis of the pattern accuracy rate, we investigate its performance over three algorithms, RMOD, MOD and *Infomap*, each of them explained next.

RMOD. This strategy randomly pick a λ within the interval $[1, 5]$ and guides the Louvain method by the parameterized modularity with this random choice.

MOD. This procedure is merely the Louvain method guided by the pure modularity, i.e., with the value 1 for the λ parameter.

Infomap. It is a multi-level strategy based on the compression information on the graph structure [13]. This strategy was chosen for comparison due to a comprehensive comparative analysis in [6] which concluded *Infomap* outperformed the other algorithms.

For the first neural network, we set up the artificial graphs by employing the following parameters for the software defined in [7]: number of vertices: 1000, 2000, 3000, 4000 and 5000; average degree: 20; maximum degree: 50; mixing parameter (μ): $\{0.1, 0.2, \ldots, 0.8\}$ (the more defined the clusters, the lower this parameter); minimum for the community sizes: 10; maximum for the community sizes: 50;

For each possible combination of parameters we generated ten different graphs. By applying the Louvain method guided by the parameterized modularity, we could observe the following distribution of graphs among the 8 classes of λ: 152 from class 1, 95 from class 2, 66 from class 3, 26 from class 4, 13 from class 5, 21 from class 6, 10 from class 7 and 17 from class 8. Concerning the learning of MLP, its training patterns included the minimum between 10 and the total number of graphs of each of the classes. This choice was motivated by the desire of covering as good as possible the graph topologies for accurately designing the network during the learning process.

We employed almost the same configuration for generating the large-sized artificial graphs as for small-sized clusters from the first MLP. The difference between them is the average and maximum degrees which, in the latter graphs, are, respectively, 20 and 50. Additionally, the graphs are distributed among the classes according to their λ as follows: 294 from class 1, 23 from class 2, 20 from class 3, 18 from class 4, 15 from class 5, 9 from class 6, 14 from class 7 and 7 from class 8. Again, regarding the learning of MLP, its training patterns included the minimum between 10 and the total number of graphs of each of the classes.

It is sensible to keep records that the first class is much denser than the other classes. This behavior was expected, since modularity works pretty well for medium to large-sized cluster graphs.

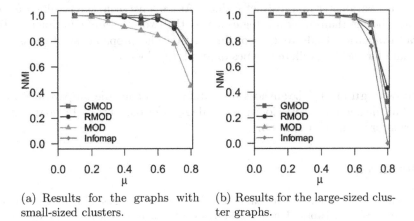

(a) Results for the graphs with small-sized clusters.

(b) Results for the large-sized cluster graphs.

Fig. 2. Performance of the proposed strategy in comparison to three other strategies

The main conclusion drawn with the results of network was that, even though it presented a reasonable pattern accuracy rate of 48% on average, it clearly outperformed RMOD and MOD. To firmly establish its superior performance regarding these other strategies, we summarized the average results in Figure 2a. This graphic shows the relation between the average NMI and mixing parameter. This means that we considered every graph of a certain mixing parameter (of every sizes), calculated the average NMI of them and plotted these results.

Additionally, still in Figure 2a, we observe that, except for the graphs with mixing parameter of 0.5, GMOD outperformed the other strategies. Nevertheless, with mixing parameter of 0.5, GMOD still produced good quality results, achieving the second best ones on average. Moreover, GMOD was very competitive with *Infomap*, being slightly worse than it for $\mu = 0.5$, but slightly better for $\mu = 0.8$.

Regarding the second neural network, it achieved a satisfactory 50% pattern accuracy rate for this class of instances. Nevertheless, the results displayed in Figure 2b show a much better accomplishment of GMOD since it outperformed the other strategies on average considering every different type of graph we designed, except for $\mu_t = 0.8$ which it took the second best result.

6 Final Remarks

In this paper, we proposed an efficient strategy based on the MLP to automatically proportionate values for an important parameter of a community assessment measure [12]. For attesting its superior quality over other methods, we guided the Louvain method with this measure for detecting communities and compared its results with the partitions provided by Infomap and by the same method, however, guided by two different objective functions. The first objective function is a measure extensively adopted for detecting communities, the modularity. The other is the same as our method, however, by adjusting the mentioned parameter in a distinct fashion. As we have seen, the results are consistently better than other strategies across the two separate experiments. The overall evaluation leads to the conclusion that the proposed strategy is very adequate for tackling different types of graph topologies.

Acknowledgments. The authors are grateful to Fundação de Amparo à Pesquisa do Estado de São Paulo and Coordenação de Aperfeiçoamento de Pessoal de Nível Superior for the financial support.

References

1. Blondel, V.D., Guillaume, J., Lambiotte, R., Lefebvre, E.: Fast unfolding of communities in large network. Journal of Statistical Mechanics: Theory and Experiment, P10008 (2008)
2. Brandes, U., Delling, D., Gaertler, M., Görke, R., Hoefer, M., Nikolosk, Z., Wagner, D.: On modularity clustering. IEEE Transactions on Knowledge and Data Engineering 20, 172–188 (2008)
3. Danon, L., Díaz-Guilera, A., Duch, J., Arenas, A.: Comparing community structure identification. Journal of Statistical Mechanics: Theory and Experiment 2005(09), P09008 (2005)
4. Fortunato, S., Barthélemy, M.: Resolution limit in community detection. Proc. Natl. Acad. Sci. USA 104, 36–41 (2007)
5. Girvan, M., Newman, M.: Community structure in social and biological networks. Proc. Natl. Acad. Sci. USA 99, 7821–7826 (2002)
6. Lancichinetti, A., Fortunato, S.: Community detection algorithms: a comparative analysis. Phys. Rev. E 80, 056117 (2009)
7. Lancichinetti, A., Fortunato, S., Radicchi, F.: Benchmark graphs for testing community detection algorithms. Phys. Rev. E 78, 046110 (2008)
8. Meira, L.A.A., Máximo, V.R., Fazenda, A.L., da Conceição, A.F.: Acc-motif: Accelerated motif detection using combinatorial techniques. In: 2012 Eighth International Conference on Signal Image Technology and Internet Based Systems (SITIS), pp. 744–753 (2012)
9. Milo, R., Shen-Orr, S., Itzkovitz, S., Kashtan, N., Chklovskii, D., Alon, U.: Network motifs: Simple building blocks of complex networks. Science 298(5594), 824–827 (2002)
10. Newman, D.J.: Communities, modules and large scale structure in networks. Nature Physics (2011)
11. Newman, M.E.J., Girvan, M.: Finding and evaluating community structure in networks. Physical Review E 69, 026113–026127 (2004)
12. Reichardt, J., Bornholdt, S.: Statistical mechanics of community detection. Phys. Rev. E 74, 016110 (2006)
13. Rosvall, M., Bergstrom, C.T.: Maps of random walks on complex networks reveal community structure. Proceedings of the National Academy of Sciences U.S.A. 105, 1118–1123 (2007)
14. Rumelhart, D.E., Hinton, G.E., Williams, R.J.: Learning representations by back-propagation errors. Nature (1986)

A Dynamic Pruning Strategy
for Incremental Learning on a Budget⋆

Yusuke Kondo and Koichiro Yamauchi

Department of Computer Science Chubu University Kasugai Aichi Matsumoto 1200
yamauchi@cs.chubu.ac.jp
http://sakura.cs.chubu.ac.jp/

Abstract. Several kernel-based perceptron learning methods on a budget have been proposed. In the early steps of learning, such methods record a new instance by allocating it a new kernel. In the later steps, however, useless memory must be forgotten to make space for recording important and new instances once the number of kernels reaches an upper bound. In such cases, it is important to find a way to determine what memory should be forgotten. This is an important process for yielding a high generalization capability. In this paper, we propose a new method that selects between one of two forgetting strategies, depending on the redundancy of the memory in the learning machine. If there is redundant memory, the learner replaces the most redundant memory with a new instance. If there is less redundant memory, the learner replaces the least recently used / least frequently used memory. Experimental results suggest that this proposed method is superior to existing learning methods on a budget.

Keywords: learning on a budget, regression, forgetting, virtual concept drifting environments.

1 Introduction

Several researchers have recently developed kernel-based perceptron learning methods on a budget [1] [2] [3]. These learning methods facilitate online learning with a set of kernels, whose number is limited to a certain upper bound. Upon reaching this upper bound, the most ineffective kernel is replaced with a new one, and the new instance is recorded. This ability is measured by implementing a cumulative error, which is the sum of errors in the presented samples.

Dekel et al. proposed the Forgetron, an algorithm for pruning the oldest kernel [1] . Furthermore, the Forgetron shrinks the parameters associated with each kernel to reduce the adverse effects caused by pruning the oldest kernel. Orabona et. al. proposed the projectron, which projects current new instance on to the space spanned by the existing kernels[2] . According to theoretical analysis, an

---

⋆ This research was supported by JST Adaptable and Seamless Technology Transfer Program through Target-driven R&D Exploratory Research AS221Z01499A.

C.K. Loo et al. (Eds.): ICONIP 2014, Part I, LNCS 8834, pp. 295–303, 2014.

extended version of the Projectron, namely, Projectron++, outperforms existing methods. Wenwu et al. proposed a kernel-based perceptron method with dynamic memory (PDM). PDM is an improvement to the Projectron. It is a method for both projecting the new instance onto the space spanned by the existing kernels and replacing the most ineffective kernel. To reduce the adverse effects from the pruning process, decremental projection is performed prior to pruning the kernel[3].

The kernel-based perceptron method, however, is nonetheless greatly affected by pruning the existing kernel. To correct this problem, we have in our prior work proposed a limited general regression neural network (LGRNN) [4] [5]. The LGRNN's output function is virtually the same as that of the original general regression neural network, and its behavior is similar to algorithms using nearest neighbors. With this output function, the learner reduces the adverse effects from replacing one of the kernels. To do so, the LGRNN manipulates four learning options, including projection and the option to replace the kernel using decremental projection.

Thus, existing methods, including our previous method, employ one of two existing pruning strategies. One is to focus on how recently each kernel has been activated. The other is to focus on redundant kernels.

In this paper, we propose an extended LGRNN method that is able to exploit both pruning strategies. If there is redundancy, the extended LGRNN replaces the redundant kernel with a new one, and if no redundant kernels are found, a new kernel replaces the kernel judged to be old and infrequently used.

2   LGRNN Using Dynamic Pruning Policy

In our previous work, we proposed a limited general regression neural network [4] [5] [6] . This LGRNN chooses the most ineffective kernel by estimating the approximated linear dependency (ALD) of each kernel. The new LGRNN proposed in this paper uses a dynamic policy, which is a combination of the ALD and LRFU policies. In what follows, Section 2.1 provides an outline of the LGRNN, and Section 2.3 describes the policy for choosing the most ineffective kernel.

2.1   Outline

The LGRNN is an extended GRNN. A GRNN normally allocates a new kernel to the memory for learning a new instance. The LGRNN, however, continues to learn new instances even with a limited memory capacity.

Although the LGRNN's output function resembles that of the GRNN [7], we represent the LGRNN's output function using a vector in Hilbert space. Given this, the output value $y(\boldsymbol{x})$ is

$$y(\boldsymbol{x}) = \frac{\langle f_t, K(\boldsymbol{x}, \cdot) \rangle}{\langle g_t, K(\boldsymbol{x}, \cdot) \rangle}, \quad f_t = \sum_{j \in I_t} w_j K(\boldsymbol{u}_j, \cdot), \quad g_t = \sum_{j \in I_t} R_j K(\boldsymbol{u}_j, \cdot), \tag{1}$$

where w_j and R_j denote the output connection strength and the number of learned samples of the j-th hidden unit. Let f_t, g_t be the functions after the t-th iterations. I_t denotes the size of the support sets. We can regard each hidden unit as a Gaussian kernel function such that $\langle K(u_j, \cdot), K(x, \cdot) \rangle = \exp\left(-\frac{\|x - u_j\|^2}{2\sigma^2}\right)$. In its initial state, LGRNN contains no kernels. When a new instance (x_t, y_t) is presented, the LGRNN appends a new kernel to record the instance. If the number of kernels reaches an upper bound, the LGRNN replaces the most ineffective kernel with a new kernel, whose center position is equivalent to the new current input vector.

Unfortunately, there are cases in which the replacement process destroys a part of the learned knowledge, degrading its generalization capability. To correct this problem, the LGRNN adopts one of four learning options, according to the predicted error, when the upper bound for the number of kernels is reached. The four learning options are explained in the following sections.

2.2 Projection Based Learning

When the number of kernels reaches the upper bound, the LGRNN selects one of four learning options either to modify, ignore, replace, or replace with a substitution according to the error expected post-learning. When replacement is necessary, the LGRNN replaces the most ineffective kernel with a new instance. In the next section, we describe the process for establishing the most ineffective kernel. This section describes the modification option. Modification involves projecting the current new instance onto a space spanned by the existing kernels. Thus, no kernels are consequently pruned by implementing this option. Therefore,

$$f_t = f_{t-1} + y_t P_{t-1} K(x_t, \cdot), \quad g_t = g_{t-1} + P_{t-1} K(x_t, \cdot), \tag{2}$$

where $P_{t-1} K(x_t, \cdot)$ is the projected vector, which is described by a linear combination of existing kernels: $P_{t-1} K(x_t, \cdot) = \sum_{i=1}^{B} a_i K(u_i, \cdot)$, where a_i is a coefficient for the i-th kernel. Let $a = [a_1 \ a_2 \ \cdots a_N]^T$, then a is

$$a = K_{t-1}^{-1} k_{t-1}(x_t), \quad \delta = \left\{1 - k^T(x_t)a\right\}, \tag{3}$$

K_{t-1} denotes the kernel matrix at round $t-1$ and $[K_{t-1}]_{p,q} = K(u_p, u_q)$, where $p \neq q$, $p \in I_{t-1}$ and $q \in I_{t-1}$.

2.3 Pruning with Replacement

Although projecting a new instance updates the combination of existing kernels to reduce the error, this option is sometimes insufficient for completely eliminating the error. In such cases, the LGRNN prunes one of the existing kernels, replacing it with a new kernel, whose center position is the new instance.

$$f_t = f_{t-1-i} + \tau w_i P_{t-1-i} K(u_i, \cdot) + y_t K(x_t, \cdot),$$
$$g_t = g_{t-1-i} + \tau R_i P_{t-1-i} K(u_i, \cdot) + K(x_t, \cdot), \tag{4}$$

where $\tau \in \{1, 0\}$ is a coefficient for switching two replacement options: $\tau = 1$ denotes selecting the "substitution and pruning with replacement" option, while $\tau = 0$ denotes selecting the "pruning with replacement" option. These two options are selected according to the expected error for each modes, as described in 2.4. The option for "substitution and pruning with replacement" involves projecting the kernel to be pruned onto the space spanned by the remaining kernels. In Eq(4), $P_{t-1-i}K(\boldsymbol{u}_i, \cdot)$ denotes the projected vector defined in Eq(5).

The target kernel to be replaced is determined by two methods: redundancy-based selection and frequency-based selection.

Finding Ineffective Kernel in Terms of Redundancy. If a kernel is redundant, the LGRNN effectively replaces it, because, the LGRNN has the ability to prune such kernels without making large changes to the output function of the LGRNN. Let us assume that one of the vectors $K(\boldsymbol{u}_i, \cdot)$ can be written as a linear combination of $K(\boldsymbol{u}_j, \cdot)$ $(j \neq i)$. This suggests that $K(\boldsymbol{u}_j, \cdot)$ is redundant in this function approximation. Therefore, the system chooses the i-th kernel that has the smallest value δ_i:

$$\delta_i = \min_{\boldsymbol{a}_i} \|K(\boldsymbol{u}_i, \cdot) - P_{t-1-i}k(\boldsymbol{u}_i, \cdot)\|^2, \quad P_{t-1-i}k(\boldsymbol{u}_i, \cdot) \equiv \sum_{j \neq i} a_{ij} K(\boldsymbol{u}_j, \cdot) \quad (5)$$

The kernel having the minimum δ_i value is suitable for being relieved of its duty because the adverse effects from its substitution are minimal. The optimal values of $\boldsymbol{a}_i$ and δ_i are obtained from:

$$\boldsymbol{a}_i = K_{t-i}^{-1}k(\boldsymbol{u}_i), \quad \delta_i = \left\{1 - k_{t-i}^T(\boldsymbol{u}_i)\boldsymbol{a}_i\right\}, \quad (6)$$

where K_{t-i} is a kernel matrix at round t, whose i-th row and column are removed: $[K_{t-i}]_{pq} = K(\boldsymbol{u}_p, \boldsymbol{u}_q)$ and $k_{t-i}(\boldsymbol{x}) = [K(\boldsymbol{u}_1, \boldsymbol{x}), \cdots, K(\boldsymbol{u}_{i-1}, \boldsymbol{x}), K(\boldsymbol{u}_{i+1}, \boldsymbol{x}), \cdots]^T$.

The system chooses the i-th kernel where $i = \arg\min_j\{\delta_j\}$, and $\min_j\{\delta_j\} < \theta$. Therefore, if a redundant kernel exists, the system replaces it with the new kernel.

Finding Ineffective Kernel in Terms of Activation-Frequency. If there are no redundant kernels ($\min_j\{\delta_j\} \geq \epsilon$), the LGRNN cannot prune one of the kernels only by using a redundancy based pruning strategy. Without the pruning process, the LGRNN yields large errors in cases where the distribution of new instances differs from the distribution of past instances. To correct this problem, the modified LGRNN uses a frequency-based pruning strategy: the LRFU policy.

In operating systems, an LRFU policy [8] is a combination of least recently and least frequently used policies for the page replacement algorithm. Hence, once the main-memory has reached its capacity, the operating system moves the most useless memory page into the swap space on the hard-disk drive. LRFU is a useful policy for selecting the most useless memory page.

In this study, we apply an LRFU policy for choosing the most ineffective kernel in LGRNN. The LRFU policy is useful for detecting the least activated and oldest kernel. It is probable that such kernels will not activate again for an extended time, indicating that the cumulative error for the learner remains low even if the kernel is pruned.

LRFU estimates the recency of each kernel's activation time by calculating the time difference between the time of activation, when the kernel center is the nearest to the input, and the current time. Let t_c be the current time and t_j be the j-th kernel's activation time. The kernel's effectiveness is measured by the following weighting function $f_j[.]$.

$$C_j(t_c) \equiv \sum_{k=1}^{w} f[t_c - t_{jk}], \quad where \quad f[t_c - t_j] \equiv \left(\frac{1}{2}\right)^{\lambda(t_c - t_j)} \tag{7}$$

Note that $0 < f[n_j] \leq 1$ and $0 \leq \lambda \leq 1$. From this equation, we can obtain the following recurrence formula.

$$C[j] := \begin{cases} 1 + \left(\frac{1}{2}\right)^{\lambda} C[j] & \text{if } j\text{-th kernel center is the nearest to the input} \\ \left(\frac{1}{2}\right)^{\lambda} C[j] & \text{otherwise} \end{cases} \tag{8}$$

Thus, we can detect the most ineffective kernel in terms of its recency as $i = \arg\min_j C[j]$. λ was set 0.0005 in the experiments.

Unification of the Two Evaluation Objectives. The LGRNN manages the two methods for finding ineffective kernels by using the rule here explained. Because ALD and LRFU are effective policies for finding the most redundant and the least frequently/recently used kernels, respectively, we use each policy for a different purpose. On the one hand, where redundant kernels remain in the LGRNN, ALD is the appropriate method for finding the target kernel, that is, the kernel to be replaced. On the other hand, if there are no redundant kernels in the LGRNN, LRFU is useful for determining the target kernel.

Nevertheless, the LGRNN cannot beforehand establish whether any redundant kernels remain. Thus, the LGRNN applies ALD based kernel replacement option at first, and after that, the LRFU based kernel replacement option is applied (see "ignore option" in 2.4).

2.4   The Best Learning Option to Be Executed

When the number of kernels reaches the upper bound, the LGRNN selects the best learning option of the four available: "modification," "replacement" or "replacement with substitution," and "ignore." The best learning option is selected in accordance with the expected loss for each of the four learning options.

To estimate predicted loss, each kernel also counts the number of learned samples, N_i. N_i is typically a natural number but in cases where the i-th unit substitutes other ineffective units, it is a real number . The LGRNN selects learning option with the least expected loss.

- **Pruning with Substitution and Replacement:** The expected loss under this option is the sum of the loss due to projection and pruning. Therefore,

$$e_{substitute} \equiv \sum_{j \neq i} \left\{ \frac{a_{ji}(w_i^* - w_j^*)}{R_j(t) + R_i(t)a_{ji}} \right\}^2 N_j + N_i(w_{Nearest(i)}^* - y_t)^2 \delta_i \quad (9)$$

where i denotes the index of the most ineffective kernel and w_i^* is $w_i^* \equiv w_i/R_i$. N_i denotes the number of samples, recorded by the i-th kernel. By default, the number of N_i is one, but if the i-th kernel substitutes another kernel, N_i increases: $N_i := N_i + N_j|a_{ji}|/\{\sum_k |a_{jk}|\}$. $Nearest(i)$ denotes the kernel nearest to the i-th kernel.

However, if $R_j + R_i a_{ij} < 0$ for $j \neq i$, this option is passed over for the candidate R_j to prevent it from being a negative value.

- **Pruning with Replacement:** A "Pruning with replacement" is the option to replace the kernel without a projection process such that $\tau = 0$. However, the number of learned samples for the i-th unit is added to $N_{Nearest(i)}$ prior to replacement, where the latter is the number of samples learned by the nearest kernel. $N_{Nearest(i)} := N_{Nearest(i)} + N_i$ The expected loss from this option is the loss due to pruning. Therefore,

$$e_{prune} \equiv (w_{Nearest(i)}^* - w_i^*)^2 N_i, \quad (10)$$

After the replacement, N_i is reset to 1.

- **Modification:** The expected loss with this option is the sum of the losses due to projection and pruning. Therefore,

$$e_{modify} \equiv \sum_j \left\{ \frac{a_{jnew}(y_{new} - w_j^*)}{R_j(t) + a_{jnew}} \right\}^2 N_j + N_{new}(w_{Nearest(New)}^* - y_t)^2 \delta_{new} \quad (11)$$

where $Nearest(New)$ denotes the kernel nearest to the new instance and $\delta_{new} = \|K(\boldsymbol{x}_t, \cdot) - P_{t-1}K(\boldsymbol{x}_t, \cdot)\|^2$. N_i is updated by $N_i := N_i + |a_{inew}|/\{\sum_j |a_{jnew}|\}$.

- **Ignore:** Under this option, the expected loss is the loss caused by doing nothing.

$$e_{ignore} \equiv N_{new}(y_t - y(\boldsymbol{x}_t))^2 \quad (12)$$

The LGRNN does nothing if e_{ignore} is the smallest of all. However, if e_{ignore} is larger than a certain threshold ϵ, the LGRNN applies the pruning with replacement option to the target kernel, selected using the LRFU evaluation method. In the experiment, ϵ was 0.1.

Subsequent to the four estimation procedures, the option with the least expected loss is selected for incremental learning.

3 Experiments

The proposed LGRNN with dynamic pruning policy was compared with three other kernel-based perceptron learning methods: Projectron++ [2], PDM [3] and the original LGRNN [4]. These models were evaluated using the servo, housing, heartal, concrete, cpu-performance and mpg datasets for regression, stored in the UCI machine-learning repository. A benchmark test was repeated 50 times by changing the dataset sequence. The resulting cumulative errors were averaged over 50 trials and 95% confidence intervals were estimated. Note that the projectron++ and PDM are were modified to solve the regression problems since they are originally designed to solve the clustering problems. Therefore, the label for each instance y_t was replaced with the residual error of the kernel perceptron $e_t = y_t - \langle f_{t-1}, K(x_t, \cdot) \rangle$, where f_t is the function of the kernel perceptron at the t-th step.

The learner's performances under a virtual concept drifting environments was also investigated. To assess this performance, the datasets were divided into several clustered groups using an EM-algorithm and re-arranged them into the order of the clusters. The method proposed and its competitors then learned the re-arranged samples. Figure 1 provides an example of the performances for heartal and concrete datasets, where the left and right parts are the results in the cases of identically and independently distributed (i.i.d) samples and virtual concept drifting samples, respectively. The cumulative errors after finishing the learning are also listed in Table 1.

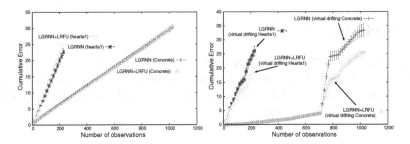

Fig. 1. Cumulative errors for heartal and Concrete datasets. Upper bound on the number of kernels: 10, vertical bar denotes 95% confidence interval. i.i.d data: left, Virtual Drifting data: right.

In this example, the number of kernels was restricted to 10. The results in Figure1 demonstrate that the proposed LGRNN with LRFU have smaller cumulative errors than those for the original LGRNN in the case of virtual concept drifting datasets, meanwhile these two models yield almost the same cumulative errors under i.i.d. dataset. This result confirms that the proposed method is stronger than the original LGRNN under virtual concept drifting environments.

Table 1 also suggest that LGRNN with LRFU performances were statistically better than those of the other methods under the virtual drifting environments.

Although LGRNN with LRFU performances were nearly the same as those of the other methods under i.i.d. samples, the averaged performances were slightly better than those of the others under the seven datasets in the twelve datasets. Therefore, LGRNN with LRFU performances are better than or equal to the other methods.

Table 1. Cumulative errors after the learning with 95% confidence interval: i.i.d. and virtual concept drifting samples

Dataset Name	Projectron++	PDM	LGRNN	LGRNN+LRFU
servo(i.i.d.)	13.5±0.1	13.5±0.1	_11.0±0.6_	11.3±0.4
housing(i.i.d.)	102.5±1.9	86.5±1.1	_12.0±0.3_	12.8±0.4
concrete(i.i.d.)	1212.3±1303.2	106.5±2.8	_30.5±0.8_	30.7±0.8
cpu_performance(i.i.d.)	5.1±0.1	3.3±0.1	_1.4±0.1_	1.5±0.1
he!arta1(i.i.d.)	120.9±0.328	120.4 ±0.4	_23.0±1.1_	24.9±0.9
mpg(i.i.d.)	89.093±23.8	10.0±0.3	_4.0±0.1_	4.2±0.1
servo(drifting)	14.3±0.04	13.6±0.1	_7.8±0.1_	7.9±0.2
housing(drifting)	105.8±0.04	65.4±0.8	15.6±1.0	_12.8±0.5_
concrete(drifting)	201.7 ± 5.1	60.6±1.2	33.6±2.3	_25.7±1.4_
cpu_performance(drifting)	3.826±0.176	2.326±0.134	1.85±0.081	_1.8±0.07_
heart!a1(drifting)	121.6±0.03	118.1±0.2	26.4±1.8	_20.8±0.5_
mpg(drifting)	133.6±125.8	7.9±0.1	10.9±0.8	_7.7±0.7_

4   Conclusion

In this paper, an extended LGRNN with a dynamic pruning strategy was proposed and evaluated. This method is similar to that of the original LGRNN; it differs in that it uses an LRFU-based pruning with replacement strategy when adopting the option to "ignore."

References

1. Dekel, O., Shalev-Shwartz, S., Singer, Y.: The forgetron: A kernel-based perceptron on a fixed budget. Technical report (2005), http://www.pascal-network.org/
2. Orabona, F., Keshet, J., Caputo, B.: The projectron: A bounded kernel-based perceptron. In: ICML 2008, pp. 720–727 (2008)
3. He, W., Wu, S.: A kernel-based perceptron with dynamic memory. Neural Networks 25, 105–113 (2011)
4. Yamauchi, K.: Pruning with replacement and automatic distance metric detection in limited general regression neural networks. In: IJCNN 2011, pp. 899–906. IEEE (July 2011)

5. Yamauchi, K.: Incremental learning on a budget and its application to quick maximum power point tracking of photovoltaic systems. In: The 6th International Conference on Soft Computing and Intelligent Systems, pp. 71–78. IEEE (November 2012)
6. Yamauchi, K., Kondo, Y., Maeda, A., Nakano, K., Kato, A.: Incremental learning on a budget and its application to power electronics. In: Lee, M., Hirose, A., Hou, Z.-G., Kil, R.M. (eds.) ICONIP 2013, Part II. LNCS, vol. 8227, pp. 341–351. Springer, Heidelberg (2013)
7. Specht, D.F.: A general regression neural network. IEEE Transactions on Neural Networks 2(6), 568–576 (1991)
8. Lee, D., Noh, S.H., Min, S.L., Choi, J., Kim, J.H., Cho, Y., Sang, K.C.: Lrfu: A spectrum of policies that subsumes the least recently used and least frequently used policies. IEEE Transaction on Computers 50(12), 1352–1361 (2001)

Neural Computing with Concurrent Synchrony

Victor Parque, Masakazu Kobayashi, and Masatake Higashi

Toyota Technological Institute,
2-12-1 Hisakata, Tempaku-ku, Nagoya 468-8511, Japan

Abstract. Neural networks are important modeling tools to implement intelligent behaviour in a wide variety of phenomena. We introduce the concept of concurrent synchrony in spikes to enable the efficient representation of neural networks to process sensory stimuli. Using different sensory modalities, we show that information processing from stimuli can be represented compactly. This approach aims at introducing homeostasis into the behavior of neural populations in order to construct diverse and sophisticated control rules without increasing network complexity.

Keywords: synchrony, concurrency, spiking networks, neural representation.

1 Introduction

Neurons inspire natural models for intelligent behaviour in many applications. Over the last century, the focused selection and inhibition of potentially competing motor programs in the basal ganglia has been described using spikes, and this view has induced important developments in computing, from the first neuron of McCulloh and Pitts in the 1940's[1] to more recent neural architectures for pattern recognition and decision making[2,3]. Some widely known architectures include the Feed Forward Networks[1], the Multilayer Perceptrons[4], the Recurrent Networks[5,6], the Radial Basis Function Networks[7], the Time-delay Neural Networks[8], the Universal Learning Networks[9] and the Spiking Neural Networks[10,11].

Broadly speaking, the *firing rates* mechanism has been widely adopted as a filtering and mapping method; and a plethora of high performing algorithms for classification and computer vision have been developed over the last decade. Nevertheless, there still exists gaps that undermine the applicability to more general problems. First, it is unable to describe memory formation within the scope of Hebbian learning theories, for which efficient learning algorithms and experimental observations exist. Second, neural populations with the *firing rates* mechanism lack of plasticity to handle noise. Third, it is unclear the role of synchrony and concurrency in computing with *firing rates*; instead, asynchrony is a desired feature for speeding up the paralleling processing of signals, where the heterogeneity of neurons, the spike timing and the concurrency of neural pathways have a null contribution for detecting the invariants in sensor stimuli; as results, the unit of computation is the neural population, not the neuron itself, and the heterogeneity of neurons is reduced to ease the complexity of learning. Finally, the representations using *firing rates* lack of specificity as to model and explain natural phenomena such as hyperkinesia.

C.K. Loo et al. (Eds.): ICONIP 2014, Part I, LNCS 8834, pp. 304–311, 2014.

To tackle the above issues, we propose in this paper a new model to compute with spikes: using *concurrency*[12,13] and *synchrony*[14], where *concurrency* has the role of activating multiple neural pathways, while *synchrony* has the role of detecting invariants and structure in sensory stimuli. The unique point in our model is that action selection is not only the net result of input stimuli, but also it emerges from the focused and the concurrent activation of both excitatory and inhibitory pathways, enabling to depart from causality principles and to consider counterfactuals[15] when designing neural representations.

The organization of this paper is as follows. Section 2 describes and exemplifies the main tenets in our model. Section 3 discusses simulation results on different sensory modalities. Section 4 concludes the paper.

2   Neural Computing with Concurrent Synchrony

Let an agent, as shown in Fig. 1, evaluate complementary stimuli from sensors A and B to decide on the execution of the motor program *Go* by using *direct* (excitatory, solid line) and *indirect* (inhibitory, dashed line) pathways.

2.1   Concurrent Pathways

A *concurrent pathway* represents the set of conjunctive rules that activate to *facilitate* the desired motor program, $A \to promote$ *Go*, and to *inhibit* potentially competing motor programs, $A \to inhibit$ *NoGo*, as shown in Fig. 1, right column. Note that action selection implies using both the *direct* and the *indirect* pathways, instead of the independent and disjunctive rules as shown in Fig. 1, left column, thus defining rules for both

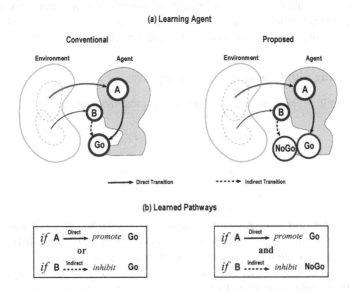

Fig. 1. Basic comparison between the conventional and the proposed pathways

why to *Go* and why not *NoGo* conjunctively. In this scheme, the receptors in B have a different role: instead of being the logical negation of the receptors in A, they help inhibit the actions that would normally interfere with the desired action. Physiologically, the receptors in $A(B)$ represent the $D_1(D_2)$ receptors in the striatum[16]. In terms of a propositional system, **concurrent pathways** help modelling *counterfactuality* in knowledge representation, where go/no-go decisions are triggered by conjunctive/concurrent pathways coordinating to encode a meaningful rule set.

2.2   Decoding Synchrony

Consider a postsynaptic neuron *Go* receiving inputs from different presynaptic sensor stimuli A and B as Fig. 1 shows. A **synchrony receptive field** of the neuron *Go* is the set of stimuli inducing synchronous firing in the neuron[14], where synchrony can be easily decoded by a *coincidence detector* in a noisy integrate-and-fire model:

$$\tau_n \frac{dv}{dt} = -v + n \tag{1}$$

$$\tau_n \frac{dn}{dt} = -n + \sigma \xi(t) \sqrt{\frac{2}{\tau_n}} \tag{2}$$

where v is the membrane potential of the postsynaptic neuron, τ_n is the membrane time constant, $n(t)$ is a noise filter with standard deviation σ, and $\xi(t)$ is a white noise.

In a population of heterogeneous neurons, a **synchrony group** is the group of neurons that fire within a small time interval (they fire more when the input stimulus is coincident). Thus, a given stimuli divides the population into different synchrony groups, where neurons in the same group project to a unique (single) postsynaptic neuron. This feature brings benefits to encode information of stimuli that may not be represented in individual receptive fields, since synchrony reveals sensory invariants, and the more heterogeneity of neurons, the better (finer) precision to encode synchrony.

2.3   Computing with Concurrent Synchrony

Let a presynaptic neuron with *rebound spiking* be modeled with:

$$\tau_0 \frac{dv}{dt} = E_l - v + g_{max} g_1 (E_k - v) \tag{3}$$

$$\tau_1 \frac{dg_1}{dt} = \left[1 + exp\left(\frac{V_a - v}{k_a} \right) \right]^{-1} - g_1 \tag{4}$$

where τ_0 and τ_1 are membrane time constants; E_l is the leak reversal potential; E_k is the reversal potential; g_1 is the low-threshold conductance; V_a is the half activation voltage; k_a is the activation factor and g_{max} is the maximal conductance. For simplicity, we set $V_a = -70$, $k_a = 5$, $E_l = -35$, $E_k = -90$, and $\tau_0 = U(10, 50)$ and $\tau_1 = U(100, 400)$. A spike is produced when $V_t > -55$, then the membrane potential is reset at -70mV.

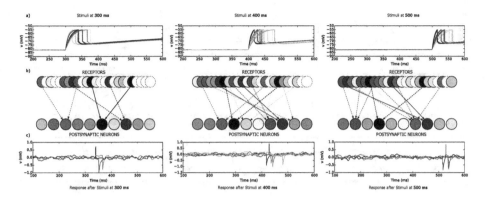

Fig. 2. Example of a spiking neural network with concurrent synchrony

Let a population of postsynaptic neurons (*coincidence detectors* with $\tau_n = 5$ms, $\sigma = 2$, and $V_t = 0.6$) receive input from a population of presynaptic neurons with *rebound spiking* as shown by Fig. 2. Also, let the presynaptic neurons receive inhibitory stimuli for 300ms, 400ms and 500ms, as shown by Fig. 2(a). Let the *synchrony groups* be formed by the ensembles of neurons that fire within 2ms, each of which is colored differently in Fig. 2 (a)-(b) (neurons that do not fire are white-dashed circumferences). Then, presynaptic neurons in the same *synchrony group*, i.e., same color in Fig. 2 (a), make pathways (synapses) onto the same postsynaptic neurons. For simplicity we omit the synapses of synchrony groups with one neuron (grey color) and set the weight on every synapse to $1/N$, where $N > 1$ is the number of presynaptic neurons in the corresponding synchrony group. Note that each stimuli duration induce different *synchrony groups*: the two neurons colored in red for the 300ms stimulus are not synchronous for the 400ms stimulus, thus each stimulus duration is associated with different groups of postsynaptic neurons.

Let the postsynaptic neurons represent (possibly competing) motor programs. Also, let one of them arbitrarily be the *desired* motor program[1] (black for the 300ms stimulus, blue for the 400ms stimulus and blue for the 500ms stimulus) and the rest be *competing* motor programs. Pathways projecting to the *desired* (*competing*) motor program are *direct* (*indirect*) and have *positive* (*negative*) synapse weight. From Fig. 2(b), we can easily note that for any kind of stimuli duration and when all pathways are **concurrent** the selected action is the net result from *direct* pathways facilitating the *desired* motor program, i. e., solid-lined arrows in Fig. 2(b), and *indirect* pathways inhibiting competing motor programs i. e., dash-lined arrows in Fig. 2(b). The resulting membrane potential of the postsynaptic neurons with concurrent synchrony is shown in Fig 2(c).

In the scheme of concurrent synchrony, it is natural to include synaptic plasticity: positive (negative) rewards increase (decrease) synapse connections (weights of the pathways projecting from the presynaptic neurons), being consistent with Hebbian learning theories and observations in vivo of memory formation[17]. Also, modeling with concurrent synchrony can explain the emergence of a number of natural phenom-

---

[1] Representing a previously rewarding course of action.

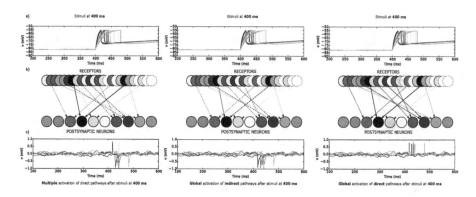

Fig. 3. Examples of natural phenomena in neural networks with concurrent synchrony

ena in real world neural populations. For example, assuming stimuli durations of the same interval (400ms), the multiple activation of only direct pathways can lead to parallel activation of postsynaptic neurons representing different (possibly complementary) motor programs, as shown by the membrane potentials of postsynaptic neurons in the left column of Fig. 3. Also, ablating or inhibiting all direct (indirect) pathways induces a global and non-selective activation of indirect (direct) pathways, inhibiting (promoting) most motor programs and not only undesired (desired) ones, thus leading to bradykinesia (hyperkinesia), as shown in the membrane potentials of postsynaptic neurons in the middle (right) column of Fig. 3. All spiking neurons were modeled with the Brian simulator[18].

3   Computational Experiments

3.1   Odor Recognition

The purpose of these experiments is to show the applicability the proposed scheme for classification (recognition) problems. For this purpose, we model odor concentration by a half-wave rectified low-pass filtered noise (Ornstein-Uhlenbeck process)[14]:

$$\tau_x \frac{dx}{dt} = -x + \xi(t)\sqrt{2\tau_x} \tag{5}$$

with membrane time constant $\tau_x = 75$ms, and odor concentrations proportional to $[x]^+$ and binding coefficients in $U(10^{-7}, 10^{-1})$.

Let $N = 5000$ presynaptic neuron receptors receive stimuli from odor fluctuations transformed into spikes by the integrate-and-fire model:

$$\tau_i \frac{dv}{dt} = -v + I(c) \tag{6}$$

with $\tau_x = 20$ms, c is the odor concentration, and $I(c)$ is the transduction current:

$$I(c) = \frac{I_{max}c^n}{c^n + K^n} \tag{7}$$

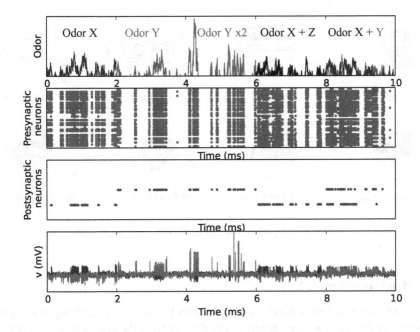

Fig. 4. Computing with concurrent synchrony in olfaction

where $I_{max} = [1 - exp(-(F_{max}\tau_x)^{-1})]^{-1}$ is the input current to produce a maximum firing rate $F_{max} = 40$Hz, $n = 3$ is the Hill function constant, and $K = 1$ is the half activation constant for relative odor concentrations.

Let $M = 20$ postsynaptic neurons be *coincidence detectors* with $\tau_n = 8$ms, $\sigma = 0.15$, and voltage threshold $V_t = 0.8$. As in our previous example, the **synchrony groups** are formed by ensembles of neurons that fire within 2ms; and the synapse weights are set to $1/N_g$, where N_g is the number of neurons in g-th synchrony group.

Let odors X, Y and Z be modeled by Eq. 5. Also, assume two odors X, Y be previously learned odors by two (arbitrarily chosen) postsynaptic neurons. Odors are presented in sequence every 2ms.: X alone, Y alone, Y alone with higher intensity, X and distracting odor Z, and both X and Y. Fig. 4 shows different situations for odor concentration (top), the activity of the 100 receptor neurons (out of the 5000), and both the activation behaviour and the membrane potential of the postsynaptic neurons. Note that when either (or both) X-Y odor is presented, the corresponding postsynaptic neurons are activated, and odor intensity increases spiking behaviour. In contrast, tolerance to a distracting odor Z (noise) is seen as the absence of spiking, thus no synchrony group formation, instead of false spiking rates. In difference with previous results[14], the use of both direct and indirect pathways within a synchrony detector, brings the efficiency of using only one postsynaptic neuron for classification and not a population of neurons.

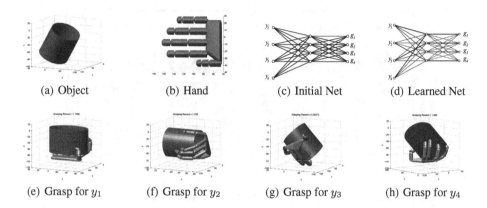

(a) Object   (b) Hand   (c) Initial Net   (d) Learned Net

(e) Grasp for y_1 (f) Grasp for y_2 (g) Grasp for y_3 (h) Grasp for y_4

Fig. 5. Grasping actions for each object condition

3.2 Hand Grasping

We consider the problem of hand grasping in order to show the learnability of computing with concurrent synchrony neurons. For this purpose, the goal is to learn a neural net to hold a rotated object (cylinder) in Y-axis by $y = \{\pi, \frac{\pi}{2}, \frac{\pi}{4}, \frac{-\pi}{4}\}$ with a 5-fingered robotic hand (20 junctions in active state)[19] and a reward function r defined by the minimum singular value of the grasping matrix[19].

Grasping actions $g = \{g_1, ..., g_4\}$ rotate the hand by $\frac{-\pi}{2}, \frac{\pi}{2}, \frac{-\pi}{2}, \pi$ in X, Y, Z and X-axis, respectively, and then close it until fingers and palm hold the object. Synaptic weights are updated by:

$$s(a)_i^D \leftarrow s(a)_i^D + \delta.\Delta s_i.r \qquad (8)$$

$$s(a)_i^I \leftarrow s(a)_i^I - \delta.\Delta s_i.r \qquad (9)$$

where a is a grasping action, $s(a)_i^D$ and $s(a)_i^I$ is the synaptic strength related to action a in the *direct* and *indirect* pathways, respectively; $\delta = sgn(r)$ is the desirability of the action, and Δs_i is the updating step. For $N = 4$ actions, the initial values for $s(a)_i^D$ and $s(a)_i^I$ are set to $\frac{1}{N}$ and Δs_i is set to $\frac{1}{N^2}$. The spiking neural net considers a integrate-and-fire model with refractory time set to 10ms, resting potential set to -1, membrane threshold set to 2, and maximum potential threshold set to 2.

Fig. 5 shows (a) the object, (b) the initial position of the hand, (c) the studied topology of the network, (d) the final weights and configuration of the learned network[2], and (e-h) the grasping actions for the object in the four rotation contexts y. We note that for every input case, the network compactly encodes the best possible action and, at the same time, which actions not to take. For example, when the context is y_3, the network recommends to perform action g_2 and not g_1, g_3 or g_4. Here, which pathway to execute is dependant on the long-term learned strength on both direct and indirect pathways.

---

[2] Solid lines represent direct pathways, dashed lines represent indirect pathways.

4 Conclusion

We have proposed a new approach to model control rules using spiking neural networks using concurrency and synchrony during pathway activation. Computational experiments show the robustness to noise signals, the efficiency in network representation in an odor recognition problem; and the compactness of networks in a hand grasping problem. Further development and comparison with the conventional learning strategies and evaluations on diverse planning problems are part of our future agenda.

References

1. McCulloch, W.S., Pitts, W.: A logical calculus of the ideas immanent in nervous activity. Bull. Math. Biophysics 5, 115–133 (1943)
2. Rumelhart, D.E., et al.: Parallel Distributed Processing. MIT Press, Cambridge (1986)
3. Hopfield, J.J.: Neural Networks and physical systems with emergent computational abilities. Proc. Natl. Acad. Sci., 2554–2558 (1982)
4. Rosenblat, F.: The perceptrons: A probabilistic model for information storage and organization in the brain. Psychol. Rev. 65, 386–408 (1958)
5. Williams, R.J., Zipser, D.: A learning algorithm for continually running fully recurrent neural networks. Neural Computation 1(2), 270–280 (1989)
6. Williams, R.J., Zipser, D.: Gradient based learning algorithm for recurrent connectist networks. College of Computer Science, Northeastern University, Boston, MA, Tech. Rep., NU-CCS-90-9 (1990)
7. Moody, R.J.J., Darken, C.J.: Fast learning networks iof locally-tunned processing units. Neural Computation 1, 281–294 (1989)
8. Lin, E.T., Dayhoff, J.E., Ligomenides, P.A.: Trajectory production with the adaptive time delay neural network. Neural Networks 8(3), 447–461 (1995)
9. Hirasawa, K., Murata, J., Hu, J., Jin, C.: Universal learning network and its application to robust control. IEEE. Trans. Man and Cybernetics, Part B: Cybernetics 30(3), 419–430 (1995)
10. Maass, W.: Networks of spiking neurons: The third generation of neural network models. In: Australian Conference on Neural Networks (1996)
11. Izhikevich, E.M.: Hybrid spiking models. Philosophcial Transactions of the Royal Society A 368, 5061–5070 (2010)
12. Mink, J.W.: The Basal Ganglia and involuntary movements: Impaired inhibition of competing motor patterns. Archives of Neurology 60, 1365–1368 (2003)
13. Cui, G., et al.: Concurrent activation of striatal direct and indirect pathways during action initiation. Nature 494, 238–242 (2012)
14. Brette, R.: Computing with Neural Synchrony. PLOS Computational Biology 8(6), e1002561 (2012), doi:10.1371/journal.pcbi.1002561
15. Pearl, J.: The algorithmization of counterfactuals. Annals for Mathematica and Artifical Intelligence 61, 29–39 (2011)
16. Gerfen, C.R., Surmeier, D.J.: Modulation of striatal projection systems by dopamine. Annu. Rev. Neurosci. 34, 441–466 (2011)
17. Navabi, S., et al.: Engineering a memory with LTD and LTP. Nature, doi:10.1038/nature13294
18. Goodman, D.F., Brette, R.: Brian: A simulator for spiking neural networks in Python. Front. Neuroinform., doi:10.3389/neuro.11.005.2008
19. Malvezzi, M., Gioioso, G., Salvietti, G., Prattichizzo, D., Bicchi, A.: SynGrasp: A MATLAB Toolbox for Grasp Analysis of Human and Robotic Hands. In: Proc. IEEE International Conference on Robotics and Automation, Karlsruhe, Germany, pp. 1088–1093 (2013)

A Line-Partitioned Heteroassociative Memory for Storing Binary Fresnel Hologram

Peter Wai Ming Tsang and Chi-Sing Leung

Dept. of Electronic Engineering, City University of Hong Kong, Hong Kong
{eewmtsan,eeleungc}@cityu.edu.hk

Abstract. The Hopfield network is a classical and interesting type of artificial neural networks that can be employed for storing multiple patterns. As such, the Hopfield Network can serve as a content-addressable, auto-associative memory (AAM) whereby a stored pattern can be retrieved from an incomplete or damaged copy of itself. Recently, it has been demonstrated that a Hopfield Network can be applied as an AAM for storing complex holograms. Despite the success, the structure of the AAM is rather cumbersome as a pair of networks is required for storing the real and the imaginary components of the complex hologram. In this paper, we proposed an enhanced method to alleviate this shortcoming with a single line-partitioned hetero-associative memory (LP-HAM). Briefly, the LP-HAM is trained to memorize the imaginary components of a collection of binary holograms, so that each of them can be recalled from its corresponding real component. Subsequently, the input (real component) and the output (recalled imaginary component) images will be combined to form the complex hologram.

Keywords: Auto-associative Memory, hetero-associative memory, hologram.

1 Introduction

Over the years, numerous research works have been conducted in different disciplines to model and simulate the various functions of human brain. Amongst different schemes, the Hopfield network invented by J. Hopfield [1] has provided an effective means for understanding, as well as modelling the mechanism of engrams. The Hopfield network is composed of a group of interconnected McCulloch-Pitts nodes (threshold logic units) which mimics a simplified version of the massive interconnected neurons of the brain. As such, the network can effectuate the content-addressable, autoassociative memory (AAM) function whereby multiple patterns can be recorded by suitably adjusting the connection weight (a process commonly known as learning or training) between each pair of nodes. A stored pattern can be recalled by presenting a complete, or partial version of itself.

Later, the AAM has been extended to the heteroassociative memory (HAM) [2–4]. Being different from an AAM, the HAM is capable of associating an input pattern with a different store pattern. Training of the AAM can be conveniently conducted with the Hebbian learning, with which the connection weight between a pair of neurons is increased if they have similar activations, and vice versa. The emergence of the AAM has instigated a lot of interesting applications in the area of 2-D image storage and recognition (such as, but not limited to [5–7]).

C.K. Loo et al. (Eds.): ICONIP 2014, Part I, LNCS 8834, pp. 312–318, 2014.
© Springer International Publishing Switzerland 2014

Recently, it has also been demonstrated that the AAM can be extended to the storage of complex holograms representing 3-D object scenes [8]. In this approach, the real and the imaginary images of the hologram to be recorded are first converted into binary 2D arrays through error diffusion. Next, a pair of separate AAM is employed to record the binarized 2D arrays with each element having a value of ± 1. The AAM is formed by an array of sub-AAM, each storing a row of the input binary 2D array. It should be noticed that the reconstruction images from the binary 2D array is still a grey level image.

As illustrated in [8] the decomposition of the AAM into sub-AAMs, which is referred to as the line-partitioned AAM (LP-AAM), is over 4 order of magnitude smaller in size as compare with the straightforward implementation, and also capable of recalling the stored hologram with a noise contaminated or damaged copy of itself. The training process is computationally efficient, as only a one-shot learning process is involved in the storage of each hologram. Apparently, the fidelity of the holograms retrieved from the LP-AAM is also superior to that obtained with existing approach based on the back-propagated (BP) neural network [9].

On the downside, a pair of LP-AAMs is required to store the 2 orthogonal components of the complex holograms. Recalling a hologram is cumbersome, as both the real and the imaginary 2D arrays of the hologram have to be input to their respective AAMs. In this paper, we presented a method to overcome the above-mentioned problems. Briefly, a single line-partitioned hetero-associative memory (LP-HAM) is employed to record the imaginary 2D array of the binarized holograms. Upon presenting the real part of a hologram to the LP-HAM, the imaginary part will be recalled. Subsequently, the input and the output images will be combined into a complex hologram.

Organization of the paper is given as follows. In Section 2, the concept of hologram is reviewed. Section 3 presents our proposed method. Section 4 presents the experimental evaluation. The paper is then concluded in Section 5

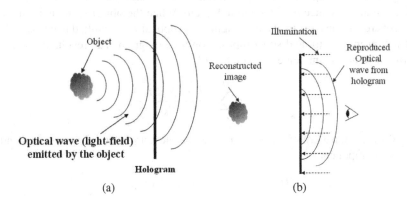

(a)   (b)

Fig. 1. Concept of a hologram. (a) Recording of the object wave with a hologram. (b) Optical reconstruction of object image from hologram.

2   Holography and Computer Generation Holography

Figure 1 illustrates the mechanism of holography. As shown in Figure 1(a), when a 3D scene is illuminated by a coherent light source, the scene scatters light field to the surrounding environment. A hologram is a 2D media which records the scattered light field. It records the fringe patterns that are resulted from the interference between the object wavefront and a reference optical beam. In the traditional approach, the 2D media is a high resolution photographic film.

A virtual image of the original 3D scene can be reproduced by illuminating the hologram with a coherent optical wave, as shown in Figure 2(b). With the holography technique, we can store the 3-D information of the scene in a 2D media.

Nowadays, with the advancement of computing and optical technologies, the holography technique can be implemented in a digital way. That is, we compute the amplitude and phase of the optical wave that hits on each point on the hologram plane (2D media) in a digital way. As a result, we have a 2-D array of complex terms (each referred to as a hologram pixel). It is referred as the digital holograms.

The virtual image can be displayed on a high resolution device, such as a spatial light modulator (SLM). From the light field emitted from the SLM, the user is able to visualize the 3-D scene recorded in the digital hologram. As the technique of generating a digital hologram is in the numerical way, we call the technique computer generated holography (CGH) [10–12].

3   Proposed Line-Partitioned Hetero-Associative Memory for Storing of Binary Hologram

For the sake of completion, we would like to outline the process of generating a binary Fresnel hologram from a 3-D scene. Suppose the latter is comprised of a collection of M object points $P = \{p_0, \cdots, p_{M-1}\}$, each having an intensity I_k $(k = 0, \cdots, M - 1)$, and located at a distance d_k from the hologram. When the object scene is illuminated with a coherent optical beam of wavelength λ, the optical wave emitted from each object point will propagate to, and superimposed on the hologram. The resultant diffraction pattern on the hologram plane is given by [12]

$$H(x, y) = \sum_{k=0}^{M-1} \frac{I_k \exp(j 2\pi \lambda)}{d_k} . \tag{1}$$

As the diffraction pattern on the hologram is complex, it can be split into a real and an imaginary component as

$$H(x, y) = H_R(x, y) + H_I(x, y). \tag{2}$$

Direct storing a Q-bits complex hologram into an associative memory involves huge number of connection weights. To reduce the the number of connection weights, we use the Floyd-Steinberg error diffusion [13] to convert the real and the imaginary components of a complex hologram into two binary images $H_R^B(x, y)$ and $H_I^B(x, y)$, respectively.

We will only describe the conversion of the real component $H_R(x, y)$ since an identical process is performed on the imaginary component $H_I(x, y)$. In the error diffusion process, a hologram is scanned along the top-to-bottom, and left-to-right manner. Each visiting pixel is assigned a value of '1' or '0', given by

$$H_R^B(x, y) = \begin{cases} 1 & H_R(x, y) \geq 0.5 \\ 0 & \text{otherwise} \end{cases} . \tag{3}$$

The conversion of a visited pixel to a binary value results in an error, given by

$$e(x, y) = H_R(x, y) - H_R^B(x, y) . \tag{4}$$

The error of the current pixel is then distributed to the neighboring unvisited pixels, given by

$$H_R(x, y + 1) \longleftarrow H_R(x, y + 1) + w_1 e(x, y) \tag{5a}$$
$$H_R(x + 1, y - 1) \longleftarrow H_R(x + 1, y - 1) + w_2 e(x, y) \tag{5b}$$
$$H_R(x + 1, y) \longleftarrow H_R(x + 1, y) + w_3 e(x, y) \tag{5c}$$
$$H_R(x + 1, y + 1) \longleftarrow H_R(x + 1, y + 1) + w_4 e(x, y) \tag{5d}$$

where the operator "$\longleftarrow$' means to replace the term on the left hand side with the value derived from the right hand side. The weighting coefficients w_i's are equal to $\{w_1 = 716, w_2 = 316, w_3 = 516, w_4 = 116\}$ [13]. Since the input format of associative memories is ± 1, at the end of the binarization process, value '0' is changed to '-1'.

Now, we describe our proposed LP-HAM for storing the binary holograms that are generated with the previous steps. To begin with, the following notation are adopted. The holograms are with resolution $N \times N$. The number of holograms is denoted as N_h.

An LP-HAM is composed of a number of HAMs. Each HAM store the association between a particular row of the imaginary images and the corresponding row of the real images of the binary holograms. The pixels at the k-th row of the real and the imaginary images of the k-th hologram are represented with the column vectors $A_{i;k} = [a_{i;k;0}, a_{i;k;1}, \cdots, a_{i;k;N-1}]^T$, and $B_{i;k} = [b_{i;k;0}, b_{i;k;1}, \cdots, b_{i;k;N-1}]^T$, respectively, where is the number of pixels in a row. Each LP-HAM is composed of two layers of neurons, $f_{A,k}$ and $f_{B,k}$. The two layers are connected with a $N \times N$ matrix. In the training stage, the weight matrix is trained with Hebbian rule to memorize a pattern pair. Mathematically, the connection weight matrix of the k-th HAM is given by

$$W_k = \sum_{i=1}^{N_h} B_{i;k} A_{i;k}^T . \tag{6}$$

The retrieval process is an iterative process that starts with a stimulus vector $A_k^{(0)}$ in $f_{A,k}$. The vector $B_k^{(1)}$ in $f_{B,k}$ is generated based on (the superscript (n) is the iteration index):

$$B_k^{(n)} = \text{sgn}\left[W_k A_k^{(n-1)}\right], \tag{7}$$

$$\text{where } \text{sgn}(x) = \begin{cases} +1 & x > 0 \\ -1 & x < 0 \\ \text{state unchanged (during recalling)} & x = 0 \\ 0 \text{ (during training)} & x = 0 \end{cases}.$$

The new state $B_k^{(n)}$ in $f_{B,k}$ is then fed backward to generate the state in $f_{A,k}$:

$$A_k^{(n)} = \text{sgn}\left[W_k^T B_k^{(n)}\right]. \tag{8}$$

Kosko [2] proved that the sequence $\{(A_k^{(n)}, B_k^{(n)})\}$ converges to one of the fixed points in a finite number of iterations for any real connection matrix.

Table 1. Optical setting for generating the hologram

Size of the images	512×512
Size of the hologram	512×512
Pixel size of the hologram	$12\mu m$
Wavelength of optical beam	650 nm

4 Experiment

The performance of our proposed method is evaluated with a set of test images, shown in Figure 2. For each image, the steps described in Section III are applied to generate a binary complex hologram based on the optical setting listed in Table 1. The hologram is parallel to the source image and located at a distance of 0.3m from the latter. For the sake of simplicity, we have only selected the test image Mandrill in Figure 3(a) in our illustration.

Fig. 2. The ten test images

The real and the imaginary images of the binary hologram representing the test image, together with the numerical reconstructed image, are shown in Figures 3(b) to 3(d), respectively.

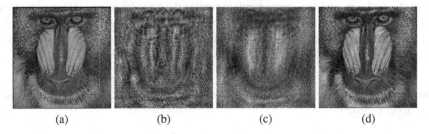

Fig. 3. The hologram of the test image Mandrill. (a) test image Mandrill. (b) Real image of the hologram. (c) Imaginary image of the hologram. (d) Reconstructed image of the hologram from presenting real part.

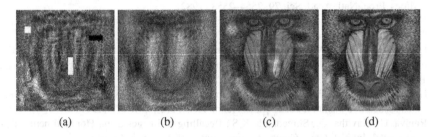

Fig. 4. Recall of the test image Mandrill. (a) Damaged real image of the hologram $A^{(n)}$. (b) Imaginary image of the hologram, recalled from the LP-HAM after one iteration $B^{(1)}$. (c) Reconstructed image (after one iteration). (d) final reconstructed image.

Next, the binarized holograms are taken to train the LP-HAMs, with the real images as the input and the imaginary images as the output of the network. To test the effectiveness of our proposed method, we present the real part of the hologram representing the test image Mandrill to the trained LP-HAM. As shown in Figure 4(a), the image has been damaged in several regions, and added with patches of random noise patterns. After one iteration, the recalled imaginary image of the hologram is shown in Figure 4(b). Subsequently, the initial input and the output images of the LP-HAM are combined to recover the complex hologram, and the numerical reconstructed image is shown in Figure 4(c). Comparing with the reconstructed image of the original hologram, we observed that apart from some degradation around the defective parts of the real image, the fidelity of the reconstructed image is preserved favorably. In addition, after one more recall cycle, the original image is recovered.

5 Conclusion

This paper reported a method for storing a collection of complex Fresnel holograms into a line partitioned HAM. First, a complex hologram is first reduced in size by binarizing it with error diffusion. Next, instead of using a classical associative memory with $(2 \times N^2 \times Q)^2$ interconnections to record the entire holograms. In our approach, we use N HAMs to store the hologram. Each HAM stores the associative between a particular row of the real part and the corresponding row of the imaginary part. Hence, the total number of interconnections can be reduced to N^3. Using our experiment as an example, we can reduce the number of interconnections from $4 \times 512^4 \times 8^2$ (8 bit resolution) to 512^3.

Acknowledgement. The work was supported by a grant from RGC Hong Kong (Project No.: CityU 115612).

References

1. Hopfield, J.J.: Neural networks and physical systems with emergent collective computational abilities. Proc. Natl. Acad. Sci. 79, 2554–2558 (1982)
2. Kosko, B.: Adaptive bidirectional associative memories. IEEE Trans. on Systems, Man, and Cybernetics 18(1), 49–60 (1988)
3. Leung, C.S., Chan, L.W., Lai, E.: Stability, capacity, and statistical dynamics of second-order bidirectional associative memory. IEEE Trans. on Systems, Man and Cybernetics 25(10), 1414–1424 (1995)
4. Leung, C.S., Chan, L.W., Lai, E.: Stability and statistical properties of second-order bidirectional associative memory. IEEE Trans. Neural Netw. 8(2), 267–277 (1997)
5. Ramya, C., Kavitha, G., Shreedhara, K.S.: Recalling of images using Hopfield neural network model. In: Nat. Conf. Comp., Comm. Cont., vol. 11, pp. 2–4 (2011)
6. Costantini, G., Casali, D., Perfetti, R.: Neural associative memory storing Gray-coded grayscale images. IEEE Trans. Neural Netw. 14(3), 703–707 (2003)
7. Singh, Y.P., et al.: Analysis of Hopfield autoassociative memory in the character recognition. Int'l. J. Comp. Sci. Engg. 2(3), 500–503 (2010)
8. Tsang, P.W.M., Ng, K.T.: Efficient recording and retrieval of complex digital fresnel holograms based on the line partitioned autoassociative memory. Neurocomputing (accepted)
9. Yang, G., Zhang, C., Xie, H.: Information Compression of Computer-Generated Hologram Using BP Neural Network. In: Biomedical Optics and 3-D Imaging, OSA Technical Digest (CD) (Optical Society of America) (2010)
10. Shimobaba, T., Nakayama, H., Masuda, N., Ito, T.: Rapid calculation algorithm of Fresnel computer-generated-hologram using look-up table and wavefront-recording plane methods for three-dimensional display. Opt. Express 18, 19504–19509 (2010)
11. Tsang, P.W.M., Cheung, W., Poon, T.C., Zhou, C.: Holographic video at 40 frames per second for 4-million object points. Opt. Express 19, 15205–15211 (2011)
12. Poon, T.C., Liu, J.P.: Introduction to Modern Digital Holography: With Matlab. Cambridge University Press (2014)
13. Floyd, R.W., Steinberg, L.: An adaptive algorithm for spatial grey scale. In: Proc. Soc. Info. Disp., pp. 75–77 (1976)

A Unified Framework
for Privacy Preserving Data Clustering

Wenye Li

Macao Polytechnic Institute,
Rua de Luís Gonzaga, Macao SAR, China
wyli@ipm.edu.mo
http://staff.ipm.edu.mo/~wyli

Abstract. We study the problem of publishing a data table containing personal information, while ensuring individual privacy and maintaining data integrity to the possible extent. One popular technique in literature is through k-anonymization. A release is considered to preserve k-anonymity if the record corresponding to any person cannot be distinguished from that of at least $k - 1$ other individuals whose information also appears in the release. In order to achieve k-anonymity, we propose an unsupervised learning framework. We further show an instantiation of the framework, which leads to an exemplar-based clustering algorithm for practical applications, and report promising results.

Keywords: k-Anonymity, Clustering, Linear Programming.

1 Introduction

Given the advances of intelligent data processing, there is increasing demand to make data publicly available so that analytical methods might yield important new discoveries. Unfortunately much personal information is inevitably involved in such data. The disclosure of personal data obviously raises serious concerns about privacy. The risks are real, as have been demonstrated by recent successful re-identifications of individuals in published data sets [1, 2].

The threaten of disclosing personal privacy data is becoming even more notable recently, mostly due to the rapid development of world wide web, cloud computing and big data processing [3]. More and more data are published in Internet, which certainly facilitates numerous applications while at the same time posing nontrivial danger of tracking back to specific persons from the published data.

Increasing efforts are being devoted to avoid the risk of releasing privacy information. Technically, a number of techniques have been developed for privacy preserving data publishing. A central idea is to remove personally-identifying information that could be used for backtracking, based on the principle of k-anonymity [4]. That is, reduce the granularity of data representation by mapping each record to an equivalence class of at least $k - 1$ other records in the data

C.K. Loo et al. (Eds.): ICONIP 2014, Part I, LNCS 8834, pp. 319–326, 2014.

collection. In this way, the probability of determining the identity of an individual (re-identification probability) is decreased. Unfortunately, for a given data set, producing the optimal k-anonymized version is NP-hard [5, 6]. In practice, therefore, researchers have resorted to approximating this by heuristic methods. Common implementations of k-anonymity use transformation techniques such as suppression and generalization [2, 4, 7].

In this paper, a unified framework is proposed to achieve k-anonymity in given data. We model the task as a clustering problem [8, 9], which maps each data record from a table to an element in a representation set. Each released element has at least k associated records to ensure k-anonymity in the resulting data representation. The framework is flexible. We are free to exert constraints for different applications. Another nice property is that the required computation is much less dependent on the anonymity parameter k than most other approaches, which is often appealing in practice.

2   Background and Related Work

2.1 k-anonymity

The model of k-anonymity is closely related to the concept of *quasi-identifiers*. Let $X(A_1, \cdots, A_d)$ be a table with a finite set of attributes $\{A_1, \cdots, A_d\}$ and a finite number of records $\{x_1, \cdots, x_m\}$. Each record x_i is a sequence of values with attributes $\{A_1, \cdots, A_d\}$. A *quasi-identifier* of X, written Q_X, is a subset of attributes $\{A_1, \cdots, A_d\}$ that can be used to identify individuals within $\{x_1, \cdots, x_m\}$.

A released data set is considered to preserve k-anonymity when for any quasi-identifier, a record is indistinguishable from $k-1$ others whose information also appears in the release. An k-anonymized table protects individual privacy in the sense that, even if an adversary has access to all the quasi-identifying attributes of all individuals represented in the table, he would not be able to track down an individual's record further than a set of at least k records, in the worst case. Without other information, the probability of back-tracing an individual (*re-identification probability*) is kept no larger than $\xi = \frac{1}{k}$ as each individual is hidden in a crowd with $k-1$ other people. Thus, releasing a table with k-anonymity prevents definitive record linkages with publicly available databases.

Formally, the k-anonymity is defined as follows:

Definition 1. *Let $X(A_1, \cdots, A_d)$ be a table and Q_X be any quasi-identifier associated with it. X is said to preserve k-anonymity if and only if each sequence of values in $X[Q_X]$ appears with at least k occurrences in $X[Q_X]$, where $X[Q_X]$ denotes the projection, maintaining duplicate tuples, of attributes Q_X in X.*

Beyond k-anonymity, ℓ-diversity [10] and a number of other models were further developed, which are omitted here to simplify the discussion. The method proposed in this paper, however, can be extended to work with ℓ-diversity and related models trivially.

2.2 Suppression and Generalization

The k-anonymity is typically achieved through *suppression* and *generalization*. The idea is to suppress/generalize some of the entries in the table so as to ensure that for each tuple in the modified table, there are at least $k-1$ other tuples in the modified table that are identical to it along the quasi-identifiers. The objective is to minimize the extent of suppression and generalization.

In suppression, each record x_i in a table is mapped to $\tilde{x}_i$ by hiding some components of x_i, so that each $\tilde{x}_i$ is identical to at least $k-1$ other $\tilde{x}_j$'s. In generalization, in addition to suppressing entry values, it is also allowed to replace them with less specific but semantically consistent values. For example, a date can be made less specific by omitting the day and only revealing the month and year. In doing so, it is assumed that for each attribute, a generalization hierarchy is provided as part of the input.

The two procedures provide a natural way in preserving k-anonymity. Unfortunately the computation for optimal suppressions/generalzations is demanding for large-scale problems and people have to seek approximations, due to the following result [5, 6].

Theorem 1. *Obtaining k-anonymity with suppression is NP-hard even for a ternary alphabet, i.e., each attribute A_i has a value within $\{0, 1, 2\}$.*

2.3 Clustering Algorithms

A number of clustering algorithms have been designed for privacy preserving data publishing. Specifically, the work of [11] linked the concept of k-anonymity to clustering in a metric space. The objective is to minimize the maximal radius among the clusters. The work of [12] studied a *k-member* clustering problem, with the objective of minimizing the intra-cluster pair-wise (point-to-point) distances. The work of [13] studied a bipartite object-feature graph partition problem with *sub-modular* objectives and developed a flow-based algorithm to balance the number of nodes in partitions.

All these algorithms are designed for specific applications other than a generic data anonymization procedure, and are not suitable for producing k-anonymized data in general situations.

3 A Clustering Framework to Preserve k-anonymity

3.1 Model

We propose an unsupervised learning framework in preserving k-anonymity when releasing the data. Given a table X with a finite set of attributes $\{A_1, \cdots, A_d\}$ and a finite set of records $\{x_1, \cdots, x_m\}$, let V_i denote the set of possible values A_i has. For another set of attributes $\{A'_1, \cdots, A'_{d'}\}$, similarly let V'_i to denote the set of possible values A'_i has. A k-anonymity perturbation function f maps each x_i in X to an element $f(x_i)$ in a given *representation set* $Y \subseteq V'_1 \times \cdots \times V'_{d'}$

so that every $f(x_i)$ is identical to at least $k-1$ other $f(x_j)$'s. Among all such functions $\mathcal{F}$, we choose the one that minimizes the total perturbation cost, that is,

$$\min_{f \in \mathcal{F}} \sum_{i=1}^{m} d_{x_i, f(x_i)}, \tag{1}$$

where $d_{x_i, f(x_i)}$ is a positive value and denotes a known cost in mapping x_i to $f(x_i)$.

The freedom in choosing the representation set Y permits us to release the k-anonymized data in different ways. For example, we can suppress the data by taking $A_i' = A_i$, $V_i' = V_i \cup \{*\}$ and $Y = V_1' \times \cdots \times V_d'$. Here the "$*$" value has the effect of hiding the corresponding attribute value, which actually provides a mechanism for suppression. If we further supply V_i' with the generalization hierarchies of the attribute value for A_i, we can also publish the data with generalization.

3.2   ILP Instantiation

Next we show an instantiation of the framework, which leads to a clustering algorithm developed in [14]. The framework in (1) can be expressed by an integer linear program (ILP) when the representation set Y has a finite number of entries $\{y_1, \cdots, y_n\}$. Here let d_{ij} denote the cost of mapping x_i to y_j. Now the objective is to seek a set of decision variables $P = \{p_{ij}\}$ to

$$\min_{P} \sum_{i=1}^{m} \sum_{j=1}^{n} d_{ij} \times p_{ij} \tag{2}$$

satisfying

$$p_{ij} \in \{1, 0\}, \; for \; all \; i, j \tag{3}$$

$$\sum_{j'=1}^{n} p_{ij'} = 1, \; for \; all \; i \tag{4}$$

$$p_{ij} \leq \frac{1}{k} \sum_{i'=1}^{m} p_{i'j}, \; for \; all \; i, j \tag{5}$$

Each decision variable p_{ij} takes a value 1 or 0, and indicates whether a record x_i should be mapped to y_j. Constraint (4) enforces that a record is mapped to exactly one element in Y. Constraints (5) guarantees that a record is identical to at least $k-1$ other records after mapping.

This is a simple yet flexible model. Slight variations of the ILP make it possible to capture other objective functions and constraint equations on decision variables. Any objective function that is a linear combination of the decision variables can be used. By adding further linear constraints, it is also feasible to implement other privacy models, such as ℓ-diversity, without difficulty.

3.3   LP Relaxation

To overcome the NP-hardness inherent in the ILP formulation, we resort to a linear program (LP) relaxation, which can be solved efficiently. Each decision variable is relaxed to a real number between 0 and 1, which gives

$$\min \sum_{i=1}^{m} \sum_{j=1}^{n} d_{ij} \times p_{ij} \qquad (6)$$

subject to

$$0 \le p_{ij} \le 1, \; for \; all \; i,j. \qquad (7)$$

$$\sum_{j'=1}^{n} p_{ij'} = 1, \; for \; all \; i \qquad (8)$$

$$p_{ij} \le \frac{1}{k} \sum_{i'=1}^{m} p_{i'j}, \; for \; all \; i,j \qquad (9)$$

For the LP, we remark:

Lemma 1. *If no solution exists for the LP relaxation (6), then there is no solution to the problem (2).*

Lemma 2. *If the solutions $\{p_{ij}\}$ to the LP relaxation (6) are all integers, that is, $p_{ij} \in \{1,0\}$, then $\{p_{ij}\}$ is the optimal assignment to the problem (2).*

With the first remark, if people could not find a solution to a relaxed instance, there will be no solution to the original ILP problem.

The relaxed problem can be solved efficiently by modern mathematical optimization packages. When the fractional solution is ready, the binary result can be made by rounding it into binary decisions. In practice, an iterative rounding strategy often reports excellent results for related problems [14–16].

3.4   Extension

Another instantiation example of the framework is to consider the case that Y has an infinite number of elements. Each record $x_i \in R^d$ and the representation set $Y = R^d$. The cost of mapping x to y is measured by their squared distance $\|x - y\|^2$. Now the problem becomes partitioning the points $x_1, \cdots, x_m$. Let c_i denote the center of the group (the mean of all points in the group) that data point x_i belongs to. The objective becomes

$$\min \sum_{i=1}^{m} \|x_i - c_i\|^2, \qquad (10)$$

such that each group has at least k points.

Unlike the classical clustering [8, 9], this model implicitly confines the maximal cluster number by constraining the number of points in each cluster. Theoretically this problem is again NP-hard, but in practice can be solved efficiently through a convex relaxation approach, which will be reported in our separate work.

Table 1. Comparison of different clustering algorithms on UCI data sets. Each item represents a maximal re-identification risk.

DATASET	m	k	k-ANONYMIZED	k-MEANS	k-CENTERS
PYRIM	74	10	.100	.333	.250
		20	.0500	.0588	.050
IRIS	150	10	.100	.333	.333
		20	.0500	.0833	.100
WINE	178	10	.100	.500	.500
		20	.0500	.250	.0770
TRIAZINES	186	10	.100	.100	.333
		20	.0500	.200	.200
SONAR	208	10	.100	.500	1.00
		20	.0500	.167	.167
GLASS	214	10	.100	1.00	1.00
		20	.0500	1.00	.500

4 Evaluation

We carried out preliminary experiments to show the necessity of the work. Especially, we hope to show previous popular clustering algorithms do not keep k-anonymity. So specialized clustering algorithms are desired if preserving privacy is a concern.

Specifically, we compared the re-identification probabilities among k-means, k-centers and our k-anonymity clustering algorithm (ref. Section 3.3)[1]. The k-means algorithm requires that the data is from a vector space. It clusters the objects into k partitions with the objective of minimizing the total distances between each object to its center. The center of a cluster is *averaged* over the objects that belong to the cluster. The k-centers algorithm differs from k-means in that the centers are required to be real objects, and the data objects are not required to be within a vector space [17–19].

We used UCI data sets [20]. The sizes (m) of the data sets vary from 74 to 214. Different privacy parameters k from 10 to 20 were used. For each k, the number of clusters was set to be $\lfloor \frac{m}{k} \rfloor$ for k-means and k-centers.

The results are summarized in table (1). It shows the maximal re-identification probability among the clusters. From the results it can be seen that, although k-means and k-centers algorithms have achieved lower costs, the two algorithms cannot provide any guarantee on the privacy re-identification risk. Given k, the risk is often far beyond the preferred threshold $\frac{1}{k}$. Thus the two algorithms cannot be applied directly where the privacy issue is a concern.

---

[1] Here note the meanings of k are different for different algorithms.

5 Conclusion

Publishing data without releasing sensitive information is important. To achieve k-anonymity when releasing privacy data, we proposed a general unsupervised learning framework for a variety of applications. We demonstrated algorithmic approaches that perform efficiently. Without other information to help identify individuals, our method provides strong guarantees that the re-identification risk is below or equal to a user-specified threshold.

This paper focuses on the discussion of the framework itself. Preliminary experimental results were reported to demonstrate the necessity of the work. In the future, extensive verification of the work is expected.

Acknowledgment. The work is partially supported by The Science and Technology Development Fund, Macao SAR, China.

References

1. Sweeney, L.: Uniqueness of simple demographics in the U.S. population (2000), http://privacy.cs.cmu.edu/
2. Aggarwal, C., Yu, P.: Privacy-Preserving Data Mining: Models and Algorithms. Springer (2008)
3. Armbrust, M., Fox, A., Griffith, R., Joseph, A.D., Katz, R.H., Konwinski, A., Lee, G., Patterson, D.A., Rabkin, A., Stoica, I., Zaharia, M.: A view of cloud computing. Communications of the ACM 53, 50–58 (2010)
4. Sweeney, L.: k-anonymity: A model for protecting privacy. Int. J. Uncertainty Fuzziness Knowledge Based Syst. 10 (2002)
5. Meyerson, A., Williams, R.: On the complexity of optimal K-anonymity. In: Proceedings of PODS 2004. ACM (2004)
6. Aggarwal, G., Feder, T., Kenthapadi, K., Motwani, R., Panigrahy, R., Thomas, D., Zhu, A.: Approximation algorithms for k-anonymity. Journal of Privacy Technology (2005)
7. Yu, T., Jajodia, S. (eds.): Secure Data Management in Decentralized Systems, vol. 33. Springer (2007)
8. MacQueen, J.: Some methods for classification and analysis of multivariate observations. In: Proceedings of the Fifth Berkeley Symposium on Mathematical Statistics and Probability, vol. 1. University of California Press (1967)
9. Jain, A., Murty, M., Flynn, P.: Data clustering: A review. ACM Computing Surveys 31, 264–323 (1999)
10. Machanavajjhala, A., Gehrke, J., Kifer, D., Venkitasubramaniam, M.: l-diversity: Privacy beyond k-anonymity. In: Proceedings of ICDE 2006 (2006)
11. Aggarwal, G., Feder, T., Kenthapadi, K., Khuller, S., Panigrahy, R., Thomas, D., Zhu, A.: Achieving anonymity via clustering. In: Proceedings of PODS 2006. ACM (2006)
12. Byun, J.-W., Kamra, A., Bertino, E., Li, N.: Efficient k-anonymization using clustering techniques. In: Kotagiri, R., Radha Krishna, P., Mohania, M., Nantajeewarawat, E. (eds.) DASFAA 2007. LNCS, vol. 4443, pp. 188–200. Springer, Heidelberg (2007)

13. Narasimhan, M., Bilmes, J.: Local search for balanced submodular clusterings. In: Proceedings of the 20th International Joint Conference on Artificial Intelligence (2007)
14. Li, W.: r-anonymized clustering. In: Huang, T., Zeng, Z., Li, C., Leung, C.S. (eds.) ICONIP 2012, Part I. LNCS, vol. 7663, pp. 455–464. Springer, Heidelberg (2012)
15. Li, W., Schuurmans, D.: Modular community detection in networks. In: Proceedings of the 22nd International Joint Conference on Artificial Intelligence, AAAI, pp. 1366–1371 (2011)
16. Li, W.: Modularity segmentation. In: Lee, M., Hirose, A., Hou, Z.-G., Kil, R.M. (eds.) ICONIP 2013, Part II. LNCS, vol. 8227, pp. 100–107. Springer, Heidelberg (2013)
17. Frey, B.J., Dueck, D.: Clustering by passing messages between data points. Science 315 (2007)
18. Li, W.: Clustering with uncertainties: An affinity propagation-based approach. In: Huang, T., Zeng, Z., Li, C., Leung, C.S. (eds.) ICONIP 2012, Part V. LNCS, vol. 7667, pp. 437–446. Springer, Heidelberg (2012)
19. Li, W., Xu, L., Schuurmans, D.: Facility locations revisited: An efficient belief propagation approach. In: 2010 IEEE International Conference on Automation and Logistics, pp. 408–413. IEEE (2010)
20. Bache, K., Lichman, M.: UCI machine learning repository (2013)

Spiking Neural Network with Lateral Inhibition for Reward-Based Associative Learning

Nooraini Yusoff and Farzana Kabir Ahmad

School of Computing, College of Arts and Sciences,
Universiti Utara Malaysia, 06010 UUM Sintok, Kedah, Malaysia
nooraini@uum.edu.my

Abstract. In this paper we propose a lateral inhibitory spiking neural network for reward-based associative learning with correlation in spike patterns for conflicting responses. The network has random and sparse connectivity, and we introduce a lateral inhibition via an anatomical constraint and synapse reinforcement. The spiking dynamic follows the properties of Izhikevich spiking model. The learning involves association of a delayed stimulus pair to a response using reward modulated spike-time dependent plasticity (STDP). The proposed learning scheme has improved our initial work by allowing learning in a more dynamic and competitive environment.

Keywords: Lateral inhibition, Spiking neural network, Associative Learning, Spike-time dependent plasticity.

1   Introduction

It has been evidently known that, in many parts of the brain, networks are recurrent in nature with sparse connectivity, e.g., [6],[7]. In the systems with sparse representation, neurons cooperate and compete with each other to accomplish a task. It has also been proposed that lateral inhibition plays a key role in many of the brain's fundamental computational abilities. Nevertheless, the underlying mechanism in a neural system with such sparse representation still remains intriguing. In a dynamic and competitive environment, not much is known how a lateral inhibition acts as a filtering apparatus in information processing to provide more intense representation of stimuli.

In this study, we show how a lateral inhibition between neuronal groups can be solved via synapse reinforcement based on reward modulated learning. Given a learning setting with some degrees of correlation in spike patterns, during a response interval time, the proposed learning scheme first triggers the network inhibitory response groups to depress activations of their competitors, and then strengthens the connectivity to its target excitatory response groups. The reinforcement signal is dependent on activation rate (i.e. firing activity) in response groups. The lateral inhibition results in stronger synapses in both target inhibitory and excitatory pathways.

C.K. Loo et al. (Eds.): ICONIP 2014, Part I, LNCS 8834, pp. 327–334, 2014.
© Springer International Publishing Switzerland 2014

1.1   Initial Work

From our preliminary work [3], we introduced a pair-associate learning for stimulus-stimulus-response (S-S-R) association. The learning scheme trains a spiking neural network to associate a delayed stimulus pair to a response. The first stimulus is presented to the network, followed by the second stimulus after a delay, the activity of the response subpopulations is then observed within an interval. The response group with highest activation rate is considered to be the winner.

The simulation model was a spiking neural network with random and sparse connectivity (probability $p=0.1$) consisting of 1000 neurons (80% of excitatory and 20% inhibitory neurons). The network has random synaptic transmission delays between 1 to 20 ms [5],[8]. The spiking dynamics of a neuron follow the properties of Izhikevich model [4]. The excitatory synapses are plastic whilst, the inhibitory synapses are not plastic. The excitatory neurons population is divided into subpopulations of m stimulus groups S, n response groups R and non-selective neurons NS. In the initial model, the inhibitory subpopulation IH acts as global inhibition (Fig. 1).

A.

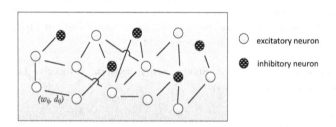

B.

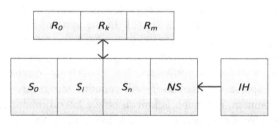

Fig. 1. (A) Schematic view of a recurrent spiking neural network consisting of 80% excitatory (N_E) neurons and 20% of inhibitory (N_I) neurons, with sparse and random connectivity, $p = 0.1$ (no self-feedback),   i.e. $N_E \rightarrow \{N_E, N_I\}$ and $N_I \rightarrow N_E$. Each synaptic transmission has random delay $d \in [1, 20]$. (B) Neurons are divided into subpopulations of stimulus groups (S), response groups (R), non-selective neurons (NS) and inhibitory pool (IH). S and R are composed of 50, and 100 excitatory neurons, respectively.

With a simple network structure in learning we implemented the winner-take-all (WTA) strategy via application of random excitatory bias signals to the winner of target response groups. With the WTA method, it could increase the probability of activation of some neurons in a target response group that had not been fired. This would consequently result in higher activation in the target response group compared to its competitors. However, the simplicity of the structure has some limitations for learning with high competition. For learning with high correlation in spike patterns, the model performance decreased due to undesired causal firings, e.g. when the network was trained to associate $(S_0, S_1) \rightarrow R_A$ and $(S_0, S_2) \rightarrow R_B$, with two competing responses, i.e. neural subpopulations R_A and R_B. Furthermore, strengthening of synaptic strength between $S_i \rightarrow R_A$ could also lead to activation of neurons in response group R_B due to triggering of synapses $R_A \rightarrow R_B$, i.e. firings of postsynaptic neurons of R_A in R_B.

2   Network with Lateral Inhibition

To improve the discrimination rate in a competitive learning, we suggest a modified network topology with a lateral inhibition mechanism (see Fig. 2). In the network (consisting of 1000 neurons) with lateral inhibition, we eliminate the excitatory synaptic connections between response groups. Excitatory neurons in each response group, e.g. R_{+m}, are connected to their inhibitory pool, e.g., R_{-m}. The inhibitory pool provides inhibition to its competitor group(s) through negative synaptic connections.

The synaptic strength from an inhibitory pool of a response group to the excitatory neurons in its competitor is set to -4.0 (a strong inhibition). Generally, each neuron has connectivity of 0.1 (i.e. 100 out of 1000 neurons). Each excitatory neuron in the response groups has 50 postsynaptic neurons from its inhibitory pool, and 50 postsynaptic neurons consisting of neurons from the same excitatory response group and/or excitatory neurons in the input module. Meanwhile each inhibitory neuron in the response groups is connected to other 100 excitatory neurons of its competitor groups. By having such anatomical constraint in the response module, activation of any neuron in a response group will invoke its inhibitory pool that eventually sends out some amount of inhibitory postsynaptic potentials to its competitor(s).

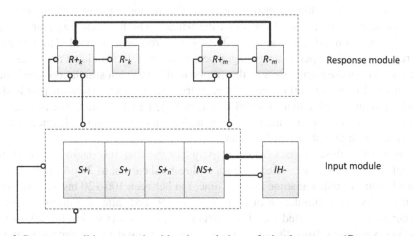

Fig. 2. Recurrent spiking network with subpopulations of stimulus groups (*S*), response groups (*R;R$_+$ and R$_-$*), non-selective neurons (*NS*) and inhibitory pool (*IH*). Lines end with open circle show excitatory connections, and lines end with solid circle indicate inhibitory connections.

For synaptic connections in stimulus neurons population (i.e. S and NS), each excitatory neuron has random connections to 100 neurons from the whole populations (from 1000 neurons), and each inhibitory neuron in this module is connected to 100 excitatory neurons from the whole population as in the network without lateral inhibition in our earlier model.

3   Stimulus-Stimulus-Response Associative Learning

All training simulations presented in this paper were implemented in C++ and testing or probe trials were performed in MATLAB.

3.1   Simulation Method

For stimulus representation, we randomly select 50 neurons from each group to deliver a superthreshold current of 20 pA, for example in group S_0 consisting of 100 neurons, 50 neurons are selected to be paired with 50 neurons from group S_1 (out of 100 neurons, chosen randomly). Hence for two stimulus pairs, e.g. $(S_0, S_1) \rightarrow R_A$ and $(S_0, S_2) \rightarrow R_B$, the stimulus S_0 might have a number of overlapping neurons.

In a 20-minute simulated time, we implement an association of a set of stimulus pairs to their target responses. The learning is initialised with a random background activity for 100 ms. During the initialisation phase, we stimulate an arbitrary neuron with 20-pA (strong) current for every ms. With the same random background activity, we present to the network a pair of stimulus (S_i, S_j), selected randomly, via intensification of 1-pulse current (i.e. 20 pA) to all neurons in the selected stimulus groups. After that, a group S_i is stimulated, followed by its associated pair S_j after an inter-stimulus interval (ISI). An optimal ISI is chosen from a range of $10 - 50$ ms based on a preliminary experiment.

From the onset of the second stimulus, we count the number of activations in the response groups, R_k, within 20 mstime interval. The response group with the highest number of activations is considered to be the winner. The next learning pair is presented after a 100-ms delay from the offset of each response interval. The learning result reported in this paper, is an averaged performance of 10 simulated networks.

For a testing phase also known as the "probe trial", we run a simulation consisting of a number of trials for 200 ms each. In each trial, we present a stimulus pair to the network randomly with equal probability for each pair to be tested. We also apply some degree of distortion via smaller random activation of neurons in a learned stimulus group i.e. with probability of less than 1.0.

The network with some background activity (for the first 100 ms in each trial) as described before is then intensified with super threshold current of 20 pA applied onto the tested prime stimulus at some random time, t in between 100-120 ms, i.e. after the random activity. The stimulation of its pair group proceeds after the prime stimulus group depending on the tested ISI. The number of spike counts within the 20-ms response interval (starts from the onset of the *choice*) is used to compute a winning response. The testing result expresses the averaged percentage of performance over a number of trials, i.e. performance = (number of correct recall/number of trials)*100.

3.2   Learning Rules

The synaptic efficacy is dependent on the reward signal $r(t)$ (2). The signal modulates the synaptic changes read from a spike-timing dependent plasticity (STDP) function (as in 1).

$$\Delta w_{stdp} = \Theta \{ A_+e^{-\Delta t/\tau+}, \Delta t \geq 0; A_-e^{\Delta t/\tau-}, \Delta t < 0 \} \tag{1}$$

From (1), the synapse is potentiated if the difference in firing times (Δt) between a postsynaptic neuron and its presynaptic neuron (i.e. t_{post}-t_{pre}) is ≥ 0, otherwise the synapse is depreciated. The magnitude of potentiation (depression) is given by $A_+e^{-\Delta t/\tau+}$ ($A_-e^{\Delta t/\tau-}$), where A represents the maximal change when the spike timing difference Δt is approaching 0, and τ is the time constant (in ms). For our STDP curve, $\tau_+ = \tau_- = 20$ ms, $A_+ = 0.1$, and $A_- = 0.15$ [2].

The reward signal $r(t)$ determines the amount of modulation to the summation of Δw_{stdp}. Therefore, the reward modulated STDP learning holds [1], [2]:

$$\Delta w(t) = [\alpha + r(t)] z(t) \tag{2}$$

where α is the activity-independent increase of synaptic weight, and $z(t)$ represents the summation of Δw_{stdp} obtained from (1). Excitatory and inhibitory weights are initialised to 1.0 and -1.0, respectively. To avoid infinite growth, weights are kept to be in the range between 0 and 4 mV.

3.3   Synapse Reinforcement

Synapse reinforcement is implemented based on a reward policy. The reward policy determines the amount of synapse potentiation (i.e. strong or weak potentiation) or depression. The network is given a strong positive reward, $r(t-1) + 0.5$, if a target response group, e.g. R_A is the winner having neuron firing rate (F) in the group greater or equal than twice of its closest competitor, e.g. R_B, or a weak reinforcement signal, $1-(F_{R\_A}/ F_{R\_B})$ if the neuron firing rate is greater than (and less twice of) its closest competitor. Meanwhile, the network receives a negative reward signal -0.1 if $F_{R\_A} < F_{R\_B}$.

Synapse reinforcement is implemented in two phases. In the first phase, within the 20-ms interval, we reward the network based on the number of activations in the response inhibitory groups within the first 10 ms. This is to strengthen the synapses for connectivity between a stimulus and the target response inhibitory group for preventing the activation of response competitor groups. In the second phase, we reward the network for the number of activations in the response excitatory groups within the 20-ms response interval for synapse reinforcement from the stimulus group to the target response excitatory group.

4   Results

4.1   Correlation in Spike Pattern

As discussed in Section 1.1, we initially trained a network with fully sparse and random connectivity. We trained the network with a set of learning pairs consisting of exclusive stimulus neurons groups, $Pair\text{-}Response = \{(S_0, S_1) \rightarrow R_A, (S_2, S_3) \rightarrow R_B, (S_4, S_5) \rightarrow R_A, (S_6, S_7) \rightarrow R_B\}$. As a result of learning, the average number of spikes for target response groups is 9.98, when compared with the non-reinforced groups with 7.18 and the negatively rewarded groups with 3.15. The correct memory recall was achieved at 99.9%.

We further experimented the learning with non-exclusive stimulus groups to see the effect of spike correlation for three conditions of learning pairs, condition I – shared the first stimulus, $Pair\text{-}Response = \{(S_0, S_1) \rightarrow RA, (S_0, S_2) \rightarrow B\}$, condition II – interference from non-exclusivity with identical orthogonality, $Pair\text{-}Response = \{(S_0, S_1) \rightarrow R_A, (S_0, S_2) \rightarrow R_B, (S_1, S_0) \rightarrow R_B\}$, and condition III - non-exclusivity with asymmetrical difference, $Pair\text{-}Response = \{(S_0, S_1) \rightarrow R_A, (S_0, S_2) \rightarrow R_B, (S_2, S_1) \rightarrow R_A\}$. To create more interference effects due to neural spike train correlation, the ISI was set to 10 ms as the average of synaptic delays in the range of 1 to 20 ms. The results are exhibited in Table 1. From Table 1, the results demonstrate the level of interference that could disrupt the stability of a pattern due to conflicting responses. The effect of non-exclusivity could be observed when any of learning pairs shared the first or second stimulus.

4.2   Learning with Lateral Inhibition

We repeated the learning experiment with non-exclusive groups for the network with lateral inhibition as described in Section 2.In all the three conditions, learning performance could be improved through implementation of our proposed lateral inhibition (see Table 1). For training, the averaged discrimination rates in conditions I, II and III are 86.56%, 76.99% and 93.79%, respectively, in comparison with learning without the lateral inhibition, 53.89%, 46.30% and 78.26% for conditions I, II and III, respectively.

Table 1. Correct memory recall to target response for condition I – shared the first stimulus, condition II – interference from non-exclusivity with identical orthogonality, and condition III - non-exclusivity with asymmetrical difference

Condition	Correct memory recall (%)	
	No lateral inhibition	With lateral inhibition
I	50.25	85.40
II	47.33	73.73
III	83.60	96.00

We also ran memory recalls for noisy stimuli with only a fraction of neurons stimulated randomly with $0.7 \leq p \leq 1.0$. The results were as follows for the distorted test pairs, 70.80%, 70.07% and 81.60% respectively (without lateral inhibition: 51.13%, 44.40% and 70.13%). Furthermore, the network was also trained with non-exclusive stimulus groups with stimulus pairs as follows (multiple responses): *Pair-Response* = $\{(S_0,S_2){\rightarrow}R_A,\ (S_0,S_3){\rightarrow}R_B,\ (S_1,S_2){\rightarrow}R_C,\ (S_1,S_3){\rightarrow}R_D\}$. The correct recall rate was achieved at 78.47% and 78.70% for training and testing, respectively.

5   Conclusion

Initially, learning tasks only involved association to two response groups, R_A and R_B. In such cases, neurons in both groups act as the dopamine neurons whose activation within its group in an interval time could be a behavioural action in anticipation of the reward.

There was a limitation due to high correlation of spike patterns that might cause instability of learning pairs. We have analysed several levels of interference that can lead to high competition of responses. Even the performance in some cases was above chance, non-inclusivity in learning pairs could somehow affect discrimination of temporal sequences. For example for a system with shared stimulus, e.g. *Pair-Response* $\in \{(S_0,S_1){\rightarrow}R_A,\ (S_2,S_1){\rightarrow}R_B\}$, any of the stimulus pairs could be dragged to an undesired response. As an immediate solution, we introduced some anatomical constraints on the current network model by eliminating the excitatory connections and inserting inhibitory connections between neurons in response groups. This provides a solution to enhance the discrimination rate for some learning conditions with non-exclusive stimulus groups. As learning progresses, reinforcement of synapses is achieved not only to target response groups but also to its inhibitory pool from a triggered stimulus pair. Strengthening of synaptic connections to an inhibitory pool could facilitate discrimination of a target group as neurons in the competitor groups will be suppressed.

We have improvised the excitatory-inhibitory network as proposed in [2] and [11] by adding lateral inhibiton connections that can prevent activations of non-desired responses. Even though the biological interpretation of such an inhibition mechanism is not well defined in our model, this serves as an initial attempt for understanding the synapses of the anterior cingulate cortex (ACC) triggered on events related with conflict or error detection, e.g., [9], [10].

We have also tested learning in environments with higher competition of responses. We extended the training to discriminate paired stimuli for four responses, i.e. R_A, R_B, R_C and R_D. Moreover, using the real images data, we have also performed learning for visual recognition task. The training result achieved at 89.46% and all image pairs were correctly discerned with 100.00% accuracy in probe trials [12].

The performance indicates some potential of our model in learning multiple input-output mappings with high competition of outputs. Nevertheless, the increase in the number of responses requires greater number of spikes from the input neurons with minimum of 80% activation from each stimulus group.

Acknowledgements. This research has been funded by the Ministry of Higher Education (Malaysia).

References

1. Florian, R.V.: Reinforcement learning through modulation of spike-timing dependent synaptic plasticity. Neural Comput. 6, 1468–1502 (2007)
2. Izhikevich, E.M.: Solving the distal reward problem through linkage of STDP and dopamine signaling. Cereb. Cortex 17, 2443–2452 (2007)
3. Yusoff, N., Grüning, A.: Learning Anticipation through Priming in Spatio-temporal Neural Networks. In: Huang, T., Zeng, Z., Li, C., Leung, C.S. (eds.) ICONIP 2012, Part I. LNCS, vol. 7663, pp. 168–175. Springer, Heidelberg (2012)
4. Izhikevich, E.M.: Simple Model of Spiking Neurons. IEEE Trans. Neural Networks 14(6), 1569–1572 (2003)
5. Izhikevich, E.M.: Polychronization: Computation with Spikes. Neural Computatio. 18, 245–282 (2006)
6. Jones, E.G.: Microcolumns in the cerebral cortex. Proceedings of the National Academy of Sciences 97(10), 5019–5021 (2000)
7. Cutsuridis, V., Wennekers, T.: Hippocampus, microcircuits and associative memory. Neural Networks 22, 1120–1128 (2009)
8. Paugam-Moisy, H., Martinez, R., Bengio, S.: Delay learning and polychnization for reservoir computing. Neurocomputing 71(7-9), 1143–1158 (2008)
9. Botvinick, M., Braver, T., Barch, D., Carter, C., Cohen, J.: Conflict monitoring and cognitive control. Psychological Review 108(3), 624–652 (2001)
10. Kaplan, G.B., Sengor, N.S., Gurvit, S., Guzelis, C.: Modelling The Stroop Effect: A Connectionist Approach. Neurocomputing 70, 1414–1423 (2007)
11. Brunel, N., Lavigne, F.: Semantic Priming in a Cortical Network Model. Journal of Cognitive Neuroscience 21(12), 2300–2319 (2009)
12. Yusoff, N., Grüning, A.: Biologically Inspired Temporal Sequence Learning. Journal of Procedia Engineering 41, 319–325 (2012)

Fuzzy Signature Neural Networks
for Classification: Optimising the Structure

Tom Gedeon, Xuanying Zhu, Kun He, and Leana Copeland

Research School of Computer Science, College of Engineering and Computer Science,
The Australian National University, Canberra, ACT 0200, Australia
{tom.gedeon,u5251881,u5058161,leana.copeland}@anu.edu.au

Abstract. We construct fuzzy signature neural networks where fuzzy signatures replace hidden neurons in a neural network similar to a radial basis function neural network. We investigated the properties of a naïve and a principled approach to fuzzy signature construction. The naïve approach provides very good results on benchmark datasets, but is outperformed by the principled approach when we approximate the noisy nature of real world datasets by randomly eliminating 20% of the data. The major benefit of the principled approach is to substantially improve robustness of the fuzzy signature neural networks we produce.

Keywords: fuzzy signature, neural network, RBF, clustering, salary selection, diabetes.

1   Introduction

Given k inputs, and at most T linguistic terms per dimension of X for the α-cover, the number of fuzzy rules covering X at least to α is $|R| = O(T^k)$ which is very high, unless k is very small. The exponential explosion in rules is a major problem hindering the application of fuzzy techniques beyond control systems. We have previously developed and adapted techniques which partially address decreasing T, and of k (by rule interpolation and hierarchical rule bases respectively [1-3]. Encouraging results of our previous research give further motivation to continue along these lines, and to develop further new techniques that will be suitable for the solution of even harder problems.

We have developed two kinds of approaches to the exponential explosion. First, sparse hierarchical fuzzy systems reduce both T and k simultaneously by finding (top-down) sub-spaces in the data, which allows some dimensions and rules to be ignored. The use of sparse rule base allows proper fuzzy reasoning even if the rule set contains "gaps" [4]. For various technical reasons, eliminating dimensions and rules at the same time gives the best results from the produced rule bases and is at the same time the most efficient [5].

The second is fuzzy signatures – constructing characteristic fuzzy structures, modelling the complex structure of the data points (bottom up) in a hierarchical manner.

C.K. Loo et al. (Eds.): ICONIP 2014, Part I, LNCS 8834, pp. 335–341, 2014.
© Springer International Publishing Switzerland 2014

Fuzzy signatures result in a much reduced order of complexity, at the cost of more complex aggregation techniques.

General fuzzy signatures are vector valued fuzzy sets, where each vector component can be a further vector valued fuzzy set, and use aggregations to propagate the fuzzy values from low levels to high levels in the structures, resulting in effective and efficient fuzzy inference. Such aggregations encompass the simple classic fuzzy conjunction and union operations, but still maintain transparency of fuzzy reasoning.

A Fuzzy Signature can be graphically represented by two structures, hierarchical structure, which is a vector, and a tree. Figure 1 below illustrates these two Fuzzy Signature structures for a SARS patient [6], conveying the same information.

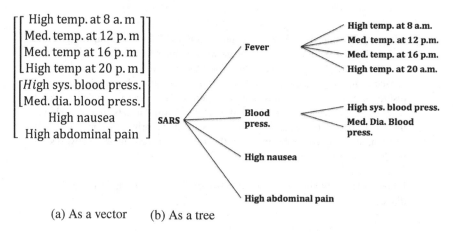

(a) As a vector   (b) As a tree

Fig. 1. Example of two Fuzzy Signature structures for a SARS patient

2 Fuzzy Signature Neural Networks

A fuzzy signature neural network (FZSIGNN) is similar to a radial basis function (RBF) neural network, where the RBF neurons are replaced by fuzzy signature neurons, where each neuron is a fuzzy signature. RBF neural networks are faster to train than back-propagation neural networks of a similar topology, as only the output layer of weights is trained.

We would expect a fuzzy version to be faster, since we would use fuzzy membership regions which are rectilinear as opposed to the (usually Euclidean) distance calculations required by RBF networks.

Each hidden neuron has a particular fuzzy signature associated with it, and outputs the similarity between input vector and fuzzy signature. This means that the inputs to the hidden layer are unmodified values read directly from input data. Hence, the weights connected between input layers and hidden layers are constants. Therefore, all the learning is done via the weights connected between hidden layer and output layer. Thus, the time to train these neural networks should be reduced due to decreasing size of weight matrix, and eliminating the need to propagate errors to slowly train layers of weights further from the output. The construction of the fuzzy signatures may take some time, which is investigated in this paper, in comparing a naïve but fast approach with a principled but slower approach.

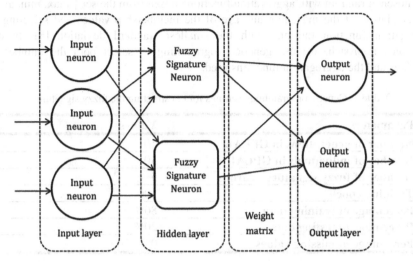

Fig. 2. Fuzzy Signature Neural Network structure

3   Fuzzy Signature Construction

In both of our techniques we use hierarchical agglomerative clustering since it does not require users to specify the number of clusters and contains more information than flat clustering [7].

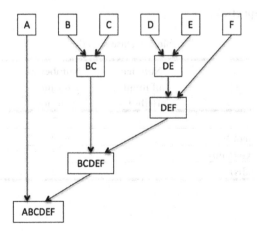

Fig. 3. Example of Agglomerative hierarchical clustering

We use various heuristics to restrict the number of clusters to 5, and use the cluster centroids to generate fuzzy signatures. In the naïve approach, 2 to n-1 input variables are chosen at random with aggregation function chosen from the set { max, min, ave } by calculation of the max, min and ave of the membership values and selecting as aggregation function the one with the smallest standard deviation [8]. In our principled approach, we use genetic programming to select both the number of variables and the aggregation functions used.

Table 1. Genetic programming settings for constructing the fuzzy signatures

Parameters	Values
Number of generations in GPLAB	10
Number of individuals in GPLAB	50
Number of fuzzy signature neurons	5
Training epoch	100
Percentage of training data	80%
Percentage of testing data	20%
Percentage of missing values	0% / 20%

4   Results

Four datasets are used for evaluation, two of which are Salary problems from Gedeon [9], one is Salary problem from Mendis [10], and the rest are Cancer and Diabetes problems from University of California Irvine (UCI) [11]. Table 2 below shows the general information about these datasets.

4.1   Dataset Properties

Table 2. Datasets used

Dataset	Number of input attributes	Number of output columns	Number of observations
Diabetes	8	2	768
High Salary (Gedeon)	3	2	200
Medium Salary (Gedeon)	3	2	200
Low Salary (Mendis)	3	2	135

4.2   Clean Data

We first used our two approaches on the clean datasets with no changes.

Table 3. Using principled approach to construct fuzzy signatures

Dataset	Testing		
	Mean (%)	StDv	MSE
Diabetes	68.5	0.0184	0.2406
High Salary (Gedeon)	81.5	0.0141	0.1903
Medium Salary (Gedeon)	83.0	0.0283	0.1423
Low Salary (Mendis)	80.1	0.0406	0.1607
Average	78.3	0.0254	0.1835

We provide average prediction performance results only for rough comparison between our two techniques, as such averages do not make sense in general.

Table 4. Using naïve approach to construct fuzzy signatures

Dataset	Testing		
	Mean (%)	StDv	MSE
Diabetes	68.2	0.0359	0.1987
High Salary (Gedeon)	78.0	0.0622	0.1160
Medium Salary (Gedeon)	88.5	0.0379	0.0876
Low Salary (Mendis)	68.9	0.1425	0.1934
Average	75.9	0.0696	0.1489

We can see that the naïve approach is worse on average, though on the simplest dataset (medium salary) it performs the best. On the hardest dataset (low salary) it performs the worst.

4.3 Damaged Data

We then used our two approaches on the datasets with 20% of the data omitted at random.

Table 5. Principled approach on damaged data

Dataset	Testing		
	Mean (%)	StDv	MSE
Diabetes	65.1	0.0147	0.2438
High Salary (Gedeon)	83.0	0.0694	0.1428
Medium Salary (Gedeon)	87.5	0.0530	0.1077
Low Salary (Mendis)	74.8	0.1154	0.1805
Average	77.6	0.0631	0.1687

On the hardest dataset (low salary) and on the diabetes dataset, the performance has decreased but on the other two dataset including the easiest dataset, the performance has actually increased. We believe this is due to elimination of some outliers from a dataset which otherwise has redundant data. We have seen such results in our previous work in heuristic pattern reduction and bimodal distribution removal. Overall, the principled approach is quite robust to the damaged data.

Table 6. Naïve approach on damaged data

Dataset	Testing		
	Mean (%)	StDv	MSE
Diabetes	64.6	0.0461	0.2180
High Salary (Gedeon)	79.5	0.0209	0.1485
Medium Salary (Gedeon)	77.5	0.0467	0.1614
Low Salary (Mendis)	55.6	0.0642	0.2388
Average	69.3	0.0445	0.1917

The naïve approach has reduced performance quite significantly overall, but has retained its performance on one of the datasets.

5   Conclusion and Future Work

We have shown our proposed approach to produce fuzzy signature neural networks, with both a naïve and a principled approach to construction of the fuzzy signature neurons.

The principled approach is more expensive computationally but is robust in the face of significant damage to the datasets of 20% deletion at random.

On the other hand, the naïve approach is computationally cheap, and performed well on at least one dataset even with damage.

Our future work will include identifying the applicability conditions where the naïve approach can be expected to perform well, and further improvements of our principled approach including particularly reduction of the computational complexity – it may be possible in certain circumstances which we intend to delineate to replace the global search of the evolutionary approach we used with a local gradient descent search.

Acknowledgements. The authors would like to express their appreciation to Pranay Chandra and Wei Fan who contributed to early parts of this work.

References

1. Tikk, D., Baranyi, P., Gedeon, T.D., Muresan, L.: Generalization of the rule interpolation method resulting always in acceptable conclusion. Tatra Mountains Math. Publ. 21, 73–91 (2001)

2. Mendis, B.S.U., Gedeon, T.D., Kóczy, L.T.: Investigation of aggregation in fuzzy signatures. In: 3rd International Conference on Computational Intelligence, Robotics and Autonomous Systems, vol. 406 (2005)
3. Mendis, B.S.U., Gedeon, T.D., Botzheim, J., Kóczy, L.T.: Generalised weighted relevance aggregation operators for hierarchical fuzzy signatures. In: 2006 International Conference on Computational Intelligence for Modelling, Control and Automation and International Conference on Intelligent Agents, Web Technologies and Internet Commerce, pp. 198. IEEE (2006)
4. Tikk, D., Biró, G., Gedeon, T.D., Kóczy, L.T., Yang, J.D.: Improvements and critique on Sugeno's and Yasukawa's qualitative modeling. IEEE Transactions on Fuzzy Systems 10(5), 596–606 (2002)
5. Koczy, L.T., Hirota, K., Gedeon, T.D.: Fuzzy rule interpolation by the conservation of relative fuzziness. International Journal of Advanced Computational Intelligence 4(1), 95–101 (2000)
6. Wong, K.W., Gedeon, T., Kóczy, L.F.: Construction of fuzzy signature from data: An example of SARS pre-clinical diagnosis system. In. In: Proceedings of the IEEE International Conference on Fuzzy Systems, vol. 3, pp. 1649–1654. IEEE (2004)
7. Tevor, H., Robert, T., Jerome, F.: The Elements of Statistical Learning, 2nd edn., pp. 520–528. Springer, New York (2009)
8. Mendis, B.S.U., Gedeon, T.D.: Complex Structured Decision Making Model: A hierarchical frame work for complex structured data. Information Sciences 194, 85–106 (2011)
9. Gedeon, T., Wong, K., Tikk, D.: Constructing hierarchical fuzzy rule bases for classification. In: Proceedings of IEEE International Conference on Fuzzy Systems, pp. 1388–1391 (2001)
10. Mendis, B.S.U.: Fuzzy Signatures: Hierarchical Fuzzy Systems and Applications, PhD thesis, Department of Computer Science, The Australian National University (March 2008)
11. Bache, K., Lichman, M.: The UCI Machine Learning Repository. Center for Machine Learning and Intelligent Systems, University of California, Irvine (1987), http://archive.ics.uci.edu/ml/datasets.html (viewed September 15, 2013)

Self-organizing Map-Based Probabilistic Associative Memory

Yuko Osana

Tokyo University of Technology,
1401-1 Katakura, Hachioji, Tokyo, Japan
osana@stf.teu.ac.jp

Abstract. In this paper, we propose a Self-Organizing Map-based Probabilistic Associative Memory (SOMPAM). The proposed SOMPAM is based on Self-Organizing Map and it is composed of the Input/Output Layer and the Map Layer. In this model, stored pattern sets are memorized with its own brief degree, and probabilistic associations based on brief degree for analog pattern sets including one-to-many relations can be realized. And it can also realize additional learning. We carried out a series of computer experiments and confirmed that the proposed SOMPAM can realize probabilistic associations and additional learning.

1 Introduction

As the model which can realize probabilistic associations, the Boltzmann machine[1] has been proposed. However, the learning of the Boltzmann machine needs much time. Moreover, it cannot realize additional learning. On the other hand, as the model which can realize one-to-many associations, we have proposed the Kohonen feature map associative memory with area representation[2] which is based on the Self-Organizing Map[3] and the Kohonen feature map associative memory[4]. This model can realize not only one-to-many associations of binary patterns but also one-to-many associations of analog patterns. However, this model cannot realize probabilistic associations and additional learning. In contrast, the variable-sized KFM associative memory with refractoriness based on area representation can realize additional learning. And, the variable-sized Kohonen feature map probabilistic associative memory can realize probabilistic associations and additional learning. However, the learning process of this model is very complex.

In this paper, we propose the Self-Organizing Map-based Probabilistic Associative Memory (SOMPAM). The proposed SOMPAM is based on Self-Organizing Map and it is composed of the Input/Output Layer and the Map Layer. In this model, stored pattern sets are memorized with its own brief degree, and probabilistic associations based on brief degree for analog pattern sets including one-to-many relations can be realized. And it can also realize additional learning with simple process.

C.K. Loo et al. (Eds.): ICONIP 2014, Part I, LNCS 8834, pp. 342–349, 2014.
© Springer International Publishing Switzerland 2014

2 Self-organizing Map-Based Probabilistic Associative Memory

Here, we explain the proposed Self-Organizing Map-based Probabilistic Associative Memory (SOMPAM).

2.1 Structure

Figure 1 shows the structure of the proposed SOMPAM. As shown in Fig. 1, the proposed SOMPAM is composed of (1) Input/Output Layer and (2) Map Layer. The Input/Output Layer is divided into M parts corresponding to M-tuple pattern and one neuron corresponding to the brief degree. In the proposed model, probabilistic association based on the brief degree can be realized.

2.2 Learning

The learning process of the proposed SOMPAM is based on that of the original self-organizing map[3]. The learning procedure is as follows.

Step 1 : The connection weights are initialized randomly.
Step 2 : The Euclidean distance between the input vector $\boldsymbol{x}^{(p)}$ and the connection weights of the neuron i in the Map Layer $\boldsymbol{w}_i$ is calculated. The Euclidean distance $d^L(\boldsymbol{x}^{(p)}, \boldsymbol{w}_i)$ is given by

$$d^L(\boldsymbol{x}^{(p)}, \boldsymbol{w}_i) = \sqrt{\sum_{j=1}^{N^{IO}-1} (x_j - w_{ij})^2} \tag{1}$$

where N^{IO} is the number of neurons in the Input/Output Layer. In this calculation, the brief degree corresponding to the neuron N^{IO} in the Input/Output Layer is not considered.

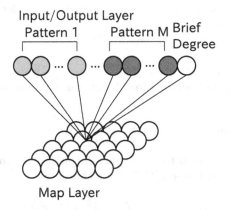

Fig. 1. Structure of SOMPAM

Step 3 : The winner neuron is decided based on the Euclidean distance calculated in **Step 2**. The winner neuron c is given by

$$c = \underset{i}{\operatorname{argmin}}\, d^L(\boldsymbol{x}^{(p)}, \boldsymbol{w}_i). \tag{2}$$

Step 4 : The connection weights $\boldsymbol{w}_i$ are updated by

$$\boldsymbol{w}_i \leftarrow \boldsymbol{w}_i + \eta \cdot \exp\left(\frac{\sqrt{(x_i - x_c)^2 + (y_i - y_c)^2}}{\delta(t)^2} \right) (\boldsymbol{x}^{(p)} - \boldsymbol{w}_i) \tag{3}$$

where η is the learning rate, x_i, y_i are the coordinates of the neuron i in the Map Layer and x_c, y_c are the coordinates of the winner neuron in the Map Layer. $\delta(t)$ is the neighborhood function at the time t, and it is given by

$$\delta(t) = \delta^{ini} \cdot \left(\frac{\delta^{fin}}{\delta^{ini}} \right)^{\frac{t}{T}} \tag{4}$$

where δ^{ini} is the initial size of neighborhood area, δ^{fin} is the final size of neighborhood area and T is the maximum learning time.

Step 5 : Steps 2∼4 are repeated T times for all training patterns to be stored T times.

Step 6 : The Euclidean distance between the input vector $\boldsymbol{x}^{(p)}$ and the connection weights of the neuron i in the Map Layer $\boldsymbol{w}_i$, $d^L(\boldsymbol{x}^{(p)}, \boldsymbol{w}_i)$ is calculated by Eq.(1).

Step 7 : The winner neuron c is determined by

$$c = \underset{i \notin C^F}{\operatorname{argmin}}\, d^L(\boldsymbol{x}^{(p)}, \boldsymbol{w}_i) \tag{5}$$

where C^F is the set of neurons in the Map Layer whose connection weights are fixed.

Step 8 : The connection weights of the winner neuron c is determined as follows:

$$\boldsymbol{w}_c = \boldsymbol{x}^{(p)} \tag{6}$$

and those connection weights are fixed.

Step 9 : Steps 6∼8 are repeated for all patterns to be stored.

2.3　Recall Process

When the pattern $\boldsymbol{x}$ is given to the Input/Output Layer, the internal state of the neuron i in the Map Layer u_i^{MAP} is calculated by

$$u_i^{MAP} = w_{iN^{IO}} g\left(1 - \frac{d^R(\boldsymbol{x}^{IN}, \boldsymbol{w}_i)}{\sqrt{N^{IN'}}} \right) \tag{7}$$

where $N^{IN'}$ is the number of neurons which receives input in the Input/Output Layer. And, $d^R(\boldsymbol{x}, \boldsymbol{w}_i)$ is the Euclidean distance between the input vector $\boldsymbol{x}^{IN}$ and the connection weights $\boldsymbol{w}_i$ and it is given by

$$d^R(\boldsymbol{x}^{IN}, \boldsymbol{w}_i) = \sqrt{\sum_{\substack{j=1 \\ j:x_j \neq -1}}^{N^{IO}} (x_j^{IN} - w_{ij})^2} \tag{8}$$

In the recall process of the proposed SOMPAM, only a part of the Input/Output Layer receives input and the part of the input which is not given to the Input/Output Layer is set to -1. And the Euclidean distance is calculated only considering the part where receives the input. And $g(\cdot)$ is given by

$$g(a) = \begin{cases} a, & (a \geq \theta^R) \\ 0, & (a < \theta^R) \end{cases} \tag{9}$$

where θ^R is the threshold.

And one winner neuron is selected based on the following probability:

$$P(x_i^{MAP} = 1) = \frac{u_i^{MAP}}{\sum\limits_{i'=1}^{N^{MAP}} u_{i'}^{MAP}} \tag{10}$$

The output of the neuron j in the Input/Output Layer is determined as

$$\boldsymbol{x}^{IO} = \boldsymbol{w}_c \tag{11}$$

where $\boldsymbol{w}_c$ is the connections weights of the winner neuron c.

2.4   Additional Learning Process

In the proposed SOMPAM, when the new pattern is given after the learning, the pattern can be stored additionally.

Step 1 : When the pattern $\boldsymbol{x}^{new}$ is given to the Input/Output Layer and all neurons in the Input/Output Layer receives input, the Euclidean distance between the input $\boldsymbol{x}^{new}$ and the connection weights $\boldsymbol{w}_i$, $d^L(\boldsymbol{x}^{new}, \boldsymbol{w}_i)$ is calculated for all neurons in the Map Layer.

Step 2 : If

$$\min_i(d^L(\boldsymbol{x}^{new}, \boldsymbol{w}_i)) > \theta^L \tag{12}$$

is satisfied, the input pattern is regarded as a new pattern. If $\boldsymbol{x}^{new}$ is regarded as a new pattern, go to **Step 3**.

Step 3 : The neuron which learns the new pattern $\boldsymbol{x}^{new}$ is determined.

(a) When the connection weights of all neurons in the Map Layer are fixed

If there is no neurons whose connection weights are not fixed, one neuron is added in the Map Layer. The coordinate of the new neuron (x_n, y_n) is determined as follows:

$$x_n = (x_c + x_c^*)/2 \tag{13}$$
$$y_n = (y_c + y_c^*)/2 \tag{14}$$

where (x_c, y_c) is the coordinate of the neuron c, and c is given by

$$c = \underset{i}{\operatorname{argmin}}\, d^L(\boldsymbol{x}^{new}, \boldsymbol{w}_i). \tag{15}$$

And, (x_c^*, y_c^*) is the coordinate of the neuron c^*, and c^* is given by

$$c^* = \underset{i\,:\,d_{ic}\,=\,\underset{j\neq c}{\min}\{d_{jc}\}}{\operatorname{argmin}} d^L(\boldsymbol{x}^{new}, \boldsymbol{w}_i). \tag{16}$$

where d_{ic} is the distance between the neuron i and the neuron c in the Map Layer, and it is given by

$$d_{ic} = \sqrt{(x_i - x_c)^2 + (y_i - y_c)^2}. \tag{17}$$

The connection weights of the new neuron $\boldsymbol{w}_n$ is set to

$$\boldsymbol{w}_n = \boldsymbol{x}^{new}. \tag{18}$$

(b) When some neurons whose connection weights are not fixed exist

If some neurons whose connection weights are not fixed exist, the neuron c whose connection weights is most similar to the input $\boldsymbol{x}^{new}$ is selected from all neurons.

$$c = \underset{i}{\operatorname{argmin}}\, d^L(\boldsymbol{x}^{new}, \boldsymbol{w}_i) \tag{19}$$

And, the neuron c^{UF} whose connection weights is most similar to the input $\boldsymbol{x}$ is selected from the neurons whose connection weights are not fixed.

$$c^{UF} = \underset{i\in C^{UF}}{\operatorname{argmin}}\, d^L(\boldsymbol{x}^{new}, \boldsymbol{w}_i) \tag{20}$$

where C^{UF} is the set of the neurons whose connection weights are not fixed. And, the neuron b whose connection weights are most different from the input $\boldsymbol{x}^{new}$ is selected from all neurons.

$$b = \underset{i}{\operatorname{argmax}}\, d^L(\boldsymbol{x}^{new}, \boldsymbol{w}_i) \tag{21}$$

If the neuron c^{UF} satisfies

$$\frac{d^L(\boldsymbol{x}^{new}, \boldsymbol{w}_{c^{UF}}) - d^L(\boldsymbol{x}^{new}, \boldsymbol{w}_c)}{d^L(\boldsymbol{x}^{new}, \boldsymbol{w}_b) - d^L(\boldsymbol{x}^{new}, \boldsymbol{w}_c)} < \theta^L, \tag{22}$$

the neuron c^{UF} is selected as the neuron which learns the new input.

The connection weights of the neuron c^{UF} ($\boldsymbol{w}_{c^{UF}}$) is set to

$$\boldsymbol{w}_{c^{UF}} = \boldsymbol{x}^{new}. \tag{23}$$

If the neuron c^{UF} does not satisfy Eq.(22), new neuron added to the Map Layer in the same way in (a).

3   Computer Experiment Results

Here, we show the computer experiment results to demonstrate the effectiveness of the proposed SOMPAM.

3.1   Probabilistic Association

In this experiment, the analog patterns including one-to-many relations shown in Fig.2(a) were memorized in the proposed SOMPAM composed of 801 neurons in the Input/Output Layer and 9 neurons in the Map Layer. Figure 2(b) and (c) show a part of the association results. From these results, we can confirm that the proposed model can recall the corresponding plural patterns correctly.

Figure 3 shows the Map Layer after the pattern pairs shown in Fig.2 (a) were memorized.

Table 1 shows the recall times of each pattern in the trial of Fig.2 (b) (t=1∼1000) and Fig.2 (c) (t=1∼1000). In this table, normalized values are also shown in ( ). From these results, we can confirm that the proposed SOMPAM can recall each pattern by the probability according to the brief degree.

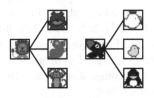

(a) Stored Patterns

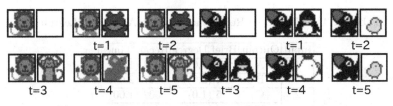

(b) *lion* was given   (c) *crow* was given

Fig. 2. One-to-Many Associations

Fig. 3. Stored Patterns in Map Layer

Table 1. The Number of Recall Times

Input	Output	Brief Degree	Recall Times
lion	bear	5 (1.0)	333 (1.00)
	monkey	5 (1.0)	330 (1.01)
	mouse	5 (1.0)	337 (1.02)
crow	hen	1 (1.0)	167 (1.00)
	chick	2 (2.0)	333 (1.99)
	penguin	3 (3.0)	500 (2.99)

```
  2-C      3-E      3-F            2-C      3-E  3-O  3-F
   o        o        o              o        o    o    o
                                   2-K             o 3-N
                     1-B           2-J             o 3-M
   o        o        o              o        1-H      1-B
                                             o        o
  2-D               1-A           2-D 1-I  1-G      1-A
   o        o        o             o  o  o   o        o
                                      2-L

      (a) Before                        (b) After
```

Fig. 4. Map Layer of Before and After Additional Learning

3.2   Additional Learning

In this experiment, the six pattern sets (1-A, 1-B, 2-C, 2-D, 3-E, 3-F) were memorized first in the proposed SOMPAM which has 9 (=3×3) neurons in the Map Layer. Figure 4(a) shows the Map Layer after these six pattern sets are memorized. As shown in this figure, the pattern sets which has common term (for example, 1-A and 1-B) are assigned to similar position. And then, nine pattern sets (1-G, 1-H, 1-I, 2-J, 2-K, 2-L, 3-M, 3-N, 3-O) were memorized additionally and Fig.4(b) shows the Map Layer after all pattern sets are memorized. In Fig.4(b), red circles shows new added neurons.

Tables 2 and 3 show the recall times of each pattern ($t=1\sim1000$). In this table, normalized values are also shown in ( ).

Table 2. The Number of Recall Times before Additional Learning

Input	Output	Brief Degree	Recall Times
1	A	1 (1.0)	333 (1.00)
	B	2 (2.0)	667 (2.00)
2	C	5 (1.0)	502 (1.00)
	D	5 (1.0)	498 (1.00)
3	E	1 (1.0)	499 (1.00)
	F	1 (1.0)	501 (1.00)

Table 3. The Number of Recall Times after Additional Learning

Input	Output	Brief Degree	Recall Times
1	A	1 (1.0)	60 (1.00)
	B	2 (2.0)	120 (2.00)
	G	3 (3.0)	199 (3.32)
	H	4 (4.0)	265 (4.42)
	I	5 (5.0)	355 (5.92)
2	C	5 (1.0)	196 (1.07)
	D	5 (1.0)	193 (1.05)
	J	5 (1.0)	202 (1.10)
	K	5 (1.0)	183 (1.00)
	L	5 (1.0)	225 (1.23)
3	E	1 (1.0)	66 (1.00)
	F	1 (1.0)	68 (1.00)
	M	5 (5.0)	300 (4.55)
0	N	5 (5.0)	280 (4.32)
	O	5 (5.0)	285(4.24)

4 Conclusion

In this paper, we have proposed the Self-Organizing Map-based Probabilistic Associative Memory (SOMPAM). The proposed SOMPAM is based on Self-Organizing Map and it is composed of the Input/Output Layer and the Map Layer. In this model, stored pattern sets are memorized with its own brief degree, and probabilistic associations based on brief degree for analog pattern sets including one-to-many relations can be realized. And it can also realize additional learning. We carried out a series of computer experiments and confirmed that the proposed SOMPAM can realize probabilistic associations and additional learning.

References

1. Hinton, G.E., Sejnowski, T.J.: Learning and Relearning in Boltzmann Machines. In: Rumelhart, D.E., McClelland, J.L., the PDP Research Group (eds.) Parallel Distributed Processing: Explorations in the Microstructure of Cognition. Foundations, vol. 1, pp. 282–317. MIT Press, Cambridge (1986)
2. Abe, H., Osana, Y.: Kohonen feature map associative memory with area representation. In: Proceedings of IASTED Artificial Intelligence and Applications, Innsbruck (2006)
3. Kohonen, T.: Self-Organizing Map. Springer (1994)
4. Ichiki, H., Hagiwara, M., Nakagawa, M.: Kohonen feature map as a supervised learning machine. In: Proceedings of IEEE International Conference on Neural Networks, pp. 1944–1948 (1993)
5. Imabayashi, T., Osana, Y.: Variable-sized KFM associative memory with refractoriness based on area representation. In: Proceedings of IEEE International Conference on System, Man and Cybernetics, San Antonio (2009)
6. Sato, H., Osana, Y.: Variable-sized kohonen feature map probabilistic associative memory. In: Villa, A.E.P., Duch, W., Érdi, P., Masulli, F., Palm, G. (eds.) ICANN 2012, Part II. LNCS, vol. 7553, pp. 371–378. Springer, Heidelberg (2012)

A Causal Model for Disease Pathway Discovery

Ruichu Cai[1,2], Chang Yuan[1], Zhifeng Hao[1,3], Wen Wen[1],
Lijuan Wang[1], Weiqi Chen[2], and Zhihao Li[1]

[1] School of Computer Science and Technology,
Guangdong University of Technology, Guangzhou, P.R. China
[2] State Key Lab. for Novel Software Technology, Nanjing University, P.R. China

Abstract. Pathway provides a deep insight into the mechanism of the biological process. With the increasing of high-throughout gene expression monitoring technology, a lot of data driven methods have been proposed to reconstruct the pathways from the observation data. Low reliability of the discovered results, especially the direction of the regulatory relation, is the main challenge of the existing methods. In this work, a level-wise causal search (LWCS) based disease pathway discovery method is proposed. The following three steps are conducted in each searching level of LWCS to locate the causal variables: firstly, in the parents and children (PC) discovery step, structure learning approach is employed to discover the candidate causal genes; then, in the casual direction learning step, additive noise models are explored to determine the direction of the edges, finally, the trivial causal candidates are pruned and not contained in the further level search. The proposed method is tested and verified on real life gene expression data sets. The success of the proposed method reflects that the causality is a proper model to present the regulatory relations among the genes and phenotypes.

Keywords: Causality, Additive Noise Model, Gene Expression Data.

1 Introduction

The development of genomic techniques make it is possible to measure the entire genome's expression level in one experiment. Microarray and HiSeq are such techniques [18], which have been widely used in diagnosing disease, discovering the regulatory network among the genes, and so on applications [8]. Understanding the behind mechanism of certain disease is one of the ultimate goal of gene expression data analysis [2].

Identifying the differentially expressed genes is a traditional preprocessing for the disease-related gene discovery method [11]. Typically methods includes, the statistical based approaches [11] and machine learning based approaches [13]. Though the relations among the genes are already considered to reduce the redundancy of the discovered, these methods only return a set of genes and the critical problem, how the genes effects the genes, is still not answered [4].

Different from discovering of differentially expressed genes, disease pathway reconstruction provides a convenient way to distill the information from the data

C.K. Loo et al. (Eds.): ICONIP 2014, Part I, LNCS 8834, pp. 350–357, 2014.

set and an intuitive way to understanding the mechanism of the disease process [17]. Bayesian network [12] and its extensions, such as dynamic Bayesian network [9] are typical used models to learning the pathway skeleton from the data. The main drawbacks of the Bayesian network based model includes, the large samples sizes needed to find a reliable model, and the Markov Equivalent class problems of the Bayesian network related model [7]. For example, the following three local structures, $g_1 \rightarrow g_2 \rightarrow g_3$ and $g_1 \leftarrow g_2 \rightarrow g_3$ belong to the same Markov equivalent class, which means we can't use the Bayesian network model to determine the regulatory direction between g_1 and g_2.

Generally, there is no efficient computational model for the disease oriented pathway reconstruction method. Fortunately, the recently developed causal discovery method provides an inspiration for the problem. In detail, the gene regulatory relation and the gene-disease all can be considered as special cases of causality. The recently developed Additive Noise Model provides a good solution to determine the direction of the causalities, here is the regulatory direction of the genes. Thus, the disease related pathway is modeled as a causal network reconstruction problem on the gene expression level and the disease state, with some significance constraint. Moreover, to reduce the complexity of the problem, we only consider the partial pathway that is closely related to the diseases. Considering the example given in Figure 1, the abnormal express of genes g_2 and g_3 is the cause of the disease, and the partial pathway containing the dash nodes is the concerns of the disease.

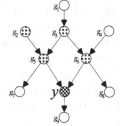

Fig. 1. An example of Disease Pathway

Thus, a level-wise search causal pathway discovery method is proposed, which search around level by level. Each level contains three main steps, in the PC step, we use structure learning approach to discovery the candidate causal genes, then in the casual learning step, we use additive noise model to determine the direction of the edges, finally the candidate which is not differentially expressed is pruned and not contained in the further level search. This level-wise search procedure ensures the important genes are discovered in the early stage, and facilitates the further discovery of the casual relation among the genes.

2   Related Work

Gene expression data analysis and causality model are two main backgrounds of this work. Gene selection is one of most basic tasks and an important preprocessing used in the gene expression data analysis, for example discovery of the differential expression genes across health v.s. disease samples [11]. The focus of this task is controlling the falsely discovery rate and the redundancy of the results [22]. Though some work has considered the interactions among genes to reduce the redundancy, for example, the causal gene selection method [6], the dynamic properties and the complex interactions are still not captured.

Different from the gene selection methods, the pathway reconstruction tries to model the complex interaction among the genes and related entities. Data driven and knowledge-driven are two main categories of pathway reconstruction [27]. Knowledge driven methods construct a pathway by integrating prior knowledge from particular domains of interest. Data-driven pathway construction usually generates relationship information of genes or proteins from the experimental data, for example the gene expression data. Bayesian network based methods [12,9], co-expression network [25], are typical used models to learn the pathway skeleton from the data. Recently, a lot of integrated methods are also proposed by considering the observed samples and prior knowledge in one model, for example, the predictive network [14].

The work is also high related to causal discovery models. Causal Bayesian network and Additive Noise Model are two main approaches. Pearl's Inductive Causality [21] provides the fundamental of causal Bayesian network based approaches. A large number of extensions are proposed, focuses on its applications on the real problems, for example the work on the small sample size problems [3], the large scale problem [16]. However, all these approaches are based on independence conditional testing, and cannot distinguish two causality structures if they come from a so-called Markov equivalence class [20]. Additive Noise Model can break the limitation of Markov equivalence class, by exploiting the asymmetric property of the noises in the generative progress, for example, the nonlinear non-Gaussian method [15] works when the data generating process is nonlinear, the discrete model works on the discrete data [23]. The regression model [19] and kernel independence test [28] are core steps of the additive noise model.

3   Method

Assume a gene expression sample consists of the expression level of m different genes, i.e. $G = \{g_1, g_2, \cdots, g_m\}$, and the disease state of the sample y. Let $D = \{x_1, x_2, \cdots, x_n\}$ denote the complete sample set. Each sample x_i can be denoted by a vector $x_i = (x_{i1}, x_{i2}, ..., x_{im}, y_i)$, where x_{ij} is the expression level of the sample x_i on gene g_j, y_i is the corresponding disease state. The disease pathway is defined on the full gene set and the state of disease, i.e., $V = G \cup \{y\}$.

The disease pathway discovery problem can be defined as finding a partial causal graph closely related to the disease. For each gene in the pathway, the following two conditions should be satisfied, 1) the genes appear in the pathway should be direct or indirect causes of the disease; 2) the genes' expression level should be significantly differentially expressed across disease states. These two conditions dramatically reduce the complexity of the original problem, and make it is possible to provide a concise disease related pathway.

Thus, the following Level-Wise Causal Search (LWCS) based disease pathway discovery method is proposed. In LWCS, firstly the direct causes of the disease are discovered. If the direct cause is differentially expressed, then its direct causes are discovered similarly. There are several benefits of this level-wise search method, firstly the important genes can be discovered in the early stages; secondly, the early discovered causal genes facilitate the further discovery of the pathway; finally, the level-wise search method can accelerate the mining process for it only works on a constrained search space.

Figure 2 gives a running example of the level-wise search framework. Firstly, the causal node of the target y, $\{g_5, g_6\}$, are discovered in Fig.2(a). Secondly, because both g_5 and g_6 are differentially expressed, the causal node of g_5, g_6 are discovered respectively as shown Fig.2(b) and Fig.2(c).

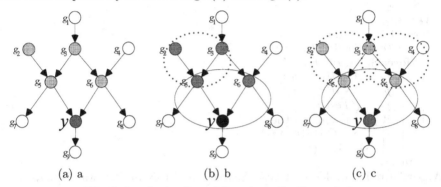

(a) a   (b) b   (c) c

Fig. 2. Level-wise Causal Search on the Example 1

In the algorithm, the causal structure is contained in the CS in the form of direct edges and the candidate causal nodes are maintained and updated in V_t. Here, the candidate causal nodes refer to the variables that are the causes of the target variable and differentially expressed across different disease state.

In each iteration, the algorithm picks up one candidate causes v_t, and expands from this node. Given a target node v_t, the following three main steps are used to find the causal nodes of v_t.

Firstly, in the PC discovery step (line 4), we use structure learning approach to discovery the parents and children nodes of the target variables, for the parents and children nose are the candidates of the causal genes. In this work, the PC discovery framework proposed in is used [5]. For the gene expression data is continuous, thus the kernel conditional independence test is used in the PC framework [28]. For the binary disease state is mapped to -1, 1.

Secondly, in the causal direction learning step, we use additive noise model (ANM) to determine the direction of the edges (line 6-7, 9), for its strong discovery ability on causal discovery. In detail, given a target variable and its PC set $PC(v_t)$, each variables in the $PC(v_t)$ is considered as the causal candidate of vt and the following additive noise models are constructed to determine the direction. Considering the additive noise model $ANM(v \rightarrow v_t)$, the following regression model $v = f(v_t) + n$ is constructed using Gaussian process regression model, and the p value is obtained using the kernel independence test, $n \perp v_t$.

Finally, in the pruning step (line 8, 9), the candidate genes which are not differentially expressed across the different diseases state are pruned and not contained in the further level search. Different from the previous work on differentially expressed genes, we focus on pruning the trivial causal candidates and some redundancy is acceptable. Thus, the traditional χ^2-test is used in this work. This step will remove the suspicious causal candidates and improve the algorithm's performance on the large scale problem.

The full map of the algorithm is summarized in the Algorithm 1.

CS=LWCS (D, V, y)
Input: D: the data set, V: the variable index, y: the target variable index
Output: CS: the causal structure
$CS = \phi$;
$V_t = \{y\}$;
foreach $v_t \in V_t$ **do**
$\quad PC(v_t)=\text{FindPC}(D, V, v_t)$;
$\quad$ **foreach** $v \in PC(v_t)$ **do**
$\quad\quad p_1 = ANM(v \to v_t)$;
$\quad\quad p_2 = ANM(v_t \to v)$;
$\quad\quad p_3 = \chi^2 Test(v, y)$;
$\quad\quad$ **if** $p_1 > p_2 \& p_3 > \alpha$ **then**
$\quad\quad\quad CS = CS \cup \{v \to v_t\}$;
$\quad\quad\quad V_t = V_t \cup \{v_t\}$;
return CS;

Algorithm 1. Level-Wise Causal Search based Disease Pathway Discovery

---

4   Experiments

The effectiveness of LWCS is experimentally studied on real gene expression data. All the experiments are conducted in Matlab 2011 environment, and several existing open source packages are used in the experiments. In detail, the Rasmussen and Williams's implementation of the Gaussian process regression[1] is used as the regression method in the ANM step, and Zhang's implementation of kernel independence test[2] are used to detect the conditional independence relations in both PC and ANM step.

Data Preparation. We test LWCS on two real gene expression data and verify the results on the previous literatures. In detail, we run our analysis on *DLBCL* dataset [26] and *Colon* data set [1]. The task of DLBCL is to find the causal pathway for the difference of lymphoma, diffuse large B-cell lymphoma(DLBCL) and follicular lymphoma(FL). The aim of Colon dataset is to find the causal pathway for the colorectal cancer. Please refer to the original references for the details of the datasets.

---

[1] http://www.gaussianprocess.org/gpml/code/matlab/doc/
[2] http://people.tuebingen.mpg.de/kzhang/KCI-test.zip

Results. The discovered pathway on the DLBCL data set is shown in Figure 3(a). Most of the genes have been reported in the previous research of DLBCL or FL, for example, Z21966_at is considered as cause gene of FL in [26]; U68030_at and L42324_at are considered as the causal genes of DLBCL in [24]. One of the interesting discovery of our work is the gene, M14328_s_at. Though M14328_s_at has already been reported differentially expressed in DLBCL but not on FL, the detail mechanism is still not clear. The discovered pathway shows that the M14328_s_at effects the disease through the gene U68030_at, which provides a new explanation of the confusing observation. The detail descriptions of the related genes are listed in Table 1.

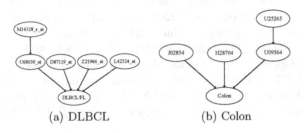

(a) DLBCL   (b) Colon

Fig. 3. Disease Pathway Discovered on DLBCL and Colon

Table 1. Results on DLBCL Data Set

Probe ID	Description
U68030_at	Human G protein-coupled receptor (STRL22) mRNA
D87119_at	Human cancellous bone osteoblast mRNA for GS3955
L42324_at	Homo sapiens G protein-linked receptor gene (GPCR) gene
Z21966_at	H.sapiens mPOU homeobox protein mRNA
M14328_s_at	Human alpha enolase mRNA

Figure 3(b) gives the pathway discovered on the Colon dataset. In the pathway, the gene J02854 and H28704 have been reported as the cause gene of colorectal cancer [1]. MEK5 (U25265) is the upstream activator of ERK5 in colon epithelial cells [10], and ERK5 is a member of the mitogen-activated protein kinase family (U09564), all of them are high related to colorectal cancer.

Table 2. Results on Colon Data Set

Probe ID	Description
J02854	myosin regulatory light chain 2, smooth muscle isoform
H28704	retinoic acid receptor RXR-beta isoform 1 (Homo sapiens)
U25265	Human MEK5 mRNA
U09564	Human serine kinase mRNA

5   Conclusions and Discussion

We propose a new causality model to present and discover the disease pathway from the high dimensional and small sample size gene expression data. The proposed method is tested on real life gene expression data sets. The success of the proposed method reflects that causality is a proper model to present the regulatory relations among the genes, and the interactions among the genes and the phenotype. The proposed approach provides a good causal inference method for the high dimensional and small sample size data.

Acknowledgements. This work is financially supported by Natural Science Foundation of China (61100148, 61202269), Key Technology Research and Development Programs of Guangdong Province (2012B01010029), Science and Technology Plan Project of Guangzhou City(12C42111607, 201200000031, 2013Y2-00034, 2014Y2-00027), Opening Project of the State Key Laboratory for Novel Software Technology (KFKT2014B03), Specialized Research Fund for the Doctoral Program of Higher Education (20134420110010), Discipline Construction and Quality Engineering of Higher Education in Guangdong Province(PT2011JSJ).

References

1. Alon, U., Barkai, N., Notterman, D.A., Gish, K., Ybarra, S., Mack, D., Levine, A.J.: Broad patterns of gene expression revealed by clustering analysis of tumor and normal colon tissues probed by oligonucleotide arrays. Proceedings of the National Academy of Sciences 96(12), 6745–6750 (1999)
2. Barabási, A.-L., Gulbahce, N., Loscalzo, J.: Network medicine: A network-based approach to human disease. Nature Reviews Genetics 12(1), 56–68 (2011)
3. Bromberg, F., Margaritis, D.: Improving the reliability of causal discovery from small data sets using argumentation. Journal of Machine Learning Research, 301–340 (2009)
4. Cai, R., Hao, Z., Yang, X., Wen, W.: An efficient gene selection algorithm based on mutual information. Neurocomputing 72(4), 991–999 (2009)
5. Cai, R., Zhang, Z., Hao, Z.: Bassum: A bayesian semi-supervised method for classification feature selection. Pattern Recognition 44(4), 811–820 (2011)
6. Cai, R., Zhang, Z., Hao, Z.: Causal gene identification using combinatorial v-structure search. Neural Networks 43, 63–71 (2013)
7. Cai, R., Zhang, Z., Hao, Z.: Sada: A general framework to support robust causation discovery. In: ICML, pp. 208–216 (2013)
8. Cookson, W., Liang, L., Abecasis, G., Moffatt, M., Lathrop, M.: Mapping complex disease traits with global gene expression. Nature Reviews Genetics 10(3), 184–194 (2009)
9. Dondelinger, F., Lèbre, S., Husmeier, D.: Non-homogeneous dynamic bayesian networks with bayesian regularization for inferring gene regulatory networks with gradually time-varying structure. Machine Learning 90(2), 191–230 (2013)
10. Drew, B.A., Burow, M.E., Beckman, B.S.: Mek5/erk5 pathway: the first fifteen years. Biochimica et Biophysica Acta (BBA)-Reviews on Cancer 1825(1), 37–48 (2012)

11. Dudoit, S., Yang, Y.H., Callow, M.J., Speed, T.P.: Statistical methods for identifying differentially expressed genes in replicated cdna microarray experiments. Statistica Sinica 12(1), 111–140 (2002)
12. Friedman, N., Linial, M., Nachman, I., Pe'er, D.: Using bayesian networks to analyze expression data. Journal of Computational Biology 7(3-4), 601–620 (2000)
13. Guyon, I., Weston, J., Barnhill, S., Vapnik, V.: Gene selection for cancer classification using support vector machines. Machine Learning 46(1-3), 389–422 (2002)
14. Haibe-Kains, B., Olsen, C., Djebbari, A., Bontempi, G., Correll, M., Bouton, C., Quackenbush, J.: Predictive networks: A flexible, open source, web application for integration and analysis of human gene networks. Nucleic Acids Research 40(D1), 866–875 (2012)
15. Hoyer, P.O., Janzing, D., Mooij, J.M., Peters, J., Schölkopf, B.: Nonlinear causal discovery with additive noise models. In: Neural Information Processing Systems, pp. 689–696 (2008)
16. Kalisch, M., Bühlmann, P.: Estimating high-dimensional directed acyclic graphs with the pc-algorithm. The Journal of Machine Learning Research 8, 613–636 (2007)
17. Kotera, M., Yamanishi, Y., Moriya, Y., Kanehisa, M., Goto, S.: Genies: Gene network inference engine based on supervised analysis. Nucleic Acids Research 40(W1), W162–W167 (2012)
18. Minoche, A.E., Dohm, J.C., Himmelbauer, H., et al.: Evaluation of genomic high-throughput sequencing data generated on illumina hiseq and genome analyzer systems. Genome Biol. 12(11), R112 (2011)
19. Mooij, J., Janzing, D., Peters, J., Schölkopf, B.: Regression by dependence minimization and its application to causal inference in additive noise models. In: ICML, pp. 745–752. ACM (2009)
20. Pearl, J.: Causality: models, reasoning and inference, vol. 29. Cambridge University Press (2000)
21. Pearl, J., Verma, T.S.: A theory of inferred causation. Studies in Logic and the Foundations of Mathematics 134, 789–811 (1995)
22. Peng, H., Long, F., Ding, C.: Feature selection based on mutual information criteria of max-dependency, max-relevance, and min-redundancy. IEEE Transactions on Pattern Analysis and Machine Intelligence 27(8), 1226–1238 (2005)
23. Peters, J., Janzing, D., Scholkopf, B.: Causal inference on discrete data using additive noise models. IEEE Transactions on Pattern Analysis and Machine Intelligence 33(12), 2436–2450 (2011)
24. Piersanti, S., Martina, Y., Cherubini, G., Avitabile, D., Saggio, I.: Use of dna microarrays to monitor host response to virus and virus-derived gene therapy vectors. American Journal of Pharmacogenomics 4(6), 345–356 (2004)
25. Ruan, J., Dean, A.K., Zhang, W.: A general co-expression network-based approach to gene expression analysis: Comparison and applications. BMC Systems Biology 4(1), 8 (2010)
26. Shipp, M.A., Ross, K.N., Tamayo, P., Weng, A.P., Kutok, J.L., Aguiar, R.C., Gaasenbeek, M., Angelo, M., Reich, M., Pinkus, G.S., et al.: Diffuse large b-cell lymphoma outcome prediction by gene-expression profiling and supervised machine learning. Nature Medicine 8(1), 68–74 (2002)
27. Viswanathan, G.A., Seto, J., Patil, S., Nudelman, G., Sealfon, S.C.: Getting started in biological pathway construction and analysis. PLoS Computational Biology 4(2), e16 (2008)
28. Zhang, K., Peters, J., Janzing, D., Schölkopf, B.: Kernel-based conditional independence test and application in causal discovery. In: UAI, pp. 804–813 (2011)

Enhanced Non-linear Features for On-line Handwriting Recognition Using Deep Learning

Qing Zhang*, Minhua Wu**, Zhenbo Luo, and Youxin Chen

Handwriting Group,
Samsung Research Center Beijing,
18/F Taiyanggong Building, ChaoYang District, Beijing, 100028, China
{qing18.zhang,zb.luo,youxin.chen}@samsung.com, mwupangpang722@gmail.com

Abstract. Conventionally, a deep neural network (DNN) is trained to predict probabilities of class labels. Recently, DNN has shown great success in many pattern recognition tasks such as speech recognition and handwritten digit recognition. To take advantage of its great learning power, we propose and build an on-line handwriting recognition system for French strings, which applies a DNN for non-linear feature transformation before training the character models. When a DNN is predicting class labels, its hidden layer outputs can be regarded as a better representation of the original features extracted from handwriting trajectory data. In this paper, we demonstrate that the proposed system can achieve a relative character error rate (CER) reduction of about 28.5% when being compared to a conventional system without feature transformation. We also notice that the CER could be further reduced by 3.3% relatively when the transformed features are used along with the original features.

Keywords: Deep neural network (DNN), On-line handwriting recognition, Feature transformation, Hidden Markov model (HMM), Gaussian mixture model (GMM).

1 Introduction

In the age of information, the pen-based handwriting recognition is becoming a new form of interaction between humans and computers. Being compared with the traditional input devices of the keyboard and the mouse, a pen could not only manage the basic operations of clicking and dragging, but also act as a more natural writing tool. To achieve more reliable recognition rate, research on handwriting recognition has been conducted for about thirty years [1][2]. The problem of handwriting recognition can be divided into two categories which are on-line and off-line. In the on-line recognition system, a sequence of feature vectors representing movement of the pen tip is transformed into meaningful text, while in an off-line recognition task, only the image of the text is available. In this paper, we focus on

---

* Corresponding author.
** This work was done while Minhua Wu was an intern at Samsung Research Center, Beijing.

C.K. Loo et al. (Eds.): ICONIP 2014, Part I, LNCS 8834, pp. 358–365, 2014.
© Springer International Publishing Switzerland 2014

the problem of on-line recognition and propose a more advanced system to recognize handwritten strings by applying deep neural network (DNN).

Neural network has shown its great advances in many pattern recognition tasks. Many complex approaches based on neural network have been proposed for handwriting recognition and have acquired great improvement. The on-line handwriting recognition system NPen++ is based on a multi-state time delay neural network (MS-TDNN) [3]. Convolutional neural network (CNN) [4] and recurrent neural network (RNN) [5] have also been applied on different handwriting recognition tasks. Apart from those complex approaches, the simpler architecture of feedforward deep neural network has also achieved great success. In the field of automatic speech recognition (ASR), DNN has been successfully applied to predict posteriors of class labels and obtained remarkable accuracy gain [6][7]. DNN has also been proved to perform well on handwritten digit recognition even without unsupervised pre-training [8]. Recently, graphics processing units (GPU) technology has been widely adopted in scientific computation and it can dramatically speed up intensive computational task which can be parallelized. The training speed of a DNN can be effectively accelerated with the aid of GPU. However, higher time consumption during the decoding stage remains to be a problem, especially when using a very deep neural network for class modeling. To make use of the strong modeling power of DNN without increasing too much computational load for decoding, we propose to use DNN for exploiting more useful feature representation in an on-line handwriting recognition system. Our goal is to investigate if we could find a better representation of the input observations while applying a deep learning scheme.

The rest of the paper is organized as follows. Section 2 describes how a conventional on-line handwriting recognition system work and how to integrate DNN in such a system for more effective feature extraction. Experimental results are presented and analyzed in section 3. Finally, section 4 ends up with conclusions and discusses possible future work.

2   Methods

2.1   Overview of Baseline Handwriting Recognition System

There are four main steps being involved in building a conventional system, and we use it as the baseline of our work for comparison. The first stage is data pre-processing which usually contains noise reduction and normalization, which normalizes the size of the handwriting according to the slant corrected baseline. The next stage is feature extraction. The function of feature extraction is to extract the dynamic and static information of handwriting data. The features are computed form the normalized sequence of captured coordinates (x, y). The dynamic features consist of the information extracted from each point considering the neighbors with respect to the time. The static features take the off-line structure representation of the handwriting into account. This step converts each frame of the input data into a multi-dimensional feature vector. We then implement genetic algorithm (GA) [9] as search heuristic that mimics the process of natural selection helping us to find an optimal subset of features as well as achieving dimension reduction. After obtaining feature vectors, we then use these features and their corresponding transcriptions for model training. Normally, each character is modeled as an Hidden Markov model (HMM). The emission probability for each state is modeled with Gaussian mixture model (GMM). For such a GMM-HMM based system, the training scheme is the same as that applied in a traditional automatic speech recognition (ASR) system using minimum phone error (MPE) criterion [10] .

This set of trained HMMs are then integrated with a language model (a character loop for this task) to build a decoding network. Given an unknown sequence of feature vectors, we can find the most probable word sequence in the network through standard Viterbi search [11].

2.2 Deep Neural Network for Non-linear Feature Enhancement

Typically, a DNN models posterior probabilities of different classes given a feature vector o_t. The input to the network is the frame at t with several of its neighboring frames. The outputs are posterior probabilities of different classes $p(s|o_t)$.

Estimation of the posterior probabilities in a DNN could be considered as a two step process . Suppose there is a DNN with $L+1$ layer, in the first step, the observation vector o_t is transformed into a feature vector v^L through L layers of non-linear transform. This step is illustrated in Eq. 1, where $\sigma(z^l(v^l))$ is output vectors at layer l given input vector v^l (with sigmoid activation σ) and it could be used as input vector for layer $l+1$. W^l is weight matrix at layer l and a^l stands for biases for neurons at layer l.

$$v^{l+1} = \sigma(z^l(v^l)) = \sigma((W^l)^T v^l + a^l), 0 \leq l < L \tag{1}$$

In the second step, the top layer models the desired posterior probabilities, which is summarized in Eq. 2.

$$P(s|o_t) = \frac{e^{z_s^L(v^L)}}{\sum_{s'} e^{z_{s'}^L(v^L)}} = softmax_s(z^L(v^L)) \tag{2}$$

As intermediate products, output vectors at hidden layer l could be regarded as a non-linear transform of the original features. In this paper, we propose to use the non-linear transformed features calculated at the hidden layer for traditional GMM-HMM system training. Furthermore, we can concatenate the original features and the transformed ones when building a GMM-HMM system. Figure 1 displays the proposed system using non-linear transformed features. At first, a DNN with hidden layers is trained. We set a middle hidden layer with fewer units than the others (We set the number of neurons at this layer close to the dimensionality of original feature vector so that it will not cost more time for decoding). The trained network can take the features of the training data as the inputs and we select linear outputs of the middle hidden layer as new transformed features. These transformed features are then applied to train a new GMM-HMM based system as mentioned in section 2.1.

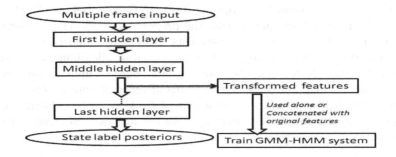

Fig. 1. Proposed system using non-linear transformed features

2.3 Training of DNN

DNNs are often trained with error back-propagation (BP) which is based on stochastic gradient descent to minimize an objective function E with a learning rate ϵ.

$$(W^l, a^l) \leftarrow (W^l, a^l) - \epsilon \frac{\partial E}{\partial(W^l, a^l)} \tag{3}$$

For the task described in this paper, the training criterion of minimum cross-entropy error (MCE) is used as in Eq. 4. The objective function represents the entropy of the target values, where $y_i(o_t)$ denotes actual output at node i in output layer (softmax output), $tar_{ti} = 1$ indicates a frame o_t belongs to class i, and vice versa. K is the number of classes and N is the number of sample frames for training.

$$E = -\sum_{t=1}^{N} \sum_{i=1}^{K} tar_{ti} log(y_i(o_t)) \tag{4}$$

The partial derivative of the cross-entropy error is measured with respect to each weight or bias in the network (back-propagate error if weight is not at the output layer). One specifically obtained partial derivative is then used to shift the corresponding weight and bias along their optimal direction as indicated in Eq. 3. Weights and biases of DNN are learned in this way epoch by epoch until the modeling accuracy converges. In practice, the gradient descents are done in mini-batches for each epoch. The learning rate normally starts with a fast rate and slows down as epoch goes.

3 Experiment

3.1 Corpus

In this paper, an on-line handwriting recognition system for single character and strings are established. Experimental results reported in this paper are based on handwriting database owned by Samsung Research Center, Beijing. The French corpus is used for training and testing the proposed system. The handwriting data was collected from French local residents. They were employed to write the prepared transcripts with stylus pen on a big screen mobile phone (e.g. Galaxy NoteII and Galaxy Note10.1). The training data could be divided into several main categories which are single letter, words of different cursive levels, single symbol, single digit and strings set containing letters, symbols and digits. Data used for testing also covers these writing styles but it does not overlap with the training set.

3.2 Baseline

The baseline established in this task is a traditional GMM-HMM based system. Each of the 116 characters in French was trained as a left-to-right HMM. The emission probability for each state was modeled with 48 GMMs. Models were trained using MPE criteria. A simple character loop was used as the language model. We applied a standard Viterbi decoder to find recognized results. The grammar scale factor and the insertion penalty were both set to 0.0.

When evaluating the baseline, the trained HMMs were at first applied on 4 validation sets containing single letter (780 characters), single symbol (8741 characters), single digit (10233 characters) and strings combined by letters, symbols and digits (8505 strings consisting of 27462 characters). After that, models were then tested on a new unlearned set containing all these writing styles (13582 characters) mentioned above. We evaluate the performance of the system using character error rate (CER).

3.3 Experiments on Transformed Features

The non-linear transformed features were obtained by applying a DNN on the original features used in the baseline. When establishing the new system on the French corpus, the DNN with 800, 40 and 800 neuron at each hidden layer was trained at first. Specifically, we used the current frame and its 10 neighboring frames of features as the input to the DNN. The whole DNN predicted posteriors of 834 state classes in the system. We used a learning rate of 0.1 for the first 10 epochs and 0.03 for the next 5 epochs. After that, the learning rate would decrease by a constant factor for each epoch (Experimentally, the factor ranges from 0.90 to 0.95). Training were stopped when the decrement of validation error got smaller than a threshold. We adopted stochastic gradient descent with mini-batch for BP. The first two hidden layer of the trained DNN were then used for extracting the new non-linear transformed features by passing the original features through. The extracted new features was then used alone (40-dimension) or concatenated with the original features (80-dimension) to train a new GMM-HMM based recognition system. We reduced the number of each HMM state mixtures to 32 GMMs in the new system for computational efficiency. Evaluation criteria of the new system were the same as that for the baseline.

The performance of systems using different feature schemes is summarized in figure 2, where CER% is defined as the sum of deletions, substitutions and insertions over the number of characters in the reference. Detailed information is recorded in Table 1. For the task of French handwriting recognition, after applying the new non-linear transformed features, no matter used alone or with concatenation, CER on different types of validation sets will all decrease even with fewer Gaussian mixtures than the baseline, which means the modeling power of the new HMMs based on the new non-linear transformed features has been effectively improved. When applying the new models trained from the transformed features alone on the unseen test data, the CER decreases relatively by about 28.5% and reach 10.69%. If the new models were trained using the concatenated features. the CER could be further reduced by about 3.3% relatively and reach 10.34%.

Table 1. Performance of systems using different features schemes
(N: Words in the reference; H: Correctly recognized words; D: Deletions; S: Substitutions; I: Insertions)

Set	N	Baseline					Transformed features					Concatenated features				
		H	D	S	I	CER%	H	D	S	I	CER%	H	D	S	I	CER%
1	780	558	0	222	29	32.18	564	0	216	19	30.13	570	0	210	15	28.85
2	8741	7044	0	1697	193	21.62	7297	0	1462	137	18.29	7567	0	1174	94	14.51
3	10233	8897	0	1336	844	21.30	9034	0	1199	663	18.20	9144	0	1089	665	17.74
4	27462	22525	193	4744	4240	33.42	23296	159	4007	3037	26.23	23621	177	3664	2782	24.12
Test	13582	11916	276	1390	313	14.57	12328	176	1078	198	10.69	12361	194	1027	183	10.34

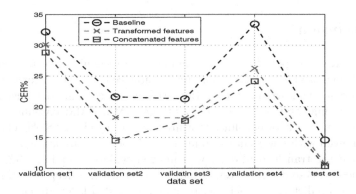

Fig. 2. CER% of systems using different feature schemes. The 4 validation sets contain single letter, single symbol, single digit and strings respectively. The unlearned test set contains all these writing styles.

The improvements on the CER of the new systems mainly result from the DNN. To model posteriors of different states from the original features, a DNN would automatically try to find a better representation of the observation vectors during the processing at the hidden layers. These intermediate outputs at the hidden layer can thus be regarded as a set of more representative features with higher robustness for GMM-HMM modeling.

To investigate more about the three systems with different feature schemes, Table 2 summarizes approximate number of parameters involved for each system. The GMM-HMM system using the transformed feature alone is more advanced than the traditional baseline since not only its CER is reduced sharply by 28.5% but also the quantity of parameters involved for system modeling decreases by about 21.2%. The main drawback for such a system is that its additional feature transform step will cost more computational time and make the recognition process slower. For transforming each frame of features, we need extra large matrix multiplication and non-linear activation function compared to the baseline. When using the higher dimensional concatenated features, though system performance would be further improved by about 3.3%, such a system is not economical since the size of the resultant model is almost doubled and the recognizing process becomes slower.

Table 2. Comparison for number of parameters involved in different systems

Feature scheme	#params	CER% on test set
Original features	3.21M	14.57
Transformed features	2.53M	10.69
Concatenated features	4.66M	10.34

4   Conclusion

In this paper, we proposed an on-line handwriting recognition system which integrates deep neural network for more representative feature transformation. On a task of French handwriting recognition, we have successfully demonstrated its advantage. Being compared to the conventional GMM-HMM baseline, applying a deep neural network for non-linear feature transform yields a 28.5% relative improvement even with fewer Gaussian mixtures. When concatenating the original features and the transformed ones, we can even achieve a further 3.3% relative improvement. These exciting results indicate that when a deep neural network is trained to predict class labels, outputs at the hidden layer are characterized with a more proper representation of the original handwriting features.

The main challenge for such a system is its higher time consumption resulted from feed-forward process of deep neural network. Thus, it is necessary to apply more efficient algorithms of matrix computation in the future. Meanwhile, it is worthwhile to spend time extending this method on handwriting recognition tasks based on other languages. Another valuable extension would be building a general feature extractor for multiple languages, since those different languages sharing the same origination may be similar in curve and shape characteristics such as English, French and some other European languages. Such a general feature extractor would be much more efficient while being applied on an embedded system.

Acknowledgements. We would like to thank all colleagues of Handwriting Group in Samsung Research Center, Beijing for valuable supports and discussions.

References

1. Plamondon, R., Srihari, S.N.: Online and off-line handwriting recognition: A comprehensive survey. IEEE Transactions on Pattern Analysis and Machine Intelligence 22(1), 63–84 (2000)
2. Santosh, K., Nattee, C., et al.: A comprehensive survey on on-line handwriting recognition technology and its real application to the nepalese natural handwriting. Kathmandu University Journal of Science, Engineering, and Technology 5(I), 31–55 (2009)
3. Jaeger, S., Manke, S., Reichert, J., Waibel, A.: Online handwriting recognition: The npen++ recognizer. International Journal on Document Analysis and Recognition 3(3), 169–180 (2001)
4. Ciresan, D.C., Meier, U., Gambardella, L.M., Schmidhuber, J.: Convolutional neural network committees for handwritten character classification. In: 2011 IEEE International Conference on Document Analysis and Recognition (ICDAR), pp. 1135–1139 (2011)
5. Liwicki, M., Graves, A., Bunke, H., Schmidhuber, J.: A novel approach to on-line handwriting recognition based on bidirectional long short-term memory networks. In: Proc. 9th Int. Conf. on Document Analysis and Recognition, vol. 1, pp. 367–371 (2007)
6. Seide, F., Li, G., Yu, D.: Conversational speech transcription using context- dependent deep neural networks. In: Interspeech, pp. 437–440 (2011)
7. Deng, L., Yu, D.: DEEP LEARNING: Methods and Applications. NOW Publishers (2014)
8. Ciresan, D.C., Meier, U., Gambardella, L.M., Schmidhuber, J.: Deep, big, simple neural nets for handwritten digit recognition. Neural Computation 22(12), 3207–3220 (2010)
9. Mitchell, M.: An introduction to genetic algorithms. MIT Press (1998)
10. Povey, D., Woodland, P.C.: Minimum phone error and I-smoothing for improved discriminative training. In: 2002 IEEE International Conference on Acoustics, Speech, and Signal Processing (ICASSP), vol. 1, pp. 105–108 (2002)
11. Forney Jr., G.D.: The viterbi algorithm. Proceedings of the IEEE 61(3), 268–278 (1973)

Recognizing Human Actions by Using the Evolving Remote Supervised Method of Spiking Neural Networks

Xiurui Xie, Hong Qu*, Guisong Liu, and Lingshuang Liu

School of Computer Science and Engineering, University of Electronic Science and
Technology of China, Chengdu 611731, P.R. China
hongqu@uestc.edu.cn

Abstract. This paper proposes a novel approach based on spiking neural networks to recognize human actions in videos. In our method, a star
skeleton detector is designed to extract spatial features of input videos,
and a classifier using evolving ReSuMe algorithm is proposed, with scale
and shift invariance, to recognize input patterns. In learning algorithm,
the remote supervised learning method(ReSuMe) is improved by the particle swarm optimization(PSO) algorithm. Experimental results on KTH
and Weizmann dataset prove that our method achieves a significant improvement in performance compared with traditional ReSuMe and other
method based on neural networks.

Keywords: Human actions recognition, Spiking neural networks, Particle swarm optimization, Remote supervised learning method.

1 Introduction

Recognizing human actions in videos is an important task in computer vision
which has been applied to various fields, such as surveillance system, man-
machine interface and medical monitoring. In recent years, plenty of approaches
have been developed, including engineering method and bionics method. Unlike
engineering method that has achieved lots of success, the bionics method is still
in its infancy, and developed rapidly in recent years. Since bionics method imitates mechanisms of human's visual system and human brain's learning rule, it
has more potential to develop especially in neural networks. Recently, there are
some researches using neural networks to recognize human actions [1]-[3], but
their performance is not good enough.

In this paper, a novel human action recognition method is proposed to solve
this problem with two contributions: (1) A new star skeleton detector based on
spiking neural networks is presented, and it is easier to adjust precision than
traditional approaches. (2) A classifier using the evolving ReSuMe algorithm is

* This work was supported by National Science Foundation of China under Grant
61273308 and the Fundamental Research Funds for the Central Universities under
Grant ZYGX2012J068.

C.K. Loo et al. (Eds.): ICONIP 2014, Part I, LNCS 8834, pp. 366–373, 2014.

devised, that improves the performance of ReSuMe algorithm. Besides, the scale of our network neurons is independent with the size of pictures, which reduces computational complexity compared with [2].

2   The Proposed Method

2.1   Star Skeleton Detector

In this section, a contour's star skeleton detector is presented. In our method, a point in a contour is represented by its corresponding angle and distance to its centroid. To achieve scale invariance, the distance of every contour is normalized to 1-10.

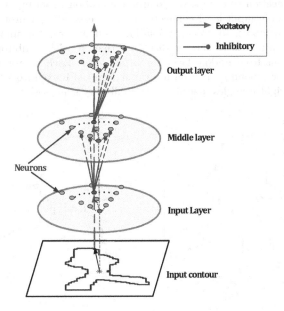

Fig. 1. Framework of star skeleton detector

As shown in Fig.1, the star skeleton detector consists of three layers, each layer is a circle with 360 degrees, and each degree has 10 neurons. Its detailed neuron connections are shown in Fig.2. The connection depicted in Fig.2(a) enable our detector to locate the effective star points and in Fig.2(b) to filter out adjacent points. Firing neurons in output layer corresponding to star points of input contour. Obviously, this detector can be implemented with high efficiency since it does not requires weight adjustments.

Different with traditional methods used in [1] and [4], star skeletons with different precisions can be obtained easily by modifying constant a in our method. Fig.3 shows simulation results with different a.

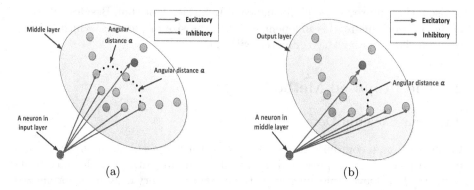

Fig. 2. Star detector connection. The centroid neuron in our method is a reference without any connection to others. (a) Neuron connections from input layer to middle layer. Each neuron in input layer connects to its corresponding neuron in middle layer with excitatory synapses, to its corresponding's left and right a angles'(from 1 to a angle's) neurons which have shorter distances than it with inhibitory synapses.(b) Neuron connections from middle layer to output layer. Each neuron in middle layer connects to its corresponding neuron in output layer with excitatory synapses, to its corresponding's right a angles' neurons with inhibitory synapses.

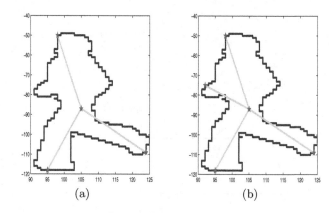

Fig. 3. Star skeletons obtained with different constant a. (a) $a = 15$; (b) $a = 5$;

2.2   Our Evolving-ReSuMe Classifier

In this section, an evolving ReSuMe classifier is proposed. As shown in Fig.4(a), the classifier with multi-synapses from middle layer to output neuron, and a target signal neuron is connected to supervise its learning. The detailed connections of our classifier are shown in Fig.4(b). In our classifier, the most often used spiking model, SRM_0 [5] is employed, with parameters set as follows.

External voltage of each neuron in input and middle layer are set to 0, while in the output neuron, it is set by the following equation,

$$U^{ext} = \begin{cases} U/5, \text{ if } t \in [t_i + (C_{t_i} - 1) * 0.1, t_i + (C_{ti} - 1)] \\ \\ 0, \quad \text{otherwise}, \end{cases} \tag{1}$$

where U is the voltage of the output neuron, and t_i is the firing time of a video, which is obtained by a linear one-to-one mapping from object's velocity to its firing time. C_{t_i} denotes class labels that the firing time t_i is related to. Every class with samples firing at time t_i is related to t_i.

The threshold is set to 0 except in output layer, in which the threshold is changed dynamically: suppose there are C_n intervals after t_i, each interval has one delayed sub-synapse potential and one threshold Θ_i ($i = 1, 2, \ldots, C_n$). The threshold Θ_i is changed by the following equation with $\Theta_1 = 0$,

$$\Theta_i = \begin{cases} Umax_i + 0.1, \text{ if output neuron fires at interval i} \\ \Theta_i, \quad\quad\quad\quad \text{otherwise}, \end{cases} \tag{2}$$

where $Umax_i$ is the maximum neuron voltage that a video produced in interval i. Then the last interval in which the output neuron emit spike has the biggest voltage, which indicates its corresponding class label.

In learning algorithm, the ReSuMe [6] learning rule is employed and improved. The weight change of ReSuMe can be expressed by the following equation,

$$\frac{d}{dt}w_{oi}(t) = [S_d^{(t)} - S_o^{(t)}][a_d + \int_0^\infty a_{di}(s)S_i(t - s)ds], \tag{3}$$

where $w_{oi}(t)$ is the weight from input neuron to the output one, a_d is a constant and a_{di} is a learning window. $S_o^{(t)}$ denotes the spike trains of the output neuron.

The value of the parameter a_d in Eq.3 influences weight distribution significantly, but there is no existing method to assign an appropriate value to it. In our study, the Particle Swarm Optimization(PSO) algorithm [7] is applied to optimize this parameter, which can be described as follows,

$$\nu_i(k + 1) = \nu_i(k) + \gamma_{1i}(b_i - a_i(k)) + \gamma_{2i}(g_i - a_i(k)), \tag{4}$$

$$a_i(k + 1) = a_i(k) + \nu_i(k + 1), \tag{5}$$

where i is the particle index, γ_{1i} and γ_{2i} are random number from 0 to 1, k is the discrete time index, v_i is the velocity of ith particle and a_i is the position of ith particle. b_i denotes the best position found by ith particle and g_i means the best position found by all particles. In our method, the recognition accuracy is regarded as the fitness of particles.

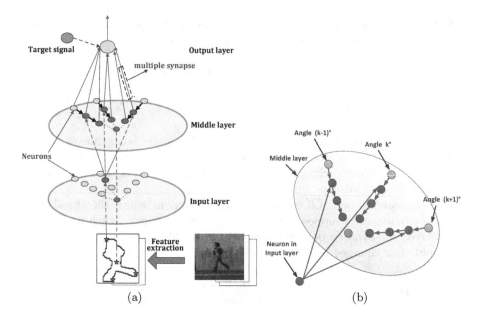

Fig. 4. The classifier framework with multiple synapses. (a) The framework of our classifier. (b) Neuron connections from input layer to middle layer. Each input neuron connects with excitatory synapses to its corresponding neuron(with distance d and angle k in middle layer), its corresponding neuron's left neuron(with distance d and angle $k - 1$ in middle layer) and right neuron (with distance d and angle $k + 1$ in middle layer). A neuron in middle layer connects to its neighbouring neuron which has the same angle but shorter distance than it with excitatory synapses.

3   Experimental Results

3.1   On Weizmann Dataset

The Weizmann database is one of benchmark datasets in human action recognition which contains ten actions, each action has nine videos which are provided by different people, these video sequences can be found at http://www.wisdom. weizmann.ac.il/~vision/SpaceTimeActions.html. In our simulations, human are detected by the motion detection method proposed by Viola, P. [10]. The PSO evolving results are shown in Fig.5. It indicates that our method is quickly converged within 20 epoches and obtains effective results.

Table 1 shows the recognition accuracy of each action, where N is the overall number of input videos, N_R is the number of videos which are correctly classified, and the accuracy R is the percentage of N_R from R. As shown in Table 1, most actions can be recognized correctly with the average recognition accuracy 95.56%.

Experimental results compared with different methods are listed in Table 2. It indicates that our method outperform the method proposed by Meng Y. [2] even using traditional ReSuMe algorithm. Besides, with the improvement of PSO, the

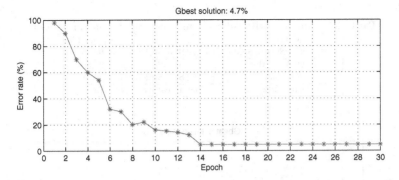

Fig. 5. Evolving results on Weizmann dataset, global best solution is 0.047 and $a =$ 0.0193. In our simulations, the population size is 25, with the maximum number of epochs 30, and the initial inertia weight 0.9, the final inertia weight 0.4. The range of particle is [-50,50], maximum velocity is 20% of the maximum speed range, and the minimum global error gradient is 10^{-25}.

Table 1. Results of action recognition on Weizmann dataset

Actions	N	N_R	$R(\%)$
walking	9	9	100
running	9	8	88.89
waving2	9	9	100
pjumping	9	9	100
siding	9	8	88.89
jacking	9	9	100
skipping	9	7	77.78
waving1	9	9	100
bending	9	9	100
jumping	9	9	100
overall	90	86	95.56

accuracy of our ReSuMe classifier is elevated by 6.28%. Since in this dataset, samples for each action are insufficient, the adaptive boosting (AdaBoost)[8] is applied to improve the performance of our method on PSO+ReSuMe. It combines several weak classifiers to obtain a better performance, and in this simulation, 5 same classifiers are employed. As shown in Table 2, AdaBoost improves our performance of PSO+ReSuMe by 3.33%.

3.2 On KTH Dataset

The KTH dataset is another benchmark dataset that is tested to investigate performance of our method. It includes 6 kinds of actions, each one is provided by 25 people under four different scenarios [9]. As shows in Fig.6, our method is quickly converged with result of error rate 0.125.

Table 2. Compared with different methods on Weizmann dataset

Mehtod	feature	accuracy (%)
AdaBoost+PSO+ReSuMe	Velocity+Star Skeletons	98.89
PSO+ReSuMe	Velocity+Star Skeletons	95.56
ReSuMe	Velocity+Star Skeletons	89.28
Meng Y. [2]	Spatial feature	75.55

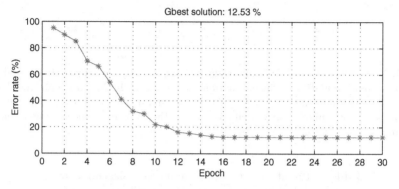

Fig. 6. Evolving result in KTH dataset, global best solution is 12.53 and $a = 0.0012$

The accuracy of each action is listed in Table 3, which indicates that most of actions can be recognized successfully with the overall performance of 87.47%.

Table 4 shows the comparable results using different approaches. Similar to previous results, our PSO+ReSuMe method outperform the BCM-GRN method presented by Meng Y. [2] and the traditional ReSuMe algorithm. In this dataset, AdaBoost has no help for our performance, since there are sufficient samples in each action(about 100 samples for each action), and there are much input noises. In this situation, Adaboost can not take its advantage.

Table 3. Results of action recognition on KTH dataset

Actions	N	N_R	$R(\%)$
walking	100	92	92.00
jogging	99	88	88.89
running	100	93	93.00
boxing	100	86	86.00
hand-waving	100	80	80.00
hand-clapping	100	85	90.00
overall	599	524	87.47

Table 4. Comparison of different methods on KTH dataset

Mehtod	feature	accuracy (%)
PSO+ReSuMe	Velocity+Skeletons	87.47
AdaBoost+PSO+ReSuMe	Velocity+Skeletons	85.26
ReSuMe	Velocity+Skeletons	84.33
Meng Y. [2]	Spatial feature	82.80

4   Conclusions

A common framework for star skeleton detection and human action classification was proposed in this paper, and experiments on a number of benchmark datasets demonstrated its performance. It is applicable even when contours are not extracted accurately, or have regular noises. Future researches will address the problem of recognize actions in dynamic backgrounds and with in-regularity noises. Besides, a more efficient learning algorithm is required.

References

1. Petridis, V., Deb, B., et al.: Detection and Identification of Human Actions Using Predictive Modular Neural Networks. In: 17th Mediterranean Conference on Control & Automation, pp. 406–411. IEEE Press, Thessaloniki (2009)
2. Meng, Y., Jin, Y., et al.: Modeling Activity-Dependent Plasticity in BCM Spiking Neural Networks with Application to Human Behavior Recognition. IEEE Transactions on Neural Networks 22, 1952–1966 (2011)
3. Escobar, M.-J., Masson, G.S., et al.: Action Recognition Using a Bio-Inspired Feedforward Spiking Network. International Journal of Computer Vision 82, 284–301 (2009)
4. Chen, H.S., Chen, H.T., et al.: Human Action Recognition using star skeleton. In: 4th ACM International Workshop on Video Surveillance and Sensor Networks, pp. 171–178. ACM Press, Santa Barbara (2006)
5. Gerstner, W., Kistler, W.M.: Spiking Nerual Models: Single Neurons, Populations, Plasticity. Cambridge University Press, Cambridge (2002)
6. Ponulak, F., Kasiński, A.: Supervised Learning in Spiking Nerual Networks with ReSuMe: Sequence Learning, Classification, and Spike Shifting. Neural Computation 22, 467–510 (2010)
7. Poli, R.: Analysis of the Publications on the Applications of Particle Swarm Optimisation. Journal of Artificial Evolution and Applications 2008, 1–10 (2008)
8. Han, J., Kamber, M.: Data Mining: Concepts and Techniques. Morgan Kaufmann, San Francisco (2011)
9. Schüldt, C., Laptev, I., et al.: Recognizing Human Actions: A Local SVM Approach. In: 17th International Conference on Pattern Recognition, pp. 32–36. IEEE Press, Cambridge (2004)
10. Viola, P., Jones, M.J., et al.: Detecting pedestrians using patterns of motion and appearance. In: 9th IEEE International Conference in Computer Vision, pp. 734–741. IEEE Press, Cambridge (2003)

A Neural Networks Committee
for the Contextual Bandit Problem

Robin Allesiardo[1,2], Raphaël Féraud[2], and Djallel Bouneffouf[2]

[1] TAO - INRIA, CNRS, University of Paris-Sud, 91405 Orsay, France
[2] Orange Labs, 2 av. Pierre Marzin, 22300 Lannion, France

Abstract. This paper presents a new contextual bandit algorithm, NeuralBandit, which does not need hypothesis on stationarity of contexts and rewards. Several neural networks are trained to modelize the value of rewards knowing the context. Two variants, based on multi-experts approach, are proposed to choose online the parameters of multi-layer perceptrons. The proposed algorithms are successfully tested on a large dataset with and without stationarity of rewards.

1 Introduction

In online decision problems such as online advertising or marketing optimization, a decision algorithm must select amoung several actions. Each of these options is associated with side information (profile or context) and the reward feedback is limited to the chosen option. For example, in online advertising, a visitor queries a web page; a request with the context (web page address, cookies, customer profile, etc.) is send to the server; the server sends an ad which is displayed on the page. If the visitor clicks on the ad the server receives a reward. The server must trade-off between the explorations of new ads and the exploitation of known ads. Moreover, in an actual applications, both rewards and data distributions can change with time. For instance, the display of a new ad can change the probability of clicks of all ads, the content of a web page can change over time. Robustness to non-stationarity is thus strongly recommended.

2 Previous Work

The multi-armed bandit problem is a model of exploration and exploitation where one player gets to pick within a finite set of decisions the one which maximizes the cumulated reward. This problem has been extensively studied. Optimal solutions have been provided using a stochastic formulation [1,2], a Bayesian formulation [3,4], or using an adversarial formulation [5,6]. Variants of the initial problem were introduced due to practical constraints (appearance of a new advertisement after the beginning of learning [7,8], fixed number of contractual page views [9,10]). However these approaches do not take into account the context while the arm's performance may be correlated therewith.

A naive solution to the contextual bandit problem is to allocate one bandit problem for each context. A tree structured bandit such as X-armed bandits [11] or a UCT variant [12] can be used to explore and exploit a tree structure of contextual variables to

C.K. Loo et al. (Eds.): ICONIP 2014, Part I, LNCS 8834, pp. 374–381, 2014.
© Springer International Publishing Switzerland 2014

find the leaves which provide the highest rewards [13]. The combinatorial aspect of these approaches limits their use to small context size. Seldin et al [14] modelize the contexts by state sets, which are associated with bandit problems. Without prior knowledge of the contexts, it is necessary to use a state per context, which is equivalent to the naive approach. Dudík et al [15] propose an algorithm of policies elimination. The performance of this algorithm depends on the presence of a good policy in the set. The epoch-greedy algorithm [16] alternates exploration then exploitation. During exploration, the arms to play are randomly drawn to collect an unbiased training set. Then, this set is used to train a classifier which will be used for exploitation during the next cycle. The nature of the classifier remains to be defined for concrete use. In LINUCB [17,18] and in Contextual Thompson Sampling (CTS) [19], the authors assume a linear dependency between the expected reward of an action and its context and model the representation space using a set of linear predictors. Banditron [20] uses a perceptron per action to recognize rewarded contexts. Furthermore, these algorithms assume that the data and the rewards are drawn from stationary distribution, which limits their practical use. The EXP4 algorithm [6] selects the best arm from N experts advices (probability vectors). Exploring the link between the rewards of each arm and the context is delegated to experts. Unlike the previous algorithms, the rewards are assumed to be chosen in advance by an adversary. Thus, this algorithm can be applied to non-stationnary data.

At first we will formalize the contextual bandit problem and will propose a first algorithm: NeuralBandit1. Inspired by Banditron [20] it estimates the probabilities of rewards by using neural networks in order to be free of the hypothesis of linear separability of the data. Neural networks are universal approximators [21]. They are used in reinforcement learning [22,23] and can estimate accurately the probabilities of rewards within actual and complex problems. In addition, the stochastic gradient achieves good performances in terms of convergence to the point of best generalization [24] and has the advantage of learning online. In seeking to reach a local minimum, the stochastic gradient can deal with non-stationarity. This will result in a change of the cost function landscape over the time. If this landscape changes at a reasonable speed, the algorithm will continue the descent to a new local minimum. The main issue raised by the use of multilayer perceptrons remains the online setting of various parameters such as the number of hidden neurons, the value of the learning step or the initalization of the weight seed. We propose two advanced versions of the algorithm NeuralBandit1 to adjust these settings using adversarial bandits that seek, among several models initialized with different parameters, the best one. We conclude by comparing these different approaches to the state-of-the-art on stationnary and on non-stationnary data.

3 Our Algorithm: NeuralBandit1

Definition 1 (Contextual bandit). *Let $x_t \in X$ be a contextual vector and $(y_{t,1}, ..., y_{t,K})$ a vector of rewards associated with the arm $k \in [K] = \{1, ..., K\}$ and $((x_1, \mathbf{y}_1), ..., (x_T, \mathbf{y}_T))$ the sequence of contexts and rewards. The sequence can be drawn from a stochastic process or chosen in advance by an adversary. At each round $t < T$, the context x_t is announced. The player, who aims to maximize his cumulated rewards, chooses an arm k_t. The reward $y_{t,k}$ of the played arm, and only this one, is revealed.*

Definition 2 (Cumulated regret). *Let* $H : X \rightarrow [K]$ *be a set of hypothesis*, $h_t \in H$ *a hypothesis computed by the algorithm* A *at the round* t *and* $h_t^* = \underset{h_t \in H}{\operatorname{argmax}}\, y_{t,h_t(x_t)}$ *the optimal hypothesis at the same round. The cumulated regret is:*

$$R(A) = \sum_{t=1}^{T} y_{t,h_t^*(x_t)} - y_{t,h_t(x_t)}$$

The purpose of a contextual bandit algorithm is to minimize the cumulative regret.

Each action k is associated with a neural network with one hidden layer which learns the probability of reward for an action knowing its context. We choose this modelization rather than one neural network with as many outputs as actions to be able to add or remove actions easier.

Let K be the number of actions, C the number of neurons of each hidden layers and $\mathbf{N}_t^k : X \rightarrow Y$ the function associating a context x_t to the output of the neural network corresponding to the action k at the round t. N denotes the number of connections of each network with $N = dim(X)C + C$. To simplify the notations, we place the set of connections in the matrix W_t of size $K \times N$. Thus, each row of the matrix W_t contains the weight of a network. Δ_t is the matrix of size $K \times N$ containing the update of each weight between rounds t and $t + 1$. The update equation is:

$$W_{t+1} = W_t + \Delta_t$$

The backpropagation algorithm [25] allows calculating the gradient of the error for each neuron from the last to the first layer by minimizing a cost function. Here, we use the quadratic error function and a sigmoid activation function.

Let λ be the learning step, $\hat{x}_t^{n,k}$ the input associated with the connection n in the network k, $\delta_t^{n,k}$ the gradient of the cost function at round t for the neuron having as input the connection n in the network k and $\Delta_t^{n,k}$ the value corresponding to the index (n, k) of the matrix Δ_t. When the reward of an arm is known, we can compute:

$$\Delta_t^{n,k} = \lambda \hat{x}_t^{n,k} \delta_t^{n,k}$$

In the case of partial information, only the reward of the arm k_t is available. To learn the best action to play, an approach consists of a first exploration phase, where each action is played the same number of times in order to train a model, and then an exploitation phase where the obtained model is used. Thus, the estimator would not be biased on the most played action. However, this approach would have abysmal performances in case of non-stationary data. We choose to use an exploration factor γ, constant over time, allowing continuing the update of the model in the case of non-stationary data. The probability of playing the action k at round t knowing that $\hat{k}_t$ is the arm with the highest reward prediction is:

$$\mathbf{P}_t(k) = (1 - \gamma)\mathbf{1}[k = \hat{k}_t] + \frac{\gamma}{K}$$

We propose a new update rule taking into account the exploration factor:

$$W_{t+1} = W_t + \tilde{\Delta}_t \,, \tag{1}$$

$$\text{with } \tilde{\Delta}_t^{n,k} = \frac{\lambda \hat{x}_t^{n,k} \delta_t^{n,k} \mathbf{1}[\hat{k}_t = k]}{\mathbf{P}_t(k)}$$

Theorem 1. *The expected value of $\tilde{\Delta}_t^{n,k}$ is $\Delta_t^{n,k}$.*

The proof is straightforward:

$$\mathbf{E}[\tilde{\Delta}_t^{n,k}] = \sum_{k=1}^{K} \mathbf{P}_t(k)\left(\frac{\lambda \hat{x}_t^{n,k} \delta_t^{n,k} \mathbf{1}[\hat{k}_t = k]}{\mathbf{P}_t(k)}\right)$$
$$= \lambda \hat{x}_t^{n,k} \delta_t^{n,k}$$
$$= \Delta_t^{n,k}$$

□

The proposed algorithm, NeuralBandit1, can adapt to non-stationarity by continuing to learn over time, while achieving the same expected result (in the case of stationary data) as a model trained in a first phase of exploration.

Algorithm 1. NeuralBandit1

Data: $\gamma \in [0, 0.5]$ et $\lambda \in]0, 1]$
begin

  Initialize $W_1 \in]-0.5, 0.5[^{N \times K}$
  for $t = 1, 2, ..., T$ **do**
    Context x_t is revealed
    $\hat{k}_t = \arg\max_{k \in [K]} \mathbf{N}_t^k(x_t)$
    $\forall k \in [K]$ on a $\mathbf{P}_t(k) = (1 - \gamma)\mathbf{1}[k = \hat{k}_t] + \frac{\gamma}{K}$
    $\tilde{k}_t$ is drawn from $\mathbf{P}_t$
    $\tilde{k}_t$ is predicted and $y_{t,\tilde{k}_t}$ is revealed
    Compute $\tilde{\Delta}_t$ such as $\tilde{\Delta}_t^{n,k} = \frac{\lambda \hat{x}_t^{n,k} \delta_t^{n,k} \mathbf{1}[k_t = k]}{\mathbf{P}_t(k)}$
    $W_{t+1} = W_t + \tilde{\Delta}_t$

---

4   Models Selection with Adversarial Bandit

Performances of neural networks are influenced by several parameters such as the learning step, the number of hidden layers, their size, and the initalization of weights. The multi-layer perceptron corresponding to a set of parameters is called model. Using batch learning, the models selection is done with a validation set. Using online learning, we propose to train the models in parallel and to use the adversarial bandit algorithm EXP3 [5,6] to choose the best model. The choice of an adversarial bandit algorithm is justified by the fact that the performance of each model changes overtime due to the learning itself or due to the non-stationarity of the data.

Exp3. Let $\gamma_{\text{model}} \in [0,1]$ be an exploration parameter, $w_t = (w_t^1, ..., w_t^M)$ a weight vector, where each of its coordinate is initialized to 1, and M the number of models. Let m be the model chosen at time t, and $y_{t,m}$ be the obtained reward. The probability to choose m at round t is:

$$\mathbf{P}_{\text{model}_t}(m) = (1 - \gamma_{\text{model}}) \frac{w_t^m}{\sum_{i=1}^M w_t^i} + \frac{\gamma_{\text{model}}}{M} \qquad (2)$$

The weight update equation is:

$$w_{t+1}^i = w_t^i \exp\left(\frac{\gamma_{\text{model}} \mathbf{1}[i = m] y_{t,m}}{\mathbf{P}_{\text{model}_t}(i) M} \right) \qquad (3)$$

NeuralBandit2 (see Algorithm 2). If we consider that a model is an arm, a run of this model can be considered as a sequence of rewards. The algorithm takes as inputs a list of M models and a model exploration parameter γ_{model}. For each element of the list, one NeuralBandit1 instance is initialized. Each instance corresponds to an arm. EXP3 algorithm is used to choose the arms over time. After receiving a reward, each neural network corresponding to the played arm is updated, and the weights of EXP3 are updated.

NeuralBandit3 (see Algorithm 3). The use of the algorithm NeuralBandit2 corresponds to the assumption that there is a model NeuralBandit1 which is the best for all actions. The algorithm NeuralBandit3 lifts this limitation by associating one EXP3 per action.

NeuralBandit3 has greater capacity of expression than NeuralBandit2 as each action can be associated with different models. However if the best model in NeuralBandit3 exists in NeuralBandit2, then NeuralBandit2 should find this model faster than Neural-Bandit3 because it has only one instance of EXP3 with less possibility to update.

5   Experiments

The Forest Cover Type dataset from the UCI Machine Learning Repository is used. It contains 581.000 instances and it is shuffled. We have recoded each continuous variable using equal frequencies into five binary variables and we have recoded each categorical variable into binary variables. We have obtained 94 binary variables for the context, and we have used the 7 target classes as the set of actions. In order to simulate a datastream, the dataset is played in loop. At each round, if the algorithm chooses the right class the reward is 1 or else 0. The cumulated regret is computed from the rewards of an offline algorithm fitting the data with 93% of classification. The plot of the curves (Figure 1) are produced by averaging 10 runs of each algorithms with $\gamma = 0.005, \gamma_{model} = 0.1$. Each run began at a random position in the dataset. The parameters of each model (NeuralBandit1) are the combination of different sizes of hidden layer $(1, 5, 25, 50, 100)$ and different values of λ $(0.01, 0.1, 1)$.

Algorithm 2. NeuralBandit2

Data: $\gamma_{\text{model}} \in [0, 0.5]$ and a list of M models parameters
begin

    Initialize M NeuralBandit1 ;
    Initialize the EXP3 weight vector w_0 with $\forall m \in [M]\ w_0^m = 1$;
    for $t = 1, 2, ..., T$ **do**

        Context x_t is revealed;
        m_t is drawn from $\mathbf{P}_{\text{model}\,t}$ (2);
        The model m_t choose action $\tilde{k}_t$;
        $\tilde{k}_t$ is predicted and $y_{t,\tilde{k}_t}$ is revealed;
        Update of each network corresponding to the action $\tilde{k}_t$ for each model with (1);
        Update of the EXP3 weight vector w_t with (3);

Algorithm 3. NeuralBandit3

Data: $\gamma \in [0, 0.5]$, $\gamma_{\text{model}} \in [0, 0.5]$ and a list of M model parameters
begin

    Initialize K neural networks per model m ;
    Initialize K instance of EXP3;
    for $t = 1, 2, ..., T$ **do**

        Context x_t is revealed;
        for $k = 1, 2, ..., K$ **do**

            m_t^k is drawn from $\mathbf{P}_{\text{model}_t^k}$ (2);
            Action k is scored $s_t^k = \mathbf{N}_t^{m_t^k, k}(x_t)$;

        $\hat{k}_t = \arg\max_{k \in [K]} s_t^k$;
        $\forall k \in [K]$ on a $\mathbf{P}_t(k) = (1 - \gamma)\mathbf{1}[k = \hat{k}_t] + \frac{\gamma}{K}$;
        $\tilde{k}_t$ is drawn from $\mathbf{P}_t$;
        $\tilde{k}_t$ is predicted and $y_{t,\tilde{k}_t}$ is revealed;
        Update of each network corresponding to the action $\tilde{k}_t$ for each model with (1);
        Update of each EXP3 weight vectors with (3);

On Stationary Data (left part of Figure 1). Banditron achieves a high cumulated regret (57% of classification computed on the 100.000 last predictions) and is outperformed by all other contextual bandit algorithms on this dataset. The cumulated regret and classification rate of LinUCB (72%), CTS (73%) and NeuralBandit3 (73%) are similar and their curves tend to be the same. NeuralBandit2 has the fastest convergence rate and achieves a smaller cumulated regret with 76% of classification.

On Non-Stationary Data (right part of Figure 1). Non-stationarity is simulated by swapping classes with a circular cycle $(1 \rightarrow 2, 2 \rightarrow 3, ..., 7 \rightarrow 1)$ every 500.000 iterations. At each concept drift, curves increase then stabilize. On stationary data LinUCB and CTS achieve a lowest regret than Banditron but can't deal with non-stationarity thus are outperformed by it after the first drift. Banditron converges again near instantly.

Models selection algorithms need between 75.000 and 350.000 in the worst case to stabilize at each drift. NeuralBandit2 and NeuralBandit3 are more complex models than Banditron but achieve better performances on this non-stationary datastream. Neural-Bandit3 seem to be more robust to nonstationarity than NeuralBandit2 on this dataset. This can be explained by the fact that sometime, one neural network can stay on a bad local minimum. If this append in NeuralBandit2, the entire model is penalized while in NeuralBandit3 the algorithm can still use all the other networks.

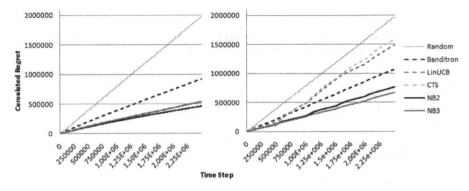

Fig. 1. The cumulated regret of different contextual bandit algorithms over time on Forest Cover Type. The left part is on a stationary datastream and the right part on a non-stationary datastream.

6   Conclusion

We introduced a new contextual bandit algorithm NeuralBandit1. Two variants with models selection NeuralBandit2 and NeuralBandit3 used an adversarial bandit algorithm to find the best parameters of neural networks. We confronted them to stationary and non-stationary datastream. They achieve a smaller cumulated regret than Banditron, LinUCB and CTS. This approach is successful and has the advantage of being trivially parallelizable. Models differentiation show a significant gain on the Forest Covert Type dataset with nonstationarity. We also showed empirically that our two models selection algorithms are robust to concept drift. These experimental results suggest that neural networks are serious candidates for addressing the issue of contextual bandit. However, they are freed from the constraint of linearity at the expense of the bound on the regret, the cost function being not convex.

References

1. Lai, T.L., Robbins, H.: Asymptotically efficient adaptive allocation rules. Advances in Applied Mathematics 6(1), 4–22 (1985)
2. Auer, P., Cesa-Bianchi, N., Fischer, P.: Finite-time analysis of the multiarmed bandit problem. Machine Learning 47(2-3), 235–256 (2002)
3. Thompson, W.: On the likelihood that one unknown probability exceeds another in view of the evidence of two samples. Biometrika 25, 285–294 (1933)

4. Kaufmann, E., Korda, N., Munos, R.: Thompson Sampling: An Asymptotically Optimal Finite-Time Analysis. In: Bshouty, N.H., Stoltz, G., Vayatis, N., Zeugmann, T. (eds.) ALT 2012. LNCS (LNAI), vol. 7568, pp. 199–213. Springer, Heidelberg (2012)
5. Auer, P., Cesa-Bianchi, N.: On-line learning with malicious noise and the closure algorithm. Ann. Math. Artif. Intell. 23(1-2), 83–99 (1998)
6. Auer, P., Cesa-Bianchi, N., Freund, Y., Schapire, R.E.: The nonstochastic multiarmed bandit problem. SIAM J. Comput. 32(1), 48–77 (2002)
7. Kleinberg, R.D., Niculescu-Mizil, A., Sharma, Y.: Regret bounds for sleeping experts and bandits. In: COLT, pp. 425–436 (2008)
8. Chakrabarti, D., Kumar, R., Radlinski, F., Upfal, E.: Mortal multi-armed bandits. In: NIPS, pp. 273–280 (2008)
9. Feraud, R., Urvoy, T.: A stochastic bandit algorithm for scratch games. In: Hoi, S.C.H., Buntine, W.L. (eds.) ACML. JMLR Proceedings, vol. 25, pp. 129–143. JMLR.org. (2012)
10. Feraud, R., Urvoy, T.: Exploration and exploitation of scratch games. Machine Learning 92(2-3), 377–401 (2013)
11. Bubeck, S., Munos, R., Stoltz, G., Szepesvári, C.: Online optimization in x-armed bandits. In: Koller, D., Schuurmans, D., Bengio, Y., Bottou, L. (eds.) NIPS, pp. 201–208. Curran Associates, Inc. (2008)
12. Kocsis, L., Szepesvári, C.: Bandit based monte-carlo planning. In: Fürnkranz, J., Scheffer, T., Spiliopoulou, M. (eds.) ECML 2006. LNCS (LNAI), vol. 4212, pp. 282–293. Springer, Heidelberg (2006)
13. Gaudel, R., Sebag, M.: Feature selection as a one-player game. In: Fürnkranz, J., Joachims, T. (eds.) ICML, pp. 359–366. Omnipress (2010)
14. Seldin, Y., Auer, P., Laviolette, F., Shawe-Taylor, J., Ortner, R.: Pac-bayesian analysis of contextual bandits. In: Shawe-Taylor, J., Zemel, R.S., Bartlett, P.L., Pereira, F.C.N., Weinberger, K.Q. (eds.) NIPS, pp. 1683–1691 (2011)
15. Dudík, M., Hsu, D., Kale, S., Karampatziakis, N., Langford, J., Reyzin, L., Zhang, T.: Efficient optimal learning for contextual bandits. CoRR (2011)
16. Langford, J., Zhang, T.: The epoch-greedy algorithm for multi-armed bandits with side information. In: Platt, J.C., Koller, D., Singer, Y., Roweis, S.T. (eds.) NIPS. Curran Associates, Inc. (2007)
17. Li, L., Chu, W., Langford, J., Schapire, R.E.: A contextual-bandit approach to personalized news article recommendation. CoRR (2010)
18. Chu, W., Li, L., Reyzin, L., Schapire, R.E.: Contextual bandits with linear payoff functions. In: Gordon, G.J., Dunson, D.B., Dudk, M. (eds.) AISTATS. JMLR Proceedings, vol. 15, pp. 208–214. JMLRorg. (2011)
19. Agrawal, S., Goyal, N.: Thompson sampling for contextual bandits with linear payoffs. CoRR (2012)
20. Kakade, S.M., Shalev-Shwartz, S., Tewari, A.: Efficient bandit algorithms for online multiclass prediction. In: Proceedings of the 25th International Conference on Machine Learning, ICML 2008, pp. 440–447. ACM, New York (2008)
21. Hornik, K., Stinchcombe, M., White, H.: Multilayer feedforward networks are universal approximators. Neural Netw. 2(5), 359–366 (1989)
22. Tesauro, G.: Programming backgammon using self-teaching neural nets. Artificial Intelligence 134, 181–199 (2002)
23. Mnih, V., Kavukcuoglu, K., Silver, D., Graves, A., Antonoglou, I., Wierstra, D., Riedmiller, M.: Playing atari with deep reinforcement learning. CoRR (2013)
24. Bottou, L., LeCun, Y.: On-line learning for very large datasets. Applied Stochastic Models in Business and Industry 21(2), 137–151 (2005)
25. Rumelhart, D.E., Hinton, G.E., Williams, R.J.: Parallel distributed processing: Explorations in the microstructure of cognition, vol. 1, pp. 318–362. MIT Press, Cambridge (1986)

Multi-step Predictions of Landslide Displacements Based on Echo State Network

Wei Yao[1,2], Zhigang Zeng[1], Cheng Lian[1], Huiming Tang[3], and Tingwen Huang[4]

[1] School of Automation, Huazhong University of Science and Technology, Wuhan, China
zgzeng_hust@163.com
[2] School of Computer Science, South-Central University for Nationalities, Wuhan, China
hevigreen@gmail.com
[3] Faculty of Engineering, China University of Geosciences, Wuhan, China
[4] Texas A&M University at Qatar, Doha, Qatar

Abstract. Time series prediction theory and methods can be applied to many practical problems, such as the early warning of landslide hazard. Most already existing time series prediction methods cannot be effectively applied on landslide displacement prediction tasks, mainly for two problems. Firstly, the underlying dynamics of landslides cannot be properly modeled; secondly, it is difficult to perform effective long term predictions. Considering these problems, a dynamic predictor is proposed in our paper. The predictor is established on a recurrent network structure and trained by a newly proposed learning algorithm, namely echo state network. Furthermore, multi-step predictors are built based on echo state network, following different predicting strategies. Experimental results show that, the dynamic predictors perform better than static predictors, and can produce reliable multi-step ahead predictions of landslide displacements.

Keywords: Time series prediction, landslide, recurrent neural network, echo state network, multi-step prediction.

1 Introduction

For a long period, Landslides have always been big threats to the life and property safety of residents in mountainous areas and river basins. The Three Gorges area is a well-known landslide-prone area in China. In recent years, the Three Gorges dam and Reservoir project dramatically changed the hydro-geological condition of the entire region [1]. The stability of the 5900 km river shore in this area is deteriorated by the rapid and significant increase of water level in the reservoir. More than 2500 instable slopes have already been located [2], while the significant environmental changes may further increase the landslide hazard in this area. Therefore, it's more and more an urgent issue to establish an effective early warning system for landslide disasters in this area.

C.K. Loo et al. (Eds.): ICONIP 2014, Part I, LNCS 8834, pp. 382–388, 2014.
© Springer International Publishing Switzerland 2014

Mechanisms of landslides have been studied. However, sufficiently precise mechanism models can hardly be obtained. Since the slope failure can be forecasted from the trend of displacement, lots of effort has been dedicated to the prediction of landslide displacement [1], [3, 4], and data-driven time series prediction approaches are usually resorted to. State of the art time series prediction techniques mainly fall into two categories, namely phase space reconstruction approaches and dynamic modeling approaches. Since the evolutions of landslides are in essence dynamic processes [5], dynamic predictors could be more suitable than feed forward neural networks and SVMs. It has been argued that in principle, an artificial recurrent neural network (RNN) can learn to mimic a target system with arbitrary accuracy [6]. In our research, we build dynamic models of landslides, using echo state network (ESN) [7], which is a novel learning paradigm for RNNs. Furthermore, following quite different predicting strategies, multi-step predictors based on ESN have also been proposed in this paper.

2   Multi-step ESN Predictors

The landslide displacement predictor, using a RNN for dynamic modeling, can be built as

$$X(t+1) = f[W_x \cdot X(t) + W_i \cdot d_r(t)]$$
$$d_p(t+1) = g[W_o \cdot X(t+1)] \tag{1}$$

where d_r and d_p denote the recorded and the predicted displacements, respectively. W_x is the recurrent internal connection matrix; W_i and W_o are the input and output connection matrix. $f[]$ and $g[]$ are the active functions of the internal and output neurons in the network, which represent the nonlinearity of the system. In ESN, the network is randomly created and remains unchanged during training, while only the synaptic connections from the internal neurons to the output readout neurons are modified, as illustrated in Fig. 1. The training process of the predictor is aimed at minimizing the difference between the realistic displacement $d_r(t+1)$ and the predicted value $d_p(t+1)$, by tuning W_o. The training can therefore be simplified into a linear regression problem, which is much easier to solve than gradient-descent problems. And the shortcomings of gradient based trainings of RNNs can also be avoided.

The predictor of (1) is a one-step ESN predictor. Following different strategies [8], four kinds of multi-step ESN predictors can be built as follows:

- Iterative multi-step ESN predictor

The multi-step prediction is produced by a one-step predictor iteratively. The longer term predictions will be made based on previous predictions. Therefore, after training, the function of the predictor expands into

$$X(t+1) = f[W_x \cdot X(t) + W_i \cdot d_p(t)]$$
$$d_p(t+1) = g[W_o \cdot X(t+1)] \tag{2}$$

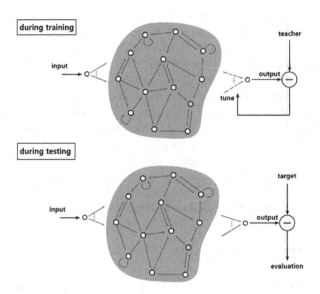

Fig. 1. Training and testing processes of the ESN predictor

This iterative multi-step predictor is in fact still a one-step predictor, and it needs to run multiple times to produce multi-step predictions.

- Direct multi-step ESN predictor

This one is trained as a multi-step ahead predictor directly. And the function of the network during both the training and the after-training phases can be expressed as

$$X(t+1) = f[W_x \cdot X(t) + W_i \cdot d_r(t)]$$
$$d_p(t+T) = g[W_o \cdot X(t+1)]$$

(3)

where T denote the prediction horizon. As compared to the iterative ESN predictor, this direct predictor won't suffer from the accumulative error problem. However, the larger prediction horizon makes it more difficult to obtain accurate predictions.

- Multi-output ESN predictor

The direct ESN predictor can produce only one prediction each time, while the multi-output predictor can give multiple predictions simultaneously. Its function is

$$X(t+1) = f[W_x \cdot X(t) + W_i \cdot d_r(t)]$$
$$[d_p(t+1), d_p(t+2), ..., d_p(t+T)] = g[W_o \cdot X(t+1)]$$

(4)

In fact, the main advantage of this multi-output ESN predictor is that it can preserve the dependencies between future values of landslide displacements. From such a respect, it is not just equivalence to the simple combination of T direct predictors with different prediction horizons.

- Direct Multi-output ESN predictor

The multi-output ESN predictor constrains all the horizons to be forecasted with the same model structure. This drawback could reduce the flexibility of the predictor. Therefore, a new multi-output prediction strategy is proposed [8], which divide one large multi-output predictor into some smaller multi-output predictors. The following function is an example.

$$X(t+1) = f[W_x \cdot X(t) + W_i \cdot d_r(t)]$$
$$[d_p(t+T/2+1),...,d_p(t+T)] = g[W_{o1} \cdot X(t+1)] \tag{5}$$
$$[d_p(t+1),...,d_p(t+T/2)] = g[W_{o2} \cdot X(t+1)]$$

As above, two multi-step predictors are in charge of different prediction horizon ranges respectively. This predictor is a tradeoff between the direct predictor and the multi-output predictor.

3   Experiments

Predictors for accumulative displacements of landslides located in the Three Gorges area are established. The experiments are two-fold. First of all, one-step-ahead predictions produced by the ESN predictor are tested and compared to the predictions of static predictors. Then multi-step predictions of ESN predictors are also produced.

3.1   Data Description

In our study, displacement datasets are recorded monthly for Yuhuangge landslide, Huangtupo landslide and Baishuihe landslide. These datasets are denoted as YHG, HTP and BSH hereafter. Systematical monitoring of all the three landslides started at around 2003 to 2004. The lengths of the three recordings are 101 months, 61 months and 101 months respectively. The developing trends of these displacement curves are quite different, corresponding to three typical landslide types.

3.2   One Step Ahead Predictions

For comparison, extreme learning machine (ELM) [9] and support vector machine (SVM) [10, 11], are employed to construct static predictors. ELM and SVM are two high quality learning algorithms which can produce highly accurate results in both classification and regression tasks [12, 13]. Since there is randomness involved in the training of the ESN predictor and the ELM predictor, experiments concerning these two predictors are repeated for ten times and the averaged accuracies are reported. During the training of the SVM predictor, cross validation is implemented to determine the kernel parameters [14] as well as the embedding dimension for prediction. Prediction errors as regard to root mean square error (RMSE) and mean absolute percentage error (MAPE) are reported in Table 1.

Table 1. Comparisons between predictors based on ESN, ELM and SVM

Measures	Data Set	ESN	ELM	SVM
	YHG	1.44mm	2.59mm	2.32mm
RMSE	HTP	1.21mm	4.59mm	17.23mm
	BSH	30.38mm	39.96mm	13.70mm
	YHG	6.31%	7.78%	7.05%
MAPE	HTP	0.68%	2.53%	10.33%
	BSH	1.36%	1.66%	0.50%

The best predictions on both the YHG dataset and the HTP dataset are produced by the ESN predictor. And the ESN predictor performs better than the ELM predictor on all the three datasets. Although the SVM predictor outperforms the other two on the BSH dataset, it tends to be quite instable as much bigger error has been made on the HTP dataset. The experimental results are consistent with our arguments that, a dynamic predictor based on ESN will be more proper for the prediction of landslide displacements, as compared to static models. Prediction curves are presented in Fig. 2. The overall consistencies between predictions of the ESN predictor and the actual displacement recordings are better than the other two predictors.

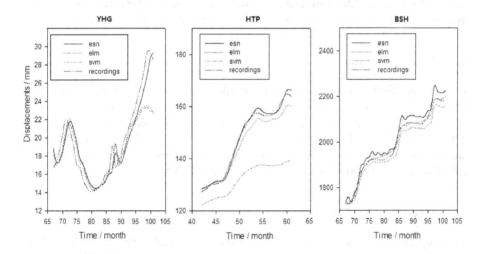

Fig. 2. Prediction made by the ESN, ELM and SVM predictors on different datasets

3.3   Multi-step Predictions of ESN

Multi-step ahead predictions are produced by the aforementioned four different ESN predictors, which will be denoted as Iterative, Direct, MO and DMO hereafter. MAPE of the predictions produced by these four predictors are reported in Table 2. The results are concerning nine multi-step prediction tasks, namely the 2-step, the 3-step and the 4-step predictions for the datasets of YHG, HTP and BSH. As for the MO predictor, although predictions for different horizons are produced at the same time, only the longest term predictions are considered when evaluating the predictor's performance. For the DMO predictor, only the 4-step prediction is implemented, in which two double-output predictors are established. Because when dealing with the 2-step and 3-step prediction tasks, the DMO predictor and the Direct predictor are the same. As shown by the values in Table 2, none of the four predictors is overwhelming to the others. The MO predictor and the Direct predictor respectively achieved the best results in three out of the nine tasks. The Iterative predictor seems to be more suitable for the BSH dataset. The DMO predictor cannot bring sufficient improvements in our landslide displacement tasks.

Table 2. Performances of different multi-step ESN predictors on practical landslide datasets

	2-Step			3-Step			4-Step		
	YHG	HTP	BSH	YHG	HTP	BSH	YHG	HTP	BSH
Iterative	11.11%	2.11%	4.07%	15.10%	3.56%	6.10%	18.25%	4.25%	8.73%
Direct	10.59%	2.11%	3.64%	15.30%	3.19%	7.46%	17.56%	3.86%	11.53%
MO	11.10%	2.08%	3.57%	14.30%	3.47%	8.84%	17.99%	4.22%	9.32%
DMO	-	-	-	-	-	-	17.55%	4.36%	11.47%

4   Conclusions

For time series prediction tasks such as landslide displacement prediction discussed in this paper, it's important to properly handle the dynamic nature of the underlying system. When static predictors are used, dynamic processes are simplified as static mappings. A more proper approach is to build predictors which are in essence dynamic, and that is what has been done in our study. Dynamic predictors are established based on ESN and extended into multi-step predictors. The effective of the proposed dynamic predictors are verified in experiments implemented on realistic landslide displacement datasets. Furthermore, four multi-step ESN predictors are also built. And it will be beneficial to select the proper predictor out of these four for a specific prediction tasks.

Acknowledgement. This work was supported by the Natural Science Foundation of China under Grant 61203286 and 61125303, National Basic Research Program of China (973 Program) under Grant 2011CB710606, the Program for Science and Technology in Wuhan of China under Grant 2014010101010004, the Program for

Changjiang Scholars and Innovative Research Team in University of China under Grant IRT1245, the Qatar National Research Fund under Grant #NPRP 4-1162-1-181.

References

1. Du, J., Yin, K., Lacasse, S.: Displacement prediction in colluvial landslides, Three Gorges Reservoir, China. Landslides 10, 1–16 (2012)
2. Bai, S.B., Wang, J., Lü, G.-N., Zhou, P.-G., Hou, S.-S., Xu, S.-N.: GIS-based logistic regression for landslide susceptibility mapping of the Zhongxian segment in the Three Gorges area, China. Geomorphology 115, 23–31 (2010)
3. Petley, D.N., Bulmer, M.H., Murphy, W.: Patterns of movement in rotational and translational landslides. Geology 30, 719–722 (2002)
4. Ran, Y., Xiong, G., Li, S., Ye, L.: Study on deformation prediction of landslide based on genetic algorithm and improved BP neural network. Kybernetes 39, 1245–1254 (2010)
5. Principe, J.C., Rathie, A., Kuo, J.M.: Prediction of chaotic time series with neural networks and the issue of dynamic modeling. International Journal of Bifurcation and Chaos 2, 989–996 (1992)
6. Príncipe, J.C., Kuo, J.M.: Dynamic modeling of chaotic time series with neural networks. In: Advances in Neural Information Processing Systems, vol. 7, pp. 311–318 (1995)
7. Jaeger, H., Haas, H.: Harnessing Nonlinearity: Predicting Chaotic Systems and Saving Energy in Wireless Communication. Science 304, 78–80 (2004)
8. Taieb, S.B., Bontempi, G., Atiya, A.F., Sorjamaa, A.: A review and comparison of strategies for multi-step ahead time series forecasting based on the NN5 forecasting competition. Expert Systems with Applications 39, 7067–7083 (2012)
9. Huang, G.B., Wang, D., Lan, Y.: Extreme learning machines: A survey. International Journal of Machine Learning & Cybernetics 2, 107–122 (2011)
10. Vapnik, V.N.: An overview of statistical learning theory. IEEE Transactions on Neural Networks 10, 988–999 (1999)
11. Chang, C.-C., Lin, C.-J.: LIBSVM: A library for support vector machines. ACM Transactions on Intelligent Systems and Technology 2, 1–27 (2011)
12. Yi, W., Wei, G.: Local prediction of the chaotic fh-code based on LS-SVM. Journal of Systems Engineering and Electronics 19, 65–70 (2008)
13. Ma, Y., Wang, X., An, L.: Fatigue life prediction of ductile iron based on DE-SVM algorithm. Physics Procedia 33, 1309–1315 (2012)
14. Huang, C.-L., Wang, C.-J.: A GA-based feature selection and parameters optimization for support vector machines. Expert Systems with Applications 31, 231–240 (2006)

Discrete-Time Nonlinear Generalized Policy Iteration for Optimal Control Using Neural Networks*

Qinglai Wei, Derong Liu, and Xiong Yang

The State Key Laboratory of Management and Control for Complex Systems
Institute of Automation, Chinese Academy of Sciences, Beijing 100190, China
{qinglai.wei,derong.liu,xiong.yang}@ia.ac.cn

Abstract. In this paper, a new generalized policy iteration (GPI) based adaptive dynamic programming (ADP) algorithm is developed to solve optimal control problems for infinite horizon discrete-time nonlinear systems. The GPI algorithm is a general idea of interacting policy and value iteration algorithms of ADP. There are two iteration indices, which iterate for policy improvement and policy evaluation, respectively, in the GPI algorithm. The convergence properties of the GPI algorithm are developed. Finally, simulation results are presented to illustrate the performance of the developed algorithm.

Keywords: Adaptive critic designs, adaptive dynamic programming, approximate dynamic programming, neuro-dynamic programming, nonlinear systems, optimal control, generalized policy iteration, reinforcement learning.

1 Introduction

Optimal control of nonlinear systems has been the focus of control fields for many decades [7, 11]. Dynamic programming has been a useful technique in handling optimal control problems for many years, while it is often computationally untenable to perform it to obtain the optimal solutions. Characterized by strong abilities of self-learning and adaptivity, adaptive dynamic programming (ADP), proposed by Werbos [17,18], has demonstrated the capability to find the optimal control policy and solve the Hamilton-Jacobi-Bellman (HJB) equation in a piratical way [5, 10, 12, 13]. There were several synonyms used for ADP, including "adaptive critic designs", "adaptive dynamic programming", "approximate dynamic programming" [1], "neuro-dynamic programming", "neural dynamic programming", and "reinforcement learning". In [8], ADP approaches were classified into several main schemes which include heuristic dynamic programming (HDP), action-dependent HDP (ADHDP), also known as Q-learning, dual heuristic dynamic programming (DHP), action-dependent DHP (ADDHP), globalized DHP (GDHP), and action-dependent GDHP (ADGDHP). Iterative methods are primary tools in ADP to obtain the solution of HJB equation indirectly and have received more and more attentions [6, 14–16]. In [3, 6], iterative ADP algorithms were

---

* This work was supported in part by the National Natural Science Foundation of China under Grants 61034002, 61233001, 61273140, 61304086, and 61374105, and in part by Beijing Natural Science Foundation under Grant 4132078.

C.K. Loo et al. (Eds.): ICONIP 2014, Part I, LNCS 8834, pp. 389–396, 2014.

classified into two main schemes which were based on policy iteration [4, 20] and value iteration [19], respectively.

In [9], a generalized policy iteration (GPI) algorithm, which is a general idea of interacting policy and value iteration algorithms, was constructed as a new iterative ADP algorithm to solve optimal control problems. There are two revolving iteration procedures for GPI algorithms, which are policy evaluation, making the performance index function in critic consistent with the current policy, and policy improvement, making the policy in actor greedy with respect to the current performance index function [9]. In this paper, for the first time the GPI algorithm is developed to solve optimal control problems for DT nonlinear systems, where the properties of convergence are analyzed. First, the detailed iteration procedure of the GPI algorithm for DT nonlinear systems is presented. Initialized by an arbitrary positive semi-definite function, two iteration procedures, which are policy evaluation and policy improvement, are developed. Second, the properties of the GPI algorithm are analyzed. The convergence criteria will be obtained. It will be shown that the iterative performance index function converges to the optimum if the convergence criteria are satisfied. Simulation results will illustrate the effectiveness of the developed algorithm.

2   Problem Statement

In this paper, we will study the following DT nonlinear system

$$x_{k+1} = F(x_k, u_k), \ k = 0, 1, 2, \ldots, \tag{1}$$

where $x_k \in \mathbb{R}^n$ is the state vector and $u_k \in \mathbb{R}^m$ is the control vector. Let x_0 be the initial state and $F(x_k, u_k)$ be the system function.

Let $\underline{u}_k = (u_k, u_{k+1}, \ldots)$ be a sequence of controls from k to ∞. The performance index function for state x_0 under the control sequence $\underline{u}_0 = (u_0, u_1, \ldots)$ is defined as

$$J(x_0, \underline{u}_0) = \sum_{k=0}^{\infty} U(x_k, u_k), \tag{2}$$

where $U(x_k, u_k) > 0$, for $\forall x_k \neq 0$, and $\forall u_k \neq 0$, is the utility function. The goal of this paper is to find an optimal control scheme which stabilizes system (1) and simultaneously minimizes the performance index function (2). Define the control sequence set as $\underline{\mathfrak{U}}_k = \{\underline{u}_k : \underline{u}_k = (u_k, u_{k+1}, \ldots), \forall u_{k+i} \in \mathbb{R}^m, i = 0, 1, \ldots\}$. Then, for a control sequence $\underline{u}_k \in \underline{\mathfrak{U}}_k$, the optimal performance index function can be defined as $J^*(x_k) = \inf_{\underline{u}_k} \{J(x_k, \underline{u}_k) : \underline{u}_k \in \underline{\mathfrak{U}}_k\}$. According to Bellman's principle of optimality, $J^*(x_k)$ satisfies the following HJB equation $J^*(x_k) = \inf_{u_k} \{U(x_k, u_k) + J^*(F(x_k, u_k))\}$. Generally speaking, $J^*(x_k)$ is difficult to obtain directly by HJB equation. To overcome this difficulty, a new iterative algorithm based on ADP is developed.

3   Generalized Policy Iteration Algorithm

In this section, the GPI algorithm is developed to obtain the optimal control law for DT nonlinear systems.

3.1 Derivation of the GPI Algorithm

The developed GPI algorithm contains two iteration procedures, which are i-iteration and j-iteration, respectively. Both of the iteration indices increase from 0. Let $\Psi(x_k)$ be a positive semi-definite function. Let

$$V_0(x_k) = \Psi(x_k) \tag{3}$$

be the initial iterative performance index function. Let $\{N_1, N_2, \ldots\}$ be a sequence, where $N_i \geq 0, i = 1, 2, \ldots$, is a nonnegative integer. For $i = 1, 2, \ldots$, the GPI algorithm can be expressed by the following two iteration procedures.

i-iteration:

$$v_i(x_k) = \arg\min_{u_k} \{U(x_k, u_k) + V_{i-1}(x_{k+1})\} \tag{4}$$

j-iteration:

$$V_{i,j_i+1}(x_k) = U(x_k, v_i(x_k)) + V_{i,j_i}(F(x_k, v_i(x_k))), \tag{5}$$

where the iteration index j_i increases from 0 to N_i and

$$V_{i,0}(x_k) = \min_{u_k} \{U(x_k, u_k) + V_{i-1}(x_{k+1})\}. \tag{6}$$

Define the iterative performance index function as $V_i(x_k) = V_{i,N_i}(x_k)$.

3.2 Properties of the GPI Algorithm

In this subsection, the properties of the GPI algorithm are analyzed. The convergence analysis will also be presented in this subsection.

Theorem 1. *For $i = 1, 2, \ldots$, let $V_{i,j_i}(x_k)$ and $v_i(x_k)$ be obtained by (3)–(6). Let $0 < \gamma_i < \infty$ and $1 \leq \delta_i < \infty$ be constants that satisfy*

$$\gamma_i U(x_k, u_k) \geq V_{i-1}(x_k), \text{ and } V_{i,0}(x_k) \leq \delta_i V_{i-1}(x_k), \tag{7}$$

respectively. Then, for $\forall i = 1, 2, \ldots$ and $j_i = 0, 1, \ldots, N_i$, we have

$$V_{i,j_i}(x_k) \leq \left(1 + \sum_{\rho=1}^{j_i} \frac{\gamma_i^\rho \delta_i^{\rho-1}(\delta_i - 1)}{(1 + \gamma_i)^\rho}\right) V_{i,0}(x_k), \tag{8}$$

where we define $\sum_{\alpha}^{\beta} (\cdot) = 0$ for $\alpha > \beta$.

Proof. The theorem can be proven by mathematical induction. Obviously, inequality (8) holds for $j_i = 0$. For $j_i = 1$, we have

$$
\begin{aligned}
V_{i,1}(x_k) &= U(x_k, v_i(x_k)) + V_{i,0}(x_{k+1}) \leq U(x_k, u_k) + \delta_i V_{i-1}(x_{k+1}) \\
&\leq \left(1 + \gamma_i \frac{\delta_i - 1}{1 + \gamma_i}\right) U(x_k, u_k) + \left(\delta_i - \frac{\delta_i - 1}{1 + \gamma_i}\right) V_{i-1}(x_{k+1}) \\
&\leq \left(1 + \gamma_i \frac{\delta_i - 1}{1 + \gamma_i}\right) V_{i,0}(x_k).
\end{aligned}
\tag{9}
$$

Thus, (8) holds for $j_i = 1$. Assume the conclusion holds for $j_i = l-1, l = 1, 2, \ldots, N_i$. Then for $j_i = l$, we have

$$
\begin{aligned}
V_{i,l}(x_k) &= U(x_k, v_i(x_k)) + V_{i,l-1}(x_{k+1}) \\
&\leq \left(1 + \frac{\gamma_i}{1+\gamma_i} \left(\delta_i - 1 + \sum_{\rho=1}^{l-1} \frac{\gamma_i^\rho \delta_i^\rho (\delta_i - 1)}{(1+\gamma_i)^\rho} \right) \right) U(x_k, v_i(x_k)) \\
&\quad + \left(\delta_i + \sum_{\rho=1}^{l-1} \frac{\gamma_i^\rho \delta_i^\rho (\delta_i - 1)}{(1+\gamma_i)^\rho} - \left(\frac{\delta_i - 1}{1+\gamma_i} + \sum_{\rho=1}^{l-1} \frac{\gamma_i^\rho \delta_i^\rho (\delta_i - 1)}{(1+\gamma_i)^{\rho+1}} \right) \right) V_{i-1}(x_{k+1}) \\
&= \left(1 + \sum_{\rho=1}^{l} \frac{\gamma_i^\rho \delta_i^{\rho-1} (\delta_i - 1)}{(1+\gamma_i)^\rho} \right) V_{i,0}(x_k).
\end{aligned}
\tag{10}
$$

Hence, (8) holds for $j = l$. The mathematical induction is completed.

Theorem 2. (Local convergence criterion) *Let* $0 < \gamma_i < \infty$ *and* $1 \leq \delta_i < \infty$ *be constants that satisfy* (7). *If for* $i = 1, 2, \ldots$ *and* $j_i = 0, 1, \ldots,$ *the constant* δ_i *satisfies* $\delta_i < \frac{1+\gamma_i}{\gamma_i}$, *then the iterative performance index function* $V_{i,j_i}(x_k)$ *is convergent as* $j_i \to \infty$, *i.e.,* $\lim_{j_i \to \infty} V_{i,j_i}(x_k) \leq \frac{1}{1 + \gamma_i - \gamma_i \delta_i} V_{i,0}(x_k)$.

Theorem 3. *For* $i = 0, 1, \ldots,$ *let* $V_{i,j_i}(x_k)$ *and* $v_i(x_k)$ *be obtained by* (3)–(6). *Let* $0 < \gamma < \infty$ *and* $1 \leq \delta < \infty$ *be constants that satisfy the following inequalities*

$$
J^*(F(x_k, u_k)) \leq \gamma U(x_k, u_k) \text{ and } V_0(x_k) \leq \delta J^*(x_k). \tag{11}
$$

For $\forall i = 0, 1, \ldots,$ *if* $\delta_i < \frac{1+\gamma_i}{\gamma_i}$, *then* $V_{i,j_i}(x_k)$ *satisfies*

$$
\begin{aligned}
V_{i,j_i}(x_k) &\leq \left(1 + \sum_{\rho=1}^{j_i} \frac{\gamma_i^\rho \delta_i^{\rho-1}(\delta_i - 1)}{(1+\gamma_i)^\rho} \right) \times \left(1 + \sum_{l=1}^{i-1} \frac{\gamma^{i-l} \gamma_l (\delta_l - 1)}{(1+\gamma)^{i-l} \prod\limits_{\eta=l}^{i-1} (1 - \gamma_\eta (\delta_\eta - 1))} \right. \\
&\quad \left. + \frac{\gamma^i (\delta - 1)}{(1+\gamma)^i \prod\limits_{\eta=1}^{i-1} (1 - \gamma_\eta (\delta_\eta - 1))} \right) J^*(x_k).
\end{aligned}
\tag{12}
$$

Proof. The statement can be proven by mathematical induction. First, the conclusion is obviously true for $i = 0$. Let $i = 1$. For $\forall j_1 = 0, 1, \ldots,$ we have

$$
\begin{aligned}
V_{1,j_1}(x_k) &\leq \left(1 + \sum_{\rho=1}^{j_1} \frac{\gamma_i^\rho \delta_i^{\rho-1}(\delta_i - 1)}{(1+\gamma_i)^\rho} \right) V_{1,0}(x_k) \\
&\leq \left(1 + \sum_{\rho=1}^{j_1} \frac{\gamma_i^\rho \delta_i^{\rho-1}(\delta_i - 1)}{(1+\gamma_i)^\rho} \right) \min_{u_k} \left\{ \left(1 + \gamma \frac{\delta - 1}{1+\gamma} \right) \right. \\
&\quad \times U(x_k, u_k) \\
&\quad \left. + \left(\delta - \frac{\delta - 1}{1+\gamma} \right) J^*(x_{k+1}) \right\}
\end{aligned}
$$

$$\leq \left(1 + \sum_{\rho=1}^{j_1} \frac{\gamma_i^\rho \delta_i^{\rho-1}(\delta_i - 1)}{(1 + \gamma_i)^\rho}\right)\left(1 + \frac{\gamma(\delta - 1)}{(1 + \gamma)}\right) J^*(x_k), \tag{13}$$

which proves inequality (12) for $i = 1$. Assume that the conclusion holds for $i = \vartheta - 1$, $\vartheta = 1, 2, \dots$ we can get

$$V_{\vartheta-1,j_{\vartheta-1}}(x_k) \leq \frac{1}{1 - \gamma_{\vartheta-1}(\delta_{\vartheta-1} - 1)} \left(1 + \sum_{l=1}^{\vartheta-2} \frac{\gamma^{i-\vartheta-1}\gamma_l(\delta_l - 1)}{(1 + \gamma)^{i-\vartheta-1}\prod\limits_{\eta=l}^{\vartheta-2}(1 - \gamma_\eta(\delta_\eta - 1))}\right.$$
$$\left. + \frac{\gamma^{\vartheta-1}(\delta - 1)}{(1 + \gamma)^{\vartheta-1}\prod\limits_{\eta=l}^{\vartheta-2}(1 - \gamma_\eta(\delta_\eta - 1))}\right) J^*(x_k). \tag{14}$$

For $i = \vartheta$, we can get inequality

$$V_{\vartheta,j_\vartheta}(x_k) \leq \left(1 + \sum_{\rho=1}^{j_\vartheta} \frac{\gamma_\vartheta^\rho \delta_\vartheta^{\rho-1}(\delta_\vartheta - 1)}{(1 + \gamma_\vartheta)^\rho}\right) V_{\vartheta,0}(x_{k+1})$$

$$\leq \left(1 + \sum_{\rho=1}^{j_\vartheta} \frac{\gamma_\vartheta^\rho \delta_\vartheta^{\rho-1}(\delta_\vartheta - 1)}{(1 + \gamma_\vartheta)^\rho}\right) \min_{u_k}\left\{U(x_k, u_k) + \frac{1}{1 - \gamma_{\vartheta-1}(\delta_{\vartheta-1} - 1)}\right.$$

$$\times \left(1 + \sum_{l=1}^{\vartheta-2} \frac{\gamma^{\vartheta-l-1}\gamma_l(\delta_l - 1)}{(1 + \gamma)^{\vartheta-l-1}\prod\limits_{\eta=l}^{\vartheta-2}(1 - \gamma_\eta(\delta_\eta - 1))}\right.$$

$$\left.\left. + \frac{\gamma^{\vartheta-1}(\delta - 1)}{(1 + \gamma)^{\vartheta-1}\prod\limits_{\eta=l}^{\vartheta-2}(1 - \gamma_\eta(\delta_\eta - 1))}\right) J^*(x_{k+1})\right\}, \tag{15}$$

which obtains (12).

Theorem 4. (Global convergence criterion) *If for* $\forall i = 0, 1, \dots$, δ_i *satisfies*

$$\delta_i < q_i \frac{1}{\gamma_i(1 + \gamma)} + 1, \tag{16}$$

where $0 < q_i < 1$ *is a constant, then for* $\forall j_i = 0, 1, \dots, N_i$, *the iterative performance index function* $V_{i,j_i}(x_k)$ *is convergent as* $i \to \infty$.

Proof. Define an iteration index set Ω_{IN} as $\Omega_{IN} = \{i \mid i = 1, 2, \dots\}$. For $\forall i \in \Omega_{IN}$, if we let $q = \max\limits_{i \in \Omega_{IN}} \{q_i\}$, then we have $0 < q < 1$. According to (12), we can get that

$$\sum_{l=1}^{i-1} \frac{\gamma^{i-l}\gamma_l(\delta_l - 1)}{(1 + \gamma)^{i-l}\prod\limits_{\eta=l}^{i-1}(1 - \gamma_\eta(\delta_\eta - 1))} < \lim_{i \to \infty} \sum_{l=1}^{i-1} \frac{\gamma^{i-l}\frac{q}{1+\gamma}}{(1 + \gamma)^{i-l}\left(1 - \frac{q}{1+\gamma}\right)^{i-l}}$$

$$= \frac{q}{1 - q}\frac{\gamma}{1 + \gamma}. \tag{17}$$

If δ_i satisfies (16), we can get $\lim\limits_{i \to \infty} \dfrac{\gamma^i (\delta-1)}{(1+\gamma)^i \prod\limits_{\eta=l}^{i-1}(1-\gamma_\eta(\delta_\eta-1))} = 0$, and $q_i \dfrac{1}{\gamma_i(1+\gamma)} + 1 < $

$\dfrac{1+\gamma_i}{\gamma_i}$. Hence, we have

$$\sum_{\rho=1}^{j_i} \frac{\gamma_i^\rho \delta_i^{\rho-1}(\delta_i - 1)}{(1+\gamma_i)^\rho} < \lim_{j_i \to \infty} \sum_{\rho=1}^{j_i} \frac{\gamma_i^\rho \delta_i^{\rho-1}(\delta_i - 1)}{(1+\gamma_i)^\rho} = \frac{q}{\gamma+1-q}. \tag{18}$$

According to (17)–(18), we can obtain that $\lim\limits_{i \to \infty} V_{i,j_i}(x_k) \leq \left(1 + \dfrac{q}{\gamma+1-q}\right)\left(1 + \right.$

$\left. \dfrac{q}{1-q} \dfrac{\gamma}{1+\gamma}\right)$. The the iterative performance index function is convergent to optimum, as $i \to \infty$. The proof is completed.

4   Simulation Example

We now examine the performance of the developed algorithm in an inverted pendulum system [2]. The dynamics of the pendulum is expressed as

$$\begin{bmatrix} x_{1(k+1)} \\ x_{2(k+1)} \end{bmatrix} = \begin{bmatrix} x_{1k} + \Delta t x_{2k} \\ \frac{g}{\ell}\Delta t \sin(x_{1k}) + (1 - \kappa\ell\Delta t)x_{2k} \end{bmatrix} + \begin{bmatrix} 0 \\ \frac{\Delta t}{m\ell^2} \end{bmatrix} u_k. \tag{19}$$

where $m = 1/2\,\mathrm{kg}$ and $\ell = 1/3\,\mathrm{m}$ are the mass and length of the pendulum bar, re-

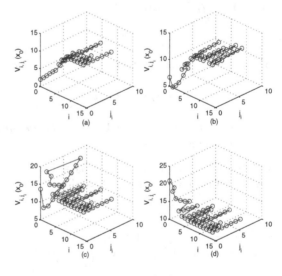

Fig. 1. Iterative performance index function $V_{i,j_i}(x_k)$ for $x_k = x_0$ and different $\bar{\Psi}(x_k)$'s. (a) $\bar{\Psi}^1(x_k)$. (b) $\bar{\Psi}^2(x_k)$. (c) $\bar{\Psi}^3(x_k)$. (d) $\bar{\Psi}^4(x_k)$.

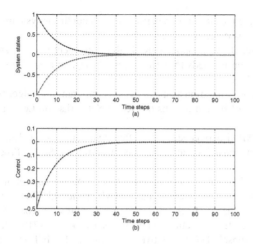

Fig. 2. Optimal trajectories. (a) Optimal states. (b) Optimal control.

spectively. Let $\kappa = 0.2$ and $g = 9.8\,\mathrm{m/s^2}$ be the frictional factor and the gravitational acceleration, respectively. Let the initial state be $x_0 = [1, -1]^T$. Let the structures of the critic and action networks be 2–12–1 and 2–12–1. We choose $p = 10000$ states to train the NNs. To illustrate the effectiveness of the algorithm, we also choose four different initial performance index functions which are expressed by $\bar{\Psi}^\varsigma(x_k) = x_k^T \bar{P}_\varsigma x_k$, $\varsigma = 1, \ldots, 4$. Let $\bar{P}_1 = 0$. Let $\bar{P}_2$–$\bar{P}_4$ be initialized by positive definite matrices given by $\bar{P}_2 = [2.98, \ 1.05; 1.05, \ 5.78]$, $\bar{P}_3 = [6.47, \ -0.33; -0.33, \ 6.55]$, and $\bar{P}_4 = [22.33, \ 4.26; 4.26, \ 7.18]$, respectively. For $\forall i = 0, 1, \ldots$, let $q_i = 0.9999$. Let the iteration sequence be $\{N_i^\varsigma\}$, where $N_i^\varsigma \in [0, 10]$ be a random nonnegative integer. Then, initialized by $\bar{\Psi}^\varsigma(x_k)$, $\varsigma = 1, \ldots, 4$, the GPI algorithm is implemented for $i = 15$ iterations. Train the the critic and the action networks under the learning rate 0.01 and set the NN training errors as 10^{-6}. The curves of the iterative performance index functions $V_i(x_k)$ are shown in Fig. 1. Let the execution time $T_f = 100$ time steps. The optimal trajectories of optimal control laws and system states are shown in Fig. 2.

5   Conclusions

In this paper, an effective GPI algorithm is developed to solve infinite horizon optimal control problems for DT nonlinear systems. The iterative ADP algorithm can be implemented by an arbitrary positive semi-definite function. It has been proven that the performance index function for the GPI algorithm converges to the optimum. Convergence criteria of the algorithm are obtained. Finally, simulation results are given to illustrate the effectiveness of the developed algorithm.

References

1. Al-Tamimi, A., Abu-Khalaf, M., Lewis, F.L.: Adaptive critic designs for discrete-time zero-sum games with application to H_∞ control. IEEE Transactions on Systems, Man, and Cybernetics-Part B: Cybernetics 37, 240–247 (2007)

2. Beard, R.: Improving the closed-loop performance of nonlinear systems, Ph.D. Thesis, Rensselaer Polytechnic Institute, Troy, NY (1995)
3. Lewis, F.L., Vrabie, D., Vamvoudakis, K.G.: Reinforcement learning and feedback control: using natural decision methods to design optimal adaptive controllers. IEEE Control Systems 32, 76–105 (2012)
4. Liu, D., Wei, Q.: Policy iteration adaptive dynamic programming algorithm for discrete-time nonlinear systems. IEEE Transactions on Neural Networks and Learning Systems 25, 621–634 (2014)
5. Liu, D., Wei, Q.: Multi-person zero-sum differential games for a class of uncertain nonlinear systems. International Journal of Adaptive Control and Signal Processing 28, 205–231 (2014)
6. Liu, D., Wei, Q.: Finite-approximation-error based optimal control approach for discrete-time nonlinear systems. IEEE Transactions on Cybernetics 43, 779–789 (2013)
7. Liu, D., Zhang, Y., Zhang, H.: A self-learning call admission control scheme for CDMA cellular networks. IEEE Transactions on Neural Networks 16, 1219–1228 (2005)
8. Prokhorov, D.V., Wunsch, D.C.: Adaptive critic designs. IEEE Transactions on Neural Networks 8, 997–1007 (1997)
9. Sutton, R.S., Barto, A.G.: Reinforcement Learning: An Introduction. MIT Press, Cambridge (1998)
10. Wang, F., Jin, N., Liu, D., Wei, Q.: Adaptive dynamic programming for finite-horizon optimal control of discrete-time nonlinear systems with ϵ-error bound. IEEE Transactions on Neural Networks 22, 24–36 (2011)
11. Wei, Q., Liu, D.: Stable iterative adaptive dynamic programming algorithm with approximation errors for discrete-time nonlinear systems. Neural Computing & Applications 24, 1355–1367 (2014)
12. Wei, Q., Wang, D., Zhang, D.: Dual iterative adaptive dynamic programming for a class of discrete-time nonlinear systems with time-delays. Neural Computing & Applications 23, 1851–1863 (2013)
13. Wei, Q., Liu, D.: Numerically adaptive learning control scheme for discrete-time nonlinear systems. IET Control Theory & Applications 7, 1472–1486 (2013)
14. Wei, Q., Liu, D.: An iterative ϵ-optimal control scheme for a class of discrete-time nonlinear systems with unfixed initial state. Neural Networks 32, 236–244 (2012)
15. Wei, Q., Zhang, H., Dai, J.: Model-free multiobjective approximate dynamic programming for discrete-time nonlinear systems with general performance index functions. Neurocomputing 72, 1839–1848 (2009)
16. Wei, Q., Zhang, H., Liu, D., Zhao, Y.: An optimal control scheme for a class of discrete-time nonlinear systems with time delays using adaptive critic programming. ACTA Automatica Sinica 36, 121–129 (2010)
17. Werbos, P.J.: Advanced forecasting methods for global crisis warning and models of intelligence. General Systems Yearbook 22, 25–38 (1977)
18. Werbos, P.J.: A menu of designs for reinforcement learning over time. In: Miller, W.T., Sutton, R.S., Werbos, P.J. (eds.) Neural Networks for Control, pp. 67–95. MIT Press, Cambridge (1991)
19. Zhang, H., Wei, Q., Luo, Y.: A novel infinite-time optimal tracking control scheme for a class of discrete-time nonlinear systems via the greedy HDP iteration algorithm. IEEE Transactions on System, Man, and cybernetics-Part B: Cybernetics 38, 937–942 (2008)
20. Zhang, H., Wei, Q., Liu, D.: An iterative adaptive dynamic programming method for solving a class of nonlinear zero-sum differential games. Automatica 47, 207–214 (2011)

ANFIS-Based Model for Improved Paraphrase Rating Prediction

El-Sayed M. El-Alfy*

College of Computer Sciences and Engineering
King Fahd University of Petroleum and Minerals, Dhahran 31261, Saudi Arabia
alfy@kfupm.edu.sa

Abstract. Paraphrase rating is an important problem with very interesting applications in plagiarism detection, language translation, text summarization, question answering, web search and information retrieval. In this paper, we present an adaptive neuro-fuzzy inference system (ANFIS) based model for automatic rating of semantic equivalence of pairs of sentences. Using a corpus of human-judged sentence pairs, lexical similarity metrics are first computed. Then, a model is constructed for predicting the mean of the rates assigned by a number of human beings. The correlation with the actual ratings and the prediction errors are studied for individual metrics as well as the model output using a non-linear logistic regression function. The experimental results showed that much higher correlations and low error rates can be achieved with the proposed method compared to those obtained with individual metrics.

Keywords: Neural networks, Fuzzy inference, Adaptive neuro-fuzzy inference system, Lexical similarity scores, Prediction, Paraphrase rating.

1   Introduction

There are many potential forms for expressing the essential meaning of a text or passage using different words and/or sentence structures. This process is known as paraphrasing, bidirectional text entailment or sematic equivalence [3]; which can occur in a variety of ways. Paraphrase rating depends on the level of inference that can be made about one sentence from another. For instance, on a scale of 5, a pair of identical sentences will be rated 5 (since one is completely inferred from the other or completely equivalent to the other) whereas a pair of totally different sentences will be rated 0 (since no inference can be made).

The difficulty of this problem can be attributed to the non-existence of a very precise definition of paraphrasing and the subjectivity of human judgements. Nonetheless, rating automation has several interesting applications in processing large volumes of available texts and documents, e.g. document summarization, plagiarism detection, language translation, question answering, web search and information retrieval. In the past decade, there has been increasing interest on

---

* On leave from the College of Engineering, Tanta University, Egypt.

C.K. Loo et al. (Eds.): ICONIP 2014, Part I, LNCS 8834, pp. 397–404, 2014.

problems related to paraphrasing and text entailment [2]. For text similarity detection, a number of approaches have been suggested in the literature using lexical similarity metrics. Among these approaches is the work described in [6] using a support vector machine and a set of machine translation related metrics. In [11], string similarity measures were combined to recognize paraphrases among sentences. Another approach is presented in [7] to detect similarity between pairs of short textual units by combining primitive and composite linguistic features with a rule-based machine learning algorithm.

Although various similarity metrics have been proposed in the literature, each has its own drawbacks and, as we will show, does not correlate well with the human judgements. The main goal of this paper is to propose and evaluate a novel composite metric for paraphrase rating of pairs of sentences. The idea of the proposed approach is based on combining five lexical similarity metrics using an adaptive neuro-fuzzy inference system. The output of the inference system is found to have much better correlations with the rates assigned by human judges.

The rest of the paper is organized as follows. Section 2 briefly describes the similarity metrics considered in this study. Section 3 describes the nonlinear logistic function used for studying the dependance relationship between each metric and the human rates. Section 4 presents the proposed prediction model. Empirical evaluation and results are discussed in Section 5. Finally, Section 6 concludes the paper.

2   Textual Similarity Metrics

2.1   Word Edit Distance

This is a word-level modified version of the well-know Levenshtein edit distance between two character strings [13,5]. It is computed from the number of deletion, insertion, and substitution operations of words that are needed to convert one sentence into the other. The computation of edit distance is performed using a dynamic programming algorithm which has a complexity of $O(|s_1| \times |s_2|)$ where $|s_i|$ is the length of sentence i [12]. Mathematically, $ed(i, 0) = i$, $ed(0, j) = j$, and recursively $\forall i = 1 : |s_1|, \forall j = 1 : |s_2|$, $ed(i, j) = min\{ed(i - 1, j - 1) + (s_1[i] = s_2[j]?0 : 1), ed(i - 1, j) + 1, ed(i, j - 1) + 1\}$, where the expression $(s_1[i] = s_2[j]?0 : 1)$ is equal to 0 if $s_1[i] = s_2[j]$; otherwise it is 1. When there are a few changes, i.e. very similar pair of sentences, the value of this metric will be small. As more changes are introduced, the edit distance will likely increase. The effectiveness of this metric depends on the level of paraphrasing but it fails to detect paraphrases with too many lexical changes.

2.2   Simple Word N-Gram Overlap

This metric uses the word n-gram overlap for $n = 1, 2, \ldots, N$ and computes the average ratio as follows,

$$sim_o(S_1, S_2; N) = \frac{1}{N} \times \sum_{n=1}^{N} \frac{S_1^n \cap S_2^n}{min(S_1^n, S_2^n)} \tag{1}$$

where $S_1^n \cap S_2^n$ is the number of common n-grams in sentences S_1 and S_2, and $min(S_1^n, S_2^n)$ is the number of n-grams in the shorter sentence.

2.3 Exclusive LCP N-Gram Overlap

This metric is similar to sim_o but it counts the prefix overlapping of uni-gram, bi-gram, ... N−gram considering the Longest Common Prefix (LCP) [4],

$$sim_{exc}(S_1, S_2; N) = \max_{n \in \{1,2,...,N\}} \frac{S_1^{n,exc} \cap S_2^{n,exc}}{min(S_1^{n,exc}, S_2^{n,exc})} \tag{2}$$

2.4 BLEU Metric

BLEU is an abbreviation of BiLingual Evaluation Understudy and has been one of the first devised methods for automatic quality evaluation of machine translation of a natural language in terms of the semantic equivalence to reference translations made by professional humans [14]. It is widely used for evaluating machine translation and benchmarking other related metrics. Later, it has been used in some studies for paraphrase generation and identification as well, e.g. [6], [11], [10]. The computation of BLEU is based on counting matches of word n-grams for the input sentences (for $n \leq N$ where N is frequently preset to 4).

2.5 Sumo Metric

This metric is proposed in [4] to alleviate some of the limitations faced by other metrics. It is based on exclusive lexical links between the two sentences. It penalizes equal and almost equal sentences and considers pairs with different syntactic structure and high degree of lexical reordering.

3 Logistic Regression

After extracting the various similarity metrics, we study their abilities in predicting the paraphrase rate. In order to achieve this, we fit a nonlinear logistic regression prediction model between the human assigned paraphrase score and each similarity metric. This nonlinear logistic regression model is described by five parameters which must be estimated from the historical data. Mathematically, the predicted score, q, is expressed as a function of a given metric, u, by the following equation: $q(u) = \beta_1(1/2 - (1 + e^{\beta_2(u-\beta_3)})^{-1}) + \beta_4 u + \beta_5$, where β_i are the parameters to be estimated to minimize the RMSE.

4 ANFIS-Based Rating Model

Fuzzy logic based systems allow mapping of prior human knowledge or experience into the inference process using linguistic variables which is an advantage but a cumbersome task. The adaptive neuro-fuzzy inference system augments a fuzzy inference system with the capability of neural networks to learn from the data [8]. Thus, ANFIS combines the advantages of both fuzzy inference and neural network and has been successively applied to a wide spectrum of domains [9].

The construction of the model starts with a training dataset and a defined structure such as the one shown in Fig. 1. In our case, there are five inputs representing the lexical similarity metrics and one output representing the expected rate. The fuzzy inference maps inputs to outputs using a set of $if - then$ rules with sets of linguistic variables to handle inherent uncertainty in the problem. The input variables are denoted a_1, a_2, a_3, a_4, and a_5 representing word edit distance (ed), simple word N-gram overlap (sim_o), exclusive LCP N-gram overlap (sim_{exc}), BLEU, and Sumo metrics, respectively. The output, z, is a score in the range of 0 (complete inequivalence) to 5 (complete equivalence).

Each input variable is normalized to fall in the range from 0 to 1, Then, using three fuzzy sets (terms) initially defined for each input variable (denoted as small, medium, and large) and generalized bell-shaped membership functions, the first layer (fuzification) converts crisp values into membership values. In the second layer, the firing strength for each rule is computed as the multiplication of its inputs which are the membership values from the previous layer; these are denoted as w_m and computed from,

$$w_m = \prod_{k=1}^{5} \mu_{A_{km_k}}(a_k) \tag{3}$$

where A_{km_k} are the terms associated with rule m. In the third layer, weights are normalized using,

$$\bar{w}_m = \frac{w_m}{\sum_{k=1}^{R} w_k} \tag{4}$$

where R is the number of rules. For the first-order Sugeno-type ANFIS system, the consequent part of each rule is expressed as a linear combination of the inputs. The fourth layer has square-shaped nodes with node functions given as,

$$f_m = \sum_{k=1}^{5} c_k^{(m)}.a_k + c_0^{(m)} \tag{5}$$

where $c_k^{(m)}$ for $k = 0$ to 5 are the coefficients of the first-order equation of rule m. Finally, the last layer node conducts summation of all incoming signals to generate the output as weighted sum,

$$\tilde{z} = \sum_{k=1}^{R} \bar{w}_k f_k \tag{6}$$

The objective of the training algorithm is to update the consequent and premise parameters in order to achieve the least error between the predicted and the desired target outputs. A hybrid training algorithm composed of a least square method and a gradient descend back-propagation algorithm is applied to tune the parameters of the ANFIS model.

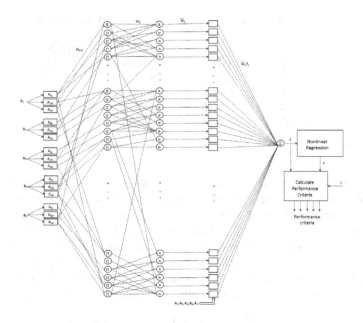

Fig. 1. ANFIS-based rating model architecture

5 Experimental Results

5.1 Dataset

The dataset used in this paper is the one published as part of SemEval-2012 workshop under Task 6 for Semantic Textual Similarity (STS) [1]. This dataset is drawn from the Microsoft Research Paraphrase Corpus. It consists of 750 sentence pairs rated on a continuous scale from 0 to 5 (known as golden standard rating). This rating represents the degree of graded bidirectional semantic equivalence for each sentence pair (with 0 means different sentences and 5 means fully semantically equivalent sentences). To reduce subjectivity among human judgments, the golden standard rating is an average of the scores assigned by five human judges.

5.2 Performance Measures

For the purpose of evaluating the performance of the proposed model, we used the following performance measures: Pearson's linear correlation coefficient (ρ), Spearman's rank correlation coefficient (ρ_s), Kendall's τ-rank correlation coefficient (τ), Root Mean Square Error ($RMSE$) and Mean Absolute Error (MAE). These performance criteria are common in prior studies, e.g. [15]. The Pearson's linear correlation coefficient, ρ, is given by:

$$\rho = \frac{\sum_i (z_i - \bar{z})(q_i - \bar{q})}{\sqrt{\sum_i (z_i - \bar{z})^2 \sum_i (q_i - \bar{q})^2}} \tag{7}$$

Table 1. Performance comparison (approx. to 4 decimal places)

	ρ	ρ_s	τ	$RMSE$	MAE
ed	0.1537	0.0947	0.0659	0.1932	0.1508
sim_o	0.3695	0.3156	0.2194	0.1817	0.1427
sim_{ex}	0.5021	0.4311	0.3043	0.1691	0.1351
$bleu$	0.2843	0.2448	0.1771	0.1875	0.1467
$sumo$	0.4505	0.4134	0.2895	0.1746	0.1368
Proposed	0.9966	0.9933	0.9588	0.0159	0.0068

where z_i and q_i are the actual and predicted scores, and $\bar{z}$ and $\bar{q}$ are the mean values of actual and predicted scores. To assess the monotonicity relationship between predicted value and actual value for a particular model, we used Spearman's rank order coefficient ρ_s. When $\rho_s = 0$, there is no tendency that q increases or decreases with the increase of z. In contrast, $\rho_s = \pm 1$, when q and z are perfectly monotonically related. This measure is computed using the same equation for Pearson's coefficient but replacing the raw scores by their ranks zr_i and qr_i. The computation ρ_s and Kendall's correlation, τ are as follow:

$$\rho_s = 1 - \frac{6\sum_i (zr_i - qr_i)}{N(N^2 - 1)}, \text{ and } \tau = \frac{2(n_c - n_d)}{N(N - 1)} \tag{8}$$

where n_c and n_d are the number of concordant and the number of discordant ranking pairs. With a value of $\tau = 0$ when the two variables are independent.

We also used two error measures: mean absolute error (MAE) and root mean square error (RMSE) which are calculated as follows:

$$MAE = \frac{1}{N}\sum_{i=1}^{N} |z_i - \hat{z}_i|, RMSE = \sqrt{\frac{1}{N}\sum_{i=1}^{N} (z_i - \hat{z}_i)^2} \tag{9}$$

where N is the number of sentence pairs.

5.3   Results

Using MATLAB Release 2014a, we developed the described model. The dataset is first processed to compute the lexical similarity metrics. The metrics and the rate values are normalized to be in the range from 0 to 1. We studied the dependance between the paraphrase rating and each of the lexical metrics. As depicted in Table 1 and Figure 2, there is a little correlation between each lexical metric and the paraphrase rate (GS, i.e. Gold Standard). Hence, the prediction errors, when using nonlinear logistic regression, are relatively high.

Applying the proposed model to combine all lexical metrics to predict the paraphrase rate has demonstrated a significant improvement as shown in the last row of Table 1. The correlation coefficients are close to 1 and the errors are much lower than those for individual lexical metrics. This result is also supported by the very clear linear relationship between the predicted and the actual rates which is shown in the last plot in Figure 2. The membership function for the input variables after training the model are as shown in Figure 3 with three linguistic terms for each input variable, a.k.a. lexical metric, and generalized bell-shaped membership functions. The parameters of the membership functions as well as the other Sugeno-type ANFIS model parameters are adjusted by a hybrid learning algorithm, as explained previously, to better fit the dataset.

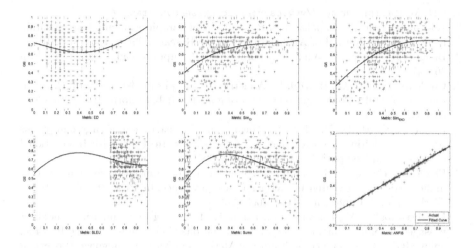

Fig. 2. Scatter diagrams and regression curves comparisons

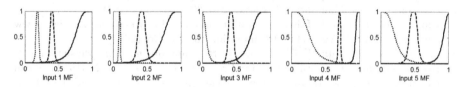

Fig. 3. Membership functions of input variables after training

6   Conclusion

In this paper, we proposed an automatic paraphrase rating model based on the Sugeno-type ANFIS system with hybrid learning to adjust the model parameters. We showed that the traditional lexical metrics have poor correlation with the human judgments when used individually with nonlinear logistic regression. However, with the proposed model for combining them much better correlation and low errors have been achieved. The experimental results confirms the validity of this conclusion using a benchmark dataset and five performance measures.

Acknowledgments. The author would like to acknowledge the support provided by the Deanship of Scientific Research (DSR) at King Fahd University of Petroleum & Minerals (KFUPM) for funding this work under Project No. RG1113-1&2.

References

1. STS MSR corpus, http://www.cs.york.ac.uk/semeval/task6/
2. Androutsopoulos, I., Malakasiotis, P.: A survey of paraphrasing and textual entailment methods. Journal of Artificial Intelligence Research 38(1), 135–187 (2010)
3. Bhagat, R., Hovy, E.: What is a paraphrase? Computational Linguistics 39(3), 462–472 (2013)
4. Cordeiro, J., Dias, G., Brazdil, P.: A metric for paraphrase detection. In: Proc. International Multi-Conference on Computing in the Global Information Technology (ICCGI) (2007)
5. Dolan, B., Quirk, C., Brockett, C.: Unsupervised construction of large paraphrase corpora: Exploiting massively parallel news sources. In: Proc. 20th International Conference on Computational Linguistics (2004)
6. Finch, A., Hwang, Y.S., Sumita, E.: Using machine translation evaluation techniques to determine sentence-level semantic equivalence. In: Proc. 3rd International Workshop on Paraphrasing, IWP 2005 (2005)
7. Hatzivassiloglou, V., Klavans, J.L., Eskin, E.: Detecting text similarity over short passages: Exploring linguistic feature combinations via machine learning. In: Proc. Joint SIGDAT Conference on Empirical Methods in Natural Language Processing and Very Large Corpora (1999)
8. Jang, J.S.: Anfis: Adaptive-network-based fuzzy inference system. IEEE Transactions on Systems, Man and Cybernetics 23(5/6), 665–685 (1993)
9. Kar, S., Das, S., Ghosh, P.K.: Applications of neuro fuzzy systems: A brief review and future outline. Applied Soft Computing 15, 243–259 (2014)
10. Madnani, N., Tetreault, J., Chodorow, M.: Re-examining machine translation metrics for paraphrase identification. In: Proc. Conference of the North American Chapter of the Association for Computational Linguistics: Human Language Technologies (2012)
11. Malakasiotis, P.: Paraphrase recognition using machine learning to combine similarity measures. In: Proc. of ACL-IJCNLP Student Research Workshop (2009)
12. Manning, C.D., Raghavan, P., Schütze, H.: Introduction to Information Retrieval. Cambridge University Press (2009)
13. Navarro, G.: A guided tour to approximate string matching. ACM Computing Surveys 33(1), 31–88 (2001)
14. Papineni, K., Roukos, S., Ward, T., Zhu, W.J.: BLEU: A method for automatic evaluation of machine translation. In: Proc. 40th Annual Meeting on Association for Computational Linguistics (2002)
15. Pelan, A., Steinhaeuser, K., Chawla, N.V., de Alwis Pitts, D.A., Ganguly, A.R.: Empirical comparison of correlation measures and pruning levels in complex networks representing the global climate system. In: Proc. IEEE Symposium Series on Computational Intelligence and Data Mining (CIDM) (2011)

Contextual Bandit for Active Learning:
Active Thompson Sampling

Djallel Bouneffouf, Romain Laroche, Tanguy Urvoy,
Raphael Feraud, and Robin Allesiardo

Orange Labs, 2, Avenue Pierre Marzin, 22307 Lannion, France
{djallel.bouneffouf,romain.laroche,tanguy.urvoy,
raphael.feraud,robin.allesiardo}@Orange.com

Abstract. The labelling of training examples is a costly task in a supervised classification. Active learning strategies answer this problem by selecting the most useful unlabelled examples to train a predictive model. The choice of examples to label can be seen as a dilemma between the exploration and the exploitation over the data space representation. In this paper, a novel active learning strategy manages this compromise by modelling the active learning problem as a contextual bandit problem. We propose a sequential algorithm named Active Thompson Sampling (ATS), which, in each round, assigns a sampling distribution on the pool, samples one point from this distribution, and queries the oracle for this sample point label. Experimental comparison to previously proposed active learning algorithms show superior performance on a real application dataset.

Keywords: Contextual Bandits, Active learning, Thompson sampling.

1 Introduction

Active Learning (AL) has emerged as a popular approach for solving machine learning problems with limited labelled data [8]. In this approach the learning algorithm is "active", and is allowed to query, an oracle O, for the label of points that are maximally informative for the learning process. The result is that, by using few but well chosen labels, the active learning algorithm is able to learn as well as a passive learning algorithm that has access to more labelled data.

In selective sampling, the choice of examples to be labelled can be seen as the dilemma between exploration and exploitation (exr/exp) on the training data. On one hand, an active learning strategy that just exploits the data will be specialized in certain areas of the input space X but will be very poor in generalization. On the other hand, a strategy which uses that exploring data does not focus on regions where X is known to improve the predictive model. These two situations illustrate the need for an active learning to find a compromise between exr/exp of labelling data strategy.

In [5], a similar analysis of the problem led the authors to model the active learning problem as a multi-armed bandit problem. They suppose that the different hypotheses of distribution $h \in H$ of the data are the arms and use an

C.K. Loo et al. (Eds.): ICONIP 2014, Part I, LNCS 8834, pp. 405–412, 2014.

adapted UCB (upper confident bound) to select the most promising hypothesis of distribution to select the points to be labelled. The drawback of this approach is in the non consideration of the context or the different features that characterize the points. For example the number of points in a area space, the proportion of labelled points, the ratio of the classes in the area or the density of points in the area can be useful to determine the most interesting to label.

To tackle this problem, we propose to model the active learning as a contextual bandit problem, where we have at first clustered the input space: each cluster is considered as an arm and the different features of the cluster are the context of the arm. Then, we implement a novel algorithm named Active Thompson Sampling (ATS) adapting the Thompson Sampling to the active learning problem. Finally, we evaluate ATS on actual data and find out that ATS outperforms all other algorithms in our panel.

The remaining of the paper is organized as follows. Section 2 reviews related works. Section 3 describes our multi armed contextual bandit model and the ATS algorithm. The experimental evaluation is illustrated in Section 4. The last section concludes the paper and points out possible directions for future works.

2   Related Work

We refer, in the following, recent works that address Active Learning problem and the exr/exp trade-off (bandit algorithm).

Active Learning. A variety of AL algorithms have been proposed in the literature employing various query strategies. One of the most popular strategy is called uncertainty sampling (US), where the active learner queries the point whose label is the most uncertain [6]. Usually the uncertainty in the label is calculated using variance of the label distribution [8]. The authors in [9] introduced the query-by-committee (QBC) strategy where a committee of potential heterogeneous models, is learnt from the labelled data, and used to select for querying, the point where most committee members disagree. Other strategies include the maximum expected reduction in error [11] or variance reducing query strategies [10] to querying the optimal point. All above proposed approaches only exploit the data.

Contextual Multi-Armed Bandit. Multi-armed bandit (MAB) problems model the exr/exp trade-off inherent in many sequential decision problems. A particularly useful version is the contextual multi-armed bandit problem. In this problem, in each iteration, an agent has to choose between arms. Before making the choice, the agent sees a d-dimensional feature vector (context vector), associated with each arm. The learner uses these context vectors along with the rewards of the arms played in the past to make the choice of the arm to play in the current iteration. Overtime, the learner's aim is to collect enough information about how the context vectors and rewards relate to each other, so that it can predict the next best arm to play by looking at the feature vectors.

Recently, the contextual bandit has been used in different domains such as recommender system (RS) and information retrieval. For example, in [3, 2], authors model RS as a contextual bandit problem. The authors propose an algorithm called Contextual-ϵ-greedy which sequentially recommends documents based on contextual information about the users. In [4], authors analyse the Thompson Sampling (TS) in contextual bandit problem. The study demonstrates that it has a better empirical performance compared to the state-of-art methods. The TS is one of the oldest heuristics for multi-armed bandit problems and it is a randomized algorithm based on Bayesian ideas. Authors in [4, 3] describe a smart way to balance exr/exp, but do not study the contextual bandit in the active learning problem.

Multi-Armed Bandit for Active Learning. To our knowledge there has been only two papers bridging the world of active learning and MAB. [7] adapted the EXP4 algorithm which is a MAB algorithm with expert advice, where the different active learning algorithms are the various experts and the different points in the pool are the arms of the MAB. An each iteration, every expert provides a sampling distribution on the pool. EXP4 maintains an estimation of the error rate for each expert, and uses exponential weight to select the optimal sampling distribution on the pool. Authors in [5] propose an adaptation of UCB called LCB algorithm, the authors suggested minimizing an unbiased estimator of risk of h, and a sampling distribution that was in proportion to the entropy of the prediction on the pool. The authors consider the arms of the bandit as the different hypothesis, and querying a data point, as the process of improving their estimation of the risk of the different hypothesis.

Our Contributions. As it is observed above, none of the described works has dressed the active learning problem from a contextual bandits view, although the consideration of the pool context might be a very informative feature for an active learning algorithm. This is precisely what we intend to do by exploiting the following new features: (1) We model the active learning as a contextual bandit problem, where each cluster of points in the space is an arm and the different features of the cluster are the context of the arms. (2) We propose a new algorithm named Active Thompson Sampling (ATS), that adapts the TS to the active learning problem. (3) We evaluate it against other methods form the state-of-the-art.

3   Key Notions and Proposed Model

This section focuses on the proposed model, beginning by introducing the key notions used in this paper.

   In pool based AL we are provided with a pool $U_0 = \{x_1, ...x_n\}$ of unlabelled points, an empty set of labelled points $L_0 = \{\}$ and a labelling oracle O, which when queried for the label of x, returns y. Algorithms in the pool based setting have to decide which points to query by looking at the entire pool.

Definition (Contextual Bandit Problem with Linear Payoffs). In a contextual bandits problem with linear payoffs, there are N arms. At time $t = 1, 2, ...$, a context vector $b_i(t) \in R^d$, is revealed for every arm i. History

$Q_{t-1} = \{a_\tau, r_\tau, b_i(\tau), i = 1, ..., N, \tau = 1, ..., t-1\}$ where a_τ denotes the arm played at time τ and that triggered reward r_τ. Given $b_i(t)$, the reward for arm i at time t is generated from an (unknown) distribution with mean $b_i(t)^\top \mu$, where $\mu \in R^d$ is a fixed but unknown parameter. An algorithm for the contextual bandit problem needs to choose, at every time step t, an arm a_t to play, using history Q_{t-1} and current contexts $b_i(t), i = 1, ..., N$.

Let a_t^* denote the optimal arm at time t, i.e. $a_t^* = argmax_i b_i(t)^\top \mu$. And let $\Delta_i(t)$ be the difference between the mean rewards of the optimal arm and the arm i played at time t, i.e., $\Delta_i(t) = b_{a_t^*}(t)^\top \mu - b_i(t)^\top \mu$. Then, the regret at time t is defined as $regret(t) = \Delta_{a_t}(t)$. The objective is to minimize the total regret $R(T) = \sum_{t=1}^{T} regret(t)$. The time horizon T is finite and known in our case.

To model the active learning problem as a contextual bandit with linear payoffs we need to define the arms of the bandit, the rewards of the environment and the context of each arms.

Construction of the Arms. We cluster corpus points $\{x_1, x_2, ..., x_n\}$. The resulted clusters $c \subset U$ are considered as the arms of the bandit.

Context of the Arms. We consider a context vector $b_i(t)$ that describes the the arms (the clusters), and contains the features that characterise the clusters.

Reward. A metric is used to measure the variation of the hypothesis learned by the model between two iterations. More the hypothesis learned by the model varies more is the received reward. We now define the function $d(h_{t-1}, h_t)$ that we use to get the variation of the model. Let $U_0 = \{x_1, ..., x_n\} = L_t \cup U_t$ be the set of labelled and unlabelled training examples that we have. Then for each of the two real-valued hypotheses $h_{t-1}(.), h_t(.)$, we define the vectors $H_{t-1} = (h_{t-1}(x_1), h_{t-1}(x_2), ..., h_{t-1}(x_n))$ and $H_t = (h_t(x_1), h_t(x_2), ..., h_t(x_n))$, i.e. vectors of the real-valued predictions of h_{t-1} and h_t.

$$d(h_{t-1}, h_t) = \frac{H_{t-1}.H_t}{||H_{t-1}||.||H_t||} \tag{1}$$

In Eq. 1, we compute the cosine similarity between the two vectors H_{t-1} and H_t. Thus $d(h_{t-1}, h_t) \in [-1, +1]$ is the cosine of the angle between H_{t-1} and H_t, and we normalise the result in the interval $[0, 1]$ using $y(t) = \frac{2.cos^{-1}(d(h_{t-1}|h_t))}{\pi}$.

Stationarity of the Reward. We have observed that, more an area is sampled by the model less is the received reward (nonstationarity of the rewards). We have confirmed this common sense idea from an off-line evaluation (see Fig. 2). In order to circumvent this nonstationarity we assume that the reward function $y(t) = r_t \cdot D(t)$, where $D(t)$ is a decreasing function that follows the decreasing reward given by the environment and r_t is the stationary reward. The process for obtaining the function $D(t)$ is described in Section 4.

Contextual Bandit Algorithm. A Contextual bandits algorithm determines a cluster $c \subset U_t$ to be sampled at each time step t, based on the previous observation sequence $Q_{t-1} = \{c_\tau, r_\tau, b_c(\tau), c = 1, ..., N, \tau = 1, ..., t-1\}$, and its current context $b_c(t)$.

3.1 Active Thompson Sampling

Thompson sampling is understood in a Bayesian setting as follows. The set of past observations Q is made of triplets $(c_t, r_t, b_c(t))$ and are modelled using a parametric likelihood function $Pr(r_t|\tilde{\mu})$ depending on some parameters $\tilde{\mu}$. Given some prior distribution $Pr(\tilde{u})$ on these parameters, the posterior distribution of these parameters is given by the Bayes rule, $Pr(\tilde{\mu}|r_t) \propto Pr(r_t|\tilde{\mu})Pr(\tilde{\mu})$.

From [1], we can say that the posterior distribution at time $t+1$, $Pr(\tilde{\mu}|r_t) \propto Pr(r_t|\tilde{\mu})Pr(\tilde{\mu})$ were given by a multivariate Gaussian distribution $\mathcal{N}(\hat{\mu}(t+1), v^2 B(t+1)^{-1})$, where $B(t) = I_d + \sum_{\tau=1}^{t-1} b_{c_\tau}(\tau)b_{c_\tau}(\tau)^\top$ with d the size of the context vectors, $v^2 \in]0,1]$ is a constant fixed to 0.25 according to [4] and $\hat{\mu} = B(t)^{-1}(\sum_{\tau=1}^{t-1} b_{c_\tau}(\tau)b_{c_\tau}(\tau))$. Every step t consists of generating a d-dimensional sample $\tilde{\mu}$ from $\mathcal{N}(\hat{\mu}(t), v^2 B(t)^{-1})$, and solving the problem $\underset{c \subset U_t \wedge |c|>0}{argmax} \, b_c(t)^\top \tilde{\mu}$ (select the cluster c that maximizes $b_c(t)^\top \tilde{\mu}$. After that the algorithm selects randomly an individual $x \in c$, requests a labelling from the oracle O and observes reward $y(t)$.

Algorithm 1. The Active Thompson Sampling algorithm

1: **Require.** $B = I_d$ set $\hat{\mu} = 0_d, f = 0_d$.
2: **Foreach** $t = 1, 2, ..., T$ **do**
3: Sample $\tilde{\mu}$ from the $N(\hat{\mu}, v^2 B^{-1})$ distribution.
4: Select cluster $c_t = \underset{c \subset U_t \wedge |c|>0}{argmax} \, b_c(t)^\top \tilde{\mu}$
5: $x_t = Random(c_t)$.
6: Query O for label y_t of x_t
7: Observe $y(t)$ and compute r_t
8: $B = B + b_{c_t}(t)b_{c_t}(t)^T, f = f + b_{c_t}(t).r_t$ **else** $\hat{\mu} = B^{-1}f$
9: **End**

---

4 Experimental Evaluation

To conduct our evaluation, we have got from our company a corpus containing French utterances collected from a commercial spoken dialogue system. There are 7 765 utterances annotated by human experts. The unannotated part consists of 3 911 695 utterances.

We use a corporate supervised algorithm (rule-based algorithm), being a part of a spoken language dialogue system. We simulate in the experiments an expert (oracle) on the unannotated corpus by using the rule-based algorithm which was designed using the 7 765 annotated utterances. In our experiments, the clustering algorithm (k-means in our case) uses the cosine similarity as a similarity metric between utterances. We have considered different features in the context vector of the clusters $b_c(t) = (Mdis_c, Vdis_c, |c|, plb_{c,t}, MixRate_{c,t})$, where $Mdis_c$ and $Vdis_c$ are respectively the average distance between individuals in the cluster, and its variance. $|c|$ gives the number of points in the cluster. $plb_{c,t}$ gives the proportion of labelled individuals in the cluster at time t and $MixRate_{c,t}$ gives the ratio of the classes in the cluster at time t (the proportion of examples labelled in each class in the cluster). To obtain the decreasing function $D(t)$, we assume $D(t) = \alpha e^{-\beta t}$, and we compute the parameters α and β of $D(t)$ by sampling uniformly the different clusters and drawing the rewards in Fig.1, then we lit $D(t)$ on the reward curve. We obtain $\alpha = 0.61$ and $\beta = 0.12$.

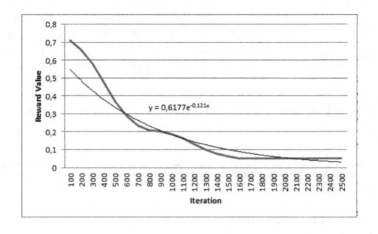

Fig. 1. Reward function

To evaluate the active learning algorithms, we have considered a version of the rule based algorithm without training. At each iteration the active learning selects from the unannotated corpus the relevant utterances to annotate and integrates it in the training set of the rule based algorithm. By relating the results to the newer versions, one can verify the usefulness of the proposed approach. We average the regret over 1000 times with a time horizon of 2000 sentences to label which correspond to our budget in term of labelling. To compute the regret, we have supposed that the optimal policy is given by the oracle. In addition to the random (baseline), we have compared our algorithm to the ones described in the related work (Sec. 2). QBC, US, and the different approaches that consider the bandit algorithms in the active learning as EXP4 used in [7] and LCB used in [5]. In Fig. 2, the horizontal axis represents the number of iterations and the vertical axis gives the cumulative regret (performance metric) which is the sum of the regrets from the first iteration to the current iteration.

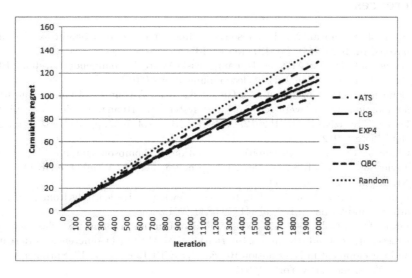

Fig. 2. Cumulative Regret for Active learning algorithms

From the Fig. 2 we observe that over all strategies gives better result than a random selection. Neither QBC nor US gives a good results. This confirms that a pure exploitation is not efficient, and it confirms the need of the exr/exp trade-off. While EXP4 algorithm gets a low cumulative regret, its overall performance is not as good as LCB and ATS. ATS and LCB indeed have the best cumulative regrets, ATS decreases the cumulative regret with 29% over the baseline and LCB, with 23%. The improvement comes from a dynamic exr/exp. These algorithms take full advantage of exploration from the beginning of the exploration rather than other strategies like uncertainty sampling or request by committee that need enough iteration to construct their models. Finally, as expected, ATS outperforms LCB, which is explained by the consideration of the context, and also that TS performs better exr/exp trade-off than UCB.

5   Conclusion

In this paper, we study the active learning problem from the side of contextual bandit and propose a new approach that adaptively balances exr/exp regarding the context of the cluster (arms). We have validated our work with data from real-world application and shown that the proposed algorithm offered promising results. This study yields to the conclusion that considering the contextual bandit model for the active learning significantly increases the results. Considering these results, we plan to study the theoretical regret of the proposed algorithm.

References

[1] Agrawal, S., Goyal, N.: Thompson sampling for contextual bandits with linear payoffs. In: ICML (3), pp. 127–135 (2013)

[2] Bouneffouf, D.: DRARS, A Dynamic Risk-Aware Recommender System. PhD thesis, Institut National des Télécommunications (2013)

[3] Bouneffouf, D., Bouzeghoub, A., Gançarski, A.L.: A contextual-bandit algorithm for mobile context-aware recommender system. In: Huang, T., Zeng, Z., Li, C., Leung, C.S. (eds.) ICONIP 2012, Part III. LNCS, vol. 7665, pp. 324–331. Springer, Heidelberg (2012)

[4] Chapelle, O., Li, L.: An empirical evaluation of thompson sampling. In: Shawe-Taylor, J., Zemel, R.S., Bartlett, P.L., Pereira, F.C.N., Weinberger, K.Q. (eds.) NIPS, pp. 2249–2257 (2011)

[5] Ganti, R., Gray, A.G.: Building bridges: Viewing active learning from the multi-armed bandit lens. CoRR, abs/1309.6830 (2013)

[6] Lewis, D.D., Gale, W.A.: A sequential algorithm for training text classifiers. In: Proceedings of the 17th Annual International ACM SIGIR Conference on Research and Development in Information Retrieval, SIGIR 1994, pp. 3–12. Springer-Verlag New York, Inc., New York (1994)

[7] Osugi, T., Kim, D., Scott, S.: Balancing exploration and exploitation: A new algorithm for active machine learning. In: Fifth IEEE International Conference on Data Mining, p. 8 (November 2005)

[8] Settles, B.: Active Learning Literature Survey. Technical Report 1648, University of Wisconsin–Madison (2009)

[9] Seung, H.S., Opper, M., Sompolinsky, H.: Query by committee. In: Proceedings of the Fifth Annual Workshop on Computational Learning Theory, COLT 1992, pp. 287–294. ACM, New York (1992)

[10] Zhang, T., Oles, F.J.: A probability analysis on the value of unlabeled data for classification problems. In: 17th International Conference on Machine Learning (2000)

[11] Zhu, X., Lafferty, J., Ghahramani, Z.: Combining active learning and semi-supervised learning using gaussian fields and harmonic functions. In: ICML 2003 Workshop on The Continuum from Labeled to Unlabeled Data in Machine Learning and Data Mining, pp. 58–65 (2003)

Choosing the Best Auto-Encoder-Based Bagging Classifier: An Empirical Study

Yifan Nie[1], Wenge Rong[2,3], Yikang Shen[1], Chao Li[2,3], and Zhang Xiong[2,3]

[1] Sino-French Enginnering School, Beihang University, Beijing, China
[2] School of Computer Science and Engineering, Beihang University, Beijing, China
[3] Research Institute of Beihang University in Shenzhen, Shenzhen, China
{yifan.nie,yikang.shen}@ecpk, {w.rong,licc,xiongz@}buaa.edu.cn

Abstract. Feature learning plays an important role in many machine learning tasks. As a common implementation for feature learning, the auto-encoder has shown excellent performance. However it also faces several challenges among which a notable one is how to reduce its generalization error. Different approaches have been proposed to solve this problem and Bagging is lauded as a possible one since it is easily implemented while can also expect outstanding performance. This paper studies the problem of integrating different prediction models by bagging auto-encoder-based classifiers in order to reduce generalization error and improve prediction performance. Furthermore, experimental study on different datasets from different domains is conducted. Several integration schemas are empirically evaluated to analyse their pros and cons. It is believed that this work will offer researchers in this field insight in bagging auto-encoder-based classifiers.

Keywords: bagging, auto-encoders, feature learning, neural networks.

1 Introduction

When it comes to feature learning, there are two commonly used approaches, i.e., handcrafting features learning and unsupervised features learning [15]. Compared with manually selecting features, which is arduous and requires several validation tests to determine which features are the most representative [10], unsupervised features learning employs learning models to obtain appropriate features. In recent years neural network based feature learning model has been attached much importance as it allows users to perform unsupervised feature learning with unlabelled data [12] and one of the popular learning models is stacked auto-encoder [7].

To build a stacked auto-encoder-based feature learning model, a common approach is to add one classification layer on the top of stacked auto-encoder and then take the learned features as input for the classification layer [9]. Though the auto-encoder-based classification model has many advantages, it still faces several challenges among which a notable one is how to reduce generalization error [4]. To meet this challenge, a lot of approaches have been proposed among which widely adopted ones are ensemble methods [13].

C.K. Loo et al. (Eds.): ICONIP 2014, Part I, LNCS 8834, pp. 413–420, 2014.

Bagging (Bootstrap Aggregation) is a popular ensemble method which consists in bootstrapping several copies of the training set and then employing them to train several separate models. Afterwards it combines the individual predictions together by a voting scheme for classification applications [5]. As each bootstrapped training set is slightly different from each other, each model trained on theses training sets has different weights and different focus, thereby having different generalization error. By combining them together, the overall generalization error is expected to decrease to some extent.

Previous works have shown that bagging works well for unstable predictors [14]. Considering neural-network-based models are also unstable predictors [19], it is intuitive to assume that applying bagging methods to feature learning based models could also probably improve classification performance. Therefore a fundamental question is arisen accordingly: is it possible to integrate feature learning with ensemble method such as bagging? If so, how can we effectively integrate them? In this research, we thoroughly investigate the possibility of integrating feature learning with bagging ensemble method and further provide different integration schemas and empirically evaluate their pros and cons.

The remainder of this paper is organised as follow. In section 2 we introduce the background about auto-encoder and bagging techniques. Section 3 presents the proposed evaluation architecture. In section 4 we will present the experimental settings and also discuss the experimental results. Finally section 5 concludes the paper and points possible future work.

2 Background

2.1 Auto-Encoder

Auto-encoder has been widely used as an unsupervised feature learning tool [1]. Typically, a basic auto-encoder consists of three layers, i.e, input layer, hidden layer (or code layer), and output layer [17] which has the same number of units as the input layer. This structure can achieve unsupervised feature learning by feeding the unlabelled data into the input layer and forcing the output layer to reconstruct the input. The encode phase is implemented by mapping the input $x \in \mathbb{R}^m$ to a hidden representation (code) $h \in \mathbb{R}^n$, which has the form:

$$h = f(x) = s(Wx + b_h) \tag{1}$$

where s is the sigmoid activation function $s(z) = \frac{1}{1+e^{-z}}$, m is the number of input units, n is the number of hidden units, x is the input feature vector, h is the extracted code and the encoder is parametrised by the $n \times m$ weight matrix W, and b_h is the bias vector. Afterwards the decoder maps the hidden representation back to a reconstruction $\hat{x} \in \mathbb{R}^m$:

$$\hat{x} = g(h) = s(W^T h + b_{\hat{x}}) \tag{2}$$

where s is the sigmoid activation function $s(z) = \frac{1}{1+e^{-z}}$, m is the number of output units, n is the number of hidden units, h is the extracted code, $\hat{x}$ is the reconstructed input feature vector, and the parameters are a bias vector $b_{\hat{x}}$ and W^T is the tied weight matrix. The training process consists in finding the parameters $\theta = \{W, b_h, b_{\hat{x}}\}$, which aim to minimise the cost function:

$$J(\theta) = \sum_{x \in trainset} L(x, g(f(x)))$$ (3)

where x is a feature vector and f, g are encode and decode function in Eq. 1 and Eq. 2. The squared error is often used as the reconstruction error L, i.e., $L(x, \hat{x}) = \|x - \hat{x}\|$. Once the training process is completed, it is able to take the code in the hidden layer as unsupervised learned features.

2.2 Bagging

Bagging is a widely used integration technique to combine several prediction models in order to improve accuracy [6]. Firstly, bootstrapped training sets are generated from the original training set $X = \{x_1, x_2, ..., x_m\}$, and this procedure is conducted by sampling with replacement m elements with equal probability from the original training set X. Because the sampling is conducted with replacement, there will be repeated examples in the bootstrapped training set. As a result if m is large, asymptotically the fraction of unique examples is $lim_{m \to \infty} 1 - (1 - \frac{1}{m})^m = 63.2\%$. As there are repeated data, the unique data in each bootstrapped training set are randomly different. As such the models trained on these bootstrapped training sets have different focuses. Suppose x is a test example, y is one possible label, Y is the set containing all possible labels, h_i is a basic prediction model and h^* is the combined prediction function, the combining process will combine each trained classifier in the following way [2]:

$$h^*(x) = \arg\max_{y \in Y} \sum_{i:h_i(x)=y} 1$$ (4)

3 Evaluation Architecture

In order to validate the possibility and evaluate the performance of bagging different feature learning models, in this paper an evaluation architecture is proposed and its architecture is presented in Fig. 1(b).

In this architecture, each model is an auto-encoder-based classification model which is chosen to integrate unsupervised feature learning into prediction models. The model is composed by two parts: 1) stacked auto-encoder-based feature learning layers and 2) a classification layer. The stacked auto-encoder-based feature learning part has multiple layers in order to obtain higher-level representation of the data and performs feature learning [3]. The classification layer is on top of the feature learning layers, and it takes the learned features as input and performs classification, as shown in Fig. 1(a).

The stacked auto-encoder is configured with 0 sparsity penalty and no denoising treatment. All activation functions are sigmoid function and for every

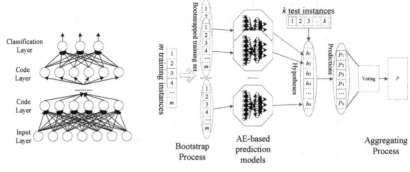

(a) Auto-encoder Structure   (b) Evaluation Architecture

Fig. 1. Auto-encoder-based Bagging Prediction Architecture

dataset, we try 1, 2, 3 code layers in the proposed model, respectively, as indicated in [18]. As for training process, a two-stage training approach is adopted. Firstly the proposed model performs a layer-wise unsupervised pre-training with unlabelled data [7] . This procedure aims to obtain unsupervised learned features in code layers. Once each code layer is pre-trained, the whole network can be further fine-tuned to include the classification layer with labelled data by traditional back-propagation algorithm [16].

Afterwards n bootstrapped training sets are generated and fed into n auto-encoder-based classification models. Finally through bagging the n basic models together, the overall performance will be evaluated. In this research the number n is set to vary from 1 to 20. The decision of limiting the number of trained basic models is both important in order to complete the experimental study in a reasonable time and for practical applications as indicated in [2].

4   Experimental Study

4.1   Dataset and Evaluation Metrics

In this research, five commonly used datasets[1,2] are employed and listed in Table 1(a). And the raw performance improvement and the relative performance improvement [2,5] over the average single model are utilised to evaluate the bagging auto-encoder performance:

$$raw\ improvement = (ACC_{bag} - ACC_{avg}) * 100\% \tag{5}$$

$$relative\ improvement = \frac{ACC_{bag} - ACC_{avg}}{100\% - ACC_{avg}} * 100\% \tag{6}$$

---

[1] http://yann.lecun.com/exdb/mnist/
[2] http://archive.ics.uci.edu/ml/

where ACC_{bag} stands for the accuracy after bagging, and ACC_{avg} stands for the average accuracy of single model.

Table 1. Datasets and Configuration

(a) Datasets

Dataset	#training instances	#test instances	#features	#classes
MNIST	60000	10000	784	10
Optdigit	3822	1000	64	10
Yeast	1000	484	8	10
CNAE9	990	90	856	9
Semeion	1400	193	256	10

(b) Layer Size Configuration

Dataset	1 Code Layer	2 Code Layers	3 Code Layers
MNIST	784-100-10	784-100-100-10	784-100-100-50-10
Optdigit	64-32-10	64-50-25-10	64-50-40-30-10
Yeast	8-50-10	8-50-25-10	8-100-80-40-10
CNAE9	856-100-9	856-428-214-9	856-428-214-107
Semeion	256-128-10	256-128-64-10	256-128-64-32-10

4.2 Experimental Result and Discussion

In this research, the number of units in each code layer is chosen by empirical evaluation to achieve best single model accuracy for each dataset [8]. The layer size configuration and experimental results on the five different datasets are summarised in Table 1(b) and Fig. 2. The performance in the best case is presented in Table 2.

Table 2. Performance in the Best Configuration

datasets	single avg	bagging	raw improvement	relative improvement	number of models
MNIST-1L	97.107%	97.840%	0.733%	25.346%	9
MNIST-2L	97.173%	97.990%	0.817%	28.909%	19
MNIST-3L	97.108%	98.080%	0.972%	33.599%	13
Optdigit-1L	95.650%	96.272%	0.621%	14.286%	6
Optdigit-2L	96.034%	96.828%	0.795%	20.031%	18
Optdigit-3L	95.727%	96.494%	0.767%	17.958%	19
Yeast-1L	48.347%	51.033%	2.686%	5.200%	2
Yeast-2L	49.403%	54.959%	5.556%	10.980%	9
Yeast-3L	51.033%	55.372%	4.339%	8.861%	8
CNAE9-1L	92.222%	94.444%	2.222%	28.571%	7
CNAE9-2L	97.259%	97.778%	0.519%	18.919%	15
CNAE9-3L	97.556%	97.778%	0.222%	9.091%	10
Semeion-1L	90.271%	92.746%	2.476%	25.444%	9
Semeion-2L	90.587%	94.301%	3.713%	39.450%	12
Semeion-3L	90.748%	94.301%	3.553%	38.400%	14

The experimental result has illustrated that it is possible to integrate feature learning with ensemble method such as bagging and the relative improvement in accuracy is satisfactory for big training set with many input features. Even for small training set with fewer input features, bagging still works and yields reasonable boost in the best configuration. Particularly the study on five datasets with different configurations reveal several lessons, which may give some insight for researchers in the domain for future bagging auto-encoder-based classifiers.

1. Bagging too few models (e.g., $n < 3$) is fruitless and sometimes the overall performance is even worse than the single model performance. It is observed

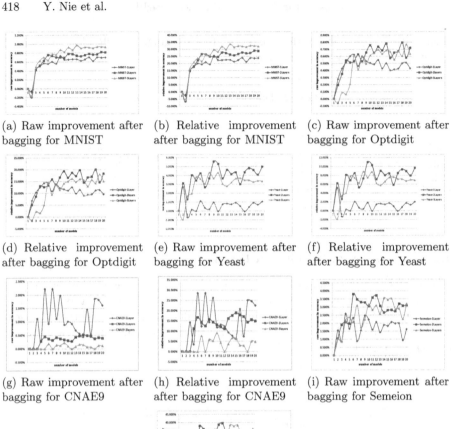

(a) Raw improvement after bagging for MNIST

(b) Relative improvement after bagging for MNIST

(c) Raw improvement after bagging for Optdigit

(d) Relative improvement after bagging for Optdigit

(e) Raw improvement after bagging for Yeast

(f) Relative improvement after bagging for Yeast

(g) Raw improvement after bagging for CNAE9

(h) Relative improvement after bagging for CNAE9

(i) Raw improvement after bagging for Semeion

(j) Relative improvement after bagging for Semeion

Fig. 2. Bagging Improvement for Five Datasets

from the experimental study that for some datasets (MNIST, Optdigit and Yeast) this will cause a negative boost and for other datasets (CNAE9 and Semeion), the boost is not quite significant compared with the performance of a single model. One possible explanation for this phenomenon is that each basic model trained on different bootstrapped training set has different focus. And given the too small number of models, when a controversial case comes, the high disagreement during bagging between the models will probably result in finally worse performance.

2. As the number of models increases, bagging begins to show its promising capability. The boost increases and seems to oscillate around a limit value. The possible reason for this phenomenon is that when the number of models

becomes bigger, there must be some overlaps between each bootstrapped training set. Since each trained model has different focus, the difference of theses focuses could not always be novel and might be repeated. As a result there could be a limit for the bagging performance. Bagging too many models does not help neither.

3. For text-based dataset (CNAE9), multiple code layers does not work well, as shown in Fig 2(g), single code layer configuration outperforms multiple code layers configuration. Whereas for image-based datasets (MNIST, Optdigit, and Semeion), multiple code layers configuration outperforms single code layer configuration. This may be due to the fact that texts themselves are already in a highly abstract form, more code layers could not extract further abstract features. However, for image-based training examples, the neighbouring features represent neighbouring pixels which are continuous and correlated. Multiple code layers can help extract more abstract features and thus increase performance.

4. For datasets which have too few features (Yeast with only 8 features), the boost with multiple code layers compared to single code layer configuration is much significant than datasets with more features (MNIST, Optdigit and Seimeion). This phenomenon might be due to the fact that for datasets with few features, more code layers can extract higher-level features and salient patterns which will improve the discriminative power of the classifier.

5 Conclusion and Future Work

The auto-encoder-based classification model has many advantages, but still faces several challenges. One of the challenges is how to reduce the model's generalization error. There are many ways to achieve this goal among which bagging is one of possible solutions. This study empirically studied the mechanism of integrating feature learning based classification models through bagging. By analysing the experimental result on different datasets with different configuration, some lessons are also summarised to show the pros and cons of the bagging oriented auto-encoder classifications. Besides auto-encoder, there exists other common feature learning structures among which DBN [11] is gaining more and more popularity. In the future, it is worthwhile to investigate bagging on these DBN based prediction models.

Acknowledgements. This work was partially supported by the National Natural Science Foundation of China (No. 61332018), the National High Technology Research and Development Program of China (No. 2013AA01A601), and the Fundamental Research Funds for the Central Universities. We are grateful to Shenzhen Key Laboratory of Data Vitalization (Smart City) for supporting this research.

References

1. Baldi, P.: Autoencoders, unsupervised learning, and deep architectures. Journal of Machine Learning Research-Proceedings Track 27, 37–50 (2012)

2. Bauer, E., Kohavi, R.: An empirical comparison of voting classification algorithms: Bagging, boosting, and variants. Machine Learning 36(1-2), 105–139 (1999)
3. Bengio, Y.: Learning deep architectures for ai. Foundations and Trends® in Machine Learning 2(1), 1–127 (2009)
4. Bengio, Y.: Deep learning of representations: Looking forward. In: Dediu, A.-H., Martín-Vide, C., Mitkov, R., Truthe, B. (eds.) SLSP 2013. LNCS (LNAI), vol. 7978, pp. 1–37. Springer, Heidelberg (2013)
5. Breiman, L.: Bagging predictors. Machine Learning 24(2), 123–140 (1996)
6. Bühlmann, P.: Bagging, boosting and ensemble methods. In: Handbook of Computational Statistics, pp. 985–1022. Springer (2012)
7. Erhan, D., Bengio, Y., Courville, A., Manzagol, P.A., Vincent, P., Bengio, S.: Why does unsupervised pre-training help deep learning? Journal of Machine Learning Research 11, 581–616 (2010)
8. Erhan, D., Manzagol, P.A., Bengio, Y., Bengio, S., Vincent, P.: The difficulty of training deep architectures and the effect of unsupervised pre-training. In: Proceedings of the 12th International Conference on Artificial Intelligence and Statistics, pp. 153–160 (2009)
9. Glorot, X., Bordes, A., Bengio, Y.: Domain adaptation for large-scale sentiment classification: A deep learning approach. In: Proceedings of the 28th International Conference on Machine Learning, pp. 513–520 (2011)
10. Guyon, I., Elisseeff, A.: An introduction to variable and feature selection. The Journal of Machine Learning Research 3, 1157–1182 (2003)
11. Kang, S., Qian, X., Meng, H.: Multi-distribution deep belief network for speech synthesis. In: ICASSP, pp. 8012–8016 (2013)
12. Le, Q.V., Ranzato, M., Monga, R., Devin, M., Chen, K., Corrado, G.S., Dean, J., Ng, A.Y.: Building high-level features using large scale unsupervised learning. arXiv preprint arXiv:1112.6209 (2011)
13. Li, L., Hu, Q., Wu, X., Yu, D.: Exploration of classification confidence in ensemble learning. Pattern Recognition 47(9), 3120–3131 (2014)
14. Liang, G., Zhu, X., Zhang, C.: An empirical study of bagging predictors for different learning algorithms. In: Proceedings of the Twenty-Fifth AAAI Conference on Artificial Intelligence (2011)
15. Netzer, Y., Wang, T., Coates, A., Bissacco, A., Wu, B., Ng, A.Y.: Reading digits in natural images with unsupervised feature learning. In: NIPS Workshop on Deep Learning and Unsupervised Feature Learning, vol. 2011 (2011)
16. Örkcü, H.H., Bal, H.: Comparing performances of backpropagation and genetic algorithms in the data classification. Expert Systems with Applications 38(4), 3703–3709 (2011)
17. Rifai, S., Mesnil, G., Vincent, P., Muller, X., Bengio, Y., Dauphin, Y., Glorot, X.: Higher order contractive auto-encoder. In: Gunopulos, D., Hofmann, T., Malerba, D., Vazirgiannis, M. (eds.) ECML PKDD 2011, Part II. LNCS (LNAI), vol. 6912, pp. 645–660. Springer, Heidelberg (2011)
18. Vincent, P., Larochelle, H., Lajoie, I., Bengio, Y., Manzagol, P.A.: Stacked denoising autoencoders: Learning useful representations in a deep network with a local denoising criterion. Journal of Machine Learning Research 11, 3371–3408 (2010)
19. Yu, L., Wang, S., Lai, K.K.: A neural-network-based nonlinear metamodeling approach to financial time series forecasting. Applied Soft Computing 9(2), 563–574 (2009)

Classification of fMRI Data in the NeuCube Evolving Spiking Neural Network Architecture

Norhanifah Murli[1,2], Nikola Kasabov[1], and Bana Handaga[3]

[1] Knowledge Engineering and Discovery Research Institute,
Auckland University of Technology, Private Bag 92006, Auckland 1010, New Zealand
{nmurli,nkasabov}@aut.ac.nz
[2] UniversitiTun Hussein Onn Malaysia, Johor, Malaysia
[3] Universitas Muhammadiyah Surakarta, Indonesia
bana.handaga@ums.ac.id

Abstract. This paper presents a new method and a case study on fMRI spatio- and spectro-temporal data (SSTD) classification with the use of the recently proposed NeuCube architecture [1]. NeuCube is a three dimensional brain-like model of evolving spiking neurons that can be trained with SSTD such as fMRI, EEG and other brain data. This SSTD is mapped, analyzed, modeled and trained, and the result from these processes can be used to better understand the brain processes and to better recognize brain patterns, and thus to extract new knowledge that may reside within the SSTD. From the experimental results we can conclude that the NeuCube architecture is capable of producing significantly more accurate classification results when compared with standard machine learning methods such as SVM and MLP. Moreover, the NeuCube method facilitates deep learning of the SSTD and deeper analysis of the spatio-temporal characteristics and patterns in the fMRI SSTD.

Keywords: spatio- spectro- temporal data; functional Magnetic Resonance Imaging (fMRI); evolving spiking neural networks; NeuCube; deep learning.

1 Introduction

In recent years MRI has become one of the most powerful imaging technique for understanding brain structure and functions, compared to other techniques, because of its non-invasive nature and the ability to produce very high visualization quality of internal organs and tissues. fMRI is a special form of MRI that can be used to measure neural activity changes in the brain resulting from stimuli. It measures the ratio of oxygenated hemoglobin to deoxygenated hemoglobin in the blood, at many individual locations within the brain which is taken as an indicator of neural activity [2].

Sequence of brain images or brain slices are constructed from spatial/spectral and temporal components. Spatial components are the coordinates of the brain cuboids (voxels) where data has been measured as intensity values. The temporal component is the time of scanning the whole brain volume, which can take 0.5-4.0 seconds to complete [2]. In a typical experiment, 100 or more brain volumes are usually scanned and recorded for a single person doing a particular task.

C.K. Loo et al. (Eds.): ICONIP 2014, Part I, LNCS 8834, pp. 421–428, 2014.

Having in mind the importance of accurate analysis and study of spatio- and temporal information in any SSTD, as previously tested with SSTD stroke data [3] and proven to produce better result, we are motivated to experiment the NeuCube model with other type of SSTD, in particular fMRI data. The rest of the paper is structured as follows: Section 2 describes the methodology, continued with experiment settings (Section 3) and followed with the result and discussion (Section 4). Finally, the paper is wrapped up in Section 5 with conclusion and future works of our study.

2   NeuCube for Modelling and Classification of SSTD

Standard classifiers such as Support Vector Machine (SVM), Multilayer Perceptron (MLP) and others have been used in many successful static brain data experiments [4, 5, 6]. These methods are efficient to process only the spatial component of the data, while neglecting the temporal component that these brain data, and other complex SSTD have. Despite of the efforts to model fMRI as SSTD [7, 8] there is a need for a new architecture that could better model, analyze and facilitate understanding of this complex data. To address this issue, the recently proposed NeuCube [1] is experimented here on fMRI data with the main objective to have a better analysis of brain patterns from fMRI SSTD.

The principles and the theory of the NeuCube were proposed and illustrated in [1]. NeuCube is a learning model of evolving spiking neural networks (eSNN) that perform learning in 2 phases: unsupervised and supervised. Figure 1 shows the NeuCube architecture. SSTD are transformed into spike sequences using Address Event Representation (AER) method [9] as demonstrated in Silicon Retina [10, 11], or other encoding methods such as rank order coding (ROC) [12] and Ben's Spikes Algorithm (BSA)[13]. Encoded spikes are then processed with the use of a learning method called Spike Time Dependent Plasticity (STDP) in a 3D SNN cube (SNNc) with leaky-integrate and fire model (LIFM) of the spiking neurons. After the completion of the unsupervised learning in the SNNc, which can be just for one training iteration, the input SSTD is fed into the trained SNNc which is connected to an eSNN to classify the spatio-temporal patterns produced in the SNNr into predefined classes. Different eSNN classifiers can be applied [14] such as dynamic eSNN (deSNN) [9] and spike pattern association neurons (SPAN) [15]. The whole model that includes both a SNNc and an eSNN classifier can be optimized through several iterations by changing the parameter values in every experimental run, until maximum accuracy is achieved.

In [1] a NeuCube-based methodology for EEG SSTD modelling and classification has been also proposed and the idea of using NeuCube for fMRI data has been raised. In [3] a NeuCube-based methodology for predictive data modelling on climate SSTD and early prediction of a personal event – stroke, has been proposed. In this paper we continue the development of NeuCube–based models for another type of SSTD – fMRI.

3   A NeuCube Based Methodology for fMRI SSTD Classification Illustrated on a Case Study of StarPlus fMRI Data

3.1   The Methodological Framework

The proposed NeuCube-based framework for classification of SSTD fMRI data is presented in Figure 1. It consists of 4 major stages, each of them explained below and illustrated on the classical StarPlus fMRI dataset [16]: Spatio-temporal mapping and encoding of the fMRI data; Unsupervised learning in a SNNc; Classification in an eSNN classifier; and Parameter and model optimization.

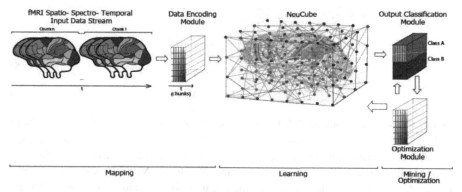

Fig. 1. A schematic representation of the NeuCube [1] based framework for fMRI data modelling and classification that consists of: 1.Spatio-temporal mapping and encoding of the fMRI data; 2. Unsupervised learning in a SNNc; 3. Classification in an eSNN classifier; 4.Parameter and system optimization

3.2   Data Mapping

Dataset. Publicly available StarPlus datasets [16] have been used to evaluate the performance of the proposed framework. The data was previously used by other researchers [4], [6, 7, 8]. The experiment was conducted into 2 set of trials in which in half of the trials, a subject was first presented with a picture stimulus for 4 sec, then a rest period for 4 sec and finally presented with a sentence stimulus for another 4 sec, and this experiment was identified as a PS trial. For half of the other trials, a subject was first presented with a sentence stimulus, followed by a rest period and then picture stimulus (SP trial). Data was measured every 500ms. We used 24 time points of PS dataset as Class 1 and another 24 time points of SP dataset as Class 2, resulting in 10 samples for each class.

Structural Mapping of fMRI SSTD into a NeuCube Structure. One of the NeuCube implementation [1] is based on Talairach brain template [17] with 1471 spiking neurons in the SNNc, so that the spatial (x,y,z) coordinates of each neuron are the same as a spatially located area of the brain according to the template. In order to map the StarPlus data into this SNNc, we have transformed the StarPlus data into 1471 integrated voxels that correspond to the 1471 Talairach template in the following way:

1. Determine mean of x and y of StarPlus coordinates, x_{meanSP} and y_{meanSP}
2. Determine stretch factor for x, y and z:

$$d_x = x_{mN}/x_{mSP} - x_{meanSP} \tag{1}$$

$$d_y = y_{mN}/y_{mSP} - y_{meanSP} \tag{2}$$

$$d_z = 10 \tag{3}$$

3. Determine new StarPlus coordinates:

$$x_{ns} = d_x * (x_{SP} - x_{meanSP}) \tag{4}$$

$$y_{ns} = d_y * (y_{SP} - y_{meanSP}) \tag{5}$$

$$z_{ns} = d_z * (z_{SP} - 4) \tag{6}$$

Where d_x, d_y and d_z are stretch factors, x_{mN} and y_{mN} are x and y minimum of Neu-Cube; x_{mSP} and y_{mSP} are x and y minimum of StarPlus; (x_{SP}, y_{SP}, z_{SP}) are original StarPlus coordinates; and $(x_{nSP}, y_{nSP}, z_{nSP})$ are new calculated StarPlus coordinates.

We define the StarPlus data points that are close to the NeuCube coordinates within a certain radius (in this case we set the radius to be 7). This is to ensure that, no single data point from the StarPlus data is located outside the NeuCube Talairach-based (x,y,z) coordinates (Figures 2a, b). fMRI input data that have already been transformed into Talairach coordinates, are encoded into spike trains using AER method.

3.3   Unsupervised Learning in the 3D SNN Cube

STDP learning method is applied in the NeuCube to initialize, modify and retain connection weights (memory) throughout the unsupervised learning stage. Experimental settings for subject s04799 as shown in Figure 3: the size of the SNNc is 1471 LIF neurons; AER threshold is 2.375:when a voxel value (integrated input voxel variables) increases above 2.375, a positive spike will be generated and entered into the SNNc at the same (x,y,z) location of the SNNc as the location of the voxel; if a voxel value decreases below 2.375, a negative spike will be generated and entered into the SNNc at the same location as the voxel co-ordinates from the data; if the voxel value is unchanged, no spike will be generated. Small world connections (SWC) regenerated probabilistically as initial connections in the SNNc reservoir (SNNcr). The radius of the SWC is chosen for this example to be 0.15. The LIFM neuron spiking threshold in the SNNc is 0.5 and a leaking parameter is chosen as 0.002. The used STDP learning rate is 0.001. In this case the number of training iterations for training the SNNc is 5. For the eSNN classifier the following parameters are used: *mod* parameter is 0. 4 (accounting for the importance of the first spikes on the neuronal inputs); and the *drift* parameter is 0.25 (accounting for the importance of the following incoming spikes to the neuronal inputs).

3.4   Classification Using deSNN Classifier

In [10] deSNN achieved fast and accurate learning of AER data for STPR as compared to eSNN and SNN that use only STDP. In the proposed framework each input neuron of the deSNN is connected to all neurons in the SNNc. The synaptic weights of the neurons in the deSNN are initialized using the rank order (RO) learning rule and the *mod* parameter, which is based on the first incoming spike that arrives on a synapse. For the following spikes, the synaptic weights are adjusted using STDP learning rule and the *drift* parameter, and the maximum of post synaptic potential and threshold value are calculated. The parameters used in the experiment are presented in Section 3.3. For the NeuCube implementation (taking subject s04799 as an example), voxel intensity values within the same trial are learned as one spatio-temporal pattern, thus producing 24 time frames, each of 1471 voxels. 20 samples are extracted from the data. Training and testing is conducted using 50:50 random split of the StarPlus data.

In the comparative analysis, the SVM experiment uses Polynomial Kernel of first degree, while the MLP model uses 20 hidden nodes and one output, with learning rate of 0.001and 500 iterations. As for standard classifiers, voxel intensity values within the same trial are concatenated and regarded as a single sample, thus producing 24 x 1471=35,304 input vector for each of the 20 samples.

4   Results and Discussion

The best accuracies obtained from the designed and trained NeuCube against standard classifiers (SVM and MLP) for 20 samples for each subject are depicted in Table 1. For illustration, the network is trained with parameters as shown in Figure 3.The classification accuracy obtained in the NeuCube model for subject s04799 was 90% (100% for class 1 and 80% for class 2, shown on Figure 3). Overall, the NeuCube models produced higher accuracy result than SVM or MLP in all experiments. Neurons connectivity before and after training are displayed in Figure 4 that can help understanding the fMRI data. The results clearly show that NeuCube model is much more accurate to handle complex fMRI data without filtering the noise from the data. This suggests that noise may carry valuable information in defining association between SSTD samples, but failed to be recognized and processed in conventional methods. The NeuCube approach does require preliminary feature extraction as it performs deep learning of the data. Additionally, classical machine learning methods are clearly not suitable for analyzing complex SSTD that have spatial and temporal information because their capability is severely limited to process only vector-based and static-based data.

Table 1. Comparative analysis of accuracy percentage for classical machine learning methods and the NeuCube model across six different subjects. The data set consists of 20 samples.

Methods / Subject	SVM	MLP	NEUCUBE
s04799	50(20,80)	35(30,40)	90(100,80)
s04820	40(30,50)	75(80,70)	90(80,100)
s04847	45(60,30)	65(70,60)	90(100,80)
s05675	60(40,80)	30(20,40)	80(100,60)
s05680	40(70,10)	50(40,60)	90(80,100)
s05710	55(60,50)	50(50,50)	90(100,80)

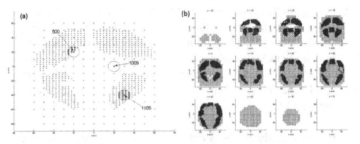

Fig. 2. (a) Blue dots are StarPlus coordinates while red dots NeuCube coordinates. As illustration three neurons are projected: the SNNc neuron 500 and its 8 neighboring StarPlus voxels represented as neurons; SNNc neuron 1105 and its 16-neighboring StarPlus voxels represented as neurons; and the SNNc neuron 1009 without StarPlus voxels around; (b) NeuCube neurons (black squares) and StarPlus voxels (blue squares) coordinates mapping in each of the z-slices; each sub-figure represents a single slice. Sub-figures which have only black squares mean that there are no StarPlus voxel coordinates mapped into that particular z-slice.

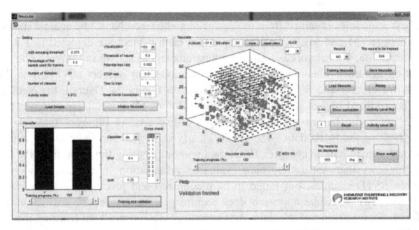

Fig. 3. A snapshot of a software implementation of the NeuCube architecture for classification of 2 class fMRI data for subject s04799. The parameter values are as in the Setting parameter box. Classifier used is deSNN with accuracy is 100% for class 1 and 80% for class 2.

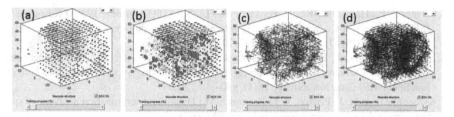

Fig. 4. Visualization of fMRI data model and connectivity between neurons of eSNN: (a) no spiking activity yet of inactive (blue) and fMRI (yellow) neurons; (b) spiking activity: active neurons (red), inactive neurons (blue), positive input neurons (magenta), negative input neurons (cyan) and zero input (yellow); (c) neurons connectivity before training (SWC): positive connections (blue) and negative connections(red); (d) neurons connectivity after training

5   Conclusion and Future Works

The experimental results produced with the proposed methodology for fMRI data modelling and classification validates its feasibility for analyzing complex fMRI data. A drawback of this approach is its high computation time particularly in determining optimal model and parameters. Multiple runs are needed to find the perfect combination of parameters in order to achieve higher accuracy. An optimizer such as Particle Swarm Optimization (PSO) for SNN [18], quantum inspired genetic algorithm [19]; and quantum inspired PSO [20] are thus planned to be developed in the future, along with NeuCube implementation on highly paralleled neuromorphic hardware [21]. We also plan to develop new techniques for the analysis of the NeuCube patterns of activity that would help to better understand the modeled fMRI data and the brain states.

Acknowledgment. This research is supported by the Knowledge Engineering and Discovery Research Institute (KEDRI, http://www.kedri.info) of the Auckland University of Technology, the Ministry of Higher Education in Malaysia and University Tun Hussein Onn Malaysia, Johor.

References

1. Kasabov, N.: NeuCube: A spiking neural network architecture for mapping, learning and understanding of spatio-temporal brain data. Neural Networks 52, 62–76 (2014)
2. Lindquist, M.A.: The statistical analysis of fMRI data. Statistical Science 23(4), 439–464 (2008)
3. Kasabov, N., Feigin, V., Hou, Z.G., Chen, Y., Liang, L., Krishnamurthy, R., Othman, M., Parmar, P.: Evolving spiking neural networks for personalised modeling, classification and prediction of spatio-temporal patterns with a case study on stroke. Neurocomputing 134, 269–279 (2014)
4. Mitchell, T.M., Hutchinson, R., Niculescu, R.S., Pereira, F., Wang, X., Just, M., Newman, S.: Learning to decode cognitive states from brain images. Machine Learning 57(1-2), 145–175 (2004)

5. Avesani, P., Hazan, H., Koilis, E., Manevitz, L., Sona, D.: Learning BOLD Response in fMRI by Reservoir Computing. In: 2011 International Workshop on Pattern Recognition in NeuroImaging (PRNI), pp. 57–60. IEEE (2011)
6. Mourão-Miranda, J., Bokde, A.L., Born, C., Hampel, H., Stetter, M.: Classifying brain states and determining the discriminating activation patterns: Support vector machine on functional MRI data. Neuroimage 28(4), 980–995 (2005)
7. Mourão-Miranda, J., Friston, K.J., Brammer, M.: Dynamic discrimination analysis: A spatial–temporal SVM. Neuroimage 36(1), 88–99 (2007)
8. Ng, B., Abugharbieh, R.: Modeling spatiotemporal structure in fMRI brain decoding using generalized sparse classifiers. In: 2011 International Workshop on Pattern Recognition in NeuroImaging (PRNI), pp. 65–68. IEEE (2011)
9. Kasabov, N., Dhoble, K., Nuntalid, N., Indiveri, G.: Dynamic evolving spiking neural networks for on-line spatio-and spectro-temporal pattern recognition. Neural Networks 41, 188–201 (2013)
10. Delbruck,   T.:   jAER   open   source   project   (2007), http://jaer.wiki.sourceforge.net
11. Lichtsteiner, P., Delbruck, T.: A 64×64 AER logarithmic temporal derivative silicon retina. Research in Microelectronics and Electronics 2, 202–205 (2005)
12. Loiselle, S., Rouat, J., Pressnitzer, D., Thorpe, S.: Exploration of rank order coding with spiking neural networks for speech recognition. In: Proceedings of the IEEE International Joint Conference on Neural Networks (IJCNN 2005), vol. 4, pp. 2076–2080 (2005)
13. Nuntalid, N., Dhoble, K., Kasabov, N.: EEG classification with BSA spike encoding algorithm and evolving probabilistic spiking neural network. In: Lu, B.-L., Zhang, L., Kwok, J. (eds.) ICONIP 2011, Part I. LNCS, vol. 7062, pp. 451–460. Springer, Heidelberg (2011)
14. Schliebs, S., Kasabov, N.: Evolving spiking neural network—A survey. Evolving Systems 4(2), 87–98 (2013)
15. Mohemmed, A., Schliebs, S., Kasabov, N.: SPAN: A neuron for precise-time spike pattern association. In: Lu, B.-L., Zhang, L., Kwok, J. (eds.) ICONIP 2011, Part II. LNCS, vol. 7063, pp. 718–725. Springer, Heidelberg (2011)
16. StarPlus fMRI data, http://www.cs.cmu.edu/afs/cs.cmu.edu/project/theo-81/www/
17. Talairach, J., Tournoux, P.: Co-planar stereotaxic atlas of the human brain. 3-Dimensional proportional system: An approach to cerebral imaging (1988)
18. Mohemmed, A., Schliebs, S., Matsuda, S., Kasabov, N.: Training spiking neural networks to associate spatio-temporal input-output spike patterns. Neurocomputing 107, 3–10 (2013), doi:10.1016/j.neucom.2012.08.034
19. Platel, M.D., Schliebs, S., Kasabov, N.: Quantum-inspired evolutionary algorithm: A multimodel EDA. IEEE Transactions on Evolutionary Computation 13(6), 1218–1232 (2009)
20. Hamed, H.N.A., Kasabov, N., Shamsuddin, S.M.: Probabilistic evolving spiking neural network optimization using dynamic quantum-inspired particle swarm optimization. Australian Journal of Intelligent Information Processing Systems 11(1) (2010)
21. Khan, M.M., Lester, D.R., Plana, L.A., Rast, A., Jin, X., Painkras, E., Furber, S.B.: SpiNNaker: Mapping neural networks onto a massively-parallel chip multiprocessor. In: IEEE International Joint Conference on Neural Networks, IJCNN 2008 (IEEE World Congress on Computational Intelligence), pp. 2849–2856. IEEE (2008)

A Hybrid Approach to Pixel Data Mining

Subana Shanmuganathan

Auckland University of Technology (AUT), New Zealand
subana.shanmuganathan@aut.ac.nz

Abstract. A hybrid approach to pixel data mining for analysing map based thematic data for segregating, identifying and characterising New Zealand's grape wine regions is elaborated. The approach consisting of self-organising map (SOM) based clustering and Top-Down Induction of Decision Tree (TDIDT) decision techniques provides a means to profiling New Zealand wine regions despite scale, resolution and extent related data analysis issues that pose constraints with traditional and even with contemporary methods, such as satellite imagery and landscape classification techniques. With the SOM-TDIDT approach viticulturist can gain further insights into existing wine regions already zoned based on traditional methods. It could also be used to evaluate the suitability of new *terroirs* for potential vineyards as the continued production of premium wines by the world famous wineries has already become a challenges due to recent climate change observed across a few wine regions in Australia and the Mediterranean.

Keywords: Self-organising maps (SOM), TDIDT, Viticulture.

1 Introduction

A hybrid approach consisting of self-organising map (SOM) based pixel clustering and Top-Down Induction of Decision Tree (TDIDT) decision techniques to segregating, identifying and characterising New Zealand (NZ) grape wine regions is presented. The conventional approaches to viticulture zoning require expertise on the environment which makes the zoning of new regions/ *terroirs* a challenging task [1]. The underlying causes relating to viticulture zoning using either single or multi-attribute spatial i.e., environmental and climatic, also combined with non-spatial attribute i.e., wine label ratings, sensory perception descriptors, as a composite index are outlined.

The independent factors used for viticulture zoning can be categorised into three main classes; 1) location, 2) wine varietal stock/ bud graft (*terroir* x *cultiva* as in the Mediterranean concept) and 3) wine quality/market price related. The viticulture zoning provides major benefits to viticulturists, resource management and the wine industry and they are; 1) gain further understanding on dominant independent factors that impact on crops to inform vineyard management for implementing mitigating operations / maximising yield i.e., cultivation practices, at the vineyard scale, 2) when selecting new sites for potential vineyards, 3) for marketing quality vintages, and 4) for the state institutions who needs to decide on the optimal use of natural resources such as land, water

C.K. Loo et al. (Eds.): ICONIP 2014, Part I, LNCS 8834, pp. 429–437, 2014.
© Springer International Publishing Switzerland 2014

[2]. However, in the good old days, viticulture zoning systems were originally set-up and implemented to protect winemaker identity and source of income [3].

Conventional approaches to viticulture zoning use one or more dominant *terroir* factors identified as the most relevant, integrated into a geographic information system (GIS) usually at the *meso* scale (covering a few/several vineyards). However, it requires extensive knowledge on the related factors for any factor discretion. For example, to classify the regions based on grapevine phenology, comprehensive knowledge on local viticulture, wine quality and taste attributes (growing degree days or GDD, frost days, berry ripening period and temperature range, and wine style/ rating/ taste attributes) is essential. Hence, characterising vineyards in the so called *new world* i.e., New Zealand, Australia, Chile or new *terroirs* is seen as a challenging task. Even for the currently world famous wine *chateaus* and vineyards gaining further understanding on local grapevine phenology has become pertinent as they are increasingly compelled to look for new sites for relocating their established vineyards due to rising temperatures that challenge the ripening of ideal grapes for continued production of premium quality wine [1][4]. For such instances, the SOM-TDIDT approach provides a means to profiling the local, environmental, viticulture and winemaker information of existing wine regions for identifying and classifying any new regions or new sites/ *terroirs* for potential vineyards using available digital map based data [5].

2   Background

2.1   Integrated Analysis of Spatial Attributes

Both simple and complex spatial data analysis methods are efficient when there is sufficient knowledge relating to the problem domain and its solutions. The simple GIS operations applied to integrated analysis of spatial attribute data can be grouped into four basic categories based on [6] [7] and they are: 1) retrieval/ classification/ measurement, 2) overlay, 3) neighbourhood and 4) connectivity of network functions [8].

2.2   Clustering in Spatial Data Mining

New algorithms are being continuously investigated for clustering spatial data to optimise the clustering efficiency [9]. The focus of recent research efforts in spatial data analysis has been; improving the cluster quality in large volumes of high dimensional data sets [10], noise removal [11], uncertainty [12], data pre-processing and reduction of clustering run time [13].

TDIDT algorithm is considered to be a powerful tool for generating classification rules for decision trees since the mid-1960s. The TDIDT algorithm [14] gives the basis for many classification systems, such as Iterative Dichotomiser 3 (ID3) and C4.5 (statistical classifier) an extension to the ID3. Using the TDIDT methods decision rules can be produced as a decision tree by repeatedly splitting the data based on the values of attributes and this is referred to as recursive partitioning. The TDIDT algorithm is based on a set of instances used for training. Each instance can be described by the values of a set of categorical attributes relating to a member of a universe of objects.

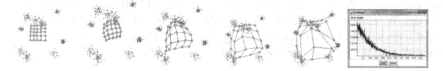

Fig. 1. SOM training that eventually covers the input data space as shown in the discrete mixture option (prob. distribution.). In graph (right), error decreases (y-axis) with training (x-axis) source: http://blog.peltarion.com/2007/06/13/the-self-organized-gene-part-2/.

2.3   Remote Sensing in Viticulture

Contemporary precision viticulture studies use increasingly high resolution aerial imagery and micro climate/ environmental data acquired using networks of wireless/ wired sensors/ probes to identify the different zones within a vineyard block [15]. Remote sensing and access to satellite imagery have led to the use of airborne multispectral and hyper spectral imagery in precision viticulture with greater flexibility especially, for yield mapping integrated with soil or disease properties [16].

3   The Methodology

3.1   SOM Clustering and WEKA's JRip Classifier

A SOM is a two-layered feed-forward artificial neural network. It uses an unsupervised learning algorithm to perform non-linear regression. During training, the network configures itself in such a way that the output gradually evolves into a display of topology preserving representation with similar input data clustered near each other (fig. 1). The topology preserving mapping of the SOM algorithm projects multidimensional data sets onto low, usually one- or two-D planes that enable the visualisation of complex data sets otherwise difficult to analyse i.e., using conventional methods [17].

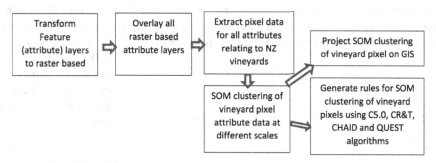

Fig. 2. A schematic diagram showing the steps adopted in the SOM-TDIDT approach

JRip classifier in WEKA is an implementation of the RIPPER rule learner created by William W. Cohen [18]. WEKA's JRip classifier model consists of collection rules and some statistics about those rules (e.g., coverage/ uncoverage, true/false positives/negatives). (WEKA functions JRIP).

3.2   Data Integration

All thematic (feature) map data in digital format (raster and vector) was processed and analysed in a GIS environment i.e., ArcGIS 10.0 (www.esri.com). The successive processing steps of the approach are outlined in fig. 2. Initially, all feature layers (polygon maps) obtained from Landcare Research web portal (www.landcareresearch. co.nz/resources/data/lris) were converted into raster maps. Secondly, pixel (point attribute) data was extracted from all raster layers for SOM clustering. Finally rule extraction was performed using TDIDT method using WEKA's JRip algorithm.

4   Results and Discussions

The 437,888 pixels extracted with their attribute values for NZ vineyards were clustered into two, five and finally into 18 clusters. The SOM clusters C1 and C2 (fig. 3), C1 (a, c and b) and C2 (a and b in fig. 4) and the final 18 clusters (fig. 4 and table 1) are analysed using their cluster profiles (charts of attribute ranges), and laid over NZ maps to see the spatial distribution of the clusters. To evaluate the use of the approach at different scales, rules generated for the SOM cluster pixels and groups of pixels at the 1) national with all NZ wine regions, 2) regional using Marlborough/ Kumeu vineyard pixels alone.

4.1   Two-Cluster SOM

In the second level SOM clustering (fig. 4 and table 1), Canterbury and coastal Otago vineyard pixels (1b) get separated (with 62.37 m average elevation, 1.09 °C average min. temperature and 4.88 average drainage) from cluster 1a & c pixels. Most of the northern North Island and Nelson vineyard pixels are in 2b with major differences in average elevation (93.84 m), average min. temperature (4.59°C), average induration (2.28), average acid soluble phosphate, age, water balance ratio and water deficiency.

In C1, 270584 pixels of the total 270,798, belong to the South Island. C2's 137,926 out of total 167,090, except for

Fig. 3. Two cluster SOM, NZ maps with C1 and C2 pixels

slope and aspect all attributes show considerable variability in the range specific to the sub clusters. C1 are at elevation 25-500 m whereas of C2 are all less than 300m. The North Island C1 pixels are in mountainous areas (centre of the island. The C2 South Island pixels are from upper North and South island's Canterbury. The pixels of the different clusters in 1395 groups (max, average and min) in the 796 vineyard polygons x 18 SOM clusters x 11 regions were used to create JRip classifier rules (fig. 5).

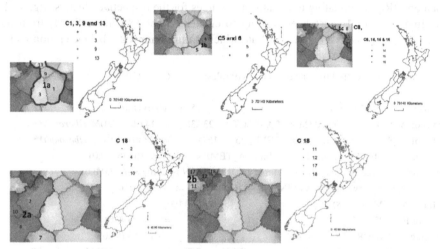

Fig. 4. SOM maps and NZ maps showing geographical distribution of the 18 SOM clusters

Table 1. Cluster profiles of 3rd level SOM.   C1a, 1b and 1c

Eighteen (18) cluster SOM (1a => 1, 3, 9 and 13) (1c=>8, 14, 15 and 16) (1b=> 5 and 6)

Clust er no	pixel count	Elevati on	A Temp	A min T	A Sol Radi ation	Indu ratio n	Exch Catio n	Acid sol Phos	Che limit	Age	Slop	Drai nage	Wat Ba R	Water deficit
C1	50313	82.33	12.39	2.40	14.9	3.32	2.00	4.19	1.00	1.97	0.06	4.81	1.40	261.4
C3	31141	111.17	12.20	2.01	14.9	2.12	2.00	3.29	1.00	2.00	0.07	2.90	1.50	247.0
C9	14330	81.52	12.40	2.40	14.9	3.32	2.00	4.20	1.00	1.98	0.04	4.80	1.40	261.7
C13	16064	108.54	12.11	1.87	14.8	2.09	2.00	3.32	1.00	2.00	0.05	2.87	1.52	239.9
C8	39187	167.00	11.90	1.00	15.2	3.90	1.90	3.90	1.00	1.60	0.03	5.00	2.10	136.2
C14	14302	167.00	11.90	1.00	15.2	3.90	1.90	3.90	1.00	1.60	0.03	5.00	2.10	136.2
C15	8945	247.64	10.20	-2.40	13.9	2.09	2.10	3.00	1.10	1.98	0.09	4.57	1.00	305.5
C16	2909	385.65	10.14	-0.78	14.2	3.37	1.62	3.10	1.00	1.99	0.25	4.49	1.64	170.1
C5	78678	54.14	11.70	1.20	14.1	3.29	2.00	3.90	1.00	1.00	0.03	4.99	1.70	213.3
C6	14929	105.75	11.24	0.52	13.9	3.38	2.09	3.65	1.00	1.99	0.05	4.34	1.73	181.6

4.2   JRip Classifier Rules for New Zealand Wine Regions

In the JRip classification, the SOM cluster no. was used as the target variable. The group values were used in this initial test as all 437,888 pixel could not be analysed with WEKA and the 27 JRip rules generated are presented in fig. 5.

Based on the first JRip rule (fig. 5), SOM cluster 15 relates to Otago vineyard pixels that have minimum water balance <=1. The main contributing attributes (1395 group summaries) for the clustering at this national scale based on the 27 JRip rules are: minimum monthly water balance ratio, annual solar radiation, elevation, annual average temperature, induration, acid soluble P, aspect and then finally annual minimum temperature. Interestingly, these are the factors used in the current system but with different terms. For instance, the monthly water balance used in this study is a surrogate for rainfall. Meanwhile, annual solar radiation and annual average temperature reflect growing degree days (GDD) used in the traditional NZ viticulture zoning system. Most of the clusters with pixels in multiple regions (fig. 5) have more than one JRip rule relating to the attribute ranges for the respective clusters (fig. 5). In the first rule, all 16 instances relating to cluster 15 (of 18-Cluster SOM) from the Otago region possess a unique attribute which is minimum water balance ration <=1.

=== Classifier model (full training set) === JRIP rules: Class: C18

(Min of W_BAL_RA <= 1)   => 15 (16.0/0.0) **Otago**
(Min of A_SOL >= 15.2) and (Min of ASPECT >= 237.31)   => 14 (26.0/1.0)*Marlborough***
(Min of A_SOL >= 15.2) and (Min of ELE_25 >= 167)   => 8 (30.0/0.0) *Marlborough***
(Average of ELE_25 >= 284.12) and (Min of A_TEMP <= 11.2) => 16 (35.0/0.0)
(Average of A_TEMP <= 10.9) and (Min of INDURATION >= 3.6)=> 16 (2.0/0.0)
(region = Canterbury) and (Max of INDURATION >= 3.6) => 6 (35.0/0.0)
(Min of A_TEMP <= 9.8)   => 6 (4.0/0.0)
(Min of ELE_25 <= 7)   => 7 (52.0/0.0)
(Average of INDURATION <= 0)   => 18 (53.0/0.0)
*(Min of ACID_S_P >= 4) and (Average of ASPECT >= 279.33) => 9 (61.0/0.0) Marlborough**
(Min of ASPECT >= 240.76) and (Min of MIN_TEMP <= 2)
and (Max of A_TEMP >= 12.2)   => 13 (64.0/0.0)
(Average of MIN_TEMP <= 0.8) and (Min of ASPECT >= 285.04) => 13 (3.0/0.0)
(Average of A_TEMP <= 11.7) and (Min of ELE_25 <= 85) => 5 (68.0/0.0)
(Max of ELE_25_2 >= 197)   => 12 (50.0/6.0)
(region = Nelson)   => 12 (42.0/18.0)
(region = Wellington) and (Average of A_TEMP <= 12.675) => 12 (14.0/6.0)
(Min of W_BAL_RA >= 3.2)   => 12 (3.0/1.0)
*(Min of ACID_S_P >= 4)   => 1 (81.0/1.0) Marlborough**
(Min of W_BAL_RA <= 1.5)   => 3 (75.0/0.0)
(Min of INDURATION >= 3.9) and (Max of ASPECT <= 127.99) => 3 (17.0/2.0)
(Min of EXCH_CAL <= 1.1) and (Average of ELE_25 >= 90.46) => 17 (104.0/0.0)
(Min of EXCH_CAL <= 1.5) and (Max of ASPECT >= 357.68) => 17 (6.0/0.0)
(Average of A_TEMP >= 15.784198)   => 17 (2.0/0.0)
(Average of DRA_25 <= 1.9) and (Min of ELE_25 <= 48)   => 10 (124.0/0.0)
(Min of A_TEMP <= 13.2)   => 2 (115.0/6.0)
(Average of ELE_25 >= 31)   => 11 (148.0/0.0)
                                                          => 4 (164.0/0.0)

Number of Rules: 27   Time taken to build model: 0.92seonds

Fig. 5. WEKA JRip rules generated for 1394 vineyard group minimum, average and maximum values for all 15 attributes using the 18 SOM cluster value as the class

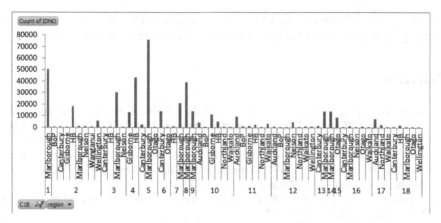

Fig. 6. Regional distribution of 18 SOM cluster pixels. SOM cluster consists of 576,333 Marlborough, 2,267 Canterbury and 78 Otago pixels. Marlborough vineyard pixels found in all but 4, 6, 10, 11, 15 and 17 clusters show the region's diverse and unique terroirs that produce word famous wine labels. Cluster 1, 8, 9 and 14 consists of only Marlborough.

The only difference between clusters 1 and 9 (* in fig. 5), both found in Marlborough alone is, aspect; cluster 9 pixels have an average aspect as 279.33°. Similarly, aspect is the only factor that differentiates the SOM clusters 8 and 14 that are again unique to the Marlborough region (** in fig. 5), for cluster 14 minimum of aspect is >=237.31. For clusters 8 and 14, all attributes were found to be same except for aspect (in the JRip rules created for Marlbrough and Kumeu alone (not included here)). The rules show the diversity and unique characters of these region's *terroirs* that give the well-known distinctive flavours of NZ appellations.

The JRip rules produced give 94.7% accuracy at 10 fold validation, meaning vineyard *terroir* attribute prediction could be made with 94.7 % accuracy.

The histogram (fig. 6) of SOM cluster and regional details plotted against the pixel counts give the different regions and their pixel counts in each SOM cluster. For instance, 263,803 pixels belonging to the Marlborough wine region can be seen in most of the SOM clusters except for 4, 6, 10, 11, 15 and 17. SOM cluster 1 consists of 50,313 pixels out of this region's pixels. Again, it shows the diverse and unique nature of the *terroirs* (or vineyards) that produce some world famous premium wine labels.

5   Conclusions

The SOM clustering and TDIDT method approach to discerning attributes for viticulture zoning using available digital data produced promising results at different scales. For Marlborough, minimum average temperature also confirms the current situation regarding frost in September which is one of the determinant factors for the crop in that region. Similarly, in [19] when Kumeu pixels alone were used to look at the success of the approach at the micro scale, age (<=1-fertile soils / 2-older less fertile soils), elevation (>92 / approx. 40m), annual average temperature <= 14/>15) along with aspect and hill shade, (the latter two are related to elevation) were found to be meaningful attributes for the characterisation of among and even within vineyards.

References

1. Shanmuganathan, S.: Viticultural Zoning for the Identification and Characterisation of New Zealand "Terroirs" Using Cartographic Data. In: Proc. of GeoCart 2010 and ICA Symposium on Cartography, September 1-3. Auckland University, Auckland (2010)
2. Riquelme, F.J., Ramos, A.B., Riquelme, F.J., Ramos, A.B.: Land and water use management in vine growing by using geographic information systems in Castilla-La Mancha, Spain. Agricultural Water Management 77, 82–95 (2005)
3. Vaudour, E., Shaw, A.B.: A Worldwide Perspective on Viticultural Zoning. S. Afr. J. Enol. Vitic. 26(2), 106–115 (2005)
4. Webb, L.B., Whetton, P.H., Barlow, E.: Impact on Australian Viticulture from Greenhouse Induced Temperature Change. In: Zerger, A., Argent, R.M. (eds.) MODSIM 2005 International Congress on Modelling and Simulation, pp. 1504–1510. Modelling and Simulation Society of Australia and New Zealand, Melbourne (2005) ISBN: 0-9758400-2-9
5. Waltman, W.J., Goddard, S., Read, P.E., Reichenb, S.E.: Digital Government: New Tools to Define Terroirs and Viticultural Areas in the Northern Great Plains. In: Proc. DG.O (2004)
6. Aronoff, S.: Geographic information systems: A management perspective, p. 294. WDL Publ., Ottawa (1989)
7. Berry, J.K.: Fundamental operations in computer-assisted map analysis. Int. J. Geogr. Inf. Syst. 1, 119–136 (1987)
8. Wei, T., Tedders, S., Tian, J.: An exploratory spatial data analysis of low birth weight prevalence in Georgia. Applied Geography 32(2), 195–207 (2012)
9. Chauhan, R., Kaur, H., Alam, M.: Data Clustering Method for Discovering Clusters in Spatial Cancer Databases. International Journal of Computer Applications (0975 – 8887) 10(6), 9–14 (2010)
10. Qian, Y., Zhang, K.: GraphZip: A Fast and Automatic Compression Method for Spatial Data Clustering. In: SAC 2004, March 14-17, Nicosia, Cyprus (2004)
11. Ester, M., Kriegel, H.-P., Jörg, S., Xu, X.: A Density-Based Algorithm for Discovering Clusters in Large Spatial Databases with Noise. In: Simoudis, E., Han, J., Fayyad, U.M. (eds.) Published in Proceedings of 2nd International Conference on Knowledge Discovery and Data Mining, KDD 1996 (1996)
12. Li, B., Shi, L., Liu, J.: Research on Spatial Data Mining Based on Uncertainty in Government GIS. In: 2010 Seventh International Conference on Fuzzy Systems and Knowledge Discovery (FSKD 2010), Yantai, China, August 10-12 (2010)
13. Walker, T.C., Miller, R.K.: Geographic information systems: an assessment of technology, applications, and products, vol. 166, p. 166. L SEAI Tech. Publ., Madison (1990)
14. Bramer, M.: Using Decision Trees for Classification. In: Principles of Data Mining, ch. 3, pp. 41–50. Springer, London (2007) ISBN-10: 1846287650 I ISBN-13: 978-1846287657
15. Taylor, J.A.: New Technologies and Opportunities for Australian Viticulture, in Precision Viticulture and Digital Terroir: Investigations into the application of information technology in Australian vineyards, ch. 3. PhD Thesis. The University of Sydney (2004), http://www.digitalterroirs.com/

16. Paoli, J.N., Tisseyre, B., Strauss, O., Roger, J.: Combination of heterogeneous data sets in Precision Viticulture. Vineyards & Sciences (2010) http://www.sferis.com/articles/050412_JnPaoli.pdf (accessed July 5, 2010)
17. Vesanto, J.: SOM-Based data visualization methods. Intelligent Data Analysis Journal (1999)
18. Cohen, W.W.: Fast effective rule induction. In: Proceedings of the Twelfth International Conference on Machine Learning. Morgan Kaufmann (1995)
19. Shanmuganathan, S., Whalley, J.: Pixel clustering in spatial data mining; An example study with Kumeu wine region in NZ. In: 20th International Congress on Modelling and Simulation, Adelaide, Australia, December 1-6, pp. 810–816 (2103), http://www.mssanz.org.au/modsim2013

A Novel SOH Prediction Framework for the Lithium-ion Battery Using Echo State Network*

Jianmin Wang, Zhe Li, Xiao Li, and Youyi Zhao

Department of Electronics Science and Technology,
Harbin University of Science and Technology, Harbin, Heilongjiang, 150080, China
wjmfuzzy@126.com

Abstract. Li-ion batteries provide lightweight, high energy density power sources for a variety of devices. Therefore, monitoring battery health in an effective way could increase the reliability and stability of the prediction system. So in this paper, we present a novel prediction framework based on Echo State Network to realize the prediction for battery state of health by training and testing battery impedance values and capacity values. To evaluate the proposed prediction approach, we have executed experiments with lithium-ion battery. Experimental results prove its effectiveness and confirm the estimation system can be effectively applied to the battery health state prediction. Moreover, the prediction system can run multiple data sets at a time to make the estimation process more efficient. Therefore, we can choose a battery which meets the requirement through the comparison between different batteries' prediction results.

Keywords: Echo State Networks, Framework, Prediction, reservoir computing, Lithium-ion Battery.

1 Introduction

The lithium-ion battery has the advantage of a low self-discharge rate, which allows the battery to remain in stock for about 12 months without additional maintenance, and requires a special circuit to control the charging and discharge process. On the other hand, the lithium-ion battery is a very important component for lots of electrical equipments and is critical to the performance of these systems. Therefore, an accurate estimation of the battery state of health (SOH) [1] in an effective approach can enhance the system reliability and stability [2]. Failure to do so can result in the performance degradation, underutilization of the equipment, and even cause catastrophic damages. Therefore, researchers have paid attention to the health monitoring and prognostics for lithium-ion battery with a variety of methods. In previous work [3] [4] [5], Bhaskar Saha using a Bayesian Framework for battery health monitoring, IL-Song Kim made the estimation through a Dual-Sliding-Mode, Datong Liu present the Gaussian process model to realize the prognostics for battery health, etc.

---

* Corresponding author.

C.K. Loo et al. (Eds.): ICONIP 2014, Part I, LNCS 8834, pp. 438–445, 2014.

The SOH which takes into account such factors as charge acceptance, internal resistance, voltage and self-discharge reflects the general condition of a battery and ability to deliver the specified performance compared with a fresh battery. SOH estimation used as a qualitative measure for the battery energy state in the system plays a major role in the battery prediction. The prediction of the SOH can be used to show the degradation of battery's performance and prevent possible accidents. There are two typical methods to calculate the SOH of the battery [6]. One method uses the battery impedance values to indicate the battery SOH. That can be determined using

$$SOH = \frac{R_i}{R_0} \times 100\% \tag{1}$$

where R_i is the ith impedance measurement varied with the cycles of charging and discharging and R_o is the battery initial impedance. On the other hand, the battery capacity C can also be used to determine the battery SOH as given in

$$SOH = \frac{C_i}{C_0} \times 100\% \tag{2}$$

where C_i is the ith capacitance value degenerated with cycles and C_0 is the initial capacity. We estimate the battery SOH using impedance and capacity values respectively

In this paper, a novel prediction framework for battery health monitoring based on ESN is proposed to innovate in the prediction system of the battery [7]. The framework contains the establishment of structures including data acquisition, data processing, parameter selection, parameter training and testing, and subsequent estimation structure with new data. We use Lithium-ion Bettery impedance or capacity values to estimate the SOH of the battery respectively. Meanwhile, We deal with the issue processing for the raw data [8]. Moreover, the prognostics system can run multiple battery data sets at a time to make the prediction process more efficient.

The paper is organized as follows. In section II. the ESN background is described. Mathematical backgrounds are given in order to define how the ESN can be used for prediction. In section III. the ESN predicrion framework is proposed formally. Also, the program of setting some parameters (reservoir design) is addressed. Before concluding, experiments are performed in order to discuss the usefulness of the ESN for prognostics purpose. Dataset used comes from experimental test of NASA Data Repository. The conclusions and future works are discussed in section V.

2   Backgrounds of ESN

As a new type of Recurrent neural networks [9]. A ESN uses the parameter α called Echo State Property (ESP) that make the reservoir dynamic. This kind of neural networks was introduced by Jaeger works about ESN in 2001 [10].

An ESN consists in a large number of neurons located in a so-called reservoir with a randomly (and fixed)connectivity between each other. Three essential elements of ESN: a. The reservoir weight matrix, called W. b. The input-reservoir weight matrix,called W^{in}. It make the the link between input and the reservoir. c. The reservoir-output weight matrix ,called W^{out}. It makes the link between the reservoir and the output. In addition, the system can add another optional matrix W^{back}, which represents the retroaction of outputs.

For the basic structure, W and W^{in} are created randomly and fixed, they do not need to be trained. Only W^{out} has to be trained. Consequently the training consists in a very simple linear regression. Compared with the traditional Recurrent neural networks, we deal with the issue the ambiguity and the complexity of training algorithm in the traditional Recurrent neural network architecture, and the training process has been simplified. Meanwhile, we also conquer the memory degradation of Recurrent neural network. A basic structure of an Echo State Network is shown in Figure 1.

2.1   Mathematical Formulation and Learning Scheme

Echo state network can be represented by state update and output equations. The update of the reservoir internal units is calculated as following:

$$x(n+1) = f\left(W^{in}u(n+1) + Wx(n) + W^{back}y(n)\right) \tag{3}$$

where $(\cdot)^T$ denotes transpose, $x(n) = (x_1(n), \cdots, x_N(n))^T$ is a state vector of the reservoir, $u(n) = (u_1(n), \cdots, u_K)^T$ is an input vector, $y(n) = (y_1(n), \cdots, y_L(n))^T$ is the output vector, $W^{in} \in R^{N \times K}, W \in R^{N \times N}, W^{back} \in R^{N \times L}$ are the internal, input and feedback connection weight matrix, respectively. $f(\cdot) = (f_1, \cdots, f_N)^T$ stands for an activation function vector. For example, $f_i(\cdot) = tanh(\cdot), i = 1, 2, \cdots, N$. Calculate the output for a basic structure ESN is:

$$y(n+1) = f^{out}\left(W^{out}(x(n+1), u(n+1), y(n)) + W^{out}_{bias}\right) \tag{4}$$

where W^{out}_{bias} is output bias or noise. The W^{out} is calculated as following:

$$(W^{out})^T = X^{-1}T \tag{5}$$

where X is a matrix created by M row vectors $(x_1(i), x_2(i), \cdots, x_N(i))(i = 1, 2, \cdots, M)$, $T = (y(1); y(2); \cdots; y(M))$ is a column vector. The learing algorithm consists in reducing the Mean Square Error (MSE) between the computed values for the training data set $y_{predicted}$ and the targets data set y_{target}

$$MSE = \frac{1}{N}\sum_{1}^{N}((y_{target}(n) - y_{predicted}(n))^2 \tag{6}$$

where N is the number of dimissed samples, due to the initial condition of the different matrix. The goal is using a linear regression approach to find the best

W^{out} weight matrix corresponding to the lowest MSE possible result. Nowadays, the nonlinear node with delayed feedback has innovated the conventional ESN. A reservoir is obtained by dividing the delay loop into N intervals and using time multiplexing. The input states are sampled and held for a delay in the feedback loop.

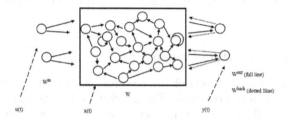

Fig. 1. Basic structure of an Echo State Network

3 ESN Prediction System

In order to obtain the SOH information of lithium-ion batteries. A framework of health prediction is proposed. We can predict different batteries SOH by a regular algorithms. The results of the prediction can be optimized by the update of the prediction system (the modulation of the data parameters).Therefore, we can get different batteries' SOH prediction information at a time with this prediction system.

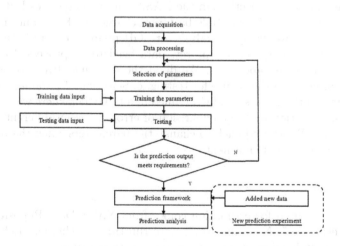

Fig. 2. The prediction framework based on ESN

The framework of the battery SOH prediction is shown in figure 2, including two process. In the process of structure establishment, The predictions we make can also with online data. The online data we may used for battery prediction including current, voltage and impedance values. Moreover, the row data need to be processed before the experiment by a variety of methods. Common methods of data processing including smoothing processing, eliminating outliers, inter-polation method, etc. In the estimation process, new data can be acquired to make the prediction for the lithium-ion battery SOH with the trained parameters directly.

3.1   Reservoir Parameters Design

Here are three major parameters of the reservoir. a. Size of the reservoir N_{res}: the number of units in the reservoir. The general wisdom is that the bigger the reservoir, the better the obtainable performance. b. c: the reservoir connectivity: the reservoir neurons are not all connected, so we have to design a connectivity parameter to get best results. It represents the percentage of non-zero weights in the reservoir and can take values between 0 and 1. c. The spectral radius: One of the most central global parameters of the reservoir connection matrix.It corresponds to maximum value of this matrix eigenvalues. In ESN the spectral radius is used to scale the non-zero elemens of W_{res}. The principle is to create the W_{res} matrix randomly with a connectivity parameter, and calculate the spectral radius of the matrix created.

4   Experiments and Analysis

In this paper, we use the impedance values and capacity values to estimate the SOH of the battery respectively in the prediction framework. a. In our experi-ment, we acquire the datasets from the NASA Ames Prognostics Data Reposi-tory. The datasets contains multiple batteries' charge, discharge and impedance information. What we want to analysis for SOH prediction here is the battery impedance and capacity. Then, the row data need to be processed by several methods (we discussed above). b. We divid the data into two groups. One is used for training, the other is use for testing. c. Selection of parameters. d. Run the program and the results will be obtained in a table file, where we can get the prediction information by the comparision of errors between different batteries in the table file. We can go back to change the paremeters when the prediction output is not meet the requirment.

4.1   Data set

The data set we used for estimation comes from NASA Data Repository. The Li-ion batteries(No.5, No.6 and No.7) were run through charge, discharge and impedance operational profiles) at room temperature. Charging was carried out at a constant current mode at 1.5A until the battery voltage reached 4.2V and then continued in a constant voltage 4.2V until the charge current dropped to 20mA. Discharge was carried out at a constant current level of 2A until the battery voltage fell to 2.7V, 2.5V, 2.2V respectively. Impedance measurement was carried out through an electrochemical impedance spectroscopy frequency sweep from 0.1Hz to 5kHz. Repeated charge and discharge cycles result in accelerated aging of the batteries. The experiments were stopped when the battery was a 30% fade in rated capacity.

Data processing: In this experiment we process the row data by deleting some abnormal points manually and using mathematical methods to reduce the signal noise. The processed battery capacity and impedance values are in figure 3.

a. Processed capacity data of battery No.5

b. Processed impedance data of battery No.5

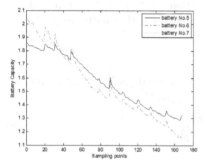 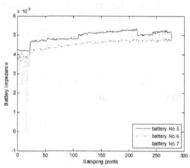

Fig. 3. Battery processed data of No.5 including processed capacity data (a) and processed impedance data (b)

4.2 Simulation Settings

The most improtant part in this experiment is parameter setting which affect the prediction results to a great extent. In this experiment. The data of each battery contains 168 cycles where we obtained the battery impedance and capacity values. The impedance values are used as inputs of an ESN. The first 84 values are used to train the network and the remaining values to test it. In the other experiment, the capacity values are also divided into two groups to train and test the network. In the prediction with capacity values of the battery, size of the reservoir N_{res} is 30, the reservoir connectivity c is $\frac{1}{3}$, the spectral radius is 0.75. In the experiment with impedance values of the battery, N_{res} is 30, c is $\frac{1}{3}$, the spectral radius is 0.5. The system could also estimate a typical battery in different operational conditions. The matrics used to check the network performance is Root Mean Square Error (RMSE). It is commonly used to quantify the difference between a predicted value and its real target.

$$RMSE = \sqrt{\frac{\sum_{t=1}^{N}((y_t - \widehat{y}_t)^2}{n}} \qquad (7)$$

a. The testing result of battery No.5
with impedance values

b. The testing result of battery No.5
with capacity values

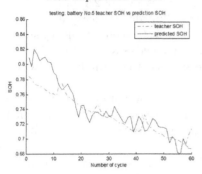 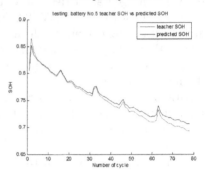

Fig. 4. Prediction result of three batteries. Prediction result of No.5 with impedance data(a). Prediction result of No.5 with capacity data(b). Other batteries prediction results are similer to battery No.5.

Mean Average Percentage Error (MAPE) is also a quantification measurement between two signals.

$$MAPE = \frac{1}{n} \sum_{t=1}^{N} \left| \frac{\widehat{y}_t - y_t}{\widehat{y}_t} \right| \tag{8}$$

4.3   Results and Discussion

Figure 4 shows the impedance prediction output and the capacity prediction output of the battery No.5. The batteries No.6 and No.7 are estimated at a time. The prediction output of the batteries No.6 and No.7 are similar to the battery No.5. The results show the degradation of different battery lifetime with the increasing of charge-discharge cycles. It also reflects the degradation of SOH well.

Table 1 shows the specific values of error. Different batteries' performance informations are displayed intuitively. The result shows the testing results is

Table 1. Three batteries' trainerror and testerror with capacity and impedance data respectively

Battery	Impedance data				Capacity data			
	RMSE		MAPE		RMSE		MAPE	
	train	test	train	test	train	test	train	test
No.5	0.028158	0.020183	0.033014	0.109070	0.001219	0.008208	0.000128	0.008391
No.6	0.036841	0.019066	0.034022	0.021449	0.000503	0.018645	0.000243	0.016310
No.7	0.018808	0.070609	0.018681	0.076688	0.000077	0.004064	0.000110	0.004603

more accurate when we use capacity values. Morever, the battery No.5 and the battery No.7 results could be more appropriate with the same set of parameters respectively.

5 Conclusion

This paper presented a framework based on ESN to evaluate the state of health. The case studies certified that the framework can be applied to different lithium-ion batteries. We can obtain different batteries' performance informations at a time. The experimental results demonstrate that the SOH estimation and prediction could obtain satisfied precision. Future work involves the optimization of the prediction of the framework for battery prediction and the development of a wider variety of methods for SOH prediction.

Acknowledgments. This work was supported by Natural Science Foundation of Heilongjiang Province of China (F201113).

References

1. Le, D., Tang, X.D.: Lithium-ion battery state of health estimation using ah-v characterization. In: Annual Conference of the Prognostics and Health Management Society, vol. 73(3), pp. 367–373 (2011)
2. Mandeleine: Development of a lifetime prediction model for lithium-ion batteries based on extended accelerated aging test data. IEEE Industrial Electronics Society 18(2), 414–423 (2012)
3. Zhou, J., Liu, D., Pang, J., Peng, Y.: Data-driven prognostics for lithium-ion battery based on gaussian process regression. In: Prognostics & System Health Management Conference, vol. 14(11), pp. 2531–2560 (2012)
4. Scott, P., Saha, B., Goebel, K., Christophersen, J.: Prognostics methods for battery health monitoring using a bayesian framework. IEEE Transactions on Instrumentation and Measurement 58(2) (February 2009)
5. Kim, I.-S.: A technique for estimating the state of health of lithium batteries through a dual-sliding-mode observer. IEEE Transactions on Power Electronics 25(4) (2010)
6. Wei, X.Z., Dai, H.F., Sun, Z.C.: A new soh prediction concept for the power lithium-ion battery used on hevs. In: Vehicle Power and Propulsion Conference, vol. 73(3), pp. 1649–1653 (August 2009)
7. Goebel, J.C.K., Saha, B., Christophersen, J.: Prognostics in battery health management. IEEE Instrumentation and Measurement Magazine 11(4), 33–40 (2008)
8. Orchard, M.E., Olivares, B.E., Cerda Munoz, M.A., Silva, J.F.: Particle- filtering-based prognosis framework for energy storage devices with a statistical characterization of state-of-health regeneration phenomena. IEEE Transactions on Instrumentation and Measurement 62(2) (February 2013)
9. Morando: Fuel cells prognostics using echo state network. IEEE Industrial Electronics Society, 265–274 (2013)
10. Jaeger, H.: The 'echo state' approach to analysing and training recurrent neural networks. Technical report GMD 148, German National Research Center Information Technology, ST., Tech. Rep. 148 (August 2001)

Significance of Non-edge Priors in Gene Regulatory Network Reconstruction

Ajay Nair[1,2,3], Madhu Chetty[4], and Pramod P. Wangikar[2]

[1] IITB-Monash Research Academy, Indian Institute of Technology Bombay, India
[2] Department of Chemical Engineering,
Indian Institute of Technology Bombay, India
[3] Faculty of Information Technology, Monash University, Australia
[4] School of Information Technology, Gippsland, Federation University, Australia
{ajaynair,wangikar}@iitb.ac.in, madhu.chetty@federation.edu.au

Abstract. It is well known that incorporating prior knowledge improves gene regulatory network reconstruction from data. Two types of prior knowledge can be given for the gene regulatory network inference - known interactions (edge priors) and known absence of interactions (non-edge priors). However, previous studies have focused mainly on edge priors. This paper shows that the edge priors give only limited improvement. Moreover, non-edge priors are crucial for better overall performance and their effect dominates edge priors at larger data samples. The studies are carried out on two real networks and a computationally tractable synthetic network, using Bayesian network framework. Further, a method to obtain large numbers of non-edge priors for real gene regulatory networks is presented.

1 Introduction

Inferring the gene regulatory network (GRN) structure from data is also known as GRN reconstruction or reverse-engineering of GRN. The availability of high through-put DNA microarray data has facilitated not only to efficiently perform classification tasks [11], but many methods have also been developed to infer the GRN. Probabilistic graphical models such as Bayesian networks (BN) and dynamic Bayesian networks (DBN) are based on the solid foundations of probability and statistics and are very popular as they can infer causality from data and integrate information from different sources, as prior knowledge during inference [5]. DBN, using time series data, can model feedback loops.

Even with BN or DBN, accurate GRN inference continues to pose a challenge, mainly due to limited experimental data [4]. Different methods have been proposed to deal with this statistical issue [5]. Using the prior knowledge of gene or protein interactions is a popular method as these interactions may be obtained from expert knowledge, experimental studies, or from literature. This prior interaction knowledge, input as initial network structure [7], [8], [13] or incorporated during sampling and evaluation [6], has shown improvement over inference from data alone. However, these studies [6], [7], [13] are limited as they consider only

C.K. Loo et al. (Eds.): ICONIP 2014, Part I, LNCS 8834, pp. 446–453, 2014.
© Springer International Publishing Switzerland 2014

known interactions. Available literature [8] considering both the presence and absence of interactions has limited scope, as the study is restricted to a single network with synthetic data and reports only the sensitivity of inference.

In this paper, two types of prior knowledge have been considered: a) presence of interaction between two genes, represented as an edge in the network and called as the edge prior and b) absence of interaction, i.e. the non-edge prior. The study reports the effect of both edge and non-edge priors - individually, in combination, at different quantities, and at different sample sizes of data. The rest of the paper is organized as follows. Section 2 explains the underlying theory and methods used. Experimental details are in section 3 and the results and discussions are given in section 4. Section 5 concludes the paper.

2   Methods

We choose globalMIT [14] for this study as it is one of the few optimal and polynomial worst-case time complex DBN structure learning algorithms available in public domain, with the source code. Its mutual information (MI) based scoring metric has been shown to have similar characteristics to other scoring metrics. Further, its MI scoring function is more efficient in searching the equivalent graph space [2], [14] than others like [16]. Since globalMIT does not support incorporation of priors, the code is modified to allow prior input and to use this prior information to restrict the search space of possible networks. The MI scoring metric is given by [2], [14].

$$S_{MIT}(G:D) = \sum_{i=1;Pa_i \neq \phi}^{n} 2N.MI(X_i, Pa_i) - \sum_{i=1;Pa_i \neq \phi}^{n} \sum_{j=1}^{s_i} \chi_{\alpha, l_{i\sigma_i(j)}} \quad (1)$$

Here, the first term to the right of equality is the MI score, equivalent to log-likelihood and measures the ability of the inferred graph to match the data. The second term is the penalizing term based on the statistical significance of the inferred network at user defined confidence level α. N is the number of samples in D, $X = \{X_1, .., X_n\}$ is the set of n nodes or variables with corresponding $\{r_1, .., r_n\}$ discrete states, $Pa_i = \{X_{i1}, .., X_{is_i}\}$ is the parent set of X_i in G with the corresponding discrete states $\{r_{i1}, .., r_{is_i}\}$, and $s_i = |Pa_i|$, is the cardinality of the parent set. $MI(X_i, Pa_i)$ is the MI between the node X_i and its parents Pa_i calculated from the data and $\chi_{\alpha, l_{i\sigma_i(j)}}$ is the value satisfying $p(\chi^2(l_{ij}) \leq \chi_{\alpha, l_{i\sigma_i(j)}}) = \alpha$. The term $l_{i\sigma_i(j)}$ is defined as

$$l_{i\sigma_i(j)} = \begin{cases} (r_i - 1)(r_{i\sigma_i(1)} - 1) & j = 1 \\ (r_i - 1)(r_{i\sigma_i(j)} - 1)\Pi_{k=1}^{j-1} r_{i\sigma_i(k)} & j = 2, .., s_i \end{cases}$$

where $\sigma_i = \{\sigma_i(1), .., \sigma_i(s_i)\}$ is any permutation of index set $\{1, .., s_i\}$ such that $\sigma_i(1)$ has the highest number of states, $\sigma_i(2)$ has second highest number of states, and so on. For further details and comparison with other scoring metrics we refer the readers to [2], [14].

Under the assumption that all variables have the same number of states (for polynomial worst-case time complexity of globalMIT), we get, $r_i = r_{ij} = r$, $l_{i\sigma_i(j)} = l_{ij} = (r-1)^2 r^{j-1}$ for $j = 1, .., s_i$, and the penalty term becomes $\sum_{i=1; Pa_i \neq \phi}^{n} \sum_{j=1}^{s_i} \chi_{\alpha, (r-1)^2 r^{j-1}}$. It can be seen that the penalty term depends only on α and s_i (r is the characteristic of input data) and as α or s_i increases the penalty also increases. The α is a user input while s_i is selected at run-time by the algorithm based on N.

From equation 1, a complex network can only be learned if the likelihood score of the network is greater than its complexity. The likelihood score increases if N or $MI(X_i, Pa_i)$ increases. MI increases if s_i increases ($MI(X, Y \cup W) \geq MI(X, Y)$) but penalty also increases at higher s_i thus overall score does not improve. Therefore, complex networks can be learned only by increasing N. However, previous studies have shown that increasing N leads to deterioration of certain performance measures such as decrease in specificity and precision (for e.g. see [8], figure 4), which is non-intuitive. We did a study on the effect of increasing sample size on the inference performance, to analyze and address this drop in performance.

Studies incorporating prior knowledge while learning from data have shown improvement in GRN inference [5]. The general scoring function for a BN/DBN learning algorithm incorporating prior knowledge is

$$S(G : D) = logP(G) + LL_D(G) - C(G)f(N) \tag{2}$$

where $S(G : D)$ is the score of the inferred graph G from the data D, $LL_D(G)$ is the log-likelihood of the data given G, N is the number of samples, $C(G)$ is the measure of network complexity, $f(N)$ is a non-negative penalty function, and $logP(G)$ is the prior probability of G based on the prior knowledge. By biasing the search towards certain graphs using prior knowledge, the total search space for G is reduced and performance can be improved. Thus, priors can be used to address the drop in inference performance. In this work, prior knowledge is given as initial structure by clamping, as this is computationally simple and efficient for good quality priors. In clamping, the prior probability of the relevant edges and non-edges are set to 1 and 0, respectively.

3   Experiments

The confidence value parameter α of algorithm is obtained using a ROC-curve analysis. For the two real networks $\alpha = 0.999$ and for the synthetic network $\alpha = 0.95$ are obtained which matches the suggested values ([2] and GlobalMIT user manual [14]). Other parameters are kept at their default values. Inference is done with no-prior knowledge (only data) and with three types of prior knowledge: a) edge priors, b) non-edge priors, and c) combined priors - containing both edge and non-edge priors. Priors are incorporated in five thresholds: 0% (corresponding to no-prior knowledge) or approximately 25%, 50%, 75%, or 100%. For each quantity of prior, the experiment is repeated for different combinations of available edge/non-edge (e.g., for 50% edge

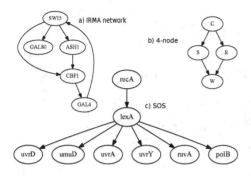

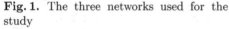

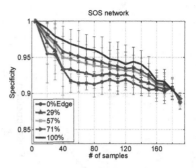

Fig. 1. The three networks used for the study

Fig. 2. Effect of edge priors on specificity

prior, different combinations of all edge priors are chosen to form the 50% input) and averaged to account for the variations. All of these are repeated at varying sample sizes of data. For each sample size, the inference is repeated at least 10 times with different data samples (if available) or by bootstrap. In each repetition, one network structure is learned and the performance parameters, i.e. true positives (TP), true negatives (TN), false positives (FP), false negatives (FN), sensitivity=TP/(TP+FN), specificity=TN/(TN+FP), precision=TP/(TP+FP), and F1-score=2*precision*recall/(precision+recall), are computed. Parameters from all the structures learned from repetition are thus calculated, averaged, and standard errors are obtained for each sample size.

Two well-studied biological networks - IRMA network [3] and *Escherichia coli* SOS network [12] are used. Further, a computationally tractable 4-node synthetic network (water-sprinkler network in Bayes Net Toolbox (BNT) [10]) is used for validating the results. Considering a synthetic network is important as the exact underlying network that produced the data is known and the quality and quantity of data samples can be ensured for the inference study. These three networks are shown in figure 1. It should be noted that the study uses smaller networks (8 nodes) due to computational constrains of structure learning using optimal algorithms rather than heuristic algorithms. For this study, atleast 7500 inference runs were performed for a single network (around 15 sample sizes, 10 bootstraps, around 10 combinations of prior knowledge, and 5 prior levels).

4   Results and Discussion

Although all performance measures mentioned in section 3 are computed and available on request, only relevant results are reported here.

4.1   Limitations of Edge Prior

A previous study [8] had reported drop in performance measures at increasing sample sizes. Our study also shows that except for sensitivity, key performance

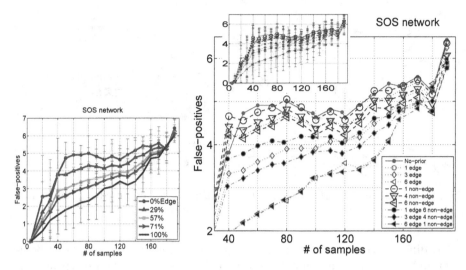

Fig. 3. Effect of edge priors on false positives   **Fig. 4.** Comparison of different types of priors for false positives

measures such as specificity, precision, and F1-score drop with increasing sample sizes. The '0%Edge' plot in figure 2 clearly shows this for the specificity of inferred SOS network. This appears counter-intuitive as we expect improved performance with increasing sample size. Our study further shows that adding different quantities of edge priors (figure 2, plots $29\%, 57\%, 71\%$, and 100%) could not prevent this performance deterioration. Moreover, it is observed that, after a certain sample size, different quantities of edge priors are not effective at all and give the same performance as no-priors (figure 2, beyond 180 samples).

Further analysis shows that this decrease in performance is due to the larger number of false edges being learned with the increasing sample sizes. Figure 3 shows this phenomena for the SOS network. Moreover, beyond a certain sample size, the networks learned without any prior and with edge-priors tend to have the same number of FPs (figure 3, beyond 180 samples). This shows the ineffectiveness of edge priors in improving the overall inference performance.

4.2   Effect of Non-edge Priors Compared to Edge Priors

As seen earlier, certain performance measures drop with increasing sample size due to the inability to discard FPs during inference. A method to avoid FPs is the incorporation of prior non-edges (equation 2). To confirm this, results with non-edge priors are compared with no-priors, edge, and combined priors. Figure 4 shows the variation of FPs with sample size for the three types of priors. To keep the comparison meaningful, only relevant plots are shown, for e.g., the combined prior plot of 3-edge-4-non-edge (3E4NE) will be shown along with plots 3-edge (3E) and 4-non-edge (4NE). Since combined prior plots in this case will have more information than either edge or non-edge plots, improvement given by combined priors will not be considered.

Figure 4 shows that the non-edge prior plots consistently show lower FPs (thus better) than no-priors. They become better than the edge prior plots beyond 180 samples for the SOS network. Similarly, at 60 and 20 samples for IRMA and 4-node networks respectively, non-edge priors become better than edge priors (figures not shown). Thus the non-edge priors give the best performance compared to no-priors or edge priors for the complete range of input samples sizes in general and particularly at higher samples sizes. In all cases, the combined prior (for e.g., see 3E4NE of SOS network in figure 4), below 180 samples is slightly better than or overlap the plots of the equivalent edge prior (3E). Above 180 samples, the combined prior (3E4NE) nearly overlaps the equivalent non-edge prior (4NE).

An interesting observation from all these results is that, at a certain sample size (180, 60, and 20 samples for SOS, IRMA, and 4-node networks respectively), the performance of edge priors drop below or become equal to no-priors and do not have any further improvement. At the same time, the performance of non-edge priors becomes superior to edge priors. We call this the 'cross-over' region since, beyond this region, the non-edge priors become consistently better than the edge priors. This phenomena can be seen in other performance parameters also (figures not shown). Thus, beyond the cross-over region only non-edge priors are effective in reducing the problem of FPs.

It should be noted that since GRNs are sparse, the number of non-edges are much larger than the edges and so considering an equal number of non-edges to edges will give only a very small percentage of total non-edge information. However, non-edge priors are always better than no-priors and their performance do not deteriorate like the edge priors.

4.3   Effect of Non-edge Priors

When a higher percentage of non-edge priors are used for inference, it is seen that specificity, precision, and F1-score improve with an increasing quantity of non-edge priors, for all the three networks. Figure 5 shows the improvement for specificity metric in SOS network, which previously showed deterioration with edge priors in figure 2. Comparing figures 2 and 5 at 180 samples shows that the non-edge priors give an improvement in specificity of $2 - 10\%$ over the edge priors at the different prior levels.

4.4   Obtaining Non-edge Priors

We have seen that non-edge priors are very important in inferring GRN. Now we show that getting many non-edge priors is not difficult even for a not well studied GRN, as the information of transcription factors (TFs) and target genes (TGs) itself can be used. Since, TGs do not have any edges going out, this will identify many of the non-edges of the network. This method is practical because GRNs are known to be sparse (i.e., there are more non-edges than edges), and

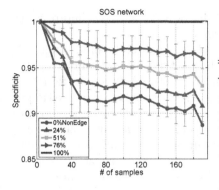

Fig. 5. Effect of non-edge priors on specificity

Table 1. SOS network gene annotation

Gene	Relevant GO-annotation	Out edges
uvrA	ATP catabolic process	No
ruvA	DNA duplex unwinding	No
polB	DNA catabolic process	No
uvrD	DNA repair	No
recA	Interacts with LexA	To LexA
uvrY	Transcription	Possible
umuD	Regulation of transcription	Possible
lexA	Transcription	Possible

have a scale-free topology (i.e., only a few genes control the majority). For small networks such as the *E. coli* SOS network (presented here as an example), the identification of TFs, TGs, and any known interactions can be done from the gene annotations. For larger networks, GO biological process annotation [1] can be used to identify the TFs and TGs.

The *E. coli* network studied here consists of 8 genes and 7 edges. Total of $8 * 8 = 64$ edges are possible, of which 8 will be edges of each gene to itself which we do not consider in the study. Thus total non-edges in the network are $64 - 7 - 8 = 49$. Since the network is small, the details about the genes are obtained from Uniprot [9] and are shown in table 1. It can be seen that only three genes (*lexA*, *umuD*, and *uvrY*) are involved in transcription and thus, can have outgoing edges. Gene *recA* interacts only with *lexA*. Thus, the 7 outgoing edges from 4 genes (*uvrA*, *ruvA*, *polB*, and *uvrD*) and outgoing edges from *recA* to all other 6 genes, can be given as non-edges. These total to $4 * 7 + 1 * 6 = 34$ non-edges, which form, $34/49 = 70\%$ of the total non-edges.

5   Conclusion

It is well known that prior knowledge improves the GRN inference from data but previous studies have not adequately addressed the significance of non-edge priors. This paper studied the effect of edge and non-edge priors separately and also their combination, to understand and establish the effect of non-edge priors. It is found that non-edge priors are complimentary to and are as important as the edge priors and both are essential in improving the overall performance of the GRN inference.

Further, it was reported previously that certain performance measures of the GRN inference deteriorate with an increasing sample size. Our study shows that this deterioration is due to the higher FPs inferred with the increasing sample size. We show that, giving prior knowledge of non-edges during inference is a good solution as the performance metrics such as specificity, precision, and

F1-score are predominantly influenced by non-edge priors than edge priors at high sample sizes. We have also shown that most non-edges of a typical GRN can be obtained easily and a simple method to find these non-edge priors is presented. Currently, our focus is on identifying different methods to improve the computational performance and extend this study to larger networks.

References

1. Ashburner, M., et al.: Gene Ontology: Tool for the Unification of Biology. Nature Genet. 25(1), 25–29 (2000)
2. de Campos, L.M.: A Scoring Function for Learning Bayesian Networks based on Mutual Information and Conditional Independence Tests. J. Mach. Learn. Res. 7, 2149–2187 (2006)
3. Cantone, I., et al.: A Yeast Synthetic Network for In-vivo Assessment of Reverse-Engineering and Modeling Approaches. Cell 137(1), 172–181 (2009)
4. De Smet, R., Marchal, K.: Advantages and Limitations of Current Network Inference Methods. Nat. Rev. Micro 8(10), 717–729 (2010)
5. Friedman, N.: Inferring Cellular Networks Using Probabilistic Graphical Models. Science 303(5659), 799–805 (2004)
6. Gao, S., Wang, X.: Quantitative Utilization of Prior Biological Knowledge in the Bayesian Network Modeling of Gene Expression Data. BMC Bioinformatics 12(1), 359 (2011)
7. Isci, S., Dogan, H., Ozturk, C., Otu, H.H.: Bayesian Network Prior: Network Analysis of Biological Data Using External Knowledge. Bioinformatics 30(6), 860–867 (2014)
8. Le Phillip, P., Bahl, A., Ungar, L.H.: Using Prior Knowledge to Improve Genetic Network Reconstruction from Microarray Data. In Silico Biology 4(3), 335–353 (2004)
9. Magrane, M.: UniProt Consortium : UniProt Knowledgebase: A Hub of Integrated Protein Data. Database (Oxford) (2011)
10. Murphy, K.P.: The Bayes Net Toolbox for Matlab. Computing Science and Statistics 33 (2001)
11. Ooi, C.H., Chetty, M., Teng, S.W.: Differential prioritization in feature selection and classifier aggregation for multiclass microarray datasets. Data Mining and Knowledge Discovery 14(3), 329–366 (2007)
12. Ronen, M., Rosenberg, R., Shraiman, B.I., Alon, U.: Assigning Numbers to the Arrows: Parameterizing a Gene Regulation Network by Using Accurate Expression Kinetics. Proc. Natl. Acad. Sci. U.S.A. 99(16), 10555–10560 (2002)
13. Steele, E., Tucker, A., 't Hoen, P.A.C., Schuemie, M.J.: Literature-based Priors for Gene Regulatory Networks. Bioinformatics 25(14), 1768–1774 (2009)
14. Vinh, N.X., Chetty, M., Coppel, R., Wangikar, P.P.: GlobalMIT: Learning Globally Optimal Dynamic Bayesian Network with the Mutual Information Test Criterion. Bioinformatics 27(19), 2765–2766 (2011)
15. Vinh, N.X., Chetty, M., Coppel, R., Wangikar, P.P.: Issues Impacting Genetic Network Reverse Engineering Algorithm Validation Using Small Networks. Biochimica et Biophysica Acta (BBA) - Proteins and Proteomics 1824(12), 1434–1441 (2012)
16. Wilczyski, B., Dojer, N.: BNFinder: Exact and Efficient Method for Learning Bayesian Networks. Bioinformatics 25(2), 286–287 (2009)

Robust Lane Detection Based On Convolutional Neural Network and Random Sample Consensus

Jihun Kim and Minho Lee

School of Electronics Engineering, Kyungpook National University
1370 Sankyuk-Dong, Puk-Gu, Taegu 702-701, South Korea
{ceuree,mholee}@gmail.com

Abstract. In this paper, we introduce a robust lane detection method based on the combined convolutional neural network (CNN) with random sample consensus (RANSAC) algorithm. At first, we calculate edges in an image using a hat shape kernel and then detect lanes using the CNN combined with the RANSAC. If the road scene is simple, we can easily detect the lane by using the RANSAC algorithm only. But if the road scene is complex and includes roadside trees, fence, or intersection etc., then it is hard to detect lanes robustly because of noisy edges. To alleviate that problem, we use CNN in the lane detection before and after applying the RANSAC algorithm. In training process of CNN, input data consist of edge images in a region of interest (ROI) and target data become the images that have only drawn real white color lane in black background. The CNN structure consists of 8 layers with 3 convolutional layers, 2 subsampling layers and multi-layer perceptron (MLP) including 3 fully-connected layers. Convolutional and subsampling layers are hierarchically arranged and their arrangement represents a deep structure in deep learning. As a result, proposed lane detection algorithm successfully eliminates noise lines and the performance is found to be better than other formal line detection algorithms such as RANSAC and hough transform.

Keywords: lane detection, neural network, deep learning, advanced driver assistance system.

1 Introduction

Road traffic accidents have become one of the most serious problems worldwide today. These accidents are caused by people, vehicle and road infrastructure. Measures to prevent these accidents can be categorized into following 3 types [1]: (1) Changing human behavior; (2) vehicle-related measures; and (3) physical road infrastructure related measures. Changing human behavior can be achieved by law enforcement, information, education and driving instructions while infrastructure measures include construction of new roads. Vehicle related measures include vehicle safety systems such as electronic stability control (ESC), anti-lock braking system (ABS) and advanced driver assistance systems (ADAS). It has been found that ESC and ABS play a crucial role in preventing accidents in crucial situations while ADAS helps to avoid

C.K. Loo et al. (Eds.): ICONIP 2014, Part I, LNCS 8834, pp. 454–461, 2014.

accidents by assisting the driver in his/her driving task continuously. Moreover, ADAS can provide more comfortable driving service.

ADAS is a common accessory in passenger and commercial vehicles these days and serves as a good solution for reducing traffic accidents [2, 3]. In general, ADAS technologies consist of adaptive cruise control (ACC), lane departure warning system, collision avoidance system, adaptive light control, automatic parking, etc. For instance, lane departure warning system [4, 5] and lateral control have been developed by detecting the lane markings of a road using forward-facing camera and computer vision techniques. Similarly lane departure system is a safety feature that informs the driver about changing lane situation. In this system and for other such systems, accurate lane detection is necessary and most important factor.

For lane detection, most common method is based on edge detection in the road scene and then application of RANSAC algorithm[6]. However, the results of RANSAC become unreliable with the growing complexity in road scenes, which may include roadside trees, fence, wall or intersection and so on. For such complex scenes, addition of CNN before and after applying the RANSAC can be a good solution. CNN is a kind of deep network and composed of multiple layers of small neuron collections, which looks at small portions of the input image. This algorithm mimics the dorsal stream of human visual system and is used for various object recognition scenarios such as face detection [7, 8], hand written characters [9], traffic signs [10], etc.

Therefore, in this paper, we propose a new robust lane detection method which combines CNN with RANSAC algorithm. When the RANSAC fails to find a load lane, the trained CNN works and provides the candidate of a road lane. The RANSAC is repeatedly applied to the candidates of road lanes that are obtained from the CNN. The proposed method is efficient to find road lanes robustly in complex real roads with noisy environment.

In Section 2, we present details of our proposed method. We present simulation results in Section 3, and followed by conclusion in Section 4.

2   Proposed Method

2.1   Overview of Proposed Method

We use two-step processing for the lane detection from real world driving videos. First step includes blurring and edge detection for removing the environment noises as a preprocessing step. Second step includes the road lane detection based on RANSAC combined with CNN processes for accurate lane detection. The whole process of proposed model is shown in Fig. 1. If road environment changes dynamically due to weather, time and objects on road, the videos contain lots of noises, which decreases lane detection accuracy in real situation. In order to alleviate that problem, we propose a new method which is combination of RANSAC and CNN. If road condition is simple and easy, we can use the RANSAC algorithm only to get lane information. But if road conditions include lots of noise factors, we can use the CNN before and after the RANSAC algorithm to get the lanes robustly.

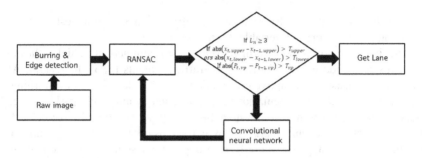

Fig. 1. Structure of proposed model

2.2 Blurred and Edge Detection

The lane is generally a white color in a gray color background road. So, an easy way to detect the lane is to use edge information. However, edge detection itself may not be sufficient way to detect the lanes because of various noises in an image such as shadow, illumination change, etc. Therefore, we use a blurring image, which are obtained by 5x5 Gaussian smoothing function before the edge detection. This blurring step reduces the environment noise, and produces more reliable information in a scene.

Moreover, for robust lane detection, the surround area of the lane needs to be suppressed. So, we use hat-shape kernel[11] to strengthen lane information, while suppressing the surroundings around the lanes in edge detection. Fig. 2.A shows the shape of hat-shape kernel. Thus, edge image calculated using convolution of lane image and this kernel. It's performance is better than other several preprocessing method. Also, result of edge detection in several situation showed in Fig. 2.B

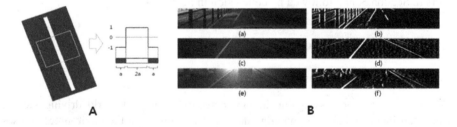

Fig. 2. The kernel of edge detection and examples. **A** : the kernel shape of edge detection. **B**: (a) fence and shadow image (b) fence and shadow image using edge detection (c) night image (d) night image using edge detection (e) changing intensity image in tunnel (f) changing intensity image in tunnel using edge detection.

2.3 Lane Detection Using RANSAC

RANSAC is an estimation technique based on the principle of hypotheses generation and verification [12, 13]. Given a model requiring a minimum of N data points to instantiate its free parameters and a set of data points P containing more than N elements, the RANSAC algorithm finds the line that has the highest possibility. We set the ROI to reduce the computation load. Since lanes look like vertical shapes from the car interior, we set the candidates of arrival and departure point on the upper and lower ROI lines. When the road scene is complex, the selected point is not likely to create a lane and accuracy of lane detection may decrease.

2.4   Reinforcement of Lane Detection Using CNN Combined with RANSAC Algorithm

The accuracy of lane detection based on RANSAC is highly dependent on the road conditions. There are three cases to fail the lanes based on the RANSAC algorithm;.(1) detecting more than two lines, (2) the shift of a lane between frames is too big and (3) the shift of the vanishing point determined by the left and right lines is too big. Fig. 3 shows the examples for three cases. In all three cases, we use the CNN before RANSAC algorithm to reinforce the lane information and suppress the surrounding noise information for stable lane detection, and then apply again the LANSAC to the candidate images obtained from CNN.

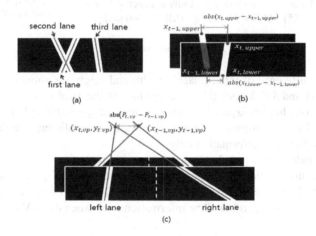

Fig. 3. Example of three cases (a) the number of lanes is three (b) position of the lane, where $x_{t,\ upper}$, $x_{t-1,\ upper}$ and $x_{t,\ lower}$, $x_{t-1,\ lower}$ in the x-axis direction are represented upper and lower points of lane in t and t-1 frame and $T_{upper}\ and\ T_{lower}$ are threshold values at upper and lower positions of lane, respectively. (c) vanishing point of lane, where $P_{t,\ vp}$ is the vanishing point at $(x_{t,vp}, y_{t,vp})$ in t frame and $P_{t-1,\ vp}$ is the vanishing point at $(x_{t-1,vp}, y_{t-1,vp})$ in t-1 frame. T_{vp} is threshold value of vanishing point.

On the other hand, the CNN is a kind of deep network which imitates the dorsal stream in human brain[14]. The dorsal stream is known as a brain area for object detection and recognition. Therefore, CNN is generally used for object detection and recognition and includes convolutional layer (simple cell) and subsampling layer (complex cell). The simple cell treats local receptive field information. In the CNN, convolutional layer works same as the simple cell. In the convolutional layer, kernels are shifted over the valid region of the input image. The subsampling layer mimics the complex cell's functions and store max-pooling or average information from convolutional layer using down sampling. After multiple convolutional and subsampling layers, a fully connected multilayer perceptron (MLP) is used to complete the CNN. In the MLP, the error back propagation is used for weight control. The output of MLP is the final image of the CNN. In this paper, we considered 2 subsampling layers, 3 convolutional layers and an MLP including 3 fully connected layers in the CNN. Fig. 4 shows its structure.

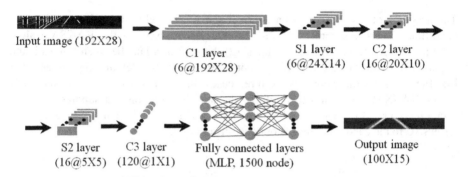

Fig. 4. Structure of CNN

Due to the difference of input image's width and height (as shown in Fig. 4), we use 8x2 kernel and 4x2 kernel for subsampling. At the final step, we use an MLP network with one hidden layer. The size of output image is 100x15. For training the CNN, we need target information of the lane. So, we manually prepare the target lane data which include lane information only.

In a test mode, when the above three cases in subsection 2.4 happen, the CNN works to find the candidate of lanes, and the output of CNN produces the 100x15 output image. Then, we apply again the RANSAC to the output images from the CNN. Finally, we can get robust lane information even when the RANSAC itself fails to lanes.

3   Experimental Results

The result of lane detection using RANSAC algorithm is shown in Fig. 5. The results highly depend on the degree of complexity of input road scenes. If there are lots of noises such as fence, wall, reflecting light in front window, lane detection based on RANSAC is very poor as shown in Fig. 5 (b).

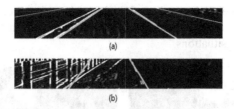

(a)

(b)

Fig. 5. Results of lane detection using RANSAC algorithm (a) well detected lane in simple image (b) wrong detected lane in complex image

To improve the accuracy of lane detection in complex road scenes, we use the RANSAC combined with CNN. The results are shown in Fig. 6. Since the CNN gives the candidate region for the lanes, and resultantly noise reduced results can be obtained as shown in Fig. 6 (b). We apply again the RANSAC to the candidate of CNN. Finally, we can get final detected actual lanes. Notice that the proposed method is robust to the road environment in lane detection as shown in Fig. 6 (c).

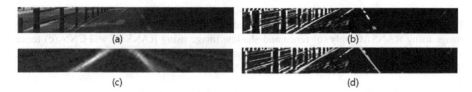

(a)   (b)

(c)   (d)

Fig. 6. Simulation results, (a) original image (b) Edge image (c) output image of CNN and (d) result of RANSAC combined with CNN

We tested our proposed RANSAC combined with CNN algorithm on complex video clips containing three different conditions. The conditions include detecting more than three lines (case 1), changing of lane position is too big (case 2) and the distance of vanishing points of left and right lanes is too big (case 3). We consider corrected detection, missed detection and false detection to evaluate the performance of these clips in Table 1. Corrected detection means that more than half region of the detected lane overlaps with a target lane. Missed detection means that less than half region of detected lane overlaps with a target lane, and false detection means that detected lane never overlaps with a target lane.

Table 1. Performance evaluation in three different conditions

Clips	Corrected detection	Missed detection	False detection
Case 1	94.7.0%	5.1%	0%
Case 2	93.9%	4.9%	1.2%
Case 3	93.2%	4.5 %	2.3%

Fig. 7 shows some experimental results of road lane detection in several complex road scenes. The experimental results show that the proposed method is appropriately performed in complex situations.

Fig. 7. Results of lane detection in different road environment conditions (a) fence and shadow image using RANSAC only (b) fence and shadow image using RANSAC and CNN (c) fence and reflecting light in front window image using RANSAC only (d) fence and reflecting light in front window image using RANSAC and CNN.

4   Conclusions

In this paper, we proposed a new method combined RANSAC with CNN for lane detection in complex scenes. As a preprocessing, we use blurring and edge detection using Gaussian smoothing and hat-type kernel. To detect the lane, we apply the RANSAC algorithm only for simple road scenes, and combined CNN with RANSAC algorithm is used for complex road scenes. The complexity of road condition is determined by the results of RANSAC algorithm. We simulated several real road environment conditions and tested the accuracy of our proposed method. Our results confirm that the proposed method has robust lane detection performance in spite of complex road conditions. In our future work, we would like to test the proposed method with larger dataset, and try to optimize it on an embedded platform.

Acknowledgments. This research was supported by the MSIP (Ministry of Science, ICT & Future Planning), Korea, under the C-ITRC (Convergence Information Technology Research Center) support program (NIPA-2014-H0401-14-1004) supervised by the NIPA (National IT Industry Promotion Agency.) and was supported by 'Software Convergence Technology Development Program', through the Ministry of Science, ICT and Future Planning(S1002-13-1014).

References

1. Lu, M., Wevers, K., Van Der Heijden, R.: Technical feasibility of advanced driver assistance systems (ADAS) for road traffic safety. Transportation Planning and Technology 28(3), 167–187 (2005)
2. Hojjati-Emami, K., Dhillon, B., Jenab, K.: Reliability prediction for the vehicles equipped with advanced driver assistance systems (ADAS) and passive safety systems (PSS). International Journal of Industrial Engineering Computations 3(5), 731–742 (2012)
3. Maag, C., Muhlbacher, D., Mark, C., Kruger, H.-P.: Studying effects of advanced driver assistance systems (ADAS) on individual and group level using multi-driver simulation. IEEE Intelligent Transportation Systems Magazine 4(3), 45–54 (2012)
4. Borkar, A., Hayes, M., Smith, M.T.: A non overlapping camera network: Calibration and application towards lane departure warning. In: International Conference on Image Processing, Computer Vision, and Pattern Recognition, IPCV 2011 (2011)
5. Tapia-Espinoza, R., Torres-Torriti, M.: Robust lane sensing and departure warning under shadows and occlusions. Sensors 13(3), 3270–3298 (2013)
6. Kim, Z.: Robust lane detection and tracking in challenging scenarios. IEEE Transactions on Intelligent Transportation Systems 9(1), 16–26 (2008)
7. Lawrence, S., Giles, C.L., Tsoi, A.C., Back, A.D.: Face recognition: A convolutional neural-network approach. IEEE Transactions on Neural Networks 8(1), 98–113 (1997)
8. Garcia, C., Delakis, M.: Convolutional face finder: A neural architecture for fast and robust face detection. IEEE Transactions on Pattern Analysis and Machine Intelligence 26(11), 1408–1423 (2004)
9. Bengio, Y., LeCun, Y., Nohl, C., Burges, C.: Lerec: A NN/HMM hybrid for on-line handwriting recognition. Neural Computation 7(6), 1289–1303 (1995)
10. Sermanet, P., LeCun, Y.: Traffic sign recognition with multi-scale convolutional networks. In: The 2011 International Joint Conference on Neural Networks (IJCNN). IEEE (2011)
11. Jiang, R., Terauchi, M., Klette, R., Wang, S., Vaudrey, T.: Low-level image processing for lane detection and tracking. In: Huang, F., Wang, R.-C. (eds.) ArtsIT 2009. LNICST, vol. 30, pp. 190–197. Springer, Heidelberg (2010)
12. Fischler, M.A., Bolles, R.C.: Random sample consensus: A paradigm for model fitting with applications to image analysis and automated cartography. Communications of the ACM 24(6), 381–395 (1981)
13. Aly, M.: Real time detection of lane markers in urban streets. In: 2008 IEEE Intelligent Vehicles Symposium. IEEE (2008)
14. Hubel, D.H., Wiesel, T.N.: Receptive fields, binocular interaction and functional architecture in the cat's visual cortex. The Journal of Physiology 160(1), 106 (1962)

Learning Local Receptive Fields in Deep Belief Networks for Visual Feature Detection

Diana Turcsany and Andrzej Bargiela

School of Computer Science, The University of Nottingham
Nottingham, United Kingdom

Abstract. Through the introduction of local receptive fields, we improve the fidelity of restricted Boltzmann machine (RBM) based representations to encodings extracted by visual processing neurons. Our biologically inspired Gaussian receptive field constraints encourage learning of localized features and can seamlessly integrate into RBMs. Moreover, we propose a method for concurrently finding advantageous receptive field centers, while training the RBM. The strength of our method to reconstruct characteristic details of facial features is demonstrated on a challenging face dataset.

Keywords: Visual information processing, neural encoding, deep belief network, receptive fields, unsupervised learning, facial feature detection.

1 Introduction

Despite strong multi-disciplinary interest the highly accurate vision system of humans and other biological systems is still not fully understood and cannot be replicated with computational methods. Important discoveries have been made regarding the morphology and functionality of neural cells and networks, however our knowledge is still far from complete. Computational models of neural circuits in the visual pathway have great importance for improving our understanding of biological visual processing. A more informed background could facilitate the design of computational visual processing units, e.g., retinal implants. Currently, with the amount of unknown details, robust computational models of biological visual processing have to account for uncertainty and unknown details. We believe flexible probabilistic models, e.g., deep belief networks (DBNs) [2] possess great potential for modeling in this uncertain environment.

Deep Networks. To learn a multi-layer generative model of the data where each higher layer corresponds to a more abstract representation of information, Hinton et al. [2] train a DBN layer by layer using unsupervised RBMs. The network parameters are subsequently fine-tuned using backpropagation. Since this efficient training method for deep networks was introduced, there has been increasing research within deep learning. The potential of deep architectures for learning meaningful features has been demonstrated on a number of visual

C.K. Loo et al. (Eds.): ICONIP 2014, Part I, LNCS 8834, pp. 462–470, 2014.

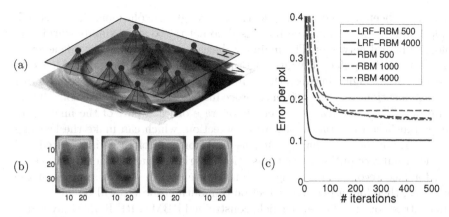

Fig. 1. (a) LRF-RBM model schematic, showing an input image in the visible layer V and local feature detector RFs (blue cones) in the hidden layer H with feature hubs around the eyes and mouth. (b) RF maps of LRF-RBMs run with different learning parameters. Automatically learned feature detector RFs are combined to show areas attracting more detectors. Darker red areas indicate higher feature density. Feature hubs emerged around average eye and mouth locations at pixels (9,15), (20,15), and (15,30). (c) Average squared reconstruction error per pixel is shown on the test set. Method names and hidden node counts are given in the graph. With the same node count LRF-RBMs give significantly lower errors than RBMs. Moreover, a 500 node LRF-RBM performs similarly to a 4000 node RBM.

tasks [3,4,5,6,7], [10], [12]. DBNs have also been shown suitable for modeling feature detection in the retina [11] and visual areas V1 and V2 [8]. Despite this success in neural modeling, primal emphasis has been given to improving performance of deep learning on visual recognition tasks, rather than increasing the fidelity of deep architectures to real neural circuits of the visual pathway. Our aim is to fill this gap by proposing deep network structures that more closely resemble biological neural networks, but still provide flexibility and great performance on visual recognition tasks. Such architectures possess high potential for modeling visual information processing in the retina and visual cortex.

Local Receptive Fields. In focus of this paper is the extension of RBMs with local receptive fields (RFs) in a way that the training process, the final architecture and the inference at test time closely resemble biological neural networks of the visual pathway. We concentrate mainly on early processing stages, the retina and V1. Our contributions are (1) a modification to the contrastive divergence [1] (CD) algorithm that introduces local receptive field constraints for hidden nodes, (2) a method for automatically identifying locations of high importance within the visual input space during RBM training, and (3) by utilizing these locations as RF centers, a compact, yet powerful encoding of visual features. We show using biologically inspired Gaussian shaped local RFs and learning advantageous RF placement improve RBM and DBN based reconstruction of face images.

RFs have been modeled in deep learning through convolutional networks [4,5, 6], [9], however the training methods used do not try to approximate learning in biological neural networks. The main emphasis is on improving the efficiency of learning to scale up deep learning algorithms to high dimensional problems. In these networks weights between the visible and hidden layers are the same across all image locations and the inference procedure can therefore utilize convolution operations. The same feature detectors operate on each part of the image providing translation invariance of feature detection, which can make the learning task easier. On the other hand, spurious detections can often be introduced and in some visual recognition tasks translation invariance may not be advantageous (e.g., for face recognition in aligned images the mouth always appears in the same area, therefore a positive detection of mouth elsewhere will be false).

In contrast, our local receptive field constrained RBM (LRF-RBM) only learns relevant feature detectors at any one image location. As opposed to a fixed grid layout of rectangular RFs [7], [12], our hidden nodes move around the visual space during training to find the best location for their Gaussian RF center. By letting the detectors move to locations of interest *"feature hubs"* can emerge in image regions where the training data has high variation, while more uniform areas will attract less detectors. The resulting network architecture extracts compact representation of visual information, provides very quick inference and by combining local features as building blocks, the network is strong at reconstructing previously unseen images. An illustration of the receptive field learning and RF distributions of our trained models are in shown in Fig. 1(a)-(b).

2   Local Receptive Field Constrained RBMs

The unsupervised phase of Hinton et al. [2]'s DBN training utilizes RBMs for learning each layer of the representation. The energy-based RBM models include a visible and a hidden layer, with connections between hidden and visible nodes but not within layers. This restriction ensures conditional independence of hidden nodes given visible nodes and vice versa, which is key for the efficiency of RBMs. In most vision tasks visible nodes correspond to pixels, while hidden nodes model visual processing neurons and detect image features.

2.1   RBM Training

The energy function of RBMs with binary visible and hidden nodes is given by:

$$E(v, h) = -a'v - b'h - h'Wv \, , \tag{1}$$

where v and h are the states of visible and hidden node, W is the weight matrix describing the symmetric connections between the visible and hidden layer, while a and b are visible and hidden biases respectively. Learning aims at reducing the energy (increasing the log probability) of the training data.

RBMs can be trained with the approximate but very efficient contrastive divergence [1] (CD) algorithm. In each step of CD, (i) visible states are initialized to a training example, then (ii) hidden states can be sampled parallel, due to the conditional independence properties, according to:

$$p(h_j = 1|v) = \frac{1}{1 + exp(-b_j - \sum_i v_i w_{ij})} \, , \tag{2}$$

followed by (iii) the reconstruction phase where visible states are sampled using:

$$p(v_i = 1|h) = \frac{1}{1 + exp(-a_i - \sum_j h_j w_{ij})} \, , \tag{3}$$

finally (iv) the weights are updated according to:

$$\Delta w_{ij} = \epsilon(< v_i h_j >_{data} - < v_i h_j >_{reconst}) \, , \tag{4}$$

where ϵ is the learning rate, correlation between the activations of v_i and h_i measured after (ii) gives $< v_i h_j >_{data}$, and the correlation after the reconstruction phase (iii) determines $< v_i h_j >_{reconst}$. A similar rule is applied to the biases.

For continuous data (e.g., the face images we studied) using Gaussian visible nodes can improve the model. In this case the energy function changes to:

$$E(v, h) = \sum_i \frac{(v_i - a_i)^2}{2\sigma_i^2} - \sum_j b_j h_j - \sum_{i,j} \frac{v_i}{\sigma_i} h_j w_{ij} \, , \tag{5}$$

where σ_i is the standard deviation at v_i. The probability of hidden node activation and the expected value of a visible node (i.e., the reconstructed value) is then given by:

$$p(h_j = 1|v) = \frac{1}{1 + exp(-b_j - \sum_i (v_i/\sigma_i)w_{ij})} \, , \tag{6}$$

$$< v_i >_{reconst} = a_i + \sigma_i \sum_j h_j w_{ij} \, . \tag{7}$$

2.2 Training with Local Receptive Fields

Neurons in early stages of the visual pathway typically only receive input from a small localized area of the previous processing layer. Moving up the layers, receptive field of neurons (the area of the photoreceptor layer in which stimulus can result in neural response) is gradually getting bigger with increasing complexity in structure. As an example, retinal ganglion cell RFs can be closely modeled by difference-of-Gaussians (DoGs), while RFs of V1 simple cells by Gabor filters.

LRF-RBMs include receptive field constraints for hidden nodes to outline the area from which the hidden node is most likely to receive input. These constraints are given in the form of RF masks, denoted by R, that operate on the RBM weights W. Each mask has a center location which corresponds to a hidden node's location in visual space. R describes the likelihood of a connection being present between a visible and a hidden node given their distance in the visual space, where the likelihood converges to 0 as the distance goes to infinity. During training to avoid the prohibitive process of sampling connections from the likelihood, we will instead use the values of R as additional weights on top of W. R thereby narrows down the scope of hidden nodes to local neighborhoods. Note that from the biological modeling point of view, R provides only a constraint or regularizer on the RF structure, the actual RFs are specified by R and W together. After training, these can show significantly different structures compared to R alone. Still, to keep the description simple we refer to R as RFs.

We found LRF-RBMs with disk or square shaped RFs efficiently learn local features, however Gaussian RFs provide smoother reconstructions with better detail and can be truncated to preserve efficiency. Gaussian RF constraints are also more adequate when modeling biological neurons in early stages of visual processing. From here on we will discuss only the Gaussian case, using fixed standard deviation (SD) for each RF, denoted by σ^{RF}.

The training algorithm described in Section 2.1 can be used for LRF-RBMs with modifications. The energy functions in Eqs. 1 and 5 and also Eqs. 2, 3, 6, 7 can be adapted by substituting w_{ij} with $r_{ij}w_{ij}$, where r_{ij} is the RF constraint on the connection between v_i and h_j. The weight update equation changes to:

$$\Delta w_{ij} = r_{ij}\epsilon(< v_i h_j >_{data} - < v_i h_j >_{reconst}).$$ (8)

Learning RF Centers. Hidden nodes can be placed at uniform distances from each other or allocated randomly, but this would not allow the network architecture to adapt to specific properties of the input data. Non-uniform feature detector distributions can be beneficial to obtain compact representations by exploiting patterns in the dataset (e.g., aligned faces have facial features at given locations, most natural images have the center of interest in the middle). When solving a task some areas of the visual space may need representation at better resolution, using many different feature detectors, while other areas do not convey much information. In the human vision system, the retina also has non-uniform ganglion cell distribution between the center (fovea) and the periphery, former being denser, thus better resolution is obtained in the center of the visual space. Our model utilizes non-uniform feature detector distribution by allowing the system to identify areas of the visual input space which need higher number of feature detectors to obtain a good data representation.

Our method learns RF centers during RBM training. In each RBM iteration, a hidden node's RF is allocated to the local area that has the strongest connections to the hidden node and thus give the most well defined feature. This is done by first (i) writing the weights of the hidden node in the shape of the input image data and (ii) applying a transformation, then (iii) filtering the weight image with a Gaussian filter (SD: σ^{RF}) on each channel, (iv) the responses over channels are combined by taking the max, finally, (v) the location with the maximum response is selected as the new RF center and (vi) a Gaussian (SD: σ^{RF}) around this center provides the updated RF. We examined element-wise transformations

Fig. 2. Samples of the test data are shown in the first row. The second and third rows show their reconstructions produced by an RBM and an LRF-RBM respectively. Note how small details, e.g., eye and mouth shapes or direction of gaze are better retained with the LRF-RBM due to the number of specialized eye and mouth detectors. Note also how images of side facing people can confuse the RBM but not so the LRF-RBM.

including identity, absolute and squared value, and found the latter two worked similarly well and superior to identity (results are shown with squared value).

LRF-DBNs. According to the DBN training procedure, we can train multiple layers of feature detectors on top of an LRF-RBM hidden layer using either RBMs or LRF-RBMs (e.g, with increasing RF sizes) with binary visible nodes. We call these models LRF-DBNs. If LRF-RBMs are used for training higher layers, hidden node locations are fixed after training a layer and RF constraints of higher layer nodes are applied when training the next layer. Although here we focus on unsupervised training, we note however for classification tasks supervised fine-tuning could subsequently be applied, analogously to DBNs.

3   Experiments

In the followings, we demonstrate how our LRF-RBM can discover important feature hubs in the deep funneled Labeled Faces in the Wild (LFW)[1] [4] face recognition dataset containing aligned faces. We have also experimented on the MNIST handwritten digit dataset using LRF-RBMs/DBNs, which successfully learned feature detectors for digit parts. When trained on the simulated photoreceptor input of [11] our LRF-RBMs were detecting local features including Gabor-like filters. Here we focus on a detailed analysis using the LFW dataset.

Dataset. LFW contains 13233 RGB images of public figures (see examples in Fig. 2 first row). The intended task on the dataset is recognizing whether two face images are taken of the same person, without having seen the person(s) during training. RBMs with rectified linear or binary hidden nodes, first trained unsupervised on single faces and subsequently fine-tuned in a supervised manner

---

[1] Available at http://vis-www.cs.umass.edu/lfw/

on pairs of images, have been shown to achieve good results on this task [10]. Applying supervised methods on pairs of faces is out of the scope of this paper, which focuses on modeling of biological vision systems. Our primal interest is to investigate the capability of LRF-RBMs to identify regions of high importance in LFW images and utilize these hubs to provide compact representation.

We applied similar pre-processing to Nair and Hinton [10] and trained RBMs with binary hidden nodes on single faces using their published settings. With this we compare our LRF-RBM model run on the same data. A separate training and test set was used, with 4038 training and 1711 test images. The central 105x153 part of the images was cropped, thereby eliminating much of the background.[2] Images were then subsampled to 27x39(x3). Finally, to simplify training, the data was normalized along each component to have zero mean and unit variance.

Training. RBMs with Gaussian visible nodes and binary hidden nodes were trained on mini-batches of size 100 for 2000 iterations, with hidden node numbers of 500, 1000 or 4000, applying a learning rate (ϵ) of 0.001 (higher learning rates failed) and momentum. LRF-RBMs were trained using the same settings, except for ϵ, where 0.1 was optimal. Results are shown when σ^{RF} of 3 and a filter size of 5 was used during RF center learning. Both RBMs and LRF-RBMs were able to learn good models within a few hundred iterations, after which performance only slightly improved. In the followings, if not stated otherwise, results are displayed for models with 4000 hidden nodes trained for 2000 iterations. We also trained LRF-DBNs to learn a second layer of feature detectors on top of our LRF-RBM features using a 1000 hidden node RBM without RF constraints.

Testing. Reconstructions are obtained by calculating the top-down activations after feeding in an image. Performance was evaluated on the test set quantitatively by comparing the squared reconstruction errors (SRE), i.e., the squared distance between the original data and its reconstruction, and qualitatively by displaying example reconstructions. In the case of LRF-RBMs, the spatial distribution of feature detectors were examined and feature hubs identified. Visualization of features learned by RBMs and LRF-RBMs are obtained by displaying their weight vector in the shape of the input images. Visualization of higher layer hidden nodes is obtained by a linear combination of the strongest connected lower level features with their weights. RBM and LRF-RBM features where compared based on the distinctiveness of their appearance and locations.

4   Results

Reconstruction. SREs are compared on normalized test data in Fig. 1(c), indicating a superior reconstruction capability for LRF-RBMs. SREs shown translate to an average 16 pixel difference on original test images for the LRF-RBM vs. 18 for the RBM. Table 1 demonstrates LRF-RBMs/DBNs give lower pixel errors than the corresponding RBMs/DBNs. The comparison of reconstructed images

---

[2] These pixels are known to unintentionally provide helpful context for recognition.

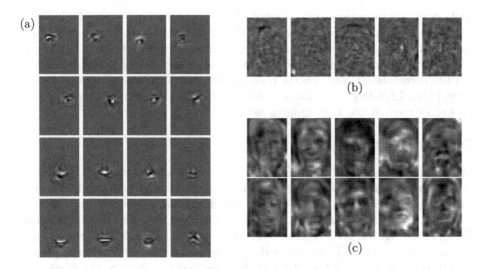

Fig. 3. (a) Distinctive looking detectors located in feature hubs within an LRF-RBM. From top to bottom row: detectors of the persons' right eye, left eye, nose and mouth can be seen. (b) RBM features having global structure. (c) Sample of second layer features learned by an LRF-DBN, corresponding to characteristic looking faces.

Table 1. Average pixel error of (LRF-)RBM reconstructions from 500 and 4000 length encodings, and (LRF-)DBN reconstructions from 1000 length encodings on top layer

hidden nodes	RBM	LRF-RBM	hidden nodes	DBN	LRF-DBN
500	22.31	**19.97**	1000-1000	22.76	**18.96**
4000	18.12	**16.06**	4000-1000	19.32	**18.19**

in Fig. 2 is even more convincing. Both models can reconstruct main features of test examples with a limited amount of nodes. However, characteristic details, especially around eye and mouth areas, are better retained using LRF-RBMs. Such details can help distinguish persons. This analysis confirms LRF-RBMs outperform RBMs for reconstructing previously unseen data.

Features. Figure 3(a) shows local facial feature detectors learned by an LRF-RBM, while Fig. 3(b) shows a sample of RBM features. All the RBMs we trained have learned features similar in nature to the ones shown, having global structure with an occasional local peak. We could not identify any clear local detector modeling a single facial feature. Our LRF-RBMs on the other hand attracted feature hubs around eye and mouth regions and by focusing on these areas have learned a number of distinctive looking eye, mouth and nose detectors. The spatial arrangement of detectors is shown in the maps of Fig. 1(b). The second map from the left belongs to the LRF-RBM that generated the local features in Fig. 3(a). Alongside these specific eye and mouth detectors, Gabor filters and

DoG detectors were also common among the learned features, especially in areas along the contour of the face. The layout of features with the emergence of feature hubs around key areas within the input space demonstrates how LRF-RBMs can identify important regions within the image data which need a higher density of feature detectors for representing their details. Features learned on the second layer in our LRF-DBN have more global receptive field structures corresponding to well defined, varied looking faces as can be seen in Fig. 3(c).

5   Conclusions

We proposed a modified unsupervised RBM training algorithm, the LRF-RBM, which poses constraint on feature detector RFs and can automatically discover advantageous placement of RF centers. We have shown how feature detectors converge to important areas within face images, e.g., eyes and mouth, forming feature hubs. We have demonstrated the superiority of LRF-RBMs to reconstruct details of test images. In future work we will incorporate RFs of varying sizes and further investigate LRF-DBNs for learning multi-layer representations.

References

1. Hinton, G.E.: Training products of experts by minimizing contrastive divergence. Neural Computation 14(8), 1771–1800 (2002)
2. Hinton, G.E., Osindero, S., Teh, Y.W.: A fast learning algorithm for deep belief nets. Neural Computation 18(7), 1527–1554 (2006)
3. Hinton, G.E., Salakhutdinov, R.R.: Reducing the dimensionality of data with neural networks. Science 313(5786), 504–507 (2006)
4. Huang, G.B., Mattar, M.A., Lee, H., Learned-Miller, E.: Learning to align from scratch. In: Neural Information Processing Systems, pp. 773–781 (2012)
5. Kavukcuoglu, K., Sermanet, P., Boureau, Y.L., Gregor, K., Mathieu, M., LeCun, Y.: Learning convolutional feature hierarchies for visual recognition. In: Neural Information Processing Systems, pp. 1090–1098 (2010)
6. Krizhevsky, A., Sutskever, I., Hinton, G.: ImageNet classification with deep convolutional neural networks. In: Neural Information Processing Systems, pp. 1106–1114 (2012)
7. Le, Q.V., Monga, R., Devin, M., Corrado, G., Chen, K., Ranzato, M.A., Dean, J., Ng, A.Y.: Building high-level features using large scale unsupervised learning. In: International Conference on Machine Learning, pp. 81–88 (2012)
8. Lee, H., Ekanadham, C., Ng, A.: Sparse deep belief net model for visual area V2. In: Neural Information Processing Systems, pp. 873–880 (2008)
9. Lee, H., Grosse, R., Ranganath, R., Ng, A.Y.: Convolutional deep belief networks for scalable unsupervised learning of hierarchical representations. In: International Conference on Machine Learning, pp. 609–616 (2009)
10. Nair, V., Hinton, G.E.: Rectified linear units improve restricted Boltzmann machines. In: International Conference on Machine Learning, pp. 807–814 (2010)
11. Turcsany, D., Bargiela, A., Maul, T.: Modelling retinal feature detection with deep belief networks in a simulated environment. In: European Conference on Modelling and Simulation, pp. 364–370 (2014)
12. Zhu, Z., Luo, P., Wang, X., Tang, X.: Deep learning identity-preserving face space. In: IEEE International Conference on Computer Vision, pp. 113–120 (2013)

Adaptive Wavelet Extreme Learning Machine (AW-ELM) for Index Finger Recognition Using Two-Channel Electromyography

Khairul Anam[1,2] and Adel Al-Jumaily[2]

[1] University of Jember, Indonesia
Khairul.Anam@student.uts.edu.au
[2] University of Technology Sydney, Australia
adel@uts.edu.au

Abstract. This paper proposes a new structure of wavelet extreme learning machine i.e. an adaptive wavelet extreme learning machine (AW-ELM) for finger motion recognition using only two EMG channels. The adaptation mechanism is performed by adjusting the wavelet shape based on the input information. The performance of the proposed method is compared to ELM using wavelet (W-ELM0 and sigmoid (Sig-ELM) activation function. The experimental results demonstrate that the proposed AW-ELM performs better than W-ELM and Sig-ELM.

Keywords: Wavelet extreme learning machine, adaptive.

1 Introduction

A wavelet neural network (WNN) is a special case of a feed-forward neural network which its activation function is wavelets [1]. A standard gradient descend can be used to train the weight of WNN. However, drawbacks of the gradient descent method such as long training time and easy trapped to local minima have hampered the implementation of WNN in the real-time application [2]. On the other hand, an extreme learning machine (ELM) was introduced to train a single-hidden layer feed-forward networks (SLFNs) resulting in a system which is fast and able to avoid a local minima [3]. Inevitably, WNN can be constructed using SLFNs.

The combination of ELM and WNN can be conducted by simply replacing the activation function of ELM with wavelets [4] [5]. This is the simplest unification of both networks as has been done in [5]. Cao et al. [6] introduced a new combination of these two algorithms by proposing a composite function of WNN with ELM. In this method, they implemented two activation functions, a wavelet function and any piecewise function which are done in order.

Another new unification of ELM and WNN was proposed by Javed et al. [7] who proposed a summation wavelet extreme learning machine (SW-ELM). Same as Cao, Javed et al. utilized two activation functions but employed them in different ways.

C.K. Loo et al. (Eds.): ICONIP 2014, Part I, LNCS 8834, pp. 471–478, 2014.

These two activation functions were done in parallel and their outputs were averaged to be the output of the hidden nodes.

This paper proposes an adaptive wavelet extreme learning machine (AW-ELM), a new unification of ELM and WNN. According to WNN structure, the proposed system utilizes a wavelet function as the activation function in the hidden node. However, the activation functions are not fixed but they are adjusted regarding to the changing in the input. The sigmoid function is used to process the input information and produce translation parameters of the wavelets in the related hidden-node. In this paper, the performance of AW-ELM will be tested to classify the finger motions from the surface Electromyography signal (EMG) extracted from two-channel sources on the forearm. In addition, its classification performance will be compared with two types of ELM, ELM with wavelet activation function (W-ELM) and sigmoid activation function (Sig-ELM).

The organization of the paper is as follows: section 2 describes the theory of W-ELM and AW-ELM, and the implementation of AW-ELM for finger motion classification. Then section 3 and 4 presents the results and the discussion. Finally section 4 will conclude this paper.

2   Methods

2.1   Wavelet Extreme Learning Machine (W-ELM)

W-ELM can be considered as a special case of extreme learning machine which its activation function is wavelets. The output function of W-ELM for arbitrary samples $(x_k, t_k) \in R^n \times R^o$ with M hidden nodes is

$$f_i^{\ k}(\mathbf{X}) = \sum_{j=1}^{M} V_{ij} \psi_{a_j b_j}(w_j, c_j, \mathbf{x}_k) \quad i = 1, 2, ..., O \tag{1}$$

where

$$\psi_{a_j b_j}(x) = \frac{1}{\sqrt{a_j}} \psi\left(\frac{x - b_j}{a_j}\right), \qquad j = 1, 2, ..., M \tag{2}$$

in which a_j and b_j are dilatation and translation parameters of the wavelets, respectively. An initialization of dilatation and translation parameters, a_j and b_j, in WNN is an important issue. The initialization should consider the input information in order to let the time domain of the wavelet covering the input domain. According to [1], suppose the input vector x_k has the domain $[x_{kmin}, x_{kmax}]$, t^* and σ^* are the centre and the radius of the mother wavelet $\psi_{a_i b_i}$, then domain of $\psi_{a_i b_i}$ is given by:

$$[b_j + a_j(t^* - \sigma^*), \ b_j + a_j(t^* + \sigma^*)]$$

Meanwhile, the input information range for ith hidden layer can be calculated as:

$$\left[\sum_{i=1}^{N} w_{ji} x_{i\,min} \quad , \quad \sum_{i=1}^{N} w_{ji} x_{i\,max}\right]$$

where w_{ji} is the weight connecting the jth hidden layer the ith input. The wavelet can cover the input space if

$$b_j + a_j\,(t^* - \sigma^*) = \sum_{i=1}^{N} w_{ji} x_{i\,min} \tag{3}$$

and

$$b_j + a_j\,(t^* + \sigma^*) = \sum_{i=1}^{N} w_{ji} x_{i\,max} \tag{4}$$

From equation (9) and (10), we can calculate a_i and b_i as:

$$a_j = \frac{1}{2\sigma^*}\left(\sum_{i=1}^{N} w_{ji} x_{i\,max} - \sum_{i=1}^{N} w_{ji} x_{i\,min}\right) \tag{5}$$

$$b_j = \frac{1}{2\sigma^*}\left(\sum_{i=1}^{N} w_{ji} x_{i\,max}\,(\sigma^* - t^*) + \sum_{i=1}^{N} w_{ji} x_{i\,min}\,(\sigma^* + t^*)\right) \tag{6}$$

2.2   Adaptive Wavelet Extreme Learning Machine (AW-ELM)

The Proposed Structure
The proposed AW-ELM is depicted by Fig. 1. If M is the number of hidden node and N is the number of input, then the input of the hidden layer P_j is given by

$$P_j(x) = \sum_{i=1}^{N} x_i\,w_{ji} + c_j \qquad j = 1, 2, ..., M \tag{7}$$

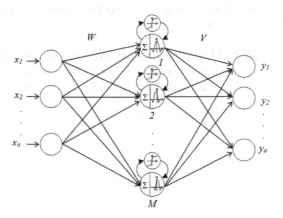

Fig. 1. The proposed adaptive wavelet extreme learning machine

where x_i are the input variables, w_{ji} are the weights of the connection between ith input and jth hidden nodes, and c_j denotes the bias of jth hidden layer. Using equation (8), the output of the hidden node is given by:

$$\psi_{a_j b_j}(P_j(x)) = \psi\left(\frac{P_j(x) - b_j}{a_j}\right), \qquad j = 1, 2, \ldots, M \tag{8}$$

In this proposed work, the Mexican Hat function [6] is used as the mother wavelet $\psi_{a_i b_i}$ as described in fig. 2a, and defined as

$$\psi(x) = e^{-x^2/2}(1 - x^2) \tag{9}$$

Therefore, the wavelet activation function of AW-ELM is:

$$\psi_{a_j b_j}(P_j) = e^{-0.5\left(\frac{P_j - b_j}{a_j}\right)^2}\left(1 - \left(\frac{P_j - b_j}{a_j}\right)^2\right) \tag{10}$$

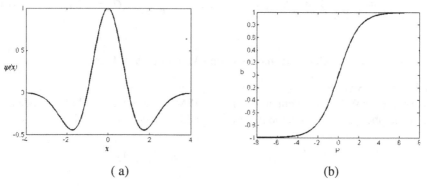

(a)   (b)

Fig. 2. Two difference functions used in this work: (a) The mother wavelet of the Mexican hat (b) A nonlinear function to produce b_j

In this proposed AW-ELM, the dilatation parameters a_j are fixed and initialized using Equation (5). As for the translation parameters b_j, they are varied according to the input information and driven by a nonlinear function f(.) as follows:

$$b_j = f(P_j) \tag{11}$$

where

$$f(P_j) = \frac{2}{1 + e^{-P_j}} - 1 \tag{12}$$

as depicted in Fig. 2b. Eventually, a new structure of an adaptive W-ELM is presented in Fig. 1. A small circle on the top of each hidden node is used to adjust the b parameters in order to change the shape of the wavelet. Thus, the output of AW-ELM is:

$$f_i^{\ k}(\mathbf{x}) = \sum_{j=1}^{M} V_{ij} \psi_{a_j b_j}(w_j, c_j, \mathbf{x}_k) = \sum_{j=1}^{M} V_{ij} \psi_{a_j b_j}(P_j(\mathbf{x}_k)) \quad i = 1, 2, ..., O \qquad (13)$$

The Learning Algorithm
For the desired output:

$$D = (d_1^T \quad d_2^T \quad ... \quad d_L^T)_{LxO} \qquad (14)$$

The AW-ELM described in (13) can be written as a linear system as follows:

$$HV = D \qquad (15)$$

where

$$H = \begin{bmatrix} \psi_{a_1 b_1}(P_1(\mathbf{x}_1)) & \cdots & \psi_{a_M b_M}(P_M(\mathbf{x}_1)) \\ \vdots & \vdots & \vdots \\ \psi_{a_1 b_1}(P_1(\mathbf{x}_L)) & \vdots & \psi_{a_M b_M}(P_M(\mathbf{x}_L)) \end{bmatrix}_{LxM} \qquad (16)$$

$$V = (v_1^T \quad v_2^T \quad \cdots \quad v_O^T)_{MxO}^T \qquad (17)$$

V can be obtained by solving the least-square solution of (15) and given by:

$$\hat{V} = H^\dagger D \qquad (18)$$

where $H^\dagger$ is the Moore-Penrose generalized inverse of the matrix H.
  The training algorithm of AW-ELM can be implemented as follows:

Algorithm of AW-ELM. Given a training set $\aleph = \{(x_k, t_k) \mid x_k \in R^n, t_k \in R^o, k = 1, 2..., L\}$, the hidden node output function $\psi_{ai, bi}(w, c; x)$ and the hidden node number M:
(1) Randomly assign vector matrix W and initialize the hidden node parameters a_j, j = 1, 2,.., M according to (5)
(2) Calculate the input hidden layer P_j in (7)
(3) Calculate $\underline{b_j}$ in (11) and (12)
(4) Calculate the hidden layer output H in (16)
(5) Calculate the output weight $\hat{V}$ in (18)

2.3   AW-ELM for Finger Motion Recognition

The proposed recognition system consists of the same stages as depicted in Figure 3. Firstly, signals from two-channel EMG located on the forearm were acquired by a data acquisition device from eight subjects. The experimental procedures for the data acquisition could be referred to [8]. Then the filtering and windowing was applied to the collected data before being extracted using a time domain (TD) and autoregressive (AR) features.

Fig. 2. The motion finger classification using AW-ELM

The features were extracted from the time domain feature set which consists of Waveform Length (WL), Slope Sign Changes (SSC), Number of Zero Crossings (ZCC), and Sample Skewness (SS). In addition, some parameters from Hjorth Time Domain Parameters (HTD) and Auto Regressive (AR) Model Parameters were included. To reduce the dimension of the features, SDRA was employed. All features were concatenated and reduced using SRDA. SRDA is an extension of LDA that can deal with singularity and a large data set. The 200 ms window length was applied to the signal to comply with the real time application along with a 25 increment.

The reduced feature set resulted in the previous stage is utilized in the classification. The objective of the classification that was performed using AW-ELM and other ELM classifiers is to recognize ten classes of the individual and combined finger movements consisting of the flexion of individuated fingers. They consisted of Thumb (T), Index (I), Middle (M), Ring (R), Little (L) and the pinching of combined Thumb–Index (T–I), Thumb–Middle (T–M), Thumb–Ring (T–R), Thumb–Little (T–L), and the hand close (HC). Finally, statistical analyses were performed to validate the result.

3   Results and Discussion

In this section, the performance of the proposed AW-ELM was compared to wavelet extreme learning machine (W-ELM) and sigmoid extreme learning machine (sig-ELM). All classifiers classified ten finger motions using EMG signal from two channel electrodes. The four-fold cross validation was used to validate the classification results. Simulation was done in the MATLAB 8.3 environment running on 2.8 GHz PC.

Table 1 shows the classification results of three classifiers in recognizing ten finger motions classes defined in 2.4. In all ELMs, the number of hidden nodes varied from 50 up to 500. The results indicate that the average accuracy the proposed of AW-ELM was higher than standard W-ELM in all cases. Likewise, the AW-ELM performance is better than Sig-ELM in all hidden node numbers except 50 and 75. In these two hidden numbers, the Sig-Elm achieved better accuracy that AW-ELM. Overall, the adaptation of wavelet shape using a sigmoid function in AW-ELM could enhance the performance of the original wavelet extreme learning machine and in several condition, could attain better performance than Sig-ELM.

Table 1. The average classification accuracy of AW-ELM across eight subjects using four-fold cross validation compared with W-ELM and Sig-ELM

# Hidden Node	Accuracy (%)		
	W-ELM	AW-ELM	Sig-ELM
50	91.07 ± 0.17	91.57 ± 0.08	**91.65 ± 0.08**
75	91.56 ± 0.14	91.93 ± 0.14	**91.97 ± 0.10**
100	91.79 ± 0.10	**92.05 ± 0.09**	92.01 ± 0.08
125	91.90 ± 0.08	**92.08 ± 0.09**	92.03 ± 0.10
150	91.94 ± 0.11	**92.06 ± 0.10**	92.04 ± 0.10
175	91.98 ± 0.09	**92.06 ± 0.09**	92.04 ± 0.08
200	91.99 ± 0.08	**92.04 ± 0.08**	92.01 ± 0.06
500	91.79 ± 0.08	**91.56 ± 0.06**	91.37 ± 0.06

Table 2. Processing time of different ELM classifiers

#Hidden Node	Training Time (s)			Testing Time (s)		
	W-ELM	AW-ELM	Sig-ELM	W-ELM	AW-ELM	Sig-ELM
50	0.16 ± 0.01	0.19 ± 0.02	0.12 ± 0.00	0.03 ± 0.00	0.06 ± 0.00	0.01 ± 0.00
75	0.28 ± 0.01	0.33 ± 0.01	0.19 ± 0.00	0.06 ± 0.01	0.07 ± 0.00	0.02 ± 0.00
100	0.38 ± 0.02	0.45 ± 0.02	0.27 ± 0.01	0.07 ± 0.00	0.10 ± 0.00	0.02 ± 0.00
125	0.59 ± 0.05	0.70 ± 0.06	0.48 ± 0.07	0.09 ± 0.00	0.14 ± 0.00	0.03 ± 0.00
150	0.71 ± 0.01	0.81 ± 0.01	0.51 ± 0.01	0.12 ± 0.00	0.19 ± 0.01	0.04 ± 0.00
175	0.93 ± 0.06	1.06 ± 0.05	0.69 ± 0.05	0.14 ± 0.00	0.22 ± 0.00	0.04 ± 0.00
200	1.07 ± 0.08	1.22 ± 0.05	0.85 ± 0.06	0.17 ± 0.00	0.26 ± 0.00	0.05 ± 0.00
500	4.20 ± 0.08	5.42 ± 0.08	2.82 ± 0.10	0.75 ± 0.01	1.28 ± 0.01	0.12 ± 0.00

In terms of processing time, the ELM using a sigmoid function (Sig-ELM) spent less training time than W-ELM and AW-ELM in as shown in Table 2. Table 2 shows that the larger the number of the hidden node, the longer the time difference between AW-ELM and Sig-ELM. Likewise, in the testing time, AW-ELM is the slowest system. The adaptive mechanism adds the processing time in both the training and testing trials.

Table 3. The p-value of anova test on the classification accuracy between AW-ELM and other tested classifiers

#Hidden Node	p-value	
	AW-ELM & W-ELM	AW-ELM & Sig-ELM
50	**0.0000**	**0.0000**
75	**0.0000**	0.1283
100	**0.0000**	**0.0006**
125	**0.0000**	0.0610
150	**0.0000**	0.3477
175	**0.0021**	0.5746
200	**0.0098**	0.0552
500	**0.0000**	**0.0000**

The one-way ANOVA test was done to evaluate the improvement significance of AW-ELM compared to W-ELM and Sig-ELM as presented in Table 4. Table 4 shows that p-values on the comparison of AW-ELM and W-ELM are less than 0.05. In other words, the performance improvement in recognizing ten finger motions by AW-ELM is significantly achieved. Furthermore, the performance of AW-ELM and Sig-ELM in some cases is significantly different in the hidden number node 50, 100 and 500 whereas it is significantly similar in other hidden node numbers. Nevertheless, the AW-ELM produced better accuracy in most trials than Sig-ELM.

4   Conclusion

This paper proposed a novel ELM i.e. an adaptive wavelet extreme learning (AW-ELM) for recognizing finger motions using two-channel EMG signals. The adaptation mechanism of the proposed method is conducted by adjusting the shape of the wavelet based on the information provided in the input. The experimental results showed that the proposed AW-ELM improved the performance of the original wavelet ELM in all cases tested and performed better than Sig-ELM in most cases observed. In the future, the performance of AW-ELM should be compared with other well-known classifiers such as support vector machine (SVM) and Linear Discriminant Analysis (LDA).

References

1. Zhou, B., Shi, A., Cai, F., Zhang, Y.-S.: Wavelet neural networks for nonlinear time series analysis. In: Yin, F.-L., Wang, J., Guo, C. (eds.) ISNN 2004. LNCS, vol. 3174, pp. 430–435. Springer, Heidelberg (2004)
2. Lin, C.-J., Tsai, H.-M.: FPGA implementation of a wavelet neural network with particle swarm optimization learning. Mathematical and Computer Modelling 47(9-10), 982–996 (2008)
3. Huang, G.B., et al.: Extreme learning machine for regression and multiclass classification. IEEE Trans. Syst. Man Cybern. B, Cybern. 42(2), 513–529 (2012)
4. Cao, J., Lin, Z., Huang, G.-B.: Composite function wavelet neural networks with extreme learning machine. Neurocomputing 73(7), 1405–1416 (2010)
5. Salih, D.M., et al.: Wavelet network based online sequential extreme learning machine for dynamic system modeling. In: 2013 9th Asian Control Conference (ASCC) (2013)
6. Ling, S.-H., et al.: Improved hybrid particle swarm optimized wavelet neural network for modeling the development of fluid dispensing for electronic packaging. IEEE Transactions on Industrial Electronics 55(9), 3447–3460 (2008)
7. Javed, K., Gouriveau, R., Zerhouni, N.: SW-ELM: A summation wavelet extreme learning machine algorithm with a priori parameter initialization. Neurocomputing 123, 299–307 (2014)
8. Anam, K., Al-Jumaily, A.A.: Swarm-based extreme learning machine for finger movement recognition. In: 2014 Middle East Conference on Biomedical Engineering (MECBME) (2014)

Text Categorization Using an Automatically Generated Labelled Dataset: An Evaluation Study

Dengya Zhu[1] and Kok Wai Wong[2]

[1] School of Information Systems, Curtin University,
GPO Box U1987, Perth, Western Australia
d.zhu@curtin.edu.au
[2] School of Engineering and Information Technology, Murdoch University
South Street, Murdoch, Western Australia, 6150
k.wong@murdoch.edu.au

Abstract. Naïve Bayes(NB), kNN and Adaboost are three commonly used text classifiers. Evaluation of these classifiers involves a variety of factors to be considered including benchmark used, feature selections, parameter settings of algorithms, and the measurement criteria employed. Researchers have demonstrated that some algorithms outperform others on some corpus, however, labeling and corpus bias are two concerns in text categorization. This paper focuses on evaluating the three commonly used text classifiers by using an automatically generated text document set which is labelled by a group of experts to alleviate subjectiveness of labelling, and at the same time to examine how the performance of the algorithms is influenced by feature selection algorithms and the number of features selected.

Keywords: Text categorization, feature selection, text classifiers.

1 Introduction

Text categorization is defined as "automated assignment of natural language texts to predefined categories based on their content" [1]. It has been widely applied in the areas such as language identification, information retrieval, opinion mining, spam filtering, and email routing [2, 3]. With the recent explosion of information on the Web, text categorization is becoming increasingly important as an approach to managing and organizing the huge volume of information on the Web. Many algorithms such as Boostexter [4] and SVMs [5] have been developed and introduced to address this problem. The evaluation of the effectiveness of the algorithms is playing an important role for both researchers and practitioners.

To evaluate a text categorization algorithm, the first consideration is the document collection to be used. Many such document collections have been developed for evaluation purposes. The widely used benchmark collections for text categorization include Reuters-21578[6], RCV1[1], UCI repository (http://archive.ics.uci.edu/ml), and OHSUMED [7].

C.K. Loo et al. (Eds.): ICONIP 2014, Part I, LNCS 8834, pp. 479–486, 2014.
© Springer International Publishing Switzerland 2014

There are issues related to the current test sets such as overfitting, restricted vocabulary [8], lack of full document text, the subjectiveness of labelling, or inconsistent or incomplete category assignments [1]. It is well known that machine learning algorithms can overfit by tuning parameters for a given dataset to make the algorithm perform extremely well on one dataset, but perform poorly on others [9]. Document labelling procedure is not only costly and labour-intensive, but also relatively subjective rather than objective and consequently results in inconsistent labelling. Another reason for the inconsistency is that some documents may have multi-labels rather than a unique one. This may cause an algorithm to perform relatively poor on one test collection but may perform better in real world and vice versa. So, there is an ongoing need of new labelled test datasets that are less subjective, have a larger vocabulary and can be generated without much expensive human-labour involved.

Dimensionality of feature space also affects the performance of a text categorization algorithm. There is a list of feature selection algorithms available to reduce the dimension of the feature space of a document set [10]. Different feature selection algorithms usually select different subspaces which consequently lead to different performance evaluation outcomes. For a given document collection, some text categorization algorithms such as kNN and LLSF are very sensitive to some feature selection algorithms (such as MI and Term Strength) and the number of features to be chosen [10].

Open Directory Project (ODP) [11], a socially constructed Web directory, provides a potential solution to alleviate the subjectiveness in document labelling. The semantic characteristics of the categories in the ODP can be used to generate a labelled document collection for the purpose of evaluating text categorization algorithms [11]. This automatically generated document collection not only can enrich the existing test document collection, but also intend to alleviate the subjectiveness of document labelling procedure because ODP document is labelled by more than one million volunteer domain experts that is kind of collective intelligence rather than by one or a small group of experts. Characteristics of the ODP dataset are introduced in section 3.

In this paper, three text categorization algorithms kNN, NB [3] and AdaBoost are evaluated on the *CategoryDocument* set, an automatically generated labelled document collection [11]. The dataset is selected because first it is labelled by over a million volunteer editors and thus alleviates subjectiveness; and two, it has more than five million categories that is the most comprehensive human edited Web directory so the corpus bias is minimized. For each of the text categorization algorithms, four feature selection algorithms, Information Gain, Mutual Information, chi-square and Odds Ratio, are employed to reduce the original high dimensional feature space into a series of subspaces. The three categorization algorithms are selected because first, as argued by [4] earlier in 2000, voluminous research has been conducted on text categorization, it would be impractical to compare all the methods; second, the three algorithms are most widely used in both research society and real world applications; and third, they are representatives of statistical-based (NB), instance-based (kNN) and ensemble-based (AdaBoost). The experimental results will enrich the list of the evaluation results of text categorization algorithms; and the influence of the number of features on Adaboost will also be explored as in [4], all terms are searched by weak learners.

2   Implementation of Categorization Algorithms

Most text categorization algorithms have their variants. For example, NB has at least two models, multinomial NB model and multinomial unigram language model [3]; AdaBoost has AdaBoost.MH and AdaBoost.MR [4]. Variations of an algorithm may produce different experimental results. Therefore, in this section, implementation details of the categorization and feature selection algorithms are introduced.

A.   *Naïve Bayes Classifier*

Let a document $\mathbf{d}_j$ is represented as a feature vector $\mathbf{d}_j = (t_1, t_2, ..., t_V)$, V is feature number in data set $D = \{\mathbf{d}_j \mid \mathbf{d}_j \in D, j = 1, ..., N\}$, N is the number of documents in the document space D. Let $C = \{c_i \mid c_i \in C, i = 1, ..., M\}$ be a set of categories where M is the total number of categories in C. If a document is assigned to one and only one category in C, the probability of $\mathbf{d}_j$ is assigned to c_i, according to Bayes rule.

To find out the most appropriate category for a document, NB classifier assigns the document to the most likely or maximum a posteriori class. This can be expressed as

$$c_{map} = \arg\max_{c_i \in C} \hat{P}(c_i \mid \mathbf{d}_j) = \arg\max_{c_i \in C} \hat{P}(c_i) \times \frac{\hat{P}(\mathbf{d}_j \mid c_i)}{\hat{P}(\mathbf{d}_j)}$$

Further assuming occurs of a particular feature t_k is statistically independent of the occurrence of other features, given that the document is of category c_i, this leads to

$$c_{map} = \arg\max_{c_i \in C} \hat{P}(c_i) \times \frac{\hat{P}(\mathbf{d}_j \mid c_i)}{\hat{P}(\mathbf{d}_j)} = \arg\max_{c_i \in C} \hat{P}(c_i) \times \frac{\prod_{k=1}^{V} \hat{P}(t_{kj} \mid c_i)}{\hat{P}(\mathbf{d}_j)}$$

Note that when estimating the $\arg\max_{c_i \in C} \hat{P}(c_i \mid \mathbf{d}_j)$, $\hat{P}(\mathbf{d}_j)$ is the same for all categories and thus can be neglected, a final estimate of c_{map} is

$$c_{map} = \arg\max_{c_i \in C} \hat{P}(c_i \mid \mathbf{d}_j) \propto \arg\max_{c_i \in C} \hat{P}(c_i) \times \prod_{k=1}^{V} \hat{P}(t_{kj} \mid c_i)$$

where $\hat{P}(t_{kj} \mid c_i)$ is the conditional probability of term t_{kj} occurring in a document of class c_i. Taking logarithms to avoid floating point underflow,

$$c_{map} \propto \arg\max_{c_i \in C}(\log \hat{P}(c_i) + \sum_{k=1}^{n_d} \log \hat{P}(t_{kj} \mid c_i)$$

The probability of $\hat{P}(c_i)$ and $\hat{P}(t_{kj} \mid c_i)$ can be estimated by the maximum likelihood estimate,

$$\hat{P}(c_i) = \frac{N_{c_i}}{N}, \quad \hat{P}(t_{kj} \mid c_i) = \frac{T_{c_i t}}{\sum_{t' \in V} T_{c_i t'}}$$

Where N_{ci} is the number of documents in class c_i, T_{cit} is the frequency of t in D from class c_i. To avoid zero estimation, Laplace smoothing is used to estimate $\hat{P}(t_{kj} \mid c_i)$,

$$\hat{P}(t_{kj} \mid c_i) = \frac{T_{c_i t} + 1}{(\sum_{t' \in V} T_{c_i t'}) + |V|}$$

This model is referred to as multinomial Bayes model, or multinomial model [3], which is implemented in the research.

B.   K-Nearest Neighbours (KNN) Classifier

The kNN [9] text categorization algorithm is a "lazy" learning approach. Without a training procedure, the similarity between a testing document and a training document is compared directly by means of approximating the distance between them. The following three factors that affect the performance of kNN are considered when the algorithm is implemented for text categorization purpose.

Similarity Estimating: There are a number of approaches to estimate the similarities between two documents such as cosine similarity, Manhattan Distance, Tanimoto Similarity, Jaccard Similarity coefficient, and Euclidean Distance. In this research, cosine similarity is used to decide the k nearest neighbours based on the estimated k top ranked similarity scores. To calculate the cosine similarity, documents are represented as a terms vector, and tf-idf term-weighting strategy is employed [3].

The number k: kNN is not an efficient algorithm and experimentally selecting an appropriate k is time consuming. In this research, k = 5, 10, and 14 are evaluated. Any k bigger than 14 is unrealistic because there are 14 categories.

Majority Voting Strategy: Supposed that each of the k member is a <category, document> tuple. Dominant Majority is defined as the majority number m > k/2 which assigns a test document to only one category. Weak majority refers to the case when m <= k/2, if the category of the top ranked member is the same as the category of the majority group, the test document is assigned to only one category which is the same as the top ranked member; otherwise, the document is assigned to two categories. One is decided by the weak group, another is decided by the top ranked member.

C.   AdaBoost Classifier

In this research, AdaBoost.MHR, the AdaBoost.MH with real-valued predictions is implemented due to the fact that it is the most effective one among the four different versions of AdaBoost [4].

The estimated value given by the final $h_t(x)$ implies a measure of "confidence" [12] to assign a category to a test document, if the two top ranked categories are both predicted with very high and very close confidence values, it is reasonable to assign the two categories to the test document.

Training round T is selected as an arbitrary number, say, 1000 in some of experiments [1, 4]. An alternative approach is to set up a terminate condition and stop the training round when the condition is satisfied [13]. One candidate terminate condition is to run a boosting algorithm until the training error reaches a predefined minimal value ε, or simply zero. Let the training round be T_0 when the terminate condition is reached, continuing the training process for another $\beta \times T_0$ rounds is suggested [13] with the intention of further reducing the testing error, because even if training error reaches zero, testing error keeps going down if the training continues.

In this research, T = 1000 is evaluated against a list of feature selection algorithms as described in the following section.

D.   *Feature Selection Algorithms*

The high dimensionality of the term space may be problematic in the field of text categorization because many sophisticated learning algorithms used for classifier induction cannot effectively handle high dimensionality [9]. Dimensionality reduction is the process to reduce the high vector space for the purpose of efficiency. In fact, dimensionality reduction not only can boost categorization efficiency, but also moderately improve the effectiveness of categorization because noise features are at the same time removed [10].

Chi-square, Information Gain (IG), Mutual Information (MI), and Odds Ratio (OR) are among the list of widely used feature selection algorithms in text categorization [10]. These will be used in this paper.

3   Experimental Results

An automatically generated labelled dataset, the Open Directory Project dataset - *CategoryDocuments* [11], is employed as the experimental dataset to evaluate the above four widely used text categorization algorithms.

The second level ODP *CategoryDocument* set is selected as the experimental dataset. The dataset contains 14 top-level ODP categories and 512 labelled documents with an average length of 47929 words.

Precision and recall are two most employed criteria of effectiveness stemmed from the area of information retrieval for text categorization [13]. To estimate the overall effectiveness of a classifier, precision and recall are usually averaged for all the categories. Micro- and Macro-averaged are the two measures that are used [1] in the paper, five-fold cross-validation is employed and precision and recall are calculated by:

Macroaveraged precision and recall

$$p = \frac{\sum_{i=1}^{|C|} p_i}{|C|}, \quad r = \frac{\sum_{i=1}^{|C|} r_i}{|C|}$$

Microaveraged precision and recall

$$p = \frac{TP}{TP + FP} = \frac{\sum_{i=1}^{|C|} TP_i}{\sum_{i=1}^{|C|} (TP_i + FP_i)}, r = \frac{TP}{TP + FN} = \frac{\sum_{i=1}^{|C|} TP_i}{\sum_{i=1}^{|C|} (TP_i + FN_i)}$$

where p_i and r_i are precision and recall for category c_i, $|C|$ is the total number of categories, TP_i, FP_i, and FN_i are true positive, false positive, and false negative for category c_i in a contingent table. Micro-averaged measure is dominated by categories that have high frequency count; whereas Macro-averaging assigns equal weight to every category and is dominated by effectiveness on low frequency categories. Since increasing precision is to certain extent at the cost of sacrificing the recall, and vice versa, the performance of a text categorization algorithms is usually evaluated by F_1 function [6] which balances the weights of precision and recall by assigning same weight to both criteria.

A. Experimental results of NB

Table 1 shows the micro-/macro-averaged F_1 for the four different feature selection algorithms.

Table 1. Results for the four feature selection algorithms for NB

#Features	50	80	100	200	300	500	1000	2000	3000	5000	10000
Mac(chi-square)	90.06	91.03	91.58	94.09	93.78	94.85	95.94	95.54	90.25	85.53	76.56
Mic(chi-suqare)	88.87	90.24	91.21	93.36	94.14	95.11	95.9	96.29	92.77	89.64	85.35
Mac (IG)	97.87	99.34	99.34	98.66	98.46	98.7	98	97.6	96.27	84.42	72.1
Mic (IG)	98.04	99.61	99.61	98.83	98.44	98.63	97.46	96.88	95.7	88.47	83.01
Mac(MI)	4.85	6.36	7.54	11.22	11.63	10.78	12.65	21.35	36.06	43.38	61.87
Mic (MI)	20.5	19.91	20.11	20.5	20.5	20.1	18.35	26.57	35.16	48.63	69.32
Mac(OR)	99.12	99.12	99.12	99.21	99.44	99.44	99	98.85	99.14	99.82	100
Mic(OR)	99.22	99.22	99.22	99.22	99.41	99.41	98.82	99.22	99.41	99.8	100

Experimental results indicated that 1) number of features affected the performance of NB significantly, 2) OR is one of the best feature selection algorithms for NB, 3) OR and IG can enable NB classifier to perform extremely well when 50 to 500 features are selected.

B. Experimental results of kNN

Table 2 illustrates the F_1 measures of kNN classifier with different feature selection algorithms. The data in Table 2 show that 1) the number of features is an essential factor which dominants the performance of kNN in terms of F_1, 2) OR is one of the best feature selection algorithms for kNN as well. It outperforms chi-square and MI, slightly better than IG, and the preferred feature range for kNN is from 300 to 500, 3) although conceptually simple, kNN, if combined with OR and IG, can perform very well for the *CategoryDocuments* set.

Table 2. Results for the four feature selection algorithms for kNN

#Features	50	80	100	200	300	500	1000	2000	3000	5000	10000
Mac(chi-square)	88.97	91.34	92.04	93.68	94.19	95.36	96.15	96.58	98.08	98.8	98.79
Mic(chi-suqare)	83.01	90.62	91.79	93.75	94.53	95.5	96.88	98.05	98.44	98.83	98.83
Mac (IG)	98.42	99.14	99.14	98.34	98.56	99.35	99.64	99.41	98.07	98.74	97.74
Mic (IG)	99.02	99.61	99.61	99.02	99.02	99.22	99.42	99.42	98.44	99.02	98.05
Mac(MI)	0.84	0.84	0.84	7.43	7.43	7.43	12.46	18.16	38.4	43.91	62.95
Mic (MI)	6.25	6.25	6.25	7.42	7.42	7.42	9.17	24.92	27.14	48.42	70.1
Mac(OR)	99.62	99.62	99.62	99.5	99.62	99.62	99.64	99.82	99.76	99.7	99.68
Mic(OR)	99.8	99.8	99.8	99.8	99.8	99.8	99.8	99.8	99.8	99.8	99.8

C. Experimental results of AdaBoost.MHR

Experimental results of AdaBoost are illustrated in Table 3. Results in Table 3 demonstrated that 1) similar to the previously evaluated two algorithms, the performance of F_1 is dominated by the number of features selected by the different feature selection algorithms, 2) as in the previous cases, OR is again the best algorithm for AdaBoost.MHR in that it yields the highest F_1 values of 99.81% and 99.78% for micro-averaging and micro-averaging measures respectively, 3) the performance of

AdaBoost.MHR is not as effective as NB and kNN in this experiment compared with the experiments of Schapire and Singer. One possible explanation for the difference is that different text categorization algorithms with different feature selection approaches on different test dataset perform differently.

Table 3. Results for the four feature selection algorithms for AdaBoost.MHR

#Features	50	80	100	200	300	500	1000	2000	3000	5000	10000
Mac(chi-square)	83.54	83.86	84.38	83.76	84.11	82.02	87.9	88.77	89.88	91.83	91.57
Mic(chi-suqare)	78.9	84.76	85.15	83.98	86.13	85.16	88.29	89.45	92.6	92.77	92.18
Mac (IG)	94.3	95.35	95.67	95.09	95.18	96.22	93.35	91.19	94.86	95.37	92.09
Mic (IG)	94.72	95.7	95.7	95.11	95.31	96.29	93.94	93.76	95.51	95.9	92.96
Mac(MI)	2.29	2.05	4.6	6.57	6.57	6.59	10.69	13	15.59	30.9	41.16
Mic (MI)	4.69	4.88	7.42	7.42	6.65	11.33	16.96	9.97	12.88	36.51	44.14
Mac(OR)	99.78	99.15	98.83	89.68	93.25	87.27	85.29	89.88	91.07	89.86	91.74
Mic(OR)	99.81	99.22	99.22	88.06	93.56	87.49	85.72	89.67	91.6	92.39	92.78

D. Comparison of experimental results

First, OR performs the best for all the three different categorization algorithms. This is confirmed by the report of [10]. The main reason may be that OR take into account both positive and negative example.

The second observation found is that Chi-square and IG are also effective feature selection algorithms and this result is similar to that of Yang and Pedersen [10]. IG and Chi-square consider not only positive examples, but also negative examples, that is why these two algorithms can also pick up informative features. It can also observe that OR gives more weight to positive examples (by ratio) than IG and Chi-square (by subtract) and this may imply why OR performs the best.

Third, MI is the worst for the three classifiers. One of the possible reasons is MI only selects features that are informative in positive examples, without considering that the same feature may also informative for negative examples.

By comparing with the experimental results from [1, 6], the three classifiers perform better on the automatically generated labeled *CategoryDocument* dataset than those on the Reuters collection. One possible reason may be that the average document length in Reuters is only about 124 words whereas the document length in *CategoryDocument* set is 47929 words. The lengthy documents have the potential to deliver more information than the shorter one, and thus more suitable to be used to test the text categorization classifiers.

Finally, the experimental results demonstrated that with few (less than 100) features select by OR, Adaboost performs extremely well. When the features increase from 100 to 1000, F_1 decreases and then begin to increase gradually with the number of features increasing. When features are selected by Chi-square, the performance of Adaboost is increasing as the number of features increasing.

4   Conclusions

In this paper, three text categorization classifiers, NB, kNN and AdaBoost were evaluated using an automatically generated labeled dataset, which is constructed by using semantic characteristics of the categories in the ODP. Features of the dataset were selected by four feature selection algorithms, Chi-square, IG, MI and OR. The experimental results demonstrated that when combined with OR or Chi-square to select a reasonable number of features, all the three classifiers can produce satisfactory results. This outcome reveals that labelling subjectiveness and bias of benchmark dataset play an essential role in evaluation of categorization algorithms, and if the two issues of a dataset can be well controlled most text categorization algorithms can perform quite well on that dataset. The experimental data also revealed that the effectiveness of the classifiers is sensitive to the feature selection algorithms and the number of features selected. The data also verified that different algorithms perform differently on different experimental document datasets.

References

1. Lewis, D.D., et al.: RCV1: A New Benchmark Collection for Text Categorization Research. Journal of Machine Learning Research 5, 361–397 (2004)
2. Aggarwal, C.C., Zhai, C.: A Survey of Text Classification Algorithms. In: Aggarwal, C.C., Zhai, C. (eds.) Mining Text Data, pp. 163–222. Springer (2012)
3. Manning, C.D., et al.: Introduction to Information Retrieval. Cambridge University Press (2008)
4. Schapire, R.E., Singer, Y.: Boostexter: A Boosting-based System for Text Categorization. Machine Learning 39(2-3), 135–168 (2000)
5. Joachims, T.: Text Categorization with Support Vector Machines: Learning with Many Relevant Features. In: Nédellec, C., Rouveirol, C. (eds.) ECML 1998. LNCS, vol. 1398, pp. 137–142. Springer, Heidelberg (1998)
6. Yang, Y.: An Evaluation of Statistical Approaches to Text Categorization. Inform. Retrieval 1(2), 69–90 (1999)
7. Hersh, W., et al.: OHSUMED: An interactive retrieval evaluation and new large test collection for research. In: Proceedings of the 19th Annual International ACM/SIGIR Conference on Research and Development in Information Retrieval, SIGIR 1994, pp. 192–201 (1994)
8. Davidov, D., et al.: Parameterized Generation of Labeled Datasets for Text Categorization Based on a Hierarchical Directory. In: Proceedings of the 27th Annual International ACM SIGIR Conference on Research and Development in Information Retrieval, pp. 250–257 (2004)
9. Sebastiani, F.: Machine Learning in Automated Text Categorization. ACM Comput. Surv. 34(1), 1–47 (2002)
10. Yang, Y., Pedersen, J.O.: A Comparative Study on Feature Selection in Text Categorization. In: Proceedings of the 14th International Conference on Machine Learning (ICML 1997), pp. 412–420 (1997)
11. Zhu, D., Dreher, H.: Characteristics and Uses of Labeled Datasets – ODP Case Study. In: Proceedings of the sixth International Conference on Semantics, Knowledge, and Grids (2010)
12. Freund, Y., Schapire, R.E.: A Short Introduction to Boosting. Journal of Society for Artificial Intelligence 14(5), 771–780 (1999)
13. Schapire, R.E., et al.: Boosting and Rocchio Applied to Text Filtering. In: Proceedings of the 21st Annual International ACM SIGIR Conference on Research and Development in Information Retrieval, SIGIR 1998, pp. 215–223 (1998)

Online Recommender System
Based on Social Network Regularization

Zhuo Wang* and Hongtao Lu

Key Laboratory of Shanghai Education Commission
for Intelligent Interaction and Cognitive Engineering
Department of Computer Science and Engineering
Shanghai Jiao Tong University, Shanghai, China
{goneera,htlu}@sjtu.edu.cn
http://www.sjtu.edu.cn

Abstract. Although model-based Collaborative Filtering approaches have been widely used in recsys in the past few years. But In practical application, where users' rating data arrives sequentially and frequently, the model-based approaches have to re-trained completely for new records. For users' social information has been succeed used in recommendation system in previous work. In this paper, we proposed several online collaborative filtering algorithms using users' social information to improve the performance of online recommender systems. The algorithms can better use the prior rating and the social network information, which compute fast and scalable in large data. The contribution of this paper are mainly two-fold: (1) We propose an online collaborative filtering algorithm which can better use the social information and prior knowledge; (2) We solve the problem of cold start and users with few ratings.

Keywords: Recommender Systems, Online Collaborative filtering, Social network, Matrix Factorization.

1 Introduction

Collaborative filtering (CF) is a method of predict users' preference by learning from other users and items ratings. It mainly divide into two schools: The memory based collaborative filtering [1], [2], [3] and model based collaborative filtering [4], [5], [6], [7] The model based CF methods have been a standard method for recommendation for a number of years. In model based CF approaches, we are assumed to have a collection of rating data, users rate the items as the indicator of their preference. Then matrix factorization methods or other model just like possibility model used on this data, find out the embed feature of users and items and make prediction.

With the emergence of large-scale data in modern recommender systems. New rating records come frequently and sometimes we have new registrants or new items. The traditional Collaborative filtering methods suffer from some

---

* Corresponding author.

C.K. Loo et al. (Eds.): ICONIP 2014, Part I, LNCS 8834, pp. 487–494, 2014.

drawbacks for its batch-trained algorithm.We have to go through all the prior ratings to re-train the whole model whenever comes a new rating record.It's non-scalable for the high expensive computing cost.Although there are several tasks [8], [9], [10], [11] investigating online learning for CF, avoid the highly expensive re-training cost of traditional batch matrix factorization algorithm. In the traditional online CF method, they simply update the user and item feature vector by the pre-defined loss function and gradient decent algorithm.

Social network information has been used in other previous works [12], [13], [14], [15] to improve the batch-train methods. For the hypothesis of one user have similar tastes with his friends, different models were proposed to improve the prediction accuracy or solve the code start problem. [15] proposed a low rank matrix factorization with social regularization, Inspired by the social regularization, we propose the methods of online collaborative filtering based on social information, to better update the user feature vector, and initialize the vector of new users and items.

In this paper, we'll proposed two different model utilizing social information on online collaborative filtering. These online social based collaborative filtering methods (OSBCF) were proven to promote the advantages of model-based collaborative filtering and improve the prediction accuracy. At the same time, avoid the highly time cost of batch-training methods.

2   Background and Related Work

In recommender Systems, collaborative filtering (CF) has been one of the most successful approaches. The CF algorithm can be classified into two categories, the memory-based algorithm and the model-based algorithm. The state-of-the-art methods for regular CF tasks is model-based algorithm, the latent factor or matrix factorization method [16], which used by the winner of the Netflix prize [18], [19].

Although the high accuracy of the batch algorithm of matrix factorization, we can't stand the high time complexity and memory cost in the real practical recommender system. Several task investigate the online algorithms from various aspects to solve the issues facing batch-trained CF algorithms. All the above-mentioned works achieved good results without using any information more.

The online learning algorithm avoid the high time complexity and memory cost but lose some accuracy at the same time. In reality, we always turn to friends we trust for movie, music or book recommendations, and our tastes can be easily affected by the company we keep. Even in some social network products, the friendship between users is build for their similar taste. Hence, [12] present a social network-based recommendation system, [15] using a social regularization to represent the social constraints on recommender systems, and design a matrix factorization objective function with social regularization.

3   Problem Definition and Notations

In this section, we describe the problem we're going to solve and the meaning of the symbols used in the paper.

In the regular collaborative filtering task, some users from a total of n users rated some items from a total of m items, and these ratings form an incomplete matrix $R \in R^{i*j}$, where r_{ij} is the rating on the j-th item given by the i-th user. The goal of collaborative filtering is to predict the unknown ratings based on the known ratings. In the basic matrix factorization algorithm, we learns the latent structure by factorizing the rating matrix to a user matrix $U \in R^{n*k}$ and item matrix $V \in R^{k*m}$. In this paper, we user U_a and V_b to denote the a-th row of U, and b-th row of V respectively.

Then, the popular model-based CF, matrix factorization algorithm can be re-formulated as the regularized optimization task as below:

$$\arg\min_{U,V} \sum_{\forall(a,b)\in S} l((U_a, V_b, r_{a,b})) + \lambda(||U_a||_F^2 + ||V_b||_F^2) \tag{1}$$

where $\lambda > 0$ is a regularization parameter, and $l(U_a, V_b, r_{a,b})$ is a loss function that defines the loss between the true $r_{a,b}$ and prediction. S is the rating set we've already known.

4   Social Based Online Collaborative Filtering

4.1   Online Collaborative Filtering(OCF)

We first present the traditional online collaborative filtering algorithm based on stochastic gradient descent, which can be adopted to reach a local minimum of the objective given in Eq.(1). Suppose the new coming rating $r_{a,b}, (a, b) \in S$, by adopt the gradient descent method, U_a, V_b move toward the average gradient descent of the loss function, by a small step controlled by η.

$$U_a = (1 - 2\eta\lambda)U_a - \eta\frac{\partial l(U_a, V_b, r_{a,b})}{\partial U_a} \tag{2}$$

$$V_b = (1 - 2\eta\lambda)V_b - \eta\frac{\partial l(U_a, V_b, r_{a,b})}{\partial V_b} \tag{3}$$

where η is the learning rate parameter, and $l(U_a, V_b, r_{a,b})$ is the loss function. To optimize the RMSE, Root Mean Square Error, in this paper, we can define the loss by the following equation:

$$l(U_a, V_b, r_{a,b}) = (r_{a,b} - U_a V_b^T)^2 \tag{4}$$

So, given an observed rating pair (a, b), the updating rule respect to the RMSE defined above can be expressed as follow:

$$U_a = (1 - 2\eta\lambda)U_a + 2\eta(V_b(r_{a,b} - U_a V_b^T)) \tag{5}$$

$$V_b = (1 - 2\eta\lambda)V_b + 2\eta(U_a(r_{a,b} - U_aV_b^T)) \tag{6}$$

4.2   Social Network Based Online Collaborative Filtering (OSBCF)

As we always turn to our friends for movie, music or book recommendations, or even in some internet products, users make friend for their similar taste, the social information seems to be significant important for the matrix factorization. In this subsection, we will interpret how the social network information used in our online models. Then, we'll give the algorithm for online learning of OSBCF.

OSBCF-I: Average-Based online CF. In this model, we assume that one user are similar to all of his friends with the same similarity, and when coming a rating record, we are not suggested to update the user feature far away from their friends. For this assumption, the first social-based matrix factorization given below:

$$\arg\min_{U,V} \sum_{\forall(a,b)\in S} l((U_a, V_b, r_{a,b})) + \lambda(||U_a||_F^2 + ||V_b||_F^2) + \alpha \sum_{f\in F} ||U_a - \frac{1}{|F|}\sum_{f\in F} U_f||_F^2 \tag{7}$$

While F is the friends set of user a, and $|F|$ is the number of set F. Then, the updating formula will turn into the following equations:

$$U_a = (1 - 2\eta\lambda)U_a + 2\eta(V_b(r_{a,b} - U_aV_b^T)) - 2\alpha(U_a - \frac{1}{|F|}\sum_{f\in F} U_f) \tag{8}$$

The update of V_b is same with the Equation 6.

OSBCF-II: Sim-Based online CF. The OSBCF-I proposed a social-based online CF, which assume users have same similarity with their friends. But it's not real in reality, while a user's friends have diverse tastes. Hence, in this model, we'll use the past ratings as the prior to define the similarity of user and his friends. Then our Sim-based online CF can be formulated as:

$$\arg\min_{U,V} \sum_{\forall(a,b)\in S} l((U_a, V_b, r_{a,b})) + \lambda(||U_a||_F^2 + ||V_b||_F^2) + \alpha \sum_{f\in F} sim(a,f)||U_a - U_f||_F^2 \tag{9}$$

The updating of item is same with the equation 6. The new updating formula of user is given below:

$$U_a = (1 - 2\eta\lambda)U_a + 2\eta(V_b(r_{a,b} - U_aV_b^T)) - 2\alpha \sum_{f\in F} sim(a,f)(U_a - U_f) \tag{10}$$

The similarity between two users can be described by the modified version of Pearson correlation coefficient employed by user-based approaches, which is given by:

$$sim(u_i, u_j) = \frac{\sum_{k \in I(c)} (r_{i,k} - \bar{r}_i)(r_{j,k} - \bar{r}_j)}{\sqrt{\sum_{k \in I(c)} (r_{i,k} - \bar{r}_i)^2} \sqrt{\sum_{k \in I(c)} (r_{j,k} - \bar{r}_j)^2}} \tag{11}$$

Where I_c is the set of items that have been rated by users U_i and U_j both. And $\bar{r}_i$ is the average rating of i-th user.

We use an additional trust matrix T to describe the similarity of user and his friends.

$$T_{i,j} = \delta_{i,j} sim(u_i, u_j) \tag{12}$$

where $\delta_{i,j}$ is an indicator function that outputs 1 when u_j is u_i's friend, otherwise 0.

As the OSBCF-II is the development of the OSBCF-I, we just give the algorithm of OSBCF-II, the sim-based online CF in algorithm 1.

Algorithm 1 OSBCF - Framework of Social Based Online Collaborative Filtering

```
Input: a sequence of ratings set S
Initialization: Initialize a random matrix U for users, and a
random matrix V for items. Import the social network In the
trust matrix T
for (u, v) in S :
    update u by equation 10
    update v by equation 6
end for
return U,V
```

Extension. In the traditional batch-training algorithm, they make an implicit assumption, both the number of users and the number of items are fixed and known before training, which is unrealistic in the online recommender system. The above **algorithm** 1, initialize the user and item matrix beforehand, can not handle the novel user or item well, when comes a novel user or item.

In this part, we aim to extend the framework of OSBCF adapt to the realistic. When a new user or item comes, a randomized vector will bring bad influence for our algorithm performance. Also, if a user or item has few ratings, the performance of the algorithm depends largely on the random processes. In order to deal with the impact of random process, and the cold start problem, we propose the following method to live initialize the user vector using the social information to replace the random initialization.

$$U_{novel} = \frac{1}{1 + |F|}(ranvec + \sum_{f \in F} U_f) \tag{13}$$

where F is the exist friends set of the novel user, and *ranvec* is a randomized vector avoid that the F is empty.

We give the detail of the extend algorithm in `algorithm` 2 below.

Algorithm 2 OSBCF - OSBCF for handling novel sample extension

```
Input: a sequence of ratings set S
Initialization: Initialize the trust matrix T, for all of its
elements set to 0; initialize two empty matrix U,V for users
and items.
for (u, v) in S :
    if u is a new user :
        randomize a vector for u
        initial u by equation 13
        extend user matrix U with u
    end if
    if v is a new item :
        randomize a vector for v
        extend item matrix V with v
    end if
    update u by equation 10
    update v by equation 6
    update the u part of trust matrix T by equation 11,12
end for
return U,V
```

5   Experiments

In this section, we conduct several experiments to compare our two social-based online collaborative filtering algorithms with the traditional online CF algorithm.

5.1   Datasets

We conduct the experiments on two publicly available datasets widely used in social recommendation.

The first data source we choose is Epinions. Epinions is a well-known general consumer review site that was established in 1999. At Epinions, members rate the product or service on a rating scale from 1 to 5 starts.The dataset consists of 49,290 users who rated a total of 139,738 different items at least once. The total number of reviews is 664,824. As to the user social trust network, the total number of issued trust statement is 487,181.

The second dataset we employ for evaluation is Flixster dataset. Flixster is a social networking service in which users can rate movies. Users can also add some users to their friend list and create a social network. Note that social relations in Flixster are undirected. The Flixster dataset consist of 1,049,511 users who have rated a total of 66,726 different items. On average, each users watched 19.5 movies, and have 8.9 friends.

Table 1. RMSE of Epinions/ K = 5

Epinons	0.2	0.4	0.6	0.8
OCF	0.9744	0.9565	0.9501	0.9478
OSBCF-I	0.9637	0.9453	0.9387	0.9366
OSBCF-II	**0.9621**	**0.9415**	**0.9379**	**0.9361**

Table 2. RMSE of Epinion / K = 10

Epinons	0.2	0.4	0.6	0.8
OCF	0.9657	0.9621	0.9491	0.9442
OSBCF-I	0.9548	0.9414	0.9361	0.9346
OSBCF-II	**0.9526**	**0.9392**	**0.9337**	**0.9287**

Table 3. RMSE of Flixster/ K = 5

Flixster	0.2	0.4	0.6	0.8
OCF	0.9745	0.9546	0.9510	0.9487
OSBCF-I	0.9555	**0.9375**	0.9349	0.9331
OSBCF-II	**0.9543**	0.9376	**0.9345**	**0.9323**

Table 4. RMSE of Flixster / K = 10

Flixster	0.2	0.4	0.6	0.8
OCF	0.9662	0.9516	0.9494	0.9472
OSBCF-I	**0.9502**	0.9370	0.9335	0.9308
OSBCF-II	0.9508	**0.9317**	**0.9301**	**0.9289**

5.2 Evaluation Metric

We user the most popular metric the Root Mean Square Error (RMSE) to measure the prediction quality of our proposed approach in comparison with traditional the-state-of-art online CF method. The metric RMSE is defined as:

$$RMSE = \sqrt{\frac{1}{T} \sum_{i,j} (R_{i,j} - \hat{R}_{i,j})^2} \tag{14}$$

Where $R_{i,j}$ denotes the rating user i gave to item j, $\hat{R}_{i,j}$ denotes the rating user i gave to item j as predicted by a method, and T denotes the number of tested ratings.

5.3 Experiment Results

We conduct experiments on the two dataset Epinons and Flixster. Table1 and Table2 are the results on the Epinions dataset and Table3 and Table4 are the results on the Flixster dataset. The columns of the table, 0.2, ..., 0.8, means we choose how many training samples of the dataset, for example, the column 0.2 suggest that we choose the 20 percent of the dataset for training and the rest for testing. K is the dimensionality of the feature vector of users and items, we also choose two different number of dimensionality.

We can find out that on the two dataset our OSBCF methods outstanding the traditional OCF algorithm on both the two dataset.We can see in the results, there is little difference between the three algorithms when K is equal to 5. When $K = 10$, we can find an obvious difference between the three algorithms.

6 Conclusion

In this paper, we focus on how the social network information can be used in the online recommender systems to improve the performance and solve the cold start problem. As the results show, we achieve good results on the two experiment datasets.

As the rapid growth of online social network sites, a variety of behavior between members are allowed in the sites, just like, communication and the forwarding of the other's review. This additional information allow us to better measure the taste similarity of members. The next step, we plan to develop similar techniques to better describe the similarity of users, and improve the recommender system.

References

1. Sarwar, B.M., Karypis, G., Konstan, J.A., Riedl, J.: Item-based collaborative filtering recommendation algorithms. In: WWW, pp. 285–295 (2001)
2. Linden, G., Smith, B., York, J.: Amazon.com recommendation: Item-to-item collaborative filterng. IEEE Internet Computing 7, 76–80 (2003)
3. Xue, G.-R., Lin, C., Yang, Q., Xi, W., Zeng, H.-J., Yu, Y., Chen, Z.: Scalable collaborative filtering using cluster-based smoothing. In: SIGIR, Salvador, Brazil, pp. 114–121 (2005)
4. Salakhutdinov, R., Mnih, A.: Probabilistic matrix factorization. In: NIPS (2007)
5. Rennie, J.D.M., Srebro, N.: Fast maximum margin matrix factorization for collaborative prediction. In: ICML 2005, Bonn, Germany, pp. 713–719 (2005)
6. Salakhutdinov, R., Mnih, A., Hinton, G.: Restricted Boltzmann Machines for collaborative filtering. In: Proc. 24th Annual International Conference on Machine Learning, pp. 791–798 (2007)
7. Hofmann, T.: Latent semantic models for collaborative filtering. ACM Transactions on Information Systems 22(1), 89–115 (2004)
8. Abernethy, J., Canini, K., Langford, J., Simma, A.: Online collaborative filtering. UC Berkeley, Tech. Rep. (2011)
9. Ling, G., Yang, H., King, I., Lyu, M.R.: Online learning for collaborative filtering. In: IJCNN, pp. 1–8 (2012)
10. Das, A., Datar, M., Garg, A., Rajaram, S.: Google news personalization: Scalable online collaborative filtering. In: WWW, pp. 271–280 (2007)
11. Liu, N.N., Zhao, M., Xiang, E.W., Yang, Q.: Online evolutionary collaborative filtering. In: RecSys, pp. 95–102 (2010)
12. Bedi, P., Kaur, H., Marwaha, S.: Trust based recommender system for semantic web. In: Proc. of IJCAI 2007, pp. 2677–2682 (2007)
13. Ma, H., King, I., Lyu, M.R.: Learning to recommend with social trust ensemble. In: Proc. of SIGIR 2009, Boston, MA, USA, pp. 203–210 (2009)
14. Massa, P., Avesani, P.: Trust-aware recommender systems. In: Proc. of RecSys 2007, Minneapolis, MN, USA, pp. 17–24 (2007)
15. Ma, H., Zhou, D., Liu, C., Lyu, M.R., King, I.: Recommender systems with social regularization. In: Proc. of WSDM 2011, Hong Kong, China (2011)
16. Koren, Y., Bell, R.M., Volinsky, C.: Matrix factorization techniques for recommender systems. IEEE Computer 42(8), 30–37 (2009)
17. Weimer, M., Karatzoglou, A., Le, Q.V., Smola, A.J.: Cofi: Rank-maximum margin matrix factorization for collaborative ranking. In: NIPS (2007)
18. Bell, R., Koren, Y., Volinsky, C.: The BellKor Solution to the Netflix Prize (2008)
19. Bennett, J., Lanning, S.: The Netflix Prize. KDD Cup and Workshop (2007), http://www.netflixprize.com

A Nonlinear Cross-Site Transfer Learning Approach for Recommender Systems

Xin Xin*, Zhirun Liu, and Heyan Huang

School of Computer Science and Technology,
Beijing Institute of Technology, Beijing, China
{xxin,zrliu,hhy63}@bit.edu.cn

Abstract. This paper targets at utilizing cross-site ratings to alleviate the data sparse problem for recommender systems. The key issue is how to bridge user features between the targeted site and the auxiliary site. In traditional transfer learning models, a linear mapping function is assumed to map the user feature in the auxiliary site into the targeted site. The limitation lies in that when the real data does not follow the linear property, such models will fail to work. Therefore, the motivation of this paper is to identify whether the rating prediction performance in recommender systems can be improved by considering the nonlinear transformation. As a primary study, we propose a nonlinear transfer learning model, and utilize the radial basis function (RBF) kernel to map user features of multiple sites. Through empirical analysis in a real-world cross-site dataset, we demonstrate that by utilizing the nonlinear mapping function RBF kernel, the rating prediction performance is consistently better than previous transfer learning models at a significant scale. It indicates that the nonlinear property does exist in real recommender systems, which has been ignored previously.

Keywords: Cross-site Recommender Systems, Collaborative Filtering, Transfer Learning, RBF Kernel.

1 Introduction

Data sparsity is a typical challenge for recommender systems. It means that the density of the user-item rating matrix is extremely low in many cases (e.g., 1.18% in the well known Netflix[1] dataset). Thus when a user/item has very few ratings, it is difficult to have accurate preference predictions. For new users/items, of which no ratings are observed, most recommendation models will fail to work. This is also known as the cold-start problem. Traditionally, researchers incorporate content features, social networks, and etc., to alleviate this problem. But most information being utilized is limited within a single social media site.

Different from previous work, in this paper, we explore information from multiple social media sites rather than a single one, to tackle the data sparsity challenge. The typical scenario is as follows. The active task is to predict an user's ratings in the targeted site, e.g., the check-in ratings in Foursquare[2]. For users who rarely have ratings in

---

* Corresponding author.

[1] https://www.netflix.com/global

[2] https://foursquare.com

C.K. Loo et al. (Eds.): ICONIP 2014, Part I, LNCS 8834, pp. 495–502, 2014.
© Springer International Publishing Switzerland 2014

Foursquare, they might have many ratings in other auxiliary sites, such as movie ratings in Netflix. Therefore, the problem is how to utilize these movie ratings in Netflix to leverage the active rating predications in Foursquare. The task belongs to the transfer learning research field in collaborative filtering (CF), but is different from many of the previous work [3,5]. In [3], user/item spaces of multiple sites are different, but in our case, we study the users who simultaneously have ratings in both sites. In [5], it has utilized items from different domains to improve the recommendation performance, but the items are still within the same site. In our case, we utilize the real cross-site data.

In considering the transfer learning technique for collaborative filtering, one limitation of previous work is that only linear transformation has been utilized in mapping user features in different sites/domains, thus when the data does not follow the linear property, most previous models will fail to work. The nonlinear property, however, might exist in real systems. For example, in the movie recommendation system, some users who like action movies are grouped into a cluster, and some other users who like romantic movies are grouped into another cluster. But in the restaurant recommender system, users who always have aggressive opinions (e.g., they always give extremely high/low ratings in all recommender systems) might be more likely to have spicy food, and form a cluster in the restaurant recommender system; and users who always have moderate opinions might be likely to have sweet food, and form another cluster. In the spicy cluster, both users who give extremely high ratings in the movie recommender system and users who give extremely low ratings are contained. This is a kind of nonlinear mapping. By previous linear mapping methods, algorithms can only choose either high-rating users or low-rating users. Thus they might fail to model the real mapping relation across different sites accurately.

In this paper, we are going to solve this problem by utilizing nonlinear mapping functions to bridge user latent features in multiple sites. As a primary empirical study, we propose an intuitive but practical nonlinear transfer learning model for collaborative filtering. Specifically, we compare linear transfer learning models and the proposed nonlinear model with the radial basis function (RBF) kernel. Through experimental verification in a real world dataset, we demonstrate that by utilizing the RBF kernel, the performance of rating prediction is consistently better than the linear transfer learning models. This indicates that nonlinear property does exist in mapping user latent features in multiple sites, and modeling the nonlinear mapping relationship across social media sites is an effective approach in alleviating the data sparsity problem for the targeted recommender system.

2   Problem Definition

The cross-site rating matrix is shown in Fig. 1 (left). The left part of the matrix denotes items in the targeted site s, and the right part of the matrix denotes items in the auxiliary site v. In the system, there are three kinds of users. Some users only have ratings in site v, such as $\{u_1, u_2, u_3\}$; some users have both rating in site s and site v, such as $\{u_4, u_5, u_6\}$; and some users only have ratings in site s, such as $\{u_7, u_8, u_9\}$. The problems to be studied in this paper are defined as follows.

1. For users who only have ratings in the auxiliary site v, e.g., $\{u_1, u_2, u_3\}$, how to predict their ratings in the targeted site?

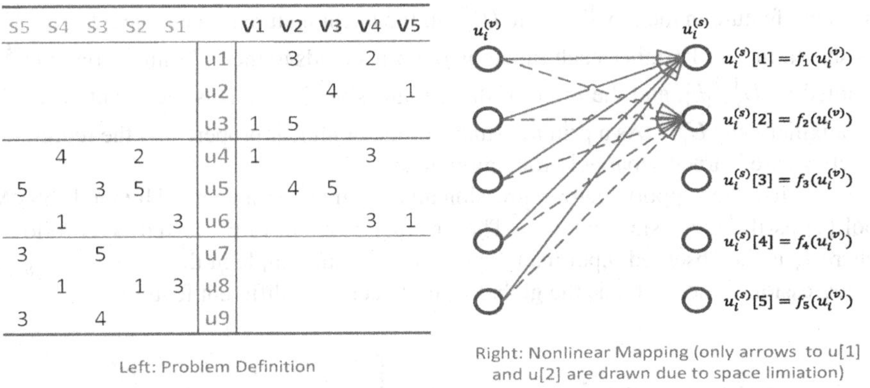

S5	S4	S3	S2	S1	user	V1	V2	V3	V4	V5
					u1		3		2	
					u2			4		1
					u3	1	5			
	4		2		u4	1			3	
5		3	5		u5		4	5		
		1		3	u6				3	1
3		5			u7					
1			1	3	u8					
3		4			u9					

$u_i^{(v)}$ $u_i^{(s)}$

$u_i^{(s)}[1] = f_1(u_i^{(v)})$

$u_i^{(s)}[2] = f_2(u_i^{(v)})$

$u_i^{(s)}[3] = f_3(u_i^{(v)})$

$u_i^{(s)}[4] = f_4(u_i^{(v)})$

$u_i^{(s)}[5] = f_5(u_i^{(v)})$

Left: Problem Definition

Right: Nonlinear Mapping (only arrows to u[1] and u[2] are drawn due to space limiation)

Fig. 1. Problem definition and nonlinear mapping for cross-site collaborative filtering

2. For users who have both ratings in site s and site v, e.g., $\{u_4, u_5, u_6\}$, how to utilize the ratings in the auxiliary site to leverage the rating prediction in the targeted site?

3 The Proposed Approach

3.1 Probabilistic Matrix Factorization

The probabilistic matrix factorization (PMF) model [6], as a kind of collaborative filtering methods, has been demonstrated successful in performing the recommendation task in large-scale datasets. It assumes a K-dimensional latent feature vector U for each user, and a K-dimensional latent feature vector V for each item. The inner product $U_i^T V_j$ is utilized to predict the rating for item j by user i. The objective function is a combination of the squared mean error and the quadratic regularization terms as follows, where N is the number of users, and M is the number of items, R_{ij} is the observed rating for item j by user i. I_{ij} is the indicator function to present whether the rating exists in the training set. The objective can be optimized effectively and efficiently by gradient descent algorithms.

$$\min_{U_i,V_j} \frac{1}{2}\sum_{i=1}^{N}\sum_{j=1}^{M} I_{ij}(R_{ij}-U_i^T V_j)^2 + \frac{\lambda_U}{2}\sum_{i=1}^{N}\|U_i\|^2 + \frac{\lambda_V}{2}\sum_{i=1}^{M}\|V_i\|^2$$

3.2 Mapping User Features Through The RBF Kernel

The mapping of user features across different sites are learned from a set of users who have a number of ratings in both sites. We utilize the PMF model to learn initial feature vectors for users/items in site s and site v separately. For the ith user, his/her latent feature in site v is denoted by $U_i^{(v)}$; and his/her latent feature in site s is denoted by $U_i^{(s)}$. We propose to utilize a group of K regression functions $f_k, (k \in 1, ..., K)$ to map the latent feature in the auxiliary site $U_i^{(v)}$ to the latent feature in the targeted site $U_i^{(s)}$. As shown in Fig. 1 (right), each function f_k corresponds to the kth dimension in $U_i^{(s)}$, denoted by $U_i^{(s)}[k]$, and the input of these regression functions is the latent vector in the auxiliary site $U_i^{(v)}$. As a primary study for the nonlinear property, all the regression functions are learned separately for simplicity.

We utilize the support vector regression algorithm [8] implemented by the LibSVM tool [1] as the regression function. The original problem to be solved is as follows, where x_i is the observed input features for the ith data sample in the training set, y_i is its regression value, and w is the global weight vector for different features.

$$\min \frac{1}{2}\|w\|^2 + C\sum_{i=1}^{l}(\xi_i + \xi_i^*), \; sub \; to \begin{cases} y_i - \langle w, x_i \rangle - b \le \varepsilon + \xi_i \\ \langle w, x_i \rangle + b - y_i \le \varepsilon + \xi_i^* \\ \xi_i, \xi_i^* \quad\quad \le \quad 0 \end{cases}$$

By solving its dual problem as

$$\max -\frac{1}{2}\sum_{i,j=1}^{l}(\alpha_i - \alpha_i^*)(\alpha_j - \alpha_j^*)\langle x_i, x_j \rangle - \varepsilon \sum_{i=1}^{l}(\alpha_i - \alpha_i^*) + \sum_{i=1}^{l} y_i(\alpha_i - \alpha_i^*)$$
$$sub \; to \; \sum_{i=1}^{l}(\alpha_i - \alpha_i^*) = 0 \; and \; \alpha_i, \; \alpha_i^* \in [0, C],$$

the kernel trick can be utilized by re-defining $\langle x_i, x_j \rangle$. In this paper, we utilize the RBF kernel as the nonlinear regression model, and its definition is

$$\langle x_i, x_j \rangle = \exp\left(-a\|x_i, x_j\|^2\right)$$

By defining the above nonlinear mapping functions, for users who only have ratings in the auxiliary site, their latent feature vector can be transferred to the targeted site. This solves the first task in the problem definition section. To avoid the latent feature permutation issue in the training process, users who only have ratings in one site are trained together with users who have ratings in both sites.

3.3 Regularization

For the second task in the problem definition, we utilize the regularization technique. The assumption is that for users who have both ratings in the auxiliary site and the targeted site, the transferred user latent feature vector from the auxiliary site should be similar with the learned user feature vector from his/her ratings in the targeted site. Therefore, we design the following strategy to leverage a user's ratings in the auxiliary site to improve his/her rating predictions in the targeted site.

The strategy is a two-step process. In the first step, feature vectors, $\{U^{(v)}, V\}$, are learned from the user-item(v) matrix, and feature vectors $\{U^{(s)}, S\}$ are learned from the user-item(s) matrix. By choosing a set of users who have a number of ratings in both sites, the mapping function $f = \{f_1, f_2, ..., f_K\}$ is learned. In the second step, for users who have ratings in both sites (the users utilized to learn f are excluded), their latent feature vectors in the targeted site are re-learned by the following objective function.

$$\min_{U_i^{(s)}} \frac{1}{2}\sum_{j=1}^{M} I_{ij}(R_{ij} - U_i^{(s)T}S_j)^2 + \frac{\lambda_U}{2}\|U_i^{(s)}\|^2 + \frac{\lambda_A}{2}\|U_i^{(s)} - f(U_i^{(v)})\|^2$$

In this objective function, only $U_i^{(s)}$ is variable, while others are fixed. Thus it can be simply solved by calculating the gradient as

$$\frac{\partial L}{\partial U_i^{(s)}} = \sum_{j=1}^{M} I_{ij}(R_{ij} - U_i^{(s)T}S_j)S_j + \lambda_U U_i^{(s)} + \lambda_A(U_i^{(s)} - f(U_i^{(v)})).$$

The second regularization term is designed to leverage auxiliary ratings to improve rating predictions in the targeted site.

4 Experiments

The experiments are targeted at justifying the following issues.

1. For users who only have ratings in the auxiliary site, to what extent can the nonlinear mapping approach outperform previous linear mapping methods for the rating prediction task in the targeted site?
2. For users who have ratings in both the targeted site and the auxiliary site, to what extent can the nonlinear mapping approach outperform previous linear mapping methods?

We compared the proposed nonlinear transfer learning approach with two state-of-the-art linear mapping methods, as well as a baseline method, in the above two configurations, respectively.

The experiments are conducted on a real-world cross-site dataset collected from two popular social media sites in China, Douban[3] and DianPing[4]. Douban is a review system for movies, and Dianping is mainly for Chinese restaurants. We first crawled ratings of from the two sites separately, and then utilized Liu et al.'s algorithm [4] to link the same users from the two sites, with the confidence value greater than 0.85. For evaluation purpose, we retain users and items that have top number of ratings. Finally, 3,924 cross-site users are selected, together with 6,842 users who only have ratings in Douban and 5,552 user who only have ratings in Dianping. Totally, these users have rated 6,885 movies, and 4,522 restaurants. There are 1,170,066 ratings in Douban (314,039 ratings are from the cross-site users), and 207,517 ratings in Dianping (20,609 ratings are from the cross-site users). We utilize Dianping as the targeted site and Douban as the auxiliary site, because Dianping's data is much more sparse than Douban's data.

We utilize two metrics, the Mean Absolute Error (MAE), and the Root Mean Square Error (RMSE) to evaluate the rating prediction performance task. Detailed definitions of these two metrics can be found in [5]. Both the two metrics measure errors. Thus a smaller MAE or RMSE value indicates a better performance. We utilize half of the cross-site users for learning the mapping function, and the other half for evaluation. For the evaluation users, we retain 0%, 25%, 50% and 75% of their ratings as training data, respectively. 0% corresponds to users who only have ratings in the auxiliary site, and the other three correspond to users who have ratings in both sites. We employ an item's

---

[3] http://www.douban.com
[4] http://www.dianping.com

Table 1. Overall performance

Dimension	Training Data	Metrics	AVG	CMF	CSVD	RBF
5	0%	MAE	0.6961	0.6892	0.6819	**0.6815**
		RMSE	0.8705	0.8664	0.8653	**0.8566**
	25%	MAE	0.7102	0.6997	0.6891	**0.6619**
		RMSE	0.8916	0.8768	0.8761	**0.8435**
	50%	MAE	0.6944	0.6880	0.6714	**0.6383**
		RMSE	0.8776	0.8770	0.8511	**0.8143**
	75%	MAE	0.6850	0.6538	0.6435	**0.6145**
		RMSE	0.8529	0.8229	0.8017	**0.7851**
10	0%	MAE	0.6961	0.6882	0.6791	**0.6694**
		RMSE	0.8705	0.8663	0.8628	**0.8481**
	25%	MAE	0.7102	0.6979	0.6857	**0.6646**
		RMSE	0.8916	0.8744	0.8699	**0.8468**
	50%	MAE	0.6944	0.6856	0.6793	**0.6551**
		RMSE	0.8776	0.8683	0.8589	**0.8317**
	75%	MAE	0.6850	0.6519	0.6392	**0.6124**
		RMSE	0.8529	0.8206	0.8017	**0.7780**

average rating as the baseline method (AVG), and two state-of-the-art transfer learning models, CMF [7] and CSVD [5] as competitive linear mapping methods.

Table. 1 shows the overall performances of different methods. In all configurations, the proposed approach (RBF) outperforms other methods consistently at a significant scale. For users who only have ratings in the auxiliary site, the proposed approach outperforms the linear mapping methods by 1.4% in MAE and 1.7% in RMSE. For users who have ratings in both sites, the proposed approach outperforms the linear mapping methods by 5.1% in MAE and 4.5% in RMSE. These results demonstrate that the nonlinear property does exist in the mapping of real recommendation systems, and modelling the transformation in the nonlinear manner is indeed effective in improving the recommendation performances. Figure. 2 shows the performances with different amount of auxiliary ratings. It demonstrates if more ratings in the auxiliary site can be obtained, the rating prediction in the targeted site will be more accurate.

5   Related Work

Probabilistic matrix factorization (PMF) [6] is one of the most competitive recommendation model, proposed by Salakhutdinov et al. Koren et al. illustrated several promising improvements on the PMF model by integrating the implicit feedback, the temporal patterns, and the confidence estimation [2]. To alleviate the data sparsity problem, one research direction is to utilize transfer learning techniques [5,7], which can be divided into two streams. In the first stream, users in the auxiliary domain and the targeted domain share the same user latent feature space, derived from matrix factorization techniques [3,7]. In the second stream, it assumes that there is a linear mapping between

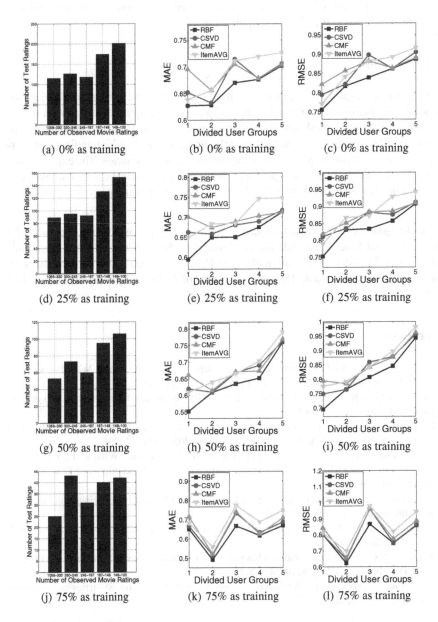

Fig. 2. Performances with different amount of auxiliary ratings. Dimension = 10. Group 1: 1058-330; Group 2: 330-245; Group 3: 245-187; Group 4: 187-148; Group 5: 148-100. The numbers of test ratings in these groups are shown in the left figure, and the results are shown in the right two figures.

user latent features across the auxiliary domain and the targeted domain [5]. The first stream is a special case of the second stream, by setting the linear mapping function among multiple sites to the identity mapping function.

6 Conclusion

In this paper, we explore how to utilize a user's real cross-site ratings to alleviate the data sparse problem in recommendation services. By an empirical comparison of linear transfer learning methods and the proposed nonlinear method, we demonstrate the nonlinear method significantly outperforms the linear methods, which indicates the existence of nonlinear property in cross-site recommendation scenarios. The experimental results also demonstrate that mapping user features by the RBF kernel is an effective manner in improving the performance of the targeted recommendation site.

Acknowledgements. The work described in this paper was mainly supported by National Natural Science Foundation of China (No. 61300076) and Ph.D. Programs Foundation of Ministry of Education of China (No. 20131101120035). It was also supported by the National Basic Research Program of China (973 Program, Grant No. 2013CB329605).

References

1. Chang, C.-C., Lin, C.-J.: Libsvm: A library for support vector machines. ACM Transactions on Intelligent Systems and Technology (TIST) 2(3), 27 (2011)
2. Koren, Y., Bell, R., Volinsky, C.: Matrix factorization techniques for recommender systems. Computer 42(8), 30–37 (2009)
3. Li, B., Yang, Q., Xue, X.: Can movies and books collaborate? cross-domain collaborative filtering for sparsity reduction. In: IJCAI, vol. 9, pp. 2052–2057 (2009)
4. Liu, J., Zhang, F., Song, X., Song, Y.-I., Lin, C.-Y., Hon, H.-W.: What's in a name?: An unsupervised approach to link users across communities. In: Proceedings of WSDM 2013, pp. 495–504. ACM (2013)
5. Pan, W., Yang, Q.: Transfer learning in heterogeneous collaborative filtering domains. Artificial Intelligence 197, 39–55 (2013)
6. Salakhutdinov, R., Mnih, A.: Probabilistic matrix factorization. In: Proc. of NIPS 2007, pp. 1257–1264 (2007)
7. Singh, A.P., Gordon, G.J.: Relational learning via collective matrix factorization. In: Proc. of SIGKDD 2008, pp. 650–658. ACM (2008)
8. Smola, A.J., Schölkopf, B.: A tutorial on support vector regression. Statistics and Computing 14(3), 199–222 (2004)

Deep Learning of Multifractal Attributes from Motor Imagery Induced EEG

Junhua Li and Andrzej Cichocki

Laboratory for Advanced Brain Signal Processing,
Brain Science Institute, RIKEN, Saitama, 351-0198, Japan
{junhua.li,a.cichocki}@riken.jp

Abstract. Electroencephalogram (EEG) is an effective metric to monitor or measure human brain activities. Another advantage for EEG utilization is non-invasive, and is not harmful to subjects. However, this leads to two drawbacks: low signal-to-noise ratio and non-stationarity. In order to extract useful features contained in the EEG, multifractal attributes were explored in this paper. A few attributes were utilized to analyze the EEG recorded during motor imagery tasks. Then, we built a deep learning model based on denoising autoencoder to recognize different motor imagery tasks. From the results, we can find that 1) Motor imagery induced EEG is of multifractal attributes, 2) multifractal spectrum D(h) and the statistics c_p based on cumulants can reflect difference between different motor imagery tasks, so they can be adopted as features for classification. 3) A deep network with initialization by denoising autoencoder is suitable to learn multifractal attributes extracted from EEG. The classification accuracies demonstrated that the proposed method is feasible.

Keywords: Multifractal Attribute, Deep Learning, Brain Computer Interface, Denoising Autoencoder, Wavelet Leader.

1 Introduction

The multifractal phenomenon widely occurs in the nature. Many objects have multi-fractal attributes, such as tree branch, coastline [1]. These objects repeatedly appear the same or equivalent attributes when they are inspected under different scales. In a task with repetitive actions, like motor imagery, the measured EEG may contain some components caused by the repetition of specific action. Besides, singularity is another component involved in EEG signal. These components can be represented by multi-fractal attributes. Up to now, multifractal analysis has been used to analyze diverse biophysiological signals and images [2-4]. For instance, Dutta et al. employed multi-fractal method to detect epileptic zones by EEG [2]. Here, we explored the feasibility of multifractal attributes for classifying motor imagery EEG. This will enrich alternative possibilities of kinds of features that represent intrinsic information related to motor imagery. The multifractal features could be integrated with power features, which are commonly used to distinguish different motor imageries [5-7], to improve the performance of decoding EEG signal [8, 9]. The performance improvement for decoding will benefit a great variety of applications of brain computer interface (BCI), such as BCI-based prosthesis [6] and entertainment [10]. In this paper, wavelet

C.K. Loo et al. (Eds.): ICONIP 2014, Part I, LNCS 8834, pp. 503–510, 2014.
© Springer International Publishing Switzerland 2014

leader multifractal formalism [11] was utilized to extract the multifractal attributes from the motor imagery EEG. Multifractal attributes (statistics c_p based on cumulants at each scale) were used as features. Unlike directly using original cumulants at each scale as in [9], we adopted higher statistical cumulants summarized from all scales. This dramatically reduces the dimension of features. After that, a deep neural network was built to learn these multifractal attributes. For the sake of effective learning, denoising autoencoder [12] was used to initialize weights of each hidden layer.

2   Methodological Framework

Figure 1 depicts the flow diagram of the proposed method. Firstly, raw EEG signal was filtered by a notch filter with band-stop at 50 Hz in order to remove the effect of electricity frequency. Then, filtered EEG was used to extract multifractal attributes. Consequently, these attribute features were fed into a deep network. To release the problem of overfitting, denoising autoencoder [12] was adopted to initialize weights of hidden layers.

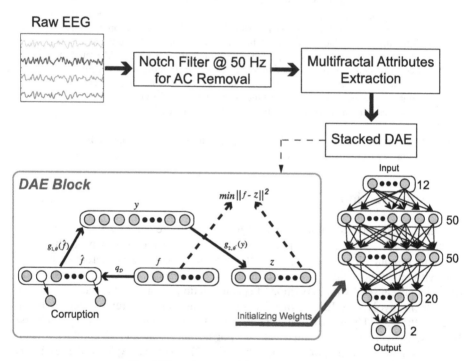

Fig. 1. Flow diagram of the proposed method

2.1   Multifractal Attributes

Let time series x(t) be a segment partitioned from a channel. After discrete wavelet transform (DWT), a time-scale representation (wavelet coefficients) of x(t) can be obtained.

Because wavelet-based multifractal analysis has two drawbacks: (1) entire multifractal spectrum of the process is not accessed under analysis, (2) not valid for all types of multifractal processes [13], we analyze multifractal attributes based on wavelet leaders [11]. Wavelet leader, denoted as $L_X(j,k)$, is the greatest value of adjacent wavelet coefficients. Adjacent region refers to time interval [k-1, k+1] at scale 2^j and all smaller scales. At a scale $a = 2^j$, the structure function is defined as the average of the q th power of $L_X(j,k)$ across time,

$$S^L(j,q) = \frac{1}{n_j}\sum_{k=1}^{n_j} L_X(j,k)^q = F_q 2^{j\zeta(q)}, \tag{1}$$

where n_j is the number of wavelet leaders $L_X(j,k)$ available at the scale 2^j, F_q is coefficient. The second equality holds in the limit of fine scales $2^j \to 0$. This indicates that structure function behaves power law behavior over scales for a set of statistical orders q (see figure 2(a) for an example).

Based on wavelet leaders, cumulants (see figure 2(b) for the first three cumulants at scales 2^1 to 2^5) at p th ($p \geq 1$) order can be defined as:

$$C^L(j,1) = E[\ln(L_X(j,\cdot))]$$
$$C^L(j,2) = E[\ln(L_X(j,\cdot)^2)] - C^L(j,1)^2$$
$$C^L(j,3) = E[\ln(L_X(j,\cdot)^3)] - 3C^L(j,2)\cdot C^L(j,1) - C^L(j,1)^3 \tag{2}$$
$$\cdots$$

If the following constraint is imposed,

$$C^L(j,p) = c_{0,p} + c_p \ln 2^j, \forall p \geq 1, \tag{3}$$

where $c_{0,p}$, c_p are coefficients, and equation (1) is transformed by the logarithm and is expressed using generating function expansion to obtain

$$\ln E e^{q \ln L_X(j,\cdot)} = \sum_{p=1}^{\infty} C^L(j,p)\frac{q^p}{p!} = \ln F_q + \zeta(q)\ln 2^j. \tag{4}$$

Then, substituting equation (3) into equation (4) to yield

$$\zeta(q) = \sum_{p=1}^{\infty} c_p \frac{q^p}{p!}. \tag{5}$$

From above equation, we can see that zeta $\zeta(q)$ is linear to q when $c_p \equiv 0, \forall p \geq 2$. In this case, it means the signal is monofractal, otherwise it is multifractal (see figure 2(c) for an example when signal is multifractal). According to equation (3), c_p can be obtained by means of linear regression in $\ln 2^j$ versus $C^L(j, p)$ coordinates,

$$c_p = (\log_2 e) \cdot \sum_{j=j_1}^{j_2} w_j C^L(j, p).$$

(6)

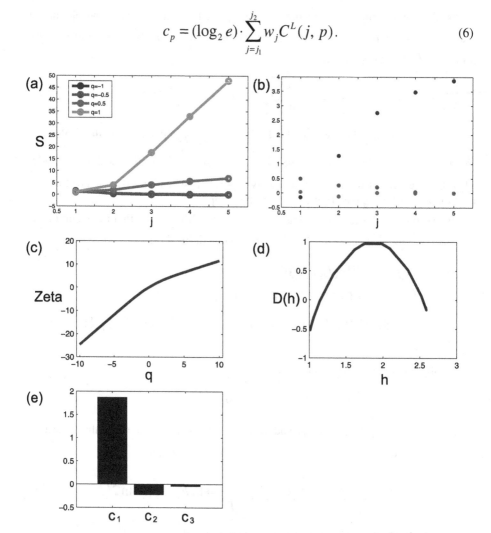

Fig. 2. An example of multifractal attributes calculated from a segment of EEG. (a) Structure function with different q at scales from 2^1 to 2^5 (note: the lines for q=-1 and q=-0.5 are overlapped). (b) Cumulants at each scale. $C^L(j,1)$, $C^L(j,2)$ and $C^L(j,3)$ are shown in blue, green, and red colors, respectively. (c) $\zeta(q)$ versus q. (d) Multifractal spectrum D(h). (e) c_p obtained from $C^L(\cdot, p)$ regression.

Figure 2(e) shows the first three c_p. The first three c_p have clearly physical meanings. c_1 is a self-similarity, which indicates the long-term temporal dependency. c_2 characterizes whether the signal has multifractality. The signal is more multifractal when c_3 is not zero. These c_p are also related to multifractal spectrum D(h) (see figure 2(d)). c_1, c_2 and c_3, respectively, correspond to the location of the maximum, width, and asymmetry of D(h) [11]. More detailed information about multifractal formalism can be referred to [11, 13-14] and the useful software can be found at http://www.irit.fr/~Herwig.Wendt/software.html.

2.2   Stacked Denoising Autoencoder

A block of denoising autoencoder (DAE) [12] is illustrated on the region enclosed by orange rectangle in figure 1. The features of multifractal attributes (i.e., c_p) were first corrupted (corrupted fraction is 0.3), denoted as $\widehat{f}$, by means of a stochastic mapping $\widehat{f} \sim q_D(\widehat{f} \mid f)$. Herein, features that were stochastically selected were enforced to be 0. Then, the corrupted features were mapped to a hidden representation by sigmoid function

$$y = g_{1,\theta}(\widehat{f}) = sigm(W \cdot \widehat{f} + b). \tag{7}$$

Consequently, we reconstructed the uncorrupted z as

$$z = g_{2,\theta'}(y). \tag{8}$$

The objective is to train parameters $\theta = \{W, b\}$ and $\theta' = \{W', b'\}$ for minimization of the average reconstruction error over training set. In other words, finding the parameters to let z as close as possible to original f, performing the following optimization:

$$
\begin{aligned}
[\theta^*, \theta'^*] &= \arg\min_{\theta, \theta'} \frac{1}{n} \sum_{i=1}^{n} \left\| f^{(i)} - z^{(i)} \right\|^2 \\
&= \arg\min_{\theta, \theta'} \frac{1}{n} \sum_{i=1}^{n} \left\| f^{(i)} - g_{2,\theta'}(g_{1,\theta}(\widehat{f}^{(i)})) \right\|^2
\end{aligned}
\tag{9}
$$

where n is the number of training samples, θ^*, θ'^* are the optimal values of θ, θ'. DAE learning rate was 0.5 with stop criterion of 30 epochs. Batch size was 75. When the training of a DAE was finished, the uncorrupted features were fed to the trained DAE. The output of previous layer of DAE was used as the input of the next DAE for training. After all DAE (stacked three DAEs) have been trained in sequence, the weights connecting between DAEs were used to initialize hidden layers of deep neural network. Finally, a top layer was added on the top of the neural network. Accordingly, the parameters were fine-tuned in a supervised way at the learning rate of 0.2 with stop criterion of 50 epochs.

3   Results

The EEG data used for evaluation is the same as used in [15]. EEG data were recorded from three subjects with fourteen electrodes mounted on sensorimotor cortex. During recording, subject was performing motor imagery tasks. There are four sessions for each subject. Each session consists of 15 trials, each of which is four seconds. A trial was divided into 25 one-second segments with an overlap of 87.5%.

3.1   Multifractal Attributes Difference between Tasks

We compared the multifractal attributes (left vs. right motor imageries) extracted from EEG based on wavelet leader multifractal formalism. Figure 3 shows the average multifractal spectrum D(h). Each subplot in the figure 3 illustrates the comparison between two tasks of motor imagery. We can clearly see that there is difference between them (i.e., the curves for different motor imageries in the figure 3 are not totally overlapped.). The larger difference of multifractal spectrum is found in subject 1. In addition, the average c_1 and c_2 respectively obtained from the regression of $C^L(\cdot,1)$ and $C^L(\cdot,2)$ are shown in figure 4. The difference between different motor imageries is observed at each session more or less. Similarly, the larger difference is appeared in the case of subject 1.

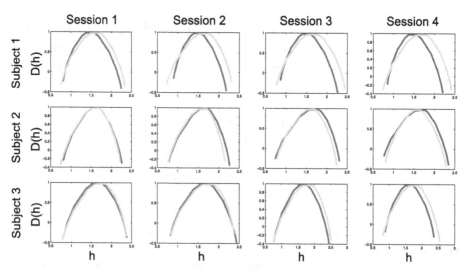

Fig. 3. Average multifractal spectrum D(h) at channel C3 for each session of all subjects. The magenta lines represent the average multifractal spectrum for left hand motor imagery over all segments within each session, and the cyan lines represent for right hand motor imagery.

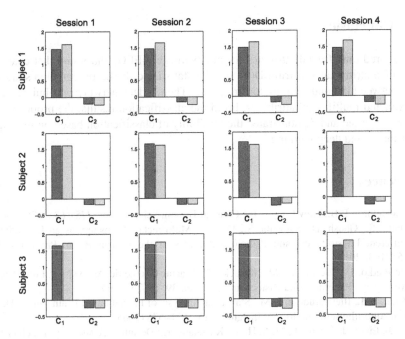

Fig. 4. Average C_1 and C_2 at channel C3 for each session of all subjects. C_1 and C_2 are obtained from the regression of corresponding cumulants. Magenta bars are for left hand motor imagery, while cyan bars are for right hand motor imagery.

3.2   Classification Accuracy

Most relevant four channels (C3, C4, Cp3, Cp4) were employed for extracting multi-fractal attributes ($c_1\ c_2\ c_3$). Preceding session data were used for training and the following session data were used for testing. Accuracies were calculated as ratios of the number of correct classification segments to the number of all segments. Table I lists all classification accuracies. Most of sessions achieved the accuracy better than chance level. Subject 1 possessed better performance, which is consistent with the analysis results in section 3.1. The more distinguished difference between tasks results in better performance in accuracy (i.e., higher accuracy).

Table 1. Classification Accuracies

Subject	1			2			3		
Session	2	3	4	2	3	4	2	3	4
Accuracy	90.93	85.87	93.07	57.60	51.47	70.93	63.20	68.27	83.47

4   Conclusion

We explored multifractal attributes for motor imagery EEG, and found that characteristics of multifractal spectrum D(h) and the statistics c_p based on cumulants are different between left and right motor imageries. Then, a deep network was built to learn the extracted multifractal attributes. From the classification results, the proposed method was worked and demonstrated the feasibility of classification based on multifractal attributes and deep learning.

References

1. Mandelbrot, B.B.: How Long Is the Coast of Britain. Science 156(3775), 636–638 (1967)
2. Dutta, S., Ghosh, D., Samanta, S., Dey, S.: Multifractal Parameters as an Indication of Different Physiological and Pathological States of the Human Brain. Physica A 396, 155–163 (2014)
3. Oczeretko, E., Juczewska, M., Kasacka, I.: Fractal Geometric Analysis of Lung Cancer Angiogenic Patterns. Folia Histochem. Cytobiol. 39, 75–76 (2000)
4. Zook, J.M., Iftekharuddin, K.M.: Statistical Analysis of Fractal-based Brain Tumor Detection Algorithms. Magn. Reson. Imaging 23(5), 671–678 (2005)
5. Li, J., Liang, J., Zhao, Q., Li, J., Hong, K., Zhang, L.: Design of Assistive Wheelchair System Directly Steered by Human Thoughts. Int. J. Neural Syst. 23(3), 1350013 (2013)
6. Pfurtscheller, G., Muller, G.R., Pfurtscheller, J., Gerner, H.J., Rupp, R.: 'Thought'-Control of Functional Electrical Stimulation to Restore Hand Grasp in a Patient with Tetraplegia. Neurosci. Lett. 351(1), 33–36 (2003)
7. Li, J., Zhang, L.: Active Training Paradigm for Motor Imagery BCI. Exp. Brain Res. 219(2), 245–254 (2012)
8. Li, J., Wang, Y., Zhang, L., Jung, T.P.: Combining ERPs and EEG Spectral Features for Decoding Intended Movement Direction. In: 34th Annual International Conference of the IEEE EMBS, San Diego, August 28-September 1, pp. 1769–1772 (2012)
9. Brodu, N., Lotte, F., Lécuyer, A.: Exploring Two Novel Features for EEG-based Brain–Computer Interfaces: Multifractal Cumulants and Predictive Complexity. Neurocomputing 79, 87–94 (2012)
10. Li, J., Liu, Y., Lu, Z., Zhang, L.: A Competitive Brain Computer Interface: Multi-person Car Racing System. In: 35th Annual International Conference of the IEEE EMBS, Osaka, Japan, July 3-7, pp. 2200–2203 (2013)
11. Wendt, H., Abry, P., Jaffard, S.: Bootstrap for Empirical Multifractal Analysis. IEEE Signal Process. Mag. 24(4), 38–48 (2007)
12. Vincent, P., Larochelle, H., Lajoie, I., Bengio, Y., Manzagol, P.A.: Stacked Denoising Autoencoders: Learning Useful Representations in a Deep Network with a Local Denoising Criterion. J. Mach. Learn. Res. 11, 3371–3408 (2010)
13. Wendt, H., Abry, P.: Multifractality Tests Using Bootstrapped Wavelet Leaders. IEEE Trans. Signal Process. 55(10), 4811–4820 (2007)
14. Lopes, R., Betrouni, N.: Fractal and Multifractal Analysis: A Review. Med. Image Anal. 13(4), 634–649 (2009)
15. Li, J., Struzik, Z., Zhang, L., Cichocki, A.: Spectral Power Estimation for Unevenly Spaced Motor Imagery Data. In: Lee, M., Hirose, A., Hou, Z.-G., Kil, R.M. (eds.) ICONIP 2013, Part I. LNCS, vol. 8226, pp. 168–175. Springer, Heidelberg (2013)

A Fast Neural-Dynamical Approach to Scale-Invariant Object Detection

Kasim Terzić, David Lobato, Mário Saleiro, and J.M.H. du Buf

Vision Laboratory (ISR-LARSyS), University of the Algarve, Faro, Portugal
{kterzic,dlobato,masaleiro,dubuf}@ualg.pt

Abstract. We present a biologically-inspired method for object detection which is capable of online and one-shot learning of object appearance. We use a computationally efficient model of V1 keypoints to select object parts with the highest information content and model their surroundings by a simple binary descriptor based on responses of cortical cells. We feed these features into a dynamical neural network which binds compatible features together by employing a Bayesian criterion and a set of previously observed object views. We demonstrate the feasibility of our algorithm for cognitive robotic scenarios by evaluating detection performance on a dataset of common household items.

1 Introduction

Reliable detection of objects in complex scenes remains one of the most challenging problems in Computer Vision, despite decades of concentrated effort. Object detection in Cognitive Robotics scenarios imposes further constraints such as real-time performance yet often with limited processing power, so efficient algorithms are needed.

In this paper, we present a fast neural approach to object detection based on cortical keypoints and neural dynamics, which can detect objects from more than 30 classes in real time. The biological foundation of our algorithm is particularly interesting for cognitive robotics based on human vision. We evaluate detection performance on a robotic vision dataset.

1.1 Related Work

Many modern object recognition algorithms begin by a keypoint extraction step to reduce the computational complexity and to discard regions which do not contain useful information. A number of keypoint detectors for extracting points of interest in images are available in the literature [1,2,3]. In biological vision, retinal input enters area V1 via the Lateral Geniculate Nucleus, and is then processed by layers of so-called simple, complex and end-stopped cells. Simple cells are usually modelled by complex Gabor filters with phases in quadrature, and complex cells by the modulus of the complex response. Simple and complex cells roughly correspond to edge-detectors in Computer Vision. End-stopped cells respond to line terminations, corners, line crossings and blobs, and can thus be

C.K. Loo et al. (Eds.): ICONIP 2014, Part I, LNCS 8834, pp. 511–518, 2014.

seen as general-purpose keypoint detectors. We base our method on fast V1 keypoints from [4], because they exhibit excellent repeatability and are biologically plausible, which makes them useful for object localisation and recognition.

There are many biologically inspired methods for object detection and recognition. Most of these are based on the Neocognitron and HMAX models, or convolutional neural networks. The Neocognitron architecture [5], originally developed for character recognition, has been successfully applied to object and face recognition [6]. Cortical simple and complex cells form the basis of the HMAX model and its derivatives [7], which alternate pooling and maximum layers to extract features of increasing complexity. However, HMAX requires an external classifier (usually an SVM) for final classification, so it is primarily a feature extraction method. Recently, deep convolutional networks have demonstrated excellent performance on a number of classification tasks, but at a considerable cost in terms of complexity and learning time [8]. The only object recognition algorithm based on neural dynamics known to us was proposed by Faubel and Schöner [9], which jointly estimates object pose and class using dynamic fields.

Concerning object detection and localisation in complex images, like in robotic scenarios, there are several main approaches: (i) sliding windows which apply a classifier at every image position and every scale [10], (ii) salience extraction followed by sequential classification [11], and (iii) voting schemes such as Generalised Hough Transform [12]. Sliding windows are computationally inefficient, while salience operators are typically based on general measures of complexity without object-specific knowledge and therefore do not reliably indicate complete objects. In contrast, our approach is based on voting and grouping, and it can be shown to maximise a Bayesian similarity criterion.

2 Method

2.1 Cortical Keypoints and Binary Descriptors

We begin by applying the fast V1 model from [4]. Given an input image, we compute the cell responses and represent them as sets of neural fields. Responses of simple cells are $R_{\lambda,\theta}$, those of complex cells are $C_{\lambda,\theta}$, and those of double-stopped cells are $D_{\lambda,\theta}$, where λ is the spatial wavelength of the Gabor filters representing simple cells, and θ their orientation. Peaks in the keypoint field $K_\lambda = \sum_\theta D_{\lambda,\theta}$ represent points in the image with high information content.

At each local maximum, we extract a binary keypoint descriptor. The descriptor is represented as a stack of neural maps B_λ^n, where $n \in 1, \ldots, N$ is the dimensionality of the descriptor. Each B_λ^n represents a comparison between the responses of two complex cells within the receptive field of the keypoint, whose size is equal λ:

$$B_\lambda^n = \mathrm{sgn} * (C_\lambda^n - C_\lambda^{\text{centre}}), \tag{1}$$

where $C_\lambda^{\text{centre}}$ is the complex cell in the middle of the receptive field, and $\mathrm{sgn} * (x)$ is 0 if $x \leq 0$ and 1 otherwise. Complex cells C_λ^n are sampled in concentric circles around the keypoint centre.

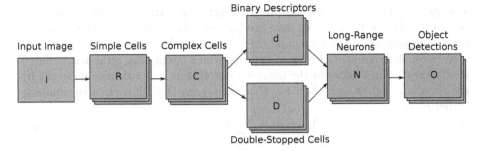

Fig. 1. Overview of the detection process. An input image I is processed by a set of retinotopic neural maps shown in the illustration as stacks of images. We begin by computing V1 responses: simple cells R, complex cells C, and double-stopped cells D, which represent keypoint activations. A local descriptor d is computed for each image location. A set of object-specific cells N responds to keypoints which are similar to those observed during training, and a set of grouping cells combines them into object heat maps O. These are fed into a stack of dynamic fields, which perform non-maximum suppression and pick the best hypothesis at each location.

2.2 Neural Object Detection Model

During training, localised objects are presented to the system and their descriptors d_i are extracted at keypoint locations x_i. For each class $c \in C$, we learn a set of neurons $N_i^{c,s}$ *at the object centre*, where each $N_i^{c,s}$ is associated with a keypoint descriptor *at keypoint location* x_i and s is the scale of the corresponding keypoint. Essentially, each neuron $N_i^{c,s}$ at the object centre has a long-range connection to its actual receptive field at x_i. During testing, keypoints and descriptors are extracted in the same way, but the the weights have been learnt such that the output of $N_i^{c,s}$ is the Hamming distance between a descriptor d_i^{test} from a novel object and a descriptor d_i^{train} observed during training. Therefore, $\sum_i N_i^{c,s}$ evaluates to zero if the system is shown one of the training objects from class c, and it grows large for very different objects.

A codebook of typical features could be learned from $N_i^{c,s}$ using a Self Organising Map, but at the moment we learn a prototype for each keypoint descriptor observed in a training image. We then threshold:

$$\hat{N}_i^{c,s} = [N_j^{c',s} - N_i^{c,s}]^+, \qquad (2)$$

with $j \neq i$ and $[\cdot]^+$ represents suppression of negative values. The introduction of the first term ensures that only neurons corresponding to small distances are activated, because large distances are not reliable for probability density estimation [13].

$\hat{N}_i^{c,s}$ are duplicated at all positions of the (subsampled) visual field. They are also duplicated at several scales by scaling the descriptor offset x_i and keypoint scale s by the same factor, resulting in a multi-scale detection framework.

If there is a keypoint in the neuron's receptive field, it is activated and its output is proportional to the thresholded Hamming distance between the observed descriptor and the training descriptor to which it was tuned. At any given position, a number of neurons $\hat{N}_i^c$ may fire for each class $c \in C$, and the largest of the active neurons are selected and the others are inhibited.

We now define a spatial map of neurons which counts the total accumulated Hamming distance between the observed descriptors and the ones expected by the object model of each class $c \in C$ and convolve it with a circular summing kernel K:

$$M_c(x, y) = \sum_{i,s} \hat{N}_i^{c,s}(x, y) , \tag{3}$$

$$O_c = M_c * K . \tag{4}$$

We assume that two objects of the same class cannot coexist at the same location in an image, so we sum over scales. This results in only one object map O_c per class.

The convolution with a circular kernel ensures that only features close to the expected position are counted towards a detection of an object, because localised features improve detection [14]. The radius of the kernel represents the maximum acceptable location error of each object part. Object detection now amounts to finding peaks in every O_C and picking the strongest peak at each position.

It can be shown that O_C actually represents at every pixel the logarithm of class likelihood conditioned on observed evidence, when using a nearest neighbour approximation of a naive Bayes classifier [15]. Selecting the class with the highest likelihood thus approximates a Maximum Likelihood classifier, which becomes a Maximum a Posteriori classifier if the prior probabilities of objects are known.

2.3   Winner Selection Using Neural Dynamics

The MAP detection model presented in this work forces a *winner takes all* decision whenever there are two competing detections at the same location. Conversely, as long as the estimate of the object likelihood is valid, picking the strongest hypothesis is equivalent to a local MAP decision between available hypotheses. We achieve this by modelling each object detection map from Eqn 4 as a dynamic field and using two inhibition schemes to force local decisions at peak locations [9].

The first inhibition scheme is global and it is applied to each field separately. It filters out the noise inherent to neural fields. In addition to global inhibition, the resting level of a field is a negative value. It acts as an activation threshold for each object class, thus removing weak and unreliable detections with little support caused by feature noise, i.e. illumination changes, occlusion, etc.

The second inhibition scheme is modelled as field interactions. Each object class is represented by a separate field and they inhibit each other: strong peaks in one field will inhibit smaller peaks at the same location in other fields. It pushes them below the detection threshold, thus forcing a winner-takes-all decision. This ensures that only one object can be detected at any one image location.

2.4   Implementation

A neuron $\hat{N}_i^{c,s}$ is only active if there is a keypoint at scale s located at offset x_i whose descriptor is similar to the one expected by $\hat{N}_i^{c,s}$. This means that the vast majority of all neurons are not active. We exploit this fact in order to improve speed. After extracting keypoints and their descriptors as described, we make each keypoint "vote" for an object centre. We do this by an efficient nearest-neighbour lookup among all descriptors learned during training and finding the k nearest neighbours. Among these, we pick the nearest descriptor for each class, and scale the offset x_i associated with this descriptor by the keypoint scale s. We then activate the neuron $\hat{N}_i^{c,s}$ corresponding to the correct class c, scale s and the scaled offset x_i/s. The distance to neighbour $k + 1$ is used to estimate the value of $N_j^{c',s}$ in Eqn 2, as suggested in [13].

The results of this alternative formulation are mathematically equivalent to using a full neural network implemention of our algorithm, but it is orders of magnitude faster and therefore usable for real-time scenarios.

Field-based dynamics were implemented using the CEDAR framework [16]. The early stages of our algorithm are implemented in C++ and OpenCV as a plug-in for CEDAR. Feature extraction and Eqns 3 and 4 were implemented on top of the public keypoint implementation from [4].

3   Evaluation

We evaluate our algorithm on the challenging IIIA30 dataset developed for robot localisation [17]. It consists of cluttered indoor images containing objects from 29 classes, with large scale and pose variance. The objects are annotated using labelled bounding boxes. As per Computer Vision convention, a detection is considered correct if it overlaps with a ground truth annotation of the same class by more than 50% (intersection over union). We compare against two standard methods used in [18]: SIFT keypoints followed by RANSAC grouping (the "classic" SIFT approach), and a Bag-of-Features method based on Vocabulary Trees built on top of SIFT descriptors. We applied our method using two descriptor types: the computational SIFT descriptor, and the biological binary descriptor based on responses of complex cells, introduced in Sec. 2.1. For brevity, we refer to our method as "NDOD": Neural-Dynamic Object Detection. The results for the standard methods were taken from [17].

Table 1 shows a summary of the results, averaged over all classes. The first row shows the best reported F1-score, averaged over all classes. The second and third rows show average recall and precision. Both values were measured separately for each class at the best F1 score for that particular class, then averaged. The fourth row shows mean Average Precision (the area under the precision-recall curve), where available. The last row shows how many classes each detector failed to detect (both recall and precision are zero). It can be seen that our full biological model compares well with the state of the art in computational vision, but does not yet match the classic SIFT approach. However, a combination of

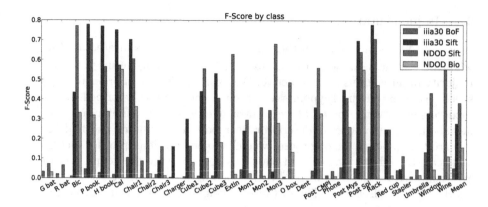

Fig. 2. Best reported F1-score for the 30 object classes from the IIIA30 dataset, compared with two state of the art methods built on SIFT descriptors. Our fully biological model achieves good performance on most classes, and the combination of our model and the SIFT descriptor outperforms all other methods.

Table 1. Common performance measures on IIIA30, averaged over all classes (see text). Our fully biological model using a cortical descriptor outperforms the Bag of Features method. It outperforms the SIFT method on many difficult classes, but is weaker on average. Our method combined with the SIFT descriptor outperforms all other methods.

	classic SIFT	SIFT+BoF	NDOD+Bio	NDOD+SIFT
Average Best F1 Score	0.281	0.054	0.159	**0.385**
Recall @ Best F1	0.260	**0.408**	0.126	0.346
Precision @ Best F1	0.372	0.032	0.301	**0.497**
Average Precision	n/a	n/a	0.076	**0.217**
% of classes failed	10	3	4	**2**

our neural object detection method and the SIFT descriptor outperforms all other methods, suggesting that a more powerful biologically plausible descriptor would significantly boost performance.

Figure 2 shows a more detailed evaluation of all four detectors. We plot the best reported F1 score for each of the 29 classes, as well as the average. The graph clearly shows that the classic SIFT method works well for some types of objects, and consistently fails with others. Both the Bag-of-Features approach and our method are more reliable with difficult classes. It can be seen that, averaged over all classes, the performance of our full biological model falls half-way between the two computational method. Our method combined with the SIFT descriptor significantly outperforms all other methods. Figure 3 shows some detections on images from the IIIA30 dataset.

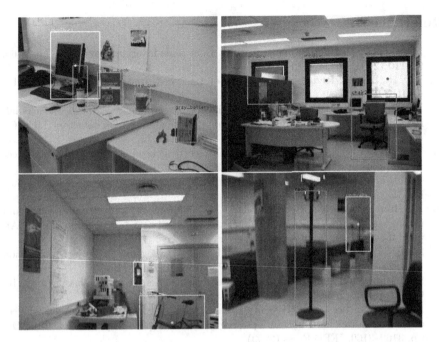

Fig. 3. Some detections from the IIIA30 dataset obtained by our method using the SIFT descriptor. We obtain high precision despite blurred images and a cluttered environment.

4   Conclusions

We have presented a real-time object detection mechanism based on cortical keypoints and neural dynamics. Our algorithm performs well on a standard dataset of household objects. To the best of our knowledge, this is the first neural object detection based on dynamic fields, and it significantly advances the state of the art in this field.

While the neural detection model is efficient and works well together with the SIFT descriptor, results show that our current biological image descriptor is holding back the performance of the complete biological model. Luckily, current research in binary image descriptors is often biologically motivated [19,20], so we expect significant progress in this area. We are currently looking into learning a powerful binary descriptor based on cortical cells, which should replace the hand-crafted one presented in this work.

Our current work focuses on using visual landmarks detected by our algorithm as localisation cues for cognitive robots, leading towards a semantic SLAM implementation.

Acknowledgements. This work was supported by the EU under the grant ICT-2009.2.1-270247 *NeuralDynamics* and the Portuguese FCT under the grant PEst-OE/EEI/LA0009/2011.

References

1. Lowe, D.G.: Distinctive image features from scale-invariant keypoints. IJCV 60, 91–110 (2004)
2. Bay, H., Ess, A., Tuytelaars, T., Van Gool, L.: Speeded-up robust features (SURF). CVIU 110, 346–359 (2008)
3. Mikolajczyk, K., Tuytelaars, T., Schmid, C., Zisserman, A., Matas, J., Schaffalitzky, F., Kadir, T., Van Gool, L.: A comparison of affine region detectors. IJCV 65, 43–72 (2005)
4. Terzić, K., Rodrigues, J., du Buf, J.: Fast cortical keypoints for real-time object recognition. In: ICIP, Melbourne, pp. 3372–3376 (2013)
5. Fukushima, K.: Neocognitron for handwritten digit recognition. Neurocomputing 51, 161–180 (2003)
6. Do Huu, N., Paquier, W., Chatila, R.: Combining structural descriptions and image-based representations for image, object, and scene recognition. In: IJCAI, pp. 1452–1457 (2005)
7. Serre, T., Wolf, L., Bileschi, S., Riesenhuber, M., Poggio, T.: Object recognition with cortex-like mechanisms. IEEE T-PAMI 29, 411–426 (2007)
8. Schmidhuber, J.: Multi-column deep neural networks for image classification. In: CVPR, pp. 3642–3649 (2012)
9. Faubel, C., Schöner, G.: A neuro-dynamic architecture for one shot learning of objects that uses both bottom-up recognition and top-down prediction. In: IROS, pp. 3162–3169. IEEE Press (2009)
10. Viola, P., Jones, M.J.: Robust real-time face detection. Int. J. Comput. Vision 57, 137–154 (2004)
11. Itti, L., Koch, C., Niebur, E.: A model of saliency-based visual attention for rapid scene analysis. IEEE T-PAMI 20, 1254–1259 (1998)
12. Leibe, B., Leonardis, A., Schiele, B.: Combined object categorization and segmentation with an implicit shape model. In: Workshop on Statistical Learning in Computer Vision, ECCV (2004)
13. McCann, S., Lowe, D.G.: Local naive bayes nearest neighbor for image classification. In: CVPR, Providence, pp. 3650–3656 (2012)
14. Mutch, J., Lowe, D.G.: Multiclass Object Recognition with Sparse, Localized Features. In: CVPR, New York, vol. 1, pp. 11–18 (2006)
15. Terzić, K., du Buf, J.: An efficient naive bayes approach to category-level object detection. In: ICIP, Paris (accepted, 2014)
16. Lomp, O., Zibner, S.K.U., Richter, M., Rañó, I., Schöner, G.: A Software Framework for Cognition, Embodiment, Dynamics, and Autonomy in Robotics: Cedar. In: Mladenov, V., Koprinkova-Hristova, P., Palm, G., Villa, A.E.P., Appollini, B., Kasabov, N. (eds.) ICANN 2013. LNCS, vol. 8131, pp. 475–482. Springer, Heidelberg (2013)
17. Ramisa, A.: IIIA30 dataset (2009), http://www.iiia.csic.es/~aramisa/datasets/iiia30.html (accessed April 30, 2014)
18. Ramisa, A.: Localization and Object Recognition for Mobile Robots. PhD thesis, Universitat Autonoma de Barcelona (2009)
19. Leutenegger, S., Chli, M., Siegwart, R.: BRISK: Binary robust invariant scalable keypoints. In: ICCV, Barcelona, pp. 2548–2555. IEEE Computer Society (2011)
20. Alahi, A., Ortiz, R., Vandergheynst, P.: FREAK: Fast retina keypoint. In: CVPR, Providence, pp. 510–517 (2012)

Improving Quantization Quality in Brain-Inspired Self-organization for Non-stationary Data Spaces

Kasun Gunawardana[1], Jayantha Rajapakse[1], and Damminda Alahakoon[2]

[1] School of Information Technology,
Monash University Malaysia
{kasun.gunawardana,jayantha.rajapakse}@monash.edu
[2] School of Information and Business Analytics, Faculty of Business & Law,
Deakin University, Australia
d.alahakoon@deakin.edu.au

Abstract. Vector quantization principles have efficiently been employed in number of clustering and classification algorithms due to its ability to estimate the probability density function of a multivariate stationary data distribution. However, application of the same principles to the algorithms which cater for non-stationary data spaces pose a massive challenge to maintain the quantization quality of learning outcomes. In order to maintain and improve the quantization quality in non-stationary data spaces, this paper presents an enhancement to an existing learning model. From experiments it has been proved that the enhanced learning model significantly improves the quantization error compared to its original version. Furthermore, the modification has resulted a less memory consuming and computationally more efficient algorithm.

Keywords: Brain-inspired, Self-organizing, Incremental Learning, Unsupervised Learning, Non-stationary Data Space.

1 Introduction

Due to the advancement of digital technology, businesses today have access to large volumes of data. These data always enfolds certain knowledge within itself. Hence, organizations demands better methods to extract hidden knowledge from their information systems. As a result interdisciplinary research in machine learning and data mining has become prominent. Especially, the discovering unseen patterns in a given data collection has been a major topic in its research.

Vector quantization principles have inspired number of learning models in this context and Self-organizing Map (SOM)[1] is a perfect example for such inspiration. SOM is a well-known neural network model which is used to uncover non-linear relationships among multivariate data points. Typically, SOM comprised with a grid like artificial neuron structure. Its learning process makes a cell or a group of cells responds to a data point at a time and as a result neuronal activation is localized and generates a spatially ordered sheet of neurons.

C.K. Loo et al. (Eds.): ICONIP 2014, Part I, LNCS 8834, pp. 519–526, 2014.

The fact that cerebral cortex of mammalian brain also consists of functionally localized sheets of neurons adds the essence of biology to SOM[1]. In fact, recognizing relationships among multivariate data in a stationary data space was a major problem at the time of SOM introduction. However with the current rate of data growth, extracting non-linear relationships in multivariate data requires new learning models which can cater for the dynamics of contemporary data spaces. Also the non-stationary nature of the problem domain demands learning models which develop its learning outcomes in an incremental manner. In order to tackle this problem in a novel approach, a learning model has been introduced in [8] and this paper proposes an enhancement to the model in [8] to improve the quality of its learning outcome.

The rest of this paper is organized as follows. Section 2 reviews handful of vector quantization inspired algorithms in the same problem domain. Section 3 describes the conceptual model in detail while section 4 presents the enhanced algorithm. Then experiments and results are discussed in section 5. Finally, the section 6 concludes the paper.

2   Related Work

In literature, it is possible to find numerous attempts that have been made to tackle the learning problem in non-stationary data spaces. However, due to the space constraints only a selected set of learning models and their key features are summarized here.

Based on local error counters and an insertion threshold of a neuron, LLCS[2] algorithm employs an adaptive insertion scheme with an adaptive learning rate to regulate the stability and the plasticity of the learning outcome. However, the number of parameters and variables which is involved with the algorithm makes its internal structure relatively a complex one. Like LLCS, SOINN[3] is also a two layered network model which employs an adaptive insertion technique and an adaptive learning rate. However in contrast to LLCS, SOINN calculates its insertion threshold by calculating min, max and average distance among neighbors. Later ESOINN[4] has been introduced to resolve and improve SOINN's inherent limitations. Unlike SOINN and LLCS, ESOINN runs on a single layer neuron structure.

IKASL[5] is another algorithm which is developed based on growing version of SOM. It runs in an incremental manner and in each increment it generalizes its learning outcome in a way that it can be used in the next increment. However the generalization process and some additionally used data structures bring extra computational cost to the algorithm. ESOM[6] is another learning model which exhibits the concept of growing network. Based on a predefined threshold, winner nodes insert new nodes to the network during the learning and at the same time inactive nodes are pruned. Conversely TASOM[7] employs time independent learning rate with neighborhood relation to accommodate dynamic data spaces. However, some of the parameters engaged with the algorithm need to be set to satisfy application's need which becomes a challenge at the execution time.The research work presented in [8] has proposed a brain inspired model to achieve biological neurons like signal propagation on a SOM like neural network. Learning model becomes more self-organizing since its neurons have autonomous *training rejection-training acceptance* mechanisms to maintain their stability and plasticity. Consequently, the algorithm has become much simpler due to these autonomous neurons. Research work presented in this paper attempts to improve the quality of quantization by enhancing its key components.

3 Learning Model

In the biological context, excitability of a neuron is decided based on the electro chemical changes which take place on the synapse with the arrival of a signal. Only the excited neurons transmit the signal further[9]. In the model [8], for each input a *best matching unit* (BMU) is elected and the BMU transmits the signal to its connected neurons. Once a signal is received, a neuron decides on its excitability. Only the excitable neurons get adapted and further transmit (see Fig. 1). Hence, it improves the similarity among excited neurons and as a result connectivity among similar neurons are strengthened. It simply reminds the Hebbian rule (*fire together - wire together*) in fundamental neuroscience. Accordingly, using the model in [8] as a basis, modified learning model can be described under three main components as follows,

3.1 *Continuity* - Sequential Presentation of Inputs

The learning model can also be seen as a continuous learning system, because the model can cater for a stream of data with temporal identities. It consumes a data point only once and reflects learning outcome just in time on its neuron structure. Unlike SOM, there is no iterative presentation of data. Hence, it enables the algorithm to accommodate any number of data points sequentially which makes it perfectly fit to a non-stationary context. Then, the learning outcome matures with the number of inputs it consumes. Interestingly the mechanism is similar to *learn by experience* in human learning; because initially formed learning outcome is revised throughout the entire lifespan of the learning system.

3.2 *Relativity* - Distance Based Learning Rate

When the probability density function is approximated on a two dimensional network structure, relative distance between two neurons represents the relative relationship of its weight vectors in input space. In fact, based on the probability of input occurrence, dense regions should be represented by more neurons compared to sparse regions. Since an input is presented only once, the learning process has to improve the density estimation in each adaptation step. Thus, the demand can be enforced by a learning rate that varies with network distance. As a result, the learning rate is a function of the network distance $d_{c,i}$, between the BMU and a given neuron such that c and i are the indexes of two neurons respectively. Hence, for an input x_t, at a given time instance t, the adjustment rule for a weight vector w, is defined as equation 1.

$$w_{t+1} = w_t + \alpha(d_{c,i}).(x_t - w_t) \qquad (1)$$

In equation 1, $\alpha(d_{c,i})$ is defined as equation 2 such that α_0 is a constant which is the maximum learning rate returned to the BMU. According to the equation 3, function $f(d_{c,i})$ controls the decay of α_0 over the network distance, $d_{c,i}$. As it can be seen in equation 3 and Fig. 2 when the $d_{c,i}$ becomes 0, $f(d_{c,i})$ becomes 1. Hence, BMU always gets the highest learning rate value α_0. Parameter l is the maximum of grid width or height which controls the overall shape of the function.

$$\alpha(d_{c,i}) = \alpha_0.f(d_{c,i}) \qquad (2)$$

$$f(d_{c,i}) = e^{\left(\frac{-(d_{c,i})^2}{l}\right)} \qquad (3)$$

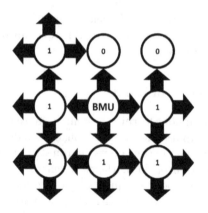

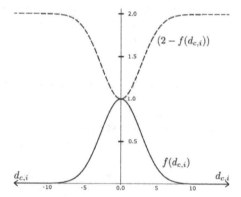

Fig. 1. Starting from a BMU, an input is transmitted by excited neurons (depicted in 1). Non-excitable neurons (depicted in 0) remain in the same state.

Fig. 2. Two functions which are employed to control learning rate decay - $f(d_{c,i})$ and receptive distance -$(2 - f(d_{c,i}))$ on a 10 x 10 network structure ($l = 10$)

3.3 *Excitability* - Adaptation Autonomy for Neurons

During the learning, SOM and SOM based learning models employ a neighborhood function to spread the impact of a single adaptation phase. In contrast, model in [8] attempts to mimic the natural way of signal transmission in biological neural networks to propagate the impact of a single adaptation phase. Hence, it introduces the notion of excitability to the artificial neuron. Technically, a neuron is excited when it receives a similar signal to the one it has already been encoded within itself. It is an autonomous decision that should be made by each neuron. If a neuron is excited then its weights will be adjusted towards the received signal and will transmit the signal to its immediate neighbors. Otherwise, the neuron stays its previous state without assisting to the signal transmission (see Fig. 1). Therefore, the signal propagation is controlled in a natural way rather than a function defined neighborhood set. In order to activate similar neurons, each neuron employs a similarity based excitability condition.

The model introduced in [8] has used an adaptation based similarity metric. In fact, a neuron in model [8] accumulates all the adaptations which take place on itself throughout the learning process. In order to decide on its excitability at any moment, a neuron checks its accumulated adaptation against the adaptation which is going to happen by the incoming signal. Since both measures are represented in vectorial form, the similarity is measured by the angle and if the cosine of the angle exceeds a certain threshold limit (*Similarity Threshold*) then the neuron decides to excite. However, this paper proposes a distinct modification to the similarity metric in order to achieve better quantization of input space. Therefore, in the modified model similarity measure is not based on the adaptation. Instead a neuron measures similarity based on the *Euclidean distance* from its weight vector to the input signal. Hence, in order to be excited, an input signal should be within certain range from the weight vector of a neuron. As a result the neuron replaces *adaptation vector* of [8] by a scalar value parameter r, which is called *receptive distance*, to keep track of its excitable distance.

At the beginning r is uninitialized and therefore, the first input is allowed to propagate over the entire network. Then at any weight adaptation, r value is also adjusted according to the equation 4. This modification makes learning model more efficient in terms of memory consumption and also for the computation.

$$r_{t+1} = \frac{r_t + (2 - f(d_{c,i})).\|x_t - w_t\|}{2} \tag{4}$$

To make BMU and its nearby neurons less susceptible to frequent modification, again network distance has to be featured in *receptive distance* calculation. Thus the actual *Euclidean distance* to the input is multiplied by an additional factor $(2 - f(d_{c,i}))$. Because the behavior of $(2 - f(d_{c,i}))$ is the complete opposite of the $f(d_{c,i})$ in equation 3 (see Fig. 2) and makes the current BMU and closer neurons less susceptible for future modifications by dissimilar signals. Consequently, as it is given in condition 5, the excitability of a neuron is assessed in accordance with the way *receptive distance* has been devised.

$$r > (2 - f(d_{c,i})).\|x_t - w_t\| \tag{5}$$

In addition, to make neurons flexible to unforeseen signals, a neuron slightly increases its *receptive distance* if it is failed to be excited. Otherwise unforeseen inputs always get rejected by the already trained neurons. In this case equation 6 has been employed to control the flexibility of r in a neuron while n denotes the number of inputs that has reached to a particular neuron.

$$r_{t+1} = \frac{(r_t \times n) + (2 - f(d_{c,i})).\|x_t - w_t\|}{n + 1} \tag{6}$$

4   Algorithm

Structurally the algorithm's functionality can be simplified into two phases; BMU election and propagation. BMU election follows the same mechanism as typical SOM (minimum Euclidean distance). In propagation phase a neuron performs two steps; first it updates weight vector and receptive distance. Then the neuron transmits the signal to its immediate neighbors. Starting from the BMU, each excited neuron executes these two steps in order to propagate the signal. The extent of input signal propagation is completely depend on the excitabilities of the neurons within the network. When all the excitable neurons are adapted, network is ready to accept the next input. Accordingly, learning continues as long as data points are available in the input spaces.

Algorithm

1. Present an input x_t, to the network (t is the time indicator)
2. Determine the BMU
3. Adjust its weight vector, w_t (equation 1)
4. Adjust its distance metric, r_t (equation 4)
5. Transmit the input signal to immediate neighbors (except the source)
6. For a neuron which receives a signal: Evaluate excitability (condition 5)
   (a) If TRUE: Go to step 3
   (b) If FALSE: Increase the value of r_t (equation 6)
7. Repeat from step 2 for a new input.

5   Experiments

In order to obtain the optimal quantization result, a learning model has to arrange its weight vectors in a way that it produces the minimum quantization error. Hence, to investigate modified algorithm's quantization error over the earlier one, four experiments were carried out. For the experiments four data sets were chosen with different complexities. Starting from Iris data set[1] which consists with data from known three classes and one class is linearly separable while other two have overlapping samples. Each sample in Iris data set is described by 4 attributes. For the second experiment Zoo data set[2] was chosen and 16 attributes have been used to describe a data point in that data set. In Flag data set[3], each sample is defined using 24 attributes and finally Urban Land Cover (ULC) data set[4] has 147 attributes to explain its single data point. Since the earlier model[8] employs a similarity threshold in its execution, that algorithm

---

[1] http://archive.ics.uci.edu/ml/datasets/Iris
[2] http://archive.ics.uci.edu/ml/datasets/Zoo
[3] https://archive.ics.uci.edu/ml/datasets/Flags
[4] http://archive.ics.uci.edu/ml/datasets/Urban+Land+Cover

was tested with four commonly used similarity thresholds 0.7, 0.75, 0.8, 0.85. Therefore, including the proposed algorithm, each experiment consisted of five tests. To conduct a fair experiment, each test was repeated 25 times for each data set. Moreover, five tests shared the same initial weight vectors at each repetition, but different randomly generated initial weights were employed in 25 repetitions. All tests were executed with $\alpha_0=0.2$ and same learning rate decay function. For Iris and Zoo data sets 4×4 network structures were used and the other two data sets were run on 6×6 network structures. After each execution quantization error was measured and results of four experiments can be summarized as in Table 1 and Table 2. As it is presented in the Table 1, modified model

Table 1. Average quantization errors and standard deviations for experiments

| | | Modified Algorithm | Initial Algorithm | | | |
			$T = 0.7$	$T = 0.75$	$T = 0.8$	$T = 0.85$
Iris	Average	0.0316	0.0330	0.0330	0.0330	0.0333
	Std. Dev.	0.0023	0.0029	0.0030	0.0028	0.0025
Zoo	Average	0.3059	0.3355	0.3291	0.3371	0.3392
	Std. Dev.	0.0190	0.0218	0.0198	0.0218	0.0263
Flag	Average	0.3694	0.4086	0.4170	0.4314	0.4803
	Std. Dev.	0.0116	0.0136	0.0159	0.0179	0.0366
ULC	Average	0.2080	0.2235	0.2321	0.2421	0.2619
	Std. Dev.	0.0093	0.0060	0.0076	0.0054	0.0113

has produced the minimum quantization error in every experiment. Further it has reported the minimum standard deviation in three cases which indicates the degree of variation of quantization error with respect to the modification. Additionally, the Table 2 portrays the error improvement as a percent of earlier model's quantization error.

Table 2. Average quantization error improvement gained by the proposed modification in each experimental setup as a percentage of the quantization error reported by earlier model in different similarity threshold values

| | Improvement (%) | | | |
	$T = 0.7$	$T = 0.75$	$T = 0.8$	$T = 0.85$
Iris	4.26	4.16	4.19	5.00
Zoo	8.81	7.04	9.25	9.80
Flag	9.59	11.42	14.36	23.08
ULC	6.94	10.39	14.07	20.58

6 Conclusion

Considering the demands of contemporary non-stationary data spaces, this paper proposes an enhancement for an existing incremental learning model[8]. The model in [8] has used a vectorial representation of total adaptation to control the stability the plasticity of a neuron over incoming signals. Instead, this paper proposes a scalar value parameter to achieve the same behaviour. From the experiments, it has been shown that the proposed algorithm achieves better quantization quality while reducing the error. Furthermore, replacement of a vector by a scalar in each neuron in the neural network causes a significant reduction in both memory consumption and computational cost. Moreover, difficulties with setting up an appropriate similarity threshold in the previous algorithm gets resolved with this enhancement since proposed model doesn't employ a threshold parameter. With all these attributes the algorithm follows a simple structure, has fewer steps and employs minimal parameters.

As a whole, the learning model consumes an input once, generates the learning outcome incrementally while improving it gradually. In the network structure, neurons autonomously organize themselves without an external involvement. Therefore, the model exhibits more self-organizing behavior. More importantly it is inspired from the characteristics of brain and mimics the natural way of signal propagation on an artificial neuron layer. Hence, the model follows intrinsic way of human learning and possesses a strong biological basis. Therefore, it is always worth to develop and make improvements on such a model.

References

1. Kohonen, T.: Self-organized formation of topologically correct feature maps. Biological Cybernetics 147, 59–69 (1982)
2. Hamker, F.H.: Life-long learning Cell Structures–continuously learning without catastrophic interference. Neural Networks 14(4-5), 551–573 (2001)
3. Shen, F., Hasegawa, O.: An incremental network for on-line unsupervised classification and topology learning. Neural Networks 19(1), 90–106 (2006)
4. Shen, F., Ogura, T., Hasegawa, O.: An enhanced self-organizing incremental neural network for online unsupervised learning. Neural Networks 20(8), 893–903 (2007)
5. De Silva, D., Alahakoon, D.: Incremental knowledge acquisition and self learning from text. In: International Joint Conference on Neural Networks, pp. 1–8 (2010)
6. Deng, D., Kasabov, N.: ESOM: An algorithm to evolve self-organizing maps from online data streams. In: Proceedings of the IEEE-INNS-ENNS International Joint Conference on Neural Networks, vol. 6, pp. 3–8 (2000)
7. Shah-Hosseini, H., Safabakhsh, R.: TASOM: A new time adaptive self-organizing map. IEEE Transactions on Systems, Man, and Cybernetics, Part B: Cybernetics 33(2), 271–282 (2003)
8. Gunawardana, K., Rajapakse, J., Alahakoon, D.: Brain-inspired self-organizing model for incremental learning. In: International Joint Conference on Neural Networks, pp. 1–8 (2013)
9. Squire, L., Berg, D., Bloom, F.E., du Lac, S., Ghosh, A., Spitzer, N.C.: Fundamental Neuroscience, 3rd edn., pp. 87–270. Academic Press (2008)

Utilizing High-Dimensional Neural Networks for Pseudo-orthogonalization of Memory Patterns

Toshifumi Minemoto[1], Teijiro Isokawa[1],
Haruhiko Nishimura[2], and Nobuyuki Matsui[1]

[1]Graduate School of Engineering, University of Hyogo, Japan
eu14n001@steng.u-hyogo.ac.jp,
{isokawa,matsui}@eng.u-hyogo.ac.jp
[2]Graduate School of Applied Informatics, University of Hyogo, Japan
haru@ai.u-hyogo.ac.jp

Abstract. Hebbian learning rule is well known as a memory storing scheme for associative memory models on neural networks. However, this rule doesn't work well in storing correlated memory patterns. Recently, a new method has been proposed based on pseudo-orthogonalization by XOR masking of original memory patterns with random patterns in order to overcome this problem. In this paper, we propose an extended method for pseudo-orthogonalization of memory patterns utilizing complex-valued and quaternionic neural networks. We demonstrate that Hebbian learning rule successfully stabilizes the correlated memory patterns, and these networks can retrieve the stored patterns corresponding to the external stimuli.

Keywords: Associative memory, Complex, Quaternion, Hopfield neural network, Pseudo-Orthogonalization, Hebbian learning.

1 Introduction

Hebbian learning rule is a well-known scheme for embedding patterns onto associative memories, such as Hopfield neural networks[1]. This scheme is simple and straightforward, however, it has a crucial issue for the embedding patterns; the patterns should be orthogonal to each other. On embedding correlated patterns by this scheme, the storage performance of the network is significantly decreased. Thus, many researches for storing correlated patterns direct to orthogonalization of these patterns, such as pseudo-inverse matrix method[2] and iterative learning scheme[3]. Though these methods enable all the correlated patterns to be stable local minima in the network, their computational costs grow with respect to the network size.

A novel scheme for embedding patterns has been proposed recently[4], which is based on orthogonalization of the stored patterns by random patterns. Called the pseudo-orthogonalization, this scheme first prepares a random pattern (masking pattern) of which size is the same as a stored pattern, and element-wise XOR

C.K. Loo et al. (Eds.): ICONIP 2014, Part I, LNCS 8834, pp. 527–534, 2014.

operation is applied between the random pattern and stored pattern. The pattern to be embedded to the network is a concatenation of XORed pattern and the corresponding random pattern, which can be embedded by Hebbian rule. The size of the pattern becomes double, but the embedded process is simple and requires lower computational cost.

Recently, the applications of complex or hypercomplex number systems to neural networks have been studied extensively[5,6]. The complex-valued or quaternionic extensions would be suitable for the pseudo-orthogonalization scheme; the pair of a pattern can be naturally embedded by utilizing imaginary part(s). However, few attempts have been accomplished for these extensions. In this paper, we extend the pseudo-orthogonalization scheme to the associative memories based on the complex-valued and quaternionic valued Hopfield networks[7]. We also investigate the performances of these extended schemes through storing binary patterns to the networks.

2   Hopfield Neural Networks

In this section, Hebbian learning rule and its network dynamics are introduced for real-valued, complex-valued, and quaternionic Hopfield neural networks.

2.1   Real-Valued Hopfield Neural Network

Let $\{\xi_{\mu,1}, \ldots, \xi_{\mu,N}\}$, $\xi_{\mu,m} \in \{+1, -1\}$ be the μth learning pattern. Hebbian learning rule for real-valued Hopfield neural network(RHNN) is represented as

$$w_{mn} = \frac{1}{N} \sum_{\mu=1}^{P} \xi_{\mu,m} \xi_{\mu,n} \qquad (1)$$

where w_{mn} is a synaptic weight between mth and nth neurons, which satisfies the conditions $w_{mm} = 0$ and $w_{mn} = w_{nm}$ for all m and n, and P is the number of the learning patterns. The dynamics of the network is given as,

$$x_m(t+1) = \text{sgn} \left(\sum_{n=1}^{N} w_{mn} x_n(t) \right) \qquad (2)$$

where $x_m(t) \in \{+1, -1\}$ represents the output of the mth neuron at the time step t. The activation function $\text{sgn}(\cdot)$ is a sign function where $\text{sgn}(u) = 1$ when $u \geq 0$, and $\text{sgn}(u) = -1$ when $u < 0$.

2.2   Complex-Valued Hopfield Neural Network

In complex-valued Hopfield neural network(CHNN), the inputs, output, and synaptic weights are encoded by complex values. The μth learning pattern is given as $\xi_{\mu,m} = \xi_{\mu,m}^{(e)} + \xi_{\mu,m}^{(i)}i$, $\xi_{\mu,m}^{(e),(i)} \in \{+1, -1\}$. Hebbian learning rule for CHNN is defined as

$$w_{mn} = \frac{1}{2N} \sum_{\mu=1}^{P} \xi_{\mu,m} \xi_{\mu,n}^{*} \qquad (3)$$

where synaptic weights satisfy the conditions $w_{mm} \geq 0$ and $w_{mn} = w_{nm}^{*}$. Here, w_{mn}^{*} denotes the complex conjugate of w_{mn}. The dynamics of the network is given as follows:

$$x_m(t+1) = \text{csgn}\left(\sum_{n=1}^{N} w_{mn} x_n(t)\right). \qquad (4)$$

The csgn$(\cdot)$ is an activate function, defined by csgn$(s) = \text{sgn}(s^{(e)}) + \text{sgn}(s^{(i)})i$.

2.3   Quaternionic Hopfield Neural Network

Quaternions form a class of hypercomplex numbers that consist of a real number and three kinds of imaginary number, i, j, k. Formally, a quaternion is defined as a vector in a four-dimensional vector space, i.e.,$x = x^{(e)} + x^{(i)}i + x^{(j)}j + x^{(k)}k$ where $x^{(e)}, x^{(i)}, x^{(j)}$ and $x^{(k)}$ are real numbers. Quaternion bases satisfy the following identities: $i^2 = j^2 = k^2 = ijk = -1$. $\mathbb{H}$, the division ring of quaternion, thus constitutes the four-dimensional vector space over the real numbers with the following bases: $1, i, j, k$. It is also written using 4-tuple or 2-tuple notations as $x = (x^{(e)}, x^{(i)}, x^{(j)}, x^{(k)}) = (x^{(e)}, \boldsymbol{x})$, where $\boldsymbol{x} = \{x^{(i)}, x^{(j)}, x^{(k)}\}$. In this representation $x^{(e)}$ is the scalar part of x, and $\boldsymbol{x}$ forms the vector part. The quaternion conjugate is defined as $x^{*} = (x^{(e)}, -\boldsymbol{x}) = x^{(e)} - x^{(i)}i - x^{(j)}j - x^{(k)}k$.

The operation between quaternions, $p = (p^{(e)}, \boldsymbol{p}) = (p^{(e)}, p^{(i)}, p^{(j)}, p^{(k)})$ and $q = (q^{(e)}, \boldsymbol{q}) = (q^{(e)}, q^{(i)}, q^{(j)}, q^{(k)})$. The addition and subtraction of quaternions are defined in the same manner as that of complex numbers or vectors by $p \pm q = (p^{(e)} \pm q^{(e)}, \boldsymbol{p} \pm \boldsymbol{q}) = (p^{(e)} \pm q^{(e)}, p^{(i)} \pm q^{(i)}, p^{(j)} \pm q^{(j)}, p^{(k)} \pm q^{(k)})$ With regard to the multiplication, the product of p and q, denoted as $p \otimes q$, is represented as follows $p \otimes q = (p^{(e)}q^{(e)} - \boldsymbol{p} \cdot \boldsymbol{q}, p^{(e)}\boldsymbol{q} + q^{(e)}\boldsymbol{p} + \boldsymbol{p} \times \boldsymbol{q})$.

In quaternionic Hopfield neural network(QHNN), all neuronal parameters in the network are encoded by quaternions. Let the μth learning pattern be $\xi_{\mu,m} = \xi_{\mu,m}^{(e)} + \xi_{\mu,m}^{(i)}i + \xi_{\mu,m}^{(j)}j + \xi_{\mu,m}^{(k)}k$, $\xi_{\mu,m}^{(e),(i),(j),(k)} \in \{+1, -1\}$. Hebbian learning rule for QHNN is represented as

$$w_{mn} = \frac{1}{4N} \sum_{\mu=1}^{P} \xi_{\mu,m} \otimes \xi_{\mu,n}^{*} \qquad (5)$$

where synaptic weights satisfy the conditions $w_{mm} \geq 0$ and $w_{mn} = w_{nm}^{*}$. The dynamics of the network is given as follows:

$$x_m(t+1) = \text{qsgn}\left(\sum_{n=1}^{N} w_{mn} \otimes x_n(t)\right). \qquad (6)$$

The qsgn($\cdot$) is an activate function that is defined by qsgn(s) = sgn($s^{(e)}$) + sgn($s^{(i)}$)$\boldsymbol{i}$ + sgn($s^{(j)}$)$\boldsymbol{j}$ + sgn($s^{(k)}$)$\boldsymbol{k}$.

3   Pseudo-orthogonalization Based on Complex Numbers and Quaternions

The purpose of the pseudo-orthogonalization method is to randomize the original patterns so that they can be stored by Hebbian learning[4]. We propose a pseudo-orthogonalization method based on complex numbers and quaternions in this section.

First, let us recapitulate the basic pseudo-orthogonalization method. Let $\{\xi_{\mu,1}, \ldots, \xi_{\mu,N}\}$ ($\xi_{\mu,m} \in \{+1, -1\}$) be the μth original pattern and $\{r_{\mu,1}, \ldots, r_{\mu,N}\}$ ($r_{\mu,m} \in \{+1, -1\}$) be a random pattern associated with the original pattern. The μth stored pattern $\{\eta_{\mu,1}, \ldots, \eta_{\mu,N'}\}$ ($\mu = 1, \ldots, P$) is generated from these patterns by

$$\eta_{\mu,m} = \begin{cases} r_{\mu,n}, & m = 2n - 1 \\ r_{\mu,n}\xi_{\mu,n}, & m = 2n \end{cases} \tag{7}$$

where $\eta_{\mu,m}$ takes either $+1$ or -1. This results in a concatenation of the random pattern and the original pattern randomized by the random pattern. The size of the stored pattern, denoted by N', becomes twice as the original one, i.e. $N' = 2N$.

Next, we introduce a scheme of complex-valued pseudo-orthogonalization. The complex-valued pseudo-orthogonalization is defined by

$$\eta_{\mu,m} = r_{\mu,m} + r_{\mu,m}\xi_{\mu,m}\boldsymbol{i}. \tag{8}$$

Original patterns can be reconstructed by

$$\xi_{\mu,m} = \eta_{\mu,m}^{(e)}\eta_{\mu,m}^{(i)}. \tag{9}$$

In this case, the sizes of the original and randomized patterns are same, i.e. $N' = N$.

Finally, the quaternionic pseudo-orthogonalization is described. A quaternionic pseudo-orthogonalization is defined by

$$\eta_{\mu,m} = r_{\mu,2m-1} + r_{\mu,2m-1}\xi_{\mu,2m-1}\boldsymbol{i} + r_{\mu,2m}\boldsymbol{j} + r_{\mu,2m}\xi_{\mu,2m}\boldsymbol{k}. \tag{10}$$

Original patterns can be reconstructed by

$$\xi_{\mu,m} = \begin{cases} \eta_{\mu,n}^{(e)}\eta_{\mu,n}^{(i)}, & m = 2n - 1 \\ \eta_{\mu,n}^{(j)}\eta_{\mu,n}^{(k)}, & m = 2n. \end{cases} \tag{11}$$

$N' = N/2$ is obtained in the quaternionic pseudo-orthogonalization method.

4   Simulations

In this section, we investigate the effectiveness of the proposed method in RHNN, CHNN and QHNN through embedding three types of stored pattern sets described as follows.

(1) Random patterns that are generated according to a uniform distribution, i.e., $\text{Prob}(\xi_{\mu,m} = \pm 1) = 1/2$.

(2) Correlated patterns that are generated by the patterns obtained by (1) and their correlations, defined by, $\text{Prob}(\xi_{\mu,m} = \pm 1) = (1 \pm \sqrt{b}\,\xi_m^0)/2$, where b is a correlation parameter, which satisfies $\text{E}[\text{Corr}(\xi_{\mu,m}, \xi_{\nu,m})] = b$, and ξ_m^0 is a random pattern.

(3) Pseudo-orthogonalized patterns, that are described in Section 3, generated from correlated patterns as the original patterns.

First, we evaluate the properties generated patterns by these three generators. For this purpose, two kinds of quantities are introduced. One is loading rate α that is defined by $\alpha = P/N$. The other is called overlap that is defined by the average of correlations among the patterns stored in the network. For RHNN, CHNN, and QHNN in this paper, we define the overlaps m_r, m_c and m_q, respectively, by

$$m_r = \frac{1}{N}\sum_{i=1}^{N}\xi_{\mu,i}x_i(t), \quad m_c = \frac{1}{2N}\sum_{i=1}^{N}\left(\xi_{\mu,i}^{(e)}x_i(t)^{(e)} + \xi_{\mu,i}^{(i)}x_i(t)^{(i)}\right),$$

$$m_q = \frac{1}{4N}\sum_{i=1}^{N}\left(\xi_{\mu,i}^{(e)}x_i(t)^{(e)} + \xi_{\mu,i}^{(i)}x_i(t)^{(i)} + \xi_{\mu,i}^{(j)}x_i(t)^{(j)} + \xi_{\mu,i}^{(k)}x_i(t)^{(k)}\right). \tag{12}$$

For the sets of the generated patterns, we calculate the overlaps for these networks. Figures 1(a), 1(b), and 1(c) show the overlaps with respect to the loading rate for RHNN, CHNN, and QHNN, respectively. The auxiliary lines $\alpha = 0.95$ and $m = 0.14$ are on these figures, showing the threshold for almost all patterns being successfully embedded. From these results, we find that the correlated patterns are hardly embedded, as compared with random patterns, and that pseudo-orthogonalization actually works for all the models. This can be also confirmed from Fig.1(d) which shows the performance comparison of pseudo-orthogonalization for the same correlated patterns, and this would imply that the stored patterns in CHNN and QHNN are more stable than those in RHNN.

Next, we show recall capabilities of the networks from an external stimulus pattern. Normally, the recall of the stored patterns in pseudo-orthogonalized method needs the random patterns used in pseudo-orthogonalization. Therefore, it is difficult to recall the stored patterns from a non-randomized pattern. However, the network dynamics can be extended for recalling from a cue pattern without the random patterns by using a simulated annealing process[4].

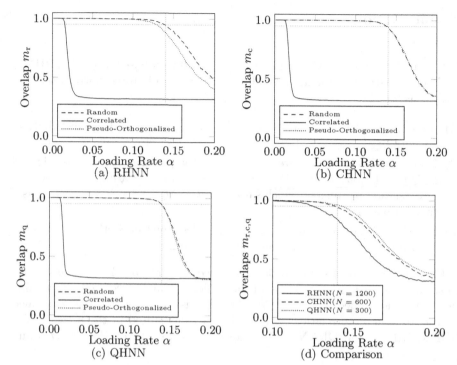

Fig. 1. Overlap m_r, m_c and m_q against loading rate α in RHNN, CHNN and QHNN. The results were obtained by averaging 1000 trials in 1000 asynchronous update($N = N' = 600, b = 0.1$). The auxiliary lines indicate $\alpha = 0.14$, $m = 0.95$.

The extended dynamics for CHNN is defined as:

$$h_m(t) = \sum_{n=1}^{N} w_{mn} x_n(t) + s z_m (x_m^{(i)}(t) + x_m^{(e)}(t)\boldsymbol{i}), \qquad (13)$$

$$\text{Prob}(x_m^{(*)} = 1) = \frac{1}{1 + \exp(-\beta(t) h_m^{(*)}(t))}, \ (*) = (e), (i), \qquad (14)$$

where $\{z_1, \ldots, z_N\}(z_m \in \{+1, -1\})$ denotes an input cue pattern, and $\beta(t+1) = \gamma\beta(t)$ is the temperature parameter which increases with time step t. Initialized randomly with $t = 0$, the state of the neuron $x_m(t)$ evolves stochastically.

Similarly, the dynamics for QHNN is defined as follows:

$$h_m(t) = \sum_{n=1}^{N} w_{mn} \otimes x_n(t) + s z_{2m-1}(x_m^{(i)}(t) + x_m^{(e)}(t)\boldsymbol{i}) + s z_{2m}(x_m^{(k)}(t)\boldsymbol{j} + x_m^{(j)}(t)\boldsymbol{k}),$$

$$(15)$$

$$\text{Prob}(x_m^{(*)} = 1) = \frac{1}{1 + \exp(-\beta(t) h_m^{(*)}(t))}, \ (*) = (e), (i), (j), (k). \qquad (16)$$

In our experiment, we use 42 learning image patterns as shown in Fig. 2. These images are strongly correlated, because they have a same local pattern on their left hand. These images are pseudo-orthogonalized and stored in RHNN, CHNN and QHNN. Figure 3(a) shows typical evolutions in recall processes for RHNN, CHNN, and QHNN for one of the stored patterns as a cue input pattern. All the networks can retrieve the pattern corresponding to the cue input, though the time steps for recalling are different each other in these networks. This can be confirmed from the result in Fig. 3(b), which shows the averaged evolutions of overlaps with different input cue patterns. At the final step($t = 1000$), the overlaps for these models become almost same. But when noisy patterns(40% elements in input pattern are flipped) are used for cue patterns, the overlaps by CHNN and QHNN are better than that by RHNN, as shown in Fig. 3(c). This suggests the robustness to the input stimuli in CHNN and QHNN.

鮪鱒鮃鮑鮗鮠鮴鰊鰤鰰鱚鱧鱸鮻
鯉鮭鯖鮫鯛鱈鮒鯱鯰鰕鰉鯧鰭鰈
鯵鮎鰯鰻鰍鰹鯨鮹鮄鮍鯑鮖鯣鯔

Fig. 2. Learning image patterns for simulation ($N = 1764, P = 42$). These patterns are Japanese kanji symbols which has same local pattern. The correlation coefficients of the pattern pairs range from 0.11 to 0.68 and its mean value is 0.41.

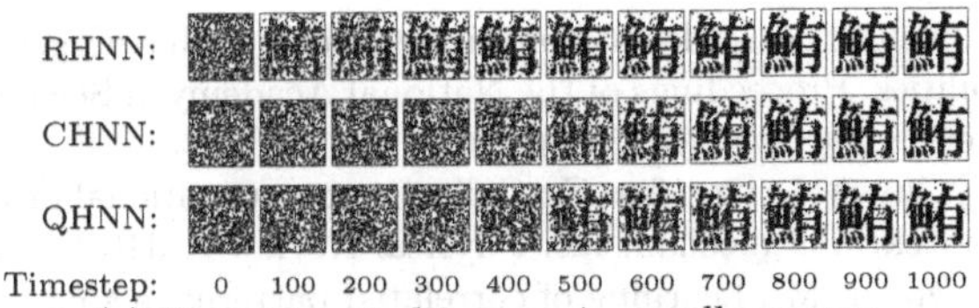

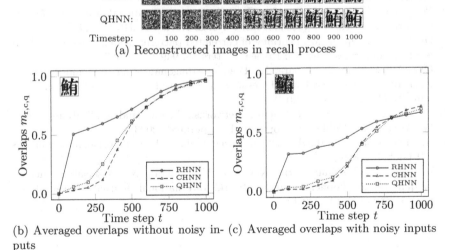

(b) Averaged overlaps without noisy in- (c) Averaged overlaps with noisy inputs
puts

Fig. 3. Results of image recall demonstrations. The parameters of the simulation are: $\beta(0) = 1.0, \gamma = 1.001$.

5 Conclusion

In this paper, we have investigated the embedding and retrieval performances for complex-valued and quaternionic extensions of the pseudo-orthogonalization scheme. In these extended schemes, the correlated patterns to be stored become orthogonalized by using random patterns associated with them, thus can be stored by Hebbian learning.

The numerical results show that the proposed scheme can embed the patterns better than the conventional (real-valued) scheme from the viewpoint of loading rates. This seems due to the improvement of orthogonalization achieved by the degree of freedom on the dimensions in complex or quaternion number systems. On retrieving the stored patterns from the input stimuli, the complex-valued and quaternionic networks can correctly retrieve the patterns regardless of noises in the input stimuli, whereas the real-valued networks often fail to retrieve ones from the noisy inputs.

Parameter dependencies for the storing and retrieval performances, such as the strength of input stimuli on the retrieval stage, should be explored in detail. Also, it is important to investigate the structure on basins of attractors for the complex-valued and quaternionic networks, as compared to the real-valued networks. These remain for our future work.

References

1. Hopfield, J.J.: Neural networks and physical systems with emergent collective computational abilities. Proceedings of the National Academy of Sciences of the United States of America 79(8), 2554–2558 (1982)
2. Personnaz, L., Guyon, I., Dreyfus, G.: Collective computational properties of neural networks: New learning mechanisms. Physical Review A 34(5), 4217 (1986)
3. Diederich, S., Opper, M.: Learning of correlated patterns in spin-glass networks by local learning rules. Physical Review Letters 58, 949–952 (1987)
4. Oku, M., Makino, T., Aihara, K.: Pseudo-orthogonalization of memory patterns for associative memory. IEEE Transactions on Neural Networks and Learning Systems 24(11), 1877–1887 (2013)
5. Hirose, A.: Complex-valued neural networks: theories and applications, vol. 5. World Scientific Publishing Company Incorporated (2003)
6. Nitta, T.: Complex-valued neural networks: Utilizing high-dimensional parameters. IGI Global (2009)
7. Isokawa, T., Nishimura, H., Kamiura, N., Matsui, N.: Associative memory in quaternionic hopfield neural network. International Journal of Neural Systems 18(02), 135–145 (2008)

Adaptive Noise Schedule for Denoising Autoencoder

B. Chandra[*] and Rajesh Kumar Sharma

Department of Mathematics, Indian Institute of Technology Delhi, New Delhi, India
bchandra104@yahoo.co.in, justrks@gmail.com

Abstract. The paper proposes an Adaptive Stacked Denoising Autoencoder (ASDA) to overcome the limitations of Stacked Denoising Autoencoder (SDA) [6] in which noise level is kept fixed during the training phase of the autoencoder. In ASDA, annealing schedule is applied on noise where the average noise level of input neurons is kept high during initial training phase and noise is slowly reduced as the training proceeds. The noise level of each input neuron is computed based on the weights connecting the input neuron to the hidden layer while keeping the average noise level of input layer to be same as that computed by annealing schedule. This enables the denoising autoencoder to learn the input manifold in greater details. As evident from results, ASDA gives better classification accuracy compared to SDA on variants of MNIST dataset [3].

Keywords: Deep Learning, Denoising, Encoder, Decoder.

1 Introduction

Deep learning algorithms generate hierarchical representation of input features. The class, to which an input belongs, depends on some intrinsic quality of input hence a good representation which captures important details of the input can be used to classify the input. More formally, a model m1(X, Y) which learns about both input X and output Y is more accurate than a model m2(Y | X) which learns only about Y given X. The goal is to learn about the input in an unsupervised manner without using any class information. Once a layer learns about the representation of input, the representation values learned by the layer are used to train the next layer in a similar unsupervised fashion. In this way, several layers are stacked to form a deep hierarchical representation of input. Finally, class information is used to fine tune the model.

In Deep Belief Network, the distribution of input is learned and restricted Boltzmann machines are stacked to form hierarchical representation [2]. Several variants of deep hierarchical representation composed of stacked autoencoders have been proposed [1], [4, 5, 6]. Autoencoders are trained in a greedy layer-wise manner to obtain deep representation of input in Stacked Autoencoders [1]. In contrastive autoencoder [4] Jacobian matrix of the hidden layer activations with respect to the input is computed and its Frobenius norm is included as regularization in the error term. Higher order contrastive autoencoder [5] computes stochastic approximation of

---

[*] Corresponding author.

C.K. Loo et al. (Eds.): ICONIP 2014, Part I, LNCS 8834, pp. 535–542, 2014.

Hessian using Jacobian matrix and extends the error term of contrastive autoencoder by including Frobenius norm of computed Hessian. Stacked denoising autoencoder [6] tries to learn the input representation by introducing noise to the input and reconstructing the original input from noisy values. In this technique, noise level during unsupervised pre-training is kept fixed.

In this paper, a novel procedure has been evolved for introducing adaptive noise to the input neurons. In the pre-training phase of denoising autoencoder, noise annealing schedule is used where average noise level of the input neurons is kept high during initial training phase. As training progresses, noise is slowly reduced. Each input neuron has different noise probability which depends on the weight of outgoing connections from that input neuron to the hidden layer. Noise probability of individual neurons is computed in such a manner that average noise level is the same as that selected by noise annealing schedule.

The intuition behind the proposed method is to impose strict restrictions on autoencoder during initial training phase which enables the autoencoder to learn general characteristics of the input. As training progresses, restriction is relaxed and the autoencoder learns about the input in more detail. This enables the autoencoder to explore the input manifold better than the case where noise is kept fixed. The idea behind varying the noise level for each input neuron is that, weights connecting input to the hidden layer indicates the extent to which the input neuron affects the activation of hidden layer.

Comparative performance evaluation of the proposed ASDA has been done with existing SDA on variants of MNIST dataset. Autoencoder is trained by using back propagation with adaptive learning rate [7]. It has been shown that ASDA performs superior compared to SDA in terms of classification accuracy. Section 2 describes SDA and Section 3 gives a detailed procedure for the proposed ASDA. Finally the results are given in Section 4.

2   Stacked Denoising Autoencoder

2.1   Autoencoder

Autoencoder is a multilayer perceptron where the output of the network is the same as original input. Transforming the input into hidden representation is done by the encoder and reconstructing the original input values is done by decoder.

Encoder: A mapping function f, that transforms input vector x into hidden representation y is called encoder and is given by:

$$y = f(x) = \sigma_e(W * x + b) \tag{1}$$

Where,
σ_e: Activation function for encoder
W: Weight matrix from input to hidden layer
b: Bias vector

Decoder: The resulting hidden vector y is mapped back to a reconstruction of x, denoted by z. The mapping function g is called decoder and is given by:

$$z = g(y) = \sigma_d(W1 * y + b1) \qquad (2)$$

Where,

σ_d: Activation function for decoder
W1: Weight matrix from hidden to output layer
b1: Bias vector from hidden to output layer.

The weights for encoder and decoder can be chosen as $W1=W^T$. Alternatively, different weights W and W1 can also be used for encoder and decoder respectively. Fig. 1 depicts a denoising autoencoder with input to hidden weights given by W and hidden to output weights given by W1.

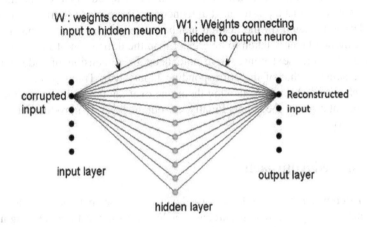

Fig. 1. Denoising Autoencoder

2.2   Overview of Stacked Denoising Autoencoder Algorithm

In order to prevent the network from learning identity and to enable it to learn useful representation, denoising is used as inherent regularization condition. After training the autoencoder with noisy input, activation values of hidden layer give a good alternate representation of input. In order to reconstruct the original input by using noisy input, hidden layer of denoising autoencoder learns a good representation of input in different dimension (same as the size of the hidden layer).

Training denoising autoencoders in a layer-wise manner by stacking one over another, is called a stacked denoising autoencoder, and creates a hierarchical representation of input [6]. Details of SDA are mentioned as follows:

While training the denoising autoencoder noise is introduced in the input. The noise level is a hyper-parameter which is kept fixed during the training. Noisy input x1 is obtained from the original input x by introducing one of the following noise:

1. Gaussian noise: x1 ~ $N(x, \sigma*I)$ where I is identity matrix and σ denotes the standard deviation. σ is a hyper-parameter which controls the level of noise in x1.
2. Masking noise: x1 is obtained by randomly making a fraction of input x to be 0. The probability of making a value of x1 to be 0 is a hyper-parameter which controls the level of noise in x1.
3. Salt and pepper noise: a fraction of input x is set as extreme values of x with certain probability.

Corrupted input x1 is used to obtain hidden representation y as given in (1). Reconstruction of input, denoted by z, is obtained using (2). Distance between reconstruction z and original uncorrupted input x is minimized by using back-propagation. For each epoch, a new x1 is generated by adding noise to x. The autoencoder learns to reconstruct the original input given the noisy input.

After training the denoising autoencoder, W1 and b1 are discarded and W and b are used to obtain hidden representation y from uncorrupted input x. The hidden representation y is used as input for next autoencoder. The process of training denoising autoencoders is repeated till the desired number of hidden layers are trained. The dimension of hidden layer is taken to be greater than the dimension of input for the first autoencoder. For the next consecutive autoencoders, dimension of hidden layer is taken to be same as that of the corresponding input layer. Training autoencoders in greedy layer-wise fashion and stacking them one over another helps to get the hierarchical representation of the input. Finally, an output layer is introduced to the pre-trained network.

3   Proposed Approach

In case of denoising autoencoders, noise level is a hyper-parameter which is kept fixed during training. Denoising autoencoder learns to reconstruct the original data based on a specific noise level. In order to improve the representation learned by the autoencoder, a novel procedure is proposed where noise level for each input neuron is different and depends on its contribution to the activation of hidden neurons. If the sum of outgoing weights from i^{th} input neuron is higher than the sum of outgoing weights from other input neurons, it implies that the ith input has higher impact on activation of hidden neurons. It also implies that the impact of other neurons is affected by the impact of ith neuron. In order to facilitate other input neurons to get better participation in activation of hidden neurons, i^{th} neuron should be turned off more often i.e. the probability of turning off i^{th} neuron should be higher. The average value of noise level (T) of the input neurons is computed at every epoch. The average noise level T is slowly reduced as the training progresses. This helps the autoencoder to learn rough approximation of the data initially and then to refine that approximation as noise is slowly reduced.

Average noise level T is slowly decreased from noise hyper-parameter A to noise hyper-parameter B during N epochs. Noise during an epoch E is given by (3).

$$T(E) = A - (A - B) * \frac{E - 1}{N - 1} \qquad (3)$$

The algorithm for Adaptive Stacked Denoising Autoencoder is given in Fig. 2.

```
// Adaptive Denoising Autoencoder
For epoch E = 1 to N
    Compute the average noise level T(E) using (3)
    For each mini-batch
        For each input neuron i
            Compute Pi, the probability of turning off input neuron i
```
$$P_i = \frac{|r_i - \mu(r)|}{K} + T(E) \qquad (4)$$
Where,
$$R_i = \sum_j W_{ij} , r_i = \frac{R_i - \min_i R_i}{\max_i R_i - \min_i R_i}$$
$$\mu(r) = \frac{\Sigma_i r_i}{n} \text{ and } K = max\left\{\frac{\max_i |r_i - \mu(r)|}{1 - T(E)}, \frac{\max_i |r_i - \mu(r)|}{T(E)}\right\}$$
```
            Introduce noise at the input neuron i using Pi.
        End
        Compute hidden representation and reconstructed input.
        Update weights using backpropagation with adaptive learning rate.
    End
End
```

Fig. 2. Algorithm for Adaptive Denoising Autoencoder

Once the autoencoder is trained, hidden representation is obtained using uncorrupted input. This hidden representation is then used as input for next adaptive denoising autoencoder. After training multiple autoencoders in a greedy fashion, output layer is added to the network and the weights are fine-tuned using backpropagation with adaptive learning rate [7].

3.1 Motivation for Using ASDA

In ASDA, the average noise is kept at a high level during initial training phase. Due to high noise, the autoencoder is forced to reconstruct the original data using only few input variables; hence it learns only general features about the data like big hills and valleys present in the manifold. As training progresses and the noise is reduced, the autoencoder encounters more input features that helps to reconstruct the original input with ease. Since more and more input features become available during each pass, the autoencoder tries to incorporate new knowledge in the existing knowledge about the data. This leads to fine tuning the curvature of hills and valleys in the manifold. Thus, the proposed method enables autoencoder to learn the manifold in a better way than fixing the noise level at pre-defined value [6]. Fig. 3 depicts the effect of reducing noise on autoencoder learning.

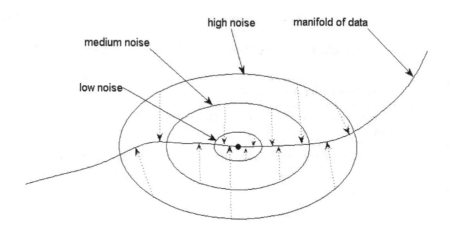

Fig. 3. Effect of noise on search neighborhood and search accuracy. For high noise in the input, reconstruction accuracy is smaller but autoencoder learns about manifold around larger neighborhood of input. This helps in getting general idea of data manifold. For smaller noise, autoencoder looks at smaller neighborhood around manifold corresponding to the input. This helps in fine tuning the knowledge about input manifold.

3.2   Implementation Details of ASDA

The deep network used for the implementation consists of an input layer, three hidden layers and an output layer with number of neurons as 784, 1000, 1000, 1000 and 10 respectively. Searching the hyper-parameter space for optimum learning rate for back-propagation algorithm is time consuming; hence backpropagation with adaptive learning rate [7] is used. Number of epochs is fixed as 200 for unsupervised pre-training of autoencoders. In ASDA, noise levels A and B are the other hyper parameters. The classification error on validation data is used for finding the values of hyper-parameters for ASDA. Although, search for hyper-parameters A and B was done using only few values, but still the performance of ASDA is found to be better than that of SDA.

4   Results

This Section gives the results of comparative performance evaluation.

4.1   Results

Performance of ASDA has been evaluated on MNIST dataset and its variants Basic, Background Images, Background Random and Rotation datasets taken from [3]. MNIST dataset consists of handwritten digit images in grey scale with 28x28 pixels.

The dataset Basic is obtained from MNIST dataset by interchanging training and testing sets. Background Random and Background Images datasets are obtained from Basic dataset by adding random noise and random images respectively to the background of Basic dataset. Rotation dataset consists of images from Basic dataset which are rotated by random angle in the range $[0, 2\pi]$. Comparison has been made with SDA. Noise levels A and B are chosen based on least validation error. The test error for various datasets are mentioned in Table 1. Corresponding noise levels are mentioned in the parenthesis. The results are obtained using backpropagation with adaptive learning rate on the same network architecture; hence a fair comparison can be made between both ASDA and SDA. For completeness, results mentioned in [6] for the same datasets are also given in Table 1.

Table 1. Classification error of various datasets (in percent)

Dataset	ASDA (adaptive learning rate)	SDA (adaptive learning rate)	SDA (optimum hyper parameters) [6]
MNIST	1.57 (0.7 -0.1)	1.98 (0.25)	1.28 (0.25)
Basic	**2.66** (0.7-0.1)	2.92 (0.1)	2.84 (0.1)
Background images	**15.42** (0.6-0.4)	16.87 (0.25)	16.68 (0.25)
Random background	12.63 (0.9-0.4)	13.46 (0.4)	10.3 (0.4)
Rotated digits	12.68 (0.7-0.4)	13.52 (0.25)	9.53 (0.25)

As evident from Table 1, for all the datasets, results obtained by the proposed ASDA gives better classification accuracy as compared to SDA when backpropagation with adaptive learning rate is used. Column 4 gives the result of SDA obtained by using optimum hyperparameters such as number of neurons in each hidden layer, learning rate, number of epochs etc. In order to come up with optimum hyperparameters, it takes enormous amount of time. The cases, where ASDA gives better accuracy than SDA obtained by optimizing hyper parameters, are marked in bold. It is seen that for the datasets Basic and Background images, the performance of ASDA with adaptive learning rate is better when compared to SDA by using optimum hyperparameters. This shows the superiority of ASDA over SDA.

5   Conclusions

In the paper, an adaptive noise schedule named as Adaptive Stacked Denoising Autoencoder (ASDA) has been proposed to train denoising autoencoders for deep learning. ASDA uses different noise level for each input neuron and slowly reduces the average noise level as training progresses. Hence, ASDA overcomes the limitations of SDA [6] in which noise level is kept fixed during the training phase of denoising autoencoder. Superiority of ASDA has been empirically shown by classification performance on variants of MNIST dataset.

Acknowledgement. The Council of Scientific and Industrial research (CSIR) India, is thankfully acknowledged for providing research fellowship to Rajesh Kumar Sharma.

References

1. Bengio, Y., et al.: Greedy layer-wise training of deep networks. In: Advances in Neural Information Processing Systems, vol. 19, p. 153 (2007)
2. Hinton, G.E., Osindero, S., Teh, Y.-W.: A fast learning algorithm for deep belief nets. Neural Computation 18(7), 1527–1554 (2006)
3. Machine Learning Laboratory, University of Montreal,
   `http://www.iro.umontreal.ca/~lisa/twiki/bin/view.cgi/Public/`
   `MnistVariations`
4. Rifai, S., et al.: Contractive auto-encoders: Explicit invariance during feature extraction. In: Proceedings of the 28th International Conference on Machine Learning, ICML 2011 (2011)
5. Rifai, S., Mesnil, G., Vincent, P., Muller, X., Bengio, Y., Dauphin, Y., Glorot, X.: Higher order contractive auto-encoder. In: Gunopulos, D., Hofmann, T., Malerba, D., Vazirgiannis, M. (eds.) ECML PKDD 2011, Part II. LNCS (LNAI), vol. 6912, pp. 645–660. Springer, Heidelberg (2011)
6. Vincent, P., et al.: Stacked denoising autoencoders: Learning useful representations in a deep network with a local denoising criterion. The Journal of Machine Learning Research 9999, 3371–3408 (2010)
7. Zeiler, M.D.: ADADELTA: An adaptive learning rate method. arXiv preprint arXiv:1212.5701 (2012)

Modeling Bi-directional Tree Contexts by Generative Transductions

Davide Bacciu[1], Alessio Micheli[1], and Alessandro Sperduti[2]

[1] Dipartimento di Informatica, Università di Pisa, Pisa, Italy
[2] Dipartimento di Matematica, Università di Padova, Padova, Italy

Abstract. We introduce an approach to integrate bi-directional contexts in a generative tree model by means of structured transductions. We show how this can be efficiently realized as the composition of a top-down and a bottom-up generative model for trees, that are trained independently within a circular encoding-decoding scheme. The resulting input-driven generative model is shown to capture information concerning bi-directional contexts within its state-space. An experimental evaluation using the Jaccard generative kernel for trees is presented, indicating that the approach can achieve state of the art performance on tree classification benchmarks.

1 Introduction

Dealing with tree data is a matter of effectively and efficiently capturing the correlation between the atomic pieces of information composing the structured sample and their context, i.e. the surrounding information bits linked through hierarchical relationships. The context of an information piece within a tree is not uniquely defined: a node label, for instance, may be evaluated in the context of either its surrounding descendants or ancestors. Generative approaches model probability distributions over spaces of trees, typically by generalizing the HMM approach for the sequential domain, through learning of an hidden generative process for labeled trees that is regulated by hidden state variables modeling the structural context of a node. By borrowing the nomenclature from HMM, these models are typically referred to as *Hidden Tree Markov Models* (HTMMs) [1,2,3]. The choice of the context ultimately determines how the structured sample is parsed and processed by such generative models, and ultimately influences their probabilistic assumptions and representational power. Generative models in literature take a uni-directional context propagation strategy that is determined by the direction of the state transition function (generative process). Top-down HTMMs (TD) [1] define a node context that captures information about the path leading to the node from the root. Bottom-up HTMMs (BU) [2,3], on the other hand, ensure that the context of a node summarizes information concerning structural properties of its descending subtrees.

In this paper, we propose an approach integrating, for the first time, both top-down and bottom-up contexts within a single generative model for trees, allowing to combine and exploit their different representational capabilities to yield

C.K. Loo et al. (Eds.): ICONIP 2014, Part I, LNCS 8834, pp. 543–550, 2014.
© Springer International Publishing Switzerland 2014

to more discriminative generative models for structured data. The intuition that knowledge from different structural contexts can be integrated to increase the computational capabilities of a learning model for tree data has been discussed in [4] for neural network approaches. A theoretical analysis on contextual models for structures within recursive incremental neural networks approaches can be found in [5]. Within the scope of generative approaches, only recently [6] has proposed a Jaccard kernel integrating information from two independent generative models by fusing their hidden state information at the kernel level: this, however, prevents the two generative models to share their contextual information. Conversely, the approach proposed here founds on an efficient composition of an homogenous top-down HTMM [1] with an input-driven bottom-up HTMM [3], realizing a structured transduction, i.e. a transformation of input tree into an output structure. We show how the realization of such transduction allows to transfer top-down context within the bottom-up model, while maintaining a contained computational complexity. An experimental assessment evaluating the discriminative power of the hidden state information of such bi-directional model is proposed within the scope of well-known tree classification benchmarks.

2   Modeling Bi-Directional Tree Contexts by Generative Model Composition

Before discussing the details of the generative models, we briefly summarize the notation used throughout the paper. Let us consider a dataset $\mathcal{D} = \{\mathbf{x}^1, \ldots, \mathbf{x}^N\}$ of N labeled rooted trees $\mathbf{x}^n$, consisting of a set of nodes $\mathcal{U}_n = \{1, \ldots, U_n\}$ with maximum finite out-degree L (i.e. the maximum number of children of a node). The term $u \in \mathcal{U}_n$ is used to denote a generic node of $\mathbf{x}^n$, whose direct ancestor, called *parent*, is denoted as $pa(u)$. A node u can have a variable number of direct descendants (*children*), such that the l-th child of node u is denoted as $ch_l(u)$. Each vertex u in the tree is associated with a label x_u which is a d-dimensional vector. The encoding tree $\mathbf{q}^n$ has the same structure of the corresponding input tree $\mathbf{x}^n$, but the labels associated to its nodes correspond to the hidden states generated by the homogenous model. Hidden state variables Q_u are indexed by the nodes u, with values over the finite set $[1, \ldots, C]$, and are associated to a state transition dynamics that follows the generative process direction.

The proposed bi-directional generative approach is realized by a circular transduction process summarized in Fig. 1. It exploits, first, a uni-directional generative model to produce an encoding of the input tree $\mathbf{x}^n$ into its hidden states, i.e. a tree $\mathbf{q}^n$ with the same structure whose nodes are labeled with the corresponding hidden states of the generative model. The encoded tree is then fed to a generative model operating on the opposite parsing direction and which is trained to transduce (i.e. decode) the encoded structure back into the original input tree $\mathbf{x}^n$. The two generative models do not differ only for the parsing direction, rather they model two different types of probabilistic distributions over trees. The former is an *homogenous* approach that models the unconditional distribution $P(\mathbf{x}^n)$ over the input trees $\mathbf{x}^n$. The latter is an *input-driven* approach

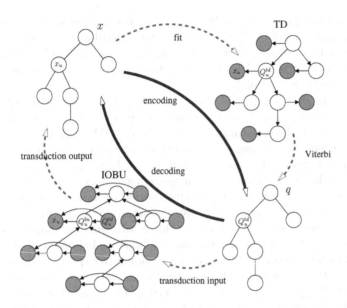

Fig. 1. Bi-directional context acquisition process: first, a TD model is fit to the input trees **x** and then the Viterbi algorithm is applied to generate the corresponding encoding trees **q**. Decoding is realized by fitting IOBU to the transduction between the encoded trees **q** (input of the transduction) to the original trees **x** (output of the transduction). The hidden states Q_u^{td}, Q_u^{bu} and original tree labels x_u are shown only for an example node u for the sake of picture clarity. Hidden and observed variables are represented as shaded and empty nodes, respectively, in the probabilistic models.

modeling the conditional distribution $P(\mathbf{x}^n|\mathbf{q}^n)$ from the encoding tree $\mathbf{q}^n$ back to the original tree $\mathbf{x}^n$. By means of the conditional distribution $P(\mathbf{x}^n|\mathbf{q}^n)$, it is possible to transfer contextual information from the first to the second generative model. The hidden state space of the input-driven model becomes dependent not only on the structural information provided by its parsing direction, but also from the contextual information coming from the opposite direction and summarized by the hidden states $\mathbf{q}^n$ of the homogeneous model.

The *encoding* step of the circular transduction is realized by an homogenous *top-down* HTMM [1] (TD in Fig. 1) implementing a generative process for all paths from the root to leaves of the trees. The direction of its generative process is modeled by the *state transition probability* $P(Q_u = j|Q_{pa(u)} = i)$. To complete the specification of the model, it is assumed that the label x_u (continuous or discrete) of a node u is completely specified by its hidden state Q_u through the *emission distribution* $P(x_u|Q_u = j)$. Following such conditional independence relations, we can factorize the joint distribution of an observed tree $\mathbf{x}^n$ with the hidden states assignment $Q_1^{td}, \ldots, Q_{U_n}^{td}$ as

$$P(\mathbf{x}^n, Q_1^{td}, \ldots, Q_{U_n}^{td}) = P(Q_1^{td})P(x_1|Q_1^{td}) \prod_{u=2}^{U_n} P(x_u|Q_u^{td})P(Q_u^{td}|Q_{pa(u)}^{td}), \quad (1)$$

where the overscript is used to highlight that the hidden states pertain to a TD model. The term $P(Q_1^{td})$ is the *prior* state distribution for the root node ($u = 1$), given that it has no parent. The likelihood for the TD is obtained from (1) by marginalizing the unknown hidden state assignment and summing across all dataset (see [1] for details).

The *decoding* step requires to learn the conditional model $P(\mathbf{x}^n|\mathbf{q}^n)$, where the encoded tree $\mathbf{q}^n$ conditions the generative process of the original tree $\mathbf{x}^n$. This is realized by the *input-output bottom-up* HTMM [3] (IOBU in Fig. 1), a recently proposed non-homogenous generative model for tree-structured data which defines a generative process propagating from the leaves to the root of the tree. The IOBU implements a generative process that composes the child subtrees of each node in a recursive bottom-up fashion. Hence its multi-nomial hidden state variables Q_u^{bu} are characterized by a children-to-parent (bottom-up) state transition dynamics. Differently from the homogenous TD, the hidden state transition and the output label emissions of IOBU are assumed to depend also on the labels on the encoding tree $\mathbf{q}^n$, that are the hidden states Q_u^{td} of the TD model. The resulting joint *state transition probability* is $P(Q_u^{bu} = i|Q_{ch_1(u)}^{bu} = j_1, \ldots, Q_{ch_L(u)}^{bu} = j_L, \mathbf{q}^n)$, assuming that each node u is conditionally independent of the rest of the tree when the joint hidden state of its direct descendants $Q_{ch_l(u)}^{bu} = j_l$ is observed together with the corresponding labels of the encoding tree $\mathbf{q}^n$. The problem with the this formulation is that it becomes computationally impractical for trees other than binary, since the size of the joint conditional transition distribution is order of C^L, where L is the node outdegree. In [3], this has been addressed by introducing a scalable *switching parent* approximation factorizing the joint transition as a mixture of L pairwise child-parent transitions: this yields the IOBU joint distribution

$$P(\mathbf{x}^n, Q_1^{bu}, \ldots, Q_{U_n}^{bu}|\mathbf{q}^n) = \prod_{u' \in \mathcal{LF}_n} P(Q_{u'}^{bu}|q_{u'})P(x_{u'}|Q_{u'}^{bu}, q_{u'})$$

$$\prod_{u \in \mathcal{U}_n \setminus \mathcal{LF}_n} P(x_u|Q_u^{bu}, q_u) \sum_{l=1}^{L} P(S_u = l)P(Q_u^{bu}|Q_{ch_l(u)}^{bu}, q_u) \tag{2}$$

where $\mathcal{LF}_n$ is the set of leaf nodes and $P(Q_u^{bu}|Q_{ch_l(u)}^{bu}, q_u)$ is the conditional contribution of the l-th child to the state transition of the parent node u having encoding label q_u. The summation term corresponds to the joint transition factorization using the switching parent $S_u \in \{1, \ldots, L\}$. This is a latent variable, independent from $Q_{ch_l(u)}^{bu}$, and whose distribution $P(S_u = l)$ measures the influence of the l-th children on a state transition to node u. The term $P(x_u|Q_u^{bu}, q_u)$ denotes the encoding-conditional emission of the original label x_u, when the node has encoding label q_u. By construction of the circular transduction, the encoding label q_u corresponds to the TD hidden state Q_u^{td} generated for the u-th node.

Training can be performed independently for the two models through an Expectation-Maximization approach applied to the log-likelihood of the models, completed with latent indicator variables that model the unknown hidden state (and switching parent) assignments. Details of the learning algorithms for the specific models can be found in the papers cited above. The resulting algorithm is a two-stage iterative procedure that, at the E-step, estimates the posterior of the indicator variables introduced in the completed log-likelihood, while, at the M-Step, it exploits such posteriors to update the model parameters θ. Posterior estimation is the most critical part of the algorithm and can be efficiently computed by message passing upwards and downwards on the structure of the nodes' dependency graph [2,7]. As shown in Fig. 1, parameter learning is first performed on the TD model to fit the unconditional distribution $P(\mathbf{x}^n)$. To generate the encoding tree $\mathbf{q}^n$, we need to determine the most likely assignment of the top-down hidden states for an input tree $\mathbf{x}^n$, i.e. $\mathbf{q}^n = \arg\max_{\mathbf{q}} P(\mathbf{X}^n = \mathbf{x}^n, \mathbf{Q}^n = \mathbf{q})$. This optimization problem can be efficiently solved for both generative models through the *Viterbi Algorithm*, a dynamic programming approach whose details can be found in [7] for the TD model. Once the encoding trees have been obtained for all the input trees, the IOBU model can be trained on the $(\mathbf{q}^n, \mathbf{x}^n)$ samples to learn the structured transduction $\tau : \mathbf{q}^n \to \mathbf{x}^n$ from the encoding back to the original tree (see the decoding step in Fig. 1). The IOBU hidden state space should now capture information from both a bottom-up context, i.e. through its own generative dynamics, and from a top-down context, i.e. through the conditioning on the TD hidden state labels in $\mathbf{q}^n$.

An effective way to mine the structural information captured by the IOBU state space is by building a *Jaccard kernel* on its hidden states multiset [8]. Roughly, each tree is represented in terms of its associated hidden states in the IOBU model; then, structure similarity is computed on the basis of the overlap in the hidden states' configurations. More specifically, given a trained IOBU, we transform a tree $\mathbf{x}^n$ (and its associated encoding $\mathbf{q}^n$) into a *bag-of-states*, that is a vector of hidden IOBU state counts, similarly to how textual documents are represented as vectors of word counts. To do so, we again need to solve a Viterbi maximization problem specific for the IOBU as described in [3]. By means of such Viterbi states, it is possible to define several bag-of-states encodings, depending on the amount of structural information that we want to introduce in the kernel feature-space representation. For instance, a *bigram* multiset measures the co-occurrence of hidden-states in a parent-children relationship: given a sample $(\mathbf{q}^n, \mathbf{x}^n)$ and associated IOBU hidden state assignments $Q_{n,u}^{bu}$ for each node u, this is represented by the (C^2)-dimensional feature-vector Φ^n, such that its i_j-th element is

$$\Phi_{i_j}^n = \sum_{u \in \mathcal{U}_n} \sum_{l \in ch(u)} \delta(Q_{n,u}^{bu}, i)\delta(Q_{n,l}^{bu}, j), \quad j = 1, \ldots, C, \; i_j = 1, \ldots, C \quad (3)$$

where $ch(u)$ is the set of children of node u and $\delta(\cdot, \cdot)$ is the Kronecker function. In this paper, following the indications in [8], we consider a *unibigram* representation corresponding to the concatenation of *unigram* and *bigram* multisets (see [8] for details). The final *Jaccard kernel* can then be computed by application of the Jaccard multiset similarity to the resulting concatenated vector as discussed in [8]. Note that here the use of the Jaccard kernel serves solely to assess the quality of the structural information captured by the bi-directional generative approach. By this means, it is possible to compare its performance with that achieved by the single uni-directional models or by other bi-directional approaches using multiple contexts only at kernel level, but not at the level of the underlying generative models. The different approaches can also be compared on the basis of the computational complexity of Jaccard kernel computation. The total complexity required for hidden state inference and kernel matrix computation is $O((N_T \cdot C^2 \cdot V) + (E + C) \cdot V)$ for uni-directional models, where N_T is the maximum tree size, V is the number of classes and $E = \min\{N_T, C^2\}$. This corresponds also to the asymptotic complexity of the bi-directional approach in [6], which differs only for constant terms. The complexity of the proposed approach is $O((N_T \cdot C^2 \cdot V) + (E + C))$.

3   Experimental Results

The experimental analysis focuses on assessing whether the tree transduction process described in Section 2 effectively captures discriminative structural information by virtue of its bi-directional context. To this end, we exploit the Jaccard kernel approach to confront the performance of the bi-directional IOBU model (BDIO, in the following) on challenging benchmarks on tree structured data classification, with that achieved by the unidirectional TD [1] and bottom-up HTMM (BU) [2]. Further, we provide an experimental comparison with the bi-directional approach realized in [6] by integrating TD and BU contexts at the kernel level (TB, in the following).

Experiments have been performed on two data sets from the 2005 and 2006 INEX XML document classification competition [9]. INEX 2005 comprises $9,361$ trees, 11 classes, with 366 possible node labels. INEX 2006 comprises $12,107$ trees, 18 classes, and 65 possible labels. Standard training and test sets are available for both datasets [9], with a 50%-50% split. The large number of classes in both data sets makes them challenging benchmarks, such that the random classifier baseline for INEX 2005 and INEX 2006 is 9% and 5.5%, respectively. Different configurations of the IOBU, TD and BU generative models have been tested by varying the number of hidden states C in $\{6, 8, 10\}$ (hidden states number follow the guidelines in [2]). Following the generative approach, we have trained a different HTMM for each class and computed the Jaccard kernel as in [8] on a total of $C \cdot V$ hidden states, where V is the number of classes. Training and test trials have been repeated 5 times for each configuration of the generative models, each time using different random initializations for the models distributions (initialization of label emission is kept fixed to the prior distribution of the multinomial labels on the training trees). SVM-based multiclass classification has been obtained by means of the LIBSVM[1] software (by a C-SVM classifier) exploiting the Jaccard Gram matrices as user defined kernel. A 3-fold cross-validation (CV) procedure has been applied to select the value of the misclassification cost parameter C_{svm} from the following set of values: 0.001, 0.01, 0.1, 1, 10, 100, 1000 using validation data external to the test set. The SVM classifier with the C_{svm} value selected on the validation set has then

---

[1] http://www.csie.ntu.edu.tw/~cjlin/libsvm

Table 1. Classification accuracy (%) on the external test set for the generative kernel configurations selected in cross-validation (variance is in brackets). The best result for each dataset is highlighted in bold.

Dataset	BDIO C	BDIO Test	BU C	BU Test	TD C	TD Test	TB C	TB Test
INEX 2005	8	95.24 (0.17)	8	94.22 (0.81)	10	93.39 (2.19)	8	**95.39** (0.14)
INEX 2006	10	**45.19** (0.12)	6	44.53 (0.09)	8	44.38 (0.06)	10	44.78 (0.02)

Table 2. Summary of the classification accuracy of non-adaptive (syntactic) tree kernels: see [10] for a quick introduction to these kernels

Dataset	ST	SST	Poly-SST	PT
INEX 2005	88.73	88.79	88.33	97.04
INEX 2006	32.02	40.41	40.12	41.13

been trained and its average performance on the 5 trials has been evaluated on the hold-out test data.

Table 1 summarizes average classification performance for the model configurations selected in CV. The results show that the bi-directional transduction approach (BDIO) achieves a classification performance that is significantly higher than that achieved by the TD and BU unidirectional approaches on both datasets, suggesting that the input-driven generative model is effectively capturing the context of two parsing directions. Compared to the bi-directional TB approach, the BDIO achieves an equivalent performance on INEX 2005 whereas it outperforms TB on INEX 2006. Note that the computational cost for computing the kernel on the top of the IOBU model is significantly lower than that required by the TB approach: the former, in fact, requires to compute the kernel on the hidden states of the IOBU model alone, while TB requires to compute and integrate two kernels built on top of the TD and BU models. Compared to results in literature, the BDIO approach shows a competitive performance, in particular as regards INEX 2006 data. Table 2 summarizes the performance of state-of-the-art syntactic (i.e. non adaptive) kernels on the two benchmarks [10]. On INEX 2005, only the computationally intensive PT kernel achieves an higher performance. Notably, BDIO achieves the best classification accuracy in literature on INEX 2006: see [11] for a recent survey of the results of several tree classification approaches on this dataset. Such performance is not obtained at the cost of computational efficiency: the average computational cost for inference and kernel computation on an INEX 2006 test tree is 1.1644sec for BDIO, 2.8732sec for TB, 1.0211sec for TD and 1.0942sec for BU.

4   Conclusion

We have introduced a novel approach to efficiently integrate bi-directional contexts in a generative tree model by means of structured transductions and a composition of top-dow and bottom-up HTMMs. This is the first approach that allows to capture information concerning bi-directional contexts within the state-space of a generative model for tree-structured data. Previously, only [6] has realized an integration of contextual information which, however, has been limited to information fusion at the kernel level. The proposed approach has shown to be effective on well-known classification benchmarks, yielding to state of the art results on the challenging INEX06 dataset. The approach allows to train bi-directional generative models in a computationally efficient way by independently training the composing top-down and bottom-up models. Further, it maintains the cost of Jaccard kernel computation comparable to that of a uni-directional approach, whereas the bi-directional kernel in [6] requires twice the effort for the same size of the hidden state space. We believe that the proposed approach is general enough to pave the way to applications on more complex classes of structured data, which will be the subject of future investigation. Further, we plan to extend the approach to allow a recursive application of the encoding-decoding scheme along the lines of [4], with the intent of capturing more articulated contextual information.

References

1. Diligenti, M., Frasconi, P., Gori, M.: Hidden tree markov models for document image classification. IEEE Trans. Pattern Anal. Mach. Intell. 25(4), 519–523 (2003)
2. Bacciu, D., Micheli, A., Sperduti, A.: Compositional generative mapping for tree-structured data - part I: Bottom-up probabilistic modeling of trees. IEEE Trans. Neural Netw. Learning Syst. 23(12), 1987–2002 (2012)
3. Bacciu, D., Micheli, A., Sperduti, A.: An input-output hidden Markov model for tree transductions. Neurocomputing 112, 34–46 (2013)
4. Micheli, A., Sona, D., Sperduti, A.: Contextual processing of structured data by recursive cascade correlation. IEEE Trans. on Neural Netw. 15(6), 1396–1410 (2004)
5. Hammer, B., Micheli, A., Sperduti, A.: Universal approximation capability of cascade correlation for structures. Neural Computation 17(5), 1109–1159 (2005)
6. Bacciu, D., Micheli, A., Sperduti, A.: Integrating bi-directional contexts in a generative kernel for trees. In: Proc. of the 2014 IEEE Int. Joint Conf. on Neural Netw., IJCNN 2014, pp. 4145–4151 (2014)
7. Durand, J., Goncalves, P., Guedon, Y., Rhone-Alpes, I., Montbonnot, F.: Computational methods for hidden Markov tree models-an application to wavelet trees. IEEE Trans. Signal Process. 52(9), 2551–2560 (2004)
8. Bacciu, D., Micheli, A., Sperduti, A.: A generative multiset kernel for structured data. In: Villa, A.E.P., Duch, W., Érdi, P., Masulli, F., Palm, G. (eds.) ICANN 2012, Part I. LNCS, vol. 7552, pp. 57–64. Springer, Heidelberg (2012)
9. Denoyer, L., Gallinari, P.: Report on the XML mining track at INEX 2005 and INEX 2006: Categorization and clustering of XML documents. SIGIR Forum 41(1), 79–90 (2007)
10. Aiolli, F., Da San Martino, G., Sperduti, A.: Route kernels for trees. In: Proc. of ICML 2009, pp. 17–24. ACM (2009)
11. Gallicchio, C., Micheli, A.: Tree echo state networks. Neurocomputing 101, 319–337 (2013)

A Novel Architecture for Capturing Discrete Sequences Using Self-organizing Maps

Manjusri Ishwara[1], Jayantha Rajapakse[1], and Damminda Alahakoon[2]

[1] Faculty of Information Technology
Monash University - Malaysia Campus
{manjusri,jayantha.rajapakse}@monash.edu
[2] School of Information and Business Analytics
Deakin University
d.alahakoon@deakin.edu.au

Abstract. This paper, proposes a layered approach to capture discrete sequences by exploiting the best matching unit (BMU) selection process of the self-organizing map (SOM) to generate a finite state machine (FSM). The FSM is used with thresholding to identify prominent subsequences available in input data.

1 Introduction

Identifying sequences and sequential behavior is an essential component of intelligence. In humans, this skill forms a fundamental part of human behavior ranging from the use of natural languages to solving complex problems. In order to make useful inferences based on sequences, learning the nature of sequences is of importance. Recently, sequence learning has gained interest in a number of diverse domains from bioinformatics to natural language processing. As a result, many approaches have been proposed for sequence learning across these domains. The proposed methodologies consist of both supervised[1,2] and unsupervised[3,4] learning algorithms. Unsupervised neural networks such as the SOM[5] has gained in popularity for sequence learning tasks(Section 2). The popularity associated with neural networks is largely due to its simplicity, generalizability and non-linearity. Apart from the three qualities mentioned above the SOM has the ability to process large amounts of data with the added capability of learning from data that is not pre-classified or pre-labeled.

Deviating from the standard norms of SOM applications to sequence data, this paper proposes a layered architecture for sequence learning by generating a finite state machine based on SOM learning(Section 3). In order to generate the FSM, SOMs' best matching unit (BMU) identification process was exploited to generate the states of the FSM in a separate layer when conditions explained in section 4 are met. The layering aids in separating the SOM learning process from the generated FSM. The FSM layer is ultimately used to identify the sequences captured. Section 5 provides an experimental evaluation of our model using mitochondrial DNA data of Homo Sapeins. Finally, we provide conclusions and possible future directions of research in section 6.

C.K. Loo et al. (Eds.): ICONIP 2014, Part I, LNCS 8834, pp. 551–558, 2014.

2   Related Work

Due to the inherent inability to involve temporal information in its learning process, the classical SOM is unable to discern meaningful information from sequential data. However, the classical SOM is able to process sequential data if supplementary mechanisms are in place to represent temporal information. In literature, there are two mechanisms to present such information to the SOM as internal time presentation[3,4,10,11] and external time presentation[7,8,6,9]. Internal mechanism represents temporal information in the form of functions or integrators attached to the network itself. This results in modifications of data representation within the network. On the contrary, the external mechanism represents time via means external to the algorithm itself such as input vector concatenation and external delay lines. Sequence learning algorithms that utilize either time representation technique mentioned above are presented next.

Kangas[6] proposed two variants of a temporal SOM that accepts temporal information externally. The first variant represents the current input to the SOM as a function of former inputs. These inputs are controlled using depth parameters that determine the intensity of influence incident on the current input by former inputs. The second variant concatenates short sequences of inputs and presents them to the network. Concatenated presentation enables the network to view the current input and a short history of previous inputs and cluster accordingly. Zehraoui et al.[7,8] introduced another concatenation based technique by modeling the matrices as a concatenation of input vectors. The technique preserves temporal information via transforming the generated matrices to their respective covariance and weighted covariance matrices. A different approach to external presentation of temporal information was adopted by the Hypermap architecture[9] proposed by Kohonen. The model generates a context vector and a pattern vector based on an external time window centered at a certain time. The generated vectors are utilized in two separate instances. The context vector nominates a set of nodes as fit for the BMU and the pattern vector identifies a node from the nominated set as the BMU for that input instance. The input initialization mechanism of the model proposed in this paper was inspired in part by three of the technique mentioned above.

In algorithms that represent temporal information internally contains additional functions attached to their inner working of the algorithm. The Temporal Kohonen Map (TKM)[3] utilizes leaky integrators to maintain activation history of neurons. The leaky potentials gradually decay overtime unless excited by an activation of a node. The recurrent SOM (RSOM)[10] improves on the TKM by associating weight adaptation as a recursive difference equation. RSOM maintains temporal information using difference vectors attached to each node in the SOM. Further algorithms of similar nature such as the recursive SOM (RecSOM)[4] and MergeSOM[11] were introduced to enhance the temporal pattern classification capability of the SOM.

Furthermore, a common observation that could be made about the techniques explained in this section is that, irrespective of the method it uses to represent temporal information, all techniques require interpretation of clustering incident on the map. Although, the interpretation of clustering is extremely important in unsupervised learning, we believe that temporal information especially transitions between instances of input vectors could be captured at the time of SOM learning to bolster the clustering results obtained by these techniques.

3   The Conceptual Model

The classical self-organizing map (SOM)[5] projects high dimensional data into a lower dimensional space through iterative presentation of inputs such that inputs with similar properties are clustered together. A given input vector x_i $(1 \le i \le n)$ could map to different neurons (winners) in the SOM throughout the SOM learning process. However, as the number of iterative presentations increase, the movements of winner neurons for a given input x_i gradually declines and becomes stable. Different inputs reach their stable winners at different epochs during the SOM training process. Once a stable winner is reached, the neuron would start accumulating hits whenever the same input or similar inputs are presented to the SOM for the remainder of the learning process. The process of reaching a stable winner for a given input has been exploited in this model to generate a state machine to represent the input data. For each unique input vector there is a corresponding winner node; these winners are modeled as states in the generated state machine. The transitions between states were modeled as the transitions between the winners for two consecutive input instances x_i and x_{i+1}. However, it is important to note that this type of modeling would only be meaningful in instances where the inputs are sequentially related. The SOM and the resulting state machine are placed on separate layers with the SOM (**S Layer**) being able to effect changes to the state machine (**V Layer**) during the SOM learning process.

In order to extract sub-sequences from the state machine generated a threshold (T) was defined based on the number of hits accumulated by neurons of a given state. Thresholding based on the hit count enables the data analyst to extract smaller state machines from the one incident on the layer V. Hit count (**H**) was used since it is a measure indicative of the stability of the winner neuron since the hit count and the duration of a winner being stable has a direct correlation. The threshold aids in identifying high frequency sub-sequences of an input sequence. The model thus far, was explained using vector inputs. This was done in order to clearly delineate the logic behind the generation of a state machine using the SOM learning phase.

Now we focus on the inputs presented to the SOM. Since this model is associated with sequential data it is expected that x_i and x_{i+1} are sequentially related. Therefore, when presenting sequential data, the temporal nature of the data needs to be preserved. This model adopts the external representation of temporal information via a two phase conversion of vector inputs. Initially the input vectors are compiled into rows of a matrix A. The number (referred to as the participant factor (pf) throughout the text) of input vectors represented by the matrix as its rows is dependent on the learning task. In order to further preserve the temporal relationship and to make inputs more reselient to noise, the weighted covariance of A was calculated and is subsequently used as the input to the SOM[1]. Modeling of sequential inputs to weighted-covariance matrices technically segments a sequence **(S)** into set of matrices representing fixed size sub-sequences of S. Therefore, according to the concept presented in this section, it is clear that the sub-sequences of S and its associated transitions represent a state machine of S over its sub-sequences.

4   Implementation of Algorithm

4.1   Input Formulation and Learning

This phase converts an input sequence S into a set of weighted covariance matrices $(WCOV)$. Lets assume a sequence S as a finite and an ordered set of n-dimensional row feature vectors (x_i) as in equation 1.

$$S = x_1 x_2 x_3 ... x_i .... x_m \; ; \; m \geq 1 \; and \; x_i \in \mathbb{R}^n \tag{1}$$

As the first step, the sequence S is segmented into a set of instance matrices M using a sliding window (ν) of size pf. The sliding window ν is slided by a single element each time the window is moved. All features vectors (x_i) that is within the perimeter of ν are set as rows of an instance matrix $(M_i \; where \; 1 \leq i \leq n-pf+1)$. For example i^{th} instance matrix (M_i^T) is presented in equation 2.

$$M_i^T = [x_i^T \; x_{i+1}^T \; ... \; x_{i+pf-1}^T] \; where M_i \in M \; ; \; M_i \in \mathbb{R}^{pf \times n} \tag{2}$$

Once the set M has been synthesized, the weighted covariance matrix for each instance matrix M_i is computed using equation 3 where q_{jk} represents the kth element of the jth row of the resulting matrix.

$$q_{jk} = \frac{\sum_{i=1}^n w_i}{(\sum_{i=1}^n w_i)^2 - \sum_{i=1}^n w_i^2} \sum_{i=1}^n w_i(x_{ij} - \mu_j)(x_{ik} - \mu_k) \tag{3}$$

In order to calculate $WCOV_i$ $(\in WCOV)$, pf weights $(0 < w_k \leq 1)$ are assigned to each row of matrix M_i such that $w_1 > w_2 > ... > w_k > ... w_n$. The decending weights ensure that the latest vector caputured by the sliding window ν has the least weight and the vector which is to be ommited due to the next slide has the highest weight. The generated set of matrices $WCOV$ is then presented to the modifed SOM that utilizes weight matrices inplace of weight vectors. This modification results in a change of policy when determining the best matching unit (BMU) of an input instance with the use of matrix distance measures. This writing assumes the Frobenius distance given in equation 4 to calculate the BMU. The remainder of the SOM algorithm remains the same with matrices participating in calculations instead of vectors.

$$F = \sqrt{tr((A - B).(A - B)^*)} \tag{4}$$

---

[1] The SOM needs to be structurally modified to support matrix based inputs.

4.2   State Machine Generation

The state machine is generated in parallel to the SOM learning process. The state machine generation process consists of four steps as state generation, state update, state transition and solidification. The first three steps are represented from steps 1(a) - 1(d) and solidification is represented from steps 1(e) - 1(g) in the following psuedo code.

Notation: $\lambda_{current/previous}$ represents tempory holders for a state. Ψ represents the set of states and ψ_i represents the state at position i in Ψ. Γ represents the set of solid sub-sequences identified during the learning process. $f(.)$ returns the sub-sequence associated with the matrix or state inserted as the parameter. If $WCOV_2$ represent sub-sequence "ABC" then $f(WCOV_2)$ returns "ABC".

1. Present each weighted covaraiance matrix $WCOV_i$ to the modified SOM and retrive the BMU for that input matrix. Then, Retrieve $\lambda_{current} = \psi_i \in \Psi$ such that $f(\psi_i) == f(WCOV_i)$
   (a) if $\lambda_{current} == null$, create a new state with the sub-sequence $f(WCOV_i)$ and position it in the V layer using the coordinates of the BMU and set the hit counter (H) to 1.
   (b) if $\lambda_{current}! = null$, verify the coordinates of the retrieved state $\lambda_{current}$ and the BMU have the same coordinates.
      i. If coordinates are identical increment H of $\lambda_{current}$ by 1.
      ii. If coordinates are not equal, relocate $\lambda_{current}$ in V layer to the location of the BMU. Reinitilize $\lambda_{current}$ such that $H = 1$ and set $\lambda_{current}$ to non-solid. Remove all sub-sequences that consist of $f(\lambda_{current})$ in set Γ.
   (c) Create transitions between $\lambda_{current}$ and the immediately preceding state $\lambda_{previous}$ (the selected state for input $WCOV_{i-1}$).
      i. If $f(\lambda_{current}) == f(\lambda_{previous})$ and X,Y coordinates of $\lambda_{current}$ and $\lambda_{previous}$ are identical, initiate repeat list R if the repeat counter $C = 0$.
      ii. If $C > 0$, retrive element C from R and increment the elements hit counter by 1.
      iii. If $C > 0$ and C is not in R, add C to R and set the hit counter of the new element to 1.
   (d) $\lambda_{previous} = \lambda_{current}$
   (e) For each $\psi_i \in \Psi$, If $H \geq threshold$ (T), mark ψ_i as a solid state.
      i. If solid state ψ_i consists of R that has a element C with $H \geq threshold(T)$, mark C as solid.
   (f) $\forall$ $\psi_i \in \Psi$ that is solid, identify sub-sequences if
      i. Rule 1 :- If two solid states A and B are linked by a transition l_1; then $f(A).appened(l_1)$ is a solid sub-sequence.
      ii. Rule 2 :- If a transition l_2 has a recursive relationship to the solid state A; then the length of the solid sub-sequences represented by A could be increased by appending the number of solid elements in A's repeat list R.
   (g) Add identified sub-sequences to Γ
2. Repeat from step 1 for all weighted covariance matrices $WCOV_i \in WCOV$
3. Repeat steps 1 - 2 for the specified number of iterations

Table 1. Solidification of subseqences in the Homo Sapiens mitochondrial genome with different T% values , number of occurances (Occ) and the ratio of the total size of sub-sequences to the size of the input string as a percentage(%)

	1%	4%	5%	6%	8.5%	Occ	(%)
AC	√	√	√	√	√	1493	9.01
CC	√	√	√	√		1770	10.68
AA	√	√	√			1602	9.67
CA	√	√	√			1529	9.23
CT	√	√	√			1441	8.70
TA	√	√	√			1374	8.29
AT	√	√				1232	7.44
TC	√	√				1203	7.26
TT	√					1003	6.05
AG	√					796	4.80
GC	√					710	4.29
GA	√					618	3.73
TG	√					513	3.10
CG	√					436	2.63
GG	√					429	2.59
GT	√					418	2.52

Table 2. Composition of the repeat lists associated with repetitive nodes GG, CC, TT and AA

	1	2	3	4	5	6
GG	3415	659	179	50	-	-
CC	30469	11808	4266	837	76	-
TT	11291	2822	661	240	60	-
AA	20500	6730	2583	1140	420	60

5   Experimental Results

In order to evaluate the proposed architectural model an experiment was performed using the mitochondrial genome of Homo sapiens[2]. The experiment was conducted using a square lattice based SOM of 10^4 neurons and a learning rate of 0.25, a participant factor (pf) of 2 and 100 iterations as network parameters.

During the execution of this experiment, the T%[3] value was varied between 1% and 8.5%. The results of solidification for the corresponding T% values are given in table 1 and the contents of R for states containing recursive transitions are given in table 2. For example, if T% value was set at 5% (82845 hits), the FSM in the V layer has six solid states. These solid states are extracted as explained

---

[2] Source:http://www.ebi.ac.uk/ Accession number: V00662.

[3] T% = (Threshold(T)/Total Number of Hits(N))*100% ; Represents the T value as a percentage.

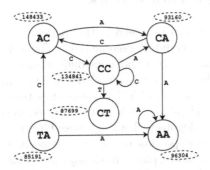

Fig. 1. The extracted FSM at P = 5%. The dotted circles represent the number of hits incured by that node after learning is complete.

Table 3. Set of identified sub-sequences origination from the state AC and the number of occurances (Occ)

TRANSITION	SUB-SEQUENCE	OCC
$AC \to CC$	ACC	515
$AC \to CA$	ACA	447
$AC \to CC \to CT$	ACCT	129
$AC \to CA \to AA$	ACAA	145
$AC \to CC \to CA \to AA$	ACCAA	49

in section 4.2 to a smaller FSM depicted in figure 1. Using the extracted FSM a set of sub-sequences (Table 3) were generated by using the state "AC" as the origin.

The generated sub-sequences in table 3 indicate that the model is able to identify sub-sequences of varying length. The length of the identified sub-sequences above could be further increased by incorporating the information available in the repeat lists. However, when considering the number of hits per repetitive element in table 2 it is clear that all the elements do not meet the T% criteria of 5%. Hence, at current value of T% the sub-sequences identified in table 3 are not extendable. Nevertheless, if the T% value was to be reduced from 5% to 0.1% (1657 hits) three additional characters of both C and A could be used to extend the sub-sequences which consist of states CC and AA. For example the sub-sequence ACCT identified above could be extended to ACCCCCCT if T% was set to 0.001% (17 hits). Also ACCCCCCT sub-sequence only occurs once throughout the entire input sequence whereas ACCT occurs 129 times. This implies our model is capable of capturing sub-sequences that has a single or a very low occurrence if the T% value is set accordingly. This draws us to an important observation. That is, the length of the sub-sequences identified is inversely proportional to the T% value.

6   Conclusion and Future Work

The sequence learning model proposed in this paper was able to identify dominant sub-sequences in a strand of genome data. The T% values that are used to differentiate between the solid and non-solid sub-sequences ensure that the data analyst is capable of identifying variable length sub-sequences from the input sequences. The versatility of T% value ensures that sub-sequences of different ranges could be identified such as the sub-sequences above or below a certain value of T% (analogous to low/high pass filtering) or within a range of two T% values. It is also important to note that experiments on larger datasets (mtDNA taxonomy generation, player modeling) were indeed performed but the page limitation enforced has disabled the inclusion of these experiments due to the inabilty to present them in sufficient detail.

Furthermore, it could be observed that the FSM generated in the V layer resembles a Markov chain. However, it is important to note that this study solely focuses on the learning aspects of the SOM when generating the FSM and the conditional probabilities are not associated with the FSM. Nevertheless, as a future work, the capability of integrating this model with the training of a Markov model is a viable path.

References

1. Graves, A., Fernández, S., Gomez, F., Schmidhuber, J.: Connectionist temporal classification: labelling unsegmented sequence data with recurrent neural networks. In: Proceedings of the 23rd International Conference on Machine Learning (ICML), pp. 369–376. ACM (2006)
2. Mozer, M.C.: Neural net architectures for temporal sequence processing. In: Santa Fe Institute Studies in the Sciences of Complexity-Proceedings, vol. 15, p. 243. Addison-Wesley Publishing Co. (1993)
3. Chappell, G.J., Taylor, J.G.: The temporal Kohønen map. Neural Networks 6(3), 441–445 (1993)
4. Voegtlin, T.: Recursive self-organizing maps. Neural Networks 15(8), 979–991 (2002)
5. Kohonen, T.: The self-organizing map. Proceeding of IEEE 78, 1464–1480 (1990)
6. Kangas, J.: Time-delayed self-organizing maps. In: International Joint Conference on Neural Networks (IJCNN), pp. 331–336. IEEE (1990)
7. Zehraoui, F., Bennani, Y.: M-SOM: Matricial Self Organizing Map for sequence clustering and classification. In: International Joint Conference on Neural Networks (IJCNN). IEEE (2004)
8. Zehraoui, F., Bennani, Y.: M-som-art: Growing self organizing map for sequences clustering and classification. In: ECAI, vol. 16 (2004)
9. Kohonen, T.: The hypermap architecture. North-Holland (1991)
10. Varsta, M., Heikkonen, J., Millan, J.: Context learning with the self-organizing map. Citeseer (1997)
11. Strickert, M., Hammer, B.: Merge SOM for temporal data. Neurocomputing 64, 39–71 (2005)

An Improved Gbest Guided Artificial Bee Colony (IGGABC) Algorithm for Classification and Prediction Tasks

Habib Shah[1], Tutut Herawan[2,3], Rozaida Ghazali[1], Rashid Naseem[1],
Maslina Abdul Aziz[4], and Jemal H. Abawajy[4]

[1] Universiti Tun Hussein Onn Malaysia
Parit Raja, 86400 Batu Pahat, Johor, Malaysia
[2] University of Malaya
50603 Lembah Pantai, Kuala Lumpur, Malaysia
[3] AMCS Research Center, Yogyakarta, Indonesia
[4] Deakin University
Waurn Ponds, Geelong, VIC, Australia
{habibshah.uthm,rnsqau}@gmail.com, tutut@um.edu.my,
rozaida@uthm.edu.my, {mabdula,jemal.abawajy}@deakin.edu.au

Abstract. Artificial Neural Networks (ANN) performance depends on network topology, activation function, behaviors of data, suitable synapse's values and learning algorithms. Many existing works used different learning algorithms to train ANN for getting high performance. Artificial Bee Colony (ABC) algorithm is one of the latest successfully Swarm Intelligence based technique for training Multilayer Perceptron (MLP). Normally Gbest Guided Artificial Bee Colony (GGABC) algorithm has strong exploitation process for solving mathematical problems, however the poor exploration creates problems like slow convergence and trapping in local minima. In this paper, the Improved Gbest Guided Artificial Bee Colony (IGGABC) algorithm is proposed for finding global optima. The proposed IGGABC algorithm has strong exploitation and exploration processes. The experimental results show that IGGABC algorithm performs better than that standard GGABC, BP and ABC algorithms for Boolean data classification and time-series prediction tasks.

Keywords: Artificial Bee Colony, Gbest Guided, Classification, Prediction.

1   Introduction

An Artificial Neural Network (ANN) is a mathematical model that tries to simulate the structure and function of human biological networks. It has the most powerful and attractive tools suitable for solving combinatorial problems such as prediction [1], forecasting [2], classification and numeric function optimization [3]. ANN have proven the outstanding performance in the area of engineering, mathematics, social computing, medical, seismic and information technology [4]. Researchers have extended ANN topologies with different connections like Multilayer Perceptron (MLP), Radial

C.K. Loo et al. (Eds.): ICONIP 2014, Part I, LNCS 8834, pp. 559–569, 2014.
© Springer International Publishing Switzerland 2014

Basis Function [5] and so on. Besides that, the ANN tools were proven to be very successful in data classification (Boolean classification) especially with the univariate, multivariate of real and time-series data. Among the examples of the ANN tools applications are weather, bankruptcy, traffic forecasting and others [6-7].

The success key point of ANN approaches are the learning algorithms which are used to train, test and validate the network for corresponding task. There are also other different learning algorithms proposed by researchers such as the supervised and unsupervised [8]. An example of the well-known supervised learning algorithm used to train ANN for different tasks is the Backpropagation (BP) algorithm. However, the BP algorithm is prone to local minima traps and has a slow convergence speed. This has motivated researchers to find suitable tools to solve these problems. There are different improvements used with the standard BP learning to overcome the drawbacks, for example introducing new parameters, learning rate, momentum, first orders and second order and others.

Swarm Intelligence (SI) has become a research interest to different domain of researchers in recent years. They take more interest to develop new learning techniques based on SI movement and thinking behaviors such as Honey Bee Colony [3], Bees Algorithm [9], Artificial Bee Colony (ABC) algorithm and others [10]. Particle Swarm Optimization (PSO) and its variations have been introduced for solving optimization problems and successfully applied to solve many real problems like classification, clustering and prediction [11]. Motivated by the foraging behavior of Honeybees [12], researchers have initially proposed ABC algorithm for solving various optimization problems.

For this research in order to balance between exploration and exploitation procedures of the Gbest Guided ABC (GGABC) algorithm and at the same time improve the effectiveness of classification and prediction tasks, an enhanced searching scheme called Improved Gbest Guided ABC (IGGABC) is proposed. The novel IGGABC algorithm and three standard learning algorithms ABC, BP and GGABC are used to train multi-layer perceptron (MLP) for Boolean functions classification and time series prediction tasks.

The rest of the paper is organized as follows. Section 2 provides preliminaries of Boolean function classification, water level height time series and rudimentary of Artificial Neural Networks (ANN). Section 3 describes the Improved Gbest-Guided Artificial Bee Colony (GGABC) Algorithm. Section 4 discusses about the experiment results. Finally, the conclusion of this work is provided in Section 5.

2   Preliminaries

In this research, the proposed and standard learning algorithms of Boolean function classification and time series prediction present two solutions. Each task is discussed in the following sub-sections.

2.1 Boolean Function Classification

There are many types of Boolean operations; however, the most commonly used are the XOR, 3-Bit Parity and Encoder/Decoder operators. XOR is not linearly separable, consequently it cannot be implemented using single layer network; a three layered network is required to solve the problem. In order to solve the XOR problem using the MLP, a hidden layer is needed with 2 or more neurons [13], where each of these neurons produces a partial solution and then the output neuron combines those solutions and solves the XOR problem [14]. The 3-bit parity or even parity problem can be considered as a generalized XOR problem but it is more difficult for classification [15]. This problem has been addressed because they are nonlinearly separable, and hence can not be solved by Single Layer Perceptron, such as the perceptron [16]. In other words, if the number of binary inputs is odd, the output is 1, otherwise it is 0. The third classification task is 4-Bit Encoder/Decoder, which is well known in computer science. The network is presented with four distinct input patterns, each having only one bit turned on. A decoder is a logic circuit which accepts a set of inputs that represents a binary number and activates only the output that corresponds to the input number. 4-Bit Encoder/Decoder is quite close to the real-world pattern classification task, where small changes in the input pattern cause small changes in the output pattern [17].

2.2 Water Level Height Time Series

Coastal flooding is primarily caused by storm surge, waves and water level height but includes many other influences. Changes in the global sea level have far reaching consequences for both humans and the natural environment. The population and the economy must be protected against flooding caused by sea water level. To achieve this, knowledge of long, medium and short-term trends of the water levels is important. Here the historical data of year 2011 of Port Aransas station PTAT2, water level height with six-minute in feet, of the water above or below Mean Lower Low Water (MLLW), offset by 10 ft [18]. Port Aransas experiences a humid subtropical climate, enjoying similar temperatures to those of other Gulf Coast regions, such as Deep South Texas and Southern Florida. The area experiences an average annual rainfall of 31.92 inches (811 mm), with prevailing winds out of the southeast from the Gulf of Mexico.

2.3 Artificial Neural Networks (ANN)

The ANN working model has different tasks such as, multiplication, summation and activation. In multiplication stage the weights are multiplied by an input signal. Initially, weights were chosen randomly but with the passage of time researchers started taking an interest to get optimal weight in weight equation. Figure 1 shows the architecture of MLP with hidden layers, output layer, and one input layer.

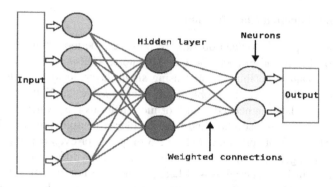

Fig. 1. Artificial Neural Networks

ANN with feed-forward topology mainly consists of one or more artificial neuron in parallel organization [19]. MLP is the well-known type of feed-forward NNs shows the flow from input to output in one direction with no back-forward firing. ANN learning is an inferred process which can not be perceived directly, but can be assumed to have happened by observing changes in performance [20]. The learning objective is to minimize the cost function, which is the difference between desired output and NN output. The networks were trained for finding optimal weights, which reduce the error until the convergence.

3   Improved Gbest-Guided Artificial Bee Colony (GGABC) Algorithm

3.1   Gbest-Guided Artificial Bee Colony (IGGABC) Algorithm

Gbest Guided Artificial Bee Colony (GGABC) algorithm as an optimization tool provides stochastic search procedure in which individuals called foods positions are adapted by the GGABC with time, and the bee's aim is to discover the places of global food sources [21]. The exploration is the famous procedure in swarm-based algorithms for finding best position within specified areas. These properties are the solution to given problems. Exploration includes things captured by terms such as search, variation, risk taking, experimentation, play, flexibility, discovery, innovation [22]. In SI, the exploitation consists of modification, alternative, invention, efficiency, collection, and execution towards getting best solution. To advance the exploitation procedure, ordinary ABC algorithm modified with Gbest guided practice called GGABC by incorporating the information of Gbest solution into the solution search equation proposed for training MLP [21]. The new candidate solution for both agents is generated by moving the old solution towards or away from another solution based on random selection from the population. Here the employed and onlookers bee section of GGABC will be modified for improving exploration procedure as:

$$v_{ij} = x_{ij} + \phi_{ij}\left(x_{ij} - x_{kj}\right) + \Psi_{ij}\left(y_j - x_{i,j}\right) \tag{1}$$

where y_j is the jth element of the global best solution, Ψ_{ij} is a uniform random number in [0,C], where C is a nonnegative constant.

3.2   An Improved Gbest Guided Artificial Bee Colony Algorithm

The exploitation and exploration procedures with balance quantity are the well-known possessions of Honey Bees based learning algorithms for competent performance. These possessions are used for solution to a given tasks. The exploitation method should refer to the ability to apply the information to find a better solution, and the exploration process leads to the ability to explore the different unknown regions in the solution space to discover the total finest [23]. Although the Gbest guided ABC algorithm has successfully increseased the poor exploration procedure of standadrd ABC in the employed and Onlooker Bees section. However, sometimes GGABC algorithm fails to overtrapped the local minima with fast convergence rate due to the similar search stretgy of employed and onlooker bees phases [21]. In other words, by using Gbest guided method if the employeed Bee phase fails to increase the exploration amount, onlooker Bees cannot reach to global optima successfully. Furthermore, the Gbest guided technique use the random strategy in scout bee phase as in standard ABC, so that exploitation procedure cannot be improved.

In addition to improve the exploration procedure and balance with exploitation, this section explains the proposed IGGABC algorithm by incorporating the information of best and guided solution into the solution search equation. The proposed IGGABC algorithm as an optimization tool that provides efficient search procedure in which individuals called foods positions are adapted by the two global guided artificial bees with time, and the bee's aim is to discover the places of global food sources. This improved technique increases the exploration procedure through the guided strategy in Scout Bee. The step by step procedures of IGGABC are given in two sections as:

3.1.1   Improved Gbest Guided Employed Bee Phase
Each global guided employed bee produces new solutions (food source positions) $V_{i,j}$ in the neighbourhood of x_{mi} for the global guided employed bee using the equation as:

$$V_{ij} = x_{ij} + \psi_{ij}\left(\left(x_j^{best} - x_{ij}\right) + \left(y_j^{best} - x_{ij}\right)\right) \tag{2}$$

Where x shows the 1st row of the foods area, i represent the 1st iteration of employed bee, j explains the fix random value in D, k explicates a solution in the neighbourhood of, i and evaluates them, x_j^{best} shows the j_{th} element of the global best solution found so far, y_j^{best} represents the j_{th} element of the best solution in the current iteration and ψ_{ij} is a uniform random number in [0,C].

3.2.2   Improved Gbest Guided Onlooker Bee Phase

An improved gbest guided onlookers bee section were produced the new solutions (new positions) v_i for the standard onlookers from the solutions x_i, selected depending on P_i, and evaluate them through equation (3) as:

$$V_{ij} = x_{ij} + \psi_{ij}(x_{ij} + x_{kj}) + \psi_{ij}(y_j best - x_{i,j}) \tag{3}$$

Where $y_j best$ represents the j_{th} element of the global best solution and ψ_{ij} is a uniform random number in $[0,C]$. The greedy selection process is used for the on-lookers between Vi of gbest guided employed section and Vj of and gbest guided onlookers bee section. The gbest guided scout bee also called unemployed bees; they choose their food source randomly. If the improved guided employed bees don't find the improved solution through a predetermined number of trial's inside limits, be-comes guided scouts and their solution abandoned.

4   Experiment Results

4.1   Parameter Setting

In this work, ABC, BP, GGABC and IGGABC algorithms are used to train MLP for Boolean classification and water level height prediction. To calculate the performance of these algorithms by Mean of MSE, NMSE, MAE, RMSE, SNR, classification and prediction accuraccy, using Matlab 2012a software. The parameters of the problems for classification and prediction are given in Tables 1 and 2, respectively.

Table 1. Parameters of the problems for boolean function classification

Problem	Colony Range	NN structure	Dimension	MCN
XOR	[10,–10]	2–2–1+ Bias(3)	13	1000
3-Bit Parity	[10, –10]	3–3–1+bias(4)	49	1000
4 BitEncoder	[10,–10]	4–2–4+Bias(6)	49	1000

Table 2. Parameters of the problems for heat waves temperature prediction

Problem	Colony Range	Hidden Nodes	Dimension	MCN
ABC	[10, –10]	[2,9]	[9, 49]	2000
GGABC	[10, –10]	[2,9]	[9, 49]	2000
IGGABC	[10, –10]	[2,9]	[9, 49]	2000

During the experimentation, 10 trials performed of training for above mentioned tasks. The sigmoid activation function used for network production for classification and prediction. During the simulation, when the number of input's signals, hidden nodes, output node and running time varies, performing training algorithms were stable, which is important for delegation ANN in the current state.

4.2   Results of Boolean Function Classification

The simulation results for the boolean classification problem using above algorithms are given in the Tables 3, 4 and 5, respectively. The above learning algorithms were tested after training the network, with trained parameters values. In Table 3, the MSE for training data set is presented from above mention algorithms. The IGGABC has outstanding performnce from others algorithms for XOR, 3 bit and 4 bit problems in terms of MSE, NMSE, MAE, convergence and accuracy. The MSE, NMSE, MAE is continuously decreasing with stable ratio by using IGGABC algorithm for boolean classification. Furthermore, the proposed IGGABC algorithm has stable convergence for XOR, 3 bit parity and 4 bit encoder / decoder problems as shown in Figure 2.

Table 3. Simulaion results for XOR classification

Algorithm	MSE	NMSE	MAE	Accuracy (%)
BP	0.018124	0.118143	0.089484	56.21
ABC	2.41E-03	1.17E-02	0.002311	97.11
GGABC	3.41E-03	1.38E-02	0.001983	99.98
IGGABC	**1.41E-04**	**5.15E-04**	**0.001721**	**99.99**

Table 4. Best single results for 3-Bit parity problem

Algorithm	MSE	NMSE	MAE	Accuracy (%)
BP	0.005334	0.018331	0.068212	76.21
ABC	0.005455	0.012245	0.018533	94.01
GGABC	0.000121	0.063811	0.010429	98.25
IGGABC	**0.000113**	**0.011217**	**0.010124**	**99.80**

Table 5. Best single results for 4-bit Enc/Dec problem

Algorithm	MSE	NMSE	MAE	Accuracy (%)
BP	0.001754	0.014466	0.022837	78.59
ABC	0.001765	6.40E-03	0.038437	92.71
GGABC	0.001543	2.00E-03	0.017211	95.25
IGGABC	**0.000872**	**1.00E-04**	**0.008423**	**97.07**

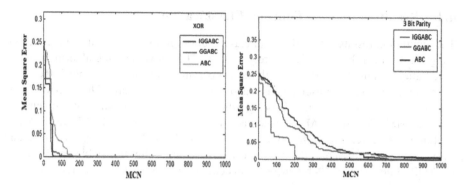

Fig. 2. Learning curves of XOR and for 3-Bit parity though learning algorithms

The simulation results of water level height by proposed IGGABC and standard algorithms are given in the Tables 6, 7, 8, 9 and Figures 3 and 4, respectively. The values of MSE, NMSE, MAE, RMSE, SNR and accuracy obtained MLP trained through IGGABC for water level height time series has efficeient performance from other algorithms. Among the standard learning algorithms for water level height prediction, the GGABC outperforms than BP and ABC algorithms.

Table 6. Average MSE for water level height prediction

MLP Structure	BP	ABC	GGABC	IGGABC
3-2-1	0.008712	0.004092	0.005098	0.001287
3-3-1	0.009634	0.002200	0.003988	0.000933
3-4-1	0.004853	0.003855	0.004833	0.001691
3-5-1	0.001562	0.000817	0.002418	0.000362
3-6-1	0.009528	0.001552	0.001565	0.000155
3-9-1	0.009631	0.023598	0.001091	0.000520

Table 7. Average NMSE for water level height prediction

MLP Structure	BP	ABC	GGABC	IGGABC
3-2-1	0.512123	0.402223	0.317421	0.228234
3-3-1	0.540935	0.302384	0.307224	0.121221
3-4-1	0.506473	1.099223	0.334234	0.113021
3-5-1	1.012634	0.323849	0.290915	0.108076
3-6-1	0.552934	0.230394	0.229117	0.024659
3-9-1	0.609298	0.027299	0.216438	0.0276794

Table 8. Average MAE for water level height prediction

NN Structure	ABC	BP	GGABC	IGGABC
2-2-1	0.005982	0.009721	0.003962	0.003281
3-3-1	0.002489	0.009978	0.002841	0.003289
4-4-1	0.005536	0.007885	0.002984	0.001629
5-5-1	0.008971	0.005624	0.008902	0.001964
7-7-1	0.002583	0.009987	0.001953	0.001190
7-9-1	0.025902	0.009531	0.001983	0.001272

Table 9. Average RMSE for heat waves temperature prediction

NN Structure	BP	ABC	GGABC	IGGABC
2-2-1	0.084764	0.096556	0.099484	0.100419
2-3-1	0.082988	0.049867	0.048969	0.064292
3-3-1	0.085343	0.047127	0.047749	0.034125
3-5-1	0.075124	0.052887	0.046915	0.081863
4-4-1	0.062225	0.091203	0.063103	0.032302
4-5-1	0.075185	0.071169	0.071197	0.031021
5-6-1	0.078237	0.032588	0.043509	0.092101
5-9-1	0.079316	0.020421	0.024224	0.021446

Table 10. Best average results for water level height prediction

Algorithm	MSE	NMSE	SNR	MAE	RMSE	Accuracy
BP	0.00024	0.12881	34.31	0.00873	0.00499	90.01
ABC	0.00018	0.20566	35.50	0.01429	0.00434	91.83
GGABC	0.00105	0.28901	36.98	0.00755	0.01027	93.93
IGGABC	**0.00092**	**0.10479**	37.11	0.00102	0.00311	**97.42**

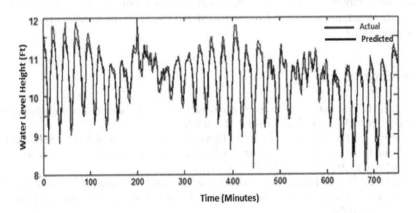

Fig. 3. Water level height prediction using GGABC algorithm

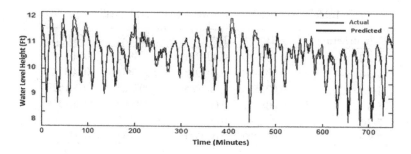

Fig. 4. Water level height prediction using IGGABC algorithm

The predicted signals of water level height time-series data are very close to the actual signals using proposed IGGABC algorithm. Furthermore, during the learning process of proposed and standard learning algorithms for water level height time series, the IGGABC algorithm has the ability for fast convergence in the collection of average simulation results.

5   Conclusion

In this paper, we have proposed the Improved Gbest Guided Artificial Bee Colony (IGGABC) algorithm to improve the existing algorithms. We also have proven that the IGGABC algorithm collected the exploration and exploitation with balance quantity successfully, through improved Gbest guided methods of employed, onlooker bees, which proves the high performance of training MLP for Boolean function classification and time series prediction. Furthermore, the proposed IGGABC algorithm achived the high accuraccy rate in classification and prediction tasks.

Acknowledgement. This work is supported by University of Malaya High Impact Research Grant no vote UM.C/625/HIR/MOHE/SC/13/2 from Ministry of Higher Education Malaysia.

References

1. Adeli, H., Panakkat, A.: A probabilistic neural network for earthquake magnitude prediction. Neural Networks 22(7), 1018–1024 (2009)
2. Ghazali, R., Hussain, A.J., Liatsis, P.: Dynamic Ridge Polynomial Neural Network: Forecasting the univariate non-stationary and stationary trading signals. Expert Systems with Applications 38(4), 3765–3776 (2011)
3. Karaboga, D.: An Idea based on Honey Bee Swarm for Numerical Optimization. Tech. Rep. TR06, Erciyes University, Engineering Faculty, Computer Engineering Departmen, pp. 1–10 (2005)
4. Romano, M., et al.: Artificial neural network for tsunami forecasting. Journal of Asian Earth Sciences 36(1), 29–37 (2009)

5. Jun Ying, C., Zheng, Q., Ji, J.: A PSO-based subtractive clustering technique for designing RBF neural networks. In: IEEE Congress on Evolutionary Computation, CEC 2008 (IEEE World Congress on Computational Intelligence) (2008)

6. Shah, H., Ghazali, R., Nawi, N.M.: Global Artificial Bee Colony Algorithm for Boolean Function Classification. In: Selamat, A., Nguyen, N.T., Haron, H. (eds.) ACIIDS 2013, Part I. LNCS (LNAI), vol. 7802, pp. 12–20. Springer, Heidelberg (2013)

7. Zhang, G., et al.: Artificial neural networks in bankruptcy prediction: General framework and cross-validation analysis. European Journal of Operational Research 116(1), 16–32 (1999)

8. Sapkal, S.D., Kakarwal, S.N., Revankar, P.S.: Analysis of Classification by Supervised and Unsupervised Learning. In: International Conference on Computational Intelligence and Multimedia Applications (2007)

9. Özbakir, L., Baykasoğlu, A., Tapkan, P.: Bees algorithm for generalized assignment problem. Applied Mathematics and Computation 215(11), 3782–3795 (2010)

10. Shah, H., Ghazali, R., Nawi, N.M.: Hybrid Ant Bee Colony Algorithm for Volcano Temperature Prediction. In: Chowdhry, B.S., Shaikh, F.K., Hussain, D.M.A., Uqaili, M.A. (eds.) IMTIC 2012. CCIS, vol. 281, pp. 453–465. Springer, Heidelberg (2012)

11. Shah, H., Ghazali, R., Mohmad Hassim, Y.M.: Honey Bees Inspired Learning Algorithm: Nature Intelligence Can Predict Natural Disaster. In: Herawan, T., Ghazali, R., Deris, M.M. (eds.) Recent Advances on Soft Computing and Data Mining SCDM 2014. AISC, vol. 287, pp. 215–225. Springer, Heidelberg (2014)

12. Karaboga, D., Akay, B.: A comparative study of Artificial Bee Colony algorithm. Applied Mathematics and Computation 214(1), 108–132 (2009)

13. Murtagh, F.: Multilayer perceptrons for classification and regression. Neurocomputing 2(5-6), 183–197 (1991)

14. Stromatias, E.: Developing a supervised training algorithm for limited precision feedforward spiking neural networks, Liverpool University (2011)

15. Stork, D.G., Allen, J.D.: How to solve the N-bit parity problem with two hidden units. Neural Networks 5(6), 923–926 (1992)

16. Minsky, M., Papert, S.A.: Perceptrons: An Introduction to Computational Geometry. MIT Press, Cambridge (1969)

17. Fahlman, S.: An empirical study of learning speed in backpropagation networks. Technical Report CMU-CS-88-162, Carnegie Mellon University, Pittsburgh, PA 15213 (September 1988)

18. United States Department of Commerce: National Oceanic and Atmospheric Administration (NOAA) (2012) (cited 2012 2012); National Data Buoy Center Bldg. 3205 Stennis Space Center, MS 39529 228-688-2805

19. Rosenblatt, F.: A Probabilistic Model for Information Storage and Organization in the Brain, pp. 386–408. Cornell Aeronautical Laboratory (1958)

20. Zurada, J.M.: Introduction to artificial neural systems, West (1992)

21. Zhu, G., Kwong, S.: Gbest-guided artificial bee colony algorithm for numerical function optimization. Applied Mathematics and Computation 217(7), 3166–3173 (2010)

22. Shah, H., Ghazali, R., Herawan, T., Khan, N., Khan, M.S.: Hybrid Guided Artificial Bee Colony Algorithm for Earthquake Time Series Data Prediction. In: Shaikh, F.K., Chowdhry, B.S., Zeadally, S., Hussain, D.M.A., Memon, A.A., Uqaili, M.A. (eds.) IMTIC 2013. CCIS, vol. 414, pp. 204–215. Springer, Heidelberg (2014)

23. Yang, S.D., Yi, Y.L., Shan, Z.Y.: Gbest-guided Artificial Chemical Reaction Algorithm for global numerical optimization. Procedia Engineering 24, 197–201 (2011)

Artifact Removal from EEG Using a Multi-objective Independent Component Analysis Model

Sim Kuan Goh[1], Hussein A. Abbass[1,2], Kay Chen Tan[1], and Abdullah Al Mamun[1]

[1] National University of Singapore, Department of Electrical and Computer Engineering,
4 Engineering Drive, 117583, Singapore
{simkuan,eletankc,eleaam}@nus.edu.sg
[2] University of New South Wales, School of Engineering and Information Technology,
Canberra, ACT 2600, Australia
h.abbass@adfa.edu.au

Abstract. Independent Component Analysis (ICA) has been widely used for separating artifacts from Electroencephalographic (EEG) signals. Still, a few challenging problems remain.

First, in real-time applications, visual inspection of components should be replaced with an automatic identification method or a heuristic for artifacts detection. Second, as we will explain more in the paper, we expect to have a clear order relationship between an electrode and a corresponding component. Third, we need to minimize the EEG information loss during artifact removal while also minimizing the residue of the artifact in the cleaned signal.

In this paper, we propose a decomposition of the independent components. This decomposition separates each component into two vectors, one - we call local vector - maintains maximum information from the unique EEG information encoded by an electrode, while the other - we call shared vector - absorbs overlapping artifact information. We present an explicit Pareto-based multi-objective optimization formulation that trade-off similarity between the local vector and the original vector on the one hand, and the uncorrelatedness of all local vectors from all components on the other hand. We demonstrate that the proposed method can automatically isolate artifacts from an EEG signal while preserving maximum EEG information.

Keywords: Electroencephalography, Independent Component Analysis, multi-objective optimization.

1 Introduction

Independent Component Analysis (ICA) is a common technique for the removal of artifact. A study in [1] shows that eye movement, eye blinking, cardiac, myographic, and respiratory artifacts can be isolated from Magnetoencephalographic (MEG) using ICA.

Previous work on the removal of eye blinks and eye movements activities relied on the substraction of the electroocculargraph (EOG) signal from the signals obtained through EEG electrodes to correct for ocular artifacts. ICA has also been demonstrated

C.K. Loo et al. (Eds.): ICONIP 2014, Part I, LNCS 8834, pp. 570–577, 2014.

to remove this type of artifact from EEG signals [2]. In both cases, however, certain level of EEG activities will still be removed; the quantification of how much of these activities have been removed and in which band remains unclear.

The wide adoption of ICA stems from the theoretical basis of the technique. However, a number of challenges remains. First, many brain computer interface (BCI) applications, such as augmented cognition applications [3,4], require real-time, or at least automatic, correction of the EEG signal. Most implementations of ICA are demonstrated on off-line correction of the signal and those used ICA for real-time applications relied on some heuristics that have not been fully validated.

A second problem is that there is no clear order on the components. In other words, we may find that one component is represented using equal weights in the mixing equation. The distribution of the mixed signal across components make sense in classical signal processing applications such as the famous cocktail party problem, which was the original motivation for ICA. In EEG recording, while we do not aim in this paper to localize the signal, it is biologically more plausible to expect more local information is captured through each electrode than non-local information.

A third problem is that there is a trade-off level between the removal of artifacts and removal of EEG information from the signal. ICA does not guarantee that all artifacts are captured by one component. Instead, some components captures most of the artifacts, but in addition, they also capture real EEG information. The more components we remove to correct the signal, the more EEG information we also remove. This level of trade-off is very difficult to evaluate.

In [5], a cut-off threshold for ICA is proposed as a preprocessing step for artifact components identification. This solves the first problem mentioned above. This paper addresses the two remaining challenges.

In this paper, we propose a decomposition of the independent components that theoretically guarantees that the first part of the decomposition captures as much local information as possible, while the second part captures joint information. Through this decomposition, we demonstrate that we are able to generate a clean signal. We rely on a multi-objective optimization approach to model this decomposition and present two metrics to quantify the amount of EEG information lost during the cleaning process. We use the pareto-based Multi-objective evolutionary optimization algorithm (NSGA-II) [6] to solve the problem.

In the remaining of this paper, we first provide background information on ICA Section 2, followed by the proposed method in Section 3, the methodology in Section 4, results in Section 5, and conclusions in Section 6.

2   Independent Component Analysis (ICA)

ICA, a blind source separation technique, has been widely used to separate signals from multiple sources. An example of a source separation problem is when multiple people talk in a cocktail party; one may pay attention to voices in one discussion while ignoring voices from other discussions. ICA can be used in such a party to separate the mixed discussions and background noise by using multiple microphones. Similarly, in BCI research, multiple electrodes measure signals from multiple brain locations. ICA would aim in this situation to separate the sources of the EEG signals.

The blind source separation problem is classically formulated as $X = AS$, Where X is the measured mixed signals, A is the mixing matrix, and S is the Sources. Mathematically, ICA performs linear transformation on mixed signals to components such that the components are statistically independent and distributed non-gaussian . Once we have A and the estimated sources $\hat{S}$, we can find W such that $\hat{S} = WX$. The cleaner signals can be reconstructed from only the useful components. $\hat{X} = W^{-1}\hat{S}$.

The fast ICA algorithm (FastICA) [7] implements the fast fixed-point algorithm for Independent Component Analysis. Two preprocessing steps of centering and whitening are implemented in the fastICA matlab package. FastICA provides a mathematical proof for the convergence of the algorithm to discover independent sources. However, it does not address any of the three problems we discussed in the introduction.

3 Proposed Method

In EEG analysis, each electrode senses local information as well as overlapping information from the rest of the scalp. If we use ICA, we will most likely obtain one component capturing the background noise. We can simply eliminate this component and reconstruct the signals. Each reconstructed signal will have the local signal without noise. However, it won't separate the local information from the other signals. The reason for this is simple, after removing the components with the noise, we are left with three components only. When we project these three components back to the four electrodes, it is not mathematically possible that we isolate the local signal of each electrode.

Now, let us assume we can isolate the local signal. This signal will be different from the local signal captured by each electrode in two aspects. First, it will exclude the background noise. Second, it will exclude the overlapping information from the other electrodes.

Let us recall the mixing equation $X = W^{-1}\hat{S}$. Remember that we used $\hat{X}$ to denote the estimated mixed signal after cleaning. Here, we use X since we do not delete any component from $\hat{S}$, therefore, it is guaranteed that we get back the original signal. Let us now rewrite this mixing equation as follows:

$$X = W^{-1}\hat{S} = W^{-1}(S^1 + S^2) = W^{-1}S^1 + W^{-1}S^2$$

Here, we decompose the independent components into two component matrices. Let us define C as the covariance matrix between X and S^1; therefore,

$$C = \frac{covar(X, S^1)}{\sigma(X) \times \sigma(S^1)}$$

In this decomposition, our aim is to make the matrix C a diagonal matrix and maximize the values on the main diagonal. This can be formulated as

$$\downarrow f_1 = \frac{1}{N} \sum_i (1 - C_{ii})^2$$

At the same time, we would like to minimize the correlation between each variables in X and all components S^1 except the one on the main diagonal corresponding to that variable.

$$\downarrow f_2 = \frac{1}{N^2 - N} \sum_i \sum_{j \neq i} C_{ij}^2$$

When values on the main diagonal are non-zero and all other off diagonal values are zeros, we have a bijective relation between the original signal and each component in S^1. In fact, since we maximize the correlations on the main diagonal, we guarantee that the matrix S^1 captures as maximum uncorrelated (notice that S^1 is a diagonal matrix with non-zero elements on the main diagonal; therefore, it has a complete rank) information as possible from the original data.

For convenience reasons, we can now formulate our multi objective optimization problem as

$$\min[f_1 \ f_2]$$

$$f_1 = \frac{1}{N} \sum_i (1 - C_{ii})^2$$

$$f_2 = \frac{1}{N^2 - N} \sum_i \sum_{j \neq i} C_{ij}^2$$

In this paper, we use the pareto-based multi-objective evolutionary optimization algorithm (NSGA-II) [6]. However, the dimension of the problem is very large. Imagine a 10-20 standard EEG cap with 19 sensors with a sampling frequency 128Hz and an epoch of 2 seconds, the genetic algorithm will attempt to optimize 19 vectors, each of them is of length 256; that is, 4864 variables. This is a very large scale optimization problem.

To overcome the dimensionality of the problem, we do the optimization in the frequency domain instead of the time domain. By using a Fast Fourier Transform (FFT) on the principal components, we at least half the dimensionality of the problem if we use a resolution of 1Hz and work with the magnitude of each spectrum. Since some bands will have close to zero values, they can also be removed from the optimization. This reduces the dimensionality even further.

More importantly, the signal in the time domain generates a sequence of interdependent values. This level of interdependency creates extensive level of linkages among the variables that makes the optimization problem very difficult. By transforming the signal to the frequency domains, amplitudes within one epoch can be assumed to be independent variables.

4   Methodology

A synthetic data set proposed by [5] will be used in this work. It assumes 6 sources of signals. Two of the sources are assumed to be EMG operating at high amplitude and high frequencies overlapping with High Beta and Gamma bands. Each source operates

Table 1. The synthesis of the six signal sources

Source ID	Band	Amplitude	Frequency	Band	Amplitude	Frequency
s_1	Delta	14	4	Beta	52	22
s_2	Theta	23	7	Beta	70	19
s_3	Delta	16	5	Alpha	43	11
s_4	Alpha	44	9	Gamma	56	47
s_5	EMG	144	31	EMG	337	51
s_6	EMG	282	28	EMG	246	49

with a mixture of two frequencies representative of classic EEG bands as shown in Table 1.

The six sources are mixed into four signals, x_1, x_2, x_3, x_4, as follows:

$$x_1 = s_1 + 0.9 * s_5$$

$$x_2 = s_2 + 0.9 * s_6$$

$$x_3 = s_3 + s_5$$

$$x_4 = s_4 + s_6$$

The signals are sampled at 256Hz. Besides, the fifth source was activated in the last 250ms of every second, and the sixth source was activated in the last 500ms of every second. All other sources were activated from time 0. The reason for delayed activation of the EMG signals is that in real-world dataset, there is no guarantee that the artifacts are synchronized with the actual sampling window. The signals are shown in Figure 3.

Given the problem structure above, the independent components contained a total of 28 unique frequencies with non-zero amplitudes. The chromosome had 60 variables in total for all components; a significantly smaller search space than the original 1024 dimensions in the time domain.

The problem in this work is a bi-objective optimization problem with a decision vector of size 60. Despite the average dimensionality, the level of non-linearity is high. Differential evolution (DE) is used in the NSGA-II framework. The following parameters are used: population size = 50, mutation rate = 0.0167, F = 0.5, Crossover Rate = 1. Generation = 1000

5   Results

The result of this work is presented by showing the Pareto front of the solutions (Figure 1).

As shown in Figure 1, the Pareto front is evenly distributed along the two objectives. It provides good tradeoffs between the two objective functions. Those Pareto solutions labelled in the figure are those solutions that were chosen for further analysis. They include the two extreme solutions and another 4 solutions selected uniformly across the Pareto frontier.

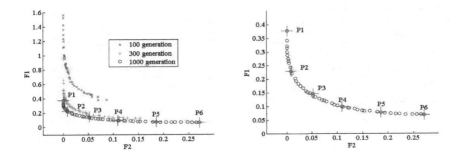

Fig. 1. The evolution of the Pareto Front From NSGA-II over selected generations (left) and the final Pareto front (right)

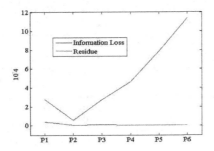

Fig. 2. Information loss and residual calculated for the 6 selecting Pareto solutions selected from the first Formulation. The blue line corresponds to artifact residue, the red line corresponds to information loss. The best solution is clearly $P2$.

The results as shown in Figure 2 demonstrates that as we move from $P1$ to $P6$ on the Pareto front, the reconstructed signals improve in terms of their EEG contents, but level of artifact also increases. P2 is found to be a good tradeoff between information loss and artifact residue.

In Figure 3, the resulting signals of the proposed method is compared to the fastICA algorithm. The information loss and residue for fastICA are 0.0103 and 0.00155 which are much higher than all solutions. The top row of the figure shows the mixed signal on the left, and the clean signal with local information alone on the right. It is evident in these plots the EMG impact at the end of the one second data.

The second row shows the output of the fastICA algorithm. The components are visualized on the left, while the reconstructed signal is visualized on the right. A closer look at the fourth variable would show that the cleaned signal carries no similarities with the original signal. This is also the case in all other signals, although the scale may mislead the eye.

The last two rows show the solutions obtained from the multi-objective optimization approach. P1-P4 are shown. The results match what we have in Figure 2. Although there

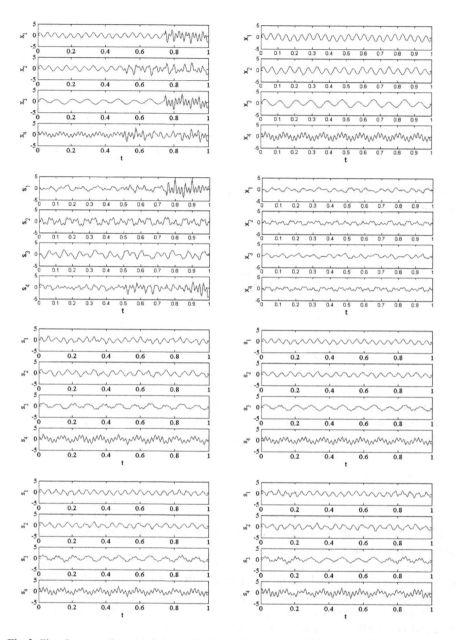

Fig. 3. First Row are the mixed signals(left) and clean signals(Right); Second Row are the independent components from ICA(left) and reconstructed signals from ICA(Right); Components from $P1$ (left) and Components from $P2$ (right) (third row); Components from $P3$ (left) and Components from $P4$ (right) (fourth row), respectively

are some distortions in the signals, the proposed method maintains the signal order, while reducing the amplitude of the artifacts.

6 Conclusion

In this paper, we presented a multi-objective formulation of the independent components analysis problem. The first objective attempts to maximize the main diagonal elements in the correlation matrix between each variable in the original mixed signal and and the corresponding components. The second objective minimizes the off diagonal elements in the same matrix; thus forcing a conflict to create a trade-off between local and joint information. Results indicate that the proposed formulation discovers components that are closer to the original clean signal with more EEG information and less artifact residual than the classical ICA.

This work raises two challenges that we will address in our future work. First, the vast amount of data in EEG experiments (big data) creates very large scale optimization problems. We will investigate efficient ways to design optimization problems for big data. Second, the role of the components is not to duplicate the original data. The mixing matrix needs to be maintained to mix the components. In this work, we discovered components that are close to the original matrix. In our future work, we will address second concern by conditioning the optimization problem with the mixing matrix.

Acknowledgement. The second author acknowledges funding from the University of New South Wales, Australia, that allowed him the time to conduct this work in Singapore.

This work was also supported by the Singapore Ministry of Education Academic Research Fund Tier 1 under the project R-263-000-A12-112.

References

1. Vigário, R., Jousmäki, V., Hämäläninen, M., Hari, R., Oja, E.: Independent component analysis for identification of artifacts in magnetoencephalographic recordings. In: Advances in Neural Information Processing Systems, pp. 229–235 (1998)
2. Vigário, R.: Extraction of ocular artefacts from EEG using independent component analysis. Electroencephalography and Clinical Neurophysiology 103(3), 395–404 (1997)
3. Abbass, H., Tang, J., Amin, R., Ellejmi, M., Kirby, S.: Augmented cognition using real-time EEG-based adaptive strategies for air traffic control. In: International Annual Meeting of the Human Factors and Ergonomic Society, HFES. SAGE (2014)
4. Abbass, H., Tang, J., Amin, R., Ellejmi, M., Kirby, S.: The computational air traffic control brain: Computational red teaming and big data for real-time seamless brain-traffic integration. Journal of Air Traffic Control 56(2), 10–17 (2014)
5. Abbass, H.A.: Calibrating independent component analysis with laplacian reference for real-time EEG artifact removal. In: Loo, C.K., Yap, K.S., Wong, K.W., Teoh, A., Huang, K. (eds.) ICONIP 2014, Part III. LNCS, vol. 8836, pp. 68–75. springer, Heidelberg (2014)
6. Deb, K., et al.: A fast and elitist multiobjective genetic algorithm: NSGA-II. IEEE Transactions on Evolutionary Computation 6(2), 182–197 (2002)
7. Hyvärinen, A., Oja, E.: Independent component analysis: algorithms and applications. Neural Networks 13(4), 411–430 (2000)

Cooperative Feature Level Data Fusion for Authentication Using Neural Networks

Mark Abernethy and Shri M. Rai

Murdoch University, Perth Australia
{mark.abernethy,s.rai}@murdoch.edu.au

Abstract. In traditional research, data fusion is referred to as multi-sensor data fusion. The theory is that data from multiple sources can be combined to provide more accurate, reliable and meaningful information than that provided by a single data source. Applications in this field of study were originally in the military domain; more recently, investigations for its application in various civilian domains (eg: computer security) have been undertaken. Multi-sensor data fusion as applied to biometric authentication is termed multi-modal biometrics. The objective of this study was to apply feature level fusion of fingerprint feature and keystroke dynamics data for authentication purposes, utilizing Artificial Neural Networks (ANNs) as a classifier. Data fusion was performed adopting the cooperative paradigm, a less researched approach. This approach necessitates feature subset selection to utilize the most discriminatory data from each source. Experimental results returned a false acceptance rate (FAR) of 0.0 and a worst case false rejection rate (FRR) of 0.0006, which were comparable to—and in some cases, slightly better than—other research using the cooperative paradigm.

Keywords: Multi-Sensor Data Fusion, Multi-Modal Biometrics, Feature Level Fusion, Cooperative Data Fusion, Fingerprint Recognition, Keystroke Dynamics, Artificial Neural Networks.

1 Introduction

Data fusion involves the combination or integration of data from multiple sensors or sources, and is traditionally referred to as multi-sensor data fusion [1]. Historically, as a formal research discipline, it was primarily the province of military research (and subsequent applications), but more recently has been applied to industrial processes, medical diagnosis, logistics, and computer security.

Using data from multiple independent sensors/sources makes a system less vulnerable to failure than a single source, because information from multiple sources (both correlated and uncorrelated) is exploited to provide a better outcome [2]. Provided the fusion method is appropriate to the types of data, and is performed correctly, the system should become less sensitive to noisy data (removing the influence of irrelevant data) and thus more robust.

Multi-modal biometrics is a term used when multi-sensor data fusion is applied in the area of biometric authentication. Though based on similar premises, and

C.K. Loo et al. (Eds.): ICONIP 2014, Part I, LNCS 8834, pp. 578–585, 2014.
© Springer International Publishing Switzerland 2014

having many similarities to formal data fusion, multi-modal biometrics has some differences in relation to data fusion levels.

Section 2 discusses the data fusion paradigms and levels as applied in multi-modal biometrics. A review of data fusion in relation to multi-modal biometrics research is presented in Sect 3. Section 4 describes the experimental methodology. Lastly, Sects 5 and 6 present the results and conclusions respectively.

2   Data Fusion Paradigms and Levels

There are three configurations that precipitate the data fusion paradigms [2]. Complementary Data Fusion is utilized where all available data from independent sources (covering a *different* region, physical attribute, or aspect of an object) are combined to provide an overall view of the region, attribute, or aspect. Competitive Data Fusion paradigm is utilized where data sources provide independent measurements of the *same* information about a region, physical attribute, or aspect of an object. They are competitive in that a fusion system must decide which sensor's data most correctly represents the region, attribute or object. Cooperative Data Fusion (the least researched) is related to the competitive paradigm. However instead of data sources competing, data are combined such that information can be derived that would be otherwise unattainable from individual data sources.

There are four data fusion levels (sensor, feature, confidence score, and decision) in a multi-modal biometric system [5,8]. Raw data represents the richest source of information (though possibly contaminated), and is fused at the sensor level. Biometric capture modules are responsible for acquisition of raw data, which can then be processed in a fusion module such that a new single vector (of integrated raw data) is obtained. An important restriction is that only samples of the same biometric trait can be used in the fusion process at this level [9].

Fusion at the feature level involves raw data (from each mode) being passed to corresponding feature extraction modules. Feature extraction requires the selection of salient features that best represent the entity and provide recognition accuracy [8]. At this stage, the features relating to each mode remain separate. Data alignment (to make the features from multiple sources compatible) and feature selection (to mitigate the 'curse of dimensionality' [5]) may be required prior to fusion. The fusion module then combines the feature vectors from each mode. The single fused feature vector is then passed to the matching module for comparison to a registered template[1]. The output from the matching module is then passed to the decision module, where the final classification decision is made. Feature level fusion results in accurate and robust authentication, because data at this level is closer to raw data—than the subsequent fusion levels—and maintains more discriminatory information than those levels [3].

Confidence score level fusion is the most commonly researched of all fusion levels in multi-modal biometrics [8,9]. The biometric capture and feature

---

[1] The registered feature vector template must have been processed in the same manner, and result in the same format, as the query feature vector.

extraction modules perform their tasks as they would for feature level fusion. However, the individual feature vectors are passed to corresponding matching modules for comparison with registered templates. For each mode, comparison between a query feature vector and its corresponding registered template generates a confidence score—in the continuous domain (interval $[0, 1]$). One confidence score is calculated in each matching module, and passed to a single fusion module where they are combined into a single scalar value in the continuous domain. This is passed to the decision module where the final classification decision is made. Fusion at this level requires less processing (than the previous levels) in order to achieve an appropriate level of system performance.

Decision level fusion is the most abstract, where accept/reject decisions—from multiple data sources—are consolidated into one final classification decision [5,8]. The biometric capture, feature extraction, and matching modules perform their tasks as they would for confidence score level fusion. The individual confidence scores are passed to corresponding decision modules, where accept/reject decisions are made for each mode. The multiple decisions are integrated (in the fusion module) to generate the final classification decision. Fusion at this level is more scalable than other levels because it requires the least amount of processing in order to achieve an appropriate level of system performance.

3 Review of Feature Level Data Fusion Research

There has been little research in the area of feature level data fusion where a cooperative approach has been applied. This review discusses some that have.

Reference [4] used two biometric characteristics (facial and iris recognition) for their experiment. They performed data fusion at the feature level by concatenating the feature sets from both sources. To reduce the high dimensionality of combined data sets, they proposed the Direct Linear Discriminant Analysis (DLDA) technique to attain a 'Reduced Joint Feature Vector' (RJFV). As the RJFV consisted of 'relevant' feature level data, their methodology achieved the same goals as a feature selection approach.

Ten samples each from 140 individuals were selected from two facial image databases. Two iris image databases were utilized; 10 high quality samples each from 100 individuals, and 10 low quality samples each from 100 individuals.

Best results were achieved when the high quality iris images were fused with the facial images. From the ROC curves presented by the authors[2], a FAR and a FRR of 0.0 were achieved.

Reference [5] used two biometric characteristics (facial recognition and hand geometry). A median normalization scheme was used to align the feature vectors. The sequential forward floating selection technique was employed to obtain a reduced feature vector.

Their experiment appears to be a combination of feature level and confidence score level fusion, which can be considered in part a cooperative data fusion approach. Five facial image samples and 5 hand geometry samples were collected

---

[2] Refer [4], Figure 5.

from 100 participants. Best results, derived from the authors ROC graph for the two fused modalities, were a FAR of 0.0001 and a FRR of 0.13[3].

Reference [6] used two biometric characteristics (facial and fingerprint recognition). A key point descriptor was defined for each minutiae extracted from a fingerprint image; this was used to align the feature sets corresponding to the two modalities. The feature sets were concatenated to give the combined feature vector and three feature reduction strategies were applied.

Five facial images and 5 fingerprint scans were collected from 50 individuals. The best results—a FAR of 0.0102 and a FRR of 0.0195—were achieved using the Delauney triangulation classifier, and their '3 points in a specific region' feature reduction strategy.

Reference [7] used two biometric characteristics (facial and palmprint recognition). Normalization was applied and a distance-based separability value—based on class membership using a Nearest Neighbour (NN) classifier—was calculated for each of the normalized data sets. Fusion was performed by concatenating normalized facial data sets with weighted normalized palm data sets. The Nearest Neighbour classifier was used to classify data by the Euclidean distance between each feature in the query and registered samples.

Twenty facial images and 20 palmprint images were collected from 119 participants. One sample (per participant) was used for training purposes (providing 119 fused training vectors); there were 2,261 (119x19) samples (per participant) used for testing purposes. Results achieved 90.73% recognition rate[4]. Though [7] did not provide values for the typical performance variables, it was of interest because fused vectors of feature level data were used when testing the fusion method.

In the next section, we describe our approach where we used feature level data fusion with a cooperative paradigm.

4   Research Method

The experiment involved the feature level fusion of fingerprint feature and keystroke dynamics data[5], adopting the cooperative approach and using Artificial Neural Networks (ANNs) for performance classification. Ninety participants were recruited from the authors' institution. The recruitment criterion required that participants typed on a standard computer keyboard on a regular basis, and could use their right index finger.

As a preliminary to the experiment described in this paper, individual experiments were conducted using keystroke dynamics and fingerprint recognition data sets [10,11]. Keystroke dynamics data collection involved participants providing 140 typed samples of a 20-character string. The keystroke duration and

---

[3] Refer [5], Figure 6.

[4] As FAR and FRR were not used, direct comparison of results is not possible.

[5] Although keystroke dynamics is widely considered a 'weak behavioural biometric', and fingerprint recognition is considered a 'strong physiological biometric', the research objective was to demonstrate the viability of the cooperative fusion process.

digraph latency metrics were extracted from recorded keystroke event time values (in milliseconds). Normality statistics (normality, kurtosis, skewness coefficients and standard deviation) were utilized for feature selection, which resulted in a 24 element array of metrics for all 140 samples for each participant [10].

Fingerprint feature data collection involved participants providing 140 scans of their right index finger. The Verifinger SDK 4.2 Visual C++ Library was utilized to extract minutiae information from each scan. Feature selection involved choosing 8 minutiae common to a participants 140 samples, and utilizing 6 attributes inherent to each minutia. This resulted in a 52 element array of metrics for all 140 samples for each participant [11].

Selected features from both data sets were normalized according to the minimax method, and written to uniquely named metrics files for each participant.

Subsequent to these initial experiments, a *complementary* data fusion experiment was conducted utilizing the same data sets [12]. As this approach utilizes all available data, the complementary fusion process concatenated (appended) a fingerprint feature sample (of 52 metrics) to a keystroke dynamics sample (of 24 metrics) resulting in 76 metrics per sample. This was performed for all 140 samples for each participant[6].

Cooperative data fusion requires feature subset selection applied to the multiple data sources [13]. The premise is that 'more relevant' features will provide improved performance, and incidentally reduce the size of each sample.

Reference [14] proposed a feature subset selection method that accumulates weights from trained ANNs. The Accumulated Relative Local Gain (ARLG) sums all weights connected to an input layer node. The ARLGs for each input layer node are ordered by magnitude, with the largest relating to the most relevant feature and the smallest to the least relevant feature. Thus a subset of the most relevant features can be determined. Equation 1 presents the method of calculation [14]:

$$LG_{ik} = \sum_j |W_{ij} \times W_{jk}| \quad (1)$$

where LG_{ik} represents the ARLG for each input layer node, W_{ij} are the weight values connecting the current i^{th} input layer node and its associated hidden layer nodes, and W_{jk} are the weight values connecting the current j^{th} hidden layer node and its associated output layer nodes.

The *cooperative* data fusion process—to select feature subsets—applied Eqn 1 to the trained ANN weights that resulted from the complementary fusion experiment [12]. Trained ANN weights had a direct correspondence to participants' complementary fused data files (which were simply concatenated samples of the data sets from both sources).

The feature subsets selected by the above process were used to extract metrics from the complementary fused data files[7]; these were subsequently used as ANN

---

[6] The experimental results for all three experiments are provided in Sect 5 for comparison with the results from the current experiment.

[7] Data sets are available upon request from the authors.

inputs in the final analysis procedure, which involved ANN training utilizing a two layered Multi-Layer Perceptron architecture (with back propagation).

Fifty participants' metrics data were randomly assigned to the training group, and the remaining 40 to the non-training group. One ANN was trained for each training group member.

For each training group member, training data was generated as follows: 30 samples were randomly chosen (and removed) from their metrics file, for the positive training case; 1 sample was randomly chosen from each of the other training group members metrics files (1 x 49), for the negative training case. This resulted in 79 samples for each member's training file. To assist training, cross validation data (for each training group member) was obtained by randomly selecting (and removing) 10 samples from their metrics file[8].

For each training group member, testing data was generated as follows: 100 samples (not used for training/validation) were used to test for Type II errors (false rejection); 100 unused samples from each of the other training group members (100 x 49), and 140 samples from each of the non-training group members (140 x 40), were used to test for Type I errors (false acceptance). This provided 10,600 samples for testing each training group member.

A dedicated ANN was trained for each training group member using their training data. The weights from each member's trained ANN were saved and used as their registered template; this meant a dedicated ANN configuration for each member. For testing, the 10,600 test samples for that particular individual were applied to their dedicated ANN and the ANN's output was used to calculate the error metrics. The same ANN architecture utilized for training was used as a pattern classifier for testing purposes.

5   Results and Discussion

Classification for a biometric authentication system involves determining the likelihood that two samples belong to the same individual [3]. This necessitates a decision based on the predicted outcome (in this case, ANN output), and involves the use of a decision threshold applied to the predicted outcome.

The two performance variables used to measure results were the False Acceptance Rate (FAR), the ratio of Type I errors to the number of samples available for testing Type I errors, and the False Rejection Rate (FRR), the ratio of Type II errors to the number of samples available for testing Type II errors. As indicated in Sect 4, there were 10,500 samples for Type I error testing and 100 samples for Type II error testing for each training group member.

The individual test results demonstrated a FAR of 0.0 for all members. This meant that no impostor samples (out of 10,500 per participant) were accepted. Results also demonstrated that only one member registered a non-zero FRR. That FRR was 0.03, which meant that (for that member only) 3 genuine samples (out of 100) were incorrectly rejected. The average FAR was 0.0 and the

---

[8] With 40 samples removed from members metrics data and used for training purposes, 100 samples remained per training group member to test for Type II errors.

average FRR was 0.0006. So no impostor samples (out of 525,000) were incorrectly accepted, and there were 3 in 5,000 genuine samples incorrectly rejected.

For comparison, Table 1 shows the average FAR and FRR figures for the previous experiments: Keystroke Dynamics only [10], Fingerprint Recognition only [11], Complementary Data Fusion approach [12]. Also included are the results for the Cooperative Data Fusion approach—i.e. the current experiment.

Table 1. Summary results for previous and current experiment

Experimental Approach	FAR	FRR
Keystroke Dynamics only	0.0277	0.0862
Fingerprint Recognition only	0.0	0.0022
Complementary Data Fusion	0.0	0.0004
Cooperative Data Fusion	**0.0**	**0.0006**

In relation to the findings presented in Sect 3, the current experiment returned FAR results comparable to [4], and slightly better than the other reviewed works. Reference [4] used 1,400 facial and 2,000 iris image samples, but did not specify how many samples were used to test Type I errors, whereas the current experiment used 10,500 samples per participant to test Type I errors.

The FRR results were comparable to the reviewed works, with [4] achieving only slightly better results. Again, the authors did not specify the number of samples used to test Type II errors; the current experiment used 100 samples per participant to test Type II errors.

6   Conclusion

Section 5 demonstrated that the cooperative feature level fusion (a less researched approach) of keystroke dynamics and fingerprint recognition data achieved results better than, or comparable to, the studies discussed in Sect 3. Feature subset selection was applied to data from both sources prior to fusion.

As demonstrated in Table 1, the cooperative data fusion approach attained improved outcomes compared to the keystroke dynamics experiment [10] and the fingerprint recognition experiment [11]. This approach also achieved an equivalent average FAR to the complementary data fusion approach [12], with a 0.0002 difference in the average FRR.

The experimental method for this study involved more participants than many published studies, and there were generally many more samples provided by participants in the current study.

Although cooperative feature level fusion has not been the subject of extensive past research, our experiment performed extremely well. The choice to utilize weights from previously trained ANNs for feature subset selection was purely to demonstrate the viability of the cooperative data fusion approach; other available feature subset selection methods may perform as well.

References

1. Hall, D.L., Llinas, J.: An Introduction to Multisensor Data Fusion. Proceedings of the IEEE 85(1), 6–23 (1997)
2. Brooks, R.R., Iyengar, S.S.: Multi-Sensor Fusion: Fundamentals and Applications with Software. Prentice Hall PTR (1998)
3. Ross, A., Jain, A.K.: Multi-modal Biometrics: An Overview. In: Proceedings of the 12th European Signal Processing Conference (EUSIPCO), pp. 1221–1224 (2004)
4. Son, B., Lee, Y.: Biometric Authentication System Using Reduced Joint Feature Vector of Iris and Face. In: Kanade, T., Jain, A., Ratha, N.K. (eds.) AVBPA 2005. LNCS, vol. 3546, pp. 513–522. Springer, Heidelberg (2005)
5. Ross, A., Govindarajan, R.: Feature Level Fusion Using Hand and Face Biometrics. In: Proceedings of the SPIE Conference on Biometric Technology for Human Identification II, pp. 196–204 (2005)
6. Rattani, A., Kisku, D.R., Bicego, M., Tistarelli, M.: Feature Level Fusion of Face and Fingerprint Biometrics. In: Proceeding of First IEEE International Conference on Biometrics: Theory, Applications, and Systems (BTAS), pp. 1–6 (2007)
7. Yao, Y.F., Jing, X.Y., Wong, H.S.: Face and Palmprint Feature Level Fusion for Single Sample Biometrics Recognition. Neurocomputing 70(7-9), 1582–1586 (2007)
8. Poh, N., Kittler, J.: Multimodal Information Fusion. In: Multimodal Signal Processing: Theory and Applications For Human-Computer Interaction, pp. 153–169. Elsevier (2010)
9. Nandakumar, K.: Multibiometric Systems: Fusion Strategies and Template Security. Michigan State University, Dept. of Computer Science and Engineering (2008)
10. Abernethy, M., Rai, S.M.: Applying Feature Selection to Reduce Variability in Keystroke Dynamics Data for Authentication Systems. In: Preceedings of the 13th Australian Information Warfare Conference, pp. 17–23 (2012)
11. Abernethy, M., Rai, S.M.: An Innovative Fingerprint Feature Representation Method to Facilitate Authentication Using Neural Networks. In: Lee, M., Hirose, A., Hou, Z.-G., Kil, R.M. (eds.) ICONIP 2013, Part II. LNCS, vol. 8227, pp. 689–696. Springer, Heidelberg (2013)
12. Abernethy, M., Rai, S.M.: Complementary Feature Level Data Fusion for Biometric Authentication Using Neural Networks. In: Proceedings of the 14th Australian Information Warfare Conference, pp. 1–8 (2013)
13. John, G.H., Kohavi, R., Pfleger, K.: Irrelevant Features and the Subset Selection Problem. In: Proceedings of the Eleventh International Conference, pp. 121–129. Morgan Kaufmann Publishers (1994)
14. Schuschel, D., Hsu, C.-N.: A Weight Analysis-Based Wrapper Approach to Neural Nets Feature Subset Selection. In: Proceedings of the 10th IEEE Conference on Tools With Artificial Intelligence, pp. 89–96 (1998)

Fuzzy Output Error as the Performance Function for Training Artificial Neural Networks to Predict Reading Comprehension from Eye Gaze

Leana Copeland, Tom Gedeon, and Sumudu Mendis

Research School of Computer Science
Australian National University
Canberra, Australia
{leana.copeland,tom.gedeon,sumudu.mendis}@anu.edu.au

Abstract. Imbalanced data sets are common in real life and can have a negative effect on classifier performance. We propose using fuzzy output error (FOE) as an alternative performance function to mean square error (MSE) for training feed forward neural networks to overcome this problem. The imbalanced data sets we use are eye gaze data recorded from reading and answering a tutorial and quiz. The goal is to predict the quiz scores for each tutorial page. We show that the use of FOE as the performance function for training neural networks provides significantly better classification of eye movements to reading comprehension scores. A neural network with three hidden layers of neurons gave the best classification results especially when FOE was used as the performance function for training. In these cases, upwards of a 19% reduction in misclassification was achieved compared to using MSE as the performance function.

Keywords: Eye tracking, reading comprehension prediction, fuzzy output error (FOE), imbalanced data sets, performance function.

1 Introduction

In this analysis we look at the practical application of predicting reading comprehension based on eye gaze recorded from participants while they read and completed a quiz. We have found no published papers on predicting reading comprehension using artificial neural networks. Current applications of eye tracking in reading analysis only take into account basic assessment of reading behavior such as using fixation time to predict when a user pauses on a word. We intend to explore the use of more complex analysis of eye gaze to make more complex prediction about the users reading behavior. We do this by investigating the use of artificial neural networks to predict these complex behaviors. However, this application poses us with several obstacles namely restricted size in the data sets that are highly imbalanced. We explore a method for improving classification performance of artificial neural networks (ANN) in this scenario. We investigate the use of fuzzy output error (FOE) [1] as the performance function for training the feed forward neural networks using back propagation training.

C.K. Loo et al. (Eds.): ICONIP 2014, Part I, LNCS 8834, pp. 586–593, 2014.
© Springer International Publishing Switzerland 2014

We assess whether the use of this performance measure is better suited to this type of problem compared to mean square error (MSE).

The intended use of reading comprehension prediction from eye gaze is in the design of adaptive online learning environments that use eye gaze to predict user reading behavior.

2   Background

2.1   Eye Movements during Reading

Eye movements can be broadly characterized as fixations and saccades. A fixation is where the eye remains relatively still to take in visual information. A saccade is a rapid movement that transports the eye to another fixation. Generally when reading English fixation duration is around 200-300 milliseconds, with a range of 100-500 milliseconds and saccadic movement is between 1 and 15 characters with an average of 7-9 characters [2]. The majority of saccades are to transport the eye forward in the text when reading English, however, a proficient reader exhibits backward saccades to previously read words or lines about 10-15% of the time [2]. Backward saccades are termed regressions. Long regressions occur due to comprehension difficulties, as the reader tends to send their eyes back to the part of the text that caused the difficulty [2]. Comprehension of the text can have significant effects on the eye movements observed [2,3]. Eye gaze patterns can be used to differentiate when individuals are reading different types of content [4]. In this application both support vector machines (SVM) and ANN were used to classify eye movement measures as either relevant or irrelevant text for answering a set of questions. ANN's have also been used to predict item difficulty in multiple choice reading comprehension tests [5]. Their analysis took into account the text structure, propositional analysis of the text, and the cognitive demand of the text, but not eye gaze.

2.2   Performance Functions for Imbalanced Data Sets

Dealing with imbalanced data sets is not a new problem. Performance functions for dealing with imbalance in data sets include increasing the weight-updating for the minority class and decreasing it for the majority class [6,7]. This error function was designed specifically for use in the back-propagation algorithm for training feed forward artificial neural networks. Many other methods have been used to overcome the problem of imbalanced data sets such as using under-sampling, over-sampling, and other forms of sampling to reduce the imbalance. An example of a cost sensitive learning algorithm is MetaCost [8] which is based on relabelling of training data with their estimated minimal cost classes. Another way of achieving cost sensitivity is to change the algorithm used to train the classifier to utilize a cost matrix, such as with neural networks [9,10].

2.3 Fuzzy Output Error (FOE)

Fuzzy Output Error (FOE) [1] is an extension of Fuzzy Classification Error (FYCLE) and Sum of Fuzzy Classification Error (SYCLE) [11]. However, FOE uses a fuzzy membership function to measure the difference between the predicted and the target values. Instead of mean square error (MSE), FOE describes the error in a fuzzy way and then sums the fuzzy errors to get the total error. FOE is defined as follows for a data set of n records with matching pairs of target and predicted values for each record 1 to n: $FOE = \sum_{i=1}^{n} 1 - \mu(\hat{y}_i - y_i)$ where $n \in \mathbb{N}$ and $\mu()$ is the membership function of a desired classification and its complement describes the error. The membership function is termed the FOE Membership Function, which we will refer to as FMF subsequently.

The FMF is used to describe the output of a fuzzy classification (or a regression) in regards to how close that output is to the target output. The membership function itself represents the fuzzy set for "good classification". The value of $\mu(x)$ gives the degree of membership of the error in the good classification fuzzy set and consequently the complement of $\mu(x)$ gives the error measure. In the case of perfect classification $\hat{y} - y = 0$ so the membership value is $\mu(\hat{y} - y) = 1$. Conversely, when $\hat{y} - y = 1$ the classification is completely wrong so the membership value is $\mu(\hat{y} - y) = 0$. FOE can represent crisp classification, i.e. the special case of $\mu(x) \in \{1,0\}$. The more $\mu(x)$ tends toward 0 the higher the error, since the difference is larger. FMFs can be created in any shape in order to describe the output of a function. It is important to note that the difference between target and predicted values is not taken as the absolute value of the difference (i.e. $|\hat{y} - y|$). Although this would make the FMF simpler as only one side of a piecewise linear function would be needed, it provides more flexibility in describing the types of error. For example, false negatives may be considered a much worse error than false positives when screening for diseases.

3 Method

A user study was conducted to collect participants' eye gaze as they read a tutorial and completed a quiz based on the tutorial's content. The tutorial and quiz were coursework from a first year computer science course at the Australian National University. The tutorial and quiz were presented to participants in two formats. The first format (denoted by A) involved presentation of the tutorial content slide followed by questions and the content slide. As there are 9 topics there are 18 slides in total displayed in this format. The second format (B) involved presentation of the questions and the content slide and so there are 9 slides in total displayed in this format. Each of the 9 slides is 400 words long with an average Flesch Kincaid Grade Level[1] of 12. All participants were university students and therefore had at least high school level education indicating that the readability of the slides should not be above their reading

---

[1] Flesch Kincaid Grade Level is an indication of the minimum level of education required to read and comprehend a piece of text. The Flesch Kincaid readability test is designed for contemporary English and United States educational system grading.

abilities. Participants answered two questions to measure their comprehension (18 questions in total); one question is multiple-choice and the other is cloze (fill-in-the-blanks). The two types of questions are to assess different forms of comprehension. The scores that the participants can receive for each question are 0, 0.5 and 1. Once the participants finished the quiz and before being shown their results, participants were asked to subjectively rate their overall comprehension on a scale of 1 to 10 with 10 being complete understanding.

Format A was presented to 15 participants (6 female, 9 male) with an average age of 22.3 years. Of these participants, 7 stated that their degree or major was related to computer science or information technology. English was not the first language for 4 participants. Format B was presented to 8 participants (1 female, 7 males), with an average age of 21.8 years. All participants stated that they had a major or degree related to computer science. English was not the first language for 3 participants.

The study was displayed on a 1280x1024 pixel monitor. Eye gaze data was recorded at 60Hz using Seeing Machines FaceLAB 5 infrared cameras mounted at the base of the monitor. The study involved a 9-point calibration sequence. EyeWorks Analyze was used to pre-process the gaze point data to give fixation points. The parameters used for this were a minimum duration of 0.06 seconds and a threshold of 5 pixels.

3.1   FMF Shapes Used to Calculate FOE

In this analysis we investigated one FMF shape used for calculating FOE. This FMF (Fig. 1) is designed to be a model of FYCLE.

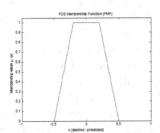

Fig. 1. FMF used to calculate FOE

3.2   Data Set Information

The raw eye gaze data consists of x,y-coordinates recorded at equal time samples (60Hz). Beyond fixation and saccade identification many other eye movement measures can be derived that reveal much about the participants' reading behavior such as; maximum fixation duration (seconds), average fixation duration (seconds), total fixation duration (seconds), and regression ratio. The number of inputs varies depending on the presentation method as the inputs are generated from the pages that the participant viewed. This means that in format A as the participants view the tutorial content page and then the questions and content page, the inputs are generated from both pages for the scores obtained from the questions and content page. Since there is a large difference in the ranges for each of the inputs we normalized the inputs to a range of [0,1]. The two outputs for all data sets are the multiple choice question score and cloze question score. The multiple-choice score can take values of 0 or 1 and the cloze score can take the values 0, 0.5 or 1. This is therefore a classification problem; a binary classification task for the multiple-choice score and a 3-class classification task for the cloze score. However, as shown in Table 1 the ratio of the number of data instances in each class for each problem is considerably imbalanced for each output.

Table 1. Properties of each data set

Properties of data set	Format A	Format B
Number of Inputs	49	26
Size	135	72
Multiple choice score class imbalance	109/26	59/13
Percentages in classes 1/0	*81%/19%*	*82%/18%*
Cloze Score class imbalance	124/11/0	69/1/2
Percentages in classes 1/0.5/0	*92%/8%/0%*	*96%/1%/3%*

4   Results and Discussion

Several ANN architectures were trained using the scaled conjugate gradient algorithm [12] and FOE used as the performance function. As a comparison the same ANN architectures were trained using MSE as the performance function and the Levenberg–Marquardt algorithm [13]. The number of inputs for each presentation format is outlined in Table 1 and all networks have 2 outputs. From initial testing it was found that a single layer network performed poorly with average misclassification rate (MCR) around the 0.5 for all both FOE and MSE. We have chosen two and three layer topologies to trial for the analysis. The following topologies were tested: [10 5], [20 10], [30 15], [50 25], [12 6 3], [16 8 4], [20 10 5], [30 20 10], and [60 30 15]. The notation [X Y Z] indicates neurons in the first hidden layer to the third hidden layer. As a baseline comparison MSE is used as one of the performance functions. Reported are the average misclassification rate (MCR) values from 10-fold cross validation with standard deviations, summarized in Table 2 and Table 3.

For format A, on average the MCR produced from using FOE as the performance function for training the neural networks to predict the question scores is lower than that from using MSE as the performance function. However, the results are not statistically different. However, there is a statistically significantly difference between the mean MCR values from 10-fold cross validation for each topology for format B (p=0.02<0.05, 2-sided, paired Student's t-test). Therefore, on average the MCR produced from using FOE, as the performance function for training the neural networks to predict the question scores is lower than that from using MSE as the performance function.

Table 2. Comparison of MCR from using FOE and MSE as the performance function for training ANNs to classify the Format A data set

| Topology | MCR | | | | Difference in MCR | % Reduction in MCR |
| | FOE Result | | MSE Result | | | |
	Mean	*St. Dev.*	*Mean*	*St. Dev.*		
[10 5]	0.32	0.29	0.25	0.10	-0.06	-25.3
[20 10]	0.42	0.28	0.35	0.09	-0.07	-19.8
[30 15]	0.33	0.16	0.38	0.15	0.04	11.9
[50 25]	0.28	0.06	0.33	0.12	0.05	14.2
[12 6 3]	0.23	0.25	0.23	0.14	0.00	-0.4
[16 8 4]	0.20	0.08	0.27	0.07	0.07	24.5
[20 10 5]	0.21	0.06	0.26	0.15	0.05	19.2
[30 20 10]	0.24	0.05	0.31	0.11	0.07	22.8
[60 30 15]	0.25	0.08	0.39	0.13	0.13	34.8
Average	*0.28*	*0.15*	*0.31*	*0.12*	*0.03*	*9.11*

Table 3. Comparison of MCR from using FOE and MSE as the performance function for training ANNs to classify the Format B data set

| Topology | MCR | | | | Difference in MCR | % Reduction in MCR |
| | FOE Result | | MSE Result | | | |
	Mean	*St. Dev.*	*Mean*	*St. Dev.*		
[10 5]	0.30	0.26	0.29	0.14	0.00	-1.5
[20 10]	0.27	0.15	0.37	0.15	0.10	26.7
[30 15]	0.33	0.21	0.46	0.13	0.13	29.1
[50 25]	0.52	0.22	0.47	0.18	-0.04	-9.3
[12 6 3]	0.16	0.07	0.24	0.13	0.07	31.1
[16 8 4]	0.16	0.06	0.30	0.14	0.14	46.4
[20 10 5]	0.16	0.07	0.24	0.10	0.07	30.6
[30 20 10]	0.26	0.11	0.39	0.15	0.13	32.6
[60 30 15]	0.35	0.22	0.36	0.10	0.01	2.5
Average	*0.28*	*0.15*	*0.35*	*0.14*	*0.07*	*20.91*

Overall, the results reflect that fact that the data sets are quite hard to classify. This could be due to several factors: class imbalance, small data sets, and too many feature inputs. However, these obstacles can be common in real world problems so it is imperative that such obstacles can be overcome. FOE has been shown to be a flexible performance function that can be tailored specifically for each problem. By defining different error membership functions (the FMF used) for FOE the outcome of training ANNs to classify the eye movement measures can be improved compared to using MSE. This is shown for both data sets where on average the use of FOE as the performance function for training the neural networks produces neural networks that are better as predicting the multiple choice and cloze scores.

Notably, for both data sets the topologies that generate the best predictions are [16 8 4], [20 10 5], and [30 20 10]. This reiterates the fact that the data set is hard to classify and contains complex relationships, as three layers of hidden neurons are needed to provide decent classification results. Furthermore, using FOE as the performance function for training generates upwards of a 19% reduction in misclassification compared to using MSE. Particularly, when using the [16 8 4] topology and FOE as the performance function for training creates a neural network that produces on average 38% and 46% reduction in misclassification, for formats A and B respectively, compared with using MSE.

5   Conclusions and Further Work

We have shown that the use of FOE as a performance function for training feed forward neural networks provides better classification of results than using MSE when the data is imbalanced. The use of FOE as the performance function for training neural networks provides significantly better classification of eye movements than reading comprehension scores. We found that the eye movement data is quite complex so it is optimal to use a neural network with three hidden layers of neurons. In these cases the use of FOE as the performance function for training gave upwards of a 19% reduction in misclassification compared to using MSE as the performance function, with a maximum of 46% reduction in misclassification, which is a significant improvement in classification. These are promising results and show that when dealing with a small data set with a large imbalance in classes MSE is not the optimal performance function to use for training neural networks. Further work will be needed to generalize to other data sets as well as with other classifiers. Additionally, we intend to extend this analysis to compare to existing techniques for handling imbalanced data sets such sampling methods and cost-sensitive learning.

One of the advantages of using FOE is that it is a flexible error function that can be tailored to the data sets and problem. Specifying the shape of the FMF used to calculate FOE does this. However, there is no simple way of constructing an FMF. In this analysis we only investigated one FMF. Other FMF shapes should be tested such as those described in [14]. However, a beneficial approach would be to learn the most appropriate FMF shape from the data set. An initial investigation on how to do this was also done in previous work but was restricted to looking at fuzzy signatures [14]. An area of further exploration is how to apply the learning of FMF shape when using other classifiers such as neural networks.

The application of predicting reading comprehension from eye gaze is in adaptive online learning environments. Prediction of comprehension would allow a system to adaptively change to a student's knowledge level making the learning process more streamlined and targeted toward their capabilities. Much is left to do in this respect. A primary area of interest is in predicting reading comprehension without questions. In both scenarios here the participants had access to the questions and the tutorial content at the same time so that they could cross-reference the text and questions to find the most appropriate answer. In a scenario where the student is shown text and no comprehension questions it would be beneficial to be able to predict their comprehension without needing to interrupt them with comprehension questions.

References

1. Gedeon, T., Copeland, L., Mendis, B.S.U.: Fuzzy Output Error. Australian Journal of Intelligent Information Processing Systems 13(2), 37–43 (2012)
2. Rayner, K.: Eye movements in reading and information processing: 20 years of research. Psychological Bulletin, 372–422 (1998)
3. Rayner, K., Chace, K.H., Slattery, T.J., Ashby, J.: Eye movements as reflections of comprehension processes in reading. Scientific Studies of Reading 10(3), 241–255 (2006)
4. Vo, T., Mendis, B.S.U., Gedeon, T.: Gaze Patterns and Reading Comprehension. In: Wong, K.W., Mendis, B.S.U., Bouzerdoum, A. (eds.) ICONIP 2010, Part II. LNCS, vol. 6444, pp. 124–131. Springer, Heidelberg (2010)
5. Perkins, K., Gupta, L., Tammana, R.: Predicting item difficulty in a reading comprehension test with an artificial neural network. Language Testing 12(1), 34–53 (1995)
6. Oh, S.-H.: Error back-propagation algorithm for classification of imbalanced data. Neurocomputing 74(6), 1058–1061 (2011)
7. Oh, S.-H.: Improving the Error Back-Propagation Algorithm for Imbalanced Data Sets. International Journal of Contents 8(2), 7–12 (2012)
8. Domingos, P.: MetaCost: a general method for making classifiers cost-sensitive. In: Proceedings of the Fifth ACM SIGKDD International Conference on Knowledge Discovery and Data Mining, pp. 155–164. ACM (1999)
9. Kukar, M., Kononenko, I.: Cost-Sensitive Learning with Neural Networks. In: 13th European Conference on Artificial Intelligence, pp. 445–449 (1998)
10. He, H., Garcia, E.A.: Learning from imbalanced data. IEEE Transactions on Knowledge and Data Engineering 21(9), 1263–1284 (2009)
11. Mendis, B.S.U., Gedeon, T.D.: A comparison: Fuzzy signatures and Choquet Integral. In: IEEE International Conference on Fuzzy Systems, FUZZ-IEEE 2008 (IEEE World Congress on Computational Intelligence), pp. 1464–1471 (2008)
12. Moller, M.F.: A scaled conjugate gradient algorithm for fast supervised learning. Neural Networks 6(4), 525–533 (1993)
13. Hagan, M.T., Menhaj, M.: Training feedforward networks with the Marquardt algorithm. IEEE Transactions on Neural Networks 5(6), 989–993 (1994)
14. Copeland, L., Gedeon, T.D., Mendis, B.S.U.: An Investigation of Fuzzy Output Error as an Error Function for Optimisation of Fuzzy Signature Parameters. RCSC TR-1 2014 (2014)

Part-Based Tracking with Appearance Learning and Structural Constrains

Wei Xiang and Yue Zhou

Institute of Image Processing and Pattern Recognition
Shanghai Jiao Tong University
Shanghai, China
{xwsjtu,yue_zhou}@126.com

Abstract. Adaptive tracking-by-detection methods are widely used in computer vision for tracking objects. Despite these methods achieve promising results, deformable targets and partial occlusions continue to represent key problem in visual tracking. In this paper, we propose a part-based visual tracking method. First, we take advantage of the existing online learning appearance model to learning the appearance of each part. Second, we propose a novel part initialization method and an affine invariant structural constrain between these parts. Third, a tracking model based on the appearance of each part and the spatial relationship between the parts is proposed. We make use of an optimization algorithm to find the best parts during tracking, update the appearance model and the structural constraints between parts simultaneously. In this paper we show our method has many advantages over the pure appearance learning based tracking model. Our method can effective solve the partial occlusion problem, and relieve the drift problems. What's more, our method achieves great result while tracking the target of which geometric appearance changes drastically over time.

Keywords: tracking, appearance model, structural constrains, part-based.

1 Introduction

Object tracking has many practical applications and has long been studied in computer vision. An approach to tracking which has become particularly popular recently is tracking-by-detection [1, 2, 3, 4, 7, 11]. It has been shown that an adaptive appearance model, which updating the model during the tracking process, is the key to good performance.

These methods train a model to separate the object from the background via a discriminative classifier can often achieve superior results. These methods often don't consider the motion model and have been termed "tracking by detection". For example, Avidan[4] used a Support Vector Machine as an off-line binary classifier to distinguish target from background. Helmut Grabner[3] used an on-line boosting method to choose discriminating features and classify the target and background. Babenko et al.[5]employ an online Multiple Instance Learning based appearance

C.K. Loo et al. (Eds.): ICONIP 2014, Part I, LNCS 8834, pp. 594–601, 2014.

model to resolve the sample ambiguity problem. All of these method delineate the tracked object by a single regular bounding box(e.g., a rectangular or circular window), which renders them sensitive to partial occlusions and shape deformations.

Deformable part-based models have been successfully applied to object detection and object recognition on numerous occasions. Part-based appearance models [6, 8, 10] have been shown to have favourable properties such as robustness to partial occlusions and articulation. In those cases, highly variable objects are represented using mixtures of deformable part-based models. But, some difficulties in extending part-based appearance models to visual tracking are still remain to solve. Firstly, it is hard to decide which parts are the "good" parts, which means these parts can well distinguish the object from background. Secondly, because the appearance model needs to be updated online, we have to adjust parts dynamically, reducing or adding parts according to the part deformation. Thirdly, when encountering partial occlusions, these parts need to be updating dynamically.

In this paper, we proposed a novel tracking method, which is based on learning the appearance of parts of the object and the spatial relationship between these parts. Our method adopts an effective appearance model to learning the appearance of each part. At the same time, we propose a novel structure description method to show spatial relationships between the parts. Our final tracking model can well handle the partial occasion and strong deformation problem.

In summary, our main contributions are: (1) we propose a novel structure description method to show spatial relationships between the parts (2) we propose a tracking model which makes full use of the appearance of each part and the spatial relationships between these parts. The appearance of each part and the structural constrains update online. We discuss the appearance model in section 2.Section 3 introduce our new tracking model, section 4 presents our experimental results, and section 5 concludes the paper.

2   Appearance Model

In this paper, we choose the compressive based features extracted appearance model [1] as our part learning model. The appearance model is generative as the object can be well represented based on the features extracted in the compressive domain. It is also discriminative because it uses these features to separate the target from the surrounding background via a naive Bayes classifier. In this appearance model, features are selected by an information-preserving and non-adaptive dimensionality reduction from the multi-scale image feature space based on compressive sensing theories.

This method assumes that the tracking window in the first frame has been determined. At each frame, sample some positive samples near the current target location and negative samples far away from the object center to update the classifier. To predict the object location in the next frame, drawing some samples around the current target location and determine the one with the maximal classification score. The fig1 illustrates the appearance model.

The appearance model is efficient and robust, but its tracking box is kept unchanged. We use the method as our part appearance model and propose the structural constrain among these parts, which make our method is robust to deformable targets.

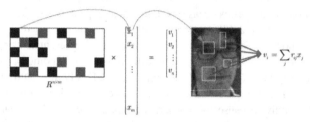

$$v_i = \sum_j r_{ij} x_j$$

Fig. 1. The illustration of the compressive based features extracted appearance model .The blue arrows illustrate that one of nonzero entries of one row of R sensing an element in x is equivalent to a rectangle filter convolving the intensity at a fixed position of an input image.

3   Parts Based Tracking

3.1   Part Initialization

Our algorithm is based on part learning. The initial position of parts has to be chosen so as to be good for image representation. It is obvious that the part which contains affluent target character information should be chosen. We propose a novel method to select the effective parts. Firstly, we use the FAST algorithm to find key points. FAST(features from accelerated segment test) is an efficient key point detecting algorithm. We assume that the area where these key points aggregating contain abundant information and should be regard as one part. Therefore, we use the K-means clustering algorithm to classify these key points into several categories. For each category, we exclude the outlier points and use a rectangle to encompass these points. These rectangles are the parts we chosen. Fig.2 show the illustration of initialization of parts. In experiment, we found that 3 or 4 parts usually is a good option.

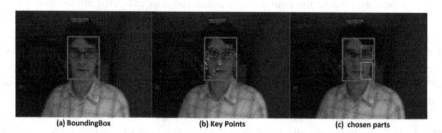

(a) BoundingBox   (b) Key Points   (c) chosen parts

Fig. 2. Example of initialization of parts in David seq. (b) displays key points detected by the FAST algorithm. (c) shows the four initialized parts.

3.2   Structure Description

Most patch-based algorithms use the distance between patch center and target center as structure description. The description is rotation-invariant. However, it cannot resist scale variance. In this paper, we propose a new method to descript the structure constrain as shown in Fig3. The position of each part is defined by $X_t = (X_t^c, X_t^1, ..., X_t^m, D_t^a, D_t^1, ..., D_t^m)$ where X_t^c denotes the center position of an object, X_t^i indicates the center position of the ith local part, we define the spatial relationship between these parts.

$$D_i = R_i / (\sum_{j=1}^{n} R_j / n), i = 1, 2, ..., n \qquad (1)$$

Where R_i is the distance between the center of ith part and target center. The description is affine invariant, Eqn.(2)proves it can resist scale variance.

$$aR_i / (\sum_{j=1}^{n} aR_j / n) = aR_i / (a\sum_{j=1}^{n} R_j / n) = R_i / (\sum_{j=1}^{n} R_j / n) = R_i / R_{avr} \qquad (2)$$

These Distances describe the structural constraints between these parts. When the target suffers rotating, scaling, translation and changes in view angle, these distances should stay the same.

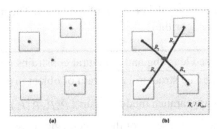

Fig. 3. Description of the structure of the target.(a) the initial part (b) illustrates the structure description of a target.

3.3   Tracking

Our tracking algorithm is based on the appearance model of each part and the structural constraints between these parts. For each part, we train the appearance model to online learning the appearance of each part. Assume four parts, P_1, P_2, P_3, P_4, we construct four trackers, T_1, T_2, T_3, T_4 based on our appearance model. During tracking, each tracker can get some candidate rectangles of each part, $L_i^1, ..., L_i^n$ with scores $S_i^1, ..., S_i^n$. These scores indicate the possibility of the rectangle to be the corresponding part. The higher score a rectangle gets, the higher possibility the rectangle can be the corresponding target. Since each contains partial information of the target, it may drift easily. We use the structural constrains between these parts to solve that problem, which lead to our final tracking model. This can be expressed by the following optimization problem.

$$\arg\max \sum_{i=1}^{n} w_i + \lambda \sum_{i=1}^{n} f(p_i) \tag{3}$$

While w_i represents the appearance similarity of each parts and $f(p_i)$ represents the structure term among each part, λ control the relative weight between the appearance term and structure term. To solve the optimization problem, we can get the appearance score of each part through our appearance model, but the structure model is hard to determine. We don't know the center of our tracking object, which should be determined by the parts. At the same time, the location of each part is still unknown. Since the problem is unsolvable, we have to find the approximate solution. We consider the fact that the variance between two serial frames in video is relative small. Thus the target is impossible to change drastically. We use the template matching method to roughly locate the object and get the center of the object. Then we can solve the Eq3 to precise positioning the location of each part. Finally, we use the minimum enclosing rectangle method to get the final tracking result. Update the appearance model of each part and the structural constrains between these parts as well.

In all, tracking procedure mainly includes three parts: (a) Using the appearance model to locate every candidate parts and get the similarity. (b) Template matching method to get the center. (c) Optimizing the Eq.2 to precisely locate each part. And finally, we use these parts to get the position of the target. The main steps of our algorithm are summarized in Algorithm1.

Algorithm 1. Part-based learning and structural constrains tracking

Input: ith video frame, the structural constrain of object in previous frame

    1. Use the each appearance model of part (P_1, P_2, P_3, P_4) to sample rectangles, and   Calculate the likelihood score of these rectangles ($W_1^{1,\ldots,n}, W_2^{1,\ldots,n}, W_3^{1,\ldots,n}, W_4^{1,\ldots,n}$).

    2. Maximize the equation 3 to find the best parts (P_1, P_2, P_3, P_4).

    3. Update the appearance model of each part and the spatial relationship between these parts.

    4. Get the final tracking result through the location of each part.

Output: Tracking location L_t and the structural constrain.

3.4   Discussion

Our tracking algorithm mainly has two advantages over some other algorithms. Firstly, it can well handle the situation that the target suffers dramatic deformation. For many tracking methods based on appearance learning, when the target or the background changes greatly, the tracking result may contain background information which may lead to the model drifting gradual. For our algorithm, due to the fact that each part contains the partial information of target object, and the structural term express the global information of the target. When the appearance of the target changes dramatically, each part can find the corresponding part. At the same time, because the structure constrain, these parts are unlikely to drift. As shown in Fig4, our tracking algorithm perform well on the ice skater video.

Fig. 4. Tracking result of ice skater

Another big advantage of our algorithm is that it can well handle the partial Occlusion problem. Each part is learned by our appearance model, and we can get some candidate rectangles with corresponding scores. When a part is occluded while other parts not, we can know this happen due to the fact that the average score of this part is much less than the other parts get. So we can dynamically reduce the learning rate of occluded part.Fig4.

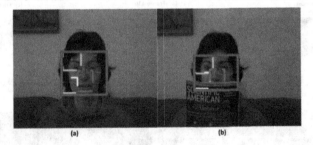

Fig. 5. (a)The target   (b) The face is partial occupied by a book. In our algorithm, we can detect the yellow rectangle and the purple rectangle is occupied by comparing the average scores they get.

4   Experiment

In experiments, we will verify the performance of our algorithm. Development environment of our algorithm is Visual Studio 2010 and Intel OpenCV library on Intel

Core(TM)2 Duo CPU E7500 and 3.00GB RAM. We evaluate our tracking algorithm with the online AdaBoost method and the compressive tracking algorithm on 4 challenging sequences.

For the David indoor sequence shown in Fig 6(1), the illumination and pose of the object both change gradually. Our method performs well but the OAB method failed and the CT approach can't locate accurately. For the Singer sequence shown in Fig 6(2) the illumination and the scale changes drastically, the proposed tracker is robust to scale and illumination change. Our method has the structural constrain term, so it can well handle this situation. For Fig 6(3), the pose of the object changes drastically, our method performs well while the other methods fail to track. For Fig 6(4), the background and the object both change drastically, our method achieved great result while the other method failed. Fig5 has demonstrated our method can well handle the partial occlusion well. Our method is very efficient, which runs at 40 frames per second (FPS) on our development environment.

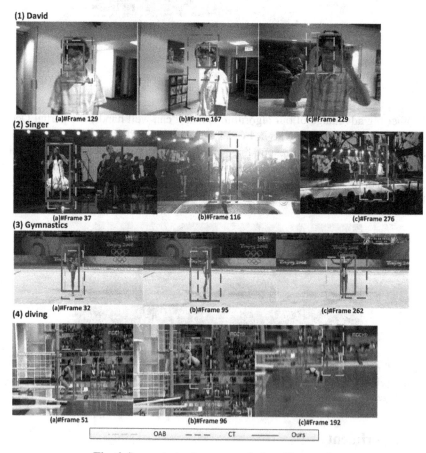

Fig. 6. Screenshots of some sampled tracking results

5 Conclusions

In this paper, we presented a novel approach to tracking based on part learning and structural constrains between these parts. Our formulation generalizes previous methods by combining the appearance learning model and the spatial relationships among parts. The appearance of parts and their spatial relationship is updating online. In contrast to other based on appearance learning model, our algorithm can well handle large variation in pose. Our approach models an object as a set of parts, each parts is trained and updated individually using the compressive based features extracted appearance model. Our algorithm can well solve the partial occlusion problem via the parts learning method. The Structural constraints between parts are rotation-invariant. So our approach can well handle the rotation and scale variation problem. When the object is very small or the resolution of target is very low, each part contains little information and may drift easily. So our approach does a relatively poor job on that situation.

References

1. Zhang, K., Zhang, L., Yang, M.-H.: Real-time compressive tracking. In: Fitzgibbon, A., Lazebnik, S., Perona, P., Sato, Y., Schmid, C. (eds.) ECCV 2012, Part III. LNCS, vol. 7574, pp. 864–877. Springer, Heidelberg (2012)
2. Yao, R., et al.: Part-based visual tracking with online latent structural learning. In: 2013 IEEE Conference on Computer Vision and Pattern Recognition (CVPR). IEEE (2013)
3. Grabner, H., Grabner, M., Bischof, H.: Real-Time Tracking via On-line Boosting. BMVC 1(5) (2006)
4. Hare, S., Saffari, A., Torr, P.H.S.: Struck: Structured output tracking with kernels. In: 2011 IEEE International Conference on Computer Vision (ICCV). IEEE (2011)
5. Babenko, B., Yang, M.-H., Belongie, S.: Visual tracking with online multiple instance learning. In: IEEE Conference on Computer Vision and Pattern Recognition, CVPR 2009. IEEE (2009)
6. Felzenszwalb, P., McAllester, D., Ramanan, D.: A discriminatively trained, multiscale, deformable part model. In: IEEE Conference on Computer Vision and Pattern Recognition, CVPR 2008. IEEE (2008)
7. Li, H., Shen, C., Shi, Q.: Real-time visual tracking using compressive sensing. In: 2011 IEEE Conference on Computer Vision and Pattern Recognition (CVPR). IEEE (2011)
8. Hare, S., Saffari, A., Torr, P.H.S.: Efficient online structured output learning for keypoint-based object tracking. In: 2012 IEEE Conference on Computer Vision and Pattern Recognition (CVPR). IEEE (2012)
9. Wu, Y., Lim, J., Yang, M.-H.: Online object tracking: A benchmark. In: 2013 IEEE Conference on Computer Vision and Pattern Recognition (CVPR). IEEE (2013)
10. Kwon, J., Lee, K.M.: Tracking of a non-rigid object via patch-based dynamic appearance modeling and adaptive basin hopping monte carlo sampling. In: IEEE Conference on Computer Vision and Pattern Recognition, CVPR 2009. IEEE (2009)
11. Kalal, Z., Mikolajczyk, K., Matas, J.: Tracking-learning-detection. IEEE Transactions on Pattern Analysis and Machine Intelligence 34(7), 1409–1422 (2012)

Estimation of Hidden Markov Chains by a Neural Network

Yoshifusa Ito[1], Hiroyuki Izumi[2], and Cidambi Srinivasan[3]

[1] School of Medicine, Aichi Medical University
Nagakute, Aichi-ken, 480-1195, Japan
ito@aichi-med-u.ac.jp
[2] Faculty of Policy Studies, Aichi-Gakuin University
Nisshin, Aichi-ken, 470-0195, Japan
hizumi@psis.agu.ac.jp
[3] Department of Statistics, University of Kentucky
Lexington, Kentucky 40506, USA
srini@ms.uky.edu

Abstract. The main theme is to show that a one-hidden-layer neural network, which has learned a Bayesian discriminant function, can be used for estimating hidden Markov chains. The crucial point of the algorithm is the use of the logistic function as the activation of the output unit of the network. The network learns a single discriminant function, but converts it to the individual discriminant functions at all the steps.

1 Introduction

We prove that a neural network, which has learned a Bayesian discriminant function, can be used for estimating a hidden Markov chain. The purpose is not to assert that the method substitutes the traditional methods of estimating hidden Markov chains but to show a neural network has this capability. Rather it is to show that the network can realize the well-known forward recursive equation.

The earlier works have dealt with neural networks which can learn Bayesian discriminant functions [3-9]. They are characterized by the logistic activation function of the output units. The activation function is used to obtain the logit transform of the output of the networks. In this article, the posterior probabilities of the hidden states are individually obtained by means of the network.

This paper treats the forward recursive algorithm of estimating two-state hidden Markov chains. Mathematical back ground of the algorithm can be seen in [2] and [10]. Conversion of the algorithm to the backward one and extension to the interpolation algorithm are its corollaries.

The network learns a Bayesian discriminant function, where the prior probabilities are the equilibrium probabilities of the states. Hence, the Markov chain can be used as a teacher sequence without modification. When the network is used for estimating a hidden Markov chain, the individual prior probabilities depend on the posterior probabilities of the previous state. Hence, the network is equipped with a module having a memory node.

C.K. Loo et al. (Eds.): ICONIP 2014, Part I, LNCS 8834, pp. 602–609, 2014.
© Springer International Publishing Switzerland 2014

2   Use of Bayesian Discriminant Functions

Let $x \in \mathbf{R}^d$ be the observables, let $S = \{s_1, s_2\}$ be the hidden states and let $\{(x^{(t)}, s^{(t)})\}_{t=1}^{T} \subset \mathbf{R}^d \times S$ be a segment of the Markov chain. We suppose that the Markov chain is irreducible and temporarily homogeneous. Hence, it is ergodic. Let $\{\pi_1, \pi_2\}$ be the equilibrium distribution of the states, and let $p_{ij}, i, j = 1, 2$ be the transition probabilities: $p_{ij} = Pr(s^{(t)} = s_j | s^{(t-1)} = s_i)$. Then,

$$\pi_1 = \frac{p_{21}}{p_{12} + p_{21}}, \qquad \pi_2 = \frac{p_{12}}{p_{12} + p_{21}}. \tag{1}$$

Denote the state-conditional probability distributions by $p(x|s_i), i = 1, 2$ and set $p(x) = \pi_1 p(x|s_1) + \pi_2 p(x|s_2)$. Then, by the Bayesian relation we have

$$P(s_i|x) = \frac{P(s_1|x)}{P(s_1|x) + P(s_2|x)} = \frac{\pi_1 p(x|s_1)}{\pi_1 p(x|s_1) + \pi_2 p(x|s_2)}. \tag{2}$$

For convenience, we set $\psi_i(x) = P(s_i|x), i = 1, 2, (\psi_2(x) = 1 - \psi_1(x))$ and sometimes write

$$\psi(x) = \psi_1(x). \tag{3}$$

Now let $\psi_i^{(t-1)}(x)$ be the posterior probabilities of the state s_i at $t - 1$. Then, the prior probabilities at t are

$$\pi_i^{(t)} = \psi_1^{(t-1)}(x^{(t-1)})p_{1i} + \psi_2^{(t-1)}(x^{(t-1)})p_{2i}, \quad i = 1, 2. \tag{4}$$

With $\pi_i^{(t)}, i = 1, 2$, the posterior probabilities of the state s_i at t are

$$\psi_i^{(t)}(x) = \frac{\pi_i^{(t)} p(x|s_1)}{\pi_1^{(t)} p(x|s_1) + \pi_2^{(t)} p(x|s_2)}. \tag{5}$$

As before, we sometimes write

$$\psi^{(t)}(x) = \psi_1^{(t)}(x). \tag{6}$$

Let σ be the logistic function: $\sigma(t) = (1 + e^{-t})^{-1}$. Then, we have

$$\psi(x) = \sigma(\log \frac{\pi_1}{\pi_2} + \log \frac{p(x|s_1)}{p(x|s_2)}), \qquad i = 1, 2, \tag{7}$$

and

$$\psi^{(t)}(x) = \sigma(\log \frac{\pi_1^{(t)}}{\pi_2^{(t)}} + \log \frac{p(x|s_1)}{p(x|s_2)}), \qquad i = 1, 2. \tag{8}$$

Hence,

$$\psi^{(t)}(x) = \sigma(\sigma^{-1}(\psi(x)) + C^{(t)}), \tag{9}$$

where

$$C^{(t)} = \log \frac{\pi_1^{(t)}}{\pi_2^{(t)}} - \log \frac{\pi_1}{\pi_2}. \tag{10}$$

Set

$$Q(x) = \log \frac{\pi_1}{\pi_2} + \log \frac{p(x|s_1)}{p(x|s_2)} \quad \text{and} \quad Q^{(t)}(x) = \log \frac{\pi_1^{(t)}}{\pi_2^{(t)}} + \log \frac{p(x|s_1)}{p(x|s_2)}. \tag{11}$$

Then, $\psi^{(t)} = \sigma(Q)$ and $\hat{\psi}^{(t)} = \sigma(\hat{Q})$. Since σ is monotone, $\psi^{(t)}, \hat{\psi}^{(t)}, Q$ and $\hat{Q}$ are all Bayesian discriminant functions [1].

When training with the teacher sequence is completed, the output of the neural network is to approximate $\psi(x)$. We denote the approximation by $\hat{\psi}(x)$. Hence, the inner potential of the output unit is $\sigma^{-1}(\hat{\psi}(x))$. The estimation $\hat{C}^{(t)}$ of $C^{(t)}$ can be obtained recursively where π_i and $\pi_i^{(t)}, i = 1, 2$, are replaced by their estimations $\hat{\pi}_i, \hat{\pi}_i^{(t)}, i, j = 1, 2$.

Consequently, the estimated Bayesian discriminant function $\hat{\psi}^{(t)}(x)$ at t can be obtained simply by shifting the inner potential of the output unit of the trained network by the constant $\hat{C}^{(t)}$ if the activation function of the output unit is σ.

3   Training of the Neural Network

Let $\xi(x, s)$ be a function on $\mathbf{R}^d \times S$, and let $E[\xi(x, \cdot)|x]$ and $V[\xi(x, \cdot)|x]$ be the conditional expectation and variance of $\xi(x, s)$ respectively. Let $F(x, w)$ be the output of the neural network, where w is the connection weight vector. Then, we have a proposition:

Proposition 1. Let

$$\mathcal{E}(w) = \int_{\mathbf{R}^d} \sum_{i=1,2} (F(x, w) - \xi(x, s_i))^2 \pi_i p(x|s_i) dx. \tag{12}$$

Then,

$$\mathcal{E}(w) = \int_{\mathbf{R}^d} (F(x, w) - E[\xi(x, \cdot)|x])^2 p(x) dx$$

$$+ \int_{\mathbf{R}^d} V[\xi(x, \cdot)|x] p(x) dx. \tag{13}$$

Since this is a slight modification of the proposition in [13], we omit the proof. It is easy to see that if $\xi(x, s_1) = 1$ and $\xi(x, s_2) = 0$, $E[\xi)(x, \cdot)|x] = \psi(x)$. Hence, when $\mathcal{E}(w)$ is minimized, the output $F(x, w)$ is expected to approximate $\psi(x)$.

Accordingly, training of the network is carried out by minimizing

$$\mathcal{E}_n(w) = \frac{1}{n} \sum_{t=1}^{T} (F(x^{(t)}, w) - \xi(x^{(t)}, s^{(t)}))^2, \tag{14}$$

where $\{(x^{(t)}, s^{(t)})\}_{t=1}^{T}$ is the teacher sequence. While training the Markov property of the chain is ignored. This algorithm of training is often used [3-11,13]. The approximation attained by this method is approximately in the sense of $L^2(\mathbf{R}^d, p), p(x) = \sum_{i=1,2} \pi_i p(x|s_i)$ [4], because the probability distribution of the observables x is $p(x)$.

4   Construction of a Neural Network

To show that the algorithm works efficiently, we present an example of simulation. Though the simulation can be with any kind of state-conditional probabilities as far as the network can approximate the discriminant functions, we suppose that they are the normal distributions:

$$p(x|s_i) = \frac{1}{\sqrt{(2\pi)^d |\Sigma_i|}} e^{-\frac{1}{2}\{(x-\mu_i)^t \Sigma_i^{-1}(x-\mu_i)\}}, \quad i = 1, 2, \tag{15}$$

where μ_i are the mean vectors and Σ_i are the covariance matrices. For simplicity, let the latter be non-degenerate. Accordingly, the Q defined by (11) is a quadratic form:

$$Q(x) = \log \frac{\pi_1 p(x|s_1)}{\pi_2 p(x|s_2)} = \log \frac{\pi_1}{\pi_2} - \frac{1}{2} \log \frac{|\Sigma_1|}{|\Sigma_2|}$$

$$-\frac{1}{2}\{(x-\mu_1)^t \Sigma_1^{-1}(x-\mu_1) - (x-\mu_2)^t \Sigma_2^{-1}(x-\mu_2)\}. \tag{16}$$

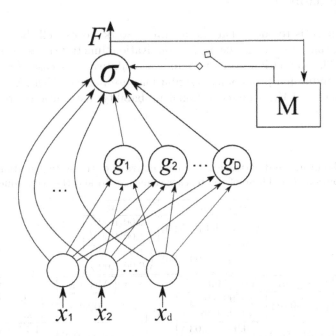

Fig. 1. The structure of the neural network used in the simulation. $D = \frac{1}{2}d(d+1)$. The box M is the module which converts the posterior probability at the previous step to the prior probabilities at the present step (see the test).

There are unit vectors v_i, $i = 1, ..., \frac{1}{2}d(d+1)$, in $\mathbf{R}^d$ such that $(v_i \cdot x)^2$ are linearly independent, where $\cdot$ stands for the inner product. We fix these vectors.

Let ν be any rapidly decreasing probability measure on $\mathbf{R}^d$ and let g be a twice continuously differentiable slowly increasing function defined on $\mathbf{R}$ such that $g^{(2)}(0) \neq 0$. It is proved in [4] that, for any quadratic form q defined on $\mathbf{R}^d$ and any $\varepsilon > 0$, there exist constants a_i, $i = 1, ..., \frac{1}{2}d(d+1)$, b_i, $i = 1, ..., d$, c and δ for which $\|q - \bar{q}\|_{L^2(\mathbf{R}^d, \nu)} < \varepsilon$, where

$$\bar{q}(x) = \sum_{i=1}^{\frac{1}{2}d(d+1)} a_i g(\delta v_i \cdot x) + \sum_{i=1}^{d} b_i x + c. \tag{17}$$

Hence, (16) can be approximated by (17) in $L^2(\mathbf{R}^d, \nu)$ with any accuracy and (17) is realized by a neural network shown in Figure 1 without the module M. Since $\psi_2^{(t-1)}(x) = 1 - \psi_1^{(t-1)}(x)$ and $\pi_2^{(t)} = 1 - \pi_1^{(t)}$, $\hat{C}^{(t)}$ is obtained if $\psi_1^{(t-1)}(x)$ is given. The function of the module M is to memorize $\psi_1^{(t-1)}(x)$ and input $\psi_1^{(t-1)}(x)$ to the output unit. If a memory unit and two units with log activation function are available, the module can be constructed.

5   Simulation

This simulation is to show that the neural network works well. It can be seen that the output unit with the logistic activation function is essential in the algorithm. The simulation is in a two-dimensional, two-state case: $d = 2, m = 2$. The state-conditional probability distributions are normal: $N(\mu_i, \Sigma_i)$, $i = 1, 2$. The parameters of the state-conditional probability distributions are in Table 1a.

Table 1. The means and variance-covariance matrices of the state-conditional probability distributions (a). Those of the teacher sequence (b) and the test sequence (c).

	State	Mean μ_i	V-C. Mat. Σ_i
a	s_1	(1.00,0.00)	$\begin{pmatrix} 2.00 & 1.00 \\ 1.00 & 1.00 \end{pmatrix}$
	s_2	(0.00,0.00)	$\begin{pmatrix} 1.00 & 0.00 \\ 0.00 & 1.00 \end{pmatrix}$

	State	Mean $\hat{\mu}_i$	V-C. Mat. $\hat{\Sigma}_i$		State	Mean $\hat{\mu}_i$	V-C. Mat. $\hat{\Sigma}_i$
b	s_1	(1.00,-0.01)	$\begin{pmatrix} 1.99 & 0.99 \\ 0.99 & 0.98 \end{pmatrix}$	**c**	s_1	(1.02,0.01)	$\begin{pmatrix} 1.96 & 0.95 \\ 0.95 & 0.95 \end{pmatrix}$
	s_2	(-0.01,-0.02)	$\begin{pmatrix} 0.93 & -0.03 \\ -0.03 & 1.05 \end{pmatrix}$		s_2	(0.08,-0.4)	$\begin{pmatrix} 1.18 & -0.09 \\ -0.09 & 1.07 \end{pmatrix}$

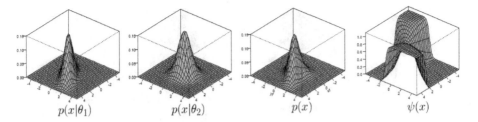

$p(x|\theta_1)$ $p(x|\theta_2)$ $p(x)$ $\psi(x)$

Fig. 2. The state-conditional probability density functions $p(x|\theta_1)$, $p(x|\theta_2)$, the probability density functions $p(x)$ of the observables and the theoretically obtained Bayesian discriminant function $\psi(x)$

The state-conditional probability density functions $p(x|\theta_1)$, $p(x|\theta_2)$, the probability density functions $p(x) = \pi_1 p(x|\theta_1) + \pi_2 p(x|\theta_2)$ of the observables are illustrated in Figure 2 with the theoretically obtained Bayesian discriminant function $\psi(x)$.

The teacher sequence $\{(x^{(t)}, s^{(t)})\}_{t=1}^{T}$ is a segment of the Markov chain from the source with $T = 5000$. The test sequence is also a segment of the Markov chain. It includes 1000 pairs. The two sequences are mutually independent. They are plotted in Figure 3. The parameters calculated from the teacher sequence are in Table 1b and those calculated from the test sequence are in Table 1c.

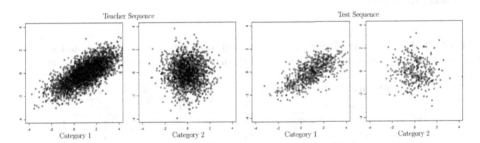

Teacher Sequence   Test Sequence

Category 1   Category 2   Category 1   Category 2

Fig. 3. Plots of the observables in the teacher sequence and test sequence

In the simulation, the activation function of the hidden layer is the logistic function, but they are shifted, $g(t) = \sigma(t+2)$, so that the condition $g^{(2)}(0) \neq 0$ is satisfied. The network is trained with the teacher sequence. The training process is shown in Figure 4 with the difference $\psi(x) - \hat{\psi}(x)$.

Both the theoretically obtained Bayesian discriminant function ψ and the simulated Bayesian discriminant function $\hat{\psi}$ are tested with the test sequence. When it was tested we supposed that $\pi_1^{(0)} = \pi_1$ and $\pi_2^{(0)} = \pi_2$. Hence, two sequences $\{\psi^{(t)}(x^{(t)})\}_{t=1}^{T}$ and $\{\hat{\psi}^{(t)}(x^{(t)})\}_{t=1}^{T}$, $T = 1000$, are obtained, where the former is obtained recursively with (4) and (5), and the latter is also obtained similarly but the discriminant function is replaced by the one obtained by simulation. Then, $x^{(t)}$ are allocated to the state s_1 or s_2 depending on whether $\psi^{(t)}(x^{(t)}) > 0.5$ ($\hat{\psi}^{(t)}(x^{(t)}) > 0.5$) or not.

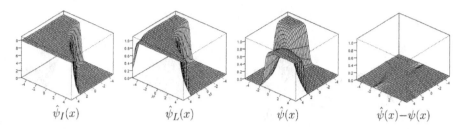

$$\hat{\psi}_I(x) \qquad\qquad \hat{\psi}_L(x) \qquad\qquad \hat{\psi}(x) \qquad\qquad \hat{\psi}(x)-\psi(x)$$

Fig. 4. The learning process and the difference $\psi(x) - \hat{\psi}(x)$. Learning proceeded from the left to the right (See text), where $\hat{\psi}_I$ is the output for the initial values, $\hat{\psi}_L$ is the output while learning, and $\hat{\psi}$ is the final output.

Among 1000 test observables, 654 are from the state s_1 and 346 from the state s_2. The allocation results are illustrated in Table 2 in the rows marked by "Forward algorithm". The number of observables correctly allocated with ψ is 765, and those with $\hat{\psi}$ is 762. In the rows marked by "With known previous state", allocation results knowing the previous states $s^{(t-1)}$ are shown. Naturally the allocations results are better. In the rows marked by "Without history", the allocation results without information on the history of the chains are shown. These are the results of the ordinary Bayesian classification. These results are worse than others. The allocations by the forward algorithm with ψ and $\hat{\psi}$ coincided for 993 observables.

Table 2. The numbers of observables allocated to the respective states s_1 and s_2, and the numbers of correctly allocated ones with the two discriminant functions ψ and $\hat{\psi}$

Theoretical ψ	Alloc. to s_1	Alloc. to s_2	Correctly alloc.
Forward algorithm	787	213	765
With known previous state	722	278	794
Without history	810	190	752

Learned $\hat{\psi}$	Alloc. to s_1	Alloc. to s_2	Correctly alloc.
Forward algorithm	788	212	762
With known previous state	724	276	798
Without history	802	198	756

6   Discussions

It is shown that a neural network having the output unit equipped with logistic function can be efficiently used for estimating the states of hidden Markov chains. The allocation results by the theoretical and simulated discriminant functions, ψ and $\hat{\psi}$, coincided for more than 99 % of observables. With the algorithm, a segment of the Markov chain can be used as a teacher sequence without modification. While training of the network, the pairs $\{(x^{(t)}, s^{(t)})\}$ are regarded as independent. The network can convert the learned discriminant function $\hat{\psi}$ to those at the respective steps $\hat{\psi}^{(t)}$, where the output unit with the logistic activation plays an essential role. As stated before, this paper is not to assert that the algorithm replaces the traditional forward methods of estimating hidden Markov chains but to show the neural network performs the work efficiently. This paper extends the range of the well-known capability of neural networks.

Acknowledgments. This work was supported by a Grant-in-Aid for Scientific Research (22500213) from the Ministry of Education, Culture, Sports, Science and Technology of Japan.

References

1. Duda, R.O., Hart, P.E.: Pattern classification and scene analysis. John Wiley & Sons, New York (1973)
2. Churchill, G.: Stochastic models for heterogeneous DNA sequences. Bull. Math. Biology 51(1), 79–94 (1989)
3. Funahashi, K.: Multilayer neural networks and Bayes decision theory. Neural Networks 11, 209–213 (1998)
4. Ito, Y.: Simultaneous approximations of polynomials and derivatives and their applications to neural networks. Neural Computation 20, 2757–2791 (2008)
5. Ito, Y., Srinivasan, C.: Multicategory Bayesian decision using a three-layer neural network. In: Kaynak, O., Alpaydın, E., Oja, E., Xu, L. (eds.) ICANN/ICONIP 2003. LNCS, vol. 2714, pp. 253–261. Springer, Heidelberg (2003)
6. Ito, Y., Srinivasan, C.: Bayesian decision theory on three-layer neural networks. Neurocomputing 63, 209–228 (2005)
7. Ito, Y., Srinivasan, C., Izumi, H.: Bayesian learning of neural networks adapted to changes of prior probabilities. In: Duch, W., Kacprzyk, J., Oja, E., Zadrożny, S. (eds.) ICANN 2005. LNCS, vol. 3697, pp. 253–259. Springer, Heidelberg (2005)
8. Ito, Y., Srinivasan, C., Izumi, H.: Discriminant analysis by a neural network with Mahalanobis distance. In: Kollias, S.D., Stafylopatis, A., Duch, W., Oja, E. (eds.) ICANN 2006, Part II. LNCS, vol. 4132, pp. 350–360. Springer, Heidelberg (2006)
9. Ito, Y., Srinivasan, C., Izumi, H.: Learning of Bayesian discriminant functions by a neural network. In: Ishikawa, M., Doya, K., Miyamoto, H., Yamakawa, T. (eds.) ICONIP 2007, Part I. LNCS, vol. 4984, pp. 238–247. Springer, Heidelberg (2008)
10. Ito, Y., Srinivasan, C., Izumi, H.: Multi-category Bayesian decision by neural networks. In: Kůrková, V., Neruda, R., Koutník, J. (eds.) ICANN 2008, Part I. LNCS, vol. 5163, pp. 21–30. Springer, Heidelberg (2008)
11. Ito, Y., Izumi, H., Srinivasan, C.: Learning of Mahalanobis discriminant functions by a neural network. In: Leung, C.S., Lee, M., Chan, J.H. (eds.) ICONIP 2009, Part I. LNCS, vol. 5863, pp. 417–424. Springer, Heidelberg (2009)
12. Khasminskii, R., Lazareva, B., Stapleton, J.: Some procedures for state estimation of a hidden Markov chain with two states. In: Gupta, S.S., Berger, J. (eds.) Statistical Decision Theory and Related Topics, pp. 477–487. Springer (1994)
13. Ruck, D.W., Rogers, S.K., Kabrisky, M., Oxley, M.E., Suter, B.W.: The multilayer perceptron as an approximation to a Bayes optimal discriminant function. IEEE Transactions on Neural Networks 1, 296–298 (1990)

Corporate Leaders Analytics and Network System (CLANS): Constructing and Mining Social Networks among Corporations and Business Elites in China

Yuanyuan Man[1], Shuai Wang[1], Tianyu Zhang[2],
T.J. Wong[2], and Irwin King[1]

[1] Department of Computer Science and Engineering
[2] School of Accountancy
The Chinese University of Hong Kong
Shatin, N.T., Hong Kong
{yyman,wangs}@cse.cuhk.edu.hk

Abstract. Social network plays a vital role in Chinese business and is highly valued by business people. However, social network analysis is difficult due to issues in data collection, natural language processing, social network detection and construction, relationship mining, etc. Thus, we develop the Corporate Leaders Analytics and Network System (CLANS) to tackle some of these problems. Our contributions are in three aspects: 1) we collect data from multiple sources and do the preprocessing to make it available to use; 2) we construct a business social network and formulate the similarity relations among individuals and corporations; 3) we conduct further data mining to discover more implicit information, including important entities finding, relation mining and shortest path finding. In this paper, we present the overview of CLANS and specifically address these three major issues. We have made an operational system and achieved basic functionalities.

Keywords: social network, business analytics, data mining, China market, business elites, corporations.

1 Introduction

Social networks are essential for business in China and many other emerging economies. Especially, relationship plays a crucial role in Chinese business model [1]. Related researches indicate that social networks among US firms benefit the debt financing [12], firm performance [6], and corporate governance [5]. However, few studies focus on corporations and elites in China. Hence, it is important to collect, investigate and analyze these business relations for corporation and elites in China.

Further, the analysis of Chinese social network is significant for business people. Investors take into account and highly value the social connecting is-

C.K. Loo et al. (Eds.): ICONIP 2014, Part I, LNCS 8834, pp. 610–618, 2014.
© Springer International Publishing Switzerland 2014

sues among Chinese firms for their investment decision. Besides, common businessman would also like to learn more about specific information for Chinese companies, senior executives and their social networks, for better or potential commercial activities.

Although the analysis of business social networks in China is important, there are a number of difficulties in data collection, natural language processing, network detection and construction, relationship mining, etc. Thus, we design and implement the Corporate Leaders Analytics and Network System (CLANS) to tackle some of these proposed problems, with the help of available computational approaches in social computing [4] [9].

The objective of CLANS is to identify and analyze social networks among corporations and business elites. Specifically, we currently focus on 2,500 Chinese listed firms and their senior managers. In this paper, we introduce the system overview of CLANS and mainly focus on addressing three issues: 1) how we collect data and make it available to use; 2) how to construct and quantify business social network; 3) how to mine more implicit information from social network.

We address the problems with our novel approaches: 1) we collect data from multiple sources and do the preprocessing to make it available to use; 2) we construct a business social network and formulate similarity relations among individuals and corporations; 3) we conduct data mining to discover more implicit information, including important entities finding, relation mining and shortest path finding.

The organization of the paper is as follows. We present the CLANS system in Section 2. Specifically, Section 2.3, 2.4, and 2.5 describe more details in addressing the three major issues. We present our system in website version in Section 3. Section 4 gives a conclusion.

2   CLANS System

2.1   System Overview

The architecture of CLANS consists of six components, shown in Fig. 1. For Data Acquisition, we collect raw data from multiple sources (Section 2.2). Then we conduct Data Preprocessing (Section 2.3) and Social Network Construction (Section 2.4) to create entities and relations respectively. Then, all entities are stored in XML files for Data Management, with an auxiliary database to store relations [13]. After that, CLANS conduct Data Mining (Section 2.5) and provide Search Services (Section 2.6) with the latest data.

2.2   Data

We collect raw data related to listed firms and senior managers from two major sources. The first source is the biography of each of these corporate leaders among all the listed firms reported in the annual reports, which are available

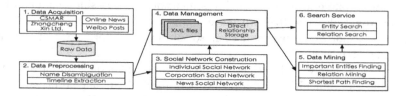

Fig. 1. Architecture of CLANS

electronically in China Securities Market and Accounting Research Database (CSMAR) and a personnel database compiled by Zhongcheng Xin Ltd. Each record in CSMAR contains company's stock code, report year, individual name, position, gender, age, title, education background, biography and payment, from year 1999 to 2012, with totally 399,216 records and about 2,500 companies. The data provided by Zhongcheng Xin Ltd. contains the structured work experiences, like company, position, start year and end year, and education experiences, like college, major, degree, start date and end date. It covers 84,859 records and 68,289 people. But it do not cover every people in CSMAR.

The second source is the online news and Sina Weibo posts related to all senior managers and listed firms. We crawled all news in "http://news.baidu.com/" by searching the name of all corporate leaders and firms. In total, we get 1,126,299 company related news and 16,374,279 people news. As Sina Weibo has become the most important micro-blogging platform in China, and news agents are more likely to utilize it to public news, we also crawled Weibo posts related to our target names. We get 2,367,619 company related posts and 19,445,929 individual posts.

2.3   Data Preprocessing

We conduct data preprocessing to create individual entities, identify related online news and Weibo posts, and extract individual detail structured timeline information.

Name Disambiguation. In this stage we encounter and tackle two major issues. Problem one is that a certain person matches multiple records, so we need to extract unique individual entities. In CSMAR, people may stay in one company for several years so he appears in annual records repeatedly, or he works in many companies so he appears in multiple companies' annual records. Meanwhile, although the data in Zhongcheng Xin Ltd. already has unique identifiers for each people, we need to map people to CSMAR. To figure out this problem, our solution is that, if two records share a high similarity of cognizable features (like name, age, gender, company and birthplace) over a defined threshold, we consider them as the same person. In this way, we identify 87,458 individual entities in CSMAR and find the common 46,130 people in two datasets.

Problem two is that a popular name in our database matches multiple people from Internet sources. A name may map multiple individual entities in our database or a name in our database may map multiple people in online news and Weibo posts. As Weibo posts are shorts and using a nonstandard language with similarities to SMS style, online news is much longer than Weibo posts, and the language styles used in news are much more standard. Then we use a two-round labeling method to identify Weibo posts and a third-round labeling method to identify related online news to a certain person. For the first round, we use the related key words (like company stock name, positions, and college) to the certain person in our database to do the filtering. Then we get a small size of reliable data set in which most posts are truly related to the people in our dataset. In the second round, we use the labeled result from the first round as training set, then transform the text into vector expression based on tf-idf [8] and calculate the cosine similarity [11] of the unlabeled posts to it. After selecting above a threshold, we get the final related Weibo posts. Same as the first two rounds in Weibo posts, in the third round for online news, we apply the Latent Semantic Allocation [2] to the labeled result from the second round, map all the documents to vectors in the lower dimensional latent semantic space, calculate documents similarities, set the threshold and label the rest news. We select a random sample of 1000 posts and news, and it shows that the problem is solved by a precision rate of 98% for Weibo posts and 86% for online news.

Timeline Extraction. For people in CSMAR, who are not covered in Zhongcheng Xin Ltd. data, we analyze personal unstructured profiles and extract structured timeline information, like education and work experience. We adopt different strategies for different parts. For education timeline, we employ rule-learning algorithm with precision rate 95.1%. For working timeline, we combine rule-learning algorithm with HMM model [3], owning to expression's diversity and complexity, and we achieve a precision rate of 83.1%.

2.4 Social Network Construction

We construct a business social network, which contains two parts of individual and corporation, and formulate similarity relations among individuals and corporations. Then we construct the news social network for individual and corporation based on Weibo posts and online news.

Individual Social Network Construction. We construct alumni and colleague social network respectively and formulate similarity relations among them, and then integrate them with weighting coefficients to construct the individual social network. We present the alumni, colleague and integrated individual social network in the website, shown in Fig. 2b.

We define the alumni relationship as the closeness of the relationship between two alumni based on the combination of four criteria, including major, degree, time of enrollment, and intersection school time. We deduce 13 types of relationships between two alumni and assign the corresponding weight empirically. For example, the closest relationship (same major, same degree and same time of enrollment) means that the two people are classmates, with a high possibility that they know each other well, so we assign the weight between them to 0.9, while the farthest relationship is 0.1 (with different major, different degree and no intersection school time).

Definition 1. *Let position rank (PS) denoted as a representation of job level by integer ranging from 0 to 9. The higher position rank has a larger value. The PS of a board chairman and a CEO are assigned to be 9 and 8 respectively, while we assign the independent director to be 1. Let value relation between two colleagues denoted as the average position rank of the two people. Let close relation between two colleagues denoted as the intersection years that they work together.*

Definition 2. *Let colleague relationship denoted as a combination of value relation and close relation. The colleague weight between person p_i and p_j is defined as*

$$\omega_{p_i,p_j} = \sum_{t \in L(p_i,p_j)} \frac{PS_{t,p_i} + PS_{t,p_j}}{2} , \tag{1}$$

where $L(p_i, p_j)$ denotes a collection of the intersection years that person p_i and p_j used to work with each other, and PS_{t,p_i} denotes the position rank of person p_i in the year t. At the end, all the weights are normalized, which is also applied in the following weight calculation.

We define the individual social network as an undirected graph $G(V, E)$. In $G(V, E)$, every edge (relationship) has weighted value, which is defined as

$$W_{i,j} = \alpha \omega_{i,j}^{al} + \beta \omega_{i,j}^{co} , \tag{2}$$

where $\omega_{i,j}^{al}$ is a weight for alumni relationship, $\omega_{i,j}^{co}$ for colleague relationship; α and β denotes the corresponding percentage that the alumni and colleague relationship contribute to the individual relation respectively. We can construct the specific individual social network according to personalized requirements by specifying different weighting coefficients.

Corporation Social Network Construction. We construct the corporation social network based on individual relations and formulate the similarity relation among corporations. We present the corporation social network in the website, shown in Fig. 2c.

Definition 3. *We define the corporation social network as an directed graph $\hat{G}(\hat{V}, \hat{E})$. In $\hat{G}(\hat{V}, \hat{E})$, every vertex (corporation) has feature set $P_i = \{p_i^1, p_i^2, \cdots, p_i^n\}$ and every direct edge (relationship) has weighted value $W_{i,j} = (\omega_{i,j}^{gp}, \omega_{i,j}^{nk})$. n is the size of the set (total number of staffs); $\omega_{i,j}^{gp}$ is a weight for group membership, $\omega_{i,j}^{nk}$ for network relationship.*
$\omega_{i,j}^{gp}, \omega_{i,j}^{nk}$ are defined as follows:

$$\omega_{i,j}^{gp} = \sum_{p_i^k \in P_i \cap P_j} PS_{p_i^k} * \omega_{p_i^k}^{gp} , \tag{3}$$

$$\omega_{i,j}^{nk} = \sum_{(p_i^k, p_j^r) \in L_2(P_i, P_j)} PS_{p_i^k} * \omega_{p_i^k, p_j^r}^{nk} . \tag{4}$$

$PS_{p_i^k}$ denotes the position rank of person p_i^k in corporation i; $\omega_{p_i^k}^{gp}$ is a weight for p_i^k connecting P_i with P_j; $L_2(P_i, P_j)$ denotes a collection of connections between $(P_i - P_i \cap P_j)$ and $(P_j - P_i \cap P_j)$; $\omega_{p_i^k, p_j^r}^{nk}$ denotes a weight between p_i^k and p_j^r calculated in the previous equation.

Thus, the corporation weight from corporation i to j is defined as $W_{i,j} = \alpha \omega_{i,j}^{gp} + \beta \omega_{i,j}^{nk}$, where α and β denotes the corresponding percentage that the two relations contribute to the corporation social network respectively.

News Social Network Construction. We construct the news social network for individual and corporation based on Weibo posts and online news. As Weibo posts and online news publish multiple news mentioned corporate leaders and firms every day, it is important to identify how they are connected with each other and provide these information to investors. We have already identified each Weibo post or online news to an individual entity (Section 2.3), so we construct their social network by identifying two individuals or corporations share the same posts or news. Using this way, we find all the relations among corporate leaders and corporations, and present their relations in the website, as shown in Fig. 2b and 2c.

2.5   Data Mining

We conduct data mining to discover implicit information in these three aspects: important entities finding, relation mining, and shortest path finding.

Important Entities Finding. We utilize two algorithms to discover important individual and corporation entities in social network respectively. We integrate this with search result, so users could find important entities when they search. For individuals finding, our algorithm is refer to the work of [14], which takes into consideration of both personal and network information. The basic idea is that commonly the person with high position level plays an important role in business social network, and if he knows someone with close relation, then that person is also important. For corporations finding, we apply PageRank [10] algorithm, which only take account of the corporations' relationships.

Relation Mining. For any specific corporation, relation mining uses a method to find out its important correlated corporations and its staffs who support those links. The corporation relations are defined as a sequence of relationships $\{\hat{e}_{i,1}, \hat{e}_{i,2}, \cdots, \hat{e}_{i,j}\}$, where i and j represents the source corporation and target corporation respectively. A clustering algorithm is utilized to group the relationships by weight, and a pre-defined threshold is used to select the relations in the group. Then we identify its important correlated corporations. Each corporation relation is defined as $\hat{e}_{i,j} = \{\tilde{e}_{p_i^k, p_j^r}^{nk}, \cdots, \tilde{e}_{p_i^d}^{gp}, \cdots\}$, where $\tilde{e}_{p_i^k, p_j^r}^{nk}$ denotes a connection between person p_i^k in corporation i and p_j^r in j, and $\tilde{e}_{p_i^d}^{gp}$ denotes person p_i^d connecting corporation i with j. We use the same method to identify the important staffs that support those corporation links.

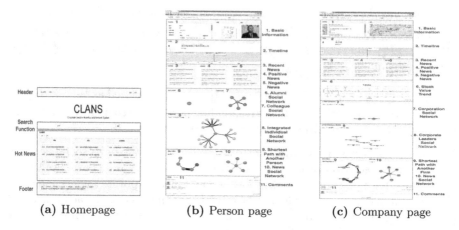

(a) Homepage   (b) Person page   (c) Company page

Fig. 2. Sample pages of websites

Shortest Path Finding. We utilize the state-of-the-art tools to identify the shortest path between three components: people-to-people, people-to-company and company-to-company. As shown in Figure 2b and 2c, users could check the shortest path among people and companies. If they have direct connection, two people have direct connection like schoolmate, family, friend or colleague, or people and company have the employment relationship, or two corporations have the cooperative relationship, the system returns the direct relation between them. For two people or two corporations, who do not have direct connection with each other, shortest path aims to find out the indirect connection between them through closest connected intermediate nodes. For people and company, shortest path aims to find out the possible link to the people who worked in the company and have a high position level. We use the state-of-the-art tools to compute the shortest path for any input person or corporation within three seconds.

2.6   Search Service

In CLANS, we provide two types of services: entity search and relation search.

Entity Search. Given any keyword, system returns a list of ranked persons and companies. Chosen a person/corporation, the system returns related information about the person/corporation.

Relation Search. Given any two keywords, the system returns shortest path between them and the corresponding intermediate nodes and link information.

3   Website Illustration

We have been establishing a website to demonstrate CLANS. Though still first version, it now can visualize basic information, temporal timeline and relations, shortest path, recent, positive and negative new (we use the sentiment analysis tools in [7]) for both companies and individuals. Fig. 2a shows the homepage, in which shows popular corporations and individuals with their news. Fig. 2b and 2c show the basic information, timeline, temporal relations, news and shortest path of a senior executive and a corporation, which are the results of Section 2.3, Section 2.4 and Section 2.5.

4 Conclusion

Social network is of great importance for Chinese business model and business people. However, the analysis is difficult due to issues in data collection, natural language processing, network detection and construction, etc. Thus, we develop CLANS to solve some of these problems. CLANS aims at constructing and mining social network among corporations and business elites. In this paper, we have described the system overview and specifically addressed three issues with our proposed novel solutions. We have established an operational system and achieved basic functionalities. We create a website to visualize information for both companies and individuals. However, it is just the first version and the development of CLANS with more powerful functions as well as a wider researched scope will be our long-term project.

Acknowledgement. This work was supported in part by a grant from the Research Grants Council of the Hong Kong Special Administrative Region, China (Project No. CUHK 413212).

References

1. Allen, F., Qian, J., Qian, M.: Law, finance, and economic growth in China. Journal of Financial Economics 77(1), 57–116 (2005)
2. Deerwester, S.C., Dumais, S.T., Landauer, T.K., Furnas, G.W., Harshman, R.A.: Indexing by latent semantic analysis. JASIS 41(6), 391–407 (1990)
3. Eddy, S.R.: Profile hidden markov models. Bioinformatics 14(9), 755–763 (1998)
4. King, I., Li, J., Chan, K.T.: A brief survey of computational approaches in social computing. In: IJCNN, pp. 1625–1632. IEEE (2009)
5. Larcker, D., Richardson, S., Seary, A., Tuna, A.: Back door links between directors and executive compensation. Back Door Links Between Directors and Executive Compensation (February 2005)
6. Larcker, D.F., So, E.C., Wang, C.C.Y.: Boardroom centrality and stock returns. Citeseer (2010)
7. Lau, T.P., Wang, S., Man, Y., Yuen, C.F., King, I.: Language technologies for enhancement of teaching and learning in writing. In: Proceedings of the Companion Publication of the 23rd International Conference on World Wide Web Companion, pp. 1097–1102. International World Wide Web Conferences Steering Committee (2014)
8. Manning, C.D., Raghavan, P., Schütze, H.: Introduction to information retrieval, vol. 1. Cambridge University Press, Cambridge (2008)

9. Mo, M., King, I.: Exploit of online social networks with community-based graph semi-supervised learning. In: Wong, K.W., Mendis, B.S.U., Bouzerdoum, A. (eds.) ICONIP 2010, Part I. LNCS, vol. 6443, pp. 669–678. Springer, Heidelberg (2010)
10. Page, L., Brin, S., Motwani, R., Winograd, T.: The pagerank citation ranking: bringing order to the web (1999)
11. Singhal, A.: Modern information retrieval: A brief overview. IEEE Data Eng. Bull. 24(4), 35–43 (2001)
12. Uzzi, B.: Embeddedness in the making of financial capital: How social relations and networks benefit firms seeking financing. American Sociological Review, 481–505 (1999)
13. Wang, S., Man, Y., Zhang, T., Wong, T.J., King, I.: Data management with flexible and extensible data schema in clans. Procedia Computer Science 24, 268–273 (2013)
14. Zhang, J., Tang, J., Li, J.: Expert finding in a social network. In: Kotagiri, R., Radha Krishna, P., Mohania, M., Nantajeewarawat, E. (eds.) DASFAA 2007. LNCS, vol. 4443, pp. 1066–1069. Springer, Heidelberg (2007)

Characteristics and Potential Developments
of Multiple-MLP Ensemble Re-RX Algorithm

Yoichi Hayashi, Yuki Tanaka, Tomoki Izawa, and Shota Fujisawa

Dept. of Computer Science, Meiji University
Tama-ku, Kawasaki, Kanagawa 214-8571, Japan
hayashiy@cs.meiji.ac.jp,
{yuuuuuki.t.124,petgoldfish123,hoiminn627}@gmail.com

Abstract. In this paper, we carefully review all our work since 2012 and establish
the concept of the Multiple-MLP Ensemble Re-RX algorithm. We first examine
the background and procedures of the Recursive-Rule Extraction (Re-RX) algorithm family and its variants, including the Multiple-MLP Ensemble Re-RX
algorithm ("Multiple-MLP Ensemble"), which uses the Re-RX algorithm as its
core to find the rules for seven kinds of mixed (i.e., discrete and continuous
attributes) datasets. We compare the accuracy against only the Re-RX algorithm
family. The Multiple-MLP Ensemble Re-RX algorithm cascades standard back-
propagation (BP) to train a multiple neural-network ensemble, where each neural
network is a Multi-Layer Perceptron (MLP). Strictly speaking, multiple neural
networks do not need to be trained simultaneously. Therefore, the Multiple-MLP
Ensemble avoids the previous complicated neural network ensemble structures and
the difficulties of rule extraction algorithms. The extremely high accuracy of the
Multiple-MLP Ensemble algorithm generally outperformed the Re-RX algorithm
and its variant. The results confirm that the Multiple-MLP Ensemble approach facilitates the migration from existing data systems toward new analytic systems and
Big data.

Keywords: Re-RX algorithm, Rule Extraction, MLP, Data Mining, Big Data,
Neural Network Ensemble, Mixed Dataset, Ensemble Concept.

1   Introduction

As Big data and business analytics become the norm, companies with existing data
warehouse architectures are concerned about implementing these new approaches into
their own systems, which are based on traditional business intelligence, to reap the benefits of the new analytic systems. Unfortunately, much current advice focuses on what is
new rather than what to do to migrate from current systems to fully integrated Big Data
and analytics. What is needed is a new information architecture that combines the best
of current data warehousing approaches and facilitates integration of new technologies.

This paper offers solutions for existing systems to move toward "Big Data mining", so that the computer scientists can meet the strong demands in our high-level
computer network society.

C.K. Loo et al. (Eds.): ICONIP 2014, Part I, LNCS 8834, pp. 619–627, 2014.
© Springer International Publishing Switzerland 2014

Real-world classification problems usually involve both discrete and continuous input attributes. All the continuous attributes must be discretized. The drawback of discretizing the continuous attributes is that the accuracy of the networks, and hence the accuracy of the rules extracted from the networks, may decrease. This is because discretization leads to a division of the input space into hyper-rectangular regions.

Setiono et al. [1] proposed a recursive algorithm for rule extraction from a neural network trained for solving a classification problem having mixed discrete and continuous input data attributes. This algorithm shares some similarities with other existing rule extraction algorithms. Setiono et al. [2] also proposed a valiant of the Recursive-Rule Extraction (Re-RX) algorithm. They [3] presented a credit card screening application of the Re-RX algorithm [1]. Utilizing the Re-RX algorithm, Hayashi et al. [4] presented a very modern marketing approach to analyze consumer heterogeneity in the context of eating-out behavior in Taiwan.

Bologna [5] proposed the Discretized Interpretable Multi-Layer Perceptron (DIMLP) model with generated rules from neural network ensembles. The DIMLP is a "special" neural network model for which symbolic rules are generated to clarify the knowledge embedded within connections and activation neurons.

Zhou et al. [6] proposed the Rule Extraction From Network Ensemble (REFNE) approach to extract symbolic rules from trained neural network ensembles that perform classification tasks.

In 2012, Hara and Hayashi [7, 8] first proposed the "Two-MLP Ensemble Re-RX algorithm" for data with mixed attributes. The accuracy of the proposed algorithm outperformed that of almost all previous algorithms. The simple and powerful idea proposed in these two papers is the impetus of the author's current paper.

In 2013, Hayashi et al. [9] presented the "Three-MLP Ensemble Re-RX algorithm" with an extremely high performance result for the German Credit dataset, which is a two-class mixed dataset.

In September 2014, Hayashi et al. [10] presented a new framework, Re-RX algorithm family, and conducted a comparative study of the accuracies in this ensemble family. In this ensemble family, each neural network is an MLP. The Re-RX algorithm is an effective rule extraction algorithm for datasets that comprise both discrete and continuous attributes, and so it is a core part of the Three-MLP Ensemble Re-RX algorithm.

The authors considered the Three-MLP Ensemble to be a "virtual" ensemble system. From an overall viewpoint, this algorithm cascades backpropagation (BP) to train the three neural-network ensemble. Thus, strictly speaking, the three neural networks do not need to be trained simultaneously. In addition, this simple and new concept of rule extraction from the neural network ensemble can avoid previous complicated neural network ensemble structures and the difficulties of rule extraction algorithms. The extracted rules maintain the high learning capabilities of neural networks while expressing highly comprehensible rules.

In this paper, we first review all our work since 2012 and establish the concept of "Multiple-MLP Ensemble Re-RX algorithm" from generalized viewpoints and also consider potential developments. This comprises Sections 2 through 5.

In Section 6, we describe the characteristics of the Multiple-MLP Ensemble Re-RX algorithm. We also consider the three major drawbacks of the Re-RX algorithm and propose some methods to resolve them. In Section 7, summarize our innovative learning method to train MLPs for extraordinarily high accuracy.

2   Recursive Rule Extraction Algorithm: Re-RX Algorithm

The Re-RX algorithm [1] is designed to generate classification rules from datasets that have both discrete and continuous attributes. The algorithm is recursive in nature and generates hierarchical rules. The rule conditions for discrete attributes are disjointed from those for continuous attributes. The continuous attributes only appear in the conditions of the rules lowest in the hierarchy. The outline of the algorithm is as follows.

Algorithm Re-RX(S, D, C)

Input: A set of data samples S having discrete attributes D and continuous attributes C.

Output: A set of classification rules.

1. Train and prune a neural network using the dataset S and all of its D and C attributes.

2. Let D' and C' be the sets of discrete and continuous attributes, respectively, still present in the network, and let S' be the set of data samples correctly classified by the pruned network.

3. If D' = ϕ, then generate a hyperplane to split the samples in S' according to the values of the continuous attributes C', and stop.

Otherwise, by using only the discrete attributes D', generate the set of classification rules R for dataset S'.

4. For each rule R_i generated:

If support(R_i)>δ_1 and error(R_i)>δ_2, then
- Let S_i be the set of data samples that satisfy the condition of rule R_i and let D_i be the set of discrete attributes that do not appear in rule condition R_i.

- If D_i = ϕ, then generate a hyperplane to split the samples in S_i according to the values of their continuous attributes C_i, and stop.

- Otherwise, call Re-RX(S_i, D_i, C_i).

The support of a rule is the percentage of samples that are covered by that rule. The support and the corresponding error rate of each rule are checked in step 4. If the error exceeds threshold δ_2 and the support meets the maximum threshold δ_1, then the subspace of this rule is further subdivided by either recursively calling Re-RX when discrete attributes are still not present in the conditions of the rule or by generating a separating hyperplane involving only the continuous attributes of the data.

3   Three-MLP Ensemble by the Re-RX Algorithm

We recently extended the Two-MLP Ensemble by the Re-RX algorithm to the Three-MLP Ensemble [9, 11] to achieve extremely high accuracy. The Three-MLP Ensemble is shown as an example of the Multiple-MLP Ensemble Re-RX algorithm.

Three-MLP Ensemble by the Re-RX algorithm

Inputs: Learning datasets LD', LDf', LDff'

Outputs: Primary rule set, secondary rule set, and tertiary rule set.

1. Randomly extract a dataset of an arbitrary proportion from learning dataset LD, and name the set of extracted data as LD'.
2. Train and prune [12] the first neural network using LD'.
3. Apply the Re-RX algorithm to the output of step 2, and output the primary rule set.
4. Based on these primary rules, create a dataset LDf from a dataset (LD) that is not correctly classified by the rules.
5. Randomly extract the dataset of an arbitrary proportion from learning dataset LDf, and name the set of extracted data as LDf'.
6. Train and prune [12] the two ensemble neural networks by using LDf'.
7. Apply the Re-RX algorithm to the output of step 6, and output the secondary rule set.
8. Integrate the primary rule set and the secondary rule set.
9. Based on the rules integrated in step 8, create a dataset LDff from a dataset (LD) that is not correctly classified by the rules.
10. Randomly extract data samples of an arbitrary proportion from learning dataset LDff, and name the set of extracted data as LDff'.
11. Train and prune [12] the ensemble of three neural networks by using LDff'.
12. Apply the Re-RX algorithm to the output of step 11, and output the tertiary rule set.
13. Integrate the primary rule set, the secondary rule set, and the tertiary rule set.

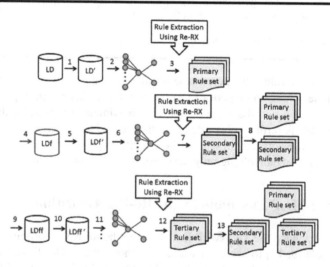

Fig. 1. Schematic diagram of Three-MLP Ensemble by the Re-RX algorithm

4 Experimental Results

We trained seven kinds of mixed datasets [13, 14] by the Re-RX algorithm family and obtained the accuracies and the number of extracted rules for each test dataset as follows. In Tables 1–4, the bold numbers are the highest test accuracies of all accuracies reported in previous papers, as of June 15, 2014. Some cells in the tables are empty, e.g., Three-MLP Ensemble of CARD1. This means that the Three-MLP Ensemble Re-RX algorithm did not provide meaningful rules, because the LDff' was too small to provide tertiary rules. The reason for this situation is described in Section 6.1. In this case, the final accuracy for the test dataset by the Two-MLP Ensemble Re-RX algorithm was the highest for each dataset.

Table 1. Performance of training and test datasets for CARD1 and CARD2

	CARD1 Training (%)	Test	No. Rules	CARD2 Training (%)	Test	No. Rules
Re-RX only	87.26	83.14	5	88.99	84.93	6
2-MLP Ensemble	95.37	**93.02**	7	97.10	**94.78**	9
3-MLP Ensemble						

Table 2. Performance of training and test datasets for CARD3 and German

	CARD3 Training (%)	Test	No. Rules	German Training (%)	Test	No. Rules
Re-RX only	86.67	80.57	7	73.00	74.40	13
2-MLP Ensemble	90.43	85.80	11	78.80	**81.80**	19
3-MLP Ensemble						

Table 3. Performance of test datasets for Bene1 and Bene2

	Bene1 Training (%)	Test	No. Rules	Bene2 Training (%)	Test	No. Rules
Re-RX only	73.02	72.07	27	72.38	72.93	42
2-MLP Ensemble	96.61	96.99	34	73.32	73.91	45
3-MLP Ensemble	97.63	**98.14**	36	81.36	**81.56**	54

Table 4. Performance of training and test datasets for Thyroid

	Thyroid Training (%)	Test	No. Rules
Re-RX only	97.50	94.89	2
2-MLP Ensemble	99.83	**99.72**	5
3-MLP Ensemble			

5   Discussion of Experimental Results

We focus only on the highlights of the experimental results. The accuracy for the test data of rules extracted from CARD1 and CARD2 by the proposed algorithm outperforms the accuracies of other previous work. We believe the approximately 10% increase in accuracies from the Re-RX only (raw Re-RX), compared to the Two-MLP Ensemble in CARD1 and CARD2, occurred for the following two reasons:

(1) After the integration of primary rules and secondary rules, we had two new rules in the CARD1 dataset. Then we conducted a sensitivity analysis of the contribution of each extracted rule for the entire test dataset. We summed the values of the two rules and obtained 93.02%. This is identical to the actual experimental value. For the CARD2 dataset, we obtained the same results.

(2) Originally, we needed the micro-analysis of the decision trees. Here, we explain only the important points for the 10% boost of the accuracies. A typical search by decision trees such as C4.5 [15] is executed many times in the Re-RX algorithm. However, the Three-MLP Ensemble Re-RX algorithm has a chance to execute a decision tree search many more times. Thus, if the decision tree search overlooks some classification rules in the Re-RX only (raw Re-RX), the search in the Two-MLP Ensemble, the Three-MLP Ensemble, or the Multiple-MLP Ensemble, could possibly find much more effective classification rules.

The accuracy for the CARD3 test dataset by Setiono et al. [3] is slightly higher than the 85.80% reported in our study.

For the Bene1 test dataset, our algorithm obtained 98.14%. The accuracy by the raw Re-RX algorithm was 73.68 % [1]. Thus, our algorithm greatly outperformed the raw Re-RX algorithm in accuracy for the Bene1 test dataset. After the sensitivity analysis of the contribution of each extracted rule for the entire test dataset, we summed the values of 9 rules and obtained 98.14%. This is identical to the actual experimental value. We believe that the reason for this value is the same as that for the CARD1 and CARD2 datasets.

For the Bene2 dataset, our algorithm obtained 81.56% for the test dataset. The number of rules extracted was 54. Setiono et al. reported an accuracy of 75.26% for the Bene2 test dataset and 67 rules extracted by the raw Re-RX algorithm for the Bene2 test dataset [1].

Clearly, our algorithm showed considerably better accuracy and a much more concise form than that of Setiono et al [1]. Since we used large test datasets, such as Bene1 and Bene2, our number of rules extracted also became large (36 rules and 54 rules for Bene1 and Bene2, respectively).

For the Thyroid dataset, our algorithm achieved 99.72% for the test data and updated the worldwide record of 99.36% by Duch [16]. The Thyroid dataset is a multiclass mixed dataset often used for benchmarking studies as a common dataset in medicine.

6 Discussion of the Multiple-MLP Ensemble Re-RX Algorithm

6.1 Characteristics of the Multiple-MLP Ensemble Re-RX Algorithm

We recently clarified the relationships between a raw Re-RX algorithm [1] and our Multiple-MLP Ensemble Re-RX algorithm [10]. Because we used an example of the Three-MLP Ensemble algorithm stated in Section 3, we can naturally verify the following four important characteristics of the proposed algorithms. [7, 8, 9, 10, 11] These four characteristics provide the most fundamental principle for the general algorithm.

(1) The most important principle in the design of our algorithm is an "attrition" system for the dataset from a learning dataset that is not correctly classified by the extracted rules. Thus, the size of the datasets for rule extraction is monotonically decreasing so that our well-designed algorithm outputs the extracted rules to be included. In addition, our algorithm uses the minimum necessary number of multiple-MLP ensembles and terminates automatically. This idea is based on the multiple use of "sieves" and is called the "French pastry chef approach".

(2) This situation can be interpreted as a kind of "saturated situation". Depending on the characteristics of the datasets, we may see the case of the saturation with only a Two-MLP Ensemble. In other cases, our algorithm may proceed to the Five-MLP Ensemble with increasing accuracies. Thus, we need not set the number of multiple MLPs in the ensemble in advance because our algorithm will terminate in the saturated situation with the minimum necessary numbers of Multiple-MLPs in the ensemble and without outputting more new extracted rules.

(3) Almost all datasets used in benchmarking or experimental comparative studies are not so huge, i.e., not Big data, so we will not be confronted with difficulties in the saturated situation.

(4) The entire Re-RX family uses C4.5 (J4.8) to generate discrete rules. Consequently, the C4.5 algorithm generates a decision tree based on the amount of information in the datasets. The branches of decision trees are not generated in the case of low information gain. Therefore, we believe that the extracted rules in smaller numbers of the Multiple-MLP Ensemble can reflect the characteristics of entire learning datasets. Accordingly, since the increased rate of accuracies decreases with the increase of Multiple-MLP Ensembles, it is a strong possibility that the final accuracy will gradually become saturated.

6.2 Potential Developments of the Multiple-MLP Ensemble Re-RX Algorithm

Here we describe the potential developments of the proposed algorithm. As an important research topic to enhance the entire performance of the proposed Multiple-MLP Ensemble Re-RX algorithm, we have the following three major issues.

(1) The Re-RX algorithm [1] includes some human interaction so currently we cannot use the Re-RX algorithm to extract rules from huge datasets like Big data. Therefore,

we have begun to implement a fully automated version of the Re-RX algorithm and will soon present the details in another paper.

(2) The Re-RX algorithm requires many iterations of BP learning to train MLP and decision tree searches such as C4.5 in the recursive nature of the Re-RX algorithm. We have heard from others that the Re-RX algorithm takes much time to extract rules, and so it is not practical. The algorithm can be used for only small academic datasets but not for Big data. Consequently, we have already implemented the Re-RX algorithm independently from Setiono et al. [1]; our program implemented in Java runs on conventional laptop computers for even Bene1 and Bene2 datasets with 3123 records and 7190 records, respectively [14].

(3) We have no reasonable or theoretical settings of parameters such as δ_1 and δ_2. These parameters are quite sensitive for the accuracy of the Re-RX algorithm stated in step 3 of Section 3.

Although we used heuristics in accordance with the properties of datasets, originally we needed a much more systematic method to determine the optimal values of δ_1 and δ_2. Needless to say, the optimal settings of values on δ_1 and δ_2 will drastically enhance the performance of the Re-RX algorithm.

For the automated version, we proposed a concrete method to find all the parameters simultaneously [11], i.e., δ_1 and δ_2 and some parameters of various backpropagation methods from the integrated viewpoints. We believe that this approach makes the entire performance of our algorithm "systematic".

7   Conclusion

In this paper, we first focused on the Re-RX algorithm family and compared the accuracies within the family. The Three-MLP Ensemble Re-RX algorithm is capable of extracting classification rules from neural network ensembles using datasets of both discrete and continuous mixed attributes. The novel characteristic of the algorithm lies in its extremely high accuracy for mixed attribute datasets.

Furthermore, since the Re-RX algorithm is a core part of the proposed algorithm, the characteristics of the Re-RX algorithm contribute to the success of the proposed algorithm._

Our major claim of the Multiple-MLP Ensemble Re-RX algorithm is its extraordinarily high accuracy, which is essential for "Big Data mining". To make this algorithm practical, we have begun to implement a fully automated version of the Re-RX algorithm and will soon present the details in another paper.

References

1. Setiono, R., et al.: Recursive neural network rule extraction for data with mixed attributes. IEEE Trans. Neural Netw. 19(2), 299–307 (2008)
2. Setiono, R., et al.: Rule extraction from minimal neural networks for credit card screening. Inter. J. of Neural Systems 21(4), 265–276 (2011)

3. Setiono, R., et al.: A note on knowledge discovery using neural networks and its application to credit card screening. European J. Operational Research 192, 326–332 (2009)
4. Hayashi, Y., et al.: Understanding consumer heterogeneity: A business intelligence application of neural networks. Knowledge-Based Systems 23(8), 856–863 (2010)
5. Bologna, G.: Is it worth generating rules from neural network ensemble? J. of Applied Logic 2, 325–348 (2004)
6. Zhou, Z.-H.: Extracting symbolic rules from trained neural network ensembles. AI Communications 16, 3–15 (2003)
7. Hara, A., Hayashi, Y.: Ensemble neural network rule extraction using Re-RX algorithm. In: Proc. WCCI (IJCNN 2012), Brisbane, Australia, June 10-15, pp. 604–609 (2012)
8. Hara, A., Hayashi, Y.: A New Neural Data Analysis Approach Using Ensemble Neural Network Rule Extraction. In: Villa, A.E.P., Duch, W., Érdi, P., Masulli, F., Palm, G. (eds.) ICANN 2012, Part I. LNCS, vol. 7552, pp. 515–522. Springer, Heidelberg (2012)
9. Hayashi, Y., et al.: A New approach to Three Ensemble neural network rule extraction using Recursive-Rule eXtraction algorithm. In: Int. Joint Conf. Neural Networks (IJCNN 2013), Dallas, August 4-9, pp. 835–841 (2013)
10. Hayashi, Y., Tanaka, Y., Fujisawa, S., Izawa, T.: Comparative Study of Accuracies on the Family of the Recursive Rule Extraction Algorithm. In: Wermter, S., Weber, C., Duch, W., Honkela, T., Koprinkova-Hristova, P., Magg, S., Palm, G., Villa, A.E.P. (eds.) ICANN 2014. LNCS, vol. 8681, pp. 491–498. Springer, Heidelberg (2014)
11. Hayashi, Y.: Neural network rule extraction by a new ensemble concept and its theoretical and historical background: A review. Int. J. of Computational Intelligence and Applications 12(4), 1340006-1–1340006-22 (2013)
12. Setiono, R., et al.: A penalty-function approach for pruning feedforward neural networks. Neural Comp. 9(1), 185–204 (1997)
13. Univ. of California, Irvine Learning Repository,
    http://archive.ics.uci.edu/m/
14. Baesens, B., et al.: Using neural network rule extraction and decision tables for credit-risk evaluation. Management Science 49(3), 312–329 (2004)
15. Quinlan, R.: C4.5: Programming for Machine Learning. Morgan Kaufman, San Mateo (1993)
16. Duch, W., et al.: A New Methodology of Extraction, Optimization and Application of Crisp and Fuzzy Logical Rules. IEEE Trans. Neural Networks 12, 277–306 (2001)

Author Index